Advances in

BOTANICAL RESEARCH

Incorporating Advances in Plant Pathology

VOLUME 44

DEVELOPMENTAL GENETICS OF THE FLOWER

Advances in

BOTANICAL RESEARCH

Incorporating Advances in Plant Pathology

Advances in

BOTANICAL RESEARCH

Incorporating Advances in Plant Pathology

DEVELOPMENTAL GENETICS OF THE FLOWER

Edited by

DOUGLAS E. SOLTIS
Department of Botany, University of Florida,
Gainesville, Florida, United States

JAMES H. LEEBENS-MACK
Department of Plant Biology,
The University of Georgia, Athens,
Georgia, United States

PAMELA S. SOLTIS
Florida Museum of Natural History,
University of Florida, Gainesville,
Florida, United States

Series Editor

J. A. CALLOW
School of Biosciences, University of Birmingham,
Birmingham, United Kingdom

VOLUME 44

2006

ELSEVIER

AMSTERDAM • BOSTON • HEIDELBERG • LONDON
NEW YORK • OXFORD • PARIS • SAN DIEGO
SAN FRANCISCO • SINGAPORE • SYDNEY • TOKYO
Academic Press is an imprint of Elsevier

Academic Press is an imprint of Elsevier
525 B Street, Suite 1900, San Diego, California 92101-4495, USA
84 Theobald's Road, London WC1X 8RR, UK

This book is printed on acid-free paper. ♾

For information on all Academic Press publications
visit our Web site at www.books.elsevier.com

ISBN-13: 978-0-12-005944-7
ISBN-10: 0-12-005944-4

Transferred to digital print, 2007
Printed and bound by CPI Antony Rowe, Eastbourne

CONTENTS

Duplication, Diversification, and Comparative Genetics of Angiosperm MADS-Box Genes

VIVIAN F. IRISH

Beyond the ABC-Model: Regulation of Floral Homeotic Genes

LAURA M. ZAHN, BAOMIN FENG AND HONG MA

Missing Links: DNA-Binding and Target Gene Specificity of Floral Homeotic Proteins

RAINER MELZER, KERSTIN KAUFMANN AND GÜNTER THEIßEN

Genetics of Floral Development in *Petunia*

ANNEKE RIJPKEMA, TOM GERATS AND MICHIEL VANDENBUSSCHE

Flower Development: The *Antirrhinum* Perspective

BRENDAN DAVIES, MARIA CARTOLANO AND ZSUZSANNA SCHWARZ-SOMMER

Floral Developmental Genetics of *Gerbera* (Asteraceae)

TEEMU H. TEERI, MIKA KOTILAINEN, ANNE UIMARI, SATU RUOKOLAINEN, YAN PENG NG, URSULA MALM, EIJA PÖLLÄNEN, SUVI BROHOLM, ROOSA LAITINEN, PAULA ELOMAA AND VICTOR A. ALBERT

Gene Duplication and Floral Developmental Genetics of Basal Eudicots

ELENA M. KRAMER AND ELIZABETH A. ZIMMER

Genetics of Grass Flower Development

CLINTON J. WHIPPLE AND ROBERT J. SCHMIDT

Developmental Gene Evolution and the Origin of Grass Inflorescence Diversity

SIMON T. MALCOMBER, JILL C. PRESTON, RENATA REINHEIMER, JESSIE KOSSUTH AND ELIZABETH A. KELLOGG

Expression of Floral Regulators in Basal Angiosperms and the Origin and Evolution of ABC-Function

PAMELA S. SOLTIS, DOUGLAS E. SOLTIS, SANGTAE KIM, ANDRE CHANDERBALI AND MATYAS BUZGO

The Molecular Evolutionary Ecology of Plant Development: Flowering Time in *Arabidopsis thaliana*

KATHLEEN ENGELMANN AND MICHAEL PURUGGANAN

A Genomics Approach to the Study of Ancient Polyploidy and Floral Developmental Genetics

JAMES H. LEEBENS-MACK, KERR WALL, JILL DUARTE, ZHENGUI ZHENG, DAVID OPPENHEIMER AND CLAUDE DEPAMPHILIS

CONTRIBUTORS TO VOLUME 44

VICTOR A. ALBERT *Natural History Museum, University of Oslo, Blindern, NO-0318 Oslo, Norway*

SUVI BROHOLM *Gerbera Laboratory, Department of Applied Biology, 00014 University of Helsinki, Finland*

MATYAS BUZGO *Florida Museum of Natural History and Department of Botany, University of Florida, Gainesville, Florida 32611*

MARIA CARTOLANO *Max Planck Institut für Züchtungsforschung, 5089 Köln, Germany*

ANDRE CHANDERBALI *Florida Museum of Natural History and Department of Botany, University of Florida, Gainesville, Florida 32611*

BRENDAN DAVIES *Centre for Plant Sciences, University of Leeds, Leeds LS2 9JT, United Kingdom*

CLAUDE DEPAMPHILIS *Department of Biology, Institute of Molecular Evolutionary Genetics, and Huck Institutes of Life Sciences, The Pennsylvania State University, University Park, Pennsylvania 16802*

JILL DUARTE *Department of Biology, Institute of Molecular Evolutionary Genetics, and Huck Institutes of Life Sciences, The Pennsylvania State University, University Park, Pennsylvania 16802*

PAULA ELOMAA *Gerbera Laboratory, Department of Applied Biology, 00014 University of Helsinki, Finland*

PETER K. ENDRESS *Institute of Systematic Botany, University of Zurich, Zollikerstrasse 107, 8008 Zurich, Switzerland*

KATHLEEN ENGELMANN *Department of Biology and Center for Comparative and Functional Genomics, New York University, New York, New York 10003*

BAOMIN FENG *Department of Biology, The Huck Institutes of the Life Sciences and the Institute of Molecular Evolutionary Genetics, The Pennsylvania State University, University Park, Pennsylvania 16802*

MICHAEL W. FROHLICH *Department of Botany, Natural History Museum, London SW7 5BD, United Kingdom*

TOM GERATS *IWWR/Plant Genetics, Toernooiveld 1, 6525ED Nijmegen, The Netherlands*

VIVIAN F. IRISH *Departments of Molecular, Cellular, and Developmental Biology and of Ecology and Evolutionary Biology, Yale University, New Haven, Connecticut 06520*

KERSTIN KAUFMANN* *Friedrich-Schiller-Universität Jena, Lehrstuhl für Genetik, Philosophenweg 12, D-07743 Jena, Germany*

ELIZABETH A. KELLOGG *Department of Biology, University of Missouri-St. Louis, One University Boulevard, Saint Louis, Missouri 63121*

SANGTAE KIM *Florida Museum of Natural History and Department of Botany, University of Florida, Gainesville, Florida 32611*

JESSIE KOSSUTH *Department of Biology, University of Missouri-St. Louis, One University Boulevard, Saint Louis, Missouri 63121*

MIKA KOTILAINEN *Gerbera Laboratory, Department of Applied Biology, 00014 University of Helsinki, Finland*

ELENA M. KRAMER *Department of Organismic and Evolutionary Biology, Harvard University, Cambridge, Massachusetts 02138*

ROOSA LAITINEN *Gerbera Laboratory, Department of Applied Biology, 00014 University of Helsinki, Finland*

JAMES H. LEEBENS-MACK† *Department of Biology, Institute of Molecular Evolutionary Genetics, and Huck Institutes of Life Sciences, The Pennsylvania State University, University Park, Pennsylvania 16802*

HONG MA *Department of Biology, The Huck Institutes of the Life Sciences and the Institute of Molecular Evolutionary Genetics, The Pennsylvania State University, University Park, Pennsylvania 16802*

SIMON T. MALCOMBER *Department of Biology, University of Missouri-St. Louis, One University Boulevard, Saint Louis, Missouri 63121*

URSULA MALM *Gerbera Laboratory, Department of Applied Biology, 00014 University of Helsinki, Finland*

RAINER MELZER *Friedrich-Schiller-Universität Jena, Lehrstuhl für Genetik, Philosophenweg 12, D-07743 Jena, Germany*

YAN PENG NG *Gerbera Laboratory, Department of Applied Biology, 00014 University of Helsinki, Finland*

DAVID OPPENHEIMER *Department of Botany and the Genetics Institute, University of Florida, Gainesville, Florida 32611*

EIJA PÖLLÄNEN *Gerbera Laboratory, Department of Applied Biology, 00014 University of Helsinki, Finland*

JILL C. PRESTON *Department of Biology, University of Missouri-St. Louis, One University Boulevard, Saint Louis, Missouri 63121*

MICHAEL PURUGGANAN *Department of Biology and Center for Comparative and Functional Genomics, New York University, New York, New York 10003*

* Present Address: Plant Research International, Business Unit Bioscience, Wageningen, 6700 AA, Netherlands.

† Present Address: Department of Plant Biology, University of Georgia, Athens, Georgia 30602.

RENATA REINHEIMER *Vegetal Morphology, Facultad de Ciencias Agrarias (UNL), Kreder 2805, 3080 Esperanza, Santa Fe, Argentina*

ANNEKE RIJPKEMA *IWWR/Plant Genetics, Toernooiveld 1, 6525ED Nijmegen, The Netherlands*

SATU RUOKOLAINEN *Gerbera Laboratory, Department of Applied Biology, 00014 University of Helsinki, Finland*

ROBERT J. SCHMIDT *Section of Cell and Developmental Biology, Division of Biological Sciences, University of California at San Diego, La Jolla, California 92093*

ZSUZSANNA SCHWARZ-SOMMER *Max Planck Institut für Züchtungsforschung, 5089 Köln, Germany*

DOUGLAS E. SOLTIS *Department of Botany, University of Florida, Gainesville, Florida 32611*

PAMELA S. SOLTIS *Florida Museum of Natural History, University of Florida, Gainesville, Florida 32611*

TEEMU H. TEERI *Gerbera Laboratory, Department of Applied Biology, 00014 University of Helsinki, Finland; Department of Biology, University of Tromsø, NO-9037 Tromsø, Norway*

GÜNTER THEIßEN *Friedrich-Schiller-Universität Jena, Lehrstuhl für Genetik, Philosophenweg 12, D-07743 Jena, Germany*

ANNE UIMARI *Gerbera Laboratory, Department of Applied Biology, 00014 University of Helsinki, Finland*

MICHIEL VANDENBUSSCHE *IWWR/Plant Genetics, Toernooiveld 1, 6525ED Nijmegen, The Netherlands*

KERR WALL *Department of Biology, Institute of Molecular Evolutionary Genetics, and Huck Institutes of Life Sciences, The Pennsylvania State University, University Park, Pennsylvania 16802*

CLINTON J. WHIPPLE *Section of Cell and Developmental Biology, Division of Biological Sciences, University of California at San Diego, La Jolla, California 92093*

LAURA M. ZAHN *Department of Biology, The Huck Institutes of the Life Sciences and the Institute of Molecular Evolutionary Genetics, The Pennsylvania State University, University Park, Pennsylvania 16802*

ZHENGUI ZHENG *Department of Botany and the Genetics Institute, University of Florida, Gainesville, Florida 32611*

ELIZABETH A. ZIMMER *Department of Botany and Laboratories of Analytical Biology, National Museum of Natural History, Smithsonian Institution, Suitland, Maryland 20746*

CONTENTS OF VOLUMES 34–43

Contents of Volume 34

Contents of Volume 35

Contents of Volume 36

Contents of Volume 37

Contents of Volume 38

Contents of Volume 39

Contents of Volume 40

Contents of Volume 41

Contents of Volume 42

Contents of Volume 43

Angiosperm Floral Evolution: Morphological Developmental Framework

PETER K. ENDRESS

Institute of Systematic Botany, University of Zurich, Zollikerstrasse 107, 8008 Zurich, Switzerland

Advances in Botanical Research, Vol. 44
Incorporating Advances in Plant Pathology

0065-2296/06 $35.00
DOI: 10.1016/S0065-2296(06)44001-5

ABSTRACT

This chapter brings together aspects of floral development and evolution in angiosperms derived from the diversity of structures of extant plants. It provides a framework especially addressed to evolutionary developmental biologists. It attempts to show the spectrum of forms with special emphasis on extremes. Part of the processes discussed here have been addressed by molecular developmental geneticists, others have not yet been tackled. Topics are the contingency of the primary morphological surface of the flower, the sequence of the origin of floral organs in evolutionary history (organ categories, "organ identities"), patterns of floral phyllotaxis and symmetry, patterns of angiospermy, the significance of postgenital and congenital fusion for synorganization, the difference between autonomous and imprinted components of organ form, architectural extremes and limits discussed with the example of increased carpel number in whorled gynoecia, the decoupling of the sequence of stamen initiation from the normal centripetal floral organ initiation by ring meristems, and some potential pitfalls of generalizations extrapolated from model plants.

I. INTRODUCTION

To understand mechanisms of development it is useful to view them in an evolutionary context. This helps to ask the right questions. Since these mechanisms have evolved, glimpses of earlier evolutionary stages may open a window to better perceive the elements of development of highly derived forms.

Thus, structural patterns and developmental processes involved in the evolution of angiosperm flowers are here outlined from a diversity perspective. What patterns are common, and under which conditions do they occur? What are extreme patterns? The morphospace, the entire range of a morphological character (Hufford and McMahon, 2003), shows what has been realized in evolution but not necessarily all that is possible in principle.

Floral evolution and diversification have often been viewed as a function of floral ecology/pollination biology. This is certainly an important aspect but it is not the whole story. With the new rise of developmental biology it became obvious that it is also a function of internal, developmental limitations (constraints) and potentials for certain pathways (and propensities) of evolution, some of which had also been recognized previously in comparative morphological studies.

What has changed in our perception of flower evolution in the past 20 years? The exciting thing is that flower evolution came into the focus of and was influenced by quite different fields of biology. (1) The study of new, three-dimensionally preserved, charcoalified flower fossils, a state of preservation of fossils not known before, opened a new dimension for comparison with extant flowers (Crane *et al.*, 1989, 1995, 2004; Crepet, 1996; Crepet and Nixon, 1998; Crepet *et al.*, 2004; Friis and Endress, 1990; Friis *et al.*, 1986, 2005; Schönenberger, 2005; Schönenberger *et al.*, 2003). Fossils not only document structures of extinct plants and their diversity but also the time in which they lived, and thus they are of twofold significance for the reconstruction of phylogeny and evolution (Doyle, 2001). (2) Molecular phylogenetics (APG, 2003; Chase *et al.*, 1993; Graham and Olmstead, 2000; Mathews and Donoghue, 1999; Qiu *et al.*, 1999; Savolainen and Chase, 2003; Soltis *et al.*, 1999, 2000, 2005b), and molecular developmental genetics (Coen and Meyerowitz, 1991; Irish, 1999; Irish and Kramer, 1998; Smyth, 2005; Zik and Irish, 2003) arose as completely new fields. Molecular developmental genetics of flowers is entering a vast new open space and at present leads to more new questions than it answers old ones. (3) In addition, comparative morphological studies have been made on flowers of larger groups of angiosperms, for example, on the petals of asterids (Erbar, 1991; Erbar and Leins, 1996), the androecium of basal angiosperms (Endress and Hufford, 1989), the gynoecium through all basal angiosperm families (Endress and Igersheim, 2000a), and the flowers of legumes (Tucker, 1992, 2003a; Tucker and Douglas, 1994). The increasing resolution of phylogenetic relationships within angiosperms enables reconstructions of flower evolution (Doyle and Endress, 2000; Hufford and McMahon, 2003, 2004; Ronse De Craene *et al.*, 2003; Schönenberger and Conti, 2003; Schönenberger *et al.*, 2005; Zanis *et al.*, 2003). Detailed comparative developmental studies should also help to link the temporal sequence and relative duration of developmental steps between different plants (Bowman, 1994; Buzgo *et al.*, 2004b; Hufford, 2001a; Lord *et al.*, 1994; Vincent and Coen, 2004).

Molecular developmental genetics works with a few model species, such as *Arabidopsis thaliana*, *Antirrhinum majus*, *Oryza sativa*, and only a few others. How can these developmental aspects be connected with evolutionary aspects? There are two possibilities for approaches: (1) from the side of diversity in flowers (Cronk *et al.*, 2002; Endress, 1992; Rudall and Bateman, 2003, 2004) or (2) from an intermediate level between the entire diversity and single model species, with new semimodel species to be added to the model species (Albert *et al.*, 2005; Baum *et al.*, 2002; Becker and Theissen, 2003;

TABLE I
List of Plant Specimens and Collection Numbers Used for Original Illustrations

Annona cherimolia Mill. (Annonaceae), P. K. Endress 1375
Campanula alliariifolia Willd. (Campanulaceae), P. K. Endress 4793
Dillenia alata (DC.) Martelli (Dilleniaceae), P. K. Endress 9057
Echium vulgare L. (Boraginaceae), P. K. Endress 98–32
Geranium robertianum L. (Geraniaceae), P. K. Endress 7224
Gyrostemon brevipes Hook. ex Moq. (Gyrostemonaceae), U. Hofmann 1274
Kitaibelia vitifolia Willd. (Malvaceae), P. K. Endress 5336
Lobelia inflata L. (Campanulaceae), P. K. Endress 4824
Munroidendron racemosum (C. N. Forbes) Sherff (Araliaceae), P. K. Endress 96–10
Scaevola taccada (Gaertn.) Roxb. (Goodeniaceae), P. K. Endress 5031
Sinningia cardinalis (Lehm.) H. E. Moore (Gesneriaceae), P. K. Endress 10000
Sollya heterophylla Lindl. (Pittosporaceae), P. K. Endress 6713
Vincetoxicum nigrum Moench (Apocynaceae), P. K. Endress 4690

Cronk, 2001; Irish and Benfey, 2004; Kramer and Hall, 2005; Kramer *et al.*, 2003; Litt and Irish, 2003; Soltis *et al.*, 2002, 2005a).

In this chapter, I will concentrate on the first approach. A focus on the diversity of flowers in angiosperms can show to what extent traits are constant or variable, and what extreme features may be expected or are unlikely to evolve. Some developmental mechanisms involved in evolutionary diversification of flowers have as yet not been addressed by molecular evolutionary geneticists but would be of evolutionary interest (e.g., floral-subtending bracts, floral prophylls, floral phyllotaxis). For material used, see Table I.

II. DEVELOPMENT: FROM FLORAL PRIMORDIUM TO FLOWER—CONTINGENCY OF THE PRIMARY MORPHOLOGICAL SURFACE

Flowers develop from a floral apex. The floral organs are initiated on the flanks of the floral apex in centripetal sequence, that is, first the outermost organs, and last the innermost organs. Flower development takes place by a gradual transformation of the surface of the floral apex. The original surface of the floral apex forms the primary morphological surface of the flower throughout its development up to maturity. In a diagrammatic median longitudinal section of a flower (Fig. 1) it can be seen how the originally simple surface transforms into a complicated convoluted structure. No two points on the line from X over Y to Z change their topology, but the line becomes extremely complicated. It is especially complicated in the gynoecium. Here it also forms the inner surface including the stylar canal, locular wall, and the ovules with

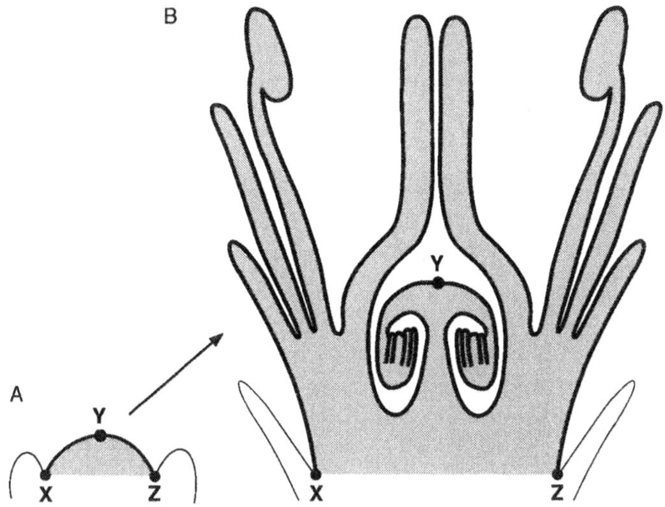

Fig. 1. Schematic LS of floral primordium and mature flower, showing contingency of primary morphological surface (thick line). The topology of the points on the line between X and Z does not change, however convoluted the line becomes. The center of the floral primordium (point Y) becomes enclosed in the ovary during flower development. The shaded region forming the floral meristem in the floral primordium will constitute the tissues of the mature flower.

nucellus and integuments. The center of the original floral apex becomes enclosed in the gynoecium (point Y in Fig. 1). Furthermore, Fig. 1 shows only a two-dimensional segment of the entire flower. The entire, three-dimensional structure is still much more complicated. In diagrammatic representations of flowers or gynoecia, this continuous surface is often not taken into account. For instance, ovary locules are confined by the primary morphological surface of the flower, and thus are always connected with the outer part of the morphological surface; in contrast, anther locules are confined by a secondary surface, which originates by tissue separation: the sporogenous tissue separates from the sterile part of the anther during or after meiosis, which leads to the formation of the locule, and thus there is no continuation with the outer surface. Thus this secondary surface of the anther locules is a mere histological surface, and not a morphological surface.

It should also be emphasized that this gradual transformation of the morphological surface is not smoothly continuous but occurs in pulses. Each time a new organ or organ part is formed, the amount of meristem in the system increases and then decreases again (Endress, 1972, 1994a). And each time a new organ or organ part is formed, the floral surface becomes more complex. A mechanism for evolutionary diversification of flower form is heterochrony during the development of organs and organ parts (Hufford,

2001a; Lord, 1991; Rudall and Bateman, 2002, 2004). Sequence models show patterns in the sequence of organ initiation and are especially interesting to establish homologies in the development of flowers with a complex and unusual disposition of organs (Hufford, 1997, 2003).

III. EVOLUTION: FROM SIMPLE TO COMPLEX FLOWERS—CONTINGENCY OF THE DEVELOPMENTAL PROGRAM

Just as one can follow the development of an individual flower from simple early stages to the complex mature form, one can trace the evolution of angiosperm flowers from simple ancestral forms to highly complicated forms of derived clades (Endress, 1994a). Successive evolutionary innovations in time gradually led to an increasing complexity of floral structure. Some innovations were highly successful and led to large radiations and thus resulted in large clades (Endress, 2001b; Ree, 2005). Other innovations were more "local" (with regard to the phylogenetic tree) and radiations were more limited and not the basis for large clades; they may be important for single genera or families. It may be practical to use the terms "key innovations" for the former and "non-key innovations" for the latter.

Key innovations are, for instance, a double perianth with sepals and petals (for eudicots), syncarpy with a compitum (see later section) (for most core eudicots and most monocots), tenuinucellate ovules (for asterids), sympetaly (for asterids), fusion of stamens with petals (for euasterids) (Endress, 2001b). Key innovations may evolve slowly and be labile at first (e.g., sympetaly in Ericales, Schönenberger *et al.*, 2005).

Non-key innovations are, for instance, a hyperstigma (see later section) (which is known only from a few genera of Monimiaceae) (Endress, 1980a), and synsepaly (in contrast to sympetaly!), which evolved many times in many genera of core eudicots, but often characterizes not even entire genera, but only subclades within genera (Endress and Matthews, in progress). Such non-key innovations are "dead ends" in the sense that although they are a novelty and have moderate evolutionary success, they do not lead to large radiations, as their potential is limited for some reason.

Some features may be key innovations in some but not in all clades in which they occur. Floral spurs represent a key innovation in *Aquilegia* (Ranunculaceae) (Hodges, 1997) but not in *Halenia* (Gentianaceae) (Kadereit and von Hagen, 2003). Floral monosymmetry was a key innovation in orchids, Lamiales, and Faboideae (Leguminosae). However, it is not a key innovation in other groups, in which it occurs again and again (e.g., many families and orders of rosids, other than Leguminosae) but did not lead to species-rich clades (see later section).

Evolution toward secondary simplification is also common. Extreme examples are present in some water plants with extreme reduction of the entire plant body, including the flowers. *Wolffia* (Araceae, monocots) has unisexual flowers consisting of a single stamen or a single carpel. *Callitriche* (Veronicaceae, core eudicots) has unisexual flowers consisting of a single stamen or two united carpels.

One of the ambitious goals in evo-devo studies of flowers will be to compare developmental programs of flowers of taxa at different levels of the phylogenetic tree and to reconstruct the evolutionary changes of these programs. Changes in the function of specific genes were shown, for example, in the development of inflorescences in different genera of grasses (Malcomber and Kellogg, 2004). It will be a big task to expand such studies to angiosperms in general and to perceive contingency in evolution, just as in individual development.

IV. FLORAL ORGANS AS MODULES, ORGAN CATEGORIES (ORGAN IDENTITY), AND THEIR EVOLUTION

A. FLORAL ORGANS AS MODULES

When working with model plants, it is not problematical to view flowers as being composed of a given, fixed number of floral organs and organ categories, and the organs always in the same position in the ensemble of the flower. Flowers of *Arabidopsis* and *Antirrhinum*, both eudicots, have 16 and 17 organs, respectively, and both have four organ categories, sepals, petals, stamens, and carpels. Ovules are an additional category. However, the situation is different if all angiosperms are taken into consideration. There is a wide range of organ numbers: between 1 and 10,000 in a flower. This entire range for the angiosperms is also represented in the single family Ranunculaceae (see Tamura, 1995). In some larger clades the number is not fixed at the genus or species level (Endress, 1990).

And even the number of organ categories becomes difficult to handle if all angiosperms are considered collectively. Stamens and carpels are present in all angiosperms. However, it appears that in basal angiosperms and monocots perianth organs do not correspond to sepals and petals in eudicots in every respect. Therefore, it makes more sense to speak of tepals and to use the modifiers sepaloid tepals and petaloid tepals (see also Warner *et al.*,

2005). Molecular pathways of evolutionary diversification of the perianth have so far been best studied in Ranunculales (Kramer *et al.*, 2003), which are sister to all other eudicots and show a great diversity of perianth conformations (Hiepko, 1965; Tamura, 1995). Another pathway of perianth evolution is present in monocots (Kanno *et al.*, 2003). Basal angiosperms are just beginning to be explored (Kim *et al.*, 2005a).

Floral organs can be seen as modules because commonly there are several of the same kind in a flower. Modules are essential units in development and evolution (Schlosser, 2004; Wagner, 1996; Wagner and Mezey, 2004). The evolutionary playing with module number is especially prominent in flowers. Whereas floral organs are the most obvious modules in flowers, there are also modules at a higher level of organization, such as fusion products of petals and stamens, or flowers can be viewed as modules of an inflorescence (see later section).

B. SEQUENCE OF EVOLUTIONARY ORIGIN OF FLORAL ORGANS

The organs of extant flowers did not all originate at the same time. Some of them have old precursors. Some organs to which we would apply the terms we use for the floral organs of extant angiosperms most likely emerged at different times in the evolutionary history of spermatophytes.

"Sporophylls" of angiosperms (stamens and carpels) today behave like phyllomes (leaf-like structures) but are not necessarily evolutionarily derived from leaves. Although in loss-of-function mutants of *Arabidopsis* all floral organs, including stamens and carpels, appear like unelaborate, simple, small leaves (phyllomes) (Meyerowitz, 1994), one must take into account that early plants may have had sporangia before there were leaves. Several very different possibilities for the evolution of sporophylls could be considered: (1) the sporangia became associated with leaves from the beginning of leaf evolution. (2) The sporangia (synangia) became secondarily transformed into organs with a dorsiventral organization, that is, leaf-like structures (for dorsiventrality see, e.g., Bowman *et al.*, 2002). (3) The sporangia became synorganized with leaves that subtended them or were otherwise topographically associated. (4) The sporangia, originally "autonomous," became combined with leaves by ectopic gene expression. (5) Leaves may be derived from sporangia by sterilization (Crane and Kenrick, 1997). If the evolutionary pathway was by combination of different organs, such as in (3) or (4), the "combined" nature is today no longer obvious. Seemingly detached structures in extant angiosperms are more likely secondary deviations from the basic pattern at the level of angiosperms, and not "atavistic" phenomena. Thus, the interpretation of "axis"-borne ovules in certain angiosperms by some earlier authors (Neumayer, 1924; Sattler, 1974; Sattler and Lacroix, 1988) would not be sensible interpretations at the level of the angiosperms (see also earlier discussions by Eckardt, 1955; Rohweder, 1963, which are still of relevance). Sattler (1974) artificially delimited carpels from their surrounding area by flat planes without considering that carpels originate not only from the surface but also from within the floral apex (the primordium encompassing several cell layers in depth). Thus the carpel is not "glued onto" the morphological surface of a flower, but rather, it is "rooted" at some depth within the primary meristem of the flower. From a developmental point of view, the carpel encompasses the entire region derived from the area that was involved in increased meristematic activity at carpel inception.

Although stamens and carpels are often said to be homologous structures, both being sporophylls, they are basically different in some respects: an ovule can also be compared with a stamen in some way, and the carpel is then more complex. An ovule is basically a megasporangium with commonly four megaspores. A stamen contains two or four microsporangia with numerous microspores.

What was the temporal sequence of evolution of the floral organ categories? It appears that the sequence of evolution is more or less the reverse of the sequence of floral development: the sequence of development is: sepals – petals – stamens – carpels – ovules. However, the probable sequence of evolution is: ovules – stamens – carpels – tepals – sepals and petals (Endress, 2001b). Thus, in the center of a flower are the evolutionarily oldest organs, the ovules, and at the periphery the evolutionarily youngest, the perianth organs (see later section). However, this question of the evolutionary sequence is perhaps not easy to answer because the definition of these structures is somewhat arbitrary. Each of these structures had its own evolutionary history. To define an evolutionary starting point for a structure depends on the wide or narrow definition of this structure, for example, the definition of a stamen may be restricted to organs that have an anther with two lateral thecae with two pollen sacs each (narrow definition), or it may also include organs that have only two lateral pollen sacs (wide definition).

C. OVULES

Ovules are perhaps the evolutionarily oldest morphological organs of flowers. They evolved in the early spermatophytes. In ancestral spermatophytes, the evolutionary precursors of the ovules, the "preovules" (going back to the Upper Devonian, 360 My, Rothwell *et al.*, 1989), basically consisted of a nucellus (megasporangium), in which female meiosis and syngamy took place, and which eventually developed into a seed. The ovules of all extant spermatophytes have an integument, which surrounds the nucellus at its flanks and forms a micropyle on top of the nucellus. In those spermatophytes in which the male gametophyte forms a pollen tube, the integument also functions in guiding the pollen tube through the micropyle to the nucellus tip, from where it can then reach the nucellus containing the female gametophyte with the egg cell. Angiosperm ovules basically have two integuments and are probably anatropous, although several basal angiosperms have orthotropous ovules (Doyle and Endress, 2000; Endress, 1994b; Endress and Igersheim, 2000a). The origin and evolution of the integuments is unknown in detail. Although there are hypotheses on their origin, this is not a topic for this chapter (because the origin of at least the first integument is preangiospermous). However, it can be assumed that in an early phase of ovule evolution in spermatophytes there was one integument, and the second (the outer) integument evolved in the ancestors of the angiosperms (and, in parallel, in a few other spermatophytes). An often discussed hypothesis for the outer integument is a derivation from the wall of a cupule, as it was present, for example, in the Mesozoic Caytoniales (Doyle, 1978, 1994; Gaussen, 1946). Thus the two integuments in angiosperms are probably of different evolutionary age and are not necessarily homologous. However, the inner integument of angiosperm ovules is probably homologous with the single integument in other spermatophytes. Dorsiventral gene expression in the outer integument was interpreted as potentially reflecting a leaf-derived origin (Meister *et al.*, 2002, Yamada *et al.*, 2003), which is, however, questioned by Sieber *et al.* (2004). Developmentally, the inner integument commonly appears slightly before the outer, more rarely both appear more or less synchronously.

D. CARPELS

Although ovules evidently preceded carpels in evolution, in modern angiosperms, ovules appear as an integral part of the carpel. The carpel contains and encloses one or several ovules (see later section). Closed carpels are unknown from fossils older than Early Cretaceous (120–130 My old) (Crane *et al.*, 1995; Friis *et al.*, 2003a). But how did ovules become combined with another structure, both of which together became a carpel? There are two main possibilities of interpretation: (1) ovules were associated with a phyllome from early on, and this association became seemingly lost in some nonangiospermous spermatophytes. A mode of derivation of a carpel from a "megasporophyll" of Caytoniales was discussed by Doyle (1978). (2) Ovules were originally not associated with a phyllome, but each ovule represented a short shoot. It became associated with either its subtending phyllome or a phyllome on the short shoot itself. The angiosperm carpel originated by synorganization of these two elements, ovule and phyllome. In the "mostly male theory of flower evolutionary origins" (Frohlich, 2001, 2003; Frohlich and Parker, 2000), it is hypothesized that the carpel originated by ectopic expression of ovules in microsporophylls (see also Meyen, 1988). Another model was proposed by Albert *et al.* (2002). There is no lead from molecular developmental genetics as to the origin of the carpel (Bowman *et al.*, 1999; Skinner *et al.*, 2004). Earlier interpretations of carpel morphology were addressed by Friis and Endress (1990).

The role of *CRABS CLAW* and *TOUSLED* in carpel development is the same in *Arabidopsis* and the basalmost extant angiosperms (*Amborella, Cabomba*), which indicates a conservative pattern of the developmental program, at least for the part in which these genes are involved (Fourquin *et al.*, 2005). This is concordant with the picture presented by a comparison of a large number of developmental studies in many angiosperm taxa by many different authors over the past decades. Carpels have a shared basic pattern of development in angiosperms but show conspicuous diversity in detail.

E. STAMENS

Stamen-like structures are present in most extant spermatophytes. However, the homology of angiosperm stamens with any such structure of nonangiospermous spermatophytes is uncertain, especially because the diversity of forms of nonangiospermous "stamens" is greater than that of ovules. In all extant spermatophytes they appear to be dorsiventral organs (for dorsiventral organs in general, see Bowman *et al.*, 2002). In the seemingly nondorsiventral male organs of *Taxus*, developmental studies suggest that they are highly reduced short shoots with small dorsiventral units (Mundry and Mundry, 2001). Whether the male organs were also dorsiventral in all extinct spermatophytes is doubtful (for surveys, see Crane, 1985; Taylor, 1988). The "stamens" of *Ginkgo*, *Gnetum*, and some conifers are the most similar organs to angiosperm stamens. In what respect are they homologous? The "stamens" of *Ginkgo*, *Gnetum*, and conifers are more simple than those in angiosperms. They have two pollen sacs (microsporangia), one on each side of the "stamen." Early fossil records of such disporangiate "stamens" are in Voltziales (Permian or Triassic, 220–250 My ago, Stewart and Rothwell, 1993). In contrast, angiosperms have two pollen sacs on each side of the stamen, which are synorganized into a theca. A theca has a single longitudinal opening slit for both sporangia, situated between the two sporangia. The most parsimonious scenario of how the angiosperm stamen could be derived from an only disporangiate nonangiosperm stamen is by subdivision of the originally one microsporangium on each side of the stamen into two, which were then organized into a theca. Such a secondary compartmentalization leading to multiplication of sporangia number would not be unique, as it has also occurred within angiosperms several times (Endress, 1994a; Endress and Stumpf, 1990). If this hypothesis is correct, the thecae of angiosperms would not really be synangia, as they are not evolutionarily derived by fusion, but just by the opposite process, by compartmentalization. Developmental studies in the multisporangiate "stamens" of cycads revealed subunits of double sporangia, which are arranged in a larger number on each side of the "stamen" (Mundry and Stützel, 2003). It is not unlikely that developmental studies in other gymnosperms will bring additional new results, which may suggest that thecae homologues are present in some extant gymnosperms. The male structures of potential ancestors of angiosperms, Caytoniales, Corystospermales, and Glossopteridales are unfortunately not known in detail (Doyle, 1978, 1994; Frohlich, 2003). Those of *Caytonanthus* (Triassic) had groups of three or four pollen sacs, which were not interpreted as dorsiventral; however, several such groups together were interpreted as forming a microsporophyll (Harris, 1941).

F. PERIANTH ORGANS

Sepals and petals are more difficult to define than stamens and carpels, which are well delimited by their male and female function and specific structures associated with these functions. Petals, although they have primarily the function of optical attraction of pollinators, may attain other functions as well. Petals commonly form the second whorl or series of floral organs in a flower. However, they may form the outer series in a few clades, in which the flowers have lost the sepals, such as probably in some Santalales; there the petals have taken over the protective function from the sepals in the basal Olacaceae, and subsequently the sepals were lost in a large part of Santalales (Endress, 1994a). Thus there is no fixed set of constant structures and functions in petals across the angiosperms or eudicots, and these organs have to be evaluated each time in the local phylogenetic surroundings of a given clade.

When working solely with model plants, this problem of sepal and petal definition is not encountered, because in the narrow framework of *Arabidopsis* and *Antirrhinum* sepals and petals are well delimited, with their characteristic position and morphological and histological differentiation. Thus sepals and petals of mutants can easily be compared with those of the wild types.

A perianth is commonly present in almost all extant angiosperms. However, it is not clear whether it has evolved at the base of the crown group or farther back in the stem lineage. Some potentially basal extant angiosperms, such as Chloranthaceae, Ceratophyllaceae, Saururaceae, and Piperaceae, do not have a perianth, and it is also lacking in the disputed Early Cretaceous angiosperm fossil *Archaefructus* (Sun *et al.*, 2002). A number of modern water plants with submerged flowers have lost their perianth. As *Archaefructus* appears to have been a water plant, it has been disputed whether the lack of its perianth was an ancestral feature or a secondary consequence of its submerged habit (Friis *et al.*, 2003a).

The presence of a perianth that is differentiated into typical sepals and petals is characteristic for eudicots; thus these organs may be ca. 150-130 My old (Wikström *et al.*, 2001) (oldest eudicot fossil reported 118 My, Magallón *et al.*, 1999). Sepal- and petal-like organs in basal angiosperms and monocots do not completely correspond to the perianth organs of eudicots. This differentiation appears to have evolved only in basal eudicots; it is especially typical for core eudicots (Lamb and Irish, 2003). Beginnings of a conspicuous differentiation can be traced in Ranunculales (see Hiepko, 1965; Kramer *et al.*, 2003). In contrast, in other basal eudicots, the perianth is only weakly differentiated, in some even lacking (*Styloceras*, Buxaceae; *Didymeles*, Didymelaceae; *Trochodendron*, Trochodendraceae) (Endress, 1986a; von Balthazar and Endress, 2002a,b; von Balthazar *et al.*, 2003). In other basal eudicots, perianth parts are well differentiated but only of one kind, which probably corresponds to a calyx (see also, Douglas and Tucker, 1996; Drinnan *et al.*, 1994; Hoot *et al.*, 1999). The structure of the perianth in *Platanus* is still enigmatic (Hoot *et al.*, 1999). And even in some basal core eudicots, the perianth is lacking or not differentiated into sepals and petals, such as in Gunnerales (Jäger-Zürn, 1967; Rutishauser *et al.*, 2004) and some Saxifragales (Endress, 1970, 1978).

Based on molecular developmental genetics, scenarios have been devised in which petals predated sepals (Albert *et al.*, 1998) or sepals predated petals (Baum, 1998; Baum and Whitlock, 1999). Studies of ABC gene expression in tepals of basal angiosperms revealed a "petal-like" pattern (Kim *et al.*, 2005a), which may support the first scenario. From morphological and functional aspects of extant angiosperms, the second scenario seems more likely. The outermost organs of eudicot flowers are sepal-like (robust, with a broad base, and often with three vascular traces) irrespective of whether there are additional perianth organs. The constitution of typical petals is functionally not suitable for being the outermost organs (they have a light construction, a narrow base, and a single vascular trace, and are retarded in development in late bud). Thus, a later appearance of petals than sepals in evolution seems more likely because: (1) the perianth organs (tepals) in basal angiosperms have more features in common with the sepals in eudicots (broad base, commonly three vascular traces at the base, rarely conspicuous retardation in bud) (Hiepko, 1965) and (2) in basal eudicots (Ranunculales) the petal homologues are often small, and are nectariferous organs, and optical attraction is often exerted by the sepals (Erbar *et al.*, 1998; Hiepko, 1965; Kosuge, 1994). Phylogenetic and molecular developmental studies indicate that petals evolved several times in basal eudicots and even in Ranunculales (Drinnan *et al.*, 1994; Kramer and Irish, 1999, 2000; Kramer *et al.*, 2003; Zanis *et al.*, 2003). However, petal evolution may have started earlier than in eudicots, as structural and molecular developmental genetic aspects in *Saruma* (Aristolochiaceae) indicate (Jaramillo and Kramer, 2004; Leins and Erbar, 1995) (and see above, Kim *et al.*, 2005a).

Nevertheless, one has to consider that in core eudicots in a number of instances petals have taken over protective functions and have become more massive, with a broader base (Matthews and Endress, 2002, 2004, 2005a,b). Also the opposite process, loss of petals and differentiation of optically attractive sepals has to be taken into account, for example, in Myrtales (Schönenberger and Conti, 2003). This switch-over and its genetic basis, which makes the definition of petals especially difficult, should be further explored.

The history of the petal is an apparent example of parallel evolution. The question arises whether such organs that have arisen separately can be treated as homologous and should be given the same name (see also Yoon and Baum, 2004). This general problem cannot be further followed here but needs to be discussed in evolutionary biology on a broader basis.

G. BRACTS AND BRACTEOLES

Bracts and bracteoles are not floral organs, but organs subtending or preceding the flower. They are involved in early floral bud protection (Endress, 1994a). Often the term bract is used for the organ subtending a flower (the subtending bract, also called pherophyll), and bracteoles for the organs that precede a flower at the floral axis (also called prophylls, in eudicots mostly two). But in many cases the usage of the terms is not precise. For clarity I prefer to use subtending bract and prophylls, respectively, for these phyllomes. One should distinguish between "subtending" and "preceding" organs. Preceding organs (one or more) are situated on the floral axis, whereas the subtending organ (only one) is situated on the axis of the next-lower order, that is, it is the organ in the axil of which the flower is situated. This distinction is based on the general rule of axillary concatenation of shoots of flowering plants and has been made in comparative morphology since the nineteenth century (Eichler, 1875, 1878). The unspecified term bract can be used for any scale-like vegetative phyllome and does not convey any positional information with regard to the branching system.

Subtending bracts and prophylls have not received much attention in developmental genetic studies because in *Arabidopsis* both organ types are lacking. That prophylls are lacking is not uncommon in flowers of eudicots. However, that subtending bracts are lacking is an idiosyncrasy of Brassicaceae. Thus, in this feature, *Arabidopsis* shows atypical behavior among eudicots. Prophylls are absent in *Antirrhinum* as well, but *inco* mutants produce prophylls (Masiero *et al.*, 2004). Absence of prophylls is not as deeply rooted in Plantaginaceae, the family to which *Antirrhinum* belongs (APG, 2003), as in Brassicaceae. This can be concluded not only from the existence of these *inco* mutants but also from the presence of a number of genera in Plantaginaceae that normally have prophylls. Upon closer inspection, subtending bracts are not completely lacking in Brassicaceae. In a few genera, the lowermost flowers of an inflorescence do have subtending bracts. Furthermore, rudiments of subtending bracts are present in youngest developmental stages of flowers also in seemingly bractless genera (Hagemann, 1963). In *Arabidopsis JAGGED* produces subtending bracts (Baum and Day, 2004; Dinneny *et al.*, 2004).

In extreme cases, plants have lost their perianth, and instead bracts (subtending bracts and prophylls) are protective organs during the entire floral development. For example, in Piperaceae (Piperales), floral-subtending bracts are the only protective organs. In *Eupomatia* (Magnoliales) a bract preceding the flower completely encloses the perianthless floral bud (Endress, 2003; Kim *et al.*, 2005b). In this case, the bract is neither a floral subtending bract nor a prophyll, because each flower is preceded by several phyllomes (foliage leaves or bracts).

V. FLORAL PHYLLOTAXIS

Each kind of floral organ occurs in various numbers in a flower, thus the organs behave as modules. These modules are formed in regular spatial patterns, called floral phyllotaxis. The two main patterns are spiral (Fig. 2A) and whorled (Fig. 2B). In spiral phyllotaxis, the organs are formed one by one, in equal time intervals (plastochrons) and equal angles in position with respect to the center of the flower (divergence angles). In whorled phyllotaxis, the organs appear in groups (whorls) and the organs within a whorl appear almost synchronously; in contrast, the whorls appear with a conspicuous plastochron (i.e., between the last formed organ of a whorl and the first formed organ of the subsequent whorl), and the organs within a whorl are in equidistant position, but the whorls alternate with each other.

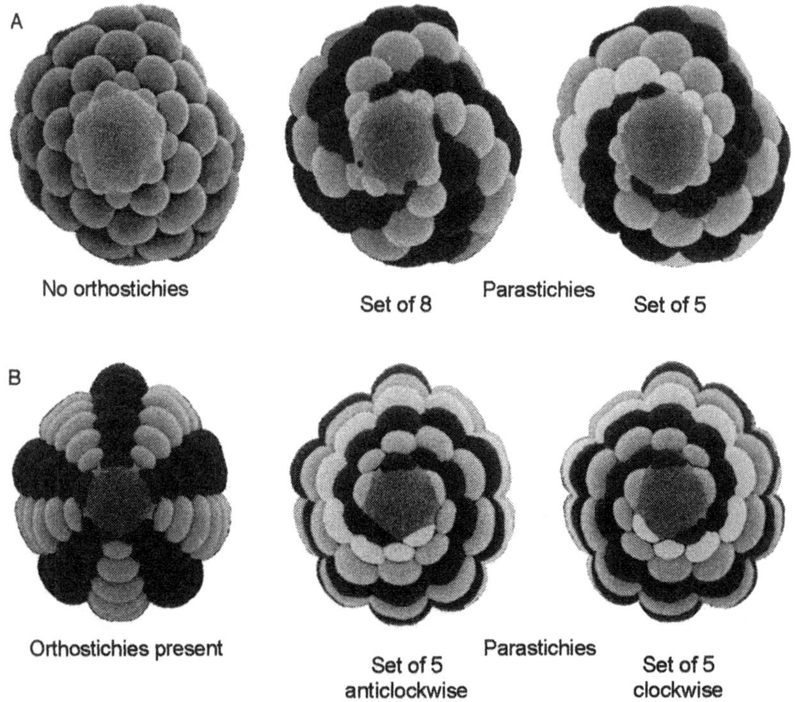

Fig. 2. Floral phyllotaxis patterns. (A) Spiral; (B) whorled. Orthostichies (radial straight lines) are only present in whorled flowers. Parastichies (spiral lines) are present in both spiral and whorled flowers. However, in whorled flowers there are always two mirror-image sets of the same number of parastichies (clockwise and anticlockwise), whereas in spiral flowers there is only one set of a given number. The illustrated spiral pattern is a "Fibonacci pattern," which is the most common spiral pattern in flowers. Parastichies and orthostichies are highlighted with different shades of gray. Modified from Endress (2004a).

Because in the most common model plants, *Arabidopsis* and *Antirrhinum*, and in cereals, the floral organs are always in whorls, molecular developmental geneticists as yet did not address the problem of how these different patterns in flowers originate. However, this problem of different phyllotaxis patterns becomes obvious when basal angiosperms are studied. In basal angiosperms, spiral and whorled phyllotaxis occur side by side at low-systematic levels (families, genera, or even species in some cases) and thus can switch relatively easily between each other (Endress, 1987a, 1990).

Endress (1987a) hypothesized that the switch between whorled and spiral floral phyllotaxis is easy in basal angiosperms because of lack of synorganization between the floral organs. Phylogenetic results that place a grade of *Amborella*, Nymphaeales, and Austrobaileyales at the base of flowering plants (Qiu *et al.*, 1999; Soltis *et al.*, 1999, 2000; Zanis *et al.*, 2003) and studies of character evolution (Doyle and Endress, 2000; Ronse De Craene *et al.*, 2003; Soltis *et al.*, 2005b) appear to confirm this, because both phyllotaxis patterns are represented in this grade. *Amborella* and Austrobaileyales have spiral flowers (Fig. 3A–F), whereas Nymphaeales have whorled flowers (except for the perianth in *Nuphar*) (Endress, 2001a, 2004b) (Fig. 3G–L). However, it has been emphasized long ago that, for example, Ranunculaceae, a family with a basal position in eudicots, also exhibits both phyllotaxis patterns (Schöffel, 1932).

How can these two patterns, spiral and whorled, be distinguished? In the literature one can find a sloppy use of these terms, and there are many mistakes of interpretation. Often young developmental stages are necessary for an interpretation, especially in flowers with a high number of organs. But also in flowers with a moderate or small number of organs, young stages of floral development show the phyllotaxis patterns more clearly than mature flowers. SEM micrographs of flowers with all organs just initiated are ideally suited to determine phyllotaxis patterns. However, this works only if the floral base is flat but not if it is pronouncedly concave or convex. In these cases, microtome section series have to be applied in addition (e.g., in Calycanthaceae, Staedler *et al.*, in progress).

The most obvious geometrical patterns are lines from the floral center to the periphery, formed by contiguous neighboring organs: spiral lines (parastichies), and radial, straight lines (orthostichies) (Fig. 2). A common mistake in the literature is that these obvious spiral lines are interpreted as indicative of spiral phyllotaxis. Spiral phyllotaxis, though, refers to the fact, that the successively initiated organs are along a spiral. This spiral, the *ontogenetic* spiral, is very flat and is not immediately obvious because the successive organs are not contiguous.

In fact, parastichies occur in both spiral and whorled patterns. However, orthostichies (straight radial lines) are present only in the whorled pattern. In whorled flowers there are always mirror image sets of parastichies, whereas this is not the case in spiral systems. For instance, in a whorled flower with 5-merous whorls, the most obvious sets of parastichies are sets of five, which are present in a clockwise and an anticlockwise version (Fig. 2B). In contrast, in a spiral flower, there is only one set of five parastichies. If this set is clockwise in a particular flower, there will be a set of three and a set of eight anticlockwise parastichies, but never a second set of five (Fig. 2A). Thus, to emphasize again, parastichies are simply a feature of all regular phyllotaxis patterns, and not restricted to spiral flowers.

In addition to spiral and whorled phyllotaxis patterns, there are also irregular patterns. Often, irregularity is connected with a high number of

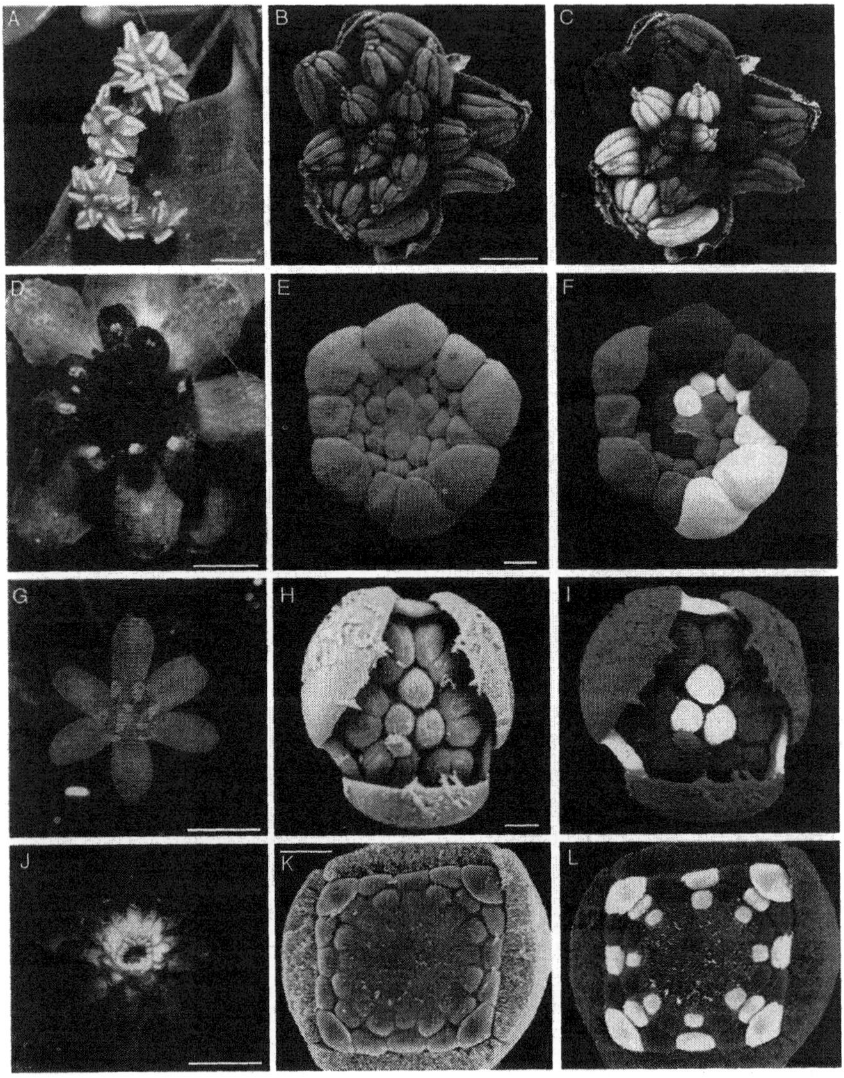

Fig. 3. Flowers of four representatives of the basalmost angiosperms (ANITA grade). Each plant depicted with a flower photograph and two SEMs, in the second SEM parastichies or orthostichies are indicated. (A–C) *Amborella trichopoda* Baill. (Amborellaceae); male flowers, spiral. (D–F) *Austrobaileya scandens* C.T. White (Austrobaileyaceae); bisexual flowers, spiral. (G–I) *Cabomba furcata* Schult. and Schult.f. (Cabombaceae); bisexual flowers, whorled. (J). *Euryale ferox* Salisb. (Nymphaeaceae); bisexual flower, whorled. (K–L) *Victoria cruziana* A.D. Orb. (Nymphaeaceae); bisexual flower, whorled. Orthostichies and parastichies are highlighted with different shades of gray. Scale bars: (A) 2 mm; (B, C) 1 mm; (D) 1 cm; (E, F, H, I, K, L) 0.2 mm; (G) 5 mm; (J) 5 cm. Modified from Endress (2004a).

organs, especially stamens (or also carpels if the gynoecium is apocarpous). In such flowers, the organ primordia are relatively small compared to the size of the entire flower at this stage. The smaller the organs are, the greater the probability that the original order decays by small local irregularities. An example is *Annona* (custard apple), which has an unordered arrangement of stamens and carpels (Fig. 4A). In the family Annonaceae, *Annona* is distinguished by its comparatively high number of stamens and carpels (Endress, 1990). Other genera of the family with lower organ numbers have a whorled phyllotaxis (*Monanthotaxis*; Ronse De Craene and Smets, 1990).

Another instance of irregular phyllotaxis is in flowers without a perianth. Apparently the formation of a perianth is a prerequisite for orderly phyllotaxis of stamens and carpels. If a perianth is absent, it appears to be more difficult to build up an order (Endress, 1978, 1989, 1990; Tucker, 1991).

It should be added that whorled floral phyllotaxis, which is so predominant in eudicots, is commonly initiated by spiral phyllotaxis in the calyx, with an equalization from the petals onwards. It should also be added that within each whorl, the developmental sequence of the organs is also often spiral (Leins and Erbar, 2004). However, this sequence is so rapid that the whorled position, with all organs of a whorl equidistant, is not affected. In monocots, which also have whorled floral phyllotaxis, the mechanism for the onset of whorls is less clear. Relatively few monocots with polysymmetric flowers have been studied in detail.

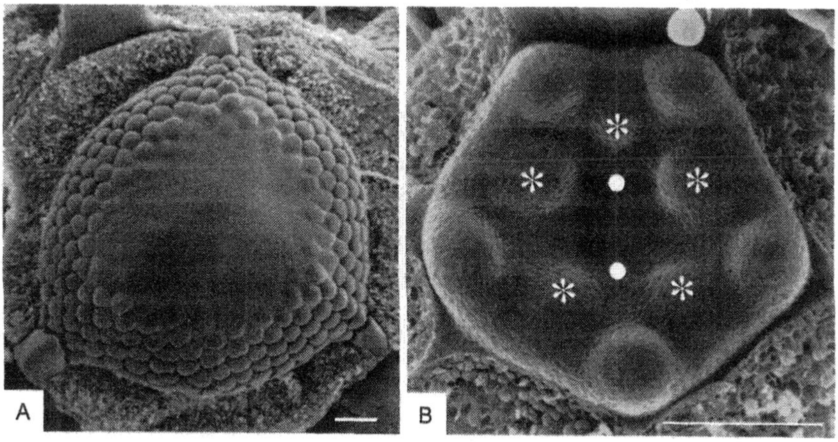

Fig. 4. Young flowers with different phyllotaxis patterns. (A) *Annona cherimolia* Mill. (Annonaceae); irregular, long time lapse between initiation of outermost stamens and carpels. (B) *Sinningia cardinalis* (Lehm.) H. E. Moore (Gesneriaceae); whorled, short time lapse between initiation of stamens and carpels. Asterisks: stamen primordia; dots: carpel primordia. Scale bars: 0.1 mm.

It has long been realized that primordia of leaves in vegetative shoots and floral organs in flowers are initiated in the "largest available space" ("Hofmeister's rule") (Hofmeister, 1868; Kirchoff, 2003; Leins, 1963). As pattern-generating mechanisms for phyllotaxis in general, molecular interactions (Meinhardt, 2003; Reinhardt *et al.*, 2003) and biophysical processes have been discussed (Green, 1999).

Floral phyllotaxis has not been addressed as yet by floral developmental genetics. Questions of interest are: What are the boundary conditions for the generation of a specific phyllotactic pattern (Rutishauser, 1998)? What is due to self-organization and what is due to genetic interaction, once a phyllotactic pattern has been initiated and is being maintained during development in a flower? Thus, what is the interdependence of genetic and epigenetic development (Müller and Olsson, 2003)?

If a flower of the basal angiosperm *Annona* (Fig. 4A) with high organ number and irregular phyllotaxis is compared with a flower of *Sinningia* (core eudicots) (Fig. 4B) with low-organ number and high regularity of phyllotaxis, the question arises: Is the positioning of the floral organs based on a "feed forward" mechanism as in vegetative phyllotaxis? Or is it completely predetermined (in a prepattern) almost at the beginning of floral development? The two contrasting examples of *Annona* and *Sinningia* suggest that both modes may occur in flowers. Perhaps the pattern is more predetermined in highly derived groups (core eudicots) than in more basal groups. But this problem needs critical developmental studies.

Another contrast between spiral and whorled flowers is the mode of the switch from one organ category to the next during development. In whorled flowers this switch takes place at the transition from one whorl to the next, resulting in a clear-cut, sudden change from one organ category to the next. In contrast, in spiral flowers the switch may be gradual in small steps encompassing several organs (e.g., *Austrobaileya*, Endress, 1980b; *Amborella*, Buzgo *et al.*, 2004a; Endress and Igersheim, 2000b; Posluszny and Tomlinson, 2003). Such spiral flowers led to the concept of sliding boundaries (Kramer *et al.*, 2003) or fading borders (Buzgo *et al.*, 2004a) for the action of floral identity genes (see also Lord, 1991).

VI. FLORAL SYMMETRY

Symmetry is a widely appreciated property of flowers. This is based in part on the regular phyllotaxis of the floral organs, but changes in symmetry are often superimposed on the patterns provided by phyllotaxis. Most flowers have either several planes of symmetry (polysymmetric, actinomorphic) or a single plane of symmetry (monosymmetric, zygomorphic) (Endress, 1999, 2001c). In developmental genetic publications sometimes the term asymmetric is used for monosymmetric. Really asymmetric flowers, i.e., flowers without a plane of symmetry, are relatively rare. Flowers with a spiral phyllotaxis are not completely polysymmetric but may be nearly so. In exceptional cases, a secondary monosymmetry may be superimposed on spiral flowers (*Aconitum*, *Delphinium*, Ranunculaceae) (Mair, 1977; Schöffel, 1932), this monosymmetry is likewise not perfect but nearly so.

Flowers of basal angiosperms are predominantly polysymmetric. The most conspicuous monosymmetric flowers in basal angiosperms are found in *Aristolochia* (Piperales). However, there the monosymmetry involves only the trimerous perianth whorl. The flowers of some other basal angiosperms are monosymmetric simply because of low organ numbers (Chloranthaceae, Piperaceae). Most monocots and eudicots have polysymmetric flowers. However, there are some large clades with elaborate monosymmetric flowers involving all organ whorls, in which monosymmetry was probably a major precondition for radiation (e.g., Orchidaceae, Leguminosae, Lamiales). A large monocot group with mostly monosymmetric flowers by reduction of organ number is the grass family (Kellogg, 2000; Rudall and Bateman, 2004). Among rosids other than Leguminosae, monosymmetric flowers have also evolved many times but without leading to large radiations, in practically all orders. Many monosymmetric flowers have an architecture with a lower and an upper lip. The lower lip is the landing platform for pollinating insects, and either the upper (in "lip flowers") or the lower lip (in "keel flowers") more or less conceals the pollinating organs (Endress, 1994a). Monosymmetry may provide higher precision in pollination than polysymmetry.

Molecular and associated comparative developmental symmetry studies were first centered around the monosymmetric flowers of *Antirrhinum* (Plantaginaceae, Lamiales) (Coen, 1996; Coen *et al.*, 1995; Luo *et al.*, 1996) but then expanded to other Lamiales and asterids (Citerne *et al.*, 2000; Cubas *et al.*, 1999; Donoghue *et al.*, 1998; Endress, 1998; Hileman and Baum, 2003; Hileman *et al.*, 2003) and other angiosperms (Citerne *et al.*, 2003; Cubas, 2004). In Lamiales, commonly the two upper petals form an upper lip and the three lower petals a lower lip (the reverse in lobelioids, Donoghue *et al.*, 1998). Of the five stamens, the median one (which is the upper one), is almost always reduced, either to a small staminode (Fig. 5A and B), or to a tiny rudiment (Fig. 5C and D), or is completely lacking (Fig. 5E and F). In some families this staminode is mostly or always present (Gesneriaceae, Bignoniaceae), in others, it is constantly lacking (Acanthaceae, Lamiaceae) (Endress, 1998, 1999). Thus there are different expressions of monosymmetry.

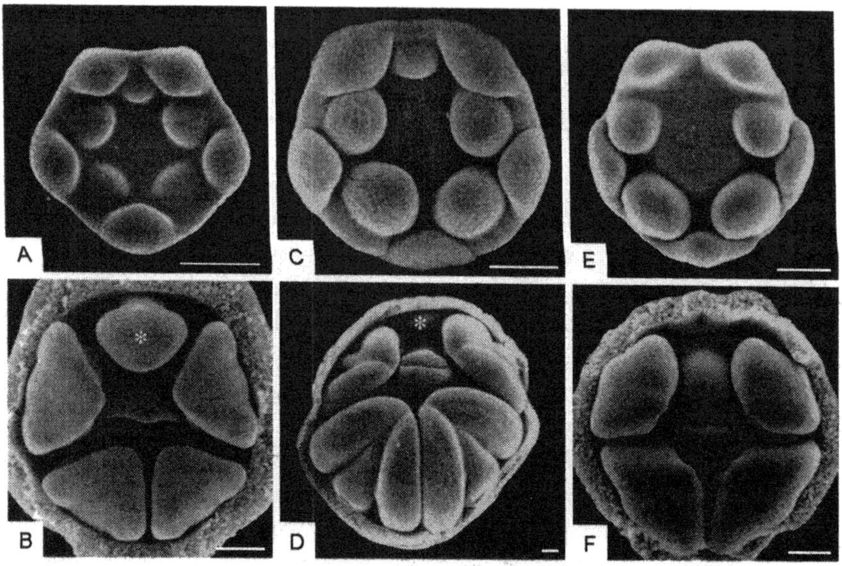

Fig. 5. Young flowers of different representatives of Lamiales, showing different expression of monosymmetry by increased reduction of upper stamen (staminode) (asterisk). Each species is represented by two developmental stages. (A, C, E) Floral organs just initiated; (B, D, F) somewhat older stages. (A, B) *Sinningia cardinalis* (Gesneriaceae). (C, D) *A. majus* L. (Plantaginaceae). (E, F) *Rehmannia angulata* Hemsl. (without family affiliation, Oxelman *et al.*, 2005), upper stamen completely absent. Scale bars: (A, B, E, F) 0.1 mm; (C, D) 0.2 mm. Modified from Endress (1998).

Several ("redundant") genes cooperate in forming such differential monosymmetry in Plantaginaceae (Hileman *et al.*, 2003). Monosymmetric flowers have given rise to less monosymmetric or nearly polysymmetric flowers many times in Lamiales (Smith *et al.*, 2004). This transition is especially prevalent in those clades of Lamiales that have retained the fifth stamen as a staminode (Baum, 1998; Endress, 1998; Smith *et al.*, 2004). However, it is difficult to calculate which of the two directions of evolutionary change was more frequent in asterids (Ree and Donoghue, 1999).

Commonly, in monosymmetric flowers, monosymmetry is less pronounced in early development than at maturity. However, in contrast, in some plants with polysymmetric flowers, early floral development is surprisingly monosymmetric by delayed development of the upper (adaxial) or lower (abaxial) portion of the flower. This may be due to an inhibitory influence of the subtending bract (mechanical or hormonal) or reflect the gradient of differentiation of the inflorescence axis from the upper to the lower end of the flower (Endress, 1999) and/or be due to the early and transient action of a monosymmetry gene (*TCP1*) (Cubas *et al.*, 2001). Symmetry may even change more than once in the development of a flower. In *Couroupita* (Lecythidaceae) development begins monosymmetric, then changes to more polysymmetric, and finally back to monosymmetric (Endress, 1994a).

VII. ANGIOSPERMY: CARPEL CLOSURE AND MODES OF SEALING

Angiospermy is the most prominent synapomorphy of angiosperms. In angiosperms, the ovules are borne on carpels, and during development the carpels become closed and the ovules hidden inside. In the process of closure the lateral carpel flanks become contiguous and begin to conceal the inner surface from the outer world (Fig. 6A and B). Carpel closure and ovule formation are not always synchronous. There is a range of different timings as seen throughout the angiosperms. In some groups the carpels close early and the ovules are only initiated in the closed carpels (e.g., in basal angiosperms) (Fig. 6A and B). In other groups the ovules begin to develop before the carpel is closed (e.g., some basal eudicots: Papaveraceae, Berberidaceae, Endress, 1995; some Caryophyllales: Amaranthaceae, Basellaceae, Plumbaginaceae, De Laet *et al.*, 1995; Hakki, 1971; Sattler and Lacroix, 1988; and some Fabaceae, Tucker and Kantz, 2001) (Fig. 6C).

Carpel closure is followed by sealing. Sealing occurs in different ways. Two extreme forms are sealing by secretion and sealing by postgenital

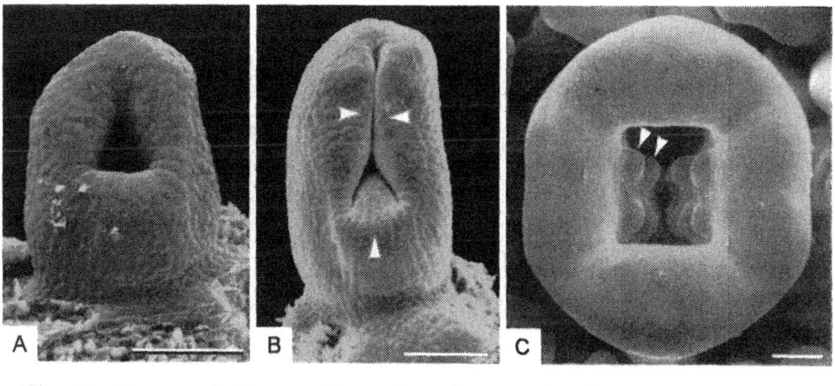

Fig. 6. Process of closure of carpels and gynoecia. (A, B) *Laurus nobilis* L. (Lauraceae); young carpels, from ventral side. (A) Carpel still open. (B) Carpel almost closed by approaching of the flanks and by a transverse ridge at the base (arrowheads). (C) *Hunnemannia fumariaefolia* Sweet (Papaveraceae); bicarpellate gynoecium still open at the time the ovules are initiated (arrowheads). Scale bars: 0.1 mm. (A, B) modified from Endress (1997a); (C) from Karrer (1991); courtesy of Andrea Fetz-Karrer.

fusion. In carpels that are sealed by secretion, the morphological surface at the site of sealing remains distinctive at anthesis, but this is not always the case in carpels sealed by postgenital fusion.

There are exceptional cases in which the carpels appear not to be sealed at anthesis. For instance, in *Tiarella* (Saxifragaceae) the carpel margins and flanks are conspicuously rolled inwards (involute) and the ovules are hidden from the surrounding world without carpel sealing; in *Reseda* (Resedaceae), the entrance into the inside of the gynoecium is covered by hairs but does not appear to be sealed. Such rare borderline cases deserve closer study with the question of how they can afford not to be sealed.

Endress and Igersheim (2000a) tentatively distinguished four types of angiospermy in their comparative study of carpel structure throughout the families of basal angiosperms: (1) sealing only by secretion (Fig. 7A), (2) mixed sealing by postgenital fusion at the periphery but with an inner canal of secretion, which reaches the outer surface, (3) mixed sealing by postgenital fusion at the entire periphery and an internal canal of secretion, which does not reach the periphery, and (4) sealing completely by postgenital

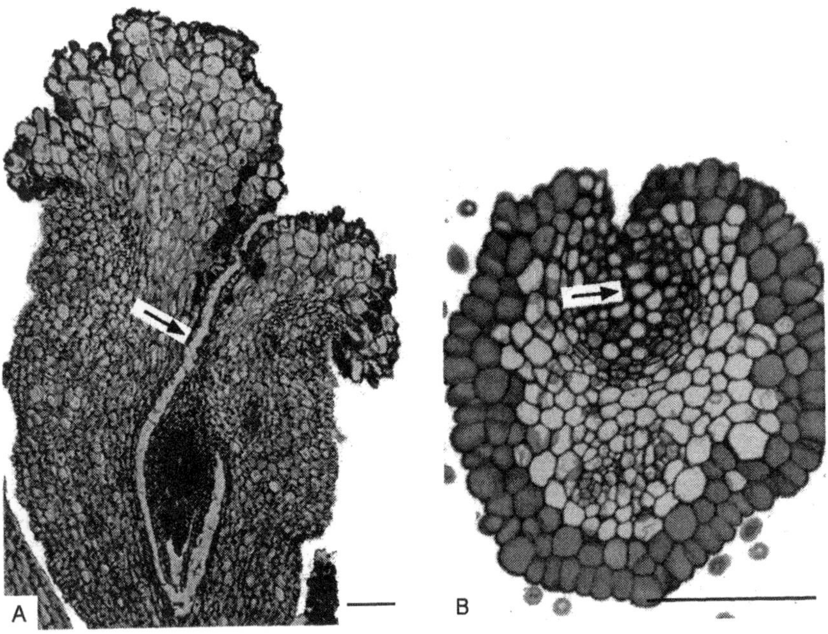

Fig. 7. Anthetic carpels with different sealing patterns of two basal angiosperms. (A) *Ascarina rubricaulis* Solms (Chloranthaceae); median LS. Sealing by secretion (arrow). (B) *Persea americana* Mill. (Lauraceae); TS of style. Sealing by postgenital fusion (arrow). Scale bars: 0.1 mm. Modified from Endress (2004a).

fusion without a canal in the center (Fig. 7B). In the basalmost angiosperms (ANITA grade, Qiu *et al.*, 1999), sealing is predominantly by secretion, thus angiospermy type 1, whereas in other basal angiosperms, postgenital fusion is more common. A functional aspect of sealing by secretion may be that basal relictual angiosperms and ancestral angiosperms are and were predominantly plants of moist habitats (Endress, 2004b; Feild *et al.*, 2004). Among the studied groups of rosids (core eudicots), types 3 and 4 are predominant: type 3 in Oxalidales and part of Crossosomatales (Matthews and Endress, 2002, 2005b), and type 4 in Celastrales, Cucurbitales, and part of Crossosomatales (Matthews and Endress, 2004, 2005a,b).

In carpels that are sealed only by secretion (type 1) there is a predominance of ascidiate (tubular) forms (Endress, 1980b, 1986b, 1987b, 2001a, 2005; Endress and Igersheim, 2000a). It is likely that this is the ancestral carpel form for angiosperms (Doyle and Endress, 2000; Endress and Igersheim, 2000a).

VIII. POSTGENITAL COHERENCE

Postgenital coherence (i.e., the coherence of parts that were free at the beginning of development) is an important mechanism in flowers (Boeke, 1973; Boeke and Vanvliet, 1979; Endress, 1997a; Raven and Weyers, 2001; van der Schoot *et al.*, 1995; Verbeke, 1992; Walker, 1975). It probably played its first role in angiospermy (see above) and was later also used for synorganization of different organs (see later section).

How do parts that are separate at first come in contact with each other, and how do their surfaces structurally interact with each other? Several genes have been found that play a role in the interaction of organ surfaces (Lolle and Pruitt, 1999; Lolle *et al.*, 1998; Pruitt *et al.*, 2000; Sinha, 2000), such as *fiddlehead* (Lolle *et al.*, 1992), *leunig* (Chen *et al.*, 2000, 2001), *aintegumenta* (Liu *et al.*, 2000), *hothead* (Krolikowski *et al.*, 2003), and *wax* (Chen *et al.*, 2003). These genes were mainly studied in teratological cases of organ coherence, although carpel sealing is so ubiquitous and fundamental in angiosperms, and postgenital coherence in other floral organs is also common. Details of how a part of an organ (a cell or a supracellular part) senses and reacts to contiguous parts of the same or other organs are poorly known. This appears to have been studied more in animals (Benjamin and Hillen, 2003) than in plants.

Postgenital fusion is an extreme form of postgenital coherence. However, the definition of postgenital fusion is somewhat arbitrary (Endress and Igersheim, 2000a). It is clear in its extreme expression, if the original morphological surfaces are no longer recognizable. However, it needs clarification if the epidermal cells are still distinct. One of the difficulties in characterizing postgenital fusion is that aspects at the structural (histological, cellular) and molecular level should be considered together. Examples of postgenital fusion are the carpel tips of Apocynaceae, as studied in detail in *Catharanthus* (this is fusion between carpels, and not within carpels) (Endress, 1994a; Verbeke, 1992; van der Schoot *et al.*, 1995; Walker, 1975). Here, interaction and coherence begin very early in development, and even plasmodesmata are secondarily formed across the two morphological surfaces (van der Schoot *et al.*, 1995). Early onset of coherence in development may be a general characteristic of postgenital fusion. In the process, epidermal cells dedifferentiate and periclinal divisions become possible (Walker, 1975). Interdigitation of epidermal cells is lacking in *Catharanthus* but is common in other cases.

In plants, postgenital coherence appears almost restricted to angiosperms. There are some isolated cases in nonangiosperms but without real fusion. For instance, the neighboring "sporophylls" of *Equisetum* show a transient coherence by interdigitation of the epidermal cells of the contiguous surfaces during development. At the time of spore dispersal the surfaces separate again (Endress, 1975). Whether it is present in reproductive organs of gymnosperms has not been worked out.

IX. SYNORGANIZATION

A. WHAT IS SYNORGANIZATION?

Based on the two issues of postgenital coherence and floral phyllotaxis, another important topic in flower evolution should now be addressed: synorganization. Organs become synorganized into new structures, "hyperorgans" (e.g., syncarpous gynoecium). The role of synorganization in floral structure and biology stands out in the most complex flowers, such as orchids (Rudall and Bateman, 2002; Vogel, 1959) and Apocynaceae (Endress, 1994a; Endress and Bruyns, 2000; Fallen, 1986; Schick, 1980, 1982). However, synorganization is also present in seemingly more simple flowers, such as, caesalpinioids and Amorpheae of Fabaceae (Endress, 1990, 1994a; McMahon and Hufford, 2002). By synorganization of organs new functions can originate, for example, with the advent of a gynostemium (congenital fusion of androecium and gynoecium) in orchids or a gynostegium (postgenital fusion of androecium and gynoecium) in Apocynaceae. Sometimes "early" evolutionary stages of synorganization that do not yet show the new function are still preserved today, such as in Apostasioideae of orchids, in which androecium and gynoecium are congenitally fused but a functional gynostemium with the production of a pollinarium is not yet present (Kocyan and Endress, 2001b).

Synorganization is a fundamental evolutionary process in flowers in general based on the presence of modules, that is, of several organs of the same kind, but it is not restricted to organs of the same kind. The advent of a new synorganization in flower evolution is sometimes a key innovation (Endress, 1997a, 2001b). Examples are syncarpy in core eudicots and monocots, sympetaly in asterids, and, in addition, fusion of petals and stamens in euasterids, fusion of stamens and carpels in orchids.

A precondition for synorganization of floral organs is whorled floral phyllotaxis, together with a moderate organ number per whorl, which is commonly not more than five. Figure 8 shows that it is much easier for whorled flowers than for spiral flowers to evolve complex synorganization of parts, because the organs of the same kind are positioned all at the same level and are equidistant from each other, and organs of different kinds are arranged in orthostichies, thus exactly behind each other. This appears why in the more highly derived angiosperms with more synorganization, spiral flowers are absent.

Synorganization of organs occurs by congenital fusion, by postgenital fusion or more loose coherence, or simply by local coordination of organs without coherence. The latter is the case, for example, with petals and stamens in Byttnerioideae-Malvaceae. Postgenital coherence is realized in different ways (sticking together, cuticular interdigitation, cellular interdigitation, postgenital fusion). Postgenital coherence is also most important in carpel sealing, that is, angiospermy, which is unique in angiosperms. Thus, these different phenomena of postgenital coherence are omnipresent in flowers, yet the developmental processes are poorly known and should be studied with molecular developmental and ultrastructural techniques (for the diversity of coherence and fusion, see later section).

Synorganization also means a transition in evolutionary emphasis from single organs to compound structures as units or modules. A sector of a syncarpous gynoecium still has the properties of a carpel. However, there are new features superimposed onto it by the congenital fusion, such as shared flanks (in gynoecia with septate ovaries) or a shared locule (in gynoecia with unilocular ovaries), a shared stigma or a compitum (Endress, 1990).

Concomitant with increased synorganization is a decrease in variability of floral organ numbers and floral phyllotaxis. Flowers with a low level of synorganization, predominantly in less derived angiosperms, may have spiral or whorled phyllotaxis, occasionally depending on the position of the flowers in the inflorescence (terminal vs lateral), for example, *Drimys* (Winteraceae,

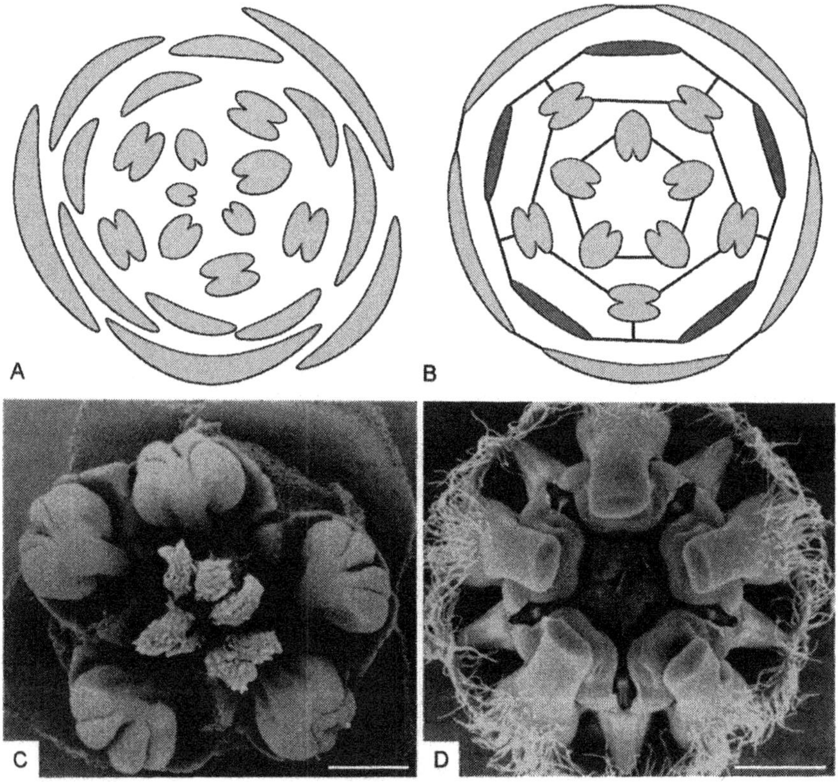

Fig. 8. Divergent potential for synorganization in flowers with spiral or whorled phyllotaxis. (A, B) Floral diagrams. (A) Flower with spiral phyllotaxis of a basal angiosperm, no potential for synorganization. (B) Flower with whorled phyllotaxis. Sites of commonly occurring synorganization indicated by lines connecting the organs (petals dark). (C, D) Examples. (C) *Hortonia angustifolia* Trimen (Monimiaceae); flower (perianth removed) with spiral phyllotaxis, without synorganization of organs. (D) *Ceropegia distincta* N. E. Br. (Apocynaceae-Asclepiadoideae); flower (perianth removed) with whorled phyllotaxis and intimate synorganization of organs by congenital and postgenital fusion. Scale bars: 0.5 mm. (C) From Endress (1992); (D) from Endress (1994a).

basal angiosperms) (Doust, 2001) and *Trochodendron* (Trochodendraceae, basal eudicots) (Endress, 1990), or different organ numbers, for example, *Distylium* (Hamamelidaceae, basal core eudicots), also depending on the position (Endress, 1970, 1978). Flowers of starved individuals (i.e., grown in nutrient-poor soil) as compared to normal individuals show different patterns of variation, depending on the degree of synorganization. Flowers with a low degree of synorganization tend to produce fewer organs in starved individuals (e.g., *Papaver*, basal eudicots) (Murbeck, 1912), whereas flowers with a high degree of synorganization tend to be smaller in size in starved individuals (e.g., *Mimulus*, asterids) (personal observation).

B. MEANS OF SYNORGANIZATION: CONGENITAL AND POSTGENITAL COHERENCE

Synorganization can take place by postgenital or congenital coherence of organs. In postgenital coherence, organs that are free at the beginning of development become secondarily coherent. In congenital coherence, there is coherence from the beginning of development, that is, morphological surfaces are not present at the site of congenital coherence. Primordia of organs are confluent so that the organs have a shared base from the beginning. Longitudinal growth is by intercalary elongation (zonal growth) of the shared basal part. Congenital fusion can also be seen as a failure of adjacent young organs to separate from each other, as for example, occurring in the presence of *FUSED FLORAL ORGANS* genes in *Arabidopsis* (Levin *et al.*, 1998).

Postgenital coherence can be realized at different structural levels: (1) at the infracellular level, either by secretion or by interdigitation of cuticular projections (e.g., sepals of Cephalotaceae); (2) at the cellular level by interdigitation of epidermal cells (e.g., tepals of Proteaceae) or by hairs (e.g., sepals of Campanulaceae); or (3) at the supracellular level by hooking together of organs (e.g., petals of Oxalidaceae and Connaraceae, Matthews and Endress, 2002).

C. SITES OF SYNORGANIZATION

As mentioned in the preceding section, synorganization is possible by congenital and postgenital coherence. I will focus here on postgenital coherence. Postgenital coherence is possible between members of all organ categories. Synorganization can take place among the organs within a whorl, which are all of the same kind, or it can take place between organs of different whorls, which are then mostly organs of different kinds. These two aspects of synorganization will be discussed here separately.

1. Synorganization among organs of the same whorl (organs of the same category)

a. Perianth. In the perianth, valvate aestivation is a precondition for postgenital coherence (Sigmond, 1929). Commonly the postgenital coherence is present in bud and disappears when the flower opens. Tepals have interdigitated epidermal cells in Proteaceae. Sepals have interdigitated epidermal cells in *Oenothera* (Onagraceae) (Chrometzka, 1967) and Vitaceae. In some Campanulaceae the sepals are reduplicative-valvate and their flanks are connected by long, curly hairs, like velcro (Fig. 9C and D). Cuticular cohesion is present in the sepals of *Cephalotus* (Cephalotaceae) (Matthews and Endress, 2002). Petals have interdigitated epidermal cells in Loranthaceae, Santalaceae, Apocynaceae-Asclepiadoideae, and a number of Asterales, such as Goodeniaceae, Menyanthaceae, Campanulaceae, and Asteraceae (Fig. 9A and B). Petals are hooked together in bud by local outgrowths of their flanks in Connaraceae and Oxalidaceae (Matthews and Endress, 2002). In complicated flowers, part of the postgenital coherence of the petals may remain at anthesis. It assists in secondary pollen presentation in *Phyteuma* (Campanulaceae) (Erbar and Leins, 1995) or in the formation of the pitcher trap flowers of *Ceropegia* (Apocynaceae-Asclepiadoideae) and, in addition, in the formation of a scent-emitting antenna in some *Ceropegia* species (Endress, 1994a; Vogel, 1961). In some Proteaceae and Loranthaceae epidermal interdigitation is the basis for the explosive opening of the flowers, when they are probed by pollinating birds.

b. Androecium. Postgenital coherence can also be present between anthers. Such postgenital union of anthers mainly occurs in two very different

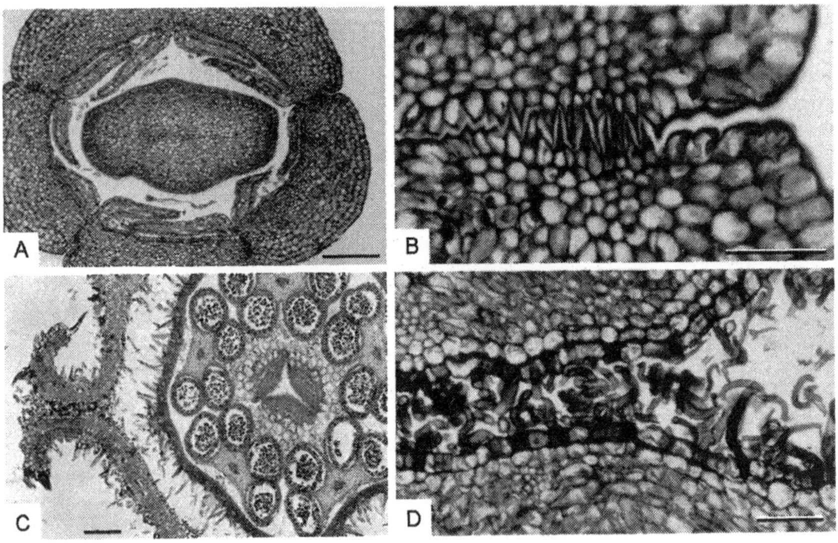

Fig. 9. Different kinds of postgenital coherence between perianth organs in floral buds of Asterales. (A, B) *Scaevola taccada* (Gaertn.) Roxb. (Goodeniaceae). (A) TS of floral bud (showing petals and gynoecium). (B) Close-up of petals showing postgenital coherence by interdigitation of epidermal cells. (C, D) *Campanula alliariifolia* Willd. (Campanulaceae). (C) TS of floral bud; (D) Close-up of sepals showing postgenital coherence by hairs. Scale bars: (A, C) 0.2 mm; (B, D) 0.05 mm.

biological situations. (1) In flowers with secondary pollen presentation: pollen is transported out of the anther tube by the elongating style and presented on the style before the stigma is receptive (Fig. 10A). This is common in Asterales, where it occurs in several variants (Erbar and Leins, 1995; Leins and Erbar, 1990). (2) In flowers adapted to vibratory pollination: in such flowers the anthers may be united into a cone by postgenital coherence (Fig. 10B). In the pollination process, the cone is embraced by pollen-collecting bees and vibratory movements cause the pollen to be expelled from the anthers onto the body of the bee. Some examples of postgenital anther coherence in vibratory-pollinated flowers are: Exaceae (Gentianaceae) (Struwe and Albert, 2002), several Rubiaceae (Puff *et al.*, 1995), *Solanum lycopersicum* (Solanaceae) (Bonner and Dickinson, 1989; Endress, 1994a; Glover *et al.*, 2004), *Curculigo capitulata* (Hypoxidaceae) (Kocyan and Endress, 2001a), *Apostasia nuda* (Orchidaceae) (Kocyan and Endress, 2001b) and others (see Endress, 1996).

c. Gynoecium. In the gynoecium, the most common union between carpels, syncarpy, is congenital, not postgenital. There are two important

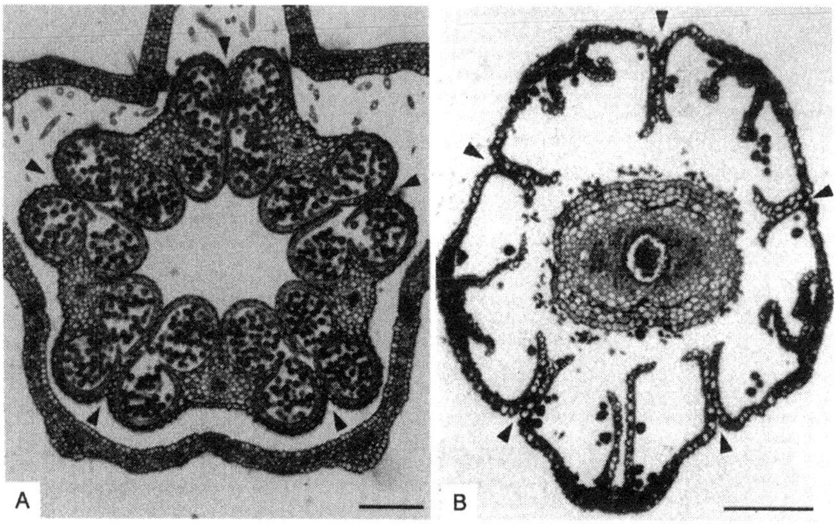

Fig. 10. Postgenital coherence between anthers in flowers of asterids. (A) *Lobelia inflata* L. (Campanulaceae); TS flower bud (showing corolla and androecium); flower with secondary pollen presentation. (B) *Sollya heterophylla* Lindl. (Pittosporaceae), TS flower (showing androecium and gynoecium); flower with vibratory pollination mechanism. Arrowheads point to areas of postgenital coherence. Scale bars: 0.2 mm.

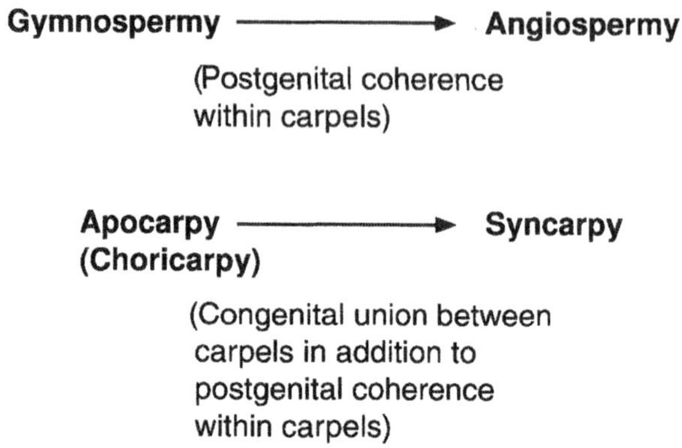

Fig. 11. The two major steps in angiosperm gynoecium evolution in which postgenital and congenital coherence are involved.

evolutionary progressions in gynoecium evolution: from gymnospermy to angiospermy and, in addition, from apocarpy to syncarpy (Fig. 11). The two phenomena, angiospermy and syncarpy, have sometimes erroneously been conflated. However, it is important to distinguish between the two. They are very different organizational/morphogenetic phenomena and are also at very different levels of the evolutionary history. Carpels and angiospermy originated at the beginning of angiosperm history. Syncarpy arrived only much later. The functional significance of angiospermy and syncarpy was discussed by Endress (1982) and Armbruster *et al.* (2002). In a syncarpous gynoecium, as is typical for core eudicots and for "core" monocots, the mostly 2–5 carpels are congenitally united. They are united from the beginning of development, and not by secondary contact of already formed morphological surfaces. In addition to congenital union between carpels there is postgenital fusion within carpels. In young developmental stages of a syncarpous gynoecium, the inner morphological surface is still visible as an often star-shaped zone in the center of the gynoecium (Fig. 12). Later this will be closed by differential growth of the carpel walls. The biologically most important aspect of syncarpy is the formation of a centralized pollen tube transmitting tract, a compitum, which allows centralized selection of all pollen tubes. There are also cases of postgenital fusion between carpels of a developmentally apocarpous gynoecium (Endress, 1982; Endress *et al.*, 1983; Jenny, 1988). This is especially present in flowers with secondary apocarpy to maintain a compitum (e.g., Malvaceae-Sterculioideae, Rutaceae, Apocynaceae), in Apocynaceae-Asclepiadoideae also to enable the formation of the devices for pollen transport (see later section) (Fig. 13).

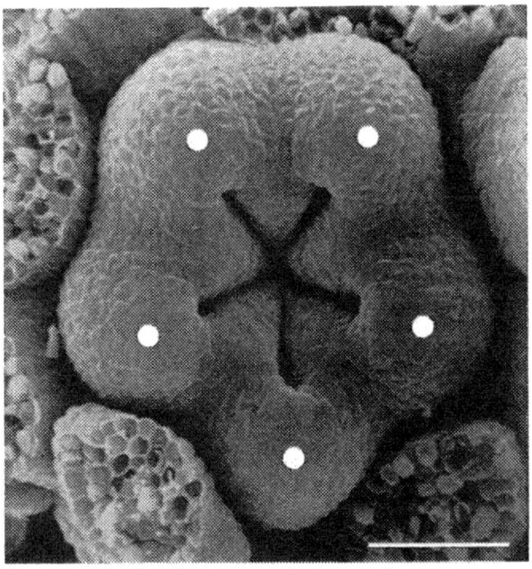

Fig. 12. *Geranium robertianum* L. (Geraniaceae); young gynoecium showing congenital union between the five carpels (syncarpy); the individual carpels are still open (angiospermy not yet established). Tip of each carpel marked with a dot. Scale bar: 0.05 mm.

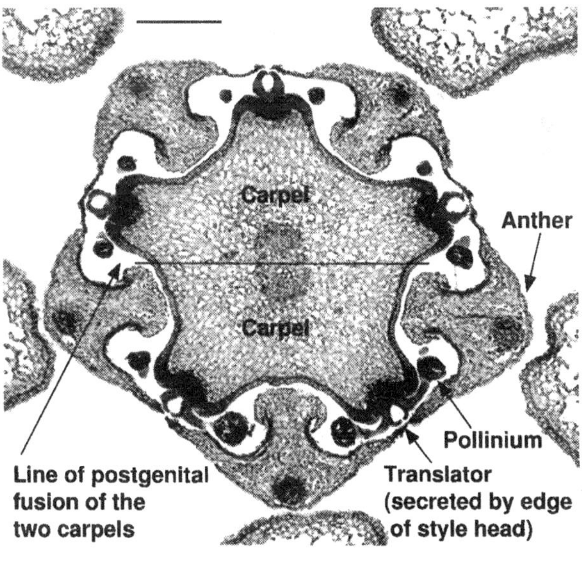

Fig. 13. *Vincetoxicum nigrum* Moench (Apocynaceae-Asclepiadoideae); TS androecium and gynoecium showing intimate synorganization among the five stamens, among the two carpels (postgenital union, indicated with line), and between the stamens and carpels, which is necessary for the formation of the five translators (that are secreted at the five edges of the style head of the gynoecium) and their connection with the pollinia. Scale bar: 0.2 mm. Modified from Endress (1999).

2. Synorganization between whorls (organs of different kinds)

Much rarer is postgenital coherence between organs of different kinds. This is present in some of the most complicated flowers of angiosperms, in the families Apocynaceae and Orchidaceae. In both families, synorganization between androecium and gynoecium results in the production of highly elaborate pollen transfer devices, with a clip or glue to be attached to pollinating animals. These devices are secreted by the gynoecium (Fig. 13), and pollen masses in the form of compact pollinia are attached to them with high precision. This is only possible by firm connection of androecium and gynoecium. In Orchidaceae this connection is congenital, in Apocynaceae it is postgenital (Fig. 14). These intimately synorganized organs influence each

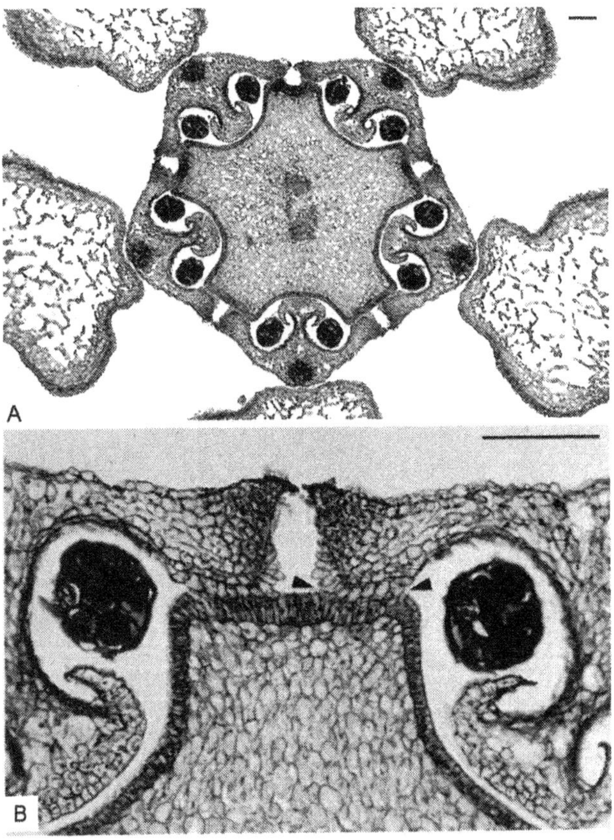

Fig. 14. *Vincetoxicum nigrum* (Apocynaceae-Asclepiadoideae). (A) TS androecium and gynoecium (section slightly lower than in Fig. 13). (B) Close-up of (A) showing postgenital fusion of anther flanks and gynoecium (arrowheads). Scale bars: 0.1 mm.

other in their shape (Endress, 1975) (see later section). An unusual formation is a hyperstigma in a few Monimiaceae (Laurales). In the female flowers the carpels are not united but they are enclosed in a floral cup consisting of several united whorls of perianth parts. The cup leaves only a narrow pore to the inside of the flower. This pore forms a unified receptive surface for pollen (Endress, 1980a).

3. Synorganization between flowers in an inflorescence

Synorganization occurs not only between organs within flowers but also between flowers within inflorescences. Flowers can be grouped together and form a superstructure (pseudanthium), which again resembles a single flower. The best-known examples of such pseudanthia are the heads of Asteraceae, in which the peripheral flowers (ray flowers) are highly monosymmetric and the united petals form conspicuous ligules, whereas the central flowers are small and polysymmetric. Thus in some way, there is a repetition of shapes at a higher structural level. In a few genera within Asteraceae this may be repeated again at a still higher level: several heads may be crowded together, and only the peripheral heads may form ray flowers at their peripheral side (e.g., *Dyssodia, Eroeda*) (Classen-Bockhoff, 1992, 1996). In addition, there are other means of making a showy periphery, for example, by enlarged and colored bracts that precede the inflorescence or by enlarged peripheral sepals or petals, a tendency present in many families, especially in asterids (Asteraceae, Dipsacaceae, Cornaceae, Apiaceae, Rubiaceae) (Albert *et al.*, 1998; Buzgo *et al.*, 2004b; Classen-Bockhoff, 1990; Uimari *et al.*, 2004).

How are such similar shapes repeatedly attained at different structural levels? Is there a similar genetic program in an asteracean head superimposed on the genetic program for the single flowers? What were the evolutionary pathways to these extreme forms?

X. PLASTICITY OF ORGAN FORM IN DEVELOPMENT SHAPED BY FLORAL ARCHITECTURE OR BY CONTIGUOUS PARTS

A. MONOSYMMETRY AND SYNORGANIZATION

In monosymmetric flowers the shape of individual organs may differ greatly from those of polysymmetric relatives. Such a "deformation" is especially prominent in stamens. One of the most impressive examples is *Salvia* (Lamiaceae), which has elaborate lip flowers with an upper and a lower lip. Only two of the originally five stamens are well developed and functional. These two stamens form an elaborate apparatus for pollen application to pollinators. They have a very short, stout filament, and the anther has a very broad connective, which is extended like a long filament. The two thecae are thus widely separated. The tip of the filament and the connective attachment form a hinge, in which the connective can be moved like a see-saw. Only one theca in each stamen is fertile. The second theca is transformed into a broadened part, which is lifted up by a visiting pollinator, resulting in a downward movement of the other side of the connective by which pollen is transfered to the dorsal side of the pollinator. Thus these stamens appear extremely "deformed" with regard to the hypothetical ancestral state in Lamiales or euasterids I (Classen-Bockhoff *et al.*, 2004). Their shape has become subordinated to the overall floral architecture or floral synorganization.

B. AUTONOMOUS SHAPES AND IMPRINTED SHAPES, POSITIONAL EFFECTS

Each floral organ has its autonomous shape but this shape may also be influenced and altered by contiguous neighboring parts by mutual pressure during development. Thus there is an autonomous component and an imprinted component. The imprinted component is conspicuous in flowers that have the organs crowded and irregularly positioned in bud. This may lead to irregular shapes of organ parts (Endress, 1975). But also crowding with regular position may affect organ shape. Petals with a contort aestivation are often asymmetric; the flank that covers its neighbor in bud may be larger than the flank that is covered by its neighbor, because it has more space to develop (e.g., some Apocynaceae). In sepals with a quincuncial aestivation often the flanks that are positioned to the outside become thick and green, whereas the flanks that are covered by neighboring flanks are thin and colorless (e.g., Warner *et al.*, 2005). Sepals or petals with valvate aestivation have thick margins, caused by mutual pressure of the two contiguous neighbors (which then allows postgenital coherence to occur).

An impressive example for imprinted shapes are highly synorganized flowers of Apocynaceae-Asclepiadoideae. Here synorganization is between anisomerous whorls. There are five stamens but only two carpels. In early development the gynoecium has two symmetry planes because of the autonomous shape of the two carpels. Then the tip of the two carpels unites postgenitally. In a next step the united tip of the gynoecium becomes five-angled by the influence of the five stamens, which act like a mold. Thus the original disymmetry (autonomous shape) is superimposed by polysymmetry (imprinted shape) (Endress, 1994a) (Fig. 15). This imprinted shape is the precondition for the development of the pollinaria in Apocynaceae-Asclepiadoideae (see earlier section). Another impressive example are gynoecia with an enlarged number of organs

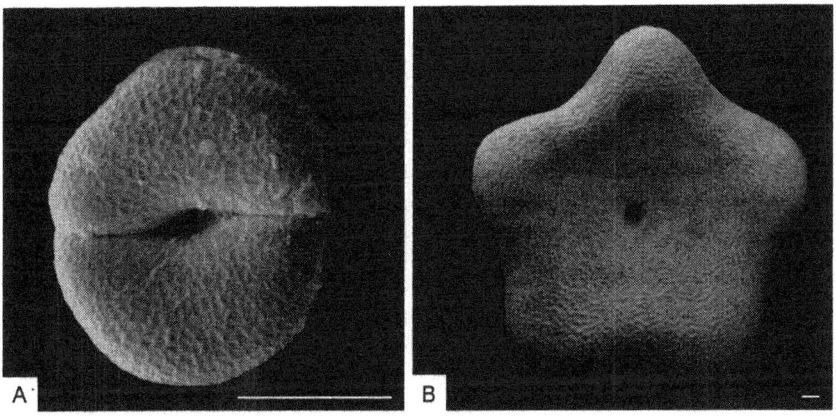

Fig. 15. *Asclepias physocarpa* Schltr. (Apocynaceae-Asclepiadoideae); gynoecium from above. (A) Young bud, the two carpels still distinct, postgenital union beginning. (B) At anthesis, the two carpels postgenitally united and with five-angular outline superimposed by the five neighboring anthers. Scale bars: 0.1 mm. Modified from Endress (1994a).

in a whorl. This leads to architectural difficulties to close the entire gynoecium in a clean manner (see later section).

XI. ARCHITECTURAL EXTREMES

Organs in a whorl occur mostly in moderate or low numbers, rarely more than five. This is especially true for the gynoecium. Architectural constraints limit the number of organs in a whorl. If there are more carpels in a whorl, it becomes difficult to maintain a regular architecture in the center of the gynoecium. This is familiar from the architecture of a cake. The more pieces in the shape of sectors one has to cut, the more difficult it is to get clean pieces, especially in the center. Thus the regular formation of the functional zones in the gynoecium becomes difficult, which is exemplified by some extreme cases that have escaped these difficulties in certain ways.

Reviews on such gynoecia with high carpel numbers in a whorl were given by Ronse De Craene and Smets (1998) for angiosperms, and Endress (2002) for asterids. However, in these studies, the architectural problems were not discussed. It is instructive to see what happens in the rare cases with increased numbers of carpels in a whorl. The more carpels there are, the more difficult it becomes to close the gynoecium in the center. Often the closure is imprecise by irregular growth of the carpel flanks, as seen, for example, in *Dillenia* (Dilleniaceae) with commonly 7 or more carpels (Endress, 1997b) (Fig. 16A and B), or, in addition, with the intervention of hairs from the carpel flanks,

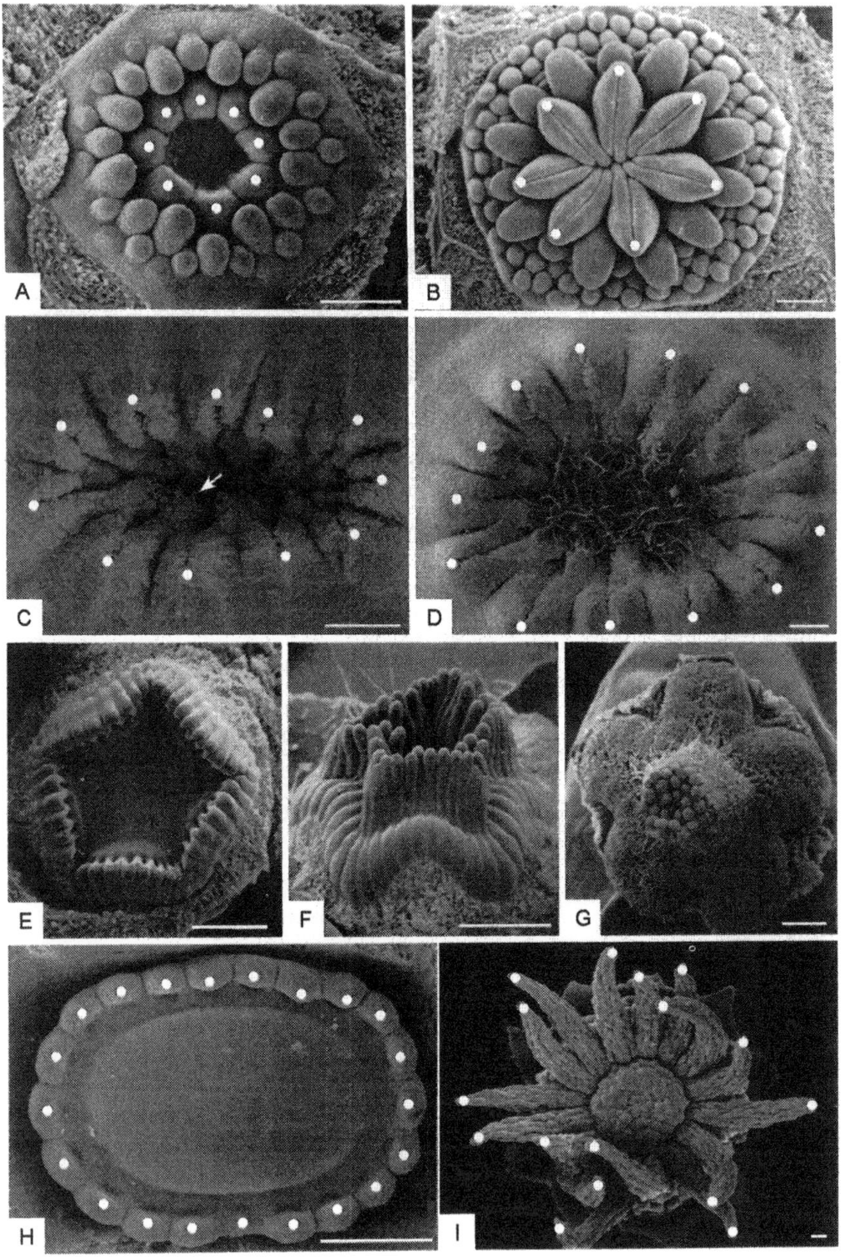

Fig. 16. Flowers with gynoecia with enlarged numbers of carpels in one whorl; inner surface of gynoecium occluded in various ways. (A, B) *Dillenia alata* (DC.) Martelli (Dilleniaceae); gynoecium closing by irregularly shaped carpel flanks. (C, D) *Munroidendron racemosum* (C. N. Forbes) Sherff (Araliaceae); center of gynoecium occluded by hairs formed by the carpel flanks (indicated by arrow in C). (E–G) *Kitaibelia vitifolia*

as in *Munroidendron* (Araliaceae) (Fig. 16C and D). Irregular closure is also present in *Decumaria* (Hydrangeaceae) with ca. 9 carpels (Hufford, 2001b), *Capparis* (Capparaceae) with up to 8 carpels (Leins and Metzenauer, 1979; Ronse De Craene and Smets, 1997b), and *Actinidia* (Actinidiaceae) with over 20 carpels (van Heel, 1987).

An extreme case is *Kitaibelia* (Malvaceae), which has about 50 carpels in a whorl (van Heel, 1995, and personal observations). Here the escape is that a large part of the floral apex is not involved in the gynoecium formation. The gynoecium forms in the shape of a five-angled ring, with completely undifferentiated center of the floral apex (Fig. 16E). Later this ring changes into a convoluted garland (Fig. 16F). And finally, by thickening of the individal styles the initial free space in the center of the flower becomes more or less occluded (Fig. 16G).

Another variant is present in *Gyrostemon* (Gyrostemonaceae). In this genus the floral apex is not even enclosed in the inner space of the gynoecium during development (Hufford, 1996). The floral apex remains completely uncovered and begins to proliferate irregularly (Fig. 16H and I). The common theme is irregular tissue proliferation, and this proliferating tissue appears to be secretory in both *Dillenia* and *Gyrostemon*. It is unknown whether this secretion is just a by-product of the overly excessive tissue proliferation or whether it is involved in some way in pollination biology.

Such architectural extremes are present in only a few erratic cases, for instance in single genera and species of larger families. They did not lead to any conspicuous evolutionary radiations. Thus they appear to be the opposite of key innovations.

XII. AN INNOVATION OF FLEXIBILITY: ESCAPE FROM CONSTRAINTS OF SYNORGANIZED FLOWERS IN CORE EUDICOTS—INCREASE OF STAMEN NUMBER AND DECOUPLING OF SEQUENCE OF INITIATION OF STAMENS AND CARPELS BY RING MERISTEMS

The normal flower development in angiosperms is with centripetal sequence of organ initiation. In spiral flowers, the single organs, and in whorled flowers, the whorls of organs appear successively from the outside to the inside of each flower (Fig. 17).

Willd. (Malvaceae); floral center occluded by thickening of styles. (H, I) *Gyrostemon brevipes* Hook. ex Moq. (Gyrostemonaceae); floral center remaining exposed and showing irregular growth. In part of the figures the carpel tips are marked with dots. Scale bars: (A–D, H, I) 0.2 mm; (E) 0.05 mm; (F, G) 0.5 mm.

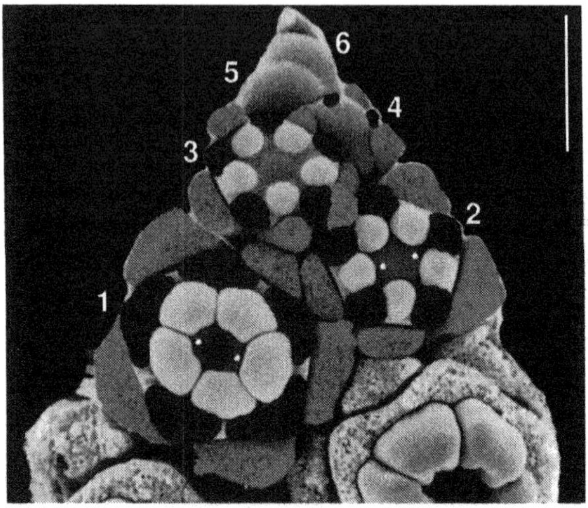

Fig. 17. *Echium vulgare* L. (Boraginaceae); tip of inflorescence showing flowers at different developmental stages, exhibiting normal acropetal initiation of organs, each organ category forming a single whorl (whorls 5-merous, but gynoecium 2-merous). Whorls highlighted by different shadings of gray. Successive flowers numbered. Scale bar: 0.2 mm.

In contrast, some derived polystemonous flowers in core eudicots deviate from a strict centripetal organ initiation. This is possible by the advent of a ring meristem in the androecium (or several sectorial meristems that are not confluent into a ring). When the carpel primordia are visible, most of the individual stamen primordia are not yet present on this ring meristem. Thus the sequence of stamen and carpel initiation is reversed. In addition, the sequence of initiation of the numerous stamens on the ring meristem has become flexible, it can be centripetal, centrifugal or bidirectional, and it can occur over an extended time, so that the last stamens are initiated after the carpels are far advanced in development. In most plant groups with a ring meristem in the androecium, stamen initiation is centrifugal. Two conspicuous examples are *Dillenia* (Dilleniaceae) (Corner, 1946; Endress, 1997b) (Fig. 18A–C) and *Couroupita* (Lecythidaceae) (Endress, 1994a; Hirmer, 1918) (Fig. 18D–F). In *Couroupita* there is even a second, crescent-shaped zone, which also produces stamens (Fig. 18F).

Thus, this decoupling of the time of initiation of stamens and carpels by intervention of a ring meristem allows extended time of stamen initiation, without holding up gynoecium formation; this is because the zone of the floral apex that is used for stamen formation is delimited early from the remaining central part of the floral apex. In some way, the ring meristem is an emancipation of the androecium from the apical floral meristem.

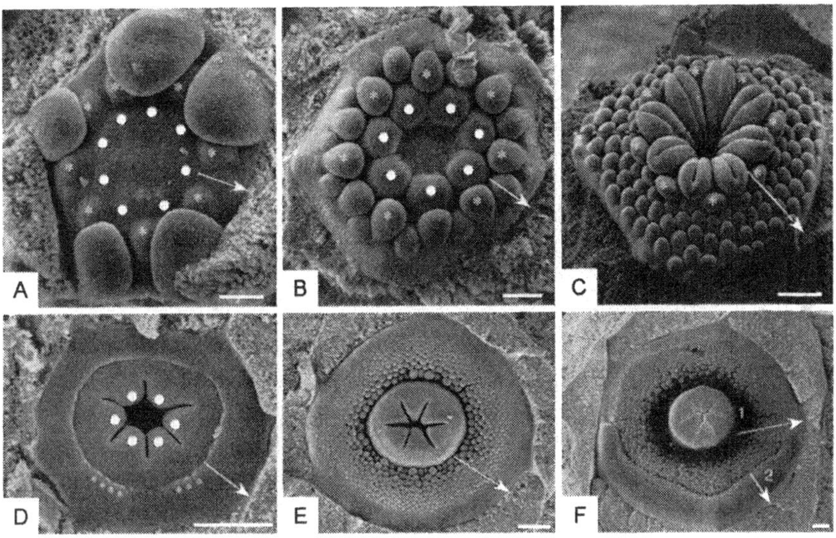

Fig. 18. Development of flowers with androecial ring meristems and centrifugal stamen initiation. (A–C) *Dillenia alata* (Dilleniaceae). (D–F) *Couroupita guianensis* Hook. (Lecythidaceae). Androecial ring meristem and direction of stamen initiation marked with arrow. In (F) arrow 1 marks the ring meristem, and arrow 2 the secondary part of the ring meristem. Individual stamens marked by asterisks, carpel tips marked by dots. Scale bars: (A, B) 0.1 mm; (C–F) 0.2 mm. (A, B) modified from Endress (1997b), (D–F) modified from Endress (1994a).

This extended time of stamen initiation, in turn, allows production of different stamen forms (called "heteranthery"), which is relevant for certain kinds of pollination biology. Heteranthery, especially in "pollen flowers," that is, flowers that offer only pollen to pollinators, allows functional specialization into "pollinating stamens" and "feeding stamens," which is the case in both *Dillenia* and *Couroupita* (Fig. 19). There are many such examples in various families of the core eudicots (e.g., Lythraceae, Leguminosae, Malvaceae) (Vogel, 1978), but heteranthery may also be present in some groups without a ring meristem and without polystemony (Commelinaceae) (Hrycan and Davis, 2005).

A decoupling of the sequence of stamen and carpel initiation in polystemonous flowers with a ring meristem has evolved several times in eudicots (with or without heteranthery). Different eudicot groups exhibit some variation of the patterns in detail, as compared with the simple initiation pattern of *Arabidopsis, Antirrhinum* and most other groups (Fig. 20). Such patterns of polystemony do not characterize one large clade (subclass "Dilleniidae") as initially assumed. However, they may be dominant at the level of some orders (Malvales), families (Lecythidaceae), or below.

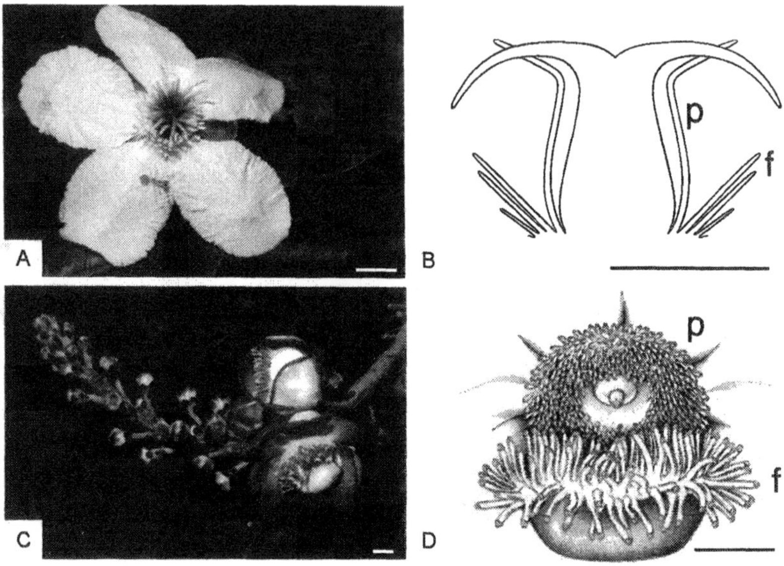

Fig. 19. Anthetic flowers with an androecial ring meristem and differentially shaped stamens (heteranthery). (A, B) *Dillenia alata* (Dilleniaceae). (A) Flower. (B) Schematic LS of androecium and gynoecium. (C, D) *Couroupita guianensis* (Lecythidaceae). (C) Inflorescence with two open flowers. (D) Androecium and gynoecium. f, Feeding stamens; p, pollinating stamens. Scale bars: ca. 1 cm. (B) From Endress (1997b); (D) from Endress (1994a).

As in *Dillenia* and *Couroupita*, stamen initiation is centrifugal in Nelumbonaceae (Hayes *et al.*, 2000), Aizoaceae and Cactaceae (Leins and Erbar, 1994), Hypericaceae (Leins, 1964), Capparaceae (Leins and Metzenauer, 1979; Ronse De Craene and Smets, 1997a), Cistaceae (Nandi, 1998; Ronse De Craene, 1992), Malvaceae (van Heel, 1966, 1995; von Balthazar *et al.*, 2004), Bixaceae (Ronse De Craene, 1989), and Theaceae (Erbar, 1986; Sugiyama, 1991; Tsou, 1998; Vishenskaya, 1980a,b). It is centripetal in *Papaver* (Karrer, 1991; Merxmüller and Leins, 1967), and bidirectional (or almost simultaneous) in *Caloncoba* and other Achariaceae (Bernhard and Endress, 1999; Endress and Voser, 1975) and in Leguminosae (Derstine and Tucker, 1991). Still another pattern was described for *Actinidia* (Actinidiaceae). It begins centrifugal but then new stamen primordia are intercalated between the older ones (van Heel, 1987).

Within the family Hamamelidaceae a centrifugal (*Fothergilla*) and a centripetal genus (*Matudaea*) was found (Endress, 1976). Seemingly disparate patterns within Loasaceae (Leins and Winhard, 1973) were found to be based on a more complicated shared pattern (Hufford, 1990). Similar complicated patterns also occur in the related family Hydrangeaceae (Hufford,

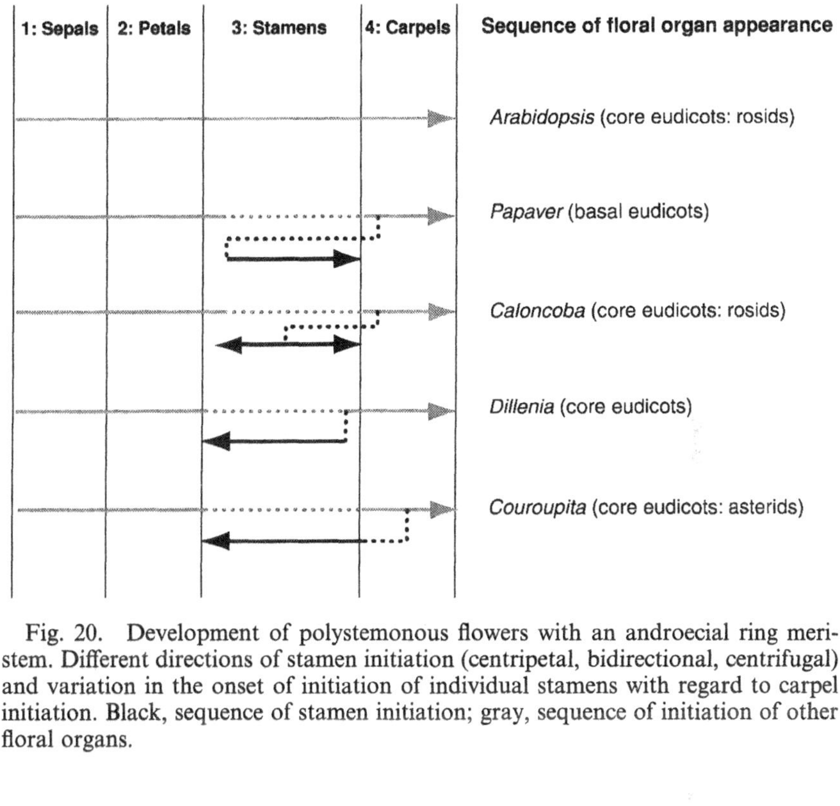

Fig. 20. Development of polystemonous flowers with an androecial ring meristem. Different directions of stamen initiation (centripetal, bidirectional, centrifugal) and variation in the onset of initiation of individual stamens with regard to carpel initiation. Black, sequence of stamen initiation; gray, sequence of initiation of other floral organs.

1998). The former Flacourtiaceae have centrifugal and bidirectional genera (Bernhard and Endress, 1999), which were found to represent two unrelated clades, now distributed to Salicaceae (centrifugal) and Achariaceae (bidirectional), within Malpighiales (Chase *et al.*, 2002). In *Swartzia* (Leguminosae) centrifugal and centripetal patterns were found in different species (also including heteranthery) (Tucker, 2003b).

Ring meristems may also occur in the perianth. They characterize plants with "early sympetaly." In early sympetaly the petals appear first as a ring before the individual tips become visible (Erbar and Leins, 1995, 2004).

XIII. POTENTIAL PITFALLS OF GENERALIZATIONS FROM MODEL PLANTS

Results derived from model species cannot necessarily be generalized for other plants. As a classic example, in their studies on periclinal chimeras of *Datura stramonium* (Solanaceae), Satina and Blakeslee (1941, 1943) found different patterns for the initiation of different organ types with regard to the cell layer of the first periclinal divisions. In foliage leaves, sepals, and petals the first periclinal cell divisions occur in the second cell layer (followed by divisons in the third layer). In contrast, in stamens, periclinal divisions occur only in the third layer. The authors concluded that "the early stages of development, therefore, suggest that the stamen is not a modified leaf, but rather a reduced axis" (Satina and Blakeslee, 1941). The carpel also develops with periclinal divisions in the third layer, and they likewise conclude that "the initiation and development of the carpel wall, septa and placentae suggest that they are axial and not foliar in origin" (Satina and Blakeslee, 1943). However, if organ initiation is compared across a broader diversity of plants, it is soon apparent that these conclusions by Satina and Blakeslee on the morphological nature of the organs in *Datura* are doubtful, because the site of the first periclinal divisions is not constant, and sometimes differs in close relatives, as discussed by Rohweder (1963). Rather, it appears that this site of periclinal cell divisions in a primordium is dependent on the future size of the organ. In small organs, they tend to be in the second cell layer, in bulky organs in a deeper layer. Initiation of stamens that are smaller than in *Datura* may be accompanied by initial periclinal cell divisions in the second, and not in the third cell layer (Rohweder, 1963). It would seem unreasonable to regard these different stamens as nonhomologous. Of course, these studies by Satina and Blakeslee are dated; however, they have led to wrong generalizations even in more recent times. In another series of classic studies it had also been shown that organs may be initiated without any periclinal cell divisions (Foard, 1971; Haber, 1962). This suggests that periclinal cell division may be merely a by-product of organ initiation.

In ovules, the integuments exhibit a similar variability. They may originate from periclinal cell divisions in the epidermis alone ("dermal" integuments) or also from deeper layers ("subdermal" integuments) (Bouman, 1984; Shamrov, 2000). According to Bouman (1984), the outer integument in angiosperm ovules is commonly subdermal, the inner dermal; however, there is considerable variation, here, too. For instance, in Piperales (basal angiosperms), the outer integument is only two-layered, and therefore dermal (Igersheim and Endress, 1998). Whether the early development is dermal or subdermal is again connected with the later thickness of the integuments but cannot be used for questions of homology or evolutionary derivation. The predominant occurrence of two integuments in angiosperms (apart from the derived asterids) indicates that the outer integuments are all homologous to each other, and the inner ones are homologous to each other, too (for different patterns of integument thickness, see the comparative studies on basal angiosperms and rosids by Endress and Igersheim, 1997, 1999, 2000a;

Igersheim and Endress, 1997, 1998; Igersheim *et al.*, 2001, Matthews and Endress, 2002, 2004, 2005a,b). Various studies have shown that properties of cells are influenced or maintained by signals from their neighborhood and not strictly by their position in a specific cell layer (discussion in Bäurle and Laux, 2003).

Organ fusion is either postgenital or congenital. To avoid confusion it is important to specify, although it is not always possible to distinguish the two types by superficial inspection. A distinction is sometimes only possible by thin microtome sections perpendicular to the two potentially postgenitally fused morphological surfaces. In a mature flower, the pattern of development can be reconstructed only by a careful analysis of the primary morphological surface.

There are some morphogenetic principles in higher plants that cannot be transcended: (1) the primary morphological surface is always continuous throughout the development. (2) Cells or organs cannot change place with each other during development. (3) Secondary surfaces may develop either by the advent of intercellular spaces (in which cell walls between contiguous cells become separated) or by dehiscence and abscission (in which cell layers disappear by cell death). The former is present, for example, inside the tissue of the septa of the gynoecium of *Arabidopsis* and other Brassicaceae, the second is ubiquitous in mature anthers as the inner surface of the anther locules, which originates by decay of the tapetal layer and one or several middle layers of each pollen sac, and is therefore not a primary, morphological surface but merely a secondary, histological surface.

XIV. CONCLUSIONS AND QUESTIONS

Research on floral development and evolution is rapidly proceeding on several different fronts. It has become increasingly difficult to keep track on all these fronts simultaneously. Therefore, it becomes ever more necessary to repeatedly review new developments and integrate them with the current knowledge. This chapter tries to bring together floral diversity and comparative floral development. There are numerous reviews on other facets of the field (Baum *et al.*, 2002; Endress, 2001b, 2002; Irish, 2003; Smyth, 2005; Soltis *et al.*, 2002). For an evolutionary synthesis it is also necessary to integrate fossil flowers (Crane *et al.*, 1995, 2004; Friis and Endress, 1990; Friis *et al.*, 2003a,b; 2005; Magallón *et al.*, 1999; Schönenberger, 2005).

Finally, some more general questions addressed in this chapter may be summarized:

What was the appearance of the first angiosperm flowers?

Floral phyllotaxis is labile in basal angiosperms, in some plants even within an individual. What is the genetic basis of changing degrees of fixation of phyllotaxis patterns in evolution? What are the boundary conditions for certain phyllotactic patterns in floral development?

To what extent is the position of all floral organs predetermined when the first organs are initiated in the development of a flower? What are differences from group to group?

How can a portion of the apical meristem be functionally decoupled from the remainder of the floral apex as a ring meristem and direction of organ initiation reversed?

Postgenital coherence plays an important role in angiospermy and in synorganization: how does it work at the molecular and the structural level?

How is morphogenesis or differentiation influenced directly by the immediate neighborhood of a part (e.g., different final structure of covering and covered edges of a perianth organ)?

How do floral organs of a category become diversified in evolution?

How to distinguish sepals and petals? This is difficult because organs, such as stamens and carpels, with a strict function, are easier to homologize throughout all angiosperms than organs that can change their functions.

How to deal with organs that apparently have evolved several times? Should "petals" in eudicots all be called petals or should they have different names depending on their origin in the phylogenetic tree?

How to deal with the problem of evolution of a predisposition for a feature but not yet the full fledged feature? And how to define the point of origin of innovations?

Floral development and evolution share two principles: contingency and punctuation. The two processes of development and evolution are gradual but not smoothly continuous. The processes are patterned. To find and understand these patterns is the common goal of developmental and evolutionary biology.

ACKNOWLEDGMENTS

I thank my scientific and technical collaborators for inspiration and support. Rosemarie Siegrist is thanked for microtome sections, Urs Jauch for support with the SEM and Alex Bernhard for support with the illustrations.

REFERENCES

Albert, V. A., Gustafsson, M. H. G. and Di Laurenzio, L. (1998). Ontogenetic systematics, molecular developmental genetics, and the angiosperm petal. *In* "Molecular Systematics of Plants II. DNA Sequencing" (D. E. Soltis, P. S. Soltis and J. J. Doyle, eds.), pp. 349–374. Kluwer, Boston.

Albert, V. A., Oppenheimer, D. G. and Lindqvist, C. (2002). Pleiotropy, redundancy and the evolution of flowers. *Trends in Plant Science* **7**, 297–301.

Albert, V. A., Soltis, D. E., Carlson, J. E., Farmerie, W. G., Wall, P. K., Ilut, D. C., Solow, T. M., Mueller, L. A., Landherr, L. L., Hu, Y., Buzgo, M. Kim, S., *et al.* (2005). Floral gene resources from basal angiosperms for comparative genomic research. *BMC Plant Biology* **5.5** (on-line).

APG (The Angiosperm Phylogeny Group) (2003). An update of the Angiosperm Phylogeny Group classification for the orders and families of flowering plants: APG II. *Botanical Journal of the Linnean Society* **141**, 399–436.

Armbruster, W. S., Debevec, E. M. and Willson, M. F. (2002). Evolution of syncarpy in angiosperms: Theoretical and phylogenetic analyses of the effects of carpel fusion on offspring quantity and quality. *Journal of Evolutionary Biology* **15**, 657–672.

Baum, D. A. (1998). The evolution of plant development. *Current Opinion in Plant Biology* **1**, 79–86.

Baum, D. A. and Day, C. D. (2004). Cryptic bracts exposed: Insights into the regulation of leaf expansion. *Developmental Cell* **6**, 318–319.

Baum, D. A. and Whitlock, B. A. (1999). Plant development: Genetic clues to petal evolution. *Current Biology* **9**, R525–R527.

Baum, D. A., Doebley, J., Irish, V. F. and Kramer, E. M. (2002). Response: Missing links: The genetic architecture of flower and floral diversification. *Trends in Plant Science* **7**, 31–34.

Bäurle, I. and Laux, T. (2003). Apical meristems: The plant's fountain of youth. *BioEssays* **25**, 961–970.

Becker, A. and Theissen, G. (2003). The major clades of MADS-box genes and their role in the development and evolution of flowering plants. *Molecular Phylogenetics and Evolution* **29**, 464–489.

Benjamin, M. and Hillen, B. (2003). Mechanical influences on cells, tissues and organs: Mechanical morphogenesis. *European Journal of Morphology* **41**, 3–7.

Bernhard, A. and Endress, P. K. (1999). Androecial development and systematics in Flacourtiaceae s. l. *Plant Systematics and Evolution* **215**, 141–155.

Boeke, J. H. (1973). Postgenital fusion in gynoecium of *Trifolium repens* L.: Light and electron microscopical aspects. *Acta Botanica Neerlandica* **22**, 503–509.

Boeke, J. H. and Vanvliet, G. J. C. M. (1979). Postgenital fusion in the gynoecium of the periclinal chimera *Laburnocytisus adamii* (Poit.) Schneid. (Papilionaceae). *Acta Botanica Neerlandica* **28**, 159–167.

Bonner, L. J. and Dickinson, H. G. (1989). Anther dehiscence in *Lycopersicon esculentum* Mill. I. Structural aspects. *New Phytologist* **113**, 97–115.

Bouman, F. (1984). The ovule. *In* "Embryology of Angiosperms" (B. M. Johri, ed.), pp. 123–157. Springer, Berlin.

Bowman, J. L. (1994). "*Arabidopsis*: An Atlas of Morphology and Development." Springer, New York.

Bowman, J. L., Baum, S. F., Eshed, Y., Putterill, J. and Alvarez, J. (1999). Molecular genetics of gynoecium development in *Arabidopsis*. *Current Topics of Developmental Biology* **45**, 155–205.

Bowman, J. L., Eshed, Y. and Baum, S. F. (2002). Establishment of polarity in angiosperm lateral organs. *Trends in Genetics* **18**, 134–141.

Buzgo, M., Soltis, P. S. and Soltis, D. E. (2004a). Floral developmental morphology of *Amborella trichopoda* (Amborellaceae). *International Journal of Plant Sciences* **165**, 925–947.

Buzgo, M., Soltis, D. E., Soltis, P. S. and Ma, H. (2004b). Towards a comprehensive integration of morphological and genetic studies of floral development. *Trends in Plant Science* **9**, 164–173.

Chase, M. W., Soltis, D. E., Olmstead, R. G., Morgan, D., Les, D. H., Mishler, B. D., Duvall, M. R., Price, R. A., Hills, H. G., Qiu, Y.-L., Kron, K. A. Rettig, J. H., *et al.* (1993). Phylogenetics of seed plants: An analysis of nucleotide sequences from the plastid gene *rbcL*. *Annals of the Missouri Botanical Garden* **80**, 528–580.

Chase, M. W., Zmarzty, S., Lledó, M. D., Wurdack, K. J., Swensen, S. M. and Fay, M. F. (2002). 'When in doubt, put it in Flacourtiaceae': A molecular phylogenetic analysis based on plastid *rbcL* DNA sequences. *Kew Bulletin* **57**, 141–181.

Chen, C. B., Wang, S. P. and Huang, H. (2000). *LEUNIG* has multiple functions in gynoecium development in *Arabidopsis*. *Genesis* **26**, 42–54.

Chen, C. B., Xu, Y., Zeng, M. H. and Huang, H. (2001). Genetic control by *Arabidopsis* genes *LEUNIG* and *FILAMENTOUS FLOWER* in gynoecium fusion. *Journal of Plant Research* **114**, 465–469.

Chen, X. B., Goodwin, S. M., Boroff, V. L., Liu, X. and Jenks, M. A. (2003). Cloning and characterization of the *WAX2* gene of *Arabidopsis* involved in cuticle membrane and wax production. *Plant Cell* **15**, 1170–1185.

Chrometzka, P. (1967). Über die Entwicklung der Zellennaht zwischen den Kelchblättern von *Oenothera*. *Österreichische Botanische Zeitschrift* **114**, 46–50.

Citerne, H. L., Moller, M. and Cronk, Q. C. B. (2000). Diversity of *cycloidea*-like genes in Gesneriaceae in relation to floral symmetry. *Annals of Botany* **86**, 167–176.

Citerne, H. L., Luo, D., Pennington, R. T., Coen, E. and Cronk, Q. C. B. (2003). A phylogenomic investigation of *CYCLOIDEA*-like TCP genes in the Leguminosae. *Plant Physiology* **131**, 1042–1053.

Classen-Bockhoff, R. (1990). Pattern analysis in pseudanthia. *Plant Systematics and Evolution* **171**, 57–88.

Classen-Bockhoff, R. (1992). Florale Differenzierung in komplex organisierten Asteraceenköpfen. *Flora* **186**, 1–22.

Classen-Bockhoff, R. (1996). Functional units beyond the level of the capitulum and cypsela in Compositae. *In* "Compositae: Biology and Utilization" (D. J. N. Hind, ed.), pp. 129–160. Royal Botanic Gardens, Kew.

Classen-Bockhoff, R., Crone, M. and Baikova, E. (2004). Stamen development in *Salvia* L.: Homology reinvestigated. *International Journal of Plant Sciences* **165**, 475–498.

Coen, E. S. (1996). Floral symmetry. *EMBO Journal* **15**, 6777–6788.

Coen, E. S. and Meyerowitz, E. M. (1991). The war of the whorls: Genetic interactions controlling flower development. *Nature* **353**, 31–37.

Coen, E. S., Nugent, J. M., Luo, D., Bradley, D., Cubas, P., Chadwick, M., Copsey, L. and Carpenter, R. (1995). Evolution of floral symmetry. *Philosophical Transactions of the Royal Society of London B* **350**, 35–38.

Corner, E. J. H. (1946). Centrifugal stamens. *Journal of the Arnold Arboretum* **27**, 423–437.

Crane, P. R. (1985). Phylogenetic analysis of seed plants and the origin of angiosperms. *Annals of the Missouri Botanical Garden* **72**, 716–793.

Crane, P. R. and Kenrick, P. (1997). Diverted development of reproductive organs: A source of morphological innovation in land plants. *Plant Systematics and Evolution* **206**, 161–174.

Crane, P. R., Friis, E. M. and Pedersen, K. R. (1989). Reproductive structure and function in Cretaceous Chloranthaceae. *Plant Systematics and Evolution* **165**, 211–226.

Crane, P. R., Friis, E. M. and Pedersen, K. R. (1995). The origin and early diversification of angiosperms. *Nature* **374**, 27–33.

Crane, P. R., Herendeen, P. and Friis, E. M. (2004). Fossils and plant phylogeny. *American Journal of Botany* **91**, 1683–1699.

Crepet, W. L. (1996). Timing in the evolution of derived floral characters: Upper Cretaceous (Turonian) taxa with tricolpate and tricolpate-derived pollen. *Review of Palaeobotany and Palynology* **90**, 339–359.

Crepet, W. L. and Nixon, K. C. (1998). Two new fossil flowers of magnoliid affinity from the Late Cretaceous of New Jersey. *American Journal of Botany* **85**, 1273–1288.

Crepet, W. L., Nixon, K. C. and Gandolfo, M. A. (2004). Fossil evidence and phylogeny: The age of major angiosperm clades based on mesofossil and macrofossil evidence from Cretaceous deposits. *American Journal of Botany* **91**, 1666–1682.

Cronk, Q. (2001). Plant evolution and development in a post-genomic context. *Nature Reviews Genetics* **2**, 607–619.

Cronk, Q. C. B., Bateman, R. M. and Hawkins, J. A. (eds.) (2002). "Developmental Genetics and Plant Evolution." Taylor & Francis, New York.

Cubas, P. (2004). Floral zygomorphy, the recurring evolution of a successful trait. *BioEssays* **26**, 1175–1184.

Cubas, P., Vincent, C. and Coen, E. (1999). An epigenetic mutation responsible for natural variation in floral symmetry. *Nature* **401**, 157–161.

Cubas, P., Coen, E. and Zapater, J. M. M. (2001). Ancient asymmetries in the evolution of flowers. *Current Biology* **11**, 1050–1052.

De Laet, J., Clinckemaillie, D., Jansen, S. and Smets, E. (1995). Floral ontogeny in the Plumbaginaceae. *Journal of Plant Research* **108**, 289–304.

Derstine, K. S. and Tucker, S. C. (1991). Organ initiation and development of inflorescences and flowers of *Acacia baileyana*. *American Journal of Botany* **78**, 816–832.

Dinneny, J. R., Yadegari, R., Fischer, R. L., Yanofsky, M. F. and Weigel, D. (2004). The role of *JAGGED* in shaping lateral organs. *Development* **131**, 1101–1110.

Donoghue, M. J., Ree, R. H. and Baum, D. A. (1998). Phylogeny and the evolution of flower symmetry in the Asteridae. *Trends in Plant Science* **3**, 311–317.

Douglas, A. W. and Tucker, S. C. (1996). Comparative floral ontogenies among Persoonioideae including *Bellendena* (Proteaceae). *American Journal of Botany* **83**, 1528–1555.

Doust, A. N. (2001). The developmental basis of floral variation in *Drimys winteri* (Winteraceae). *International Journal of Plant Sciences* **162**, 697–717.

Doyle, J. A. (1978). Origin of angiosperms. *Annual Review of Ecology and Systematics* **9**, 365–392.

Doyle, J. A. (1994). Origin of the angiosperm flower: A phylogenetic perspective. *Plant Systematics and Evolution* **8** (Suppl.), 7–29.

Doyle, J. A. (2001). Significance of molecular phylogenetic analyses for paleobotanical investigations on the origin of angiosperms. *Palaeobotanist* **50**, 167–188.

Doyle, J. A. and Endress, P. K. (2000). Morphological phylogenetic analysis of basal angiosperms: Comparison and combination with molecular data. *International Journal of Plant Sciences* **161** (Suppl.), S121–S153.

Drinnan, A. N., Crane, P. R. and Hoot, S. B. (1994). Patterns of floral evolution in the early diversification of non-magnoliid dicotyledons (eudicots). *Plant Systematics and Evolution* **8** (Suppl.), 93–122.

Eckardt, T. (1955). Nachweis der Blattbürtigkeit ("Phyllosporie") grundständiger Samenanlagen bei Centrospermen. *Berichte der Deutschen Botanischen Gesellschaft* **68**, 167–182.

Eichler, A. W. (1875). Blüthendiagramme I, II. Engelmann, Leipzig.

Endress, M. E. and Bruyns, P. V. (2000). A revised classification of the Apocynaceae s. l. *Botanical Review* **66**, 1–56.

Endress, P. K. (1970). Die Infloreszenzen der apetalen Hamamelidaceen, ihre grundsätzliche morphologische und systematische Bedeutung. *Botanische Jahrbücher für Systematik* **90**, 1–54.

Endress, P. K. (1972). Zur vergleichenden Entwicklungsmorphologie, Embryologie und Systematik bei Laurales. *Botanische Jahrbücher für Systematik* **92**, 331–428.

Endress, P. K. (1975). Nachbarliche Formbeziehungen mit Hüllfunktion im Infloreszenz- und Blütenbereich. *Botanische Jahrbücher für Systematik* **96**, 1–44.

Endress, P. K. (1976). Die Androeciumanlage bei polyandrischen Hamamelidaceen und ihre systematische Bedeutung. *Botanische Jahrbücher für Systematik* **97**, 436–457.

Endress, P. K. (1978). Blütenontogenese, Blütenabgrenzung und systematische Stellung der perianthlosen Hamamelidoideae. *Botanische Jahrbücher für Systematik* **100**, 249–317.

Endress, P. K. (1980a). Ontogeny, function and evolution of extreme floral construction in Monimiaceae. *Plant Systematics and Evolution* **134**, 79–120.

Endress, P. K. (1980b). The reproductive structures of the Austrobaileyaceae. *Botanische Jahrbücher für Systematik* **101**, 393–433.

Endress, P. K. (1982). Syncarpy and alternative modes of escaping disadvantages of apocarpy in primitive angiosperms. *Taxon* **31**, 48–52.

Endress, P. K. (1986a). Floral structure, systematics and phylogeny in Trochodendrales. *Annals of the Missouri Botanical Garden* **73**, 297–324.

Endress, P. K. (1986b). Reproductive structures and phylogenetic significance of extant primitive angiosperms. *Plant Systematics and Evolution* **152**, 1–28.

Endress, P. K. (1987a). Floral phyllotaxis and floral evolution. *Botanische Jahrbücher für Systematik* **108**, 417–438.

Endress, P. K. (1987b). The Chloranthaceae: Reproductive structures and phylogenetic position. *Botanische Jahrbücher für Systematik* **109**, 153–226.

Endress, P. K. (1989). Chaotic floral phyllotaxis and reduced perianth in *Achlys* (Berberidaceae). *Botanica Acta* **102**, 159–163.

Endress, P. K. (1990). Patterns of floral construction in ontogeny and phylogeny. *Biological Journal of the Linnean Society* **39**, 153–175.

Endress, P. K. (1992). Evolution and floral diversity: The phylogenetic surroundings of *Arabidopsis* and *Antirrhinum*. *International Journal of Plant Sciences* **153** (Suppl.), S106–S122.

Endress, P. K. (1994a). "Diversity and Evolutionary Biology of Tropical Flowers." Cambridge University Press, Cambridge.

Endress, P. K. (1994b). Floral structure and evolution in primitive angiosperms. *Plant Systematics and Evolution* **192**, 79–97.

Endress, P. K. (1995). Floral structure and evolution in Ranunculanae. *Plant Systematics and Evolution* **9** (Suppl.), 47–61.

Endress, P. K. (1996). Diversity and evolutionary trends in angiosperm anthers. *In* "The Anther: Form, Function and Phylogeny" (W. G. D'Arcy and R. C. Keating, eds.), pp. 92–117. Cambridge University Press, Cambridge.

Endress, P. K. (1997a). Evolutionary biology of flowers: Prospects for the next century. *In* "Evolution and Diversification of Land Plants" (K. Iwatsuki, P. H. Raven and P. H., eds.), pp. 99–119. Springer, Tokyo.

Endress, P. K. (1997b). Relationships between floral organization, architecture, and pollination mode in *Dillenia* (Dilleniaceae). *Plant Systematics and Evolution* **206**, 99–118.

Endress, P. K. (1998). *Antirrhinum* and Asteridae—evolutionary changes of floral symmetry. *Society of Experimental Biology Symposium Series* **51**, 133–140.

Endress, P. K. (1999). Symmetry in flowers—diversity and evolution. *International Journal of Plant Sciences* **160** (Suppl.), S3–S23.

Endress, P. K. (2001a). The flowers in basal extant angiosperms and inferences on ancestral flowers. *International Journal of Plant Sciences* **162**, 1111–1140.
Endress, P. K. (2001b). The origins of flower morphology. *Journal of Experimental Zoology (Molecular and Developmental Evolution)* **291**, 105–115.
Endress, P. K. (2001c). Evolution of floral symmetry. *Current Opinion in Plant Biology* **4**, 86–91.
Endress, P. K. (2002). Morphology and angiosperm systematics in the molecular era. *Botanical Review* **68**, 545–570.
Endress, P. K. (2003). Early floral development and the nature of the calyptra in Eupomatiaceae. *International Journal of Plant Sciences* **164**, 489–503.
Endress, P. K. (2004a). Biologie und Evolution der Blüten basaler Blütenpflanzen. *Leopoldina Jahrbuch 2003, Leopoldina*, R. 3, **49**, 467–486.
Endress, P. K. (2004b). Structure and relationships of basal relictual angiosperms. *Australian Systematic Botany* **17**, 343–366.
Endress, P. K. (2005). Carpels of *Brasenia* (Cabombaceae) are completely ascidiate despite a long stigmatic crest. *Annals of Botany* **96**, 209–215.
Endress, P. K. and Hufford, L. D. (1989). The diversity of stamen structures and dehiscence patterns among Magnoliidae. *Botanical Journal of the Linnean Society* **100**, 45–85.
Endress, P. K. and Igersheim, A. (1997). Gynoecium diversity and systematics of the Laurales. *Botanical Journal of the Linnean Society* **125**, 93–168.
Endress, P. K. and Igersheim, A. (1999). Gynoecium diversity and systematics of the basal eudicots. *Botanical Journal of the Linnean Society* **130**, 305–393.
Endress, P. K. and Igersheim, A. (2000a). Gynoecium structure and evolution in basal angiosperms. *International Journal of Plant Sciences* **161** (Suppl.), S211–S223.
Endress, P. K. and Igersheim, A. (2000b). The reproductive structures of the basal angiosperm *Amborella trichopoda* (Amborellaceae). *International Journal of Plant Sciences* **161** (Suppl.), S237–S248.
Endress, P. K. and Stumpf, S. (1990). Non-tetrasporangiate stamens in the angiosperms: Structure, systematic distribution and evolutionary aspects. *Botanische Jahrbücher für Systematik* **112**, 193–240.
Endress, P. K. and Voser, P. (1975). Zur Androeciumanlage und Antherenentwicklung bei *Caloncoba echinata* (Flacourtiaceae). *Plant Systematics and Evolution* **123**, 241–253.
Endress, P. K., Jenny, M. and Fallen, M. E. (1983). Convergent elaboration of apocarpous gynoecia in higher advanced dicotyledons (Sapindales, Malvales, Gentianales). *Nordic Journal of Botany* **3**, 293–300.
Erbar, C. (1986). Untersuchungen zur Entwicklung der spiraligen Blüte von *Stewartia pseudocamellia* (Theaceae). *Botanische Jahrbücher für Systematik* **106**, 391–407.
Erbar, C. (1991). Sympetaly—a systematic character? *Botanische Jahrbücher für Systematik* **112**, 417–451.
Erbar, C. and Leins, P. (1995). Portioned pollen release and the syndromes of secondary pollen presentation in the Campanulales-Asterales-complex. *Flora* **190**, 323–338.
Erbar, C. and Leins, P. (1996). Distribution of the character states "early sympetaly" and "Late sympetaly" within the "Sympetalae Tetracyclicae" and presumably allied groups. *Botanica Acta* **109**, 427–440.
Erbar, C. and Leins, P. (2004). Sympetaly in Apiales (Apiaceae, Araliaceae, Pittosporaceae). *South African Journal of Botany* **70**, 458–467.
Erbar, C., Kusma, S. and Leins, P. (1998). Development and interpretation of nectary organs in Ranunculaceae. *Flora* **194**, 317–332.

Fallen, M. E. (1986). Floral structure in the Apocynaceae: Morphological, functional, and evolutionary aspects. *Botanische Jahrbücher für Systematik* **106**, 245–286.

Feild, T. S., Arens, N. C., Doyle, J. A., Dawson, T. E. and Donoghue, M. J. (2004). Dark and disturbed: A new image of early angiosperm ecology. *Paleobiology* **30**, 82–107.

Foard, D. E. (1971). The initial protrusion of a leaf primordium can form without concurrent periclinal cell divisions. *Canadian Journal of Botany* **49**, 1601–1603.

Fourquin, C., Vinauger-Douard, M., Fogliani, B., Dumas, C. and Scutt, C. P. (2005). Evidence that *CRABS CLAW* and *TOUSLED* have conserved their roles in carpel development since the ancestor of the extant angiosperms. *Proceedings of the National Academy of Sciences of the United States of America* **102**, 4649–4654.

Friis, E. M. and Endress, P. K. (1990). Origin and evolution of angiosperm flowers. *Advances in Botanical Research* **17**, 99–162.

Friis, E. M., Crane, P. R. and Pedersen, K. R. (1986). Floral evidence for Cretaceous chloranthoid angiosperms. *Nature* **320**, 163–164.

Friis, E. M., Doyle, J. A., Endress, P. K. and Leng, Q. (2003a). *Archaefructus*—angiosperm precursor or specialized early angiosperm? *Trends in Plant Science* **8**, 369–373.

Friis, E. M., Pedersen, K. R. and Schönenberger, J. (2003b). *Endressianthus*, a new Normapolles-producing plant genus of fagalean affinity from the Late Cretaceous of Portugal. *International Journal of Plant Sciences* **164** (Suppl.), S201–S223.

Friis, E. M., Pedersen, K. R. and Crane, P. R. (2005). When Earth started blooming: Insights from the fossil record. *Current Opinion in Plant Biology* **8**, 5–12.

Frohlich, M. W. (2001). A detailed scenario and possible tests of the mostly male theory of flower evolutionary origins. *In* "Beyond Heterochrony: The Evolution of Development" (M. L. Zelditch, ed.), pp. 59–104. Wiley-Liss, New York.

Frohlich, M. W. (2003). An evolutionary scenario for the origin of flowers. *Nature Reviews Genetics* **4**, 559–566.

Frohlich, M. W. and Parker, D. S. (2000). The mostly male theory of flower evolutionary origins: From genes to fossils. *Systematic Botany* **25**, 155–170.

Gaussen, H. (1946). Les Gymnospermes, actuelles et fossiles. *Travaux du Laboratoire Forestier de Toulouse II*, Sect. 1, Ch. 5, **1** (3), 1–26.

Glover, B. J., Bunnewell, S. and Martin, C. (2004). Convergent evolution within the genus *Solanum*: The specialized anther cone develops through alternative pathways. *Gene* **331**, 1–7.

Graham, S. W. and Olmstead, R. G. (2000). Utility of 17 chloroplast genes for inferring the phylogeny of the basal angiosperms. *American Journal of Botany* **87**, 1712–1730.

Green, P. B. (1999). Expression of pattern in plants: Combining molecular and calculus-based biophysical paradigms. *American Journal of Botany* **86**, 1059–1076.

Haber, A. H. (1962). Nonessentiality of concurrent cell divisions for degree of polarization of leaf growth. I. Studies with radiation-induced mitotic inhibition. *American Journal of Botany* **49**, 583–589.

Hagemann, W. (1963). Weitere Untersuchungen zur Organisation des Sprossscheitelmeristems; der Vegetationspunkt traubiger Floreszenzen. *Botanische Jahrbücher für Systematik* **82**, 273–315.

Hakki, M. I. (1971). Blütenmorphologische und embryologische Untersuchungen an *Chenopodium capitatum* und *Chenopodium foliosum* sowie weiteren Chenopodiaceae. *Botanische Jahrbücher für Systematik* **92**, 178–330.

Harris, T. M. (1941). *Caytonanthus*, the microsporophyll of *Caytonia*. *Annals of Botany*, n. S., **5**, 47–58.
Hayes, V., Schneider, E. L. and Carlquist, S. (2000). Floral development of *Nelumbo nucifera* (Nelumbonaceae). *International Journal of Plant Sciences* **161** (Suppl.), S183–S191.
Hiepko, P. (1965). Vergleichend-morphologische und entwicklungsgeschichtliche Untersuchungen über das Perianth bei den Polycarpicae. *Botanische Jahrbücher für Systematik* **84**, 359–508.
Hileman, L. C. and Baum, D. A. (2003). Why did parlogs persist? Molecular evolution of *CYCLOIDEA* and related floral symmetry genes in Antirrhineae (Veronicaceae). *Molecular Biology and Evolution* **20**, 591–600.
Hileman, L. C., Kramer, E. M. and Baum, D. A. (2003). Differential regulation of symmetry genes and the evolution of floral morphologies. *Proceedings of the National Academy of Sciences of the United States of America* **100**, 12814–12819.
Hirmer, M. (1918). Beiträge zur Morphologie der polyandrischen Blüten. *Flora* **110**, 140–192.
Hodges, S. A. (1997). Rapid radiation due to a key innovation in columbines (Ranunculaceae: *Aquilegia*). *In* "Molecular Evolution and Adaptive Radiation" (T. J. Givnish and K. J. Sytsma, eds.), pp. 391–405. Cambridge University Press, Cambridge.
Hofmeister, W. (1868). Allgemeine Morphologie der Gewächse. Engelmann, Leipzig.
Hoot, S. B., Magallón, S. and Crane, P. R. (1999). Phylogeny of basal eudicots based on three molecular data sets: *atpB*, *rbcL*, and 18S nuclear ribosomal DNA sequences. *Annals of the Missouri Botanical Garden* **86**, 1–32.
Hrycan, W. C. and Davis, A. R. (2005). Comparative structure and pollen production of the stamens and pollinator-deceptive staminodes of *Commelina coelestis* and *C. dianthifolia* (Commelinaceae). *Annals of Botany* **95**, 1113–1130.
Hufford, L. (1990). Androecial development and the problem of monophyly of Loasaceae. *Canadian Journal of Botany* **68**, 402–419.
Hufford, L. (1996). Developmental morphology of female flowers of *Gyrostemon* and *Tersonia* and floral evolution among Gyrostemonaceae. *American Journal of Botany* **83**, 1471–1487.
Hufford, L. (1997). The roles of ontogenetic evolution in the origins of floral homoplasies. *International Journal of Plant Sciences* **158** (Suppl.), S65–S80.
Hufford, L. (1998). Early development of androecia in polystemonous Hydrangeaceae. *American Journal of Botany* **85**, 1057–1067.
Hufford, L. (2001a). Ontogenetic sequences: Homology, evolution, and the patterning of clade diversity. *In* "Beyond Heterochrony: The Evolution of Development" (M. L. Zelditch, ed.), pp. 27–57. Wiley-Liss, New York.
Hufford, L. (2001b). Ontogeny and morphology of the fertile flowers of *Hydrangea* and allied genera of tribe Hydrangeeae (Hydrangeaceae). *Botanical Journal of the Linnean Society* **137**, 139–187.
Hufford, L. (2003). Homology and developmental transformation: Models for the origin of the staminodes of Loasaceae subfamily Loasoideae. *International Journal of Plant Sciences* **164** (Suppl.), S409–S439.
Hufford, L. and McMahon, M. (2003). Beyond morphoclines and trends: The elements of diversity and the phylogenetic patterning of morphology. *In* "Deep Morphology: Toward a Renaissance of Morphology in Plant Systematics" (T. F. Stuessy, V. Mayer and E. Hörandl, eds.), pp. 165–186. Gantner, Ruggell.
Hufford, L. and McMahon, M. (2004). Morphological evolution and systematics of *Synthyris* and *Besseya* (Veronicaceae): A phylogenetic analysis. *Systematic Botany* **29**, 716–736.

Igersheim, A. and Endress, P. K. (1997). Gynoecium diversity and systematics of the Magnoliales and winteroids. *Botanical Journal of the Linnean Society* **124**, 213–271.

Igersheim, A. and Endress, P. K. (1998). Gynoecium diversity and systematics of the paleoherbs. *Botanical Journal of the Linnean Society* **127**, 289–370.

Igersheim, A., Buzgo, M. and Endress, P. K. (2001). Gynoecium diversity and systematics in basal monocots. *Botanical Journal of the Linnean Society* **136**, 1–65.

Irish, V. F. (1999). Patterning the flower. *Developmental Biology* **209**, 211–220.

Irish, V. F. (2003). The evolution of floral homeotic gene function. *BioEssays* **25**, 637–646.

Irish, V. F. and Benfey, P. N. (2004). Beyond *Arabidopsis*. Translation biology meets evolutionary developmental biology. *Plant Physiology* **135**, 611–614.

Irish, V. F. and Kramer, E. M. (1998). Genetic and molecular analysis of angiosperm flower development. *Advances in Botanical Research* **28**, 197–230.

Jäger-Zürn, I. (1967). Infloreszenz- und blütenmorphologische, sowie embryologische Untersuchungen an *Myrothamnus* Welw. *Beiträge zur Biologie der Pflanzen* **42**, 241–271.

Jaramillo, M. A. and Kramer, E. M. (2004). *APETALA3* and *PISTILLATA* homologs exhibit novel expression patterns in the unique perianth of *Aristolochia* (Aristolochiaceae). *Evolution and Development* **6**, 449–458.

Jenny, M. (1988). Different gynoecium types in Sterculiaceae: Ontogeny and functional aspects. *In* "Aspects of Floral Development" (P. Leins, S. C. Tucker and P. K. Endress, eds.), pp. 225–236. Cramer, Stuttgart.

Kadereit, J. W. and von Hagen, K. B. (2003). The evolution of flower morphology in Gentianaceae-Swertiinae and the roles of key innovations and niche width for the diversification of *Gentianella* and *Halenia* in South America. *International Journal of Plant Sciences* **164** (Suppl.), S441–S452.

Kanno, A., Saeki, H., Kameya, T., Saedler, H. and Theissen, G. (2003). Heterotopic expression of class B floral homeotic genes supports a modified ABC model for tulip (*Tulipa gesneriana*). *Plant Molecular Biology* **52**, 831–841.

Karrer, A. B. (1991). Blütenentwicklung und systematische Stellung der Papaveraceae und Capparaceae. Doctoral dissertation, University of Zurich.

Kellogg, E. A. (2000). The grasses: A case study in macroevolution. *Annual Review of Ecology and Systematics* **31**, 217–238.

Kim, S., Koh, J., Yoo, M.-J., Kong, H., Hu, Y., Ma, H., Soltis, P. S. and Soltis, D. E. (2005a). Expression of floral MADS-box genes in basal angiosperms: Implications for the evolution of floral regulators. *Plant Journal* **43**, 724–744.

Kim, S., Koh, J., Ma, H., Hu, Y., Endress, P. K., Hauser, B. A., Buzgo, M., Soltis, P. S. and Soltis, D. E. (2005b). Sequence and expression studies of A-, B-, and E-class MADS-box genes in *Eupomatia* (Eupomatiaceae): Support for the bracteate origin of the calyptra. *International Journal of Plant Sciences* **166**, 185–198.

Kirchoff, B. K. (2003). Shape matters: Hofmeister's rule, primordium shape, and flower orientation. *International Journal of Plant Sciences* **164**, 505–517.

Kocyan, A. and Endress, P. K. (2001a). Floral structure and development, and systematic aspects of some 'lower' Asparagales. *Plant Systematics and Evolution* **229**, 187–216.

Kocyan, A. and Endress, P. K. (2001b). Floral structure and development of *Apostasia* and *Neuwiedia* (Apostasioideae) and their relationships to other Orchidaceae. *International Journal of Plant Sciences* **162**, 847–867.

Kosuge, K. (1994). Petal evolution in Ranunculaceae. *Plant Systematics and Evolution* **8** (Suppl.), 185–191.

Kramer, E. M. and Hall, J. C. (2005). Evolutionary dynamics of genes controlling floral development. *Current Opinion in Plant Biology* **8**, 13–18.

Kramer, E. M. and Irish, V. F. (1999). Evolution of genetic mechanisms controlling petal development. *Nature* **399**, 144–148.

Kramer, E. M. and Irish, V. F. (2000). Evolution of the petal and stamen developmental programs: Evidence from comparative studies of the lower eudicots and basal angiosperms. *International Journal of Plant Sciences* **161** (Suppl.), S29–S40.

Kramer, E. M., Di Stilio, V. S. and Schluter, P. M. (2003). Complex patterns of gene duplication in the *APETALA3* and *PISTILLATA* lineages of the Ranunculaceae. *International Journal of Plant Sciences* **164**, 1–11.

Krolikowski, K. A., Victor, J. L., Wagler, T. N., Lolle, S. J. and Pruitt, R. E. (2003). Isolation and characterization of the *Arabidopsis* organ fusion gene *HOTHEAD*. *Plant Journal* **35**, 501–511.

Lamb, R. S. and Irish, V. F. (2003). Functional divergence within the *APETALA3/PISTILLATA* floral homeotic gene lineages. *Proceedings of the National Academy of Sciences of the United States of America* **100**, 6558–6563.

Leins, P. (1963). Entwicklungsgeschichtliche Studien an Ericales-Blüten. *Botanische Jahrbücher für Systematik* **83**, 57–88.

Leins, P. (1964). Die frühe Blütenentwicklung von *Hypericum hookerianum* Wight et Arn. und *H. aegypticum* L. *Berichte der Deutschen Botanischen Gesellschaft* **77**, 112–123.

Leins, P. and Erbar, C. (1990). On the mechanisms of secondary pollen presentation in the Campanulales-Asterales complex. *Botanica Acta* **103**, 87–92.

Leins, P. and Erbar, C. (1994). Putative origin and relationships of the order from the viewpoint of developmental flower morphology. *In* "Caryophyllales: Evolution and Systematics" (H.-D. Behnke and T. J. Mabry, eds.), pp. 304–316. Springer, Berlin.

Leins, P. and Erbar, C. (1995). Das frühe Differenzierungsmuster in den Blüten von *Saruma henryi* Oliv. (Aristolochiaceae). *Botanische Jahrbücher für Systematik* **117**, 365–376.

Leins, P. and Erbar, C. (2004). Floral organ sequences in Apiales (Apiaceae, Araliaceae, Pittosporaceae). *South African Journal of Botany* **70**, 468–474.

Leins, P. and Metzenauer, G. (1979). Entwicklungsgeschichtliche Untersuchungen an *Capparis*-Blüten. *Botanische Jahrbücher für Systematik* **100**, 542–554.

Leins, P. and Winhard, W. (1973). Entwicklungsgeschichtliche Studien an Loasaceen-Blüten. *Österreichische Botanische Zeitschrift* **122**, 145–165.

Levin, J. Z., Fletcher, J. C., Chen, X. and Meyerowitz, E. M. (1998). A genetic screen for modifiers of *UFO* meristem activity identifies three novel *FUSED FLORAL ORGAN* genes required for early flower development in *Arabidopsis*. *Genetics* **149**, 579–595.

Litt, A. and Irish, V. F. (2003). Duplication and diversification in the *APETALA1/FRUITFULL* floral homeotic gene lineage: Implications for the evolution of floral development. *Genetics* **165**, 821–833.

Liu, Z. C., Franks, R. G. and Klink, V. P. (2000). Regulation of gynoecium marginal tissue formation by *LEUNIG* and *AINTEGUMENTA*. *Plant Cell* **12**, 1879–1891.

Lolle, S. J. and Pruitt, R. E. (1999). Epidermal cell interactions: A case for local talk. *Trends in Plant Science* **4**, 14–20.

Lolle, S. J., Cheung, A. Y. and Sussex, I. M. (1992). *FIDDLEHEAD*—an *Arabidopsis* mutant constitutively expressing an organ fusion program that involves interactions between epidermal cells. *Developmental Biology* **152**, 383–392.

Lolle, S. J., Hsu, W. and Pruitt, R. E. (1998). Genetic analysis of organ fusion in *Arabidopsis*. *Genetics* **149**, 607–619.

Lord, E. M. (1991). The concepts of heterochrony and homeosis in the study of floral morphogenesis. *Flowering Newsletter* **11**, 4–13.

Lord, E. M., Crone, W. and Hill, J. P. (1994). Timing of events during flower organogenesis: *Arabidopsis* as a model system. *Current Topics in Developmental Biology* **29**, 325–356.

Luo, D., Carpenter, R., Vincent, C., Copsey, L. and Coen, E. (1996). Origin of floral asymmetry in *Antirrhinum*. *Nature* **383**, 794–799.

Magallón, S., Crane, P. R. and Herendeen, P. S. (1999). Phylogenetic pattern, diversity, and diversification of eudicots. *Annals of the Missouri Botanical Garden* **86**, 297–372.

Mair, O. (1977). Zur Entwicklungsgeschichte monosymmetrischer Dicotylen-Blüten. *Dissertationes Botanicae* **38**, 1–90.

Malcomber, S. T. and Kellogg, E. A. (2004). Heterogeneous expression patterns and separate roles of the *SEPALLATA* gene *LEAFY HULL STERILE1* in grasses. *Plant Cell* **16**, 1692–1706.

Masiero, S., Li, M. A., Will, I., Hartmann, U., Saedler, H., Huijser, P., Schwarz-Sommer, Z. and Sommer, H. (2004). *INCOMPOSITA*: A MADS-box gene controlling prophyll development and floral meristem identity in *Antirrhinum*. *Development* **131**, 5981–5990.

Mathews, S. and Donoghue, M. J. (1999). The root of angiosperm phylogeny inferred from duplicate phytochrome genes. *Science* **286**, 947–950.

Matthews, M. L. and Endress, P. K. (2002). Comparative floral structure and systematics in Oxalidales (Oxalidaceae, Connaraceae, Cephalotaceae, Brunelliaceae, Cunoniacaeae, Elaeocarpaceae, Tremandraceae). *Botanical Journal of the Linnean Society* **140**, 321–381.

Matthews, M. L. and Endress, P. K. (2004). Comparative floral structure and systematics in Cucurbitales (Anisophylleaceae, Begoniaceae, Coriariaceae, Corynocarpaceae, Cucurbitaceae, Datiscaceae, Tetramelaceae). *Botanical Journal of the Linnean Society* **145**, 129–185.

Matthews, M. L. and Endress, P. K. (2005a). Comparative floral structure and systematics in Celastrales (Celastraceae, Parnassiaceae, Lepidobotryaceae). *Botanical Journal of the Linnean Society* **149**, 129–194.

Matthews, M. L. and Endress, P. K. (2005b). Comparative floral structure and systematics in Crossosomatales (Crossosomataceae, Stachyuraceae, Staphyleaceae, Aphloiaceae, Geissolomataceae, Ixerbaceae, Strasburgeriaceae. *Botanical Journal of the Linnean Society* **147**, 1–46.

McMahon, M. and Hufford, L. (2002). Developmental morphology and structural homology of corolla-androecium synorganization in the tribe Amorpheae (Fabaceae: Papilionoideae). *American Journal of Botany* **89**, 1884–1898.

Meinhardt, H. (2003). Complex pattern formation by a self-destabilization of established patterns: Chemotactic orientation and phyllotaxis as examples. *Comptes Rendus Biologies* **326**, 223–237.

Meister, R. J., Kotow, L. M. and Gasser, C. S. (2002). *SUPERMAN* attenuates positive *INNER NOOUTER* autoregulation to maintain polar development of *Arabidopsis* ovule outer integuments. *Development* **129**, 4281–4289.

Merxmüller, H. and Leins, P. (1967). Die Verwandtschaftsbeziehungen der Kreuzblütler und Mohngewächse. *Botanische Jahrbücher für Systematik* **86**, 113–129.

Meyen, S. V. (1988). Origin of the angiosperm gynoecium by gamoheterotopy. *Botanical Journal of the Linnean Society* **97**, 171–178.

Meyerowitz, E. M. (1994). The genetics of flower development. *Scientific American* **271** (5), 40–47.

Müller, G. B. and Olsson, L. (2003). Epigenesis and epigenetics. *In* "Keywords and Concepts in Evolutionary Developmental Biology" (B. K. Hall and W. M. Olson, eds.), pp. 114–123. Harvard University Press, Cambridge, MA.

Mundry, I. and Mundry, M. (2001). Male cones in Taxaceae s. l.—an example of Wettstein's pseudanthium concept. *Plant Biology* **3**, 405–416.

Mundry, M. and Stützel, T. (2003). Morphogenesis of male sporangiophores of *Zamia amblyphyllidia* D. W. Stev. *Plant Biology* **5**, 297–310.

Murbeck, S. (1912). Untersuchungen über den Blütenbau der Papaveraceen. *Kungl. Svenska Vetenskapsakademiens Handlingar* **50** (1), 1–168.

Nandi, O. I. (1998). Floral development and systematics of Cistaceae. *Plant Systematics and Evolution* **212**, 107–134.

Neumayer, H. (1924). Die Geschichte der Blüte. *Abhandlungen der Zoologisch-Botanischen Gesellschaft Wien* **14**, 1–112.

Oxelman, B., Kornhall, P., Olmstead, R. G. and Bremer, B. (2005). Further disintegration of Scrophulariaceae. *Taxon* **54**, 411–425.

Posluszny, U. and Tomlinson, P. B. (2003). Aspects of inflorescence and floral development in the putative basal angiosperm *Amborella trichopoda* (Amborellaceae). *Canadian Journal of Botany* **81**, 28–39.

Pruitt, R. E., Vielle-Calzada, J. P., Ploense, S. E., Grossnikluas, U. and Lolle, S. J. (2000). *FIDDLEHEAD*, a gene required to suppress epidermal cell interactions in *Arabidopsis*, encodes a putative lipid biosynthetic enzyme. *Proceedings of the National Academy of Sciences of the United States of America* **97**, 1311–1316.

Puff, C., Igersheim, A., Buchner, R. and Rohrhofer, U. (1995). The united stamens of Rubiaceae. Morphology, anatomy; their role in pollination ecology. *Annals of the Missouri Botanical Garden* **82**, 357–382.

Qiu, Y.-L., Lee, J., Bernasconi-Quadroni, F., Soltis, D. E., Soltis, P. S., Zanis, M., Zimmer, E. A., Chen, Z., Savolainen, V. and Chase, M. W. (1999). The earliest angiosperms: Evidence from mitochondrial, plastid and nuclear genomes. *Nature* **402**, 404–407.

Raven, J. A. and Weyers, J. D. B. (2001). Significance of epidermal fusion and intercalary growth for angiosperm evolution. *Trends in Plant Science* **6**, 111–113.

Ree, R. H. (2005). Detecting the historical signature of key innovations using stochastic models of character evolution and cladogenesis. *Evolution* **59**, 257–265.

Ree, R. H. and Donoghue, M. J. (1999). Inferring rates of change in flower symmetry in asterid angiosperms. *Systematic Biology* **48**, 633–641.

Reinhardt, D., Pesce, E. R., Stieger, P., Mandel, T., Baltensperger, K., Bennett, M., Traas, J., Friml, J. and Kuhlemeier, C. (2003). Regulation of phyllotaxis by polar auxin transport. *Nature* **426**, 255–260.

Rohweder, O. (1963). Anatomische und histogenetische Untersuchungen an Laubsprossen und Blüten der Commelinaceen. *Botanische Jahrbücher für Systematik* **82**, 1–99.

Ronse De Craene, L. P. (1989). Floral development of *Cochlospermum tinctorium* and *Bixa orellana* with special emphasis on the androecium. *American Journal of Botany* **76**, 1344–1359.

Ronse De Craene, L. P. (1992). "The androecium of the Magnoliophytina: Characterisation and systematic importance." Doctoral dissertation, Catholic University of Leuven.

Ronse De Craene, L. P. and Smets, E. F. (1990). The floral development of *Popowia whitei* (Annonaceae). *Nordic Journal of Botany* **10**, 411–420.

Ronse De Craene, L. P. and Smets, E. F. (1997a). A floral ontogenetic study of some species of *Capparis* and *Boscia*, with special emphasis on the androecium. *Botanische Jahrbücher für Systematik* **119**, 231–255.

Ronse De Craene, L. P. and Smets, E. F. (1997b). Evidence for carpel multiplications in the Capparaceae. *Belgian Journal of Botany* **130**, 59–67.
Ronse De Craene, L. P. and Smets, E. F. (1998). Meristic changes in gynoecium morphology, exemplified by floral ontogeny and anatomy. *In* "Reproductive Biology in Systematics, Conservation and Economic Biology" (S. J. Owens and P. J. Rudall, eds.), pp. 85–112. Royal Botanic Gardens, Kew.
Ronse De Craene, L. P., Soltis, P. S. and Soltis, D. E. (2003). Evolution of floral structures in basal angiosperms. *International Journal of Plant Sciences* **164** (Suppl.), S329–S363.
Rothwell, G. W., Scheckler, S. E. and Gillespie, W. H. (1989). *Elkinsia* gen nov., a late Devonian gymnosperm with cupulate ovules. *Botanical Gazette (Crawfordsville)* **150**, 170–189.
Rudall, P. J. and Bateman, R. M. (2002). Roles of synorganisation, zygomorphy and heterotopy in floral evolution: The gynostemium and labellum of orchids and other lilioid monocots. *Biological Reviews* **77**, 403–441.
Rudall, P. J. and Bateman, R. M. (2003). Evolutionary change in flowers and inflorescences: Evidence from naturally occurring terata. *Trends in Plant Science* **8**, 76–82.
Rudall, P. J. and Bateman, R. M. (2004). Morphology of monocot flowers: Iterative patterns and developmental constraints. *New Phytologist* **162**, 25–44.
Rutishauser, R. (1998). Plastochrone ratio and leaf arc as parameters of a quantitative phyllotaxis analysis in vascular plants. *In* "Symmetry in Plants" (R. V. Jean and D. Barabé, eds.), pp. 171–212. World Scientific, Singapore.
Rutishauser, R., Wanntorp, L. and Pfeifer, E. (2004). *Gunnera herteri*—developmental morphology of a dwarf from Uruguay and S Brazil (Gunneraceae). *Plant Systematics and Evolution* **248**, 219–241.
Satina, S. and Blakeslee, A. F. (1941). Periclinal chimeras in *Datura stramonium* in relation to development of leaf and flower. *American Journal of Botany* **28**, 862–871.
Satina, S. and Blakeslee, A. F. (1943). Periclinal chimeras in *Datura* in relation to the development of the carpel. *American Journal of Botany* **30**, 453–462.
Sattler, R. (1974). A new approach to gynoecial morphology. *Phytomorphology* **24**, 22–34.
Sattler, R. and Lacroix, C. (1988). Development and evolution of basal cauline placentation: *Basella rubra*. *American Journal of Botany* **75**, 918–927.
Savolainen, V. and Chase, M. W. (2003). A decade of progress in plant molecular phylogenetics. *Trends in Genetics* **19**, 717–721.
Schick, B. (1980). Untersuchungen über die Biotechnik der Apocynaceenblüte. I. Morphologie und Funktion des Narbenkopfes. *Flora* **170**, 394–432.
Schick, B. (1982). Untersuchungen über die Biotechnik der Apocynaceenblüte. II. Bau und Funktion des Bestäubungsapparates. *Flora* **172**, 347–371.
Schlosser, G. (2004). The role of modules in development and evolution. *In* "Modularity in Development and Evolution" (G. Schlosser and G. P. Wagner, eds.), pp. 519–582. University of Chicago Press, Chicago.
Schöffel, K. (1932). Untersuchungen über den Blütenbau der Ranunculaceen. *Planta* **17**, 315–371.
Schönenberger, J. (2005). Rise from the ashes—reconstruction of charcoal fossil flowers. *Trends in Plant Science* **10**, 436–443.
Schönenberger, J. and Conti, E. (2003). Molecular phylogeny and floral evolution of Penaeaceae, Oliniaceae, Rhynchocalycaceae, and Alzateaceae (Myrtales). *American Journal of Botany* **90**, 293–309.

Schönenberger, J., von Balthazar, M. and Matthews, M. L. (2003). Flowers—diversity, development, and evolution: Introduction. *International Journal of Plant Sciences* **164** (Suppl.), S197–S199.

Schönenberger, J., Anderberg, A. A. and Sytsma, K. J. (2005). Molecular phylogenetics and patterns of floral evolution in the Ericales. *International Journal of Plant Sciences* **166**, 265–288.

Shamrov, I. I. (2000). The integument of flowering plants: Developmental patterns and evolutionary trends. *Acta Biologica Cracoviensia, Series Botanica* **42** (2), 9–20.

Sieber, P., Gheyselinck, J., Gross-Hardt, R., Laux, T., Grossniklaus, U. and Schneitz, K. (2004). Pattern formation during early ovule development in *Arabidopsis thaliana. Developmental Biology* **273**, 321–334.

Sigmond, H. (1929). Vergleichende Untersuchungen über die Anatomie und Morphologie von Blütenknospenverschlüssen. *Botanisches Centralblatt, Beihefte, I,* **56**, 1–67.

Sinha, N. (2000). The response of epidermal cells to contact. *Trends in Plant Science* **5**, 233–234.

Skinner, D. J., Hill, T. A. and Gasser, C. S. (2004). Regulation of ovule development. *Plant Cell* **16** (Suppl.), S32–S45.

Smith, J. F., Hileman, L. C., Powell, M. P. and Baum, D. A. (2004). Evolution of *GCYC*, a Gesneriaceae homolog of *CYCLOIDEA*, within Gesnerioideae (Gesneriaceae). *Molecular Phylogenetics and Evolution* **31**, 765–779.

Smyth, D. R. (2005). Morphogenesis of flowers: Our evolving view. *Plant Cell* **17**, 330–341.

Soltis, D. E., Soltis, P. S., Chase, M. W., Mort, M. E., Albach, D. C., Zanis, M., Savolainen, V., Hahn, W. H., Hoot, S. B., Fay, M. F., Axtell, M. Swensen, S. M., *et al.* (2000). Angiosperm phylogeny inferred from 18S rDNA, *rbcL.*, and *atpB* sequences. *Botanical Journal of the Linnean Society* **133**, 381–461.

Soltis, D. E., Soltis, P. S., Albert, V. A., Oppenheimer, D. G., dePamphilis, C. W., Ma, H., Frohlich, M. W. and Theissen, G. (2002). Missing links: The genetic architecture of flower and floral diversification. *Trends in Plant Science* **7**, 22–34.

Soltis, D. E., Albert, V. A., Kim, S., Yoo, M.-J., Soltis, P. S., Frohlich, M. W., Leebens-Mack, J., Kong, H., Wall, K., dePamphilis, C. and Ma, H. (2005a). Evolution of the flower. *In* "Plant Diversity and Evolution: Genotypic and Phenotypic Variation in Higher Plants" (R. J. Henry, ed.), pp. 165–200. CAB International, Cambridge, MA.

Soltis, D. E., Soltis, P. S., Endress, P. K. and Chase, M. W. (2005b). "Phylogeny and Evolution of Angiosperms." Sinauer, Sunderland, MA.

Soltis, P. S., Soltis, D. E. and Chase, M. W. (1999). Angiosperm phylogeny inferred from multiple genes as a tool for comparative biology. *Nature* **402**, 402–404.

Stewart, W. N. and Rothwell, G. W. (1993). "Paleobotany and the Evolution of Plants." Cambridge University Press, Cambridge.

Struwe, L. and Albert, V. A. (eds.) (2002). "Gentianaceae: Systematics and Natural History." Cambridge University Press, Cambridge.

Sugiyama, M. (1991). Scanning electron microscopy observation on early ontogeny of the flower of *Camellia japonica* L. *Journal of Japanese Botany* **66**, 295–299.

Sun, G., Ji, Q., Dilcher, D. L., Zheng, S., Nixon, K. C. and Wang, X. (2002). Archaefructaceae, a new basal angiosperm family. *Science* **296**, 899–904.

Tamura, M. (1995). Ranunculaceae. *In* "Die natürlichen Pflanzenfamilien" (A. Engler and K. Prantl, eds.), 2nd Edn., pp. 1–555. Duncker & Humblot, Berlin.

Taylor, T. N. (1988). Pollen and pollen organs of fossil gymnosperms: Phylogeny and reproductive biology. *In* "Origin and Evolution of Gymnosperms" (C. B. Beck, ed.), pp. 177–217. Columbia University Press, New York.

Tsou, C.-H. (1998). Early floral development of Camellioideae (Theaceae). *American Journal of Botany* **85**, 1531–1547.

Tucker, S. C. (1991). Helical floral organogenesis in *Gleditsia*, a primitive caesalpinioid legume. *American Journal of Botany* **78**, 1130–1149.

Tucker, S. C. (1992). The role of floral development in studies of legume evolution. *Canadian Journal of Botany* **70**, 692–700.

Tucker, S. C. (2003a). Floral development in legumes. *Plant Physiology* **131**, 911–926.

Tucker, S. C. (2003b). Floral ontogeny in *Swartzia* (Leguminosae: Papilionoideae: Swartzieae): Distribution and role of the ring meristem. *American Journal of Botany* **90**, 1271–1292.

Tucker, S. C. and Douglas, A. W. (1994). Ontogenetic evidence and phylogenetic relationships among basal taxa of legumes. *In* "Advances in Legume Systematics 6: Structural Botany" (I. K. Ferguson and S. C. Tucker, eds.), pp. 11–32. Royal Botanic Gardens, Kew.

Tucker, S. C. and Kantz, K. E. (2001). Open carpels with ovules in Fabaceae. *International Journal of Plant Sciences* **162**, 1065–1073.

Uimari, A., Kotilainen, M., Elomaa, P., Yu, D, Albert, V. A. and Teeri, T. H. (2004). Integration of reproductive meristem fates by a *SEPALLATA*-like MADS-box gene. *Proceedings of the National Academy of Sciences of the United States of America* **101**, 15817–15822.

van der Schoot, C., Dietrich, M. A., Storms, M., Verbeke, J. A. and Lucas, W. J. (1995). Establishment of a cell-to-cell communication pathway between separate carpels during gynoecium development. *Planta* **195**, 450–455.

van Heel, W. A. (1966). Morphology of the androecium in Malvales. *Blumea* **13**, 177–394.

van Heel, W. A. (1987). Androecium development in *Actinidia chinensis* and *A. melanandra* (Actinidiaceae). *Botanische Jahrbücher für Systematik* **109**, 17–23.

van Heel, W. A. (1995). Morphology of the gynoecium of *Kitaibelia vitifolia* Willd. and *Malope trifida* L. (Malvaceae-Malopeae). *Botanische Jahrbücher für Systematik* **117**, 485–493.

Verbeke, J. A. (1992). Fusion events during floral morphogenesis. *Annual Review of Plant Physiology and Plant Molecular Biology* **43**, 583–598.

Vincent, C. A. and Coen, E. S. (2004). A temporal and morphological framework for flower development in *Antirrhinum majus*. *Canadian Journal of Botany* **82**, 681–690.

Vishenskaya, T. D. (1980a). Polymerous androecium and its development in the flower of *Thea sinensis* L. (Theaceae). *Botanicheskij Zhurnal (Moscow and Leningrad)* **65**, 39–50.

Vishenskaya, T. D. (1980b). The development of the polymerous androecium in *Stuartia pseudocamellia* (Theaceae). *Botanicheskij Zhurnal (Moscow and Leningrad)* **65**, 948–957.

Vogel, S. (1959). Organographie der Blüten kapländischer Ophrydeen mit Bemerkungen zum Koaptations-Problem I/II. *Abhandlungen der Akademie der Wissenschaften und der Literatur Mainz. Mathematisch-Naturwissenschaftliche Klasse* **1959**, 265–532.

Vogel, S. (1961). Die Bestäubung der Kesselfallen-Blüten von *Ceropegia*. *Beiträge zur Biologie der Pflanzen* **36**, 159–237.

Vogel, S. (1978). Evolutionary shifts from reward to deception in pollen flowers. *In* "The Pollination of Flowers by Insects" (A. J. Richards, ed.), pp. 89–96. Academic Press, London.

von Balthazar, M. and Endress, P. K. (2002a). Development of inflorescences and flowers in Buxaceae and the problem of perianth interpretation. *International Journal of Plant Sciences* **163**, 847–876.

von Balthazar, M. and Endress, P. K. (2002b). Reproductive structures and systematics of Buxaceae. *Botanical Journal of the Linnean Society* **140**, 193–228.

von Balthazar, M., Schatz, G. E. and Endress, P. K. (2003). Female flowers and inflorescences of Didymelaceae. *Plant Systematics and Evolution* **237**, 199–208.

von Balthazar, M., Alverson, W. S., Schönenberger, J. and Baum, D. A. (2004). Comparative floral development and androecium structure in Malvoideae (Malvaceae s. l.). *International Journal of Plant Sciences* **165**, 445–473.

Wagner, G. P. (1996). Homologues, natural kinds and the evolution of modularity. *American Zoologist* **36**, 36–43.

Wagner, G. P. and Mezey, J. G. (2004). The role of genetic architecture constraints in the origin of variational modularity. *In* "Modularity in Development and Evolution" (G. Schlosser and G. P. Wagner, eds.), pp. 338–358. University of Chicago Press, Chicago.

Walker, D. B. (1975). Postgenital carpel fusion in *Catharanthus roseus* III. Fine structure of the epidermis during and after fusion. *Protoplasma* **86**, 43–63.

Warner, K., Rudall, P. J. and Frohlich, M. W. (2005). Environmental control of sepalness and petalness in waterlilies XVII International Botanical Congress, Vienna. Abstracts, 11.6.6.

Wikström, N., Savolainen, V. and Chase, M. W. (2001). Evolution of the angiosperms: Calibrating the family tree. *Proceedings of the Royal Society of London B* **268**, 2211–2220.

Yamada, T., Ito, M. and Kato, M. (2003). Expression pattern of *INNER NO OUTER* homologue in *Nymphaea* (water lily family, Nymphaeaceae). *Development, Genes and Evolution* **213**, 510–513.

Yoon, H. S. and Baum, D. A. (2004). Transgenic study of parallelism in plant morphological evolution. *Proceedings of the National Academy of Sciences of the United States of America* **101**, 6524–6529.

Zanis, M. J., Soltis, P. S., Qiu, Y.-L., Zimmer, E. A. and Soltis, D. E. (2003). Phylogenetic analyses and perianth evolution in basal angiosperms. *Annals of the Missouri Botancial Garden* **90**, 129–150.

Zik, M. and Irish, V. F. (2003). Flower development: Initiation, differentiation, and diversification. *Annual Reviews of Cell and Developmental Biology* **19**, 119–140.

Recent Developments Regarding the Evolutionary Origin of Flowers

MICHAEL W. FROHLICH

Department of Botany, Natural History Museum, London SW7 5BD, United Kingdom

Advances in Botanical Research, Vol. 44
Incorporating Advances in Plant Pathology

0065-2296/06 $35.00
DOI: 10.1016/S0065-2296(06)44002-7

ABSTRACT

The evolutionary origin of flowers has long been a famous puzzle. The problem has not been solved, but there have been advances. The Anthophyte theory was based on relationships inferred in the first cladistic analyses of seed plants done in the mid-1980s. Support for these relationships vanished when molecular phylogenetic analyses gave strong evidence that living gymnosperms are monophyletic, contradicting the anthophyte relationships. Paleobotanical cladistic analyses continue to find the anthophyte relationships (although sometimes weakly); some paleobotanists accept the molecular phylogenies but others do not. The Mostly Male theory arose from the study of the evolution of development, based on multiple sources of data. It provides a detailed scenario for flower origins, suggesting that the flower arose mostly from the male reproductive structure of the gymnosperm ancestor, and that the ovule precursors moved ectopically onto this formerly male structure. Some new data favor this theory while other data oppose it. The fossil group Corystospermales, which fits the theory well, may not figure as potential ancestors, but the Caytoniales may fill the same role. Other theories based on the evolution of development have been advanced to explain flower origins. Friedman's modular theory accounts for the angiosperm embryo sac and triploid endosperm. This paper is Floral Genome Project contribution number 56.

I. INTRODUCTION

The evolutionary origin of the flower is one of the major questions in the history of life, famously called an "abominable mystery" by Darwin (Darwin and Seward, 1903: 20–21). A multitude of theories have been proposed to account for flower origins, but it is not the objective of this chapter to summarize all theories or all previous works. This paper focuses on developments pertaining to flower origins, starting with the first cladistic analyses of seed plant morphology that led to the Anthophyte theory, which established the modern period in the study of flower origins. For broader discussions of theories on flower origins, including historical ideas, see Stebbins (1974), Beck (1976), Doyle (1978, 1996, 1998), Crane (1985), Crane *et al.* (1995), Friis *et al.* (1987), Cronquist (1988), and Taylor and Hickey (1996).

Elucidating the origin of flowers is important because of the centrality of flowers in angiosperm reproduction, in the production of human food from grains and fruits, and because of the familiarity of flowers in daily life. Flowers have traditionally provided the largest proportion of the information used to classify angiosperms. Events surrounding pollination and fertilization generate most of the isolating mechanisms involved in speciation and are commonly critical in the maintenance of species distinctions. The amazing diversity of flowers, unparalleled by any other plant organs, coupled with the intricacy of some pollination syndromes adds to their intrigue.

Historically, three interrelated problems have impeded solution to the question of flower origins: (1) the large morphological gap between angiosperms and the living and fossil gymnosperms that must include angiosperms' relatives; (2) uncertainty in the relationships among angiosperms; and (3) uncertainty in relationships among seed plants in general and gymnosperms in particular, especially fossil gymnosperms. These issues are interrelated because the large morphological gap between angiosperms and gymnosperms causes uncertainty in homology assessment between flower structures and structures of gymnosperm reproductive units, greatly contributing to uncertainties in inferring the relationships that constitute the third question. Only the second of these issues has now been largely resolved by recent molecular studies.

Angiosperms are seed plants, and share with living and fossil gymnosperms numerous complex vegetative and important reproductive characters, primarily related to the ovule and seed. Clearly angiosperms evolved from gymnospermous ancestors. Gymnosperms, as defined by the first seeds, arose in the Famennian stage of the late Devonian, 363–365 million years ago (MYA) (Hilton, J., personal communication). The earliest apparent angiosperms, based on pollen records, are in the early Cretaceous Hauterivian stage, 130–136 MYA (Friis *et al.*, 2006: 256; Hughes, 1994). Gymnosperms antedate angiosperms by ca 231 million years, longer than angiosperms have existed. Hence, considering both living and fossil members, "gymnosperms" are paraphyletic with respect to angiosperms. Along the lineage that eventually led to angiosperms, the distinction between angiosperms and gymnosperms is one of grade, not of clade. If numerous intermediate forms were known we might have difficulty in finding a natural break point at which to say that the angiosperm grade had appeared. The existence of intermediates would illuminate the sequence of evolutionary steps that led to the remarkable features of angiosperms, and there would be no uncertainty regarding angiosperm origins. The large morphological gap between angiosperms and known gymnosperms saves us from this minor nomenclatural problem, but leaves us instead the important problem of the mystery of angiosperm origins. For this chapter, I regard the angiosperms as having arisen when the first flower appeared. I define a flower as a bisexual structure with more or less recognizable carpels, that is, laminar structures largely enclosing the ovules on their upper (adaxial) surface, and with more or less recognizable stamens. The ovules also should have two integuments, like ovules in modern basal angiosperms. Petals and/or sepals (or tepals) may or may not have been present at this stage. This first flower would have appeared some time below the base of the extant angiosperm clade, that is, below the most recent common ancestor of all living angiosperms.

There is no structure among living or fossil gymnosperms that resembles a carpel. As the most complex organ of an angiosperm flower, the carpel has been the focus of most theories of angiosperm origins. An evolutionary scenario for the carpel must also account for the peculiar features of the angiosperm ovule, in particular, for the origin of the second (outer) integument of the angiosperm ovule, surrounding the inner integument and nucellus, which are homologous to the single integument and nucellus of the gymnosperm ovule (Stebbins, 1974). Angiosperm stamens also differ greatly from gymnosperm male reproductive structures (Crepet and Nixon, 1996; Hufford, 1996). Angiosperm flowers are bisexual, except for secondarily unisexual flowers. Only two gymnosperms normally show functionally bisexual reproductive units: Bennettitales, which figured prominently in the Anthophyte theory and also in the model of Meyen (1988), discussed later, and *Irania* (Schweitzer, 1977) (in the Czekanowskiales, Doyle, J. A., personal communication) which to my mind does not show closely plausible antecedents for the angiosperm carpel, ovule, and stamen, although it has been suggested as an angiosperm relative (Krassilov, 1997). Theories that derive the flower from an ancestor with separate male and female structures must also account for the origin of angiosperm bisexuality.

A full theory of flower origins must account for all the major morphological innovations of the flower (Frohlich, 2006), but older theories typically focus on only one organ or on one innovation, with little attention paid to the other novel attributes of the flower. Such theories are not only incomplete, but many are also mutually incompatible, so they cannot easily be merged to make a complete theory. Only one such theory by Gaussen (1946) is discussed later because it figures prominently in one detailed theory, the Mostly Male theory (Frohlich, 2001, 2002, 2003; Frohlich and Parker, 2000).

Vegetative morphology, while not the subject of this chapter, does provide numerous additional characters for inferring relationships among seed plants, including angiosperms (Doyle, 2006; Hilton and Bateman, 2006). There may be less difficulty in judging homology for many vegetative characters than for some reproductive characters, adding to the value of vegetative characters. In precladistic analyses, some vegetative characters were considered very strong evidence for close relationship between particular gymnosperm groups and angiosperms, such as the presence of net-veined leaves, shared by angiosperms and Caytoniales, or of the presence of vessels, found in most angiosperms and in Gnetales. However, evident homoplasy in these characters lessens their significance, so they are properly accorded only the weight of individual characters within cladistic analyses. Net-veined leaves are found in all basal angiosperms but are also present in a number of other groups, even ferns, as well as in several gymnosperm groups (although angiosperm venation is more complex than most of these) (Alvin and Chaloner, 1970; Boyce and Knoll, 2002; Glasspool *et al.*, 2004; Trivett and Pigg, 1996). Vessels are found in most but not all angiosperms, but the vesselless angiosperms are scattered among several clades, suggesting multiple gains and/or losses of vessels in disparate angiosperm lineages (Doyle and Endress, 2000; Young, 1981). Loss of vessels may be an adaptation to periodic freezing (Feild *et al.*, 2002). *Amborella*, commonly accepted as representing the basal-most clade of living angiosperms, is essentially vesselless (Feild *et al.*, 2000), and Nymphaeales, which are either the second clade diverging from the angiosperm lineage, or possibly sister to *Amborella*, have only small perforations or none at all through the pit membranes connecting adjacent vascular elements (Carlquist and Schneider, 2002). However, vessels are present in some gymnosperms, both living and extinct (Carlquist, 1996; Li and Taylor, 1999), in a number of ferns (Bhardwaha and Baijal, 1977; Carlquist and Schneider, 2001) and even in a *Selaginella*, a member of the basal-most clade of extant vascular plants (Schneider and Carlquist, 2000). This demonstrates much homoplasy in this character, if it were scored as the simple presence or absence of vessels. As in the case of many vegetative and reproductive characters representing major morphological transitions, especially those that show homoplasy, this character might be usefully resolved into multiple steps (Carlquist and Schneider, 2002), although such detailed knowledge is not yet available for all basal angiosperms. See Hilton and Bateman (2006) and Doyle (2006) for cladistic analyses of both vegetative and reproductive characters of angiosperms and gymnosperms.

Uncertainty in relationships within angiosperms has allowed broad speculation on the nature of the flower at the base of the angiosperm crown group (i.e., of the most recent common ancestor of all living angiosperms), which I will call the "ancestral flower." Attributes of the ancestral flower must be the endpoint for any reasonable theory of flower origins, as we cannot reconstruct states below that node using living taxa. At various times, angiosperms with simple, unisexual flowers consisting of only stamens or only carpels (e.g., oaks or *Ceratophyllum*) have been thought to represent the basal-most extant angiosperm group(s), suggesting that the basal angiosperm flower was very simple and unisexual (Chase *et al.*, 1993). This places very different constraints on theories of flower origin, compared to the accepted endpoint of a bisexual flower with multiple parts. Chloranthaceae and Piperaceae have also been suggested as the basal-most angiosperms (Taylor and Hickey, 1992); they each have unisexual or bisexual flowers consisting of one to three stamens and/or a single carpel (although Piperaceae may have more). Alternatively, plants with large flowers containing numerous parts, such as the woody Magnoliales or Nymphaeaceae have been widely considered characteristic of basal angiosperms, suggesting that the ancestral flower had numerous spirally arranged tepals, stamens, and carpels (Cronquist, 1988; Donoghue and Doyle, 1989).

This dispute has now been essentially resolved, with the repeated demonstration that the *Amborella*, Nymphaeales and Austrobaileyales (ANA) grade angiosperms (formerly called the ANITA grade) constitute the basal-most extant angiosperm lineages. All of them have flowers with multiple parts, and are either bisexual or clearly derived from bisexual ancestors. For flower origins the most intriguing remaining question lies at the basal node of extant angiosperms: whether *Amborella* is by itself the sister to the rests of the angiosperms, or whether *Amborella* + Nymphaeales constitute the sister of the remaining angiosperms (Barkman *et al.,* 2000; Goremykin *et al.*, 2003; Leebens-Mack *et al.*, 2005; Lockhart and Penny, 2005; Soltis and Soltis, 2004; Soltis *et al.*, 2004). I will most often assume the former, as I think the evidence favors that reconstruction. *Amborella* and Nymphaeales are the "A" and the "N" of the "ANA" acronym. Above them is a clade containing the families Schisandraceae, Trimeniaceae and Austrobaileyaceae. Illiciaceae, which provided the "I" of the old ANITA acronym, is now included in Schisandraceae (Angiosperm Phylogeny Group, 2003); the other two families provided the "T" and "A" of the old acronym. These are now all included in the Austrobaileyales (Angiosperm Phylogeny Group, 2003), providing the second "A" of the new three-letter acronym. Above the ANA grade phylogenetic relationships are in dispute, and may conservatively be described as a polytomy of Chloranthaceae, magnoliids, eudicots, and monocots, with *Ceratophyllum* most likely either sister to monocots or to the large eudicot clade, and Piperales within magnoliids (Qiu *et al.*, 2005; Soltis *et al.*, 2005; Zanis *et al.*, 2003). *Ceratophyllum* is certainly well enough above the ANA grade to have little effect on estimates of morphology at the basal node.

Discovery of a fossil angiosperm that is cladistically below the crown group (i.e., a "stem-group" angiosperm) could revolutionize studies of flower origins, by cutting the morphological long branch between crown-group angiosperms and gymnosperms. Unfortunately, no one has yet identified an angiosperm belonging to the stem group. *Archaefructus*, widely heralded as a stem-group angiosperm with elongate flowers bearing scattered carpels and stamens (Sun *et al.*, 1998, 2002), appears instead to be a member of the crown group. The elongate structures bear numerous short pedicels, which bear the carpels in pairs, and the stamens are similarly borne in pairs or larger clusters on short pedicels. This suggests that these pedicillate pairs or clusters are the true flowers, while the elongate structure is an inflorescence (Friis *et al.*, 2003). This is supported by the discovery of *Archaefructus eoflora*, which sometimes bears both carpels and stamens together in these small lateral flowers (Ji *et al.*, 2004).

The diversity of living gymnosperms is limited. They fall into only four (or five) groups: cycads, *Ginkgo*, Gnetales, and conifers, although conifers may comprise two distinct groups, Pinaceae and other (nonpinaceous) conifers. Most gymnospermous groups are extinct, so their interrelationships (and their relationships to living gymnosperms and angiosperms) must be inferred from morphological evidence. Identifying the gymnosperm group(s) that is sister to angiosperms would allow much improved inferences of the morphological start point for theories of flower evolution. The first careful cladistic studies of living and fossil seed plant relationships ushered in the modern period in the study of flower origins by allowing just such inferences (Crane, 1985; Doyle and Donoghue, 1986). They concluded that among living gymnosperms, the Gnetales are the closest relatives of angiosperms (Fig. 1A). This phylogeny has since been found repeatedly in morphological cladistic analyses. Relationships among the living gymnosperm groups and angiosperms can be addressed by molecular data, and this has yielded a great surprise: molecular studies support monophyly of the extant gymnosperms (Fig. 1B), in some studies with great strength, which conflicts with analyses of living and fossil groups based on morphology. Molecular phylogeneticists accept the molecular-derived analyses, as do some paleobotanists (Doyle, 2006), but other paleobotanists find this result surprising and they reject it, preferring the morphologically based analyses that put cycads basal among the living seed plants (Crane *et al.*, 2004; Hilton and Bateman, 2006; Rydin and Källersjö, 2002; Rydin *et al.*, 2002). Friis *et al.* (2006) consider the issue unresolved. Hence the crucial question of seed plant relationships is not resolved to everyone's satisfaction.

Theories of flower origins may be divided into two major groups based on the assumed morphological start point for flower evolution. Euanthial theories suggest that the flower arose from a shoot bearing a series of leaf homologs. Pseudanthial theories derive the flower from a compound shoot, that is, from an axis that bears leaf homologs, which in turn produce axillary shoots that bear additional leaf homologs that, in turn, bear the microsporangia and ovules. The latter theories require a great deal of fusion and/or loss of parts to achieve the morphology of a flower, while Euanthial theories require more or less elaboration to generate the observed complexity of the carpel (Crane, 1985; Doyle, 1994, 1996, 1998; Judd *et al.*, 2002).

I believe that two additional problems contribute to ongoing difficulties in resolving angiosperm relationships. In particular, the relative lack of

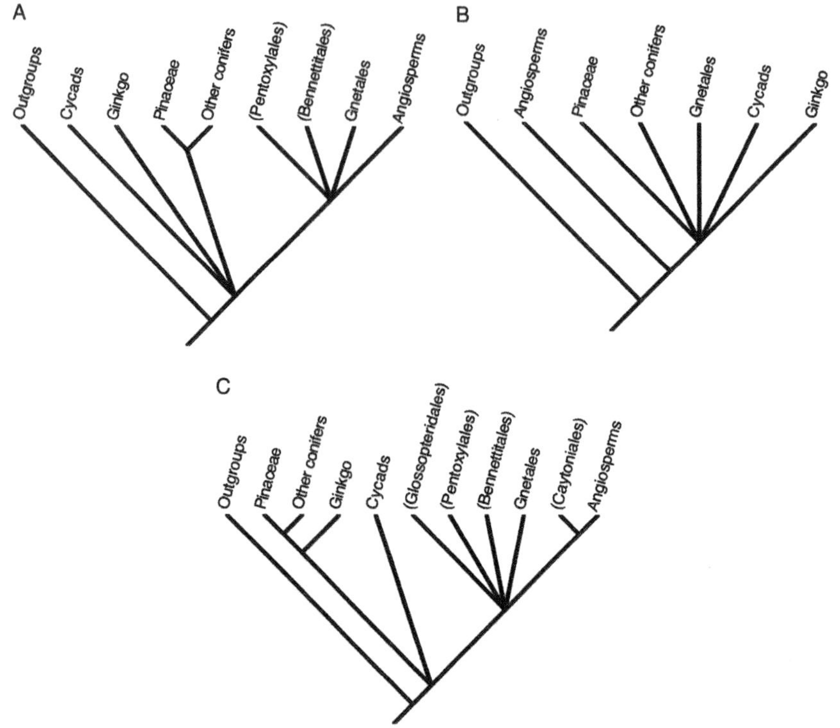

Fig. 1. (A) Simplified seed plant cladogram showing an anthophyte phylogeny. Only two extinct groups are shown, marked with parentheses. Here angiosperms are most closely related to Gnetales among living plants; along with the extinct Bennettitales and Pentoxylales they form the anthophyte group. (B) Cladogram of living seed plants showing monophyly of extant gymnosperms. Relationships within the living gymnosperms are not resolved. (C) Cladogram showing the modified anthophyte phylogeny of Doyle (1996). The fossil groups Caytoniales are sister to angiosperms, and Glossopteridales are also added to the anthophytes.

knowledge of Mesozoic pteridosperms (seed ferns), especially from the Lower Cretaceous and the Jurassic, precludes a full picture of gymnosperm diversity in the era during and immediately before the first appearance of angiosperms. This, in turn, seems partly due to the relative lack of well-preserved (especially permineralized) gymnosperm fossils known from this period, which directly reduces the knowledge that can be gained from the available fossils, and also discourages paleobotanists from intensively studying deposits of this age. The focus on studying fossil angiosperms, especially flowers, in Cretaceous and Tertiary sediments may also be contributing to the lack of work on gymnosperms of earlier periods. Study of early angiosperm fossils is crucial to understand the genesis of angiosperm diversity, but a clear understanding of angiosperm origins depends on knowledge of the gymnosperm groups of eras when close relatives of crown-group angiosperms must have been present. Not long ago, it was widely believed that significant numbers of fossil flowers would never be found, so few people looked for them. A number of localities are now known that provide excellent and diverse fossil flower material (Crane *et al.*, 1995: 30; Crepet, 2000; Friis *et al.*, 2006). Perhaps earlier deposits may exist that would provide similar excellent material to clarify gymnosperm diversity.

Fragmentary or misunderstood fossils of stem-group angiosperms may have already been collected, but may be too incomplete or poorly preserved to be recognized as such. Fossil gymnosperms that are very close outgroups to angiosperms may also have been found, but remain unrecognized for similar reasons. There are many fragmentary and/or poorly preserved fossils of uncertain relationships. Both Doyle (2006) and Hilton and Bateman (2006) restricted their published cladistic analyses to fossil plants that had been convincingly reconstructed in vegetative and male and female reproductive anatomy, to minimize missing data in their matrices, and the cladogram instability that missing data causes (although the latter, at least, did experiment with less well-understood taxa in their analyses). Much new morphological data is needed before adding these poorly understood fossils of Jurassic and Cretaceous gymnosperms to cladistic analyses.

The second problem is the deadening hand of wonderful, classic research. This phrasing is not oxymoronic. Superb studies sometimes become enthroned as classics of scientific research, to the point that no one looks at the subject again for decades, although analytic tools improve and scientific questions change. I think this has been the case in the study of the morphology of living plants, and several examples of old, long-established views that are now rejected will be mentioned later, that is, that basalmost angiosperm carpels are conduplicate (vs ascidiate), that carpels of nearly all angiosperms are sealed shut at anthesis, that double fertilization is the quintessential angiosperm feature, and that the derivation of cone scales in pine from short shoots is generalizable to all conifers (sensu lato).

II. WHAT NEEDS TO BE EXPLAINED FOR THE ORIGIN OF THE FLOWER?

In the absence of knowledge of stem-group angiosperms, the morphological endpoint for any reasonable theory of flower origins must be the "ancestral flower" as inferred at the base of the extant angiosperm clade. Such inference is based primarily on flowers of the ANA grade species, with some consideration of flower attributes of clades further up the tree, including Chloranthaceae, magnoliids, eudicots, monocots, and *Ceratophyllum*. Such inferences are made by placing character states at the terminals of a cladogram thought to reflect angiosperm phylogeny and projecting these states down the tree to the basal node. This is usually done using parsimony, as this is the simplest rule, and programs such as MacClade (Maddison and Maddison, 1992) do such reconstructions automatically. Phylogenies of the angiosperms typically contain polytomies. Reconstructing character states across polytomies is problematic; in MacClade explicit reconstructions of character states are done, assuming either: (1) the polytomy represents an actual multiple speciation event, in which case the most common descendent state is reconstructed below the polytomy or (2) the polytomy hides an unknown resolution of the dichotomizing tree in which case all character states shared by two or more descendents are placed below the polytomy. If no states are shared, then all descendent states are projected below the polytomy. This is said to be the most "favorable" reconstruction for each character (Maddison and Maddison, 1992: 92–93, 116–117). Alternatively, one can do character state reconstructions independently on each distinct dichotomous cladogram that is compatible with the polytomies, and then take the basal states reconstructed on most trees as reasonable inferences for the basal node (Frohlich, 1987). Whether *Amborella* or (*Amborella* + Nymphaeaceae) is the sister clade of the rest of the angiosperms does affect states inferred for the ancestral flower, but the morphological gap between angiosperms and gymnosperms remains so large that these have little direct effect on the question of flower origins.

Ideally, reconstruction of character states should include outgroups attaching below the node in question as well as the ingroup taxa above the node. This allows interaction and optimization in both topology and character state assignments between the gymnosperm and angiosperm groups. Doyle's (2006) cladistic analysis of seed plants includes 10 distinct angiosperm terminals in the cladogram representing each ANA grade family (combining the Nymphaeaceae and Cabombaceae, and separating the Illiciaceae from Schisandraceae), the Chloranthaceae and three magnoliid families. This is a sufficient sampling of angiosperms for reasonable reconstruction of the ancestral flower. However, there is such asymmetry between the robust, molecular-based phylogeny of living angiosperms, with the detailed knowledge of their morphologies, as compared to the much less well-known phylogeny and morphologies of the fossil gymnosperms that reconstruction of the ancestral flower should be done both with and without the gymnosperm taxa. This distinction is moot for flower characters that have no gymnosperm equivalents, as gymnosperm data will be scored as missing, but for other characters incorrect homology assessments between gymnosperms and angiosperms, or a nonhistorical tree topology among the gymnosperms could lead to erroneous state assignments at the base of the angiosperms. Furthermore, erroneous state assignments would be self-confirming for the gymnosperm portion of the tree. Different outcomes when character state reconstructions are done with and without gymnosperms would highlight characters and states that might benefit from further examination.

Excellent studies were done to reconstruct some ancestral flower character states before the ANA grade taxa were recognized as basal within angiosperms. Weller *et al.* (1995) found that gametophytic self-incompatibility systems probably did not occur at the base of the angiosperms. Bernhardt (2000) found that basal angiosperms were probably not specifically beetle-pollinated, but more likely exhibited generalist insect pollination syndromes. Albert *et al.* (1998) found that the perianth at the base of the angiosperms may well have been petaloid. Donoghue and Scheiner (1992) reconstructed endosperm evolution but with angiosperms sister to Gnetales. These results need to be checked in the light of the newer angiosperm cladograms. After recognition of the ANA grade, Bernhardt *et al.* (2003) found that *Trimenia*, of the third angiosperm clade, does have a self-incompatibility system and a generalist insect pollination syndrome, although *Illicium floridanum* apparently does not [Koehl *et al.*, 2004]. Thien *et al.* (2003) found that *Amborella* exhibits both generalist insect pollination and wind pollination, so generalist insect pollination is perhaps at the base of the angiosperms. *Amborella* is functionally dioecious, although derived from perfect-flowered ancestors (Endress and Igersheim, 2000a) so self-incompatibility is moot, and was presumably absent before the evolution of dioecy.

Doyle and Endress (2000: S146–S147) present the most-detailed reconstruction of the ancestral flower, based on angiosperm data, using a combined molecular and morphological parsimony analysis that resolves the accepted ANA topology, with *Amborella* alone specified as basal. This analysis includes many more angiosperm taxa than Doyle (2006) and also additional flower characters that are not pertinent to reconstruction of the gymnosperm phylogeny, which is central to Doyle (2006). Zanis *et al.* (2003) mapped perianth attributes onto their ANA reconstruction of basal angiosperm phylogeny. They reconstructed an undifferentiated perianth for the ancestral flower, with perianth arrangement (whorled vs spiral) and the number of parts as equivocal. Friedman and Williams (2004) reconstructed characters of embryo sac and endosperm, finding considerable differences from traditional views. Doyle (2005) reconstructed pollen attributes, also finding substantial differences from traditional views; the latter two issues are discussed later.

Parsimony ignores branch length in reconstructing character state changes on the tree. Hence, both sides of the dichotomy at the base of angiosperms have equal weight in reconstructing the states at the basal node, even although the branch to *Amborella* is long, while the internodes interconnecting the other ANA grade members near the base of the tree are much shorter (Soltis *et al.*, 2005: 38). In maximum likelihood (ML) analyses, inferred branch lengths affect the cost of placing character state changes on a particular internode (Jaramillo *et al.*, 2004; also see Pagel *et al.*, 2004 for a Bayesian approach to character mapping). Hence, state changes would tend to migrate to the longer branches compared to their reconstructed positions in a parsimony analysis. For reconstructions at the base of the angiosperms, this would have the effect of down weighting the influence of *Amborella* and increasing the input from the other near-basal taxa, especially the ANA members, but perhaps also Chloranthales, Magnoliales, eudicots, monocots, and *Ceratophyllum*. I think this is a more reasonable approach than strict parsimony because the base of the angiosperm tree is so unbalanced. For example, *Amborella* is scored as possessing an orthotropous ovule (with stalk opposite the micropyle) while the other ANA members, as well as most Magnoliales, eudicots, monocots, and most other angiosperms (but not Chloranthaceae) have anatropous ovules, with the stalk attached close to the micropyle. In a parsimony reconstruction, the basal node will be precisely equivocal for this character, because *Amborella* and the rest of angiosperms project opposite character states (Doyle and Endress, 2000). Taking into account branch length (e.g., via ML), the influence of *Amborella* would be reduced, while that of the other ANA members would be increased. Chloranthales would approximately balance Magnoliales + other higher angiosperms, so the basal node reconstruction would favor, at least to a considerable degree, anatropous over orthotropous ovules in the ancestral angiosperm flower.

III. IMPORTANT ATTRIBUTES OF THE ANCESTRAL FLOWER

A. BISEXUALITY

Although *Amborella* is functionally unisexual, the male flowers have vestigial carpels and the female flowers have vestigial stamens (Endress and Igersheim, 2000a); clearly *Amborella* had ancestors with bisexual flowers. All other basal angiosperms' groups are entirely or largely bisexual. *Amborella* is usually said to be dioecious, but some cultivated plants have been observed to switch from male to female from first year to the next (Buzgo *et al.*, 2004).

B. THE CARPEL

The carpel is the most complex organ in the flower, so it has been the prime focus of most theories of flower origin. Traditionally, the carpel has been considered a leaf homolog that is folded lengthwise (the "conduplicate" carpel), so the lateral edges meet, with at least a portion of this lateral edge becoming the stigma (Stebbins, 1974). As heralded by the name "angiosperm," the carpel is traditionally thought of as closed, that is, sealed shut by fusion where the lateral edges abut, except for a few unusual plants, such as *Degeneria* and *Drimys*. The superb work by I. W. Bailey and coworkers (Bailey and Smith, 1942; Bailey and Swamy, 1951; Swamy, 1949) is an example of classic work impeding understanding. When this issue was reexamined by Endress and Igersheim (2000b) using SEM, they showed that the carpels of basal angiosperms are generally ascidiate (urn or bucket shaped) rather than conduplicate, and that in very many basal angiosperms the carpel is not sealed by fusion of the carpel edges; instead the opening is filled with a mucilage secretion at anthesis. In many of these plants the carpels do become closed as the fruit develops, but they are not fused at anthesis. Hence, carpels of the ancestral flower are now pictured as ascidiate rather than conduplicate, with the opening occluded by mucilage.

By contrast, *Drimys* and *Degenaria* were famous as having unfused carpels at anthesis, with the abutting edges of *Degenaria* carpels connected only by interlocking trichomes, but Igersheim and Endress (1997: 224, 235, 252–253) showed that in fact their carpels are sealed shut at anthesis.

C. THE STAMEN

Angiosperm stamens are highly complex and remarkably uniform across the group (D'Arcy and Keating, 1996). Typical stamens have a filament bearing a single anther at the top. The anther consists of two thecae joined by a connective that is more or less continuous with the top of the filament. Each theca contains two embedded microsporangia that dehisce along a line joining the two locules of the microsporangia, so that after dehiscence it may not be obvious that two separate locules had been present. Dehiscence occurs by breakage of the stomium tissue that bridges the two locules of the microsporangia and extends to the external surface of the anther. In some basal angiosperms, including *Amborella*, Nymphaeaceae, and members of Austrobaileyales, the filament is laminar with the two thecae widely separated, but in Cabombaceae (the sister of Nymphaeaceae in the Nymphaeales) the filament is narrow, as in higher angiosperms. The traditional view that the ancestral flower had laminar stamens is probably correct, but there must be homoplasy in the character of laminar vs narrow filaments.

D. THE ANGIOSPERM OVULE

Angiosperm ovules typically have two integuments (coverings) and are bent over (anatropous), so the micropyle (the opening through the integuments) is near the stalk of the ovule. Gymnosperm ovules have one integument, with the micropyle at the opposite end of the ovule from the stalk. Gymnosperm ovules exude a droplet, which typically catches pollen grains and then brings them into the micropyle as the droplet is absorbed. Angiosperms make no such droplet, as the pollen is captured on the stigma of the carpel. There are numerous other differences in ovule and seed development; for example, gymnosperm seeds have a large gametophyte that serves as food storage tissue whereas angiosperms have a small gametophyte and endosperm as the typical storage tissue, which develops from the second fusion nucleus produced in double fertilization.

E. POLLEN

Typically the pollen wall of angiosperms is far more complex in structure than that of gymnosperms, probably related to the capture of pollen on the stigma and the subsequent pollen-stigma interactions, instead of capture by a gymnosperm pollination droplet (Taylor and Taylor, 1993: 738). Pollen is especially important for understanding angiosperm origins because it is so commonly represented in the fossil record, due to the resistance of sporopollenin to degradation, and because pollen grains of gymnosperms and angiosperms are clearly homologous and their complex structures provide many characters, making them highly informative. Preservation of fossil pollen is often good enough that even transmission electron microscopy (TEM) studies show useful structure. Fossil pollen has been intensively studied, in part because of the utility of spores and pollen grains to date sediments. A greater variety of extinct gymnosperms and angiosperms is surely known from their fossil pollen grains than from macrofossils, although of course fossil pollen shows a sampling bias favoring wind-pollinated plants.

Most angiosperm pollens have a columellar infratectum. That is, the middle portion of the outer layer (exine) of the wall contains vertical columns of sporopollenin, attached at their bases to a continuous layer of sporopollenin and joined at their tops by another layer that may be more or less continuous, often with perforations, or reduced to a tracery of narrow bridges between the tops of the columns. Magnoliales and Laurales commonly have a granular infratectum, without columns, which was previously thought to be the condition at the base of the angiosperms. Some living gymnosperms, such as Gnetales, also have granular walls, as do the fossil Bennettitales and *Pentoxylon* (Doyle, 2005), so this character had appeared to support a relationship between these groups and angiosperms. Other fossil pollens typically found as isolated grains also show granular walls, but are clearly gymnosperm pollens, based on other characters of the pollen, for example, *Eucommiidites* (Doyle *et al.*, 1975); this was later confirmed when associated macroscopic parts were found (Pedersen *et al.*, 1989). Based on the granular wall these had once been considered angiosperm relatives. However, molecular phylogenies show that Magnoliales and Laurales are not close to the base of angiosperms, so their granular walls must be derived from columellate. Examinations of ANA group members show that there are columnar features present in the exine of *Amborella* and also in at least some Nymphaeales. This had been considered an intermediate condition, but Doyle (2005, 2006) now considers this columellate, so Doyle reconstructs ancestral flower pollen as columellate. Hilton and Bateman (2006) regard Nymphaeales as granular, and do not include *Amborella* in their analysis, but the succeeding four subclades in their analysis are all columellate; considering only angiosperms, a parsimony analysis would make the base of angiosperms equivocal. If the genetic and biochemical mechanisms that produce these arrangements of sporopollenin could be elucidated, then a more informed delimitation of character states may be possible. The detailed analysis of basal angiosperm pollen by Doyle (2005) reconstructs many more attributes for pollen of the ancestral flower.

IV. WHAT SHOULD A THEORY DO?

A theory of flower origins should constitute a testable evolutionary scenario, accounting for all the major innovations of the flower (Frohlich, 2006). The ultimate theory should include the start point of flower evolution from known fossil gymnosperm(s) and an endpoint referable to the ancestral flower. The theory should include the series of morphological, developmental, and if possible genetic changes leading to the ancestral flower. The theory should also include suggestions regarding selective forces that could have pushed evolution in this direction. Small theories that account for only one or a few innovations in the origin of the flower are useful as components that can contribute to a grand theory, but are incomplete. Most of the early theories of flower origins dealt with only one of the flower's innovations, and typically were in conflict with each other regarding start points and end points. Such theories have limited utility. A full scenario would almost certainly contain multiple separable elements, such that some of these elements could be correct while others might be wrong. One must be careful not to discard an entire scenario if a separable component is wrong, as happened, for instance, with the theory of continental drift in the 1930s and 1940s (Frohlich, 2006).

V. THE ANTHOPHYTE THEORY

The modern period for the study of flower origins was initiated by Crane (1985), who published the first comprehensive cladistic study of the phylogeny of living and fossil seed plants, with extensive discussion on the morphology of potential angiosperm relatives. A principal objective was to identify the gymnosperm group(s) closest to angiosperms, to provide a start point for evolutionary scenarios on the origin of flowers. He found that angiosperms grouped with the living gymnosperms the Gnetales, and with two fossil groups, the Bennettitales and the Pentoxylales (Fig. 1A) (Crane, 1985; Crane *et al.*, 1995). Early in the twentieth century, Gnetales had been considered close to angiosperms, but later anatomical studies suggested that the supposedly shared features (e.g., vessels) were analogous rather than homologous (Bailey, 1944; Doyle, 1996), so Crane's result was a great surprise. Crane's work stimulated cladistic studies by Doyle and Donoghue (1986, 1987, 1992) who also found the same grouping, which they informally named the "anthophytes." Doyle and Donoghue (1986, 1987, 1992) noted that the reproductive axes of angiosperms, Gnetales, and Bennettitales all have sterile flat structures surrounding male structures with female structures in the center (although Gnetales are functionally dioecious and the ovules present in male structures are sterile, and serve only to make pollination droplets that attract pollinators). Doyle and Donoghue (1986, 1987, 1992) suggested that this overall arrangement of sterile, male and female parts might be homologous in the three groups. Doyle (1994, 1996, 1998) proposed detailed scenarios of morphological transformation, including various alternatives, starting from reproductive structures of different possible gymnosperm ancestors, most notably from idealized advanced pteridosperms, and leading to angiosperms, Gnetales and Bennettitales. However, in Doyle (1996) another fossil group, the Caytoniales, which had previously attached below anthophytes, moved up into the anthophyte clade as sister to angiosperms (Fig. 1C). Another fossil group, the Glossopteridales, also moved into the anthophyte group. Caytoniales are known only from detached megasporophylls, microsporophylls

(*Caytonanthus*), and leaves (*Sagenopteris*); there is no evidence regarding how these were borne on the intact plant, so whether they were borne in bisexual reproductive complexes or not is unknown. The placement of Caytoniales as sister to angiosperms makes homoplasy and synapomorphy equally parsimonious for the character of bisexual reproductive structures; hence this later version of the Anthophyte theory omits the inference of homology in the overall arrangement of parts in the reproductive units of the three groups (Doyle, 1996, 1998).

Morphological cladistic analyses by Nixon *et al.* (1994) found that angiosperms nested within the Gnetales, which would imply a Gnetales-style cone was the start point for the origin of the flower. Gnetales have compound cones, so this implies a pseudanthial origin of the flower, in contrast to a euanthial origin if the start point is a pteridosperm, as in the Anthophyte theory scenarios of Doyle (1994, 1996, 1998). Nixon *et al.* (1994) did not propose a detailed scenario for flower origins, but Doyle (1996, 1998) discussed various possibilities that their work would imply in comparison to the Anthophyte theory. Much discussion regarding these alternative possibilities rested on the characters and states used in the cladistic analyses to judge whether or not angiosperms really do nest within Gnetales.

VI. MOLECULAR PHYLOGENETIC ANALYSES OF SEED PLANTS UNDERMINE THE ANTHOPHYTE THEORY

Molecular phylogenies of seed plants yielded three surprises. In order of increasing controversy, they are: (1) Gnetales are not sister to angiosperms, (2) extant gymnosperms are monophyletic, and (3) Gnetales are either sister to conifers or nested within conifers and sister to the conifer family Pinaceae. For the question of angiosperm origins it is the position of Gnetales relative to angiosperms that is most important. If Gnetales are not sister to angiosperms then the underpinnings are removed from the Anthophyte theory (i.e., that angiosperms, Gnetales, Bennettitales, and Pentoxylales form a clade, with or without Caytoniales and Glossopteridales). This would not disprove the scenarios suggested for the Anthophyte theory (Doyle 1994, 1996, 1998), in particular those deriving Bennettitales and angiosperms from hypothetical pteridosperms, but Gnetales can have no place in these scenarios (Donoghue and Doyle, 2000). More important, it means Gnetales are not close angiosperm relatives, so they are not the key to understanding flower origins. Molecular studies did show unequivocally that Gnetales are monophyletic, refuting the suggestion of Nixon *et al.* (1994) that Gentales are paraphyletic.

Monophyly of the extant gymnosperms conflicts with the strongly held traditional view that cycads are basal among living seed plants. Monophyly also conflicts with the morphological cladistic studies mentioned earlier, which place angiosperms as sister to or nested within the Gnetales among living plants. If extant gymnosperms are monophyletic then no group of living gymnosperms is more closely related to angiosperms than is any other. This is of profound importance for the understanding of the origin of the angiosperms. The inference of extant gymnosperm monophyly remains controversial among paleobotanists. It is accepted by Doyle (2006), but rejected outright by Crane *et al.* (2004) and by Hilton and Bateman (2006) (although they are less concerned by the position of Gnetales than by the position of cycads, which Doyle also finds surprising (personal communication)), hence extant gymnosperm monophyly requires careful consideration.

A series of molecular studies now provide very strong evidence that angiosperms are not sister to Gnetales and that extant gymnosperms are monophyletic. Relationships among Pinaceae, Gnetales, and the remaining conifers are not directly pertinent to angiosperm origins, except that any close relationship among these groups precludes a sister-group relationship between Gnetales and angiosperms.

In the initial molecular studies of seed plant phylogeny, bootstrap support for the position of Gnetales relative to angiosperms was low (Chaw *et al.*, 1997; Goremykin *et al.*, 1996; Hansen *et al.*, 1999; Samigullin *et al.*, 1999), so these results were questioned (Doyle, 1998). However, later molecular studies using multiple genes strongly supported monophyly of living gymnosperms (Bowe *et al.*, 2000; Soltis *et al.*, 2002).

A. MOLECULAR PHYLOGENETIC STUDIES USING NONDUPLICATED GENES

Genes from all three genomic compartments have been used to address seed plant relationships, often in combinations within the same paper. These studies have the unusual outcome that different genomic compartments show significantly different results, and that adding additional genes has not made all analyses converge on similar topologies, unlike the situation in the study of angiosperm phylogeny (Soltis *et al.*, 1999). There is a trade-off in the number of genes in a particular study vs the number of taxa for which all or most sequences are available, so studies using more genes have lower taxon sampling. Most nuclear genes find extant gymnosperm monophyly and Gnetales sister to or within conifers; alternatives are weakly supported. Mitochondrial genes show extant gymnosperm monophyly with strong support. Chloroplast genes show an odd pattern, with the first and second codon positions supporting extant gymnosperm monophyly (or giving polytomies), but third positions placing Gnetales as basal within seed plants, at least in maximum parsimony (MP) analyses. For comprehensive summaries of these numerous studies see Magallón and Sanderson (2002) and Soltis *et al.* (2005). The following discussion focuses on particular studies which are central to this controversy.

Results from several studies place Gnetales several nodes away from angiosperms. For example, in Chaw *et al.* (2000), to allow Gnetales to be sister to angiosperms, the Gnetales clade would have to jump over four nodes, which have individual bootstrap supports (ML/MP) of 100/87, 100/91, 85/57, and 99/91. Although bootstraps across sequential nodes cannot be directly combined, violating so many well-supported nodes seems unlikely. (To obtain an explicit measure, one would need to check whether any bootstrap replicates that fail to support these nodes allow angiosperms to be sister to Gnetales. Considering the large number of ways that a node could lack support in a bootstrap replicate, this specific rearrangement of taxa is probably unlikely. When a specific phylogenetic placement is at issue, as here, it would be useful to report the frequency of that particular placement among the bootstrap replicates, even if—or especially—if it is rare or absent.)

Sanderson *et al.* (2000) and Magallón and Sanderson (2002) carefully analyzed two chloroplast genes, *psaA* and *psbB*. They found that the third positions had significantly different signal from the first and second positions and generated very different trees. Trees from the first and second positions of the two genes together (and also amino acid trees in the first study) support extant gymnosperm monophyly by both MP and ML, with Gnetales sister to *Pinus*, and do so with strong support, at least with MP (Fig. 2A). Trees from the third positions put Gnetales basal in seed plants and angiosperms sister to the remaining extant gymnosperms, with support at or near 100% for the critical nodes (Fig. 2B). Although very different, neither of these trees allows angiosperms to be sister to Gnetales nor supports the Anthophyte theory. When strange or strongly conflicting trees are found, long branch attraction (LBA) (Felsenstein, 1978; Hendy and Penny, 1989) is usually cited to explain the aberrant results. ML reconstructions show that, for these genes, the Gnetales are on substantially longer branches than the other seed plants, as are the outgroups. Sanderson *et al.* (2000) used simulations to study error rates and biases in these tree reconstructions. They imposed each of the three alternative trees—Gnetales basal in seed plants, Gnetales sister to angiosperms, or gymnosperms monophyletic—on the data and estimated branch lengths by ML. They then used these topologies and

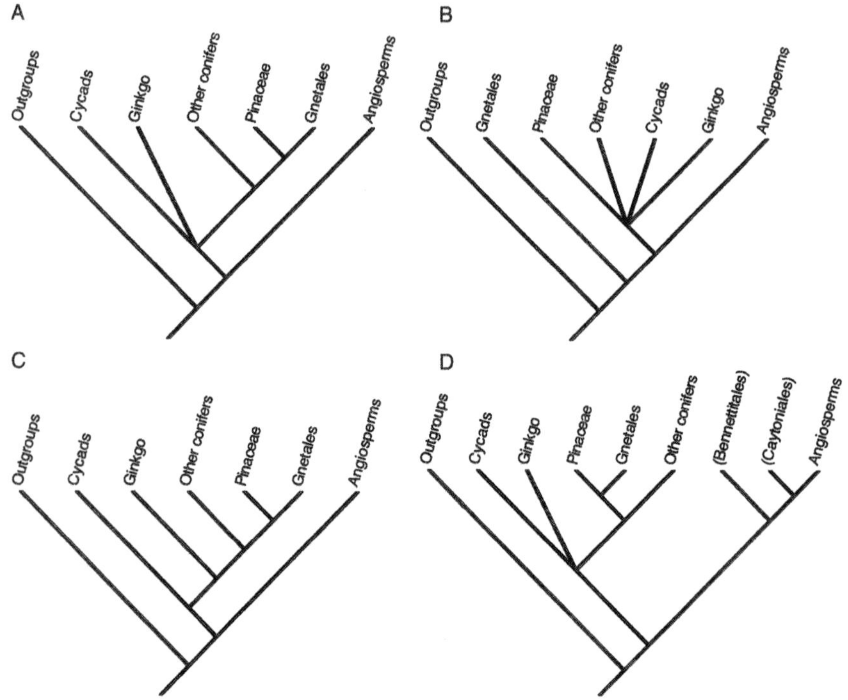

Fig. 2. (A) Cladogram showing extant gymnosperm monophyly with Gnetales sister to Pinaceae. (B) Cladogram showing Gnetales basal within seed plants, with angiosperms as the second clade. Relationships among other gymnosperm groups are not illustrated. (C) A fully resolved cladogram showing extant gymnosperm monophyly, based on Burleigh and Mathews (2004). (D) Cladogram showing positions of the extinct groups.

branch lengths to simulate sequence data, and did phylogenetic analyses to see if the imposed trees were recovered. For the third codon position there was clear bias toward the Gnetales-basal tree either if the anthophyte tree was imposed or (for *psbB*) if the monophyletic gymnosperms tree was imposed. This is due to LBA between the Gnetales sequences and the outgroup sequences, both of which are on longer branches than the other seed plant groups. They note, "At the other extreme are the first- and second-position data for both genes, which have extremely low error rates and which both point to the gymnosperm(s) [monophyletic] hypothesis as the true tree" (Sanderson *et al.*, 2000). They interpret these results as providing a "slight edge in weight of evidence" for gymnosperm monophyly. I suggest an alternative explanation for their results—that gymnosperm monophyly is the correct tree; as a result, imposing the anthophyte tree on the actual data produces aberrant branch length estimates, including a trichotomy between angiosperms, Gnetales, and the remaining gymnosperms, which is the closest the ML program is allowed to get to the monophyletic gymnosperms tree. Sanderson *et al.* (2000) note the trichotomies as strange reconstructions in the imposed ML tree, which appear for both compartments of both genes. This by itself hints that the imposed tree does not fit the data. Imposing the Gnetales-basal tree on the data has the effect of placing the two long branches together; this is also the outcome of LBA, so it is not surprising that this imposed tree is recovered by the simulated data.

Magallón and Sanderson (2002) found that the ML trees using third codon positions did not give the same results as the MP analysis using this data, but placed angiosperms sister to cycads with this clade then sister to a clade of Gnetales and conifers. This is also incompatible with the MP result from third positions that puts Gnetales sister to angiosperms. They did not report analyses using amino acid characters. They conclude that an "unambiguously supported hypothesis of phylogenetic relationships among seed plants has not yet been obtained." They note, "Adding more taxa has proven useful in solving particular phylogenetic problems because critically selected taxa may effectively break the long branches." However, they note the difficulty for adding taxa "because most seed plant diversity is extinct."

Classical LBA is not the only process that could account for Sanderson *et al.* (2000) and Magallón and Sanderson (2002) results from the third codon positions. Classical LBA arises from *random* substitutions on long branches, which, by independently inserting the same base at the same sequence position, creates homoplasious similarities (Felsenstein, 1978; Hendy and Penny, 1989). The effects of heterotachy (Lockhart *et al.*, 2006; Spencer *et al.*, 2005) might be called semirandom. Heterotachy is the variation, in different lineages (or over time) of the substitution rates at particular nucleotide positions; it is commonly observed (Spencer *et al.*, 2005). The processes that result in this variation may be nonrandom, but once established, heterotachy might allow parallel base substitutions in unrelated lineages, if both lineages exhibit elevated substitution rates at the same sites.

Nonrandom processes can also generate homoplasious similarities; in particular, differences in codon bias can strongly affect the third codon positions of affected genes (Christianson, 2005). Codon bias refers to differences in the relative frequencies of synonymous codons coding for individual amino acids. It is only partly caused by the mutational biases that affect overall GC content; it is also affected by selection on individual codons (Liu and Xue, 2005). Both factors primarily affect the third codon positions, and especially transitions, where most degeneracy in the code is found. These differences can be substantial in chloroplast genes, and they vary considerably within land plants (Wall and Herbeck, 2003). Selection on codons is thought to result from the relative abundances of tRNAs that bear different anticodons but synonymously insert the same amino acid. Protein synthesis is inhibited if scarce tRNAs are required to translate its codons, so highly expressed proteins (e.g., rbcL) are especially likely to show significant codon biases (Liu *et al.*, 2004). Codon bias could result in heterotachy appearing at many sites, if a previously severe codon bias became relaxed, allowing synonymous substitutions that were previously prevented. Alternatively, the appearance of similar, strong codon biases in different lineages could result in parallel, homoplasious substitutions at many sites, even if only one or a few amino acid codons were involved (Christianson, 2005).

The sharing of 6–10 (or perhaps fewer) character states can generate parsimony bootstrap support of 100%, so codon biases strong enough to affect phylogenetic estimation may not be apparent in standard statistical analyses of whole gene sequences. It would be interesting to use sequential character removal (Davis, 1993; Davis *et al.*, 1993) to identify which individual bases cause the Gnetales-basal trees to be favored in analyses based on the third codon position. If these character states can be explained as the result of codon bias, then trees based on the first and second positions should be accepted.

Rydin *et al.* (2002) analyzed seed plant phylogeny using plastid *rbcL* and *atpB* and nuclear 18S and 26S rDNA. They added sequences from many taxa to the previously available data, allowing much wider taxon sampling. Using all data and parsimony, Gnetales were basal within seed plants (100% jackknife support for root position); above them, angiosperms were sister to the rest of gymnosperms (100%) (Fig. 2B). The chloroplast data by themselves also placed Gnetales basal in seed plants (root position 97%), with angiosperms sister to the rest of gymnosperms (100%). The nuclear genes, by themselves, weakly favored Gnetales sister to angiosperms (56%).

Rydin *et al.* (2002) reported conflict in the chloroplast genes between transitions and transversions. They performed various analyses on the chloroplast data by itself, using all characters, or first and second positions only, or third positions only, and including or excluding transitions for the third positions or for all positions. None of these analyses placed angiosperms as sister to Gnetales. Gnetales grouped with conifers in monophyletic gymnosperms if all chloroplast positions were used, with transitions excluded only in the third position or in all positions (Fig. 1B). Gymnosperms were also monophyletic if only third positions were used with transitions excluded; however if the transitions were included then Gnetales became basal in seed plants. First and second position data gave polytomies for seed plants with or without transitions.

This clearly shows that the signal pushing Gnetales to the base of seed plants (Fig. 2B) is found among the transitions in the third codon positions.

The conflict between data partitions does not appear to be between transversions and transitions in general, but specifically with the transitions in the third codon position. This is similar to the result found by Sanderson *et al.* (2000), and might also be due to codon bias.

Rydin *et al.* (2002) reported Bremer support (decay index or branch support) in the trees they fully illustrate, in addition to jackknife support. Neither bootstrapping nor jackknifing can distinguish levels of support above 100%. Such support levels can be achieved with relatively few shared apomorphies. As data sets expand with ever longer DNA sequences, mechanisms such as LBA become ever more likely to cause sufficient spurious similarities to generate high bootstrap values, and effects of codon bias also accumulate in longer sequences. As Felsenstein (1978) stressed in his initial paper, increasing the amount of data analyzed increases apparent support for the erroneous LBA tree. Bremer support has no upper bound, so it can distinguish among all levels of support, including widely different levels that would all generate bootstraps or jackknives of 100%.

Even although anthophytes appear in only one of their trees, supported at only 56%, Rydin *et al.* (2002) conclude that sister-group status between Gnetales and angiosperms "cannot be ruled out." They note that, in the Gnetales basal trees, a change in the attachment position of the outgroup (the free-sporing plants) would make angiosperms and Gnetales sisters. Outgroup position is often labile in trees, presumably because there is often a long branch between the outgroups and the ingroup. Using all genes, and including third codon positions in the analysis, Rydin *et al.* (2002: 201, Fig. 1) found Gnetales basal within seed plants, but Bremer support for the attachment of the outgroup is 40 steps, which is one of the most strongly supported nodes in the tree. Of large groups, only monophyly of Gnetales and monophyly of Pinaceae have higher support. Instead of the outgroup being misplaced, it would be more reasonable to suggest that Gnetales are misplaced relative to angiosperms; the node-holding angiosperms above the Gnetales has only 29 steps of Bremer support. (Ideally, one should do analyses using constraint trees to impose these alternative topologies, to determine the precise numbers of additional steps required for the alternative position of interest, and to determine how such constraints affect the other portions of the trees.) However, the conflict between third position transitions and the other data is cause for worry. It strongly suggests that the distribution of character states in one or the other of these partitions reflects factors other than phylogenetic relationships. Third codon position transitions are suspect for evaluating such deep relationships, especially due to the presence of different codon biases in different groups of seed plants and outgroups (Wall and Herbeck, 2003). Coupled with the conflict between data partitions, this suggests that the third codon position transition should be discounted. Phylogenetic analyses of the coding sequences using amino acids, instead of DNA bases, would remove effects of codon biases, but are not reported. Rydin *et al.* (2002) show that analyzing nuclear and chloroplast genes together (including all positions) gives increased resolution and substantially stronger support values for many nodes compared to separate analyses (Rydin *et al.*, 2002: 201, Fig. 1 vs pp. 202, 203, Figs. 2 and 3). An analysis of all four genes, but without transitions in the two chloroplast genes might be very informative, but is not reported. I conclude that the results of Rydin *et al.* (2002) strongly oppose a sister-group relationship between Gnetales and angiosperms, and provide some additional support for extant gymnosperm monophyly.

Rydin and Källersjö (2002) evaluated taxon sampling effects with *rbcL*. They did phylogenetic analyses of multiple data sets, with 36 or 80 taxa, with various alternative taxa included. They did parsimony analyses with all data, or with transitions excluded, or with third positions excluded, and did Bayesian analysis as well. Their results are similar to those of Rydin *et al.* (2002); most parsimony analyses using all data favored Gnetales basal, but this result did not occur if transitions or third codon positions were removed. Analyses done with transitions removed favored Gnetales sister to or within conifers, or produced polytomies, as did all but one analysis with third codon positions removed. Depending on the substitution model, Bayesian analyses mostly gave either Gnetales basal or Gnetales sister to Pinaceae. This again demonstrates a conflict between transitions and transversions, and between third codon positions vs first and second positions. Third position transitions were not considered separately, but these results are consistent with them being the source of the conflict. Analyses using amino acid characters were not reported.

Rydin and Källersjö (2002) showed that the selection of taxa included in the analyses can have important effects on the outcome. One of their MP analyses, using the smaller sampling of taxa and including transitions, placed angiosperms sister to Gnetales, although with less than 50% jackknife support. One analysis using the large sampling of taxa, with third positions removed, also yielded anthophytes, but also made Pinaceae paraphyletic. Polytomies were quite common outcomes. If the conflicting signal from transitions (and perhaps just third codon transitions) is discounted, then the analyses of Rydin and Källersjö (2002) provide support for extant gymnosperm monophyly.

Rydin and Källersjö (2002) strongly advocate extensive taxon sampling. They note that increased taxon sampling can cut long branches, improving phylogenetic estimation (Graybeal, 1998). However, the rare phylogenetic outcome of Gnetales sister to angiosperms occurred once each in a large and a small taxon sample, although with low support or with the clearly aberrant inference of paraphyletic Pinaceae. It is certainly important to cut long branches, but the longest branches in the seed plant analyses are those from the base of seed plants leading to the various outgroup taxa, the Moniliformopses (ferns and fern relatives minus lycopods), the lycopods and the bryophyte groups, and also between the major seed plant groups. Adding numerous taxa within the crown groups of angiosperms, conifers or even Moniliformopses does not cut these crucial long branches. Furthermore, errors in phylogenetic estimation due to codon biases may not be resolvable simply by cutting long branches.

Burleigh and Mathews (2004) studied multiple genes in all three genomic compartments and performed detailed analyses, using ML, and MP, including ML analyses of individual genes using different exemplar taxa from each major group as well as using all taxa. They found little difference in the results, validating the use of exemplar taxa for further analyses. This reduced taxon sampling allowed calculation of ML bootstrap support. They partitioned the data into eight classes based on substitution rates, noting that this avoids lumping all third codon positions into the same class. They found that sites evolving at intermediate rates favored Gnetales sister to Pinaceae, while faster evolving sites were ambiguous, and slower sites had little phylogenetic signal. The ML analysis of all sequences using all data gave monophyletic extant gymnosperms with Gnetales sister to Pinaceae, and with all bootstrap values placing major groups at 98% or above (Fig. 2C). The MP tree with all data places Gnetales basal in seed plants, with angiosperms and the rest of gymnosperms as sisters in the node above, which is supported at 79% bootstrap (Fig. 2B). However, excluding the two fastest rate classes of data resulted in an MP tree with monophyletic extant gymnosperms and Gnetales sister to pines, with all nodes placing major groups at 100%, the same as the ML tree (Fig. 2C). Bremer support values were not reported.

Separate parsimony analyses of data from the three genomic compartments were done with all sites and also with the two highest rate classes excluded. The analyses with all sites gave differing results for each compartment, but none placed angiosperms sister to Gnetales. Nuclear genes gave monophyletic gymnosperms, with Gnetales basal within gymnosperms; chloroplast genes gave Gnetales basal within seed plants, with angiosperms and the rest of gymnosperms sisters at the node above, with 100% bootstrap support; while the mitochondrial genes gave angiosperms sister to cycads, with this pair high in the seed plant tree.

With the two fastest rate classes excluded, all three compartments gave the same tree: monophyletic gymnosperms, with cycads basal within gymnosperms and Gnetales sister to pine (Fig. 2C). In the chloroplast tree all bootstraps placing major groups were at 93% or higher. The fastest evolving sites are the ones most likely to cause LBA, and in general to exhibit noise. That the trees with all data disagree shows that there is no consistent signal in the fast sites across the three compartments. That all three compartments give the same tree when fast rate classes are excluded, minimizing the risk of LBA, provides very strong support that this is, indeed, the historically correct tree. That the chloroplast tree based on all data gives such strong support to the incompatible phylogeny placing Gnetales basal within seed plants (Fig. 2B) again demonstrates the presence of a contrary signal, which may be due to codon bias in addition to, or instead of, LBA. Analyses based on amino acid characters would have been of interest.

In sum, analyses based on nonduplicated genes repeatedly find a strong phylogenetic signal supporting extant gymnosperm monophyly. The critical alternative phylogeny that was the source of the Anthophyte theory is very strongly opposed, because it is almost never found, and in those few cases where it is found it is very weakly supported. The only serious alternative, that Gnetales might be the basal-most group within seed plants, appears only from the most rapidly changing sites, in data sets dominated by transitions or third codon positions. Sites showing high rates of change are especially vulnerable to LBA because high rates of change are the essence of a long branch. Third codon positions, and third codon transitions in particular, are subject to artifacts resulting from codon bias (Christianson, 2005); like LBA, this can result in seemingly strong support for a spurious phylogenetic arrangement. The crucial long branch, between the base of the seed plants and the free-sporing outgroups, cannot be cut by additional taxon sampling.

B. MOLECULAR PHYLOGENETIC STUDIES WITH DUPLICATED GENES

Long branches in molecular trees can be cut by gene duplications (Mathews and Donoghue, 2000), and this may be the only way to cut long branches of crucial importance in understanding plant evolution. A lineage of organisms that persists over a long geologic interval, and does not generate any surviving sister groups during that interval, cannot have its long branch cut by additional taxon sampling of unduplicated genes. Appropriate taxa are not available. But if a gene duplication occurs during this interval, and if both gene lineages survive to the present, then the long branch is cut at the point in time when the gene duplication event occurred. Duplications can occur at any point along a lineage, but duplications might occur shortly before an organismal radiation. Gene duplications are thought to be one of the motors allowing evolution to occur (Monson, 2003; Moore and Purugganan, 2003).

If the duplication contributes to the success of the organism, or if it is coopted for an important novel attribute, or merely if it accompanied a polyploidy event that contributes to evolutionary success and diversification of the organismal lineage, then the duplication event would slightly antedate the radiation. This would leave a *short* branch between the duplication event and the base of the organismal radiation—the ideal circumstance for robust phylogenetic analysis of the radiation. Each paralog of the duplicated genes can root the cladogram of the other paralog, or both together can be rooted by genes from organismal outgroups.

Studies of nuclear gene families can provide such data. The MADS-box genes are the most thoroughly studied gene family in plants, and they have undergone numerous duplications, including duplications within long branches separating major taxonomic groups (Theissen *et al.*, 2000). However, the large numbers of duplications within MADS genes, including within many subclades, presents the risk that the true sister of a gene under study might not be found, either because it has been missed among the numerous other MADS genes or because it has been lost from the genome. Gene families with fewer duplications, and therefore fewer overlooked and perhaps few lost genes would be more amenable for such studies. Even with this caveat, MADS genes have suggested that Gnetales are closer to other gymnosperms than to angiosperms. Winter *et al.* (1999) found several cases of Gnetales MADS genes sister to Pinaceae genes, sometimes with strong support, rather than sister to angiosperm MADS genes, and concluded that Gnetales were more closely related to pines than to angiosperms. At that time relatively few MADS genes were known, but current large scale expressed sequence tag (EST) projects, such as the Floral Genome Project on basal angiosperms, *Welwitschia* and *Zamia* (Albert *et al.*, 2005), other projects on *Cycas* (Brenner *et al.*, 2003) and *Ginkgo* (Brenner *et al.*, 2005), and various conifer projects will provide sufficiently dense coverage of many gene families for extremely robust phylogenetic analyses of seed plants using many duplicated genes.

A duplication in the small *FLORICAULA/LEAFY* (*LFY*) gene family is ideal for evaluating the position of angiosperms relative to Gnetales because duplications in *LFY* are rare and this particular duplication occurs below the base of extant seed plants (Frohlich and Parker, 2000). Hence, this duplication cuts the long branch between extant seed plants and their free-sporing outgroups. Frohlich and Parker (2000) did phylogenetic analyses using amino acid characters, to avoid bias due to the unusually high GC content in the *Nymphaea* sequence. In parsimony analyses we found that *LFY* supports extant gymnosperm monophyly with 17 steps of Bremer support on branches that are a maximum of ca. 100 steps long; this is enough Bremer support to provide 100% bootstrap or jackknife support several times over.

This emphatically refutes claims of close relationship between angiosperms and Gnetales. ML analyses also found extant gymnosperm monophyly.

That the longest (parsimony) branches are only ca. 100 steps means that LBA cannot account for this result. To explain away this tree by invoking LBA, one would need to pretend that chance alone could create 17 homoplasious amino acid identities uniting the extant gymnosperms, as is in fact observed, in excess of the imagined synapomorphic **plus chance** identities that unite the supposed sisters Gnetales and angiosperms. Such an excess of chance occurrences might be imagined in branches of 1000 or 10,000 steps, but branches of 100 are far too short. Note, also, that amino acid characters have 20 character states, not the four states of DNA bases. Even although substitutions are far from equiprobable among the amino acids, this does create a larger character space than for DNA, which further reduces the probability of amino acid identities arising by chance.

The tree in Frohlich and Parker (2000) has genes from few taxa, but does have representatives from conifers (sensu lato) and the other three gymnosperm groups, with genes from two Gnetales. Additional genes would not cut the branches adjacent to the nodes in question, making them of limited added value (Burleigh and Mathews, 2004). Before we had sequences from all four extant gymnosperm groups we found Gnetales closer to pine than to angiosperms in every analysis; after obtaining sequences from all four groups we found extant gymnosperm monophyly in every analysis. This has remained the universal result in more recent (unpublished) analyses, and is found in all *LFY* trees published by others.

The results with *LFY*, added to the results from nonduplicated genes, conclusively show that, among living seed plants, angiosperms and Gnetales are not sister taxa, and that the living gymnosperms form a monophyletic grouping.

C. MORPHOLOGICAL PHYLOGENETIC ANALYSES

Extant gymnosperm monophyly remains controversial among paleobotanists. Two cladistic studies on the phylogeny of living and fossil seed plants (Doyle, 2006; Hilton and Bateman, 2006) officially focus on pteridosperm ("seed fern") relationships, but include analyses of all seed plant groups that are sufficiently well known for cladistic analyses. Both studies find that morphological characters support a sister-group relationship between Gnetales and angiosperms among extant seed plants, although in Doyle (2006) it is only one step longer to put Gnetales with conifers. Among all taxa examined, living and fossil, both studies find Caytoniales as sister to angiosperms. However, both studies show that cladograms only a few steps longer would allow extant gymnosperm monophyly. If extant gymnosperm monophyly is used as a constraint, then Caytoniales remain sister to angiosperms, with Bennettitales as the next group. Results of these two analyses are broadly similar, although interpretations differ due to their differing views on extant gymnosperm monophyly.

Hilton and Bateman (2006) reject extant gymnosperm monophyly, noting disagreements among the molecular analyses, and the dangers of LBA in molecular studies that cannot sample fossil organisms. They question whether one should accept a "majority rule consensus" of molecular studies. They prefer seed plant phylogenies based on morphological analyses, although they also present analyses using the constraint of extant gymnosperm monophyly. Crane *et al.* (2004) also reject extant gymnosperm monophyly in favor of morphologically based trees.

Doyle (2006) accepts molecular analyses showing extant gymnosperm monophyly, at least provisionally to evaluate their implications for angiosperm questions (personal communication). He notes "only a few molecular analyses have linked angiosperms with Gnetales, and this with low support." He notes that "some analyses have placed Gnetales at the base of seed plants . . . but tests using likelihood and other methods suggest that this arrangement is a result of long-branch attraction, particularly affecting third codon positions . . ." Doyle (2006) therefore favors a morphological phylogeny constrained by the relationships among the living taxa deduced from molecular analyses.

As is clear from the previous section, I strongly favor the molecular analyses that show extant gymnosperm monophyly. I agree that one should not accept a "majority rule consensus" tree based on balancing the results of various studies; rather, one should examine the studies, effectively performing a meta-analysis, to determine which results are credible. As discussed above, the morphological phylogenetic result of Gnetales sister to angiosperms is thoroughly rejected. The placement of Gnetales as basal among extant seed plants arises from the least reliable data, of the most rapidly changing sites (Burleigh and Mathews, 2004) and especially from third codon transitions in chloroplast genes, which are especially subject to codon bias. When questionable data disagree with strong data one should accept the results of the strong data, as Burleigh and Mathews (2004) do. Furthermore, studies using duplicated genes can cut long branches even although no sister group survives from the interval on the long branch where the duplication occurs. If this long branch, or much of it, represents a ghost lineage (i.e., an inferred lineage without known fossil representatives), then it would remain a long branch in paleobotanical cladistic analyses. One should use extant gymnosperm monophyly to constrain cladistic analyses of morphological data. I think the best estimate of relationships among the fossil seed plants should be based on the constraint tree of extant gymnosperm monophyly.

However, whether paleobotanical authors accept or reject extant gymnosperm monophyly is not the important aspect of their work; rather, it is the careful reconstruction of fossil taxa, the analysis of characters along with their potential transformations, and the analysis of the data matrices that make both of these papers very important. It is especially valuable to have such careful analyses that are largely independent, as this is likely to speed progress in the understanding of fossil seed plant diversity, a critical issue for elucidating the origin of angiosperms. Disagreements foment progress. If there were only one such new analysis it might well become canonized as a classic of research, inhibiting further studies.

The shortest trees in both of the morphological studies show few nodes with bootstrap support over 50%, and most nodes have the smallest possible amount of Bremer support: one step (Doyle, 2006: Fig. 6; Hilton and Bateman, 2006: Fig. 4). In a molecular study this would be considered too weak to allow any conclusions to be drawn. However, I think that morphological characters count for more than individual molecular characters for two reasons: (1) major morphological characters, as in these analyses, result from the activities of many genes, and the visible character states arise through a developmental program that has been constrained by selection. Such characters are much less labile than individual DNA bases or the amino acids in a protein; hence they convey more information. However, it is crucial that the homology assessments and transformation series among such characters are correctly inferred, or the characters will support spurious relationships and (2) the characters are of interest in and of themselves. The characters in these analyses describe the remarkable evolutionary steps that have generated the diversity of gymnosperms and angiosperms through time. An understanding of the series of character state changes that led to the modern flora, as would be found on the true, historical, morphological-cladistic tree would constitute the desired account of angiosperm origins.

Hilton and Bateman (2006) did a constrained analysis, forcing extant gymnosperms to be monophyletic. The resulting trees were only four steps longer (360 steps) than their shortest unconstrained trees (356 steps). Doyle (2006) did a constrained analysis that just forced Gnetales and conifers together; this added only one step to his shortest trees of 321 steps. In that constrained analysis cycads were at the base of the lineage that included angiosperms. The stronger constraint of extant gymnosperm monophyly only added two steps. These are not very large differences; the historically correct trees might have included homoplasies that strict parsimony analysis would reject. I would certainly prefer the constrained trees consistent with the very strongly supported result of extant gymnosperm monophyly, to the unconstrained analyses that place angiosperms sister to Gnetales among living plants. The constrained trees show large polytomies. This is disappointing, but does not indicate that the trees are wrong. Doyle (2006) describes the two distinct topological islands that result from his most elaborately constrained trees, which enforce the relationships within angiosperms and within conifers to follow those derived from molecular studies. Glossopteridales + *Pentoxylon* jumps between distant locations, causing the central portion of the consensus tree to collapse into a polytomy.

Hilton and Bateman (2006) conducted many experiments with their data, by adding or removing various taxa. Excluding various of the progymnosperm outgroups led to some remarkable rearrangements of the seed plant taxa. In one case angiosperms became basal among extant seed plants, but some of the fossil hydrasperman pteridosperms moved to the top of the tree. These plants are clearly basal, as shown by stratigraphy and by morphology, so this tree is certainly erroneous. Based on this clearly erroneous result they were "inclined" to reject any tree, "based on any kind of data," that placed angiosperms below cycads (i.e., showed extant gymnosperm monophyly), "as merely demonstrating the vital importance of including sufficient well-chosen taxa to root the trees successfully." However, misplacement of some taxa does not prove that other taxa are also misplaced, and instability in a particular data set reflects attributes of that data set only, not other independent data sets. Hilton and Bateman (2006) criticized the molecular studies, saying that the long branch leading to the free-sporing outgroups could cause "spurious rooting." Yet this same phenomenon appears to be operating in this case with their morphological data set.

Both studies concur regarding some relationships of importance for the origin of angiosperms. Both studies, in all analyses, place Caytoniales as the sister group to angiosperms, with Bennettitales as the next outgroup. Also, both studies place the corystosperms close to the peltasperms at some distance from angiosperms.

Disagreements between molecular- and morphologically based analyses highlight the need for careful analyses to understand why the different data sets result in such different inferences. I favor sequential character removal (Davis, 1993; Davis *et al.*, 1993) to determine which character states are crucial for the existence of particular clades in the shortest tree(s). Morphological characters that generate the anthophyte grouping may then be examined with especial care to verify the state assignments, and to search for possible indications of homoplasy. Characters that support a clade are **not** always the characters that change state below that clade. Sometimes the crucial characters serve to pull other taxa **away** from the clade in question, ensuring that these other taxa cannot jump into the clade of interest in any bootstrap replicates, thus permitting the clade to achieve strong support. Such interactions between characters and their supported nodes are not obvious from any other sort of analysis (Davis, 1993; Davis *et al.*, 1993).

The surprise of extant gymnosperm monophyly stimulated Doyle (2006) to reexamine characters that appeared to unite Gnetales with angiosperms. This reanalysis led him to recode some characters, with the effect of reducing the support associating Gnetales with angiosperms.

A different example illustrates why individual morphological characters can appear to be conclusive evidence of relationship, yet can nevertheless suffer recoding. The surprise result that Gnetales appear as sister to Pinaceae in some molecular analyses was ridiculed because it implied convergent evolution in cone structure between Pinaceae and the conifer remnant group. Some years ago, one famous paleobotanist told me, with great vehemence, "It's stupid! It's stupid!" The evolution of Pinaceae cones was the subject of classic work by Florin (1951, 1954), who showed that the female cone scale of Pinaceae derived from a short shoot-bearing leaf homologs that, over evolutionary time, all became congenitally fused (i.e., fused from their inception as the cone develops). This cone scale is borne in the axil of a small subtending bract. The cone scale and the bract arise as clearly separate structures in development, but they usually become fused later, as they continue to develop (i.e., postgenitally fused). (Douglass fir, *Pseudotsuga menziesii*, in Pinaceae, is an exception; its bracts remain unfused, become long and thin, and protrude from between the cone scales.) Cones of other conifers look generally similar to typical Pinaceae cones (except for some highly modified cones), so their cone scales were assumed to be homologous to Pinaceae cone scales.

This provides an example of the deadening hand of classic research. No one studied the early development of nonpinaceous conifer cones until recently. Takaso and Tomlinson (1989, 1990, 1991, 1992), Farjon and Garcia (2002, 2003), and Zhang *et al.* (2004) showed that in several genera of Cupressaceae (sensu lato, including Taxodiaceae) the cone scale is the bract, whereas the short shoot in the bract's axil remains small, forming the ovule and little else. The numbers and precise positions of these axillary short shoots vary in different conifer groups. Thus, Pinaceae and these nonpinaceous conifer cones develop in different ways, suggesting that they might possibly not be homologous. However, other nonpinaceous conifers (e.g., Auracariaceae) have scales composed of both an expanded short shoot and the bract; if Auracariaceae is in the basal clade of the nonpinaceous conifers then there may have been an evolutionary progression between the two organizations. If pinaceous and nonpinaceous cones were not homologous, this would remove the strongest morphological evidence against Gnetales nesting within conifers.

Perhaps convergent evolution of similar-appearing cones could have resulted from selection to hide the pollination drop, to prevent insects from drinking the droplet as a water or nutrient source (Frohlich, 2001).

VII. THE MOSTLY MALE THEORY

The Mostly Male theory was the first to be based on evidence from genetics, developmental genetics, and the morphologies of modern and fossil plants. It is also one of the most complete theories, attempting to provide a full scenario for the origin of bisexuality, the carpel and the angiosperm ovule, with suggestions regarding possible selective forces. Subsequent theories are in part reactions to it, so this theory is discussed in detail.

The work that led to it began as an effort to test the Anthophyte theory's most accessible inference: the claimed homology in the overall arrangement of the sterile, the male and the female organs in the reproductive units of Gnetales and angiosperms. The ABC-model of Coen and Meyerowitz (1991) had already received strong support (Weigel and Meyerowitz, 1994); furthermore, the critical ABC MADS-box genes had already been cloned and they showed expression zones corresponding to the regions of the flower where they helped specify developmental fate. The initial goal was to clone the closest homologs (hopefully the orthologs) of these genes in Gnetales and determine if they had expression zones comparable to their homologs in *Arabidopsis* and *Antirrhinum*. The *LFY* gene was also studied as a back up, in case the ABC MADS-box genes turned out to be at the wrong level in the developmental control system to help elucidate the origin of flowers.

We refocused on the *LFY* gene after finding the duplication in *LFY* that antedated the split between Gnetales and Pinaceae (Frohlich and Meyerowitz, 1997). The angiosperm *LFYs* attached below the Gnetales-Pinaceae split (for both copies), hinting that Gnetales might be closer to pines than to angiosperms, but support was so weak that we did not mention this inference (although Baum, (1998) did note this implication). Subsequently, we found that *Ginkgo* and a cycad (*Zamia*) also had both duplicates; with the addition of outgroup sequences from ferns the *LFY* genes gave very strong support to extant gymnosperm monophyly (Frohlich and Parker, 2000).

A. THE GENESIS OF THE MOSTLY MALE THEORY

The trigger for the Mostly Male theory was the discovery that the *LFY* duplication antedated the split between angiosperms and extant gymnosperms (with 91% bootstrap support). Angiosperms should therefore have inherited two copies of *LFY*. (See Frohlich and Parker (2000) for a discussion of placing gene trees atop species trees, which is the source of this inference.) However, all angiosperm *LFY*s corresponded to just one of the two gymnosperm copies, the "Leaf" copy, rather than to the "Needle" copy, which was lost. We searched for the "Needle" copy in *Nymphaea*, but found only pseudogenes of the known copy of *LFY* (Frohlich and Parker, 2000). The important role of *LFY* in specifying the floral apex suggested that this duplication might be evolutionarily important. In *Arabidopsis*, where it had been studied most carefully, *LFY* is expressed in and near shoot apices in a graded manner, regulated by multiple inputs. At vegetative apices it is expressed at low to moderate levels, but if the level of *LFY* expression becomes high enough then an autoregulatory circuit is activated sending expression of its components, including *LFY* and *AP1*, to high levels, causing the apex to become specified as floral (Blázquez and Weigel, 2000; Blázquez *et al.*, 2006). *LFY* is a transcription factor that upregulates many of the major genes known to control flower development. If the *LFY* homologs of gymnosperms have similar roles then loss of the one copy might signal loss of the downstream genes it regulates.

Expression data had been reported for the two *LFY* genes in *Pinus radiata*. The Needle copy (*NEEDLY*) was cloned by Mouradov *et al.* (1998), who published *in situ* hybridizations showing low expression in both vegetative apices and in apices of male cones, but high expression in early developing female cones. In parallel with the known function in angiosperms, this suggests that *NEEDLY* might specify female cones. Mellerowicz *et al.* (1998), using northern hybridization, found low expression of the Leaf copy (*PRFLL*) in vegetative tissues and in female cones, but high expression in male cones, hinting that this gene might specify male cones.

Angiosperms retain the Leaf copy of *LFY*, suggesting that they may also have retained the downstream genes it activates, but angiosperms have lost the Needle copy, suggesting that they may have lost comparatively more of the Needle downstream genes. If the Leaf and Needle copies of *LFY* specified male and female cones respectively in gymnosperms, then the flower would have retained relatively more genes formerly active in the ancestral gymnosperm's male, rather than female cones, hence the "Mostly Male" theory. The minimum set of female genes retained could have been restricted to those required to form the ovule. The ovule would then have been ectopic (i.e., developing in a novel location on the plant) upon a formerly male reproductive unit. If the ovule formed ectopically on a flat undivided microsporophyll, and the microsporophyll subsequently lost its microsporangia and rolled around the ovule, a structure approximating the carpel would have been created. The carpel wall, then, would have derived from a microsporophyll, not from a megasporophyll.

The gymnosperm ancestor in such a scenario would have had a male reproductive unit consisting of spirally arranged simple microsporophylls. The female structure could be of any form, as long as there was an appropriate precursor for the angiosperm ovule. After elaborating the theory, I learned that the male structure of the corystosperms had been reinterpreted as a short shoot bearing spirally arranged simple microsporophylls (Yao *et al.*, 1995). The female units of corystosperms consisted of cupules containing one or two ovules of typical gymnospermous structure. The whole cupule is bent or folded, so the tip of the cupule is close to the cupule's stalk (Taylor and Taylor, 1993). Gaussen (1946) noted that a cupule of this type containing a single ovule would be similar in organization to an angiosperm ovule, homologizing the cupule wall with the outer integument of the angiosperm ovule, and the inner integument and nucellus with the single integument and nucellus of the gymnosperm ovule. Hence corystosperms' reproductive structures fit the imagined gymnosperm precursor of angiosperms, according to the Mostly Male theory (Frohlich, 2001, 2002; Frohlich and Parker, 2000).

Other classes of evidence supporting the Mostly Male theory are mentioned here only briefly; see Frohlich and Parker (2000) and Frohlich (2001, 2002, 2003) for more detailed discussions:

1. *Arabidopsis lfy* mutants completely lack stamens, but can still form carpels in their highly aberrant flowers, and also make carpels at the inflorescence terminus, suggesting that *LFY* is more directly related to formation of male structures than female, because other genes can generate carpels in the absence of *LFY*.
2. The highly uniform morphology and placement of anthers—only one per stamen—as compared to the variable insertion and numbers of ovules on carpels suggests that if either structure could have appeared ectopically within the flower, it is most likely the ovule and not the anther.
3. Ectopic ovules can appear on sepals and petals in *Petunia* through overexpression of the *Petunia* gene *FBP11* and naturally on leaves of *Ginkgo*. While ectopic stamens are not uncommon in angiosperms, ectopic anthers either do not occur or are found only on structures with at least partial staminal character, such as structures in a graded organ series with stamens. (Ectopic microsporangia do occur in *Ginkgo*, a gymnosperm (Fujii, 1896).)
4. The sterile ovules in the male structures of Gnetales apparently provide an example of ectopic ovules appearing in a male structure, indicating that this can occur in evolution. The closest outgroups of Gnetales, according to the molecular phylogenies and morphological phylogenies constrained by extant gymnosperm monophyly all have unisexual cones. Thus, the sterile ovules in Gnetales male cones represent an innovation, which is by definition a case of ectopic placement.

5. The function of the sterile ovules in Gnetales male structures is to provide pollination droplets that attract pollinating insects, so the Gnetales are insect pollinated. This provides a reasonable immediate selective advantage that might have applied to the original mutant angiosperm ancestor with ectopic ovules, allowing the ectopic ovule trait to be positively selected, and thus preserved, giving time for evolution to perfect the flower. A selective advantage is required because a mutant phenotype as drastic as ectopic ovules is unlikely to have been neutral; without a selective advantage it would have been immediately lost. This was a key component of the Mostly Male theory from its inception (Frohlich and Parker, 2000).

B. NEW EVIDENCE REGARDING THE MOSTLY MALE THEORY

The Mostly Male theory presents a highly detailed scenario for the origin of flowers, based on a variety of types of evidence; as a result, a variety of types of new evidence have a bearing on the theory. The largest quantity of new information relates to the *LFY* gene. *LFY* has now been cloned from many additional angiosperms, gymnosperms, and free-sporing plants. No additional duplication events have been found in gymnosperms, although there are a few duplications in free-sporing plants. All published *LFY* trees agree that the seed plant duplication antedates the split between angiosperms and other extant gymnosperms. No Needle copies have been found in angiosperms, although there are very many new angiosperm sequences. Some seemingly diploid angiosperms and some relatively old polyploids do contain two active *LFY* genes, which derive from duplications high in the angiosperm *LFY* tree. Presumed diploids include members of the Lamiales (Aagaard *et al.*, 2005). Old polyploids with two *LFY*s include the Malloideae (Wada *et al.*, 2002) and Maize (Bomblies *et al.*, 2003). Some relatively recent polyploids, such as *Nicotiana tabacum* have long been known to contain two copies of *LFY*. Baum *et al.* (2005) found two copies of *LFY* in some polyploids in Brassicaceae, but they also noted instances where the second copy had been lost through a speciation event. They suggested that *LFY* may be "prone to be lost by drift due to a low probability of subfunctionalization or neofunctionalization."

Albert *et al.* (2002) proposed an explanation for the failure of *LFY* to form a sizeable gene family. *LFY* is expressed in a graded manner, unlike most genes that are either expressed at high levels or are turned off. For *LFY*, small changes in expression level might have significant phenotypic effects. For example, heterozygote *lfy* mutants of *Arabidopsis* flower later than the wild type, and *Arabidopsis* plants carrying additional genomic *LFY* sequences flower earlier than wild-type plants (Blázquez *et al.*, 1997). This contrasts with the effects of most mutant genes, which are fully recessive, that is, they have no detectible phenotypic effect as heterozygotes.

Flowering time can be under very strong selection. For *LFY*, a gene duplication that results in two extra copies could be selectively deleterious, resulting in inactivation of the gene at one of the two *LFY* loci. This would not be an example of drift, as suggested by Baum *et al.* (2005), but as they suggest it would make subfunctionalization/neofunctionalization unlikely, because two copies of *LFY* would only be present briefly. The inferred brief coexistence of the two copies of *LFY* in a polyploidy has implications for the use of gene order around *LFY* in phylogenetic inference; it reduces the likelihood of parallel events that could eliminate the same copy in different organismal lineages, thus reducing the chance of homoplasy in this character, making *LFY* an especially good marker for homologous syntenic regions.

Free-sporing plants, including mosses, the lycopod *Isoetes,* and a number of Moniliformopses (ferns and fern relatives, excluding lycopods) do have two or three active copies of *LFY*. These all appear to result from duplications within these groups, suggesting that in these plants *LFY* might not have such an important role in combining inputs from various pathways resulting in graded expression levels, as in *Arabidopsis*. This might allow *LFY* duplications to persist and become subfunctionalized. No *LFY* has yet been found in an alga. It is not present in *Chlamydomonas*, which has been fully sequenced. Our attempts to find it in Characeae have not succeeded.

Shindo *et al.* (2001) provided convincing evidence that there is only one copy of *LFY* (the Leaf ortholog) in *Gnetum parviflorum* of the Gnetales. They regard this as evidence against the Mostly Male theory, but if anything it supports the theory. Both the Leaf and the Needle paralogs are found in *Welwitschia*, the sister group of *Gnetum*, and also in *Ephedra* (Frohlich, 2003), the next outgroup, and in conifers, the third outgroup. Ergo, the absence of the needle ortholog in *Gnetum* represents an independent loss of the gene in the *Gnetum* lineage (Frohlich, 2003). This provides an opportunity to study concomitant effects of the loss of the Needle copy of *LFY*. Perhaps the loss is related to the morphology of the male and female structures in *Gnetum*, which show a "collar" extending around the stem in both sexes, which subtends multiple apices that in the male can bifurcate to make numerous reproductive units similar to the axillary reproductive units of other Gnetales. Other Gnetales have a strictly decussate arrangement of bracts and axillary units in their cones (Frohlich, 2003).

C. *LFY* EXPRESSION AND INFERRED FUNCTION

The inferred function of the *LFY* gene was important in the genesis of the Mostly Male theory, although the evidence regarding *LFY* function was sparse. *LFY* mutants have now been described in a number of angiosperms, which in all cases support the role of *LFY* in helping to specify the flower, although *LFY* can acquire additional roles as well. In pea, in addition to its role of specifying the flower, *LFY* is expressed in the leaves, and is required to make the pea-leaf compound (Yaxley *et al.*, 2001). In tomato, the *LFY* gene (*FALSIFLORA*) is expressed in vegetative, inflorescence and floral meristems, and in developing leaves (Kato *et al.*, 2005; Molinero-Rosales *et al.*, 1999), yet major phenotypic effects of the *fa* mutant are confined to late flowering, loss of floral meristem specification and the reversion of inflorescences to vegetative growth. Unlike the case with pea, leaf dissection in tomato depends almost completely on Knox genes, with only a very small effect from *falsiflora* mutations on the number of intercalary leaflets (Molinero-Rosales *et al.*, 1999; Piazza *et al.*, 2005). Another gene (*SFT*) has been identified, although not cloned, that acts in concert with *LFY* in causing the floral transition in tomato (Molinero-Rosales *et al.*, 2004); perhaps *SFT* limits the ability of *LFY* to cause the floral transition in purely vegetative shoots, which would be consistent with both the genetic and expression evidence. Reversion of the inflorescence to vegetative growth in *fa* mutants may be due to loss of floral identity in floral apices (Molinero-Rosales *et al.*, 1999).

That *LFY* affects flowering time in tomato, both in terms of the number of days to flowering and also in the number of leaves produced on a vegetative shoot before it converts to an inflorescence, shows that *LFY* has effects before the flower is specified, as the plant must make an inflorescence before the first flower forms. This is similar to *Arabidopsis*, where *LFY* also affects flowering time; *lfy* heterozygotes flower later than wild type and plants with extragenomic copies of *LFY* flower earlier. Inflorescences vary so much among flowering plants that they must have arisen many times. One way to create complex structures that incorporate numerous flowers would be to expand the expression of some, but not all, flower genes to tissues outside the flower that were formerly purely vegetative. This might create novel structures, and could be a mechanism in the evolution of inflorescences (Frohlich, 2003). This might also be the explanation for the role of *LFY* in the grasses rice, *Lolium*, and maize, where *LFY* is expressed in the inflorescences, especially in spikelets (the secondary inflorescence branches) and in developing flowers (Bomblies *et al.*, 2003; Gocal *et al.*, 2001; Prasad *et al.*, 2003). In maize, double mutants for the two *LFY* genes show defects in floral organ patterning, in transition to flowering, and in inflorescence architecture; hence *LFY* in maize shows the same functions as in *Arabidopsis*, with the addition of inflorescence architecture. Inflorescences in grasses are specialized structures peculiar to that family, so it is unlikely that the inflorescence role of *LFY* derives from basal monocots.

LFY expression has now been studied in the basal eudicot *Eschscholzia californica*, the California poppy (Becker *et al.*, 2005; Busch and Gleissberg,

2003). The *Eschscholzia LFY* (*EcFLO*) is expressed in a ring at the periphery of the apical meristem, without discernable changes in expression level or zone from the seedling stage until after the plant has initiated its first flower, which terminates the primary stem axis. The *LFY* expression pattern does not change until stamen initiation begins within the flower.

Busch and Gleissberg (2003) suggested that this could be related to a function of *LFY* in the formation of compound leaves. Becker *et al.* (2005) say that this expression pattern indicates "a continuous role for this gene in meristem function," in "shoot apex maintenance"; however, the high expression of *FALSIFLORA* in tomato apices during vegetative growth, without any meristem defect at all in the *falsiflora* mutant, shows that *LFY* may have high levels of expression in an apex without any functional role in that apex. It may be, for example, that *LFY* requires a second, interacting protein to specify floral character of an apex and that this partner protein, rather than the *LFY* product, is generally the limiting factor. Mutant analysis is sorely needed to determine *EcFLO* function. *Eschscholzia* is transformable, so reverse genetics approaches may allow this question to be resolved.

Monocots and eudicots both attach to the angiosperm tree at the unresolved tetrachotomy above the ANA grade. In the only monocots where *LFY* function has been studied by mutant analysis (grasses), *LFY* shows similar functions to *Arabidopsis*, with additional functions in specifying inflorescence structures. This suggests that the function of *LFY* in basal angiosperms very well may be similar to *LFY* in *Arabidopsis*. Busch and Gleissberg (2003) suggest that *LFY* and Knox may both have been involved in leaf dissection in eudicots, and that compound leaves may have been ancestral in eudicots. However, neither ANA grade members nor Chloroanthaceae, or Magnoliids or basal monocots have compound leaves, so this role would be an innovation in the eudicots. Functional studies of *LFY* in ANA grade plants are badly needed to clarify the role of *LFY* at the base of the angiosperms.

New data on *LFY* expression in conifers clouds the issue of whether the Leaf and Needle orthologs may specify male and female reproductive structures in gymnosperms. Dornelas and Rodriguez (2005) report the expression of the Leaf ortholog from *Pinus caribaea*. Their data indicate that it is strongly expressed in the early developing female cone, and only weakly expressed in the male cone. Their *in situs* also show strong expression in the vegetative apex. This surprising result is the reverse of that of Mellerowicz *et al.* (1998), who found strong expression of the *Pinus radiata* Leaf ortholog in male cones and virtually no expression in female cones, based only on Northern blots of multiple stages of development. Dornelas and Rodriguez's (2005) results are based on Northern blots but more importantly also on *in situs* of the earliest stages in female and male cone development, which are the stages critical for specification of male or female cones. Expression at later stages could reflect secondary roles of these genes, so expression differences at the earliest stages, just as (or before) distinct male or female attributes begin to be manifest would be expected if they specify maleness and femaleness of the reproductive structures. Dornelas and Rodriguez (2005) noted that a second *LFY* gene may be present in *P. caribaea*, which would presumably be the needle ortholog, but they did not clone it. It is conceivable that they might have detected the uncloned paralog in their *in situs* (Dornelas, M., personal communication). Their RNA probe, from the entire cDNA, is similar to that of Mellerowicz *et al.* (1998), who also used RNA antisense probes made from the entire cDNA. Dornelas and Rodriguez (2005) did RNAse treatment before their high stringency wash (M. Dornelas, personal communication). Comparing the two *LFY* sequences of *P. radiata*, there seem to be enough mismatches so that RNAse treatment followed by high-stringency washing would have released the probe from the paralogous mRNA. The contrasting results of these two studies remain unexplained.

No other *in situ* experiments comparing expression of the needle and Leaf paralogs in gymnosperms have been published, although several projects are nearing completion on members of the Pinaceae and also on nonpinaceous conifers. Studies that examine expression of both paralogs would be especially helpful. If *in situ* studies show strong expression of the needle and Leaf paralogs in earliest developing female and male cones, respectively, with only comparatively weak expression in the reproductive apices of the opposite sex as well as weak expression in vegetative apices, then the inference that the *LFY* paralogs specify the sex of reproductive units in gymnosperms would be supported, but if other expression patterns are observed then it would be severely undermined (although not disproved). Conifers, including Pinaceae, have highly specialized female reproductive structures, with much fusion of parts (Florin, 1951, 1954). Expression studies are sorely needed in the other extant gymnosperm groups, such as Gnetales, which show less fusion of parts than either conifer group, and in *Ginkgo* and cycads, which attach below the clade of conifers and Gnetales. There is a very great need for a gymnosperm genetic model in which gene function may be determined by inactivating or strongly downregulating gene expression.

These results undermine the *LFY* evidence supporting the Mostly Male theory; however, the other lines of evidence remain that suggest that the ovule may have been ectopic upon a previously male structure. This is strengthened by the discovery of mutations in *Arabidopsis* that generate ectopic ovules on the cauline leaves (Kidner and Martienssen, 2005). These mutations appear unrelated to the genes whose overexpression generates ectopic ovules in *Petunia*. This shows that multiple distinct genetic mechanisms can generate ectopic ovules.

D. DORSIVENTRALITY OF THE OUTER INTEGUMENT AND OF THE CUPULE WALL

New data on dorsiventrality call into question corystosperms as possible angiosperm ancestors. This relates to the dorsiventrality of the outer integument of the angiosperm ovule, compared to that of the corystosperm cupule wall. If the corystosperm cupule gave rise to the ovule then these two structures are homologous and should have the same dorsiventral orientation (Frohlich, 2001, 2003). The orientation of vascular bundles in the permineralized fossil corystosperm cupules described by Klavins *et al.* (2002) clearly shows that the outer surface of the cupule wall is the morphological adaxial surface (facing the stem), so the ovules are borne on the abaxial surface (facing away from the stem). Orientation of the angiosperm outer integument was first inferred from expression of the YABBY gene *INO* in *Arabidopsis*, which is expressed in the outer cell layer of the outer integument (Villanueva *et al.*, 1999). The *Nymphaea* homolog of *INO* is also expressed on the outer surface of the outer integument (Yamada *et al.*, 2003). YABBY genes were thought to universally specify the abaxial surface of organs, based on expression patterns and on mutant analysis in *Arabidopsis* (Bowman, 2000). However, in *Amborella* and in maize the YABBY genes in the leaves are expressed on the adaxial surface instead, so inferences of ab- or adaxial character based on YABBY genes are ambiguous (Juarez *et al.*, 2004; Yamada *et al.*, 2004). About 80 families of angiosperms form vascular bundles in the outer integument, mostly after fertilization, as the seed develops. Literature reports rarely discuss their orientation, but in the few cases where this is reported it supports the outer side of the outer integument as the abaxial surface (Frohlich, 2003, 2006). This would eliminate corystosperms as possible angiosperm ancestors, unless the orientation of the outer integument became reversed, which might not have been impossible (Frohlich, 2003).

Other fossil gymnosperms had cupules that show the reverse orientation from the cupules of corystosperms. Weak evidence suggests that in Caytoniales cupules the outer surface was the abaxial surface. The vascular bundles of the permineralized cupule *Petriellaea* show that the outer surface was the abaxial side, so it has the same orientation as angiosperm ovules (Taylor *et al.*, 1994). Relationships of this cupule are uncertain; no other part of the plant was preserved. It might belong to the Caytoniales. Of course, far too little is known about it to be included in the analyses of Doyle (2006) or Hilton and Bateman (2006).

The phylogenies of both Doyle (2006) and Hilton and Bateman (2006) place Caytoniales as the immediate sister group of angiosperms, but corystosperms are a considerable distance away from angiosperms. This also suggests that angiosperms are not derived from corystosperms or their close relatives, but may be derived from ancestors closely related to (or within) Caytoniales.

A direct test of the central prediction of the Mostly Male theory is now underway in the Floral Genome Project; that is, that a preponderance of genes expressed in flowers (but outside the ovules) will be more closely related to genes expressed in gymnosperm male, rather than female reproductive structures. This will reveal whether the flower arose mostly from the ancestral gymnosperm male reproductive units.

VIII. ALTERNATIVES AND MODIFICATIONS OF THE MOSTLY MALE THEORY

Albert *et al.* (2002) present a complex model for the origin of bisexual flowers involving the two *LFY* paralogs in gymnosperms and the downstream MADS-box genes that they may regulate. They suggest that the Leaf copy stimulated expression of the MADS B-class gene(s) and also some expression of C-class genes, which together specify male reproductive structures. They suggest that the Needle copy only stimulated expression of the C-class gene, resulting in female reproductive structures. They further suggest that the Leaf and Needle paralog proteins had the same function, based on the similar phenotypes generated by over expressing the two *P. radiata LFY*s in *Arabidopsis*. They attribute different downstream functions of these genes exclusively to differences in their promoters. They suggest that, if changes in the promoter of the Leaf copy increased its ability to stimulate expression of the C-class gene, then selection might no longer favor maintenance of the Needle copy, leading to its loss, and to the production of bisexual reproductive units under the control of the Leaf copy. However, they do not explain how the two LFY paralogs could have different downstream effects if their proteins had the identical functions; if that were true, the information regarding which promoter had been activated would be invisible to the proteome. For the Leaf paralog to generate bisexual reproductive structures some additional spatial regulatory mechanism would be required, to generate distinct regions where B- and C-class genes were expressed, and other regions where only C-class genes were expressed.

Overexpression studies are not as sensitive to differences in function as complementation studies, such as that of Maizel *et al.* (2005), which did find differences in the ability of Leaf and Needle paralogs to complement the

Arabidopsis lfy mutation, suggesting that the proteins made by these genes could have different potential functions. The great phylogenetic and temporal distance between *Arabidopsis* and gymnosperms suggests that any differences between these proteins involving differing protein–protein interactions or differences in DNA binding may not be effective in *Arabidopsis*, so such long-distance expression studies may not be the best methods to evaluate roles of the two paralogs in gymnosperms. This again highlights the great need for methods to downregulate gymnosperm genes to allow more reliable inferences of gene function.

Theissen *et al.* (2002) and Theissen and Becker (2004) proposed the "out of male and out of female" theories for the origin of flowers. They note that in some conifers, such as in a spruce horticultural variety, bisexual cones do occur at moderate frequency (Theissen and Becker, 2004), which are male at the base and female in the terminal region. There are anecdotal reports of such cones in other conifers, such as Torrey Island Pine (*Pinus torreyana*). As with the spruce example, the male and female regions of these cones show normal structure for cones of that sex in that species. They suggest that alterations in B-gene expression, either downregulation in the terminal region of a female cone, or upregulation at the base of a male cone, could generate bisexual cones of this morphology. They suggest that such a mechanism could have generated bisexual flowers, which thus could have arisen from either the ancestral male or female reproductive units. They propose increased attractiveness to insects, resulting in insect pollination, as an immediate selective advantage that would have preserved such mutant cones. The reward could be either the pollination drop, as proposed in the Mostly Male theory (Frohlich, 2001; Frohlich and Parker, 2000), or the pollen itself.

Their theory accounts for the origin of bisexuality, but not for any of the other morphological innovations of the flower, most notably, for the carpel or carpel precursor or the ovule. There would need to be a series of unspecified evolutionary steps, either before or after the origin of bisexuality, to achieve an evolutionary grade that could be called a flower. Theories that account for only one morphological innovation are useful but are not complete.

The new results on *LFY* do suggest that some other genes might be responsible for determining the difference between male and female reproductive units in gymnosperms. The gymnosperm homologs of the B-genes certainly are involved in this specification, but it is not clear whether they are the top genes in the cascade that determine this difference in normal cones, or are effector genes active at later steps, where they may activate downstream genes needed for male development and inactivate specifically female genes. A mechanism that alters expression of an effector gene could be sufficient to create an evolutionarily important bisexual cone, and might also result in some aberrations in the misexpressed sex, if lack of the normal top gene(s) that initiate that sex results in absence of some expressed genes not controlled by the misexpressed effector gene.

For example, ectopic expression of the two B-genes and the C-gene *AGAMOUS* (*AG*) in all whorls of *Arabidopsis* results in all whorls of the flower being converted to stamens, even though the genes that normally initiate the expression of the B-genes are absent from the outer whorl.

Theissen *et al.* (2002) and Theissen and Becker (2004) suggest that the B-class MADS-box genes may be the primary-controlling genes determining male vs female cones in gymnosperms, and that changes in their expression caused the appearance of bisexuality on the lineage leading to flowers. The weakness of support for *LFY* as the gene controlling male vs female identity suggests that some other gene(s) may be responsible instead; perhaps these are the B-class genes. Alternatively, many *Arabidopsis* genes are now known that help regulate the transition to flowering (Parcy, 2005; Vijayraghavan *et al.*, 2005). Some are upstream of *LFY* and act through *LFY*, while others operate in parallel pathways. It is conceivable that homolog(s) of one or more of these transition-to-flowering genes might serve as the top genes in gymnosperms that specify male or female reproductive structures.

Sieber *et al.* (2004) show that the *PHABULOSA* (*PHB*) gene, which helps specify adaxial fate in leaves, is expressed in the inner integument of *Arabidopsis* ovules but not in the outer integument. They note that the combined expression of *INO* in the outer integument and *PHB* in the inner integument resembles the expression of these dorsiventrality genes in an individual leaf. They suggest that angiosperms might have acquired two integuments by the splitting of a single integument which had expressed one dorsiventrality gene in its outer layers and the other in its inner region. While intriguing, this ignores the dramatic differences between the two integuments, which suggest that they have different origins (Stebbins, 1974). More seriously, the lateral expansion of leaves occurs at the boundary, within the same organ, between tissues showing abaxial and adaxial fates. Splitting a single integument would eliminate this boundary. *Arabidopsis* mutants that lack one or the other integument, or have integuments of different lengths, show that the two integuments do not operate together to create such a boundary. This implies that a novel mechanism(s) stimulates expansion of the two integuments, suggesting that the two dorsiventrality genes have different roles in the integuments compared to leaves; but this undermines any inferences based on their roles in leaves. A number of other dorsiventrality genes are now known (Bowman *et al.*, 2002). Expression studies of all these genes in angiosperm and gymnosperm ovules might elucidate the role of dorsiventrality in ovule and integument development.

IX. BAUM AND HILEMAN'S SCENARIO FOR FLOWER ORIGINS

Baum and Hileman (2006) propose an intriguing, detailed scenario to account for three steps in the origin of flowers: bisexuality, the appearance of floral determinacy with axis compression, and for the origin of petals. Their scenario focuses on possible changes in gene activity, gene interactions, and gene networks, making explicit predictions that are eminently subject to testing, making their proposal a full-fledged evo-devo scenario. Bisexuality is discussed in the other current theories of flower origin, but compaction of the floral axis and floral determinacy, and regular distinction between sepals and petals are not. The latter could be considered steps that occurred after the origin of the flower, which were crucial to create the sorts of flowers familiar today.

Baum and Hileman (2006) explain the origin of bisexuality through a homeotic mechanism, similar to the out-of-male proposal of Theissen and Becker (2004), but with a much more explicit model. They note that male structures require the activity of both B-class MADS genes, whose proteins must interact in quartets with the E-function *SEPALLATA* genes. C-class gene proteins are active along with B-gene proteins in male structures, but in female structures B-function is absent while C-function is present. They suggest that the abundance of C-gene protein could have increased to very high levels in the terminal regions of a cone, complexing so much of the *SEPALLATA* protein that little *SEPALLATA* was available to interact with the B-gene protein. This would effectively eliminate B-gene activity, making the terminal region of the cone female. Explicit tests could be made by manipulating the levels of C-gene activity in different parts of an experimental flower to determine whether *SEPALLATA* genes can be titrated out of availability for other MADS genes. One would need to measure the levels of the C-gene protein and correlate this with the developmental effects. Such experiments are feasible now.

They explain floral determinacy by acquisition of the ability of the C-gene protein to downregulate the *WUSCHEL* gene, which maintains the stem cells in the central zone of the apex. As a result, the apex would stop growing, and primordia forming late in development would arise closer and closer to the center of the apical meristem. This, in turn, would allow positional rather than temporal controls to determine the developmental fate of these primordia, as occurs in flowers such as *Arabidopsis* today. They suggest the control of B-gene expression by *UNUSUAL FLORAL ORGANS* (*UFO*), as in *Arabidopsis*, could have arisen by such a series of events. This could have caused the B-gene zone to extend beyond the C-gene zone, which could have activated some but not all stamen genes, resulting in the formation of petals. They suggest that these latter evolutionary innovations may have occurred above the ANA grade, at the base of the core eudicots, creating the familiar flower with a whorl of sepals surrounding a whorl of petals.

Their scenario is especially intriguing because it is so detailed, giving it great explanatory power and allowing many explicit tests. This is the ideal for an evo-devo scenario (Frohlich, 2006). They note in their conclusions that details of their model could be wrong, even if the basic idea is right. In different words, they are referring to the separable components of their grand scenario, which need not all be exactly correct. Portions of their scenario could conceivably be combined with the other theories discussed here, to provide an even more detailed account of flower origins.

X. DOYLE'S *CAYTONIA*–GLOSSOPTERID MODEL FOR THE ORIGIN OF THE CARPEL

Doyle (2006) suggests a morphologically based model for flower origins, most notably, for the origin of the carpel. This is based on the cladistic position of Caytoniales as the immediate sister group to angiosperms in all of his (and in Hilton and Bateman (2006)) analyses, and on the position of glossopterids. Caytoniales have been a favorite starting point for scenarios on the origin of flowers since Gaussen's (1946) suggestion that the *Caytonia* cupule could give rise to the angiosperm ovule with its two integuments. Past models proposed that the narrow stalk on which the *Caytonia* cupules were borne could have become wide and enclosed the cupules to make a carpel. Doyle notes that glossopterids appear on some cladograms not much below the Caytoniales. Glossopterids are Permian, so they significantly antedate the Caytoniales. Glossopterids had flat obovate leaves, some of which bore male or female reproductive stalks in their axils or attached to the upper surface of the leaf. These stalks terminated in a wide, flat cupule that showed anatomical dorsiventrality and apparently derived from a leaf. This cupule partly enclosed the ovules, which were borne on its anatomically adaxial surface, although this surface faced the subtending leaf.

Caytonia cupule-bearing organs are found as detached fossils, so how they were borne on the plant is not known, but Doyle suggests that Caytoniales could have been derived from glossopterids, and if so then *Caytonia* reproductive stalks, like those of glossopterids, might have been axillary to leaves (or bracts), or borne on reduced axillary shoots. If this sort of cupule-bearing stalk became fused to the upper surface of the leaf, and the leaf rolled inward, then the whole structure would resemble a carpel. *Caytonia* vegetative leaves were compound with leaflets arising together at the top of the petiole, but this could be an autapomorphy of *Caytonia*.

This is a plausible scenario. Testing it depends on finding better fossils of *Caytonia* or related Mesozoic fossils. To me it is amazing that this famous and important fossil taxon has not received more extensive investigation.

Doyle's (2006) model does raise serious evo-devo questions that have not yet been sufficiently studied: the comparative attributes and mechanisms of postgenital fusion and of congenital fusion and how each might arise. An axillary reproductive stalk, as suggested for *Caytonia*, becoming fused to a subtending leaf could initially have been an example of postgenital fusion; that is, the two structures would arise as separate organs and become fused as they developed. This would certainly be the simplest way to achieve the structure that Doyle (2006) envisions, as it would allow this morphological innovation to proceed by small steps.

However, in modern carpels the ovules arise from the placenta, which may be positioned virtually anywhere within the carpel. If the carpel did arise from two separate but adjacent structures, then they are now congenitally fused. It is not clear to me how evolution would proceed from postgenital to congenital fusion, if that was the evolutionary sequence. Congenital fusion (whether or not preceded by an evolutionary stage showing postgenital fusion) may resemble the ectopic movement of ovules onto the inner surface of the carpel precursor, as in the Mostly Male theory.

There are very many cases among modern plants where congenital fusion has arisen, especially from structures that previously were separate but adjacent. Many of these cases might be explained by an overlap of developmental fields, so that the two structures that arose side-by-side in the ancestor now arise as a single fused structure (e.g., petals fused into a tube, and at least some cases of stamens borne on the corolla). The edges of primordia appear to be delimited by specific boundary genes expressed between adjacent primordia (Aida and Tasaka, 2006). If these genes are mutated then the boundary may be lost, so adjacent primordia become congenitally fused, as in mutants of the *CUP SHAPED COTYLEDON* (*CUC*) genes of *Arabidopsis* (Bruil-Broyer *et al.*, 2004; Laux *et al.*, 2004). Downregulation of boundary genes provides a mechanism for the origin of congenital fusion, but only from adjacent structures that retain their positional relationships. It is not clear whether such a mechanism could account, in *Amborella*, for the positioning of the placenta and ovule, which arise from the inner side of the "cross meristem" that spans the earlier-arising distal edges of the carpel primordium (Buzgo *et al.*, 2004); if the cross meristem expresses, from its inception, any genes specific to the ovule or placenta then evolution from an axillary structure via loss of boundary determination would be possible. However, such a mechanism would not account for the numerous ovules of other plants, such as Nymphaeaceae.

XI. MEYEN'S GAMETOHETEROTOPY THEORY (AND KRASSILOV'S (1997) BOOK)

Meyen's (1988) "gametoheterotopy" theory for flower origins has similarities to the Mostly Male theory; however, his model involves a homeotic conversion within a bennettitalean reproductive unit, rather than an ectopic transfer of ovules between separate male and female reproductive units. Typical bisexual Bennettitales reproductive units had outer sterile structures surrounding a whorl of flat or pinnate microsporophylls. Each microsporophyll bore four to many microsporangia, often in a complicated arrangement. Inserted above the microsporophylls were numerous short stalks terminating in ovules. Borne in between the ovules were numerous interseminal scales, which covered over the ovules, except for the protruding micropylar tubes that would have produced the pollination droplets. Other Bennettitales had unisexual reproductive units that were similar, except they lacked either the male or the female components (Crane, 1985; Taylor and Taylor, 1993). Based on especially well-preserved permineralized fossils, Rothwell and Stockey (2002) and Stockey and Rothwell (2003) interpreted the ovule-bearing stalks as megasporophylls and showed that the ovules were not borne in cupules. Other bennettitalians may have had cupules, however, and so might have had reasonable precursors for the angiosperm ovule (Doyle, 2006).

Meyen (1988) suggested that the developmental program of the microsporophylls became expressed in the female region, replacing the female organs with flat structures that resembled the microsporophylls, except that they bore ovules in place of microsporangia. These flat ovule-bearing structures would have then rolled inward to become carpels. Note that the female structures retain their location, but change their morphology, which is the essence of homeotic conversion.

Like the Mostly Male theory, Meyen's theory predicts that genes expressed in the carpel wall should be more closely related to genes expressed in gymnosperm male rather than female structures. In Meyen's model the early acting genes specifying the flower may not show the same relationships, providing a possible contrast with the Mostly Male theory. The gymnosperm ancestor, according to Meyen's theory, could have already been bisexual and insect pollinated, so the selective advantage invoked for ectopic ovules in the Mostly Male theory would not apply to Meyen's theory.

The two theories differ in suggested mechanism, but in 1988 neither homeotic conversion nor ectopic movement of structures were widely appreciated as possible events in evolution; so Meyen's theory was very innovative. Meyen's theory could best be tested by the discovery of appropriate fossils. If Bennettitales microsporophylls were discovered that closely resembled angiosperm stamens then a bennettitalean origin of angiosperms would be strongly supported.

In Bennettitales, the degree of diversity shown by the male and female structures is the reverse of angiosperms. Bennettitales female structures are highly uniform, while the number and placement of microsporangia on microsporophylls varies greatly. If Bennettitales reproductive units achieved bisexuality through ectopic repositioning, then the microsporangia, rather than the ovules, would be the better candidates for such movement.

Krassilov's (1997) book on the origin of angiosperms discusses a wide variety of fossils that are relatively poorly known, which is valuable. He follows an older, precladistic tradition (Crane, 1998), which I find difficult to integrate with current ideas. Krassilov includes diverse fossils in the "proangiosperm" group, but their suggested interrelationships and relationships with angiosperms are obscure. Other workers have proposed various fossils as close angiosperm relatives, but without wide acceptance, of which probably the best known is *Sanmiguelia* (Cornet, 1989). *Bevhalstia* is another (Hill, 1993).

Stuessy's (2004) "transitional-combinational" theory also appears to be fundamentally non-cladistic. Stuessy accepts a ca. 200 million years ago (MYA) estimate for phylogenetic splits among extant angiosperm lineages (perhaps between monocots and eudicots) based on molecular clocks. He attempts to reconcile this date with the much younger fossil record of the initial angiosperm radiation at ca. 130 MYA. He suggests that carpels (which he suggests are derived from cupules) evolved ca. 200 MYA, that double fertilization appeared later and that flowers appeared only shortly before the 130 MYA radiation. It is indeed likely that various angiosperm features appeared in sequence over a considerable time interval, but coupling this with early divergence dates implies extensive parallelism on multiple lineages. He notes that only carpels (and may be not even carpels (Stuessy, 2004: 10)) would have been present in the last common ancestor of extant angiosperms.

Stuessy (2004) notes that "Laurales, Magnoliales and paleoherbs, etc., may in fact represent different lineages from carpellary ancestors" (Stuessy, 2004: 9). For this to be true, the endosperm, the second ovule integument and even the flower itself would all have to be independent, parallel evolutionary innovations in about 9 lineages: the three ANA lineages (which he accepts as basal), about three clades in the near-basal polytomy (eudicots, monocots, and Chloranthaceae, depending on resolution of the polytomy) and three within the magnoliids (Magnoliales, Laurales and Piperales+Canellales (Soltis *et al.*, 2005: 78). Furthermore, each of these clades would need to have achieved flowers at about the same time. This is most unlikely. Even if the early split were limited to monocots versus eudicots there would be 6 or 5 independent origins of these crucial angiosperm features (6: monocots, eudicots, magnoliids, and the 3 ANA groups or 5: monocots, eudicots, magnoliids, Austrobaileyales and *Amborella*+Nymphaeales). Surely it is more reasonable to accept a much younger age for the split between *Amborella* and other angiosperms, still within the range (68–281 MYA) suggested by Sanderson and Doyle (2001). A younger age would be consistent with the reconstruction of the basal flower discussed earlier, and obviates the inference of massive parallel evolution implied by acceptance of the early divergence date.

XII. DOUBLE FERTILIZATION AND FRIEDMAN AND WILLIAMS' MODULAR SCENARIO

The presence of double fertilization, leading to the formation of triploid endosperm, has been considered one of the strangest and most important characters defining angiosperms (Friedman and Floyd, 2001). It has long seemed bizarre that the two sperms in the winning pollen tube would each fuse with nuclei in the embryo sac, and especially that one sperm would fuse with two nuclei to form a triploid nucleus. This has even been suggested as a prime selective advantage leading to angiosperm dominance in much of the world's vegetation. It has long been known that there is considerable variation in the structure of the embryo sac, and in the number and ploidy of the nuclei that fuse with the second sperm, and therefore in the ploidy of the endosperm (Maheshwari, 1950, 1963). However, the commonness of the standard embryo sac, with eight nuclei in seven cells leading to a triploid endosperm, fostered the view that that was the basal condition among angiosperms, and must confer a powerful selective advantage (Friedman and Floyd, 2001; Friedman and Williams, 2004).

The discovery by Friedman (1990) that both sperm nuclei undergo fusions upon entering the female gametophyte in *Ephedra*, a member of the Gnetales, caused a sensation; this appeared to be yet another feature of angiosperms that could have arisen below angiosperms and could be a synapomorphy between angiosperms and other anthophytes (Carmichael and Friedman, 1996; Donoghue and Scheiner, 1992; Friedman, 1992, 1994). The discovery that *Gnetum gnemon* also shows a form of double fertilization confirms that this is perhaps an ancestral attribute of the extant Gnetales. Details of these processes differ substantially. In *Ephedra* one sperm fuses with the egg, while the second fuses with the ventral canal cell nucleus, the sister nucleus of the egg. In *Gnetum* there is a single sperm that is binucleate. The female gameotphyte is coenocytic without a differentiated egg. Both sperm nuclei fuse with nuclei in this coenocytic gametophyte. In both *Ephedra* and *Gnetum* the products of the second fusion develop as second embryos. There is no endosperm; the storage tissue in Gnetales, as in other gymnosperm seeds, is the haploid female gametophyte (Carmichael and Friedman, 1995, 1996).

The production of an additional embryo from the second sperm in Gnetales suggested that such a second embryo could have been the evolutionary source for endosperm storage tissue. Inclusive fitness considerations would favor conversion of the second embryo to a storage function, to improve fitness of its genetically identical sibling embryo. Once a diploid storage tissue arises, conflict in resource allocation between sibling seeds with different paternity (reflecting the conflict between the pollen- and ovule-parents) results in higher female fitness if the genomic contribution of the female parent is increased. This favors the evolution and retention of a triploid endosperm storage tissue. It also suggests differential imprinting of genes from the male and female parents may arise, which has been found (Friedman and Williams, 2004). Such imprinting can also result in an advantage for triploid endosperm (Stewart-Cox *et al.*, 2004).

There are old published reports (long ignored) that in some conifers both sperms undergo fusion events (as well as similar old reports regarding Gnetales). After their discoveries in Gnetales, Friedman (1992) reevaluated the conifer reports and finds those for *Abies balsamea* (Pinaceae) and *Thuja occdentalis* (Cupressaceae) to be credible (Hutchinson, 1915; Land, 1902). The presence of double fertilization in both the Gnetales and the Pinaceae shows that this feature must have arisen at least twice, in parallel, if one accepts monophyly of extant gymnosperms. The living gymnosperms *Ginkgo* and cycads still retain the ancient swimming sperm, used by all free-sporing land plants to transport genetic material from the male gametophyte plant to the egg on the female. In *Ginkgo* and cycads the pollen tube grows most of the way toward the egg, but then releases swimming sperm which swim the remaining distance to the egg. This was long considered a strange, primitive oddity. The sperm is the only cell in seed plants that has flagellae. If the swimming sperm were lost, it would be incomprehensible for the complicated flagellar system to be maintained in the genome, for subsequent resurrection in a reversal to fertilization by a swimming sperm. Hence, loss of the swimming sperm is a Dollo character (i.e., that cannot be reversed). Even if one does not accept monophyly of extant gymnosperms, loss of swimming sperm (siphonogamy) and double fertilization would still have arisen twice; both Doyle (2006) and Hilton and Bateman (2006) show the extinct glossopterids nesting between angiosperms and conifers in all cladograms (except where these taxa are part of a polytomy), whether constrained by extant gymnosperm monophyly or not. *Glossopteris* had swimming sperm (Nishida *et al.*, 2004); as a Dollo character this would force two origins for siphonogamy and double fertilization. Hence, including fossils, siphonogamy cannot be a synapomorphy of angiosperms, conifers, and Gnetales, contrary to Crane *et al.* (2004).

In free-sporing plants, the gatekeeper that restricts fertilization to a single sperm is presumably the plasmogamy (fusion) between the sperm's cell membrane and that of the egg, rather than the fusion of their two nuclei. Perhaps in *Ginkgo* and cycads this plasmogamy also functions as the gatekeeper, and requires the swimming sperm to operate. Hence, retention of the swimming sperm would not be a strange anachronism, but a necessary mechanism to regulate correct formation of the zygote. Friedman and Williams (2004) note that the function of sperm nuclei is to fuse with other nuclei. They suggest that, if a nucleus does not find an egg to fuse with, it may fuse with any nucleus available. Perhaps, as they imply, double fertilization is a direct consequence of pollen tube delivery of the sperm nuclei, resulting from the defeat of the ancient gatekeeper. Fusion of the second nucleus with a cell other than the egg might have arisen as a mechanism to protect the zygote from fusion by the second nucleus, which would have resulted in a triploid plant that could not reproduce.

Embryo sac structure and the fertilization process in the ANA grade members have also provided surprises. Williams and Friedman (2002, 2004) and Friedman *et al.* (2003) show that in *Nuphar, Illicium*, and *Kadsura* the embryo sac contains only four haploid nuclei in four cells and that the endosperm nucleus formed by the second fertilization event is diploid, not triploid. In *Amborella* the embryo sac was thought to have the standard seven cells with eight haploid nuclei (Tobe *et al.*, 2000). By parsimony, and ignoring the outgroup, the ancestral flower would be equivocal in this character, even if Nymphaeales and *Amborella* were sisters. If branch length were considered, and if other Nymphaeales and other Austrobaileyales have four-celled embryo sacs, then the latter would be slightly favored.

Friedman and Williams (2003, 2004) propose a scenario to account for the origin of the standard eight-nucleate seven-celled (*Polygonum* type) embryo sac from a four-celled and four-nucleate precursor, which also accounts for some of the stranger embryo sacs of higher angiosperms. They suggest that the four nucleate condition constitutes a developmental module, yielding one nucleus that migrates to the central cell and later fuses to form an endosperm nucleus, while the other three organize small separate cells at the periphery of the embryo sac. At the micropylar end of the embryo sac one of these becomes the egg while the other two become synergids. Friedman and Williams (2003, 2004) suggest that this module becomes reiterated if, at early stages, nuclei become widely separated from each other into distinct cytoplasmic domains at opposite ends of the developing embryo sac. Each module then produces four nuclei, resulting in three small cells at each end of the embryo sac and two nuclei that migrate to the central cell, that is, the standard *Polygonum* type. At the end opposite the micropyle (the chalazal end) of the embryo sac they form the three antipodal cells, which do not undergo further development. This second module can be lost, as has happened in Onagraceae (an advanced eudicot) and in grasses, advanced monocots, which have four-celled embryo sacs, but are derived from ancestors with the standard *Polygonum* type. Extra modules can be gained, which can explain the strange embryo sacs of *Penaea* (a higher eudicot), which has two additional clumps of three cells each on the sides of the embryo sac, and four haploid maternal nuclei contributing to the eventual endosperm nucleus; the formation of four modules would account for this embryo sac. *Plumbago* (a higher eudicot, not close to *Penaea*) also can be described as having four modules, but instead of clumps of three cells each at the ends and sides of the embryo sac there is only a single cell in each position. Embryo sacs with other numbers of nuclei and small peripheral cells (Maheshwari, 1950: 86) would require further modifications to the basic modular system, mostly involving additional fusions and/or additional divisions added to the end of the developmental series. Note that there do not seem to be embryo sacs with small numbers of nuclei that cannot be explained by their model, that is, with five or six or seven haploid nuclei.

Friedman and Williams (2003, 2004) suggest that the ancestral flower had only a single module, so the condition of *Nuphar, Illicium*, and *Kadsura* would be a retained plesiomorphy, and that the two-module *Polygonum* embryo sac would have evolved in parallel in *Amborella* and in the lineage above the ANA group. Considering that the gymnosperm outgroups lack this character, this is the more parsimonious reconstruction.

Friedman and Williams' (2003, 2004) modular model fulfills the desired attributes of an evo-devo scenario; it accounts for the observed attribute of the diverse types of embryo sac, including some very strange types, and it makes predictions that are testable, a crucial attribute required for serious scenarios (Frohlich, 2006). For example, if a plant with the standard *Polygonum* type embryo sac has a mutation that prevents the early separation of nuclei into separate cytoplasmic domains, then a four-nucleate embryo sac should result, resembling that of *Nuphar*. Mutations that result in more than two widely separated nuclei should generate extra modules.

Mutant screens to find genes involved in female gametophyte development and function have begun (Brukhin *et al.*, 2005; Pagnussat *et al.*, 2005). The characterization of such genes should allow stringent tests of this scenario.

The modular scenario does not account for earlier evolutionary events, most notably; for the great reduction in size of the female gametophyte in angiosperms as compared to gymnosperms. Perhaps, if the Mostly Male theory is correct, and ovule precursors were ectopic on male structures, then there may have been strong selection to reduce the size of the ovule, including the female gametophyte, so the ectopic ovules would fit on a microsporophyll.

XIII. OVERALL CONCLUSIONS

The question of the evolutionary origin of the flower has not been resolved. There remain several theories in play that have a reasonable chance of being largely correct. The Mostly Male theory is one, but less-specific theories that do not involve the ectopic placement of ovules are also viable. Baum and Hileman (2006) have proposed a very detailed scenario to explain flower bisexuality and critical innovations within angiosperms—floral determinacy, the condensation of the floral axis and the origin of petals. Friedman and coworkers have proposed a reasonable scenario for the elaboration of the angiosperm embryo sac. The continuing elucidation of flower origins is likely to be an exciting story in the coming years.

ACKNOWLEDGMENTS

I thank R. M. Bateman, J. Hilton, J. Leebens-Mack, P. S. Soltis, D. E. Soltis, and C. W. dePamphilis, and especially J. A. Doyle, for useful discussions and comments on this chapter. I thank J. A. Doyle, J. Hilton, R. M. Bateman, and D. A. Baum for allowing me to use their "in press" papers. This work was funded through the NSF Plant Genome Research Program (project DBI0115684).

REFERENCES

Aagaard, J. E., Olmstead, R. G., Willis, J. H. and Phillips, P. C. (2005). Duplication of floral regulatory genes in the Lamiales. *American Journal of Botany* **92**, 1284–1293.

Aida, M. and Tasaka, M. (2006). Genetic control of shoot organ boundaries. *Current Opinion in Plant Biology* **9**, 72–77.

Albert, V. A., Gustafsson, M. H. G. and Di Laurenzio, L. (1998). Ontogenetic systematics, molecular developmental genetics, and the angiosperm petal. *In* "Molecular Systematics of Plants II DNA Sequencing" (D. E. Soltis, P. S.

Soltis and J. J. Doyle, eds.), pp. 349–374. Kluwer Academic Publishers, Boston.

Albert, V. A., Oppenheimer, D. G. and Lindqvist, C. (2002). Pleiotropy, redundancy and the evolution of flowers. *Trends in Plant Science* **7**, 297–301.

Albert, V. A., Soltis, D. E., Carlson, J. E., Farmerie, W. G., Wall, P. K., Ilut, D. C., Teri, M., Solow, T. M., Mueller, L. A., Landherr, L. L., Hu, Y. Buzgo, M., *et al.* (2005). Floral gene resources from basal angiosperms for comparative genomics research. *BMC Plant Biology* **5**, 5.

Alvin, K. and Chaloner, W. G. (1970). Parallel evolution in leaf venation: An alternative view of angiosperm origins. *Nature* **226**, 662–663.

Angiosperm Phylogeny Group. (2003). An update of the Angiosperm Phylogeny Group classification for the orders and families of flowering plants: APG II. *Botanical Journal of the Linnean Society* **141**, 399–436.

Bailey, I. W. (1944). The Development of vessels in angiosperms and its significance in morphological research. *American Journal of Botany* **31**, 421–428.

Bailey, I. W. and Smith, A. C. (1942). Degeneriaceae, a new family of flowering plants from Fiji. *Journal of the Arnold Arboretum* **23**, 356–365.

Bailey, I. W. and Swamy, B. G. L. (1951). The conduplicate carpel of dicotyledons and its initial trends of specialization. *American Journal of Botany* **38**, 373–379.

Barkman, T. J., Chenery, G., McMeal, J. R., Lyons-Weiler, J., Ellisens, W. J., Moore, G., Wolfe, A. D. and dePamphilis, C. W. (2000). Independent and combined analyses of sequences from all three genomic compartments converge on the root of flowering plant phylogeny. *Proceedings of the National Academy of Sciences of the United States of America* **97**, 13166–13171.

Baum, D. A. (1998). The evolution of plant development. *Current Opinion in Plant Biology* **1**, 79–86.

Baum, D. A. and Hileman, L. C. (2006). A developmental genetic model for the origin of the flower. *In* "Flowering and Its Manipulation" (C. Ainsworth, ed.), pp. 3–27. Blackwell Publishing, Sheffield, UK.

Baum, D. A., Yoon, H.-S. and Oldham, R. L. (2005). Molecular evolution of the transcription factor *LEAFY* in Brassicaceae. *Molecular Phylogenetics and Evolution* **37**, 1–14.

Beck, C. B. (1976). "Origin and Early Evolution of Angiosperms." Columbia University Press, New York.

Becker, A., Gleissberg, S. and Smyth, D. R. (2005). Floral and vegetative morphogenesis in California poppy (*Eschscholzia californica* Cham.). *International Journal of Plant Sciences* **166**, 537–555.

Bernhardt, P. (2000). Convergent evolution and adaptive radiation of beetle-pollinated angiosperms. *Plant Systematics and Evolution* **222**, 293–320.

Bernhardt, P., Sage, T., Weston, P., Azuma, H., Lam, M., Thien, L. B. and Bruhl, J. (2003). The pollination of *Trimenia moorei* (Trimeniaceae): Floral volatiles, insect/wind pollen vectors and stigmatic self-incompatibility in a basal angiosperm. *Annals of Botany* **92**, 445–458.

Bhardwaha, T. N. and Baijal, J. (1977). Vessels in Rhizome of *Marsilea*. *Phytomorphology* **27**, 206–208.

Blázquez, M. A., Ferrándiz, C., Madueño, F. and Parcy, F. (2006). How floral meristems are built. *Plant Molecular Biology* **60**, 855–870.

Blázquez, M. L., Soowal, I. L. and Weigel, D. (1997). *LEAFY* expression and flower initiation in *Arabidopsis*. *Development Supplement* **124**, 3835–3844.

Bomblies, K., Wang, R.-L., Ambrose, B. A., Schmidt, R. J., Meeley, R. B. and Doebley, J. (2003). Duplicate *FLORICAULA/LEAFY* homologs *zfl1* and *zfl2* control inflorescence architecture and flower patterning in maize. *Development* **130**, 2385–2395.

Bowe, L. M., Coat, G. and dePamphilis, C. W. (2000). Phylogeny of seed plants based on all three genomic compartments: Extant gymnosperms are monophyletic and Gnetales' closest relatives are conifers. *Proceedings of the National Academy of Sciences of the United States of America* **97**, 4092–4097.

Bowman, J. L. (2000). The YABBY gene family and abaxial cell fate. *Current Opinion in Plant Biology* **3**, 17–22.

Bowman, J. L., Eshed, Y. and Baum, S. F. (2002). Establishment of polarity in angiosperm lateral organs. *Trends in Genetics* **18**, 134–141.

Boyce, C. K. and Knoll, A. H. (2002). Evolution of developmental potential and the multiple independent origins of leaves in Paleozoic vascular plants. *Paleobiology* **28**, 70–100.

Brenner, E. D., Stevenson, D. W., McCombie, R. W., Katari, M. S., Rudd, S. A., Mayer, K. F., Palenchar, P. M., Runko, S. J., Twigg, R. W., Dai, G., Martienssen, R. A. Benfey, P. N., *et al.* (2003). Expressed sequence tag analysis in *Cycas*, the most primitive living seed plant. *Genome Biology* **4**, R78.

Brenner, E. D., Katari, M. S., Stevenson, D. W., Rudd, S. A., Douglas, A. W., Moss, W. N., Twigg, R. W., Runko, S. J., Stellari, G. M., McCombie, W. R. and Coruzzi, G. M. (2005). EST analysis in *Ginkgo biloba*: An assessment of conserved developmental regulators and gymnosperm specific genes. *BMC Genomics* **6**, 143.

Bruil-Broyer, S., Morel, P., de Almeida-Engler, J., Coustham, V., Negrutiu, I. and Trehin, C. (2004). High-resolution boundary analysis during *Arabidopsis thaliana* flower development. *Plant Journal* **38**, 182–192.

Brukhin, V., Curtis, M. D. and Grossniklaus, U. (2005). The angiosperm female gametophyte: No longer the forgotten generation. *Current Science* **89**, 1844–1852.

Burleigh, J. G. and Mathews, S. (2004). Phylogenetic signal in nucleotide data from seed plants: Implications for resolving the seed plant tree of life. *American Journal of Botany* **91**, 1599–1613.

Busch, A. and Gleissberg, S. (2003). *EcFLO*, a *FLORICAULA*-like gene from *Eschscholzia californica* is expressed during organogenesis at the vegetative shoot apex. *Planta* **217**, 841–848.

Buzgo, M., Soltis, P. S. and Soltis, D. E. (2004). Floral developmental morphology of *Amborella trichopoda* (Amborellaceae). *International Journal of Plant Sciences* **165**, 925–947.

Carlquist, S. (1996). Wood, bark, and stem anatomy of Gnetales: A summary. *International Journal of Plant Sciences* **157**, S58–S76.

Carlquist, S. and Schneider, E. L. (2001). Vessels in ferns: Structural, ecological, and evolutionary significance. *American Journal of Botany* **88**, 1–13.

Carlquist, S. and Schneider, E. L. (2002). The tracheid-vessel element transition in angiosperms involves multiple independent features: Cladistic consequences. *American Journal of Botany* **89**, 185–195.

Carmichael, J. S. and Friedman, W. E. (1995). Double fertilization in *Gnetum gnemon*: The relationship between the cell-cycle and sexual reproduction. *Plant Cell* **7**, 1975–1988.

Carmichael, J. S. and Friedman, W. E. (1996). Double fertilization in *Gnetum gnemon* (Gnetaceae): Its bearing on the evolution of sexual reproduction within the anthophyte clade. *American Journal of Botany* **83**, 767–780.

Chase, M. W., Soltis, D. E., Olmstead, R. G., Morgan, D., Les, D. H., Mishler, B. D., Duvall, M. R., Price, R. A., Hills, H. G., Qiu, Y.-L., Kron, K. K. Rettig, J. H., *et al.* (1993). Phylogenetics of seed plants: An analysis of nucleotide sequences from the plastid gene rbcL. *Annals of the Missouri Botanical Garden* **80**, 528–580.

Chaw, S. M., Zharkikh, A., Sung, H. M., Lau, T. C. and Li, W. H. (1997). Molecular phylogeny of extant gymnosperms and seed plant evolution: Analysis of nuclear 18s rRNA sequences. *Molecular Biology and Evolution* **14**, 56–68.

Chaw, S. M., Parkinson, C. L., Cheng, Y. C., Vincent, T. M. and Palmer, J. D. (2000). Seed plant phylogeny inferred from all three plant genomes: Monophyly of extant gymnosperms and origin of Gnetales from conifers. *Proceedings of the National Academy of Sciences of the United States of America* **97**, 4086–4091.

Christianson, M. L. (2005). Codon usage patterns distort phylogenies from or of DNA sequences. *American Journal of Botany* **92**, 1221–1233.

Coen, E. S. and Meyerowitz, E. M. (1991). The war of the whorls: Genetic interactions controlling flower development. *Nature* **353**, 31–37.

Cornet, B. (1989). The reproductive morphology and biology of *Sanmiguelia lewisii*, and its bearing on angiosperm evolution in the late Triassic. *Evolutionary Trends in Plants* **3**, 25–51.

Crane, P. R. (1985). Phylogenetic analysis of seed plants and the origin of angiosperms. *Annals of the Missouri Botanical Garden* **72**, 716–793.

Crane, P. R. (1998). Angiosperm origins: Morphological and ecological aspects (Book Review). (V. A. Krassilov, 1997). *Plant Science Bulletin* **44**, 124.

Crane, P. R., Friis, E. M. and Pedersen, K. R. (1995). The origin and early diversification of angiosperms. *Nature* **374**, 27–33.

Crane, P. R., Herendeen, P. and Friis, E. M. (2004). Fossils and plant phylogeny. *American Journal of Botany* **91**, 1683–1699.

Crepet, W. L. (2000). Progress in understanding angiosperm history, success, and relationships: Darwin's abominably "perplexing phenomenon." *Proceedings of the National Academy of Sciences of the United States of America* **97**, 12939–12941.

Crepet, W. L. and Nixon, K. C. (1996). The fossil history of stamens. *In* "The Anther Form, Function, and Phylogeny" (W. G. D'Arcy and R. C. Keating, eds.), pp. 24–57. Cambridge University Press, New York.

Cronquist, A. (1988). "The Evolution and Classification of Flowering Plants," 2nd ed., The New York Botanical Garden, Bronx, New York.

D'Arcy, W. G. and Keating, R. C. (eds.) (1996). "The Anther Form, Function, and Phylogeny." Cambridge University Press, New York.

Darwin, F. and Seward, A. C. (1903). "More Letters of Charles Darwin: A Record of his Work in a Series of Hitherto Unpublished Letters, Vol. 2." John Murray, London.

Davis, J. I. (1993). Character removal as a means for assessing stability of clades. *Cladistics* **9**, 201–210.

Davis, J. I., Frohlich, M. W. and Soreng, R. J. (1993). Cladistic characters and cladogram stability. *Systematic Botany* **18**, 188–196.

Donoghue, M. J. and Doyle, J. A. (1989). Phylogenetic analysis of the angiosperms and the relationships of the Hamamelidae. *In* "Evolution, Systematics, and Fossil History of the Hamamelidae, Vol. 1: Introduction and 'Lower' Hamamelidae" (P. R. Crane and S. Blackmore, eds.), Systematics Association special volume 40A, pp. 17–45. Clarendon Press, Oxford.

Donoghue, M. J. and Doyle, J. A. (2000). Seed plant phylogeny: Demise of the anthophyte hypothesis? *Current Biology* **10**, R106–R109.

Donoghue, M. J. and Scheiner, S. M. (1992). Evolution of endosperm: A phylogenetic account. *In* "Ecology and Evolution of Plant Reproduction: New Approaches" (R. Wyatt, ed.), pp. 356–389. Chapman and Hall, New York.

Dornelas, M. C. and Rodriguez, A. P. M. (2005). A *FLORICAULA/LEAFY* gene homolog is preferentially expressed in developing female cones of the tropical pine *Pinus caribaea* var. *caribaea*. *Genetics and Molecular Biology* **28**, 299–307.

Doyle, J. A. (1978). Origin of angiosperms. *Annual Review of Ecology and Systematics* **9**, 365–392.

Doyle, J. A. (1994). Origin of the angiosperm flower: A phylogenetic perspective. *Plant Systematics and Evolution* **8** (Suppl.), 7–29.

Doyle, J. A. (1996). Seed plant phylogeny and the relationships of Gnetales. *International Journal of Plant Sciences* **157**, S3–S39.

Doyle, J. A. (1998). Molecules, morphology, fossils, and the relationship of angiosperms and Gnetales. *Molecular Phylogenetics and Evolution* **9**, 448–462.

Doyle, J. A. (2005). Early evolution of angiosperm pollen as inferred from molecular and morphological phylogenetic analyses. *Grana* **44**, 227–251.

Doyle, J. A. (2006) Seed ferns and the origin of angiosperms. *Journal of the Torrey Botanical Society* **133**, 169–209.

Doyle, J. A. and Donoghue, M. J. (1986). Seed plant phylogeny and the origin of angiosperms: An experimental cladistic approach. *Botanical Review* **52**, 321–431.

Doyle, J. A. and Donoghue, M. J. (1987). The origin of angiosperms: A cladistic approach. *In* "The Origin of Angiosperms and their Biological Consequences" (E. M. Friis, W. G. Chaloner and P. R. Crane, eds.), pp. 17–50. Cambridge University Press, Cambridge.

Doyle, J. A. and Donoghue, M. J. (1992). Fossils and seed plant phylogeny reanalyzed. *Brittonia* **44**, 89–106.

Doyle, J. A. and Endress, P. K. (2000). Morphological phylogenetic analysis of basal angiosperms: Comparison and combination with molecular data. *International Journal of Plant Sciences* **161**, S121–S153.

Doyle, J. A., Van Campo, M. and Lugardon, B. (1975). Observations on exine structure of *Eucommiidites* and Lower Cretaceous angiosperm pollen. *Pollen Spores* **17**, 429–486.

Endress, P. K. and Igersheim, A. (2000a). The reproductive structures of the basal angiosperm *Amborella trichopoda* (Amborellaceae). *International Journal of Plant Sciences* **161**, S237–S248.

Endress, P. K. and Igersheim, A. (2000b). Gynoecium structure and evolution in basal angiosperms. *International Journal of Plant Science* **161**, S211–S223.

Farjon, A. and Garcia, S. O. (2002). Towards the minimal conifer cone: Ontogeny and trends in *Cupressus, Juniperus* and *Microbiota* (Cupressaceae s. str.). *Botanische Jahrbücher für Systematik, Pflanzengeschichte und Pflanzengeographie* **124**, 129–147.

Farjon, A. and Garcia, S. O. (2003). Cone and ovule development in *Cunninghamia* and *Taiwania* (Cupressaceae sensu lato) and its significance for conifer evolution. *American Journal of Botany* **90**, 8–16.

Feild, T. S., Zweiniecki, M. A., Brodribb, T., Jaffre, T., Donoghue, M. J. and Holbrook, N. M. (2000). Structure and function of tracheary elements in *Amborella trichopoda. International Journal of Plant Sciences* **161**, 705–712.

Feild, T. S., Brodribb, T. and Holbrook, M. (2002). Hardly a relict: Freezing and the evolution of vesselless wood in Winteraceae. *Evolution* **56**, 464–478.

Felsenstein, J. (1978). Cases in which parsimony or compatibility methods will be positively misleading. *Systematic Zoology* **27**, 401–410.

Florin, R. (1951). Evolution in Cordaites and conifers. *Acta Horti Bergiani* **15**, 285–388.

Florin, R. (1954). The female reproductive organs of conifers and taxads. *Biological Reviews* **29**, 367–389.
Friedman, W. E. (1990). Double fertilization in *Ephedra*: A nonflowering seed plant: Its bearing on the origin of angiosperms. *Science* **247**, 951–954.
Friedman, W. E. (1992). Evidence of a pre-angiosperm origin of endosperm: Implications for the evolution of flowering plants. *Science* **255**, 336–339.
Friedman, W. E. (1994). The evolution of embryogeny in seed plants and the developmental origin and early history of endosperm. *American Journal of Botany* **81**, 1468–1486.
Friedman, W. E. and Floyd, S. K. (2001). Perspective: The origin of flowering plants and their reproductive biology: A tale of two phylogenies. *Evolution* **55**, 217–231.
Friedman, W. E. and Williams, J. H. (2003). Modularity of the angiosperm female gametophyte and its bearing on the early evolution of endosperm in flowering plants. *Evolution* **57**, 216–230.
Friedman, W. E. and Williams, J. H. (2004). Developmental evolution of the sexual process in ancient flowering plant lineages. *Plant Cell* **16**, S119–S132.
Friedman, W. E., Gallup, W. N. and Williams, J. H. (2003). Female gametophyte development in *Kadsura*: Implications for Schisandraceae, Austrobaileyales, and the early evolution of flowering plants. *International Journal of Plant Sciences* **164**, S293–S305.
Friis, E. M., Chaloner, W. G. and Crane, P. R. (1987). "The Origin of Angiosperms and their Biological Consequences." Cambridge University Press, Cambridge.
Friis, E. M., Pedersen, K. R. and Crane, P. R. (2000). Reproductive structure and organization of basal angiosperms from the Early Cretaceous (Barremian or Aptian) of Western Portugal. *International Journal of Plant Sciences* **161**, S169–S182.
Friis, E. M., Pedersen, K. R. and Crane, P. R. (2006). Cretaceous angiosperm flowers: Innovation and evolution in plant reproduction. *Palaeogeography, Palaeoclimatology, Palaeoecology* **232**, 251–293.
Frohlich, M. W. (1987). Common is primitive: A partial validation by tree counting. *Systematic Botany* **12**, 217–237.
Frohlich, M. W. (2001). A detailed scenario and possible tests of the Mostly Male theory of flower evolutionary origins. *In* "Beyond Heterochrony: The Evolution of Development" (M. L. Zelditch, ed.), pp. 59–104. Wiley-Liss, New York.
Frohlich, M. W. (2002). The Mostly Male theory of flower origins: Summary and update regarding the Jurassic pteridosperm *Pteroma*. *In* "Developmental Genetics and Plant Evolution" (Q. C. B. Cronk, R. M. Bateman and J. A. Hawkins, eds.), Systematics Association special volume series 65, pp. 85–108. Taylor and Francis, London.
Frohlich, M. W. (2003). An evolutionary scenario for the origin of flowers. *Nature Reviews Genetics* **4**, 559–566.
Frohlich, M. W. (2006). Recommendations and goals for evo-devo research: Scenarios, genetic constraint, and developmental homeostasis. *In* "Monocots: Comparative Biology and Evolution, 2 Vols." (J. T. Columbus, E. A. Friar, J. M. Porter, L. M. Prince and M. G. Simpson, eds.), pp. 172–187. Rancho Santa Ana Botanic Garden, Claremont, California.
Frohlich, M. W. and Meyerowitz, E. M. (1997). The search for flower homeotic gene homologs in basal angiosperms and Gnetales: A potential new source of data on the evolutionary origin of flowers. *International Journal of Plant Sciences* **158**, S131–S142.
Frohlich, M. W. and Parker, D. S. (2000). The Mostly Male theory of flower evolutionary origins: From genes to fossils. *Systematic Botany* **25**, 155–170.

Fujii, K. (1896). On the different views hitherto proposed regarding the morphology of the flowers of *Ginkgo biloba* L. *The Botanical Magazine (Tokyo)* **10**, 7–8, 13–15, 104–110.

Gaussen, H. (1946). Les Gymnospermes actuelles et fossiles. *Travaux du laboratoire forestier de Toulouse* Tome 2; fascicule 3.

Glasspool, I. J., Hilton, J., Collinson, M. E., Want, S.-J. and Li, C.-S. (2004). Foliar physiognomy in Cathaysian gigantopterids and the potential to track Palaeozoic climates using an extinct plant group. *Palaeogeography Palaeoclimatology Palaeoecology* **205**, 69–110.

Gocal, G. F. W., King, R. W., Blundell, C. A., Schwartz, O. M., Andersen, C. H. and Weigel, D. (2001). Evolution of floral meristem identity genes. Analysis of *Lolium temulentum* genes related to *APETALA1* and *LEAFY* of *Arabidopsis*. *Plant Physiology* **125**, 1788–1801.

Goremykin, V., Bobrova, V., Pahnke, J., Troitsky, A., Antonov, A. and Martin, W. (1996). Noncoding sequences from the slowly evolving chloroplast inverted repeat in addition to rbcL data do not support gnetalean affinities of angiosperms. *Molecular Biology and Evolution* **13**, 383–396.

Goremykin, V. V., Hirsch-Ernst, K. I., Wölfl, S. and Hellwig, F. H. (2003). Analysis of the *Amborella tricopoda* chloroplast genome sequence suggests that *Amborella* is not a basal angiosperm. *Molecular Biology and Evolution* **20**, 1499–1505.

Graybeal, A. (1998). Is it better to add taxa or characters to a difficult phylogenetic problem? *Systematic Biology* **47**, 9–17.

Hansen, A., Hansmann, S., Samigullin, T., Antonov, A. and Martin, W. (1999). *Gnetum* and the angiosperms: Molecular evidence that their shared morphological characters are convergent, rather than homologous. *Molecular Biology and Evolution* **16**, 1006–1009.

Hendy, M. D. and Penny, D. (1989). A framework for the quantitative study of evolutionary trees. *Systematic Zoology* **38**, 297–309.

Hill, C. R. (1993). A plant with flower-like organs from the Wealden of the Weald (Lower Cretaceous), southern England. *Cretaceous Research* **17**, 27–38.

Hilton, J. and Bateman, R. M. (2006). Pteridosperms are the backbone of Seed-plant phylogeny. *Journal of the Torrey Botanical Society* **133**, 119–168.

Hufford, L. (1996). The origin and early evolution of angiosperm stamens. *In* "The Anther Form, Function, and Phylogeny" (W. G. D'Arcy and R. C. Keating, eds.), pp. 58–91. Cambridge University Press, Cambridge.

Hughes, N. F. (1994). "The Enigma of Angiosperm Origins." Cambridge University Press, Cambridge, UK.

Hutchinson, A. H. (1915). Fertilization in *Abies balsamea*. *Botanical Gazette* **60**, 457–472.

Igersheim, A. and Endress, P. K. (1997). Gynoecium diversity and systematics of the Magnoliales and winteroids. *Botanical Journal of the Linnean Society* **124**, 213–271.

Jaramillo, M. A., Manos, P. S. and Zimmer, E. A. (2004). Phylogenetic relationships of the perianthless Piperales: Reconstructing the evolution of floral development. *International Journal of Plant Sciences* **165**, 403–416.

Ji, Q., Li, H. Q., Bowe, L. M., Liu, Y. S. and Taylor, D. W. (2004). Early Cretaceous *Archaefructus eoflora sp nov* with bisexual flowers from Beipiao, Western Liaoning, China. *Acta Geologica Sinica-English Edition* **78**, 883–896.

Juarez, M. T., Twigg, R. W. and Timmermans, M. C. P. (2004). Specification of adaxial cell fate during maize leaf development. *Development* **131**, 4533–4544.

Judd, W. S., Campbell, C. S., Kellogg, E. A., Stevens, P. F. and Donoghue, M. J. (2002). "Plant Systematics: A Phylogenetic Approach." Sinauer, Sunderland, MA.

Kato, K., Ohta, K., Komata, Y., Araki, T., Kanahama, K. and Kanayama, Y. (2005). Morphological and molecular analyses of the tomato floral mutant leafy inflorescence, a new allele of *falsiflora*. *Plant Science* **169**, 131–138.

Kidner, C. A. and Martienssen, R. A. (2005). The role of *ARGONAUTE1 (AGO1)* in meristem formation and identity. *Developmental Biology* **280**, 504–517.

Klavins, S. D., Taylor, T. N. and Taylor, E. L. (2002). Anatomy of *Umkomasia* (Corystospermales) from the Triassic of Antarctica. *American Journal of Botany* **89**, 664–676.

Koehl, V., Thien, L. B., Heij, E. G. and Sage, T. (2004). The causes of self-sterility in natural populations of the relictual angiosperm *Illicium floridanum*. *Annals of Botany* **94**, 43–50.

Krassilov, V. A. (1997). "Angiosperm Origins: Morphological and Ecological Aspects." Pensoft Publishers, Sofia, Bulgaria.

Land, W. J. G. (1902). A morphological study of Thuja. *Botanical Gazette* **34**, 249–259.

Laux, T., Wurschum, T. and Breuninger, H. (2004). Genetic regulation of embryonic pattern formation. *Plant Cell* **16**, S190–S202.

Leebens-Mack, J., Raubeson, L. A., Cui, L. Y., Kuehl, J. V., Fourcade, M. H., Chumley, T. W., Boore, J. L., Jansen, R. K. and dePamphilis, C. W. (2005). Identifying the basal angiosperm node in chloroplast genome phylogenies: Sampling one's way out of the Felsenstein zone. *Molecular Biology and Evolution* **22**, 1948–1963.

Li, H. Q. and Taylor, D. W. (1999). Vessel-bearing stems of *Vasovinea tianii* gen. et sp. nov. (Gigantopteridales) from the Upper Permian of Guizhou Province, China. *American Journal of Botany* **86**, 1563–1575.

Liu, Q. P. and Xue, Q. Z. (2005). Comparative studies on codon usage pattern of chloroplasts and their host nuclear genes in four plant species. *Journal of Genetics* **84**, 55–62.

Liu, Q. P., Feng, Y., Zhao, X. A., Dong, H. and Xue, Q. Z. (2004). Synonymous codon usage bias in *Oryza sativa*. *Plant Science* **167**, 101–105.

Lockhart, P. J. and Penny, D. (2005). The place of *Amborella* within the radiation of angiosperms. *Trends in Plant Science* **10**, 201–202.

Lockhart, P., Novis, P., Milligan, B. G., Riden, J., Rambaut, A. and Larkum, T. (2006). Heterotachy and tree building: A case study with plastids and eubacteria. *Molecular Biology and Evolution* **23**, 40–45.

Maddison, W. P. and Maddison, D. R. (1992). "MacClade: Analysis of Phylogeny and Character Evolution, Version 3." Sinauer, Sunderland, MA.

Magall0n, S. and Sanderson, M. J. (2002). Relationships among seed plants inferred from highly conserved genes: Sorting conflicting phylogenetic signals among ancient lineages. *American Journal of Botany* **89**, 1991–2006.

Maheshwari, P. (1950). "An Introduction to the Embryology of Angiosperms." McGraw-Hill, New York.

Maheshwari, P. (1963). "Recent Advances in the Embryology of Angiosperms." International Society of Plant Morphologists, Delhi.

Maizel, A., Busch, M. A., Tanahashi, T., Perkovic, J., Kato, M., Hasebe, M. and Weigel, D. (2005). The floral regulator *LEAFY* evolves by substitutions in the DNA binding domain. *Science* **308**, 260–263.

Mathews, S. and Donoghue, M. J. (2000). Basal angiosperm phylogeny inferred from duplicate phytochromes A and C. *International Journal of Plant Sciences* **161**, S41–S55.

Mellerowicz, E. J., Horgan, K., Walden, A., Coker, A. and Walter, C. (1998). *PRFLLM*—A *Pinus radiata* homologue of *FLORICAULA* and *LEAFY* is expressed in buds containing vegetative shoot and undifferentiated male cone primordia. *Planta* **206**, 619–629.

Meyen, S. V. (1988). Origin of the angiosperm gynoecium by gametoheterotopy. *Botanical Journal of the Linnean Society* **97**, 171–178.

Molinero-Rosales, N., Jamilena, M., Zurita, S., Gomez, P., Capel, J. and Lozano, R. (1999). *FALSIFLORA*, the tomato orthologue of *FLORICAULA* and *LEAFY*, controls flowering time and floral meristem identity. *Plant Journal* **20**, 685–693.

Molinero-Rosales, N., Latorre, A., Jamilena, M. and Lozano, R. (2004). *SINGLE FLOWER TRUSS* regulates the transition and maintenance of flowering in tomato. *Planta* **218**, 427–434.

Monson, R. K. (2003). Gene duplication, neofunctionalization, and the evolution of C-4 photosynthesis. *International Journal of Plant Sciences* **164**, S43–S54.

Moore, R. C. and Purugganan, M. D. (2003). The early stages of duplicate gene evolution. *Proceedings of the National Academy of Sciences of the United States of America* **100**, 15682–15687.

Mouradov, A., Glassick, T., Murphy, L., Fowler, B., Majla, S. and Teasdale, R. D. (1998). *NEEDLY*, a *Pinus radiata* ortholog of *FLORICAULA/LEAFY* genes, expressed in both reproductive and vegetative meristems. *Proceedings of the National Academy of Sciences of the United States of America* **95**, 6537–6542.

Nishida, H., Pigg, K. B., Kudo, K. and Rigby, J. F. (2004). Zooidogamy in the late Permian genus *Glossopteris*. *Journal of Plant Research* **117**, 323–328.

Nixon, K. C., Crepet, W. L., Stevenson, D. and Friis, E. M. (1994). A reevaluation of seed plant phylogeny. *Annals of the Missouri Botanical Garden* **81**, 484–533.

Pagel, M., Meade, A. and Barker, D. (2004). Bayesian estimation of ancestral character states on phylogenies. *Systematic Biology* **53**, 673–684.

Pagnussat, G. C., Yu, H. J., Ngo, Q. A., Rajani, S., Mayalagu, S., Johnson, C. S., Capron, A., Xie, L. F., Ye, D. and Sundaresan, V. (2005). Genetic and molecular identification of genes required for female gametophyte development and function in *Arabidopsis*. *Development* **132**, 603–614.

Parcy, F. (2005). Flowering: A time for integration. *International Journal of Developmental Biology* **49**, 585–593.

Pedersen, K. R., Crane, P. R. and Friis, E. M. (1989). Pollen organs and seeds with *Eucommiidites* pollen. *Grana* **28**, 279–294.

Piazza, P., Jasinski, S. and Tsiantis, M. (2005). Evolution of leaf developmental mechanisms. *New Phytologist* **167**, 693–710.

Prasad, K., Kushalappa, K. and Vijayraghavan, U. (2003). Mechanism underlying regulated expression of *RFL*, a conserved transcription factor, in the developing rice inflorescence. *Mechanisms of Development* **120**, 491–502.

Qiu, Y. L., Dombrovska, O., Lee, J., Li, L. B., Whitlock, B. A., Bernasconi-Quadroni, F., Rest, J. S., Davis, C. C., Borsch, T., Hilu, K. W., Renner, S. S. Soltis, D. E., *et al.* (2005). Phylogenetic analyses of basal angiosperms based on nine plastid, mitochondrial, and nuclear genes. *International Journal of Plant Sciences* **166**, 815–842.

Rothwell, G. W. and Stockey, R. A. (2002). Anatomically preserved *Cycadeoidea* (Cycadeoidaceae), with a reevaluation of systematic characters for the seed cones of Bennettitales. *American Journal of Botany* **89**, 1447–1458.

Rydin, C and Källersjö, M. (2002). Taxon sampling and seed plant phylogeny. *Cladistics* **18**, 485–513.

Rydin, C., Källersjö, M. and Friis, E. M. (2002). Seed plant relationships and the systematic position of Gnetales based on nuclear and chloroplast DNA:

Conflicting data, rooting problems, and the monophyly of conifers. *International Journal of Plant Science* **163**, 197–214.

Samigullin, T. Kh., Martin, W. F., Troitsky, A. V. and Antonov, A. S. (1999). Molecular data from the chloroplast rpoC1 gene suggest a deep and distinct dichotomy of contemporary spermatophytes into two monophyla: Gymnosperms (including Gnetales) and angiosperms. *Journal of Molecular Evolution* **49**, 310–315.

Sanderson, M. J., Wojciechowski, M. F., Hu, J.-M., Khan, T. S. and Brady, S. G. (2000). Error, bias and long-branch attraction in data for two chloroplast photosystem genes in seed plants. *Molecular Biology and Evolution* **17**, 782–797.

Schneider, E. L. and Carlquist, S. (2000). SEM studies on the vessels of heterophyllous species of *Selaginella*. *Journal of the Torrey Botanical Society* **127**, 263–270.

Schweitzer, H.-J. (1977). Die räto-jurassischen Floren des Iran und Afghanistans. 4. Die rätische Zwitterblüte *Irania hermaphroditica* nov. spec. und ihre Bedeutung für die Phylogenie der Angiospermen. *Palaeontographica Abt. B* **161 S**, 98–145.

Shindo, S., Sakakibara, K., Sano, R., Ueda, K. and Hasebe, M. (2001). Characterization of a *FLORICAULA/LEAFY* homologue of *Gnetum parvifolium* and its implications for the evolution of reproductive organs in seed plants. *International Journal of Plant Science* **162**, 1199–1209.

Sieber, P., Gheyselinck, J., Gross-Hardt, R., Laux, T., Grossniklaus, U. and Schneitz, K. (2004). Pattern formation during early ovule development in *Arabidopsis thaliana*. *Developmental Biology* **273**, 321–334.

Soltis, D. E. and Soltis, P. S. (2004). *Amborella* not a "basal angiosperm?" Not so fast. *American Journal of Botany* **91**, 997–1001.

Soltis, D. E., Soltis, P. S. and Zanis, M. (2002). Phylogeny of seed plants based on evidence from eight genes. *American Journal of Botany* **89**, 1670–1681.

Soltis, D. E., Albert, V. A., Savolainen, V., Hilu, K., Qiu, Y. L., Chase, M. W., Farris, J. S., Stefanovic, S., Rice, D. W., Palmer, J. D. and Soltis, P. S. (2004). Genome-scale data, angiosperm relationships, and "ending incongruence": A cautionary tale in phylogenetics. *Trends in Plant Science* **9**, 477–483.

Soltis, P. S., Soltis, D. E. and Chase, M. W. (1999). Angiosperm phylogeny inferred from multiple genes as a tool for comparative biology. *Nature* **402**, 402–404.

Soltis, P. S., Endress, P. K., Chase, M. W. and Soltis, D. E. (2005). "Angiosperm Phylogeny and Evolution." Sinauer, Sunderland, MA.

Spencer, M., Susko, E. and Roger, A. J. (2005). Likelihood, parsimony, and heterogeneous evolution. *Molecular Biology and Evolution* **22**, 1161–1164.

Stebbins, G. L. (1974). "Flowering Plants Evolution Above the Species Level." Harvard University Press, Cambridge, MA.

Stewart-Cox, J. A., Britton, N. F. and Mogie, M. (2004). Endosperm triploidy has a selective advantage during ongoing parental conflict by imprinting. *Proceedings of the Royal Society of London series B* **271**, 1737–1743.

Stockey, R. A. and Rothwell, G. W. (2003). Anatomically preserved *Williamsonia* (Williamsoniaceae): Evidence for bennettitalean reproduction in the late Cretaceous of western North America. *International Journal of Plant Science* **164**, 251–262.

Sun, G., Dilcher, D. L., Zheng, S. and Zhou, Z. (1998). In search of the first flower: A Jurassic angiosperm, *Archaefructus*, from Northeast China. *Science* **282**, 1692–1695.

Sun, G., Ji, Q., Dilcher, D. L., Zheng, S., Nixon, K. C. and Wang, X. (2002). Arfchaefructaceae, a new basal angiosperm family. *Science* **296**, 889–904.

Swamy, B. G. L. (1949). Further contributions to the morphology of Degeneriaceae. *Journal of the Arnold Arboretum* **30**, 10–38.
Takaso, T. and Tomlinson, P. B. (1989). Aspects of cone and ovule ontogeny in *Cryptomeria* (Taxodiaceae). *American Journal of Botany* **76**, 692–705.
Takaso, T. and Tomlinson, P. B. (1990). Cone and ovule ontogeny in *Taxodium* and *Glyptostrobus* (Taxodiaceae-Coniferales). *American Journal of Botany* **77**, 1209–1221.
Takaso, T. and Tomlinson, B. P. (1991). Cone and ovule development in *Sciadopitys* (Taxodiaceae-Coniferales). *American Journal of Botany* **78**, 417–428.
Takaso, T. and Tomlinson, P. B. (1992). Seed cone and ovule ontogeny in *Metasequoia, Sequoia* and *Sequoiadendron* (Taxodiaceae-Coniferales). *Botanical Journal of the Linnean Society* **109**, 15–37.
Taylor, D. W. and Hickey, L. J. (1992). Phylogenetic evidence for the herbaceous origin of angiosperms. *Plant Systematics and Evolution* **180**, 137–156.
Taylor, D. W. and Hickey, L. J. (eds.) (1996). "Flowering plant origin, evolution and phylogeny." Chapman and Hall, New York.
Taylor, T. N. and Taylor, E. L. (1993). "The Biology and Evolution of Fossil Plants." Prentiss Hall, Englewood cliffs, NJ.
Taylor, T. N., Del Fueyo, G. M. and Taylor, E. L. (1994). Permineralized seed fern cupules from the Triassic of Antarctica: Implications for cupule and carpel evolution. *American Journal of Botany* **81**, 666–677.
Theissen, G. and Becker, A. (2004). Gymnosperm orthologues of class b floral homeotic genes and their impact on understanding flower origin. *Critical Reviews in Plant Science* **23**, 129–148.
Theissen, G., Becker, A., Di Rosa, A., Kanno, A., Kim, J. T., Munster, T., Winter, K. U. and Saedler, H. (2000). A short history of MADS-box genes in plants. *Plant Molecular Biology* **42**, 115–149.
Theissen, G., Becker, A., Winter, K.-U., Münster, T., Kirchner, C. and Saedler, H. (2002). How the land plants learned their floral ABCs: The role of MADS-box genes in the evolutionary origin of flowers. *In* "Developmental Genetics and Plant Evolution" (Q. C. B. Cronk, R. M. Bateman and J. A. Hawkins, eds.), Systematics Association special volume series 65, pp. 173–206. Taylor and Francis, London.
Thien, L. B., Sage, T. L., Jaffre, T., Bernhardt, P., Pontieri, V., Weston, P. H., Malloch, D., Azuma, H., Graham, S. W., McPherson, M. A., Rai, H. S. Sage, R. F., *et al.* (2003). The population structure and floral biology of *Amborella trichopoda* (Amborellaceae). *Annals of the Missouri Botanical Garden* **90**, 466–490.
Tobe, H., Jaffre, T. and Raven, P. H. (2000). Embryology of *Amborella* (Amborellaceae): Descriptions and polarity of character states. *Journal of Plant Research* **113**, 271–280.
Trivett, M. L. and Pigg, K. B. (1996). A survey of reticulate venation among fossil and living land plants. *In* "Flowering Plant Origin, Evolution and Phylogeny" (D. W. Taylor and L. J. Hickey, eds.), pp. 8–31. Chapman and Hall, New York.
Vijayraghavan, U., Prasad, K. and Meyerowitz, E. (2005). Specification and maintenance of the floral meristem: Interactions between positively acting promoters of flowering and negative regulators. *Current Science* **89**, 1835–1843.
Villanueva, J. M., Broadhvest, J., Hauser, B. A., Meister, R. J., Schneitz, K. and Gasser, C. S. (1999). *INNER NOOUTER* regulates abaxial-adaxial patterning in *Arabidopsis* ovules. *Genes and Development* **13**, 3160–3169.

Wada, M., Cao, Q. F., Kotoda, N., Soejima, J. and Masuda, T. (2002). Apple has two orthologues of *FLORICAULA/LEAFY* involved in flowering. *Plant Molecular Biology* **49**, 567–577.

Wall, D. P. and Herbeck, J. T. (2003). Evolutionary patterns of codon usage in the chloroplast gene *rbcL*. *Journal of Molecular Evolution* **56**, 673–688.

Weller, S. G., Donoghue, M. J. and Charlesworth, D. (1995). The evolution of self-incompatibility in flowering plants: A phylogenetic approach. *In* "Experimental and Molecular Approaches to Plant Biosystematics" (P. C. Hoch and L. J. Hickey, eds.), pp. 355–382. Monographs in Systematic Botany from the Missouri Botanical Garden **53**. Missouri Botanical Garden, St. Louis, MO.

Weigel, D. and Meyerowitz, E. M. (1994). The ABCs of floral homeotic genes. *Cell* **78**, 203–209.

Williams, J. H. and Friedman, W. E. (2002). Identification of diploid endosperm in an early angiosperm lineage. *Nature* **415**, 522–526.

Williams, J. H. and Friedman, W. E. (2004). The four-celled female gametophyte of *Illicium* (Illiciaceae; Austrobaileyales): Implications for understanding the origin and early evolution of monocots, eumagnoliids, and eudicots. *American Journal of Botany* **91**, 332–351.

Winter, K. U., Becker, A., Munster, T., Kim, J. T., Saedler, H. and Theissen, G. (1999). MADS-box genes reveal that gnetophytes are more closely related to conifers than to flowering plants. *Proceedings of the National Academy of Sciences of the United States of America* **96**, 7342–7347.

Yamada, T., Ito, M. and Kato, M. (2003). Expression pattern of *INNER NOOUTER* homologue in *Nymphaea* (water lily family, Nymphaeaceae). *Development Genes and Evolution* **213**, 510–513.

Yamada, T., Ito, M. and Kato, M. (2004). *YABBY2*: Homologue expression in lateral organs of *Amborella trichopoda*. *International Journal of Plant Science* **165**, 917–924.

Yao, X., Taylor, T. N. and Taylor, E. L. (1995). The Corystosperm pollen organ *Pteruchus* from the Triassic of Antarctica. *American Journal of Botany* **82**, 535–546.

Yaxley, J. L., Jablonski, W. and Reid, J. B. (2001). Leaf and flower development in pea (*Pisum sativum* L.): Mutants *cochleata* and *unifoliata*. *Annals of Botany* **88**, 225–234.

Young, D. A. (1981). Are the angiosperms primitively vesselless? *Systematic Botany* **6**, 313–330.

Zanis, M. J., Soltis, P. S., Qiu, Y. L., Zimmer, E. and Soltis, D. E. (2003). Phylogenetic analyses and perianth evolution in basal angiosperms. *Annals of the Missouri Botanical Garden* **90**, 129–150.

Zhang, Q, Sodmergen, Hu, Y. S. and Lin, J. X. (2004). Female cone development in *Fokienia, Cupressus, Chamaecyparis* and *Juniperus* (Cupressaceae). *Acta Botanica Sinica* **46**, 1075–1082.

Duplication, Diversification, and Comparative Genetics of Angiosperm MADS-Box Genes

VIVIAN F. IRISH

Departments of Molecular, Cellular, and Developmental Biology and of Ecology and Evolutionary Biology, Yale University, New Haven, Connecticut 06520

Advances in Botanical Research, Vol. 44
Incorporating Advances in Plant Pathology

0065-2296/06 $35.00
DOI: 10.1016/S0065-2296(06)44003-9

ABSTRACT

There is considerable morphological diversity among angiosperm flowers. What is the developmental basis of this diversity? Some of this variation appears to be due to changes in the roles of a subset of MADS-box genes, which have been shown to control various aspects of flower development in several model species, notably *Arabidopsis*. Here I focus on perianth diversification and the roles that duplication and diversification of MADS-box genes may play in this process. I review several hypotheses for the types of changes in MADS-box genes that are responsible for phenotypic variation in perianth form. To critically test such hypotheses, functional analyses in phylogenetically informative taxa are needed. I explore the possibilities for utilizing virus-induced gene silencing as a tool to functionally assess the roles of MADS-box genes in nonmodel species. By carrying out functional analyses in a phylogenetic context it will be possible to infer the direction and type of developmental shifts that have occurred. Together these approaches will provide a robust means of inferring the relative contributions of regulatory evolution, protein evolution, and duplication of the MADS-box genes for building new floral morphologies.

I. INTRODUCTION

The angiosperms, or flowering plants, are the most successful group of land plants on earth today. A large part of this success is thought to be due to their unique reproductive structures—the flowers. Flowers represent a new way to organize plant reproductive organs, and this evolutionary novelty appears to have arisen in concert with insect-mediated pollination that may have been a driving force in the rapid radiation of the angiosperms (Crepet, 2000). Insect pollination likely provided a more effective means for outcrossing, thus contributing to enhanced fitness and diversification of the angiosperms (Crane *et al.*, 1995).

Two interrelated questions have been the focus of much of modern day research in the evolution and development of flowers. First, how did the angiosperm flower arise, and what developmental mechanisms were recruited to specify this evolutionary novelty? Second, within the angiosperms, what are the underlying mechanisms responsible for the great diversity of floral forms? Here I will focus on the second issue, that of understanding the molecular and genetic underpinnings of the diversification of floral forms within the angiosperms. Flowers represent a particularly attractive system for investigating how diversification in developmental pathways has contributed to new morphologies. Flowers in general have a stereotypical architecture, yet within that context there is wide variation in morphologies (Endress, 1994; Weberling, 1989). As such, it seems likely that there will be both conservation and diversification in the developmental pathways controlling the formation of different floral forms.

II. RATIONALE FOR EXPLORING MADS-BOX GENE DIVERSIFICATION

The field of evolutionary developmental biology rests in part on dissecting how developmental pathways have diverged in different species to give rise to species-specific morphologies. This relies on first, identifying relevant developmental pathways, and second, comparing the genes and the functions encoded by such genes. By placing such changes in a phylogenetic framework, we can define the polarity of such changes, and infer the likeliest route by which shifts in developmental pathways have come about.

Most such studies rely on a candidate gene approach; characterizations of key developmental regulatory genes in one species gives us the opportunity to explore the roles of homologous genes in other species. It seems likely that the most relevant evolutionary shifts in developmental genetic pathways have come about through changes in key developmental regulatory genes—those that encode transcription factors (Carroll *et al.*, 2001; Doebley and Lukens, 1998). Since transcription factors control a whole suite of processes, simple shifts in the expression or biochemical function of the products of such genes are a likely sources of morphological variation.

The MADS-box genes are excellent candidates for investigations of flower developmental evolution. The MADS-box genes encode DNA-binding transcription factors that have been shown to play important regulatory roles in a number of developmental processes in animals, plants, and fungi (Becker and Theissen, 2003; Messenguy and Dubois, 2003; Riechmann and Meyerowitz, 1997b). These genes contain a MADS box, encoding a 58-amino acid, highly conserved DNA-binding domain, as well as several other conserved regions (Ma *et al.*, 1991; Pnueli *et al.*, 1991; Riechmann and Meyerowitz, 1997b; Shore and Sharrocks, 1995). The MIKC-type MADS-box genes appear to have arisen within the land plant lineage, and are characterized by the highly conserved MADS domain, as well as more divergent I, K, and C domains (Alvarez-Buylla *et al.*, 2000; Parenicova *et al.*, 2003). The MADS and K domains have both been shown to be required for protein dimerization (Pellegrini *et al.*, 1995; Riechmann *et al.*, 1996; Schwarz-Sommer *et al.*, 1992; Yang and Jack, 2004). In addition to the MADS and K domains, the MIKC-type MADS-box proteins contain a weakly conserved I region, which has also been implicated in protein dimerization (Riechmann and Meyerowitz, 1997b; Riechmann *et al.*, 1996), as well as a divergent carboxy terminal C domain, which in some proteins contains conserved motifs or transcriptional activation functions (Cho *et al.*, 1999; Kramer and Irish, 1999; Kramer *et al.*, 1998; Litt and Irish, 2003).

The MIKC-type MADS-box genes have duplicated extensively in plants, with 39 such genes identified in *Arabidopsis* and at least 47 in rice (Nam *et al.*,

2004; Parenicova *et al.*, 2003). Virtually all of the MADS-box genes that have been shown to play critical developmental roles in plants belong to the MIKC subclass, demonstrating the importance of this subfamily. The explosion in numbers of MIKC-type MADS-box genes, coupled with the fact that many of these genes in *Arabidopsis* play a variety of central developmental roles including regulating flowering time, flower development, and fruit formation, suggests that diversification of these transcriptional regulators may have played a central role in the evolution of novel and different angiosperm morphologies (for recent reviews, see Becker and Theissen, 2003; de Bodt *et al.*, 2003; Irish, 2003; Irish and Litt, 2005; see Melzer *et al.*, Chapter 5). In particular, a significant amount of work has been done exploring the diversification in the sublineages of MIKC-type MADS-box genes that have been shown to direct floral organ identities in *Arabidopsis.* It is likely that the duplication and diversification of these MADS-box genes paved the way for the evolution of key innovations in flower structure in different angiosperm lineages (also see Kramer and Zimmer, Chapter 9; Rijpkema *et al.*, Chapter 6; and Soltis *et al.*, Chapter 12).

III. FLOWER MORPHOLOGIES ACROSS THE ANGIOSPERMS

Flowers come in a wide variety of sizes, shapes, and forms. What are the key developmental changes that have led to differences in floral morphologies in different angiosperm lineages? Furthermore, can we define the series of events that have resulted in various developmental shifts? These questions depend on first, inferring the form of the ancestral angiosperm flower, and second, defining the types of alterations that have occurred, at both the anatomical and molecular levels, that have resulted in particular morphologies (see Endress, Chapter 1).

To begin to determine how the developmental pathways specifying floral architecture arose and diversified in the angiosperms, it is necessary to understand how angiosperm flowers differ from other plant reproductive structures. The seed plants include the extant angiosperms and the gymnosperms, which were separated about 300 million years ago (MYA) (Stewart, 1993). Gymnosperm cones differ significantly from angiosperm flowers, in that the male pollen-bearing microsporophylls and female ovule-bearing megasporophylls are physically separated on different axes in gymnosperms, while angiosperm flowers are generally bisexual and contain both male (stamen) and female (carpel) organs united on the same axis. Angiosperm flowers also generally contain sterile perianth organs surrounding the stamens and carpels, which can be differentiated into sepals and petals, or can appear morphologically similar (tepals). Within the angiosperms, certain other morphological novelties in floral structure have arisen, including multiple derivations of bilateral symmetry from an originally radially symmetric form (Cubas, 2004; Endress, 2001a) (for further discussion of angiosperm innovations, see Endress, Chapter 1).

Two competing models have been debated over the years to explain how the angiosperm single bisexual axis arose. The euanthial theory proposes that angiosperm flowers arose from the condensation of a single axis, which bore microsporophylls and megasporophylls surrounded by sterile organs (Arber and Parkin, 1907; Doyle, 1994; Endress, 1987; Endress and Friis, 1994). This model would predict that angiosperm flowers are derived from a simple single unisexual axis similar to the reproductive structures of one group of gymnosperms, the cycads, and the extinct Bennetitales (Arber and Parkin, 1907; Doyle and Donoghue, 1992; Takhtajan, 1991). This model presupposes that initially unisexual axes became bisexualized. In contrast, the pseudanthial theory proposed that flowers arose from the condensation of a series of male and female axes without a perianth (von Wettstein, 1924; Weberling, 1989). This would predict that flowers are more similar to the reproductive structures of a different group of gymnosperms, the Gnetales, which lack perianth-like organs. Defining the ancestral flower form has important implications for assessing how different angiosperm floral morphologies, such as presence or absence of a perianth, have arisen. By defining an archetypal form, one can then "polarize" characters, or in other words, define the likely evolutionary trajectory by which specific morphologies have arisen.

Three general lines of evidence have been used to try and determine which of these two scenarios more likely reflects evolutionary history, and thus predicts the form of the ancestral angiosperm flower, and in particular whether or not it may have contained a perianth. These approaches include assessing fossil evidence, defining the most basal extant angiosperm species, and defining the closest nonangiosperm relative. Each of these sources of evidence, though, has its own unique problems of interpretation.

A. FOSSIL EVIDENCE

Examination of the fossil record indicates that early angiosperms were far more diverse than extant taxa would lead us to believe (Friis *et al.*, 2005). Contained within that diversity, though, are fossil representatives that may help to shed light on some of the earliest forms of the angiosperm flower. One of the oldest known angiosperm fossils is that of *Archaefructus*

(Sun *et al.*, 1998, 2002). *Archaefructus* flowers possessed conduplicate carpels surrounded by pairs of stamens and lacked perianth organs (Sun *et al.*, 1998, 2002). However, reevaluations of these fossils suggest that the reduced floral form may have been secondarily derived, as these plants were likely to have been specialized for an aquatic lifestyle (Friis *et al.*, 2003). Other early Cretaceous angiosperm fossils have been obtained, and these also support the idea that the ancestral angiosperm had a small, few-parted flower (Crane *et al.*, 1995; Stewart, 1993). So, despite the problems with interpreting fossil data and the general paucity of early angiosperm fossils, the weight of the evidence does support the hypothesis that the ancestral angiosperm flower was small, few parted, with few or no subtending perianth organs.

B. WHAT IS THE CLOSEST EXTANT NONANGIOSPERM?

The gymnosperms and angiosperms together comprise the seed plants, and determining which extant gymnosperm group is most closely related to angiosperms is one method that can be used to make homology assessments. Morphological data have been used to argue that the Gnetales are the sister group to the angiosperms (Crane, 1985; Doyle, 1996), which would support the idea that primitive angiosperms likely had flowers similar to the Gnetalean reproductive structures. However, some recent molecular studies reject this hypothesis in favor of placing the Gnetales within a clade of extant gymnosperms (e.g., Burleigh and Mathews, 2004; Chase, 1993; Qiu *et al.*, 1999; Soltis *et al.*, 2002; Winter *et al.*, 1999). As such, these data provide little evidence to support a particular model of angiosperm flower origins.

C. WHAT ANGIOSPERM IS SISTER TO OTHER EXTANT FLOWERING PLANTS?

Great strides in phylogenetic analysis over the past few years have redefined the relationships of different angiosperm clades. As a consequence, hypotheses as to the form of the ancestral angiosperm flower and the directions of evolutionary change have been revolutionized (Soltis *et al.*, 2005). For many years, the large, multiparted *Magnolia* flower was considered the archetype of the ancestral angiosperm flower (Cronquist, 1981). However, a series of independent phylogenetic analyses have suggested that *Amborella trichopoda* together with the Nympheales (the water lilies), are the most basal extant angiosperms (Mathews and Donoghue, 1999; Parkinson *et al.*, 1999; Qiu *et al.*, 1999; Soltis *et al.*, 1999; Stefanovic *et al.*, 2004; Zanis *et al.*, 2002). *Amborella* flowers are diminutive, with spirally arranged perianth organs, surrounding numerous stamens (male flowers) or four to eight unfused carpels (female flowers). Species of Nymphaeales are aquatic, and their flowers are variable in form. They can have few or many whorled perianth organs that can display some differentiation between inner and outer organs or be uniform in morphology. Flowers of Nymphaeales can range from simple trimerous forms (as in Cabombaceae), which has been reconstructed as ancestral, to large, multiparted flowers, which appear to be derived in this clade (reviewed in Soltis *et al.*, 2005).

Together, these observations suggest that the ancestral angiosperm flower had a relatively undifferentiated perianth surrounding the reproductive organs. Furthermore, it seems likely that there was considerable evolutionary "tinkering" of floral developmental pathways, since there is extensive variation in flower morphologies among basal angiosperms. The distinctions between sepals and petals likely arose later during angiosperm evolution. However, these conclusions have to be viewed with caution, since even extant basal angiosperm lineages could have evolved floral morphologies that are considerably different from the ancestral state.

IV. PERIANTH DIVERSITY ACROSS THE MAJOR CLADES OF ANGIOSPERMS

There has been considerable diversification of floral form within the angiosperms. Many angiosperm species have a relatively stable floral architecture, but certain aspects can show a large degree of variability (see Endress, Chapter 1). Some of these variables include the degree to which inner and outer perianth organs are morphologically distinct, whether floral organs are organized into whorls or helices, the number of organs of each type, patterns of floral symmetry, and the patterns of differentiation of each type of floral organ. In this chapter, I will focus on perianth diversity as a model for understanding how new and diverse morphologies can arise by shifts in developmental pathways.

While stamens and carpels are thought to have evolved once, different perianth forms appear to have arisen multiple times during the diversification of the angiosperms. A perianth with a distinct outer whorl of sepals and inner whorl of petals is characteristic of the core eudicots (Endress, 2001c). In fact, the relatively regular architecture of the core eudicot flower, with whorls of sepals, petals, stamens, and carpels, may be associated with the tremendous success of the core eudicots, in that they comprise close to 75% of all angiosperm species. The basal eudicots, successive sisters to the core eudicots, exhibit variability in the occurrence of distinct petals and sepals, and it has been suggested that petals have arisen independently several times in this group (Drinnan *et al.*, 1994; Endress, 1994; Ronse de Craene, 2003).

Some basal angiosperm species have distinct petals and sepals, while others have a series of tepals (Ronse de Craene *et al.*, 2003). Petals may also have arisen several times independently in the basal angiosperms as well (Endress, 2001b; Ronse de Craene *et al.*, 2003; Takhtajan, 1991; Zanis *et al.*, 2002). In some cases, petals appear to have been derived from leaf-like structures like sepals, while in other lineages petals appear to have been derived as modifications of stamens. The distribution of perianth types across angiosperm species suggests that these differences arose multiple times during the radiation of the angiosperms (e.g., Soltis *et al.*, 2005).

One complication, though, in inferring where and when distinct perianth organ types arose is in actually defining "sepal" vs "petal" (Endress, 2001c; see Endress, Chapter 1). In contrast to the specific reproductive functions of the stamens and carpels, functional definition of sepals and petals is more problematic. Sepals commonly are green and leaflike and have a protective role while petals frequently serve an attractive function, although this is not always the case. Alternatively, these organ types can be defined positionally, in that sepals occupy the outer and petals the inner whorl of the perianth. However, using these criteria to define sepals vs petals can be tricky. For instance, tulips have two whorls of brightly colored organs that can be described as petaloid, despite having the positions of sepals and petals (Kanno *et al.*, 2003). The blurring of these distinctions points to the idea that the molecular genetic programs controlling the specification of these different organ types may be somewhat labile.

V. HOW DID THE GENETIC MECHANISMS CONTROLLING THESE NEW MORPHOLOGIES EVOLVE?

While many different scenarios have been postulated to describe how the developmental mechanisms controlling different flower forms may have evolved, none of these models has yet been explicitly tested. Clearly what is needed is an in-depth investigation of the genes and processes underlying the development of flowers in different species.

Two general strategies can be used to try and discern the underlying genetic changes driving the evolution of new morphologies. The first is to identify a specific morphological difference between two species, and then attempt to define the underlying genetic differences. This type of approach only works well in situations where one is comparing quite closely related species, as that makes defining the relevant molecular genetic differences tractable. Frequently many loci contribute to the phenotypic difference, and so such traits are inherited in a quantitative manner. The genomic regions controlling such traits are called quantitative trait loci (QTL). By exploiting the natural variation in phenotypes between closely related species, it is possible to define naturally occurring alleles of loci that have contributed to the morphological differences between species. For instance, QTL analyses have been very successful in defining genes that control flowering time in *Arabidopsis* (El-Din El-Assal *et al.*, 2001; Johanson *et al.*, 2000; Michaels and Amasino, 1999). QTL analyses have also been carried out in tomato and *Arabidopsis* to define loci responsible for differences in floral organ size or shape; while several QTL affecting these traits have been identified, none of the corresponding genes have yet been characterized (Frary *et al.*, 2004; Juenger *et al.*, 2000).

The other strategy is to use a "candidate gene" approach; genes important for specifying a particular aspect of morphology in one species are examined for changes in their expression or function in other species that are associated with specific morphological differences. To some degree, this is a more amenable approach, in that it is relatively straightforward to define homologs of a particular gene from widely divergent species. However, the pitfall in this approach lies with the inference that changes in expression have played a crucial role in specifying a morphological difference. Changes in gene expression can frequently be correlated with changes in morphology, but such links do not necessarily demonstrate causality. Functional analyses, while often difficult to carry out, are a necessary adjunct to such analyses. To date, though, functional analyses have only been carried out in a handful of species. In a later section I describe some of the results and approaches being used to address the roles of candidate genes in regulating perianth development in different angiosperm lineages.

A. ANALYSIS OF PERIANTH DEVELOPMENT IN *ARABIDOPSIS*, A CORE EUDICOT

Genes involved in floral development have been extensively characterized in *Arabidopsis*, and homologs of such genes are attractive contenders for playing key roles in the evolution of different flower morphologies. *Arabidopsis*, a core eudicot, has four sepals, four petals, six stamens, and two carpels that arise in a sequential fashion from a florally determined shoot apical meristem (Smyth *et al.*, 1990). One of the earliest steps in this process is the integration of floral inductive signals by the *LEAFY* (*LFY*) gene. Flowering in *Arabidopsis* is accelerated by exposure to long days, as well as by changes in the levels of the plant hormone gibberellin and by other endogenous signals (Blazquez and Weigel, 2000). *LFY* is expressed at low levels throughout the *Arabidopsis* plant, but in response to both exogenous and endogenous signals, *LFY* expression levels are upregulated and promote the transition to flowering (Blazquez and Weigel, 1999, 2000; Blazquez *et al.*, 1997, 1998). *LFY* is clearly a key regulator in this process since loss-of-function mutations in *LFY* abolish flowering, while ectopic expression of *LFY* can induce the production of flowers from normally vegetative shoots (Huala and Sussex, 1992. Schultz and Haughn, 1991; Weigel and Nilsson, 1995; Weigel *et al.*, 1992). *LFY* encodes a novel transcription factor that acts to regulate the expression of other master regulatory genes required for floral meristem and floral organ identities (Busch *et al.*, 1999; Lamb *et al.*, 2002; Parcy *et al.*, 1998; Wagner *et al.*, 1999; William *et al.*, 2004).

1. The floral homeotic genes specify organ identity

One of the consequences of *LFY* activity is the transcriptional upregulation of a number of floral homeotic genes. Analyses of mutations in these homeotic genes led to the development of the "ABC" model that posits that overlapping expression domains of these genes can specify different floral organ types (Bowman *et al.*, 1991; Coen and Meyerowitz, 1991; Meyerowitz *et al.*, 1991). In *Arabidopsis*, mutations in *APETALA1* (*AP1*) and *APETALA2* (*AP2*) affect sepal and petal development and are thus considered A group genes (Bowman *et al.*, 1989, 1993; Irish and Sussex, 1990). The *Arabidopsis* B group gene function is encoded by *APETALA3* (*AP3*) and *PISTILLATA* (*PI*); mutations in each of these genes disrupt the specification of petal and stamen identity (Hill and Lord, 1989; Bowman *et al.*, 1989). Mutations in the *Arabidopsis* C group gene *AGAMOUS* (*AG*) result in a homeotic conversion of third whorl organs to petals, while the floral meristem is indeterminate, resulting in reiterated whorls of sepals and petals (Bowman *et al.*, 1989). In addition, four *SEPALLATA* (*SEP1–4*) genes, often referred to as the "E" function genes, have redundant roles in specifying floral organ identities as well as meristem determinacy in combination with the ABC genes (Castillejo *et al.*, 2005; Ditta *et al.*, 2004; Pelaz *et al.*, 2000). The *Arabidopsis AG*, *AP3*, *PI*, *AP1*, and *SEP* genes all encode MIKC-type MADS-box proteins, and thus act as transcription factors to regulate the expression of suites of genes involved in tissue and cell-type specific processes integral to forming the different floral organ types (Ditta *et al.*, 2004; Goto and Meyerowitz, 1994; Honma and Goto, 2001; Jack *et al.*, 1992; Mandel *et al.*, 1992; Pelaz *et al.*, 2000; Yanofsky *et al.*, 1990). The *AP2* gene also has been cloned, and its predicted product shows a region of similarity to the DNA-binding domain of ethylene response element-binding proteins suggesting that *AP2* may also act as a transcription factor (Jofuku *et al.*, 1994; Magnani *et al.*, 2004; Riechmann and Meyerowitz, 1998; Weigel, 1995).

The activities of the *Arabidopsis* MIKC-type floral homeotic MADS-box genes are regulated at several different levels. First, the spatial patterns of expression of these genes help to delimit their domains of activity. Second, the encoded proteins appear to act as parts of multiprotein complexes, which may help to confer target site specificity. Third, differences in the *cis*-acting elements bound by these transcription factors likely also play a role in target site selectivity. As an example, I describe in a later section how such mechanisms operate to regulate the activity of *AP3* and *PI*.

2. Expression of AP3 *and* PI

The expression patterns of these *Arabidopsis* floral homeotic genes appear to be responsible in large part for their functional specificity, since altering these patterns has dramatic consequences in terms of floral organ identity. For instance, *AP3* and *PI* are both expressed throughout developing petal and stamen primordia, although *PI* is also transiently expressed in the fourth whorl and *AP3* is expressed in a small patch of cells at the base of the sepal (Goto and Meyerowitz, 1994; Hill *et al.*, 1998; Honma and Goto, 2000; Jack *et al.*, 1992; Tilly *et al.*, 1998). Ectopic expression of both of these genes throughout the first whorl confers petal identity, while ectopic expression of *AP3* in the fourth whorl, where *PI* is normally initially expressed, results in stamen development (Jack *et al.*, 1994; Krizek and Meyerowitz, 1996a; Weigel and Meyerowitz, 1993, 1994). Furthermore, limiting the expression of *AP3* to certain cells, or to certain phases of development, results in disruptions of petal and stamen development (Bowman *et al.*, 1989; Jenik and Irish, 2001). Together, these results indicate a continuing requirement of *AP3* in virtually all petal and stamen primordial cells throughout their development to effect normal floral organ development in *Arabidopsis*.

As described in an earlier section, *LFY* appears to have a central and direct role in transcriptionally activating floral homeotic organ identity genes, including *AP3* (Lamb *et al.*, 2002; Parcy *et al.*, 1998; Wagner *et al.*, 1999, 2004). *LFY*, however, is expressed throughout the floral meristem, so is not sufficient to determine the spatially limited domain of *AP3* expression. It appears that *LFY* acts in conjunction with *UFO*, which is transiently expressed in the second and third whorls, to activate *AP3* expression in the appropriate domain (Lamb *et al.*, 2002; Parcy *et al.*, 1998). *UFO* encodes an F-box containing protein and is a component of an SCF complex, and so may be acting by targeting an as yet unidentified negative regulator of *AP3* expression for proteolysis (Ingram *et al.*, 1995; Lee *et al.*, 1997; Samach *et al.*, 1999; Wang *et al.*, 2003). *UFO* action is dispensable at later stages of petal and stamen organogenesis, indicating that other factors play a role in the maintenance of *AP3* expression (Laufs *et al.*, 2003). Once *UFO* and *LFY* coordinately act to activate *AP3* expression, continued *AP3* expression appears to depend on binding of an AP3 and PI protein complex to the *AP3* promoter in a positive autoregulatory feedback loop (Hill *et al.*, 1998; Jack *et al.*, 1994; Riechmann *et al.*, 1996; Tilly *et al.*, 1998). However, similar direct regulation of *PI* expression by binding of an AP3/PI heterodimer has not been observed (Chen *et al.*, 2000; Honma and Goto, 2000). As such, *PI* activation clearly depends on molecular mechanisms that are distinct from that of *AP3.*

3. AP3 and PI form protein complexes

Dimerization and higher-order protein complex formation can greatly increase the numbers and specificity of distinct DNA-binding complexes. In yeast, the interaction of MADS-box proteins with a variety of cofactors confers exquisite specificity to the resulting complex (for review see Messenguy and Dubois, 2003). In *Arabidopsis*, AP3 and PI, interacting with either SEP3 or AP1, have been found to be components of multimeric protein complexes *in vitro* (Egea-Cortines *et al.*, 1999; Honma and Goto, 2001; see Melzer *et al.*, Chapter 5). AP3 and PI may also potentially interact with other, non-MADS-box accessory factors, to form transcriptional complexes. The formation of AP3/PI heterodimers or higher-order protein complexes is likely to contribute to the localization and stability of the resulting complex in the nucleus (Honma and Goto, 2001; McGonigle *et al.*, 1996). Furthermore, as a number of the MADS-box proteins lack transcriptional activation capabilities, such complexes can serve to bring an activation domain into a transcriptional complex (Honma and Goto, 2001). Although many MADS-box proteins have been shown to interact in various specific combinations *in vitro* (de Folter *et al.*, 2005), there is only limited evidence, as of yet, for the formation of such protein complexes *in vivo* (Ferrario *et al.*, 2004; Jack, 2001; Immink *et al.*, 2002; Theissen and Saedler, 2001). It will be important to demonstrate that these MADS-box protein complexes in *Arabidopsis* have different functional specificities *in vivo*, as has been postulated. Experiments in which the MADS box has been swapped between proteins indicate that variations in this DNA-binding domain have little, if any, effect on functional specificity (Krizek and Meyerowitz, 1996b; Krizek *et al.*, 1999; Riechmann and Meyerowitz, 1997a). Nonetheless, differences in protein interactions and complex composition likely have a large effect on target site specificity.

4. Downstream targets of AP3 *and* PI

In addition to differences in patterns of expression and in protein complex composition, it is also plausible that specificity can be achieved by differences in target promoter sequences. While a large number of downstream targets of *AP3* and *PI* have been identified using genomic approaches (Sablowski and Meyerowitz, 1998; Wellmer *et al.*, 2004; Zik and Irish, 2003), relatively few of these have been shown to be directly bound by AP3/PI proteins. The best analyzed downstream target of AP3/PI is the *AP3* promoter itself, and analyses of the *cis*-regulatory sequences of *AP3* indicate that there can be considerable complexity to such regulatory sequences (Hill *et al.*, 1998; Koch *et al.*, 2001; Tilly *et al.*, 1998). MADS-box proteins bind to a CC(A/T)6GG core consensus sequence termed the CArG box (Schwarz-Sommer *et al.*, 1992; Wynne and Treisman, 1992). *AP3* contains three CArG boxes in its promoter, which likely are bound by various MADS-box proteins *in vivo* (Hill *et al.*, 1998; Irish and Yamamoto, 1995; Tilly *et al.*, 1998). Despite the fact that CArG box 2 is highly conserved and matches the CArG consensus exactly, it is not bound by AP3/PI proteins, indicating that the affinity of MADS-box proteins for target sites likely depends on a variety of additional factors (Hill *et al.*, 1998; Koch *et al.*, 2001). These could include accessory proteins that modulate the activity of the MADS-box proteins, cooperative interactions of the MADS-box proteins with other transcription factors in binding to DNA, or the flanking sequence context of the binding site. The specific sequence of the binding site can influence the architecture of the transcriptional complex, affecting the degree of DNA bending which regulates the recruitment and formation of multiprotein transcription factor complexes at the promoter (Lim *et al.*, 2003; West and Sharrocks, 1999).

B. CONSERVATION AND DIVERSIFICATION IN EXPRESSION AND FUNCTION OF *AP3* AND *PI* GENES IN THE ANGIOSPERMS

The action of the *AP3* and *PI* genes in *Arabidopsis* conforms in large part to the hypotheses of the ABC model, but to what extent does such a model of overlapping gene functions in different whorls serve to explain floral development in other species? If such a model does apply to other species, then do the orthologous MADS-box genes from these other species serve the equivalent functions? Underlying these issues is an important distinction between genetic functions and orthology. It is important to point out that orthologous genes need not have equivalent functions; they can diverge significantly in their roles (Irish and Litt, 2005). As such, it is a misnomer to refer to "B function" genes in species where a "B function" (e.g., specification of petal and stamen organ identity) has not been ascribed. This issue is further complicated by the fact that additional copies of genes can arise by duplication events, which can result in a subfunctionalization or neofunctionalization of their roles (Force *et al.*, 1999; Lynch and Force, 2000; Ohno, 1970;

Teichmann and Babu, 2004; see Kramer and Zimmer, Chapter 9). These complications are highlighted by analyses of *AP3* and *PI* lineage genes in other angiosperms.

1. Gene duplication in the AP3 *and* PI *gene lineages*

The *AP3* and *PI* gene lineages appear to have arisen by a gene duplication event that predated the origin of the angiosperms (Aoki *et al.*, 2004; Kim *et al.*, 2004; Kramer *et al.*, 1998; Purugganan *et al.*, 1995; Stellari *et al.*, 2004; see Kramer and Zimmer, Chapter 9 and Soltis *et al.*, Chapter 12). This duplication appears to have occurred after the split between gymnosperms and angiosperms, since extant gymnosperms possess *AP3/PI* homologs with sequence motifs that suggest that they represent an ancestral preduplication form (Mouradov *et al.*, 1999; Sundstrom *et al.*, 1999; Winter *et al.*, 2002).

Within the angiosperms, the *AP3* lineage has undergone a gene duplication event resulting in two paralogous gene lineages, the *euAP3* and *TM6* lineages, in the core eudicots (Kramer *et al.*, 1998). While the *TM6* lineage genes have sequence motifs similar to the ancestral *paleoAP3* genes, including a C-terminal paleoAP3 motif, the *euAP3* gene lineage has undergone a frameshift mutation, resulting in a completely different C-terminal motif, the euAP3 motif (Kramer *et al.*, 1998; Vandenbussche *et al.*, 2003). Thus, the *euAP3* lineage genes represent a divergent paralogous gene lineage in the core eudicots. The *Arabidopsis AP3* gene is a member of this divergent *euAP3* lineage (Kramer *et al.*, 1998). The considerable conservation seen in the characteristic and different C-terminal motifs in each of these gene lineages has led to the suggestion that these motifs confer qualitatively different functions on the encoded proteins (Kramer *et al.*, 1998). In addition, there have also been a number of more recent gene duplications in the *AP3* and *PI* genes lineages (Kramer *et al.*, 1998, 2003; Stellari *et al.*, 2004; Vandenbussche *et al.*, 2004) as well as gene losses, such as the loss of the *TM6* representative from *Arabidopsis* (Lamb and Irish, 2003).

One of the limitations in defining where and when gene duplications and loss events have occurred is the paucity of both taxonomic and genomic sampling. Genome projects and EST databases have been developed for a number of angiosperm species (e.g., see http://www.plantgdb.org/), but these analyses are concentrated in the core eudicots and monocot grasses and miss out on many key taxa that are important from an evolutionary perspective. In addition, in-depth sampling needs to be carried out in exemplar species in order to determine the complete repertoire of a given gene family in that species. A concerted effort to obtain this type of information for the MADS-box genes from a variety of phylogenetically informative angiosperm species is underway, and should result in a robust picture of gene duplication, loss, and diversification in this gene family (Hileman *et al.*, submitted).

2. An exploration of the changing roles of the AP3 *and* PI *lineage genes*

A number of models have been proposed to explain the roles of *AP3* and *PI* lineage genes in different angiosperm taxa. These models have relied to a large extent on surveys of expression patterns of these genes, since functional analyses have been limited to only a few taxa. It seems likely that the ancestral role of the preangiosperm *AP3/PI* gene was in specifying stamen identity. This is consistent with the male reproductive organ-specific expression patterns of such genes in gymnosperms (Fukui *et al.*, 2001; Mouradov *et al.*, 1999; Sundstrom *et al.*, 1999) as well as the expression of *AP3* and *PI* lineage genes in stamens from all angiosperm taxa analyzed to date (for reviews see Irish, 2003; Irish and Kramer, 1998; Soltis *et al.*, Chapter 12; Zahn *et al.*, 2005). Furthermore, mutational analyses of *AP3* and *PI* lineage genes in a number of angiosperm taxa support the idea that these genes have a role in specifying stamen identity. However, their roles in specifying other organ or tissue types may vary in different clades of angiosperms.

While there is an ever-increasing number of expression surveys of *AP3* and *PI* homologs in various angiosperm species, there are several limitations in inferring gene function simply from expression patterns. Expression in a particular organ or tissue may reflect a homeotic role, or a very different role in elaborating a particular cell or tissue type. Expression at early vs later phases of development may also reflect differing roles for particular genes. In addition, some genes may act in a nonautonomous fashion, in which their domains of action can be at some distance from the cells in which the gene is expressed (Tokunaga, 1972). Several MADS-box genes have already been shown to act nonautonomously (Efremova *et al.*, 2001; Jenik and Irish, 2001; Sieburth *et al.*, 1998; Vincent *et al.*, 2003). Many models have been proposed to explain floral diversity based on shifts in MADS-box gene expression patterns (Bowman, 1997; Kramer *et al.*, 2003; Theissen *et al.*, 2000), but without functional data, these expression-based models should be viewed simply as attractive hypotheses for further testing. In other words, these caveats point to the need for functional analyses by which one can critically test such hypotheses. Nonetheless, there are a handful of examples where functional analyses in conjunction with expression surveys have been carried out, and these help to illuminate some of the subtleties of how these developmental programs have diversified in different species.

Within the core eudicots, the *PI* lineage genes and the *euAP3* lineage genes are generally expressed throughout developing petal and stamen primordia, although there is some slight amount of variability in this pattern in different species (for review see Irish and Kramer, 1998). In general, core eudicot *euAP3* and *PI* genes are expressed from the inception of petal and stamen primordia, in all cells of these organs, and this ubiquitous pattern of expression is maintained until very late stages of development when expression can become restricted to particular differentiating cell types. The requirement for constant and ubiquitous expression of *euAP3* and *PI* for specifying petal and stamen identities has been demonstrated experimentally in *Arabidopsis* (Bowman *et al.*, 1989; Jenik and Irish, 2001). These experiments have shown that disrupting the expression of these genes during particular times or tissues during development can result in significant defects in petal or stamen differentiation. Furthermore, even within the core eudicots, there is some diversification in the molecular mechanisms regulating the expression patterns of these genes. For instance, the initial patterns of *euAP3* and *PI* gene expression differ considerably between the core eudicot species *Arabidopsis* and *Antirrhinum* and only come to be in register later in development as a consequence of cross-regulatory interactions (Goto and Meyerowitz, 1994; Jack *et al.*, 1992; Samach *et al.*, 1997; Schwarz-Sommer *et al.*, 1992; Sommer *et al.*, 1990; Trobner *et al.*, 1992). These observations suggest that there has been strong selective pressure to maintain this specific domain of expression, which may reflect the fixed nature of floral patterning in the core eudicot clade. These expression patterns are consistent with the known functional roles of *euAP3* and *PI* genes in specifying petal and stamen identities in core eudicots (Angenent *et al.*, 1993; Goto and Meyerowitz, 1994; Immink *et al.*, 2003; Jack *et al.*, 1992; Sommer *et al.*, 1990; Trobner *et al.*, 1992; Tsuchimoto *et al.*, 2000; van der Krol *et al.*, 1993; Vandenbussche *et al.*, 2004; see Rijpkema *et al.*, Chapter 6).

Members of the *TM6* lineage, however, show a much more variable pattern of expression in the core eudicots, with expression ranging from predominantly in developing stamens and carpels [e.g., *Petunia* (Vandenbussche *et al.*, 2004) or *Populus* (Sheppard *et al.*, 2000)] to a broader and more dynamic pattern in petals, stamens, and carpels [e.g., tomato (Pnueli *et al.*, 1991)] to a very broad distribution not only in flowers but in vegetative tissues as well [e.g., *Gerbera* (Yu *et al.*, 1999)]. It is interesting to note that, within the core eudicots, it appears to be members of the divergent *euAP3* lineage that have been maintained, while *TM6* genes may be in the process of being lost from particular clades. A number of factors may influence whether a duplicate gene is lost or retained, but as of yet there is little empirical evidence to support a particular mechanism (Holland, 1999; Shimeld, 1999). The variation in *TM6* lineage gene expression patterns may potentially reflect a relaxation of the selective pressure to maintain these genes.

In turn, this may contribute to the gradual loss or degradation of *TM6* orthologs to pseudogenes in some core eudicot clades.

The functions of the *euAP3* and *PI* lineage genes have been characterized in a few core eudicot species. Mutational analyses of these genes in *Arabidopsis*, *Antirrhinum*, *Petunia*, *Gerbera*, and tomato have shown that the *euAP3* and *PI* genes in these species are responsible for specifying petal and stamen identities (Angenent *et al.*, 1993, 1995; Goto and Meyerowitz, 1994; Jack *et al.*, 1992; Sommer *et al.*, 1990; Trobner *et al.*, 1992; Tsuchimoto *et al.*, 2000; van der Krol *et al.*, 1993; Vandenbussche *et al.*, 2004; Yu *et al.*, 1999) (DeMartino, G., and V. F. I., manuscript in preparation). Disruption in the functions of the *euAP3* and *PI* orthologs in these species results in homeotic transformations of petals into sepaloid organs, while stamens are transformed into ovule-bearing carpelloid structures. In some cases, such as *Petunia*, there have been more recent gene duplication events within the *euAP3* and *PI* lineages, resulting in subfunctionalization of the resulting paralogs, but collectively these genes retain their functions in specifying petal and stamen identities (Angenent *et al.*, 1992; Immink *et al.*, 2003; van der Krol *et al.*, 1993; Vandenbussche *et al.*, 2004). Orthologs of the *TM6* lineage are being functionally characterized, and this evidence suggests that such genes may have a role in specifying stamen identity (DeMartino, G., and V. F. I., manuscript in preparation).

In basal eudicots, *paleoAP3* and *PI* lineage genes have been characterized from several species, including members of the Papaveraceae and Ranunculaceae. Multiple *paleoAP3* and *PI* gene duplication events have been identified in these groups, and may reflect modifications in the roles of these genes via sub- or neofunctionalization (Kramer and Irish, 1999, 2000; Kramer *et al.*, 1998, 2003). Consistent with this idea, the spatial and/or temporal patterns of expression of these duplicates can vary significantly from one another. Although functional analyses have not yet been carried out for any of these genes, heterologous expression studies in *Arabidopsis* support the idea that these genes retain a conserved role in stamen but not petal identity specification (Lamb and Irish, 2003). This observation is consistent with the idea that basal eudicot petals have arisen multiple times independently as modifications of stamens (Kosuge, 1994; Ronse de Craene, 2004; Takhtajan, 1991), and so may reflect independent recruitments of distinct developmental programs to specifying petal identity in each case.

The monocots, which possess *paleoAP3* and *PI* representatives, display variable patterns of expression of these genes. In some monocots, such as tulips, lilies, and *Agapanthus*, which have petaloid first whorl organs, homologs of these genes are expressed in the first, second, and third whorls (An *et al.*, 1988; Kanno *et al.*, 2003; Nakamura *et al.*, 2005; Theissen *et al.*, 2000).

These observations have led to a variety of "sliding boundary" models in which expanded expression of *paleoAP3* and *PI* genes into the first whorl is posited as sufficient to confer petaloid identity (Bowman, 1997; Kramer *et al.*, 2003; van Tunen *et al.*, 1993). However, other monocots, such as *Asparagus*, which also have petaloid first whorl organs, have *paleoAP3* and *PI* gene expression restricted to the second and third whorls (Park *et al.*, 2003, 2004). Thus, in *Asparagus*, the expression of *paleoAP3/PI* lineage genes is not correlated with petaloid identity, suggesting that petaloidy may be determined by other genes in certain monocot taxa.

The situation in grasses is more difficult to interpret since grass flowers are highly derived monocots and lack petals (Bommert *et al.*, 2005). Rice and maize *paleoAP3* representatives are expressed in stamens and in lodicules, which are grass-specific organs (Ambrose *et al.*, 2000; Kang *et al.*, 1998; Nagasawa *et al.*, 2003). Furthermore, these genes are required for stamen and lodicule development, since mutations in these genes disrupt development of these organs. Rice contains two *PI* orthologs; silencing of one of these copies using transgenic approaches disrupts stamen and lodicule development, whereas silencing of the other copy only affects lodicule formation (Kang *et al.*, 1998; Prasad and Vijayraghavan, 2003). From these results it seems clear that the functions of these genes in regulating the specification of stamen identity has been conserved in monocots and eudicots. However, it is unclear as to whether lodicules represent second whorl organs or represent modified sterile stamens (Clifford, 1988; Dahlgren *et al.*, 1985). In fact, the lodicule appears to have arisen within the grass lineage, implying that it has no counterpart in the eudicot flower (Dahlgren *et al.*, 1985). The action of grass *paleoAP3* and *PI* genes in specifying lodicule development has been used to argue that lodicules are homologous to petals (Ambrose *et al.*, 2000; Whipple *et al.*, 2004), but this does not take into account that homology implies common descent. Rather, it seems more likely that grasses and core eudicots have converged upon a similar mechanism to specify lodicules and petals, respectively. In the case of the grasses, this has entailed the recruitment of *paleoAP3* and *PI* representatives to lodicule development, while in the core eudicots, the *euAP3* and *PI* genes participate in specifying petal identity.

Expression patterns of *paleoAP3* and *PI* lineage genes have also been surveyed in some basal angiosperm species (see Kim *et al.*, 2005; see Soltis *et al.*, Chapter 12). In general, the patterns of expression in petaloid organs are somewhat variable. Variability can be seen in terms of the tissue specificity of expression, the intensity of expression, the duration of expression, as well as in the relative patterns of *paleoAP3* vs *PI* homolog gene expression (Kramer and Irish, 2000; Kim *et al.*, 2005). For instance, in different Magnoliaceae species, expression of the *paleoAP3* homolog is restricted to earlier stages of petaloid development (*Liriodendron tulipifera*), or spatially restricted to certain tissues (*Michelia figo*) (Kramer and Irish, 2000). Similarly, expression of the paleoAP3 representative in *Nuphar advena* (Nympheaceae) is restricted to tips of inner tepals, which could point to a tissue-specific or other role in later stages of tepal development (Kim *et al.*, 2005). The variation in the patterns of expression of these genes across basal angiosperms may reflect plasticity in the underlying developmental mechanisms controlling organ identity and cell- and tissue-type differentiation. It is important to point out that these hypotheses, based on expression data, can only be critically tested via functional analyses.

VI. EVOLUTION OF *AP3* AND *PI* FUNCTION

The data reviewed above provide some tantalizing glimpses of the ways in which *AP3/PI* gene functions may have evolved to give rise to new morphologies. Still, it is quite apparent that these studies are extremely limited in terms of taxonomic representation, as well as in the extent of detailed functional analyses. What, then, can we argue from these observations as to the evolutionary history of the roles of these genes? How did new functions come about?

A number of scenarios can be envisaged that are consistent with existing data. One possibility is that members of the *AP3* and *PI* gene lineages have a constant "selector gene" role in specifying organ identity. Selectors have discrete roles in regulating a network of developmental processes leading to a particular organ or tissue identity and, as such, can be distinguished from other transcription factors (Garcia-Bellido, 1975; Mann and Carroll, 2002). In this hypothesis, *AP3* and *PI* lineage genes had a presumed ancestral selector gene role in specifying stamen identity. During the radiation of the angiosperms, plasticity in organ number could have led to multiplication of stamens, and a corresponding widening of the spatial domain in which *AP3* and *PI* lineage genes acted. The outer domain of stamens then gradually became sterilized to become petals, through an accumulation of a variety of morphological and molecular changes. Such changes presumably included shifts in AP3/PI cofactor interactions (presumably facilitated by changes in C-terminal motifs), as well as shifts in the suites of target genes regulated by these MADS-box gene products.

A second possibility is that the ancestral role of *AP3/PI* lineage genes was in regulating terminal cell type differentiation. This is consistent with the roles of *AP3* and *PI* in *Arabidopsis*, in that they are required in essentially all petal and stamen cell types throughout organogenesis. This model would posit that during angiosperm evolution, *AP3* and *PI* lineage genes were first recruited to roles in the differentiation of specific cell types, and then redeployed to specify whole organs, such as petals. The transient and/or spatially restricted patterns of these genes in, for instance, basal eudicots, may reflect a cell-type specific role for *AP3/PI* lineage genes in specification of particular petal cell types. There is some evidence in animal systems that the molecular mechanisms governing the specification of particular cell types are more highly conserved than those regulating regional identity (Mann and Carroll, 2002), and, by analogy, the roles of *AP3/PI* in regulating cell type differentiation may represent the "ancestral" function. In certain clades, such as the core eudicots, gene duplication and diversification paved the way for the redeployment of such genes for a new developmental role—that of specifying petal "identity."

These models are not mutually exclusive. Furthermore, there may have been multiple recruitments of the *AP3/PI* genes to new roles and/or new spatial domains in different clades of angiosperms. These could have occurred in a similar fashion in different lineages, or have been qualitatively distinct. Distinguishing between the selector gene model and the redeployment model, at the very minimum, requires an analysis of the functional roles of these genes in a variety of angiosperm systems. Do *AP3/PI* homologs have consistent roles in specifying petal organ identity? Do they regulate homologous sets of target genes or not? Do they have other, cell specific, roles in perianth development? Or no role at all in petal development in some species? Obviously, the key to addressing these alternatives is to explicitly test the roles of these genes in phylogenetically informative species.

VII. HOW DO WE FUNCTIONALLY ANALYZE MADS-BOX GENES IN NONMODEL SPECIES?

The limitations associated with expression studies, combined with the observation of multiple occurrences of gene duplication in the *AP3/PI* lineage genes, significantly complicates interpretations of how gene functions may have diverged in different angiosperm species. Duplicate genes can diverge in function by changes in their expression patterns, or by changes in the coding region (resulting in changes in biochemical function). It is likely that many duplicate genes accrue such changes at both expression level and amino acid coding level. Assessing the loss-of-function phenotypes of such genes is a necessary prerequisite to defining the roles of such genes. However, traditional genetic approaches are not feasible in many species. Forward genetic approaches, including chemical and insertional mutagenesis, rely on the ability to outcross individuals and obtain progeny in a reasonable amount of time. Reverse genetic approaches, in which target gene expression is reduced via RNAi, in general relies on *Agrobacterium*-mediated transformation. This approach depends on transforming explanted plant tissue, and regenerating transformed plants via tissue culture. While in many species generating such transformation events is relatively straightforward, such analyses are hampered by the difficulties associated with fast and reliable means of regenerating transformed plants.

Virus-induced gene silencing (VIGS) is an attractive and effective alternative to traditional genetic or transgenic approaches (Ruiz *et al.*, 1998). VIGS is a means to transiently silence plant genes by harnessing the homology-dependent RNA-degradation defense mechanisms that plants utilize in response to viral infection. As such, viruses can be engineered to target a host gene(s) of interest for silencing (Burch-Smith *et al.*, 2004; Robertson, 2004). Thus, VIGS is a potentially powerful tool for reverse genetic analyses in a wide range of plant species.

VIGS is similar to posttranscriptional gene silencing (PTGS); both result in the degradation of endogenous RNAs with extensive sequence similarity to introduced transgenes (Fagard and Vaucheret, 2000; Robertson, 2004; Ruiz *et al.*, 1998). Both RNA viruses and DNA viruses have been used as VIGS vectors, although RNA viruses are more commonly employed (Robertson, 2004). The silencing signal is double-stranded RNA (dsRNA). In VIGS, dsRNA is produced by virally encoded RNA dependent RNA polymerase using the viral genome as a template. The host plant defense mechanism recognizes dsRNA as aberrant, resulting in the cleavage of such dsRNA into small interfering RNA (siRNA) molecules, which then target homologous endogenous RNAs for cleavage (Hamilton and Baulcombe, 1999; Zamore, 2004). VIGS is systemic, in that sequence-specific silencing can be transmitted from cell to cell (Fagard and Vaucheret, 2000). Propagation of the silencing signal to the rest of the plant likely depends upon cell-to-cell movement as well as reamplification of the signal in recipient cells.

At least two possibilities exist for adapting VIGS to nonmodel systems. One possibility is to develop specific viruses as vectors for particular hosts. For instance, barley stripe mosaic virus-based vectors have been used successfully to silence genes in the natural host, barley (Hein *et al.*, 2005; Holzberg *et al.*, 2002). A second approach is to utilize vectors based on viruses that are known to have a wide host range, relying on the likely ability of such viruses to be able to induce silencing in heterologous hosts. *Tobacco rattle virus* (TRV), a bipartite single-stranded RNA virus, is an excellent choice for such an approach. TRV has a broad host range (http://www.ncbi.nlm.nih.gov/ICTVdb/index.htm), spanning much of angiosperm diversity.

Several TRV vectors have been developed for VIGS and have been effective in silencing a wide range of endogenous genes in various Solanaceae species (Liu *et al.*, 2002; Ratcliff *et al.*, 2001). Because TRV infects the meristematic regions of plants, unlike many other viruses, it promises to have wide applicability in terms of probing developmental functions. In fact, TRV-based VIGS vectors have been effective at silencing genes expressed during floral development (Hileman *et al.*, 2005; Liu *et al.*, 2004).

TRV-VIGS has been shown to be an effective tool in silencing genes in the basal eudicot species *Papaver somniferum* (poppy) (Hileman *et al.*, 2005). Phytoene desaturase (PDS), an enzyme involved in the carotenoid biosynthetic pathway, is an excellent visual marker to follow VIGS, as silencing of the *PDS* gene results in photooxidation, and a concomitant photobleached phenotype (Demmig-Adams and Adams, 1992). Silencing of the poppy *PDS* gene can be detected by photobleaching within 2 weeks of inoculation, and affects all parts of the plant, including flowers (Hileman *et al.*, 2005). These observations indicate that gene products involved in a variety of developmental and biosynthetic processes in poppy should be amenable to functional dissection using TRV-VIGS. Using this approach, we are currently exploring the functional roles of various MADS-box genes in the poppy *Papaver somniferum*, as an exemplar basal eudicot species.

VIII. PROSPECTS

The conservation of master developmental regulators, such as the HOX genes in animals and the MADS-box genes in plants, is astonishing and rightly has captured the attention of biologists who want to understand the evolution of the underlying molecular mechanisms specifying different morphologies. However, the conservation of these genes cannot itself explain diversity. The analysis of *AP3* and *PI* lineage genes in the angiosperms has revealed layers of complexity in how members of these gene lineages operate in different species. We are just at the beginning stages of dissecting when, how, and where these genes have undergone subtle alterations—ranging in changes in their expression domains to changes in their function in terms of activating or repressing particular sets of target genes to generating "new" gene functions by duplication. Furthermore, by placing such changes in a phylogenetic context, we should be able to infer the direction and type of developmental shifts that have occurred. Through these assessments, we will eventually be in a position to weigh the relative contributions of regulatory evolution, protein evolution, and duplication of the MADS-box genes for building new floral morphologies.

ACKNOWLEDGMENTS

I thank Drs Amy Litt and Lena Hileman for their constructive comments on this manuscript. Recent work in my lab on comparative analyses of floral development has been supported by grants #IBN-0110731, DBI-0110115 and DBI-0411960 from the National Science Foundation.

REFERENCES

Alvarez-Buylla, E. R., Pelaz, S., Liljegren, S. J., Gold, S. E., Burgeff, C., Ditta, G. S., de Pouplana, L. R., Martinez-Castilla, L. and Yanofsky, M. F. (2000). An ancestral MADS-box gene duplication occurred before the divergence of plants and animals. *Proceedings of the National Academy of Sciences of the United States of America* **97**, 5328–5333.

Ambrose, B. A., Lerner, D. R., Ciceri, P., Padilla, C. M., Yanofsky, M. F. and Schmidt, R. J. (2000). Molecular and genetic analyses of the Silky1 gene reveal conservation in floral organ specification between eudicots and monocots. *Molecular Cell* **5**, 569–579.

An, G., Ebert, P. R., Mitra, A. and Ha, S. B. (1988). Binary vectors. *Plant Molecular Biology Manual* **A3**, 1–19.

Angenent, G. C., Busscher, M., Franken, J., Mol, J. N. M. and van Tunen, A. J. (1992). Differential expression of two MADS box genes in wild-type and mutant petunia flowers. *The Plant Cell* **4**, 983–993.

Angenent, G. C., Franken, J., Bussher, M., Colombo, L. and van Tunen, A. J. (1993). Petal and stamen formation in petunia is regulated by the homeotic gene fbp1. *Plant Journal* **4**, 101–112.

Angenent, G. C., Busscher, M., Franken, J., Dons, H. J. and van Tunen, A. J. (1995). Functional interaction between the homeotic genes fbp1 and pMADS1 during petunia floral organogenesis. *Plant Cell* **7**, 507–516.

Aoki, S., Uehara, K., Imafuku, M., Hasebe, M. and Ito, M. (2004). Phylogeny and divergence of basal angiosperms inferred from APETALA3- and PISTILLATA-like MADS-box genes. *Journal of Plant Research* **117**, 229–244.

Arber, E. A. N. and Parkin, J. (1907). On the origin of angiosperms. *Botanical Journal of the Linnean Society of London* **38**, 29–80.

Becker, A. and Theissen, G. (2003). The major clades of MADS-box genes and their role in the development and evolution of flowering plants. *Molecular Phylogenetics and Evolution* **29**, 464–489.

Blazquez, M. A. and Weigel, D. (1999). Independent regulation of flowering by phytochrome B and gibberellins in *Arabidopsis*. *Plant Physiology* **120**, 1025–1032.

Blazquez, M. A. and Weigel, D. (2000). Integration of floral inductive signals in *Arabidopsis*. *Nature* **404**, 889–892.

Blazquez, M. A., Soowal, L., Lee, I. and Weigel, D. (1997). LEAFY expression and flower initiation in *Arabidopsis*. *Development* **124**, 3835–3844.

Blazquez, M. A., Green, R., Nilsson, O., Sussman, M. R. and Weigel, D. (1998). Gibberellins promote flowering of *Arabidopsis* by activating the LEAFY promoter. *Plant Cell* **10**, 791–800.

Bommert, P., Satoh-Nagasawa, N., Jackson, D. and Hirano, H. Y. (2005). Genetics and evolution of inflorescence and flower development in grasses. *Plant Cell Physiology* **46**, 69–78.

Bowman, J. L. (1997). Evolutionary conservation of angiosperm flower development at the molecular and genetic levels. *Journal of Biosciences* **22**, 515–527.
Bowman, J. L., Smyth, D. R. and Meyerowitz, E. M. (1989). Genes directing flower development in *Arabidopsis*. *Plant Cell* **1**, 37–52.
Bowman, J. L., Smyth, D. R. and Meyerowitz, E. M. (1991). Genetic interactions among floral homeotic genes of *Arabidopsis*. *Development* **112**, 1–20.
Bowman, J. L., Alvarez, J., Weigel, D., Meyerowitz, E. M. and Smyth, D. R. (1993). Control of flower development in *Arabidopsis thaliana* by APETALA1 and interacting genes. *Development* **119**, 721–743.
Burch-Smith, T. M., Anderson, J. C., Martin, G. B. and Dinesh-Kumar, S. P. (2004). Applications and advantages of virus-induced gene silencing for gene function studies in plants. *Plant Journal* **39**, 734–746.
Burleigh, J. G. and Mathews, S. (2004). Phylogenetic signal in nucleotide data from seed plants: Implications for resolving the seed plant tree of life. *The American Journal of Bioethics* **91**, 1599–1613.
Busch, M. A., Bomblies, K. and Weigel, D. (1999). Activation of a floral homeotic gene in *Arabidopsis*. *Science* **285**, 585–587.
Carroll, S. B., Grenier, J. K. and Weatherbee, S. D. (2001). "From DNA to Diversity: Molecular Genetics and the Evolution of Animal Design." Blackwell Science, Malden, MA.
Castillejo, C., Romera-Branchat, M. and Pelaz, S. (2005). A new role of the *Arabidopsis* SEPALLATA3 gene revealed by its constitutive expression. *Plant Journal* **43**, 586–596.
Chase, M. W. E. A. (1993). Phylogenetics of seed plants: An analysis of nucleotide sequences from the plastid gene rbcL. *Annals of the Missouri Botanical Garden* **80**, 528–580.
Chen, X. M., Riechmann, J. L., Jia, D. X. and Meyerowitz, E. (2000). Minimal regions in the *Arabidopsis* PISTILLATA promoter responsive to the APETALA3/PISTILLATA feed back control do not contain a CArG box. *Sex Plant Reproduction* **13**, 85–94.
Cho, S., Jang, S., Chae, S., Chung, K. M., Moon, Y.-H., An, G. and Jang, S. K. (1999). Analysis of the C-terminal region of *Arabidopsis thaliana* APETALA1 as a transcription activation domain. *Plant Molecular Biology* **40**, 419–429.
Clifford, H. T. (1988). *In* "Grass Systematics and Evolution" (T. R. Soderstrom, ed.), pp. 21–30. Smithsonian Institution Press, Washington, DC.
Coen, E. S. and Meyerowitz, E. M. (1991). The war of the whorls: Genetic interactions controlling flower development. *Nature* **353**, 31–37.
Crane, P. R. (1985). Phylogenetic analysis of seed plants and the origin of angiosperms. *Annals of the Missouri Botanical Garden* **72**, 716–793.
Crane, P. R., Friis, E. M. and Pedersen, K. R. (1995). The origin and early diversification of angiosperms. *Nature* **374**, 27–33.
Crepet, W. L. (2000). Progress in understanding angiosperm history, success, and relationships: Darwin's abominably "perplexing phenomenon." *Proceedings of the National Academy of Sciences of the United States of America* **97**, 12939–12941.
Cronquist, A. (1981). "An Integrated System of Classification of Flowering Plants." Columbia University Press, New York.
Cubas, P. (2004). Floral zygomorphy, the recurring evolution of a successful trait. *Bioessays* **26**, 1175–1184.
Dahlgren, R. M. T., Clfford, H. T. and Yeo, P. F. (1985). "The Families of the Monocotyledons." Springer Verlag, New York.
de Bodt, S., Raes, J., Van de Peer, Y. and Theissen, G. (2003). And then there were many: MADS goes genomic. *Trends Plant Science* **8**, 475–483.

de Folter, S., Immink, R. G., Kieffer, M., Parenicova, L., Henz, S. R., Weigel, D., Busscher, M., Kooiker, M., Colombo, L., Kater, M. M., Davies, B. and Angenent, G. C. (2005). Comprehensive interaction map of the *Arabidopsis* MADS box transcription factors. *Plant Cell* **17**, 1424–1433.

Demmig-Adams, B. and Adams, W. W. (1992). Photoprotection and other responses of plants to high light stress. *Annual Review of Plant Physiology and Plant Molecular Biology* **43**, 599–626.

Ditta, G., Pinyopich, A., Robles, P., Pelaz, S. and Yanofsky, M. F. (2004). The SEP4 gene of *Arabidopsis thaliana* functions in floral organ and meristem identity. *Current Biology* **14**, 1935–1940.

Doebley, J. and Lukens, L. (1998). Transcriptional regulators and the evolution of plant form. *Plant Cell* **10**, 1075–1082.

Doyle, J. A. (1994). Origin of the angiosperm flower: A phylogenetic perspective. *Plant Systematics and Evolution* **8** (Suppl.), 7–29.

Doyle, J. A. (1996). Seed plant phylogeny and the relationships of the Gnetales. *International Journal of Plant Sciences* **157**, S3–S39.

Doyle, J. A. and Donoghue, M. J. (1992). Fossils and seed plant phylogeny reanalyzed. *Brittonia* **44**, 89–106.

Drinnan, A. N., Crane, P. R. and Hoot, S. B. (1994). *In* "Early Evolution of Flowers" (P. K. Endress and E. M. Friis, eds.). Springer-Verlag, New York.

Efremova, N., Perbal, M. C., Yephremov, A., Hofmann, W. A., Saedler, H. and Schwarz-Sommer, Z. (2001). Epidermal control of floral organ identity by class B homeotic genes in *Antirrhinum* and *Arabidopsis*. *Development* **128**, 2661–2671.

Egea-Cortines, M., Saedler, H. and Sommer, H. (1999). Ternary complex formation between the MADS-box proteins SQUAMOSA, DEFICIENS and GLOBOSA is involved in the control of floral architecture in *Antirrhinum majus*. *EMBO Journal* **18**, 5370–5379.

El- Din El-Assal, S., Alonso-Blanco, C., Peeters, A. J., Raz, V. and Koornneef, M. (2001). A QTL for flowering time in *Arabidopsis* reveals a novel allele of CRY2. *Nature Genetics* **29**, 435–440.

Endress, P. K. (1987). The Early Evolution of the Angiosperm Flower. *Trends in Ecology & Evolution* **2**, 300–304.

Endress, P. K. (1994). "Diversity and Evolutionary Biology of Tropical Flowers." Cambridge University Press, Cambridge.

Endress, P. K. (2001a). Evolution of floral symmetry. *Current Opinion Plant Biology* **4**, 86–91.

Endress, P. K. (2001b). The flowers in extant basal angiosperms and inferences on ancestral flowers. *International Journal of Plant Sciences* **162**, 1111–1140.

Endress, P. K. (2001c). Origins of flower morphology. *Journal of Experimental Zoology* **291**, 105–115.

Endress, P. K. and Friis, E. M. (1994). Major trends in the study of early flower evolution—introduction. *Plant Systematics and Evolution Supplement* **8**, 1–6.

Fagard, M. and Vaucheret, H. (2000). Systemic silencing signal(s). *Plant Molecular Biology* **43**, 285–293.

Ferrario, S., Busscher, J., Franken, J., Gerats, T., Vandenbussche, M., Angenent, G. C. and Immink, R. G. (2004). Ectopic expression of the petunia MADS box gene UNSHAVEN accelerates flowering and confers leaf-like characteristics to floral organs in a dominant-negative manner. *Plant Cell* **16**, 1490–1505.

Force, A., Lynch, M., Pickett, F. B., Amores, A., Yan, Y. L. and Postlethwait, J. (1999). Preservation of duplicate genes by complementary, degenerative mutations. *Genetics* **151**, 1531–1545.

Frary, A., Fritz, L. A. and Tanksley, S. D. (2004). A comparative study of the genetic bases of natural variation in tomato leaf, sepal, and petal morphology. *Theoretical and Applied Genetics* **109**, 523–533.

Friis, E. M., Doyle, J. A., Endress, P. K. and Leng, Q. (2003). Archaefructus—angiosperm precursor or specialized early angiosperm? *Trends in Plant Science* **8**, 369–373.

Friis, E. M., Pedersen, K. R. and Crane, P. R. (2005). When Earth started blooming: Insights from the fossil record. *Current Opinion in Plant Biology* **8**, 5–12.

Fukui, M., Futamura, N., Mukai, Y., Wang, Y., Nagao, A. and Shinohara, K. (2001). Ancestral MADS box genes in Sugi, *Cryptomeria japonica* D. Don (Taxodiaceae), homologous to the B function genes in angiosperms. *Plant Cell Physiology* **42**, 566–575.

Garcia-Bellido, A. (1975). Genetic control of wing disc development in *Drosophila. Ciba Found. Symp.* **29**, 161–182.

Goto, K. and Meyerowitz, E. M. (1994). Function and regulation of the *Arabidopsis* floral homeotic gene PISTILLATA. *Genes and Development* **8**, 1548–1560.

Hamilton, A. J. and Baulcombe, D. C. (1999). A species of small antisense RNA in posttranscriptional gene silencing in plants. *Science* **286**, 950–952.

Hein, I., Barciszewska-Pacak, M., Hrubikova, K., Williamson, S., Dinesen, M., Soenderby, I. E., Sundar, S., Jarmolowski, A., Shirasu, K. and Lacomme, C. (2005). Virus-induced gene silencing-based functional characterization of genes associated with powdery mildew resistance in barley. *Plant Physiology* **138** (4), 2155–2164.

Hileman, L. C., Drea, S., de Martino, G., Litt, A. and Irish, V. F. (2005). Virus induced gene silencing is an effective tool to assay gene function in the basal eudicot species *Papaver somniferum* (opium poppy). *Plant Journal* **44** (2), 334–341.

Hileman, C. L., Sundstrom, J. F., Litt, A., Chen, M., Shumba, T. and Irish, V. F. (in press). Molecular and phylogenetic analyses of the MADS-box gene family in tomato. *Molecular Biology and Evolution, Research Article.*

Hill, J. P. and Lord, E. M. (1989). Floral development in *Arabidopsis thaliana*: A comparison of the wild type and the homeotic pistillata mutant. *Canadian Journal of Botany* **67**, 2922–2936.

Hill, T. A., Day, C. D., Zondlo, S. C., Thackeray, A. G. and Irish, V. F. (1998). Discrete spatial and temporal cis-acting elements regulate transcription of the *Arabidopsis* floral homeotic gene APETALA3. *Development* **125**, 1711–1721.

Holland, P. W. (1999). Gene duplication: Past, present and future. *Seminars in Cell & Developmental Biology* **10**, 541–547.

Holzberg, S., Brosio, P., Gross, C. and Pogue, G. P. (2002). Barley stripe mosaic virus-induced gene silencing in a monocot plant. *Plant Journal* **30**, 315–327.

Honma, T. and Goto, K. (2000). The *Arabidopsis* floral homeotic gene PISTILLATA is regulated by discrete cis-elements responsive to induction and maintenance signals. *Development* **127**, 2021–2030.

Honma, T. and Goto, K. (2001). Complexes of MADS-box proteins are sufficient to convert leaves into floral organs. *Nature* **409**, 469–471.

Huala, E. and Sussex, I. M. (1992). *LEAFY* interacts with floral homeotic genes to regulate *Arabidopsis* floral development. *The Plant Cell* **4**, 901–913.

Immink, R. G., Gadella, T. W., Jr., Ferrario, S., Busscher, M. and Angenent, G. C. (2002). Analysis of MADS box protein-protein interactions in living plant cells. *Proceedings of the National Academy of Sciences of the United States of America* **99**, 2416–2421.

Immink, R. G., Ferrario, S., Busscher-Lange, J., Kooiker, M., Busscher, M. and Angenent, G. C. (2003). Analysis of the petunia MADS-box transcription factor family. *Molecular Genetics and Genomics* **268**, 598–606.

Ingram, G. C., Goodrich, J., Wilkinson, M. D., Simon, R., Haughn, G. W. and Coen, E. S. (1995). Parallels between UNUSUAL FLORAL ORGANS and FIMBRIATA, genes controlling flower development in *Arabidopsis* and *Antirrhinum*. *Plant Cell* **7**, 1501–1510.
Irish, V. F. (2003). The evolution of floral homeotic gene function. *Bioessays* **25**, 637–646.
Irish, V. F. and Kramer, E. M. (1998). Genetic and molecular analysis of Angiosperm flower development. *Advances in Botanical Research* **28**, 197–230.
Irish, V. F. and Litt, A. (2005). Flower development and evolution: Gene duplication, diversification and redeployment. *Current Opinion in Genetics & Development* **15**, 454–460.
Irish, V. F. and Sussex, I. M. (1990). Function of the apetala-1 gene during *Arabidopsis* floral development. *The Plant Cell* **2**, 741–753.
Irish, V. F. and Yamamoto, Y. T. (1995). Conservation of floral homeotic gene function between *Arabidopsis* and *Antirrhinum*. *Plant Cell* **7**, 1635–1644.
Jack, T. (2001). Relearning our ABCs: New twists on an old model. *Trends Plant Science* **6**, 310–316.
Jack, T., Brockman, L. L. and Meyerowitz, E. M. (1992). The homeotic gene APETALA3 of *Arabidopsis thaliana* encodes a MADS box and is expressed in petals and stamens. *Cell* **68**, 683–697.
Jack, T., Fox, G. L. and Meyerowitz, E. M. (1994). *Arabidopsis* homeotic gene APETALA3 ectopic expression: Transcriptional and posttranscriptional regulation determine floral organ identity. *Cell* **76**, 703–716.
Jenik, P. D. and Irish, V. F. (2001). The *Arabidopsis* floral homeotic gene APETALA3 differentially regulates intercellular signaling required for petal and stamen development. *Development* **128**, 13–23.
Jofuku, K. D., den Boer, B. G. W., Van Montague, M. and Okamuro, J. K. (1994). Control of *Arabidopsis* flower and seed development by the homeotic gene APETALA2. *Plant Cell* **6**, 1211–1225.
Johanson, U., West, J., Lister, C., Michaels, S., Amasino, R. and Dean, C. (2000). Molecular analysis of FRIGIDA, a major determinant of natural variation in *Arabidopsis* flowering time. *Science* **290**, 344–347.
Juenger, T., Purugganan, M. and Mackay, T. F. (2000). Quantitative trait loci for floral morphology in *Arabidopsis thaliana*. *Genetics* **156**, 1379–1392.
Kang, H.-G., Jeon, J.-S., Lee, S. and An, G. (1998). Identification of class B and class C floral organ identity genes from rice plants. *Plant Molecular Biology* **38**, 1021–1029.
Kanno, A., Saeki, H., Kameya, T., Saedler, H. and Theissen, G. (2003). Heterotopic expression of class B floral homeotic genes supports a modified ABC model for tulip (*Tulipa gesneriana*). *Plant Molecular Biology* **52**, 831–841.
Kim, S., Yoo, M., Albert, V. A., Farris, J. S., Soltis, P. S. and Soltis, D. E. (2004). Phylogeny and diversification of B-function genes in angiosperms: Evolutionary and functional implications of a 260-milliion year old duplication. *American Journal of Botany* **91**, 2102–2118.
Kim, S., Koh, J., Yoo, M. J., Kong, H., Hu, Y., Ma, H., Soltis, P. S. and Soltis, D. E. (2005). Expression of floral MADS-box genes in basal angiosperms: Implications for the evolution of floral regulators. *Plant Journal* **43**, 724–744.
Koch, M. A., Weisshaar, B., Kroymann, J., Haubold, B. and Mitchell-Olds, T. (2001). Comparative genomics and regulatory evolution: Conservation and function of the Chs and Apetala3 promoters. *Molecular Biology and Evolution* **18**, 1882–1891.
Kosuge, K. (1994). Petal evolution in the Ranunculaceae. *Plant Systematics and Evolution* **8**, 185–191.

Kramer, E. M. and Irish, V. F. (1999). Evolution of genetic mechanisms controlling petal development. *Nature* **399**, 144–148.

Kramer, E. M. and Irish, V. F. (2000). Evolution of petal and stamen developmental programs: Evidence from comparative studies of the lower eudicots and basal angiosperms. *International Journal of Plant Sciences* **161**, S29–S40.

Kramer, E. M., Dorit, R. L. and Irish, V. F. (1998). Molecular evolution of petal and stamen development: Gene duplication and divergence within the APETALA3 and PISTILLATA MADS-box gene lineages. *Genetics* **149**, 765–783.

Kramer, E. M., DiStilio, V. S. and Schluter, P. M. (2003). Complex patterns of gene duplication in the APETALA3 and PISTILLATA lineages of the Ranunculaceae. *International Journal of Plant Sciences* **164**, 1–11.

Krizek, B. A. and Meyerowitz, E. M. (1996a). The *Arabidopsis* homeotic genes APETALA3 and PISTILLATA are sufficient to provide the B class organ identity function. *Development* **122**, 11–22.

Krizek, B. A. and Meyerowitz, E. M. (1996b). Mapping the protein regions responsible for the functional specificities of the *Arabidopsis* MADS domain organ-identity proteins. *Proceedings of the National Academy of Sciences of the United States of America* **93**, 4063–4070.

Krizek, B. A., Riechmann, J. L. and Meyerowitz, E. M. (1999). Use of the APETALA1 promoter to assay the *in vivo* function of chimeric MADS box genes. *Sex Plant Reproduction* **12**, 14–26.

Lamb, R. S. and Irish, V. F. (2003). Functional divergence within the APETALA3/PISTILLATA floral homeotic gene lineages. *Proceedings of the National Academy of Sciences of the United States of America* **100**, 6558–6563.

Lamb, R. S., Hill, T. A., Tan, Q. K. and Irish, V. F. (2002). Regulation of APETALA3 floral homeotic gene expression by meristem identity genes. *Development* **129**, 2079–2086.

Laufs, P., Coen, E., Kronenberger, J., Traas, J. and Doonan, J. (2003). Separable roles of UFO during floral development revealed by conditional restoration of gene function. *Development* **130**, 785–796.

Lee, I., Wolfe, D. S., Nilsson, O. and Weigel, D. (1997). A LEAFY co-regulator encoded by UNUSUAL FLORAL ORGANS. *Current Biology* **7**, 95–104.

Lim, F. L., Hayes, A., West, A. G., Pic-Taylor, A., Darieva, Z., Morgan, B. A., Oliver, S. G. and Sharrocks, A. D. (2003). Mcm1p-induced DNA bending regulates the formation of ternary transcription factor complexes. *Molecular Cell Biology* **23**, 450–461.

Litt, A. and Irish, V. F. (2003). Duplication and diversification in the APETALA1/FRUITFULL floral homeotic gene lineage: Implications for the evolution of floral development. *Genetics* **165**, 821–833.

Liu, Y., Schiff, M. and Dinesh-Kumar, S. P. (2002). Virus-induced gene silencing in tomato. *Plant Journal* **31**, 777–786.

Liu, Y., Nakayama, N., Schiff, M., Litt, A., Irish, V. F. and Dinesh-Kumar, S. P. (2004). Virus induced gene silencing of a DEFICIENS ortholog in *Nicotiana benthamiana*. *Plant Molecular Biology* **54**, 701–711.

Lynch, M. and Force, A. (2000). The probability of duplicate gene preservation by subfunctionalization. *Genetics* **154**, 459–473.

Ma, H., Yanofsky, M. F. and Meyerowitz, E. M. (1991). *AGL1-AGL6*, an *Arabidopsis* gene family with similarity to floral homeotic and transcription factor genes. *Genes and Development* **5**, 484–495.

Magnani, E., Sjolander, K. and Hake, S. (2004). From endonucleases to transcription factors: Evolution of the AP2 DNA binding domain in plants. *Plant Cell* **16**, 2265–2277.

Mandel, M. A., Gustafson-Brown, C., Savidge, B. and Yanofsky, M. F. (1992). Molecular characterization of the *Arabidopsis* floral homeotic gene APETALA1. *Nature* **360**, 273–277.

Mann, R. S. and Carroll, S. B. (2002). Molecular mechanisms of selector gene function and evolution. *Current Opinion in Genetics & Development* **12**, 592–600.

Mathews, S. and Donoghue, M. J. (1999). The root of angiosperm phylogeny inferred from duplicate phytochrome genes. *Science* **286**, 947–950.

McGonigle, B., Bouhidel, K. and Irish, V. F. (1996). Nuclear localization of the *Arabidopsis* APETALA3 and PISTILLATA homeotic gene products depends on their simultaneous expression. *Genes Development* **10**, 1812–1821.

Messenguy, F. and Dubois, E. (2003). Role of MADS box proteins and their cofactors in combinatorial control of gene expression and cell development. *Gene* **316**, 1–21.

Meyerowitz, E. M., Bowman, J. L., Brockman, L. L., Drews, G. N., Jack, T., Sieburth, L. E. and Weigel, D. (1991). A genetic and molecular model for floral development in *Arabidopsis thaliana*. *Development* **1**, 157–167.

Michaels, S. D. and Amasino, R. M. (1999). FLOWERING LOCUS C encodes a novel MADS domain protein that acts as a repressor of flowering. *Plant Cell* **11**, 949–956.

Mouradov, A., Hamdorf, B., Teasdale, R. D., Kim, J., Winter, K.-U. and Theissen, G. (1999). A DEF/GLO like MADS box gene from a gymnosperm: *Pinus radiata* contains an ortholog of angiosperm B class floral homeotic genes. *Developmental Genetics* **25**, 245–252.

Nagasawa, N., Miyoshi, M., Sano, Y., Satoh, H., Hirano, H., Sakai, H. and Nagato, Y. (2003). SUPERWOMAN1 and DROOPING LEAF genes control floral organ identity in rice. *Development* **130**, 705–718.

Nakamura, T., Fukuda, T., Nakano, M., Hasebe, M., Kameya, T. and Kanno, A. (2005). The modified ABC model explains the development of the petaloid perianth of *Agapanthus praecox* ssp. *orientalis* (Agapanthaceae) flowers. *Plant Molecular Biology* **58**, 435–445.

Nam, J., Kim, J., Lee, S., An, G., Ma, H. and Nei, M. (2004). Type I MADS-box genes have experienced faster birth-and-death evolution than type II MADS-box genes in angiosperms. *Proceedings of the National Academy of Sciences of the United States of America* **101**, 1910–1915.

Ohno, S. (1970). "Evolution by Gene Duplication." Springer, New York.

Parcy, F., Nilsson, O., Busch, M. A., Lee, I. and Weigel, D. (1998). A genetic framework for floral patterning. *Nature* **395**, 561–566.

Parenicova, L., de Folter, S., Kieffer, M., Horner, D. S., Favalli, C., Busscher, J., Cook, H. E., Ingram, R. M., Kater, M. M., Davies, B., Angenent, G. C. and Colombo, L. (2003). Molecular and phylogenetic analyses of the complete MADS-box transcription factor family in *Arabidopsis*: New openings to the MADS world. *Plant Cell* **15**, 1538–1551.

Park, J. H., Ishikawa, Y., Yoshida, R., Kanno, A. and Kameya, T. (2003). Expression of AODEF, a B-functional MADS-box gene, in stamens and inner tepals of the dioecious species *Asparagus officinalis* L. *Plant Molecular Biology* **51**, 867–875.

Park, J. H., Ishikawa, Y., Ochiai, T., Kanno, A. and Kameya, T. (2004). Two GLOBOSA-like genes are expressed in second and third whorls of homochlamydeous flowers in *Asparagus officinalis* L. *Plant Cell Physiology* **45**, 325–332.

Parkinson, C. L., Adams, K. L. and Palmer, J. D. (1999). Multigene analyses identify three earliest lineages of extant flowering plants. *Current Biology* **9**, 1485–1488.

Pelaz, S., Ditta, G. S., Baumann, E., Wisman, E. and Yanofsky, M. F. (2000). B and C floral organ identity functions require SEPALLATA MADS-box genes. *Nature* **405**, 200–203.

Pellegrini, L., Tan, S. and Richmond, T. J. (1995). Structure of the serum response factor core bound to DNA. *Nature* **376**, 490–498.

Pnueli, L., Abu-Abeid, M., Zamir, D., Nacken, W., Schwarz-Sommer, Z. and Lifschitz, E. (1991). The MADS box gene family in tomato: Temporal expression during floral development, conserved secondary structures and homology with homeotic genes from *Antirrhinum* and *Arabidopsis*. *The Plant Journal* **1**, 255–266.

Prasad, K. and Vijayraghavan, U. (2003). Double-stranded RNA interference of a rice PI/GLO paralog, OsMADS2, uncovers its second-whorl-specific function in floral organ patterning. *Genetics* **165**, 2301–2305.

Purugganan, M. D., Rounsley, S. D., Schmidt, R. J. and Yanofsky, M. F. (1995). Molecular evolution of flower development: Diversification of the plant MADS box regulatory gene family. *Genetics* **140**, 345–356.

Qiu, Y.-L., Lee, J., Bernasconi-Quadroni, F., Soltis, D. E., Soltis, P. S., Zanis, M., Zimmer, E. A., Chen, Z., Savolainen, V. and Chase, M. W. (1999). The earliest angiosperms: Evidence from mitochondrial, plastid and nuclear genomes. *Nature* **402** (6760), 404–407.

Ratcliff, F., Martin-Hernandez, A. M. and Baulcombe, D. C. (2001). Technical Advance. Tobacco rattle virus as a vector for analysis of gene function by silencing. *Plant Journal* **25**, 237–245.

Riechmann, J. L. and Meyerowitz, E. M. (1997a). Determination of floral organ identity by *Arabidopsis* MADS domain homeotic proteins AP1, AP3, PI, and AG is independent of their DNA-binding specificity. *Molecular Biology of the Cell* **8**, 1243–1259.

Riechmann, J. L. and Meyerowitz, E. M. (1997b). MADS domain proteins in plant development. *Biological Chemistry* **378**, 1079–1101.

Riechmann, J. L. and Meyerowitz, E. M. (1998). The AP2/EREBP family of plant transcription factors. *Biological Chemistry* **379**, 633–646.

Riechmann, J. L., Krizek, B. A. and Meyerowitz, E. M. (1996). Dimerization specificity of *Arabidopsis* MADS domain homeotic proteins APETALA1, APETALA3, PISTILLATA, and AGAMOUS. *Proceedings of the National Academy of Sciences of the United States of America* **93**, 4793–4798.

Robertson, D. (2004). VIGS vectors for gene silencing: Many targets, many tools. Annual. *Review of Plant Biology* **55**, 495–519.

Ronse de Craene, L. P. (2003). The evolutionary significance of homeosis in flowers: A morphological perspective. *International Journal of Plant Sciences* **164**, S225–S235.

Ronse de Craene, L. P. (2004). Floral development of *Berberidopsis corallina*: A crucial link in the evolution of flowers in the core eudicots. *Annals of Botany* **94**, 741–751.

Ronse de Craene, L. P., Soltis, P. S. and Soltis, D. E. (2003). Evolution of floral structures in basal angiosperms. *International Journal of Plant Sciences* **164**, S329–S363.

Ruiz, M. T., Voinnet, O. and Baulcombe, D. C. (1998). Initiation and maintenance of virus-induced gene silencing. *Plant Cell* **10**, 937–946.

Sablowski, R. W. M. and Meyerowitz, E. M. (1998). A homolog of NO APICAL MERISTEM is an immediate target of the floral homeotic genes APETALA3/PISTILLATA. *Cell* **92**, 93–103.

Samach, A., Kohalmi, S. E., Motte, P., Datla, R. and Haughn, G. W. (1997). Divergence of function and regulation of class B floral organ identity genes. *Plant Cell* **9**, 559–570.

Samach, A., Klenz, J. E., Kohalmi, S. E., Risseeuw, E., Haughn, G. W. and Crosby, W. L. (1999). The UNUSUAL FLORAL ORGANS gene of *Arabidopsis thaliana* is an F-box protein required for normal patterning and growth in the floral meristem. *Plant Journal* **20**, 433–445.

Schultz, E. A. and Haughn, G. W. (1991). LEAFY, a homeotic gene that regulates inflorescence development in *Arabidopsis*. *Plant Cell* **3**, 771–781.

Schwarz-Sommer, Z., Hue, I., Huijser, P., Flor, P. J., Hansen, R., Tetens, F., Lonnig, W.-E., Saedler, H. and Sommer, H. (1992). Characterization of the *Antirrhinum* floral homeotic MADS-box gene *deficiens*: Evidence for DNA binding and autoregulation of its persistent expression throughout flower development. *EMBO Journal* **11**, 251–263.

Sheppard, L. A., Brunner, A. M., Krutovskii, K. V., Rottmann, W. H., Skinner, J. S., Vollmer, S. S. and Strauss, S. H. (2000). A DEFICIENS homolog from the dioecious tree black cottonwood is expressed in female and male floral meristems of the two-whorled, unisexual flowers. *Plant Physiology* **124**, 627–640.

Shimeld, S. M. (1999). Gene function, gene networks and the fate of duplicated genes. *Seminars in Cell & Developmental Biology* **10**, 549–553.

Shore, P. and Sharrocks, A. D. (1995). The MADS-box family of transcription factors. *European Journal of Biochemistry* **229**, 1–13.

Sieburth, L. E., Drews, G. N. and Meyerowitz, E. M. (1998). Non-autonomy of AGAMOUS function in flower development: Use of a Cre/loxP method for mosaic analysis in *Arabidopsis*. *Development* **125**, 4303–4312.

Smyth, D. R., Bowman, J. L. and Meyerowitz, E. M. (1990). Early flower development in *Arabidopsis*. *Plant Cell* **2**, 755–767.

Soltis, D. E., Soltis, P. S. and Zanis, M. J. (2002). Phylogeny of seed plants based on evidence from eight genes. *American Journal of Botany* **89**, 1670–1681.

Soltis, D. E., Soltis, P. S., Endress, P. K. and Chase, M. W. (2005). "Phylogeny and Evolution of Angiosperms." Sinauer Assoc, Sunderland, MA.

Soltis, P. S., Soltis, D. E. and Chase, M. W. (1999). Angiosperm phylogeny inferred from multiple genes as a tool for comparative biology. *Nature* **402**, 402–404.

Sommer, H., Beltran, J.-P., Huijser, P., Pape, H., Lonnig, W.-E., Saedler, H. and Schwarz-Sommer, Z. (1990). *Deficiens*, a homeotic gene involved in the control of flower morphogenesis in *Antirrhinum majus*: The protein shows homology to transcription factors. *EMBO Journal* **9**, 605–613.

Stefanovic, S., Rice, D. W. and Palmer, J. D. (2004). Long branch attraction, taxon sampling, and the earliest angiosperms: Amborella or monocots? *BMC Evolutionary Biology* **4**, 35.

Stellari, G. M., Jaramillo, M. A. and Kramer, E. M. (2004). Evolution of the APETALA3 and PISTILLATA lineages of MADS-box-containing genes in the basal angiosperms. *Molecular Biology and Evolution* **21**, 506–519.

Stewart, W. N. and Rothwell, G. W. (1993). "Paleobotany and the Evolution of Plants." Cambridge University Press, Cambridge.

Sun, G., Dilcher, D. L., Shaoling, Z. and Zhekun, Z. (1998). In search of the first flower: A Jurassic angiosperm, Archefructus, from Northeast China. *Science* **282**, 1692–1695.

Sun, G., Ji, Q., Dilcher, D. L., Zheng, S., Nixon, K. C. and Wang, X. (2002). Archaefructaceae, a new basal angiosperm family. *Science* **296**, 899–904.

Sundstrom, J., Carlsbecker, A., Svenson, M., Svensson, M. E. and Engstrom, P. (1999). MADS box genes active in developing pollen cones of Norway Spruce are homologous to the B class floral homeotic genes in angiosperms. *Developmental Genetics* **25**, 253–266.

Takhtajan, A. (1991). "Evolutionary Trends in Flowering Plants." Columbia University Press, New York.

Teichmann, S. A. and Babu, M. M. (2004). Gene regulatory network growth by duplication. *Nature Genetics* **36**, 492–496.

Theissen, G. and Saedler, H. (2001). Plant biology. Floral quartets. *Nature* **409**, 469–471.

Theissen, G., Becker, A., Di Rosa, A., Kanno, A., Kim, J. T., Muenster, T., Winter, K.-W and Saedler, H. (2000). A short history of MADS-box genes in plants. *Plant Molecular Biology* **42**, 115–149.

Tilly, J., Allen, D. W. and Jack, T. (1998). The CArG boxes in the promoter of the *Arabidopsis* floral organ identity gene *APETALA3* mediate diverse regulatory effects. *Development* **125**, 1647–1657.

Tokunaga, C. (1972). Autonomy or nonautonomy of gene effects in mosaics. *Proceedings of the National Academy of Sciences of the United States of America* **69**, 3283–3286.

Trobner, W., Ramirez, L., Motte, P., Hue, I., Huijser, P., Lonnig, W. E., Saedler, H., Sommer, H. and Schwarz-Sommer, Z. (1992). Globosa—a homeotic gene which interacts with Deficiens in the control of *Antirrhinum* floral organogenesis. *EMBO Journal* **11**, 4693–4704.

Tsuchimoto, S., Mayama, T., van der Krol, A. and Ohtsubo, E. (2000). The whorl-specific action of a petunia class B floral homeotic gene. *Genes to Cells* **5**, 89–99.

van der Krol, A. R., Brunelle, A., Tsuchimoto, S. and Chua, N. H. (1993). Functional analysis of petunia floral homeotic MADS box gene pMADS1. *Genes and Development* **7**, 1214–1228.

van Tunen, A. J., Eikelboom, W. and Angenent, G. C. (1993). Floral organogenesis in Tulipa. *Flowering Newsletter* **16**, 33–38.

Vandenbussche, M., Theissen, G., Van de Peer, Y. and Gerats, T. (2003). Structural diversification and neo-functionalization during floral MADS-box gene evolution by C-terminal frameshift mutations. *Nucleic Acids Research* **31**, 4401–4409.

Vandenbussche, M., Zethof, J., Royaert, S., Weterings, K. and Gerats, T. (2004). The duplicated B-class heterodimer model: Whorl-specific effects and complex genetic interactions in *Petunia hybrida* flower development. *Plant Cell* **16**, 741–754.

Vincent, C. A., Carpenter, R. and Coen, E. S. (2003). Interactions between gene activity and cell layers during floral development. *Plant Journal* **33**, 765–774.

von Wettstein, R. (1924). Handbuch der systematischen Botanik Franz Deutige, Leipzig.

Wagner, D., Sablowski, R. W. M. and Meyerowitz, E. M. (1999). Transcriptional activation of APETALA1 by LEAFY. *Science* **285**, 582–584.

Wagner, D., Wellmer, F., Dilks, K., William, D., Smith, M. R., Kumar, P. P., Riechmann, J. L., Greenland, A. J. and Meyerowitz, E. M. (2004). Floral induction in tissue culture: A system for the analysis of LEAFY-dependent gene regulation. *Plant Journal* **39**, 273–282.

Wang, X., Feng, S., Nakayama, N., Crosby, W. L., Irish, V., Deng, X. W. and Wei, N. (2003). The COP9 signalosome interacts with SCF(UFO) and participates in *Arabidopsis* flower development. *Plant Cell* **15**, 1071–1082.

Weberling, F. (1989). "Morphology of Flowers and Inflorescences." Cambridge University Press, Cambridge.

Weigel, D. (1995). The APETALA2 domain is related to a novel type of DNA binding domain. *Plant Cell* **7**, 388–389.

Weigel, D. and Meyerowitz, E. M. (1993). Activation of floral homeotic genes in *Arabidopsis*. *Science* **261**, 1723–1726.

Weigel, D. and Meyerowitz, E. M. (1994). The ABCs of floral homeotic genes. *Cell* **78**, 203–209.

Weigel, D. and Nilsson, O. (1995). A developmental switch sufficient for flower initiation in diverse plants. *Nature* **377**, 495–500.

Weigel, D., Alvarez, J., Smyth, D. R., Yanofsky, M. F. and Meyerowitz, E. M. (1992). *LEAFY* controls floral meristem identity in *Arabidopsis*. *Cell* **69**, 843–859.

Wellmer, F., Riechmann, J. L., Alves-Ferreira, M. and Meyerowitz, E. M. (2004). Genome-wide analysis of spatial gene expression in *Arabidopsis* flowers. *Plant Cell* **16**, 1314–1326.

West, A. G. and Sharrocks, A. D. (1999). MADS-box transcription factors adopt alternative mechanisms for bending DNA. *Journal of Molecular Biology* **286**, 1311–1323.

Whipple, C. J., Ciceri, P., Padilla, C. M., Ambrose, B. A., Bandong, S. L. and Schmidt, R. J. (2004). Conservation of B-class floral homeotic gene function between maize and *Arabidopsis*. *Development* **131**, 6083–6091.

William, D. A., Su, Y., Smith, M. R., Lu, M., Baldwin, D. A. and Wagner, D. (2004). Genomic identification of direct target genes of LEAFY. *Proceedings of the National Academy of Sciences of the United States of America* **101**, 1775–1780.

Winter, K.-U., Becker, A., Munster, T., Kim, J. T., Saedler, H. and Theissen, G. (1999). MADS box genes reveal that gnetophytes are more closely related to conifers than to flowering plants. *Proceedings of the National Academy of Sciences of the United States of America* **96**, 7342–7347.

Winter, K. U., Saedler, H. and Theissen, G. (2002). On the origin of class B floral homeotic genes: Functional substitution and dominant inhibition in *Arabidopsis* by expression of an orthologue from the gymnosperm Gnetum. *Plant Journal* **31**, 457–475.

Wynne, J. and Treisman, R. (1992). SRF and MCM1 have related but distinct DNA binding specificities. *Nucleic Acids Research* **20**, 3297–3303.

Yang, Y. and Jack, T. (2004). Defining subdomains of the K domain important for protein-protein interactions of plant MADS proteins. *Plant Molecular Biology* **55**, 45–59.

Yanofsky, M. F., Ma, H., Bowman, J. L., Drews, G. N., Feldmann, K. A. and Meyerowitz, E. M. (1990). The protein encoded by the *Arabidopsis* homeotic gene *agamous* resembles transcription factors. *Nature* **346**, 35–39.

Yu, D., Kotilainen, M., Pollanen, E., Mehto, M., Elomaa, P., Helariutta, Y., Albert, V. and Teeri, T. (1999). Organ identity genes and modified patterns of flower development in *Gerbera hybrida* (Asteraceae). *Plant Journal* **17**, 51–62.

Zahn, L. M., Leebens-Mack, J., DePamphilis, C. W., Ma, H. and Theissen, G. (2005). To B or Not to B a flower: The role of DEFICIENS and GLOBOSA orthologs in the evolution of the angiosperms. *The Journal of Heredity* **96**, 225–240.

Zamore, P. D. (2004). Plant RNAi: How a viral silencing suppressor inactivates siRNA. *Current Biology* **14**, R198–R200.

Zanis, M. J., Soltis, D. E., Soltis, P. S., Mathews, S. and Donoghue, M. J. (2002). The root of the angiosperms revisited. *Proceedings of the National Academy of Sciences of the United States of America* **99**, 6848–6853.

Zik, M. and Irish, V. F. (2003). Global identification of target genes regulated by APETALA3 and PISTILLATA floral homeotic gene action. *Plant Cell* **15**, 207–222.

Beyond the ABC-Model: Regulation of Floral Homeotic Genes

LAURA M. ZAHN, BAOMIN FENG AND HONG MA

Department of Biology, The Huck Institutes of the Life Sciences and the Institute of Molecular Evolutionary Genetics, The Pennsylvania State University, University Park, Pennsylvania 16802

Advances in Botanical Research, Vol. 44
Incorporating Advances in Plant Pathology

0065-2296/06 $35.00
DOI: 10.1016/S0065-2296(06)44004-0

ABSTRACT

The ABC-model explains the genetic basis for the specification of floral organ identities, stating that A-function is required for the sepal, (A + B)-functions are necessary for the petal, (B + C)-functions are needed for the stamen, and C-function is essential for the carpel. Molecular cloning revealed that all ABC-genes encode transcriptional regulators and most are MADS-box genes. This chapter focuses on molecular genetic studies that have identified numerous genes that affect A-, B-, and C-functions by regulating specific genes that are required for A-, B-, or C-function. These genes act at both the transcriptional and posttranscriptional levels and encode a variety of regulatory proteins. Discoveries and analyses of these genes provide further understanding of the molecular mechanisms underlying floral development and lay a foundation for greater advances in this fertile field of plant development. Future studies using genomic and evolutionary approaches promise to reveal an expanding knowledge of the floral molecular machinery.

I. INTRODUCTION: THE ABC-MODEL

The ABC-model presents a genetic mechanism for the specification of floral organ identity on the basis of mutant studies in *Arabidopsis thaliana* and *Antirrhinum majus*, representing the core eudicot rosid and asterid lineages, respectively (Coen and Meyerowitz, 1991; Ma, 1994). This model states that the four whorls of organs found in most flowering plants and exemplified by *Arabidopsis* and *Antirrhinum*, the sepals, petals, stamens, and carpels, respectively, are due to specific spatially defined activities of homeotic genes. In the ABC-model, A-function alone specifies the sepals, both A and B functions control the petals, (B + C)-functions determine the stamens and C-function alone directs the carpels (Fig. 1). In addition, A- and C-functions are mutually antagonistic (Bowman *et al.*, 1991b).

Genes that are required for one of the ABC-functions are homeotic genes such that when their function is lost there is a change in the floral organ type in two adjacent floral whorls. Specifically, in *Arabidopsis* the loss of A-function results in a flower that is composed of whorl one carpels, whorl two stamens, whorl three stamens, and whorl four carpels. Similarly, when B-function is lost, the flower consists of whorls of sepals, sepals, carpels, carpels and, when C-function is lost, the flower has sepals, petals, petals, sepals in the first four whorls (Bowman *et al.*, 1989, 1991a; Coen and Meyerowitz, 1991; Irish and Sussex, 1990; Ma, 1994; Weigel and Meyerowitz, 1994) (Fig. 1).

Besides the A-, B-, and C-functions in specifying floral organ identity, two additional functions have been proposed: D- and E-functions. D-function specifies the ovule identity, a reproductive function that is separate from the development of the carpels and acts later than the ABC-functions (Angenent *et al.*, 1995; Colombo *et al.*, 1995). E-function refers to a floral-specific

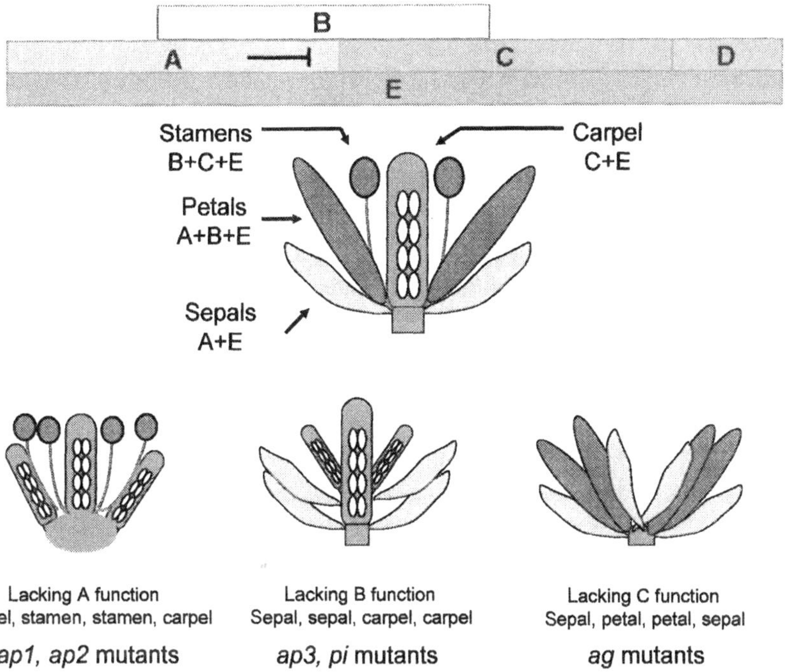

Fig. 1. An illustration of the ABC-model. (See color Plate 1.)

function required for all four whorls of floral organs. When first introduced, "E-function *sensu stricto*" specified the development of the floral organs in the inner three whorls (Pelaz *et al.*, 2000; Theissen, 2001; Theissen and Saedler, 2001). However, studies suggest that "E-function *sensu lato*" specifies floral organ development in all four floral whorls (Ditta *et al.*, 2004; Ma, 2005; Zahn *et al.*, 2005a). When D-function is lost, the ovules develop into leaf-like structures (Favaro *et al.*, 2003). When E-function *sensu stricto* is lost, the floral organs of the inner three whorls become sepals (Pelaz *et al.*, 2000), whereas when E-function *sensu lato* is lost, all floral organs become leaf-like (Ditta *et al.*, 2004). As described in more detail later, most of the known ABC-function genes were found to be MADS-box genes, which encode a group of (putative) transcription factors found in all major lineages of eukaryotes (Ma, 1994; Ng and Yanofsky, 2001; Riechmann and Meyerowitz, 1997; Sommer *et al.*, 1990; see also Davies *et al.*, Chapter 7; Irish, Chapter 3; Kramer and Zimmer, Chapter 9; Melzer *et al.*, Chapter 5; and Rijpkema *et al.*, Chapter 6). Furthermore, some D- and all E-function genes identified to date are also MADS-box genes (Ma, 2005; Pelaz *et al.*, 2000; Pinyopich *et al.*, 2003; Zahn *et al.*, 2005a).

In *Arabidopsis*, A-function is specified primarily by two genes, *APETALA1* (*AP1*) and *APETALA2* (*AP2*) (Bowman *et al.*, 1989, 1991b, 1993; Haughn and Somerville, 1988; Irish and Sussex, 1990; Jofuku *et al.*, 1994; Kunst *et al.*, 1989). *AP1* is a MADS-box gene (Mandel *et al.*, 1992), whereas *AP2* is the founding member of the AP2-domain gene family encoding plant-specific transcription factors (Jofuku *et al.*, 1994). In addition to the control of sepal and petal identity, *AP1* also functions in regulating the floral meristem (Bowman *et al.*, 1993; Mandel *et al.*, 1992). In strong *ap1* flowers, the outermost organs are converted to bracts and the second-whorl organs are missing. Weak *ap1* mutant flowers have defects in outer whorl organs (Bowman *et al.*, 1993; Mandel *et al.*, 1992). In flowers of strong *ap2* alleles, the outer two whorl organs undergo a homeotic conversion to carpelloid and staminoid organs, respectively (Bowman *et al.*, 1989, 1991b; Kunst *et al.*, 1989). *AP2* is expressed in all four floral whorls, the developing ovules and the leaves (Jofuku *et al.*, 1994), indicating both posttranscriptional control of its role in A-function (see Section II) and a possible role outside of A-function (see Section IV).

The conservation of A-function has not been extensively tested. The *Antirrhinum* ortholog of *AP1, SQUAMOSA* (*SQUA*), functions primarily in floral meristem identity, as any flowers that occur in *squa* mutants have only minor defects in the petals (Huijser *et al.*, 1992). The *Antirrhinum* homologs of *AP2, LIPLESS1*, and *LIPLESS2* (*LIP1, LIP2*), are functionally redundant and play a role in sepal, petal, and ovule development (Keck *et al.*, 2003). The *lip1 lip2* double mutants are defective in all floral organs, but their mutant phenotypes are significantly different from those of the *ap2* mutants in *Arabidopsis* (Keck *et al.*, 2003). Instead, in *lip1 lip2* plants, the sepals are transformed into bract-like structures unlike sepals and the lips of the petals were either severely reduced or lost (Keck *et al.*, 2003). These findings suggest that sepal identity may represent an independently evolved genetic program, varying between bracts and petals, rather than one that is controlled by a single-conserved function (Buzgo *et al.*, 2004). This idea is further supported by the fact that the expression patterns of homologs of floral genes in the outer perianth appear to be less conserved than expression patterns in the well-studied petals, stamens, and carpels (Kim *et al.*, 2005).

As with A-function, B-function in *Arabidopsis* is primarily conferred by two genes, *APETALA3* (*AP3*) and *PISTILLATA* (*PI*), both of which are MADS-box genes. Single mutations of either gene, as well as those in their orthologs in *Antirrhinum, DEFICIENS* (*DEF*) and *GLOBOSA* (*GLO*), result in a homeotic conversion of the petals into sepals and stamens into carpelloid organs (Bowman *et al.*, 1989; Goto and Meyerowitz, 1994; Jack *et al.*, 1992; Sommer *et al.*, 1990; Trobner *et al.*, 1992). AP3 and PI, or DEF and GLO, form heterodimers that bind to specific DNA sequences, and mutations affecting this heterodimerization disrupt gene function (Goto and Meyerowitz, 1994; Jack *et al.*, 1994; Riechmann *et al.*, 1996b; Samach *et al.*, 1997; Schwarz-Sommer *et al.*, 1992; Trobner *et al.*, 1992; Yang *et al.*, 2003a; see Melzer *et al.*, Chapter 5). Among the angiosperms, the function of *AP3* and *PI* homologs appears to be highly conserved, although it is important to note that sub- and/or neofunctionalization has apparently occurred within these gene lineages such that B-function can be partitioned among multiple copies of *AP3* and/or *PI* homologs and may not be conserved among orthologs (Ma and dePamphilis, 2000; Zahn *et al.*, 2005b).

C-function in *Arabidopsis* is primarily conferred by a single MADS-box gene, *AGAMOUS* (*AG*). *AG* functions in stamen and carpel organ identity as well as floral meristem determinacy (Bowman *et al.*, 1989; Mizukami and Ma, 1992; Okamuro *et al.*, 1996). In *ag* mutants, flowers lack stamens and carpels and instead produce petals and sepals in their place (Bowman *et al.*, 1989). In addition, the growth at the center of the flower is indeterminate, resulting in repeating whorls of sepals, petals, and petals (Bowman *et al.*, 1989; Mizukami and Ma, 1992; Okamuro *et al.*, 1996; Yanofsky *et al.*, 1990). Ectopic expression of *AG* results in sepals with carpelloid features and staminoid petals (Mizukami and Ma, 1992). Although less than the *AP3* and *PI* lineages, the genes in the *AG* subfamily have also apparently undergone sub- and/or neofunctionalization that has partitioned C-function among different gene copies (Zahn *et al.*, 2006). The MADS-box genes that confer D-function include *SHATTERPROOF1* and *2* (*SHP1, 2*) and *SEEDSTICK* (*STK*) (Favaro *et al.*, 2003; Pinyopich *et al.*, 2003), although these genes have also demonstrated function in fruit development and seed dispersal (Liljegren *et al.*, 2000; Pinyopich *et al.*, 2003). The *SHP1, 2* and *STK* genes are all members of the *AG* subfamily (Kramer *et al.*, 2004; Zahn *et al.*, 2006).

Both C- and D-functions are conserved among members of the *AG* subfamily in a number of plants (Kramer *et al.*, 2004; Zahn *et al.*, 2006). However, like that observed in B-function genes, phylogenetically strict orthologs do not always have conserved functions (Kramer *et al.*, 2004; Zahn *et al.*, 2006; see Davies *et al.*, Chapter 7). In particular, the *Arabidopsis AG* and the *Antirrhinum PLENA* (*PLE*) genes are functionally most similar, as both are required for C-function, yet they are not orthologous. In contrast, the *AG* ortholog in *Antirrhinum, FARENELLI* (*FAR*), has a more specialized function in anther and pollen development (Davies *et al.*, 1999; see Davies *et al.*, Chapter 7). Conversely, the closest homologs of *PLE* in *Arabidopsis* are *SHP1* and *SHP2*, which are the result of a duplication event in the Brassicaceae (Blanc and Wolfe, 2004; Blanc *et al.*, 2003; Moore *et al.*, 2005;

Simillion *et al.*, 2002; Vision *et al.*, 2000). Unlike *PLE, SHP1/2* do not have early function in controlling both male and female reproductive organ identities, rather these genes are important for fruit and ovule development (Liljegren *et al.*, 2000). Because *AG* (and perhaps *PLE*) have anther, pollen and ovule development functions (Ito *et al.*, 2004; Pinyopich *et al.*, 2003), the specialized functions of *FAR* and *SHP1/2* may represent examples of sub- and/or neofunctionalization (Zahn *et al.*, 2006).

The four *SEPALLATA* genes (*SEP1,2,3,4*; formerly *AGL2,3,9,4*) were initially isolated as *AG* homologs (Ma *et al.*, 1991; Mandel and Yanofsky, 1998); they confer E-function and are highly redundant (Ditta *et al.*, 2004; Malcomber and Kellogg, 2005; Pelaz *et al.*, 2000; Zahn *et al.*, 2005a). It was observed that the flowers of triple *sep1 sep2 sep3* plants are composed solely of sepal-like structures, leading to the idea that the *SEP1,2,3* genes are required for both B- and C-functions (Pelaz *et al.*, 2000). Quadruple *sep1 sep2 sep3 sep4* plants produce leaf-like organs in place of floral organs (Ditta *et al.*, 2004), indicating that the four *SEP* genes together are responsible for the floral state in *Arabidopsis*. Additionally, triple and quadruple *sep* mutant plants show a loss of meristem determinacy (Ditta *et al.*, 2004; Pelaz *et al.*, 2000). The structure of the *SEP* subfamily was demonstrated to be highly similar to the *DEF/GLO* and *AG* subfamilies, including an early duplication before the divergence of the extant angiosperms and a second duplication early in the origin of the eudicots (Zahn *et al.*, 2005a). In addition, the fact that E-function is the only function that is both floral specific and floral-organ nonspecific and that it is required for the conversion of leaves into floral organs in *Arabidopsis* (Honma and Goto, 2001; Pelaz *et al.*, 2000, 2001) indicates that it is an excellent candidate function that distinguishes angiosperms from gymnosperms. The idea that E-function might have been critical for the emergence of angiosperms is further supported by the apparent origin of the *SEP* subfamily near the origin of the flowering plants (Zahn *et al.*, 2005a).

Further extending the genetic ABC + (DE)-model is the quartet model which couples genetic and molecular data to explain how ABC + (DE)-genes work together to specify floral organ identity (Theissen, 2001; Theissen and Saedler, 2001; see Melzer *et al.*, Chapter 5). The quartet model specifies that MADS-box proteins form heterotetramers that direct organ-specific development. It is hypothesized that SEP-AP1-AP3-PI specifies petals, SEP-AG-AP3-PI specifies stamens, SEP-SEP-AG-AG specifies carpels, and SEP-SEP (or AG)-SHP-STK specifies the ovules (Kaufmann *et al.*, 2005; Zahn *et al.*, 2005b). At the time when the quartet model was proposed, E-function (*SEP* genes) was not known to be required for the sepal identity. We now know that *SEP* genes are also required for sepals (Ditta *et al.*, 2004). Additionally,

AP1 has a close homolog, *CAULIFLOWER* (*CAL*), that was formed by a duplication at the base of the Brassicaceae (Blanc and Wolfe, 2004; Blanc *et al.*, 2003; Moore *et al.*, 2005; Simillion *et al.*, 2002; Vision *et al.*, 2000). Although the *cal* single mutant has no detectable phenotype, *ap1 cal* double mutants display a more severe phenotype than that of the *ap1* single mutant (Kempin *et al.*, 1995), strongly suggesting that CAL may also play a role in A-function and floral meristem identity. Data suggest that it is plausible to extend the quartet model to include the hypothesis that SEP-SEP-CAL-AP1 complexes specify sepals (Castillejo *et al.*, 2005; Ma, 2005; Zahn *et al.*, 2005b). These hypotheses have been supported by protein complexes in yeast-two and yeast-three hybrid studies (Davies *et al.*, 1996; Fan *et al.*, 1997; Honma and Goto, 2001; Immink *et al.*, 2003; Moon *et al.*, 1999), along with the observation that the ectopic expression of SEP-AP1-PI-AP3 converts leaves into petals and the ectopic expression of SEP-PI-AP3-AG converts leaves into stamens (Honma and Goto, 2001; Pelaz *et al.*, 2001). Phylogenetic studies (Kramer *et al.*, 2004; Litt and Irish, 2003; Zahn *et al.*, 2005a,b, 2006) support the hypotheses that gene duplicates in the *AP1, AP3/PI, AG*, and *SEP* MADS-box subfamilies may have allowed for the formation of novel protein–protein interactions among these important floral proteins, and that selective pressure for such novel interactions may have contributed to the maintenance of these duplicated genes (see Kramer and Zimmer, Chapter 9).

In summary, a number of floral regulatory genes critical for the ABCDE-functions have been identified using both forward and reverse genetic studies. With the exception of the A-function *AP2* gene (Jofuku *et al.*, 1994), these floral genes are all MADS-box genes encoding putative transcription factors (Ma, 1994, 2005; Ng and Yanofsky, 2001; Pelaz *et al.*, 2000; Pinyopich *et al.*, 2003; Riechmann and Meyerowitz, 1997; Sommer *et al.*, 1990; Zahn *et al.*, 2005b). Several MADS-box subfamilies include members that have evolutionarily conserved function in floral development. Phylogenetic analyses of the subfamilies containing the A-, B-, C-, D- and E-genes have found that the *Arabidopsis* genes within these subfamilies, and their closest homologs, are often the result of relatively ancient gene duplication events. In each of the B (*DEF/GLO*), C + D (*AG*), and E (*SEP*) subfamilies, a gene duplication event occurred between the time of the divergence of the extant gymnosperms and the extant angiosperms ~300 million years ago (MYA) and the time of the diversification of extant angiosperms at least 210–150 MYA (Aoki *et al.*, 2004; De Bodt *et al.*, 2005; Kim *et al.*, 2004, 2005; Kramer *et al.*, 2004; Soltis *et al.*, 2002; Stellari *et al.*, 2004; Zahn *et al.*, 2005a,b, 2006). Furthermore, a second gene duplication occurred in specific clades containing genes with A-, B-, C-, and E-functions close to the time of the diversification of early eudicots ~125 MYA (Aoki *et al.*, 2004; Crane *et al.*, 1995; Kim *et al.*, 2004; Kramer *et al.*, 2004; Litt and Irish, 2003; Stellari *et al.*, 2004; Zahn *et al.*, 2005a,b, 2006). Within the Brassicaceae, there have been further gene duplications estimated to have occurred ~30 MYA (Blanc and Wolfe, 2004; Blanc *et al.*, 2003; Moore *et al.*, 2005; Simillion *et al.*, 2002; Vision *et al.*, 2000) and the retention of paralogs has resulted in (partial) functional redundancy in *Arabidopsis* in three of the four subfamilies. Therefore, gene duplications, followed by either functional divergence or persistent functional conservation, particularly at the sequence level, have played critical roles in the evolution of the members of the MADS-box gene family required for the ABC-functions (Davies *et al.*, 1996; Malcomber and Kellogg, 2005; Whipple *et al.*, 2004; Zahn *et al.*, 2005a, 2006). Other chapters of this volume provide more detailed accounts of such evolutionary changes in specific subfamilies and in particular groups of plants (Davies *et al.*, Chapter 7; Irish, Chapter 3; Kramer and Zimmer, Chapter 9; Melzer *et al.*, Chapter 5; and Rijpkema *et al.*, Chapter 6). In this chapter, we focus on genes that regulate the expression and/or function of ABC-genes, primarily from the analyses in *Arabidopsis*.

II. REGULATORS OF A-FUNCTION

A. REGULATION OF *AP1* BY MERISTEM IDENTITY GENES

The A-function *AP1* gene is expressed beginning at floral stage 1, when the floral meristem has emerged from the inflorescence meristem. By stage 2, *AP1* is expressed throughout the spherical floral meristem (Mandel *et al.*, 1992). Subsequently, *AP1* expression is restricted to the outer region of the stage-3 floral primordium, but is excluded from the central region that gives rise to the reproductive organs occupying the inner two whorls (Mandel *et al.*, 1992). Early *AP1* expression is positively regulated by the floral meristem gene *LFY*, which encodes a plant-specific transcription factor and is generally found as a single copy in the angiosperms (Frohlich and Meyerowitz, 1997) (Fig. 2). In *lfy* mutant flowers, there is a delay and reduction in *AP1* expression (Weigel and Meyerowitz, 1993). In addition, *AP1* expression is accelerated in transgenic plants constitutively expressing *LFY* from the 35S promoter (Weigel and Nilsson, 1995). Moreover, the LFY protein can bind to the *AP1* promoter and transgenic plants with an inducible form of LFY also show inducible activation of *AP1* expression (Wagner *et al.*, 1999; William *et al.*, 2004). Specifically, *AP1* expression was activated after the induction of a fusion protein between LFY and the hormone receptor portion of the glucocorticoid receptor (Wagner *et al.*, 1999; William *et al.*,

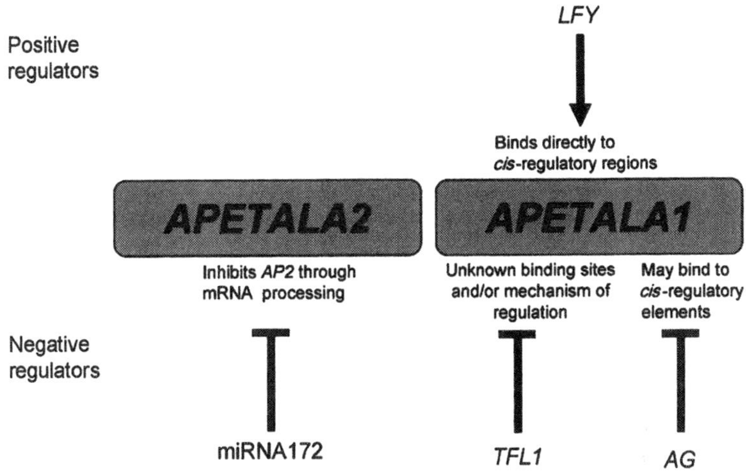

Fig. 2. Regulation of A-function genes.

2004). Furthermore, chromatin immunoprecipitation experiments indicate that LFY binds to the *AP1* promoter region *in vivo* (William *et al.*, 2004). Consistent with these observations are the phenotypes of *ap1* and *lfy* single and double mutants (Bowman *et al.*, 1993; Huala and Sussex, 1992; Weigel *et al.*, 1992). Similar to *ap1* mutants, *lfy* single mutants have defects in the formation of the perianth organs, including an absence of petals in strong alleles and abnormal petals in weak alleles (Huala and Sussex, 1992; Weigel *et al.*, 1992). Also, *ap1 lfy* double mutants exhibit more severe phenotypes than either single mutant (Bowman *et al.*, 1993; Weigel and Meyerowitz, 1993), suggesting that these genes have partially overlapping functions and that LFY positively and directly regulates *AP1* expression.

In addition to *LFY*, the *CAL* and *FRUITFULL* (*FUL*) genes positively interact with *AP1*. *CAL* and *FUL* are both MADS-box genes in the *AP1* subfamily (Gu *et al.*, 1998; Kempin *et al.*, 1995; Litt and Irish, 2003; Mandel and Yanofsky, 1995). The functional relatedness of *AP1, CAL*, and *FUL* is best illustrated by the analysis of the *ap1 cal* double mutant and the *ap1 cal ful* triple mutant. Although *cal* and *ful* single mutants do not have defects in the earliest stages of flower development, they enhance *ap1* mutant phenotypes (Bowman *et al.*, 1993; Ferrandiz *et al.*, 2000). The *ap1* single-mutant flowers lack normal perianth organs, but produce functional reproductive organs, indicating that the mutant still retains some floral functions. In contrast, *ap1 cal* double mutants are arrested at the early floral meristem stage and cannot produce any floral organs during much of reproductive development (Bowman *et al.*, 1993). The *ap1 cal* double mutant eventually produces some flowers and is fertile. However, the *ap1 cal ful* triple mutant cannot produce flowers at all (Bowman *et al.*, 1993; Ferrandiz *et al.*, 2000). Furthermore, the *cal lfy* double mutant has a lower level of *AP1* expression than the *lfy* single mutant, indicating that *CAL* plays a role in promoting *AP1* expression (Bowman *et al.*, 1993). These results suggest that A-function is related to a ground floral state. This is also supported by the observation that the simultaneous ectopic expression of A- and B-function genes in the leaves can partially convert them into petals, although the conversion is not as strong as that when one of the *SEP* genes is also ectopically expressed (Honma and Goto, 2001; Pelaz *et al.*, 2001).

Besides being regulated positively by LFY and others, *AP1* expression is also negatively regulated by *TERMINAL FLOWER1* (*TFL1*) (Fig. 2), which acts to promote the inflorescence meristem (Gustafson-Brown *et al.*, 1994). *AP1* is not normally expressed at the apical center of the inflorescence meristem, where indeterminacy is maintained (Mandel *et al.*, 1992). In the *tfl1* mutant, *AP1* is ectopically expressed in the inflorescence meristem, as is *LFY* (Bradley *et al.*, 1997; Gustafson-Brown *et al.*, 1994; Weigel *et al.*, 1992). This ectopic expression of *AP1* contributes to the formation of the terminal flower(s) in the *tfl1* mutant. Conversely, overexpression of *AP1* represses *TFL1* expression, and leads to the production of terminal flowers in transgenic plants (Liljegren *et al.*, 1999; Ratcliffe *et al.*, 1999). *TFL1* encodes a protein that has sequence similarity (Bradley *et al.*, 1997; Ohshima *et al.*, 1997) to animal phosphatidylethanolamine-binding proteins involved in cell signaling (Keller *et al.*, 2005), suggesting that it may play a role in signal transduction, but it is not known how TFL1 may regulate *AP1* expression.

B. REGULATION OF *AP1* BY *AG*

As a part of the antagonism between A- and C-functions, *AP1* expression is negatively regulated by *AG* (Fig. 2). At stage 3 of floral development in the wild-type flower, when *AG* expression is initiated in the center of the floral meristem, *AP1* is repressed in the same region (Gustafson-Brown *et al.*, 1994). In contrast, in *ag* mutant flowers, *AP1* expression is expanded into the central region of the floral meristem, indicating that *AG* is required for the exclusion of *AP1* expression at the center (Gustafson-Brown *et al.*, 1994). Although the *AP1* promoter has putative *cis* sites that resemble MADS-box protein-binding sites (Parcy *et al.*, 1998), it has not yet been demonstrated that AG directly represses *AP1* expression. In yeast, it is known that the MCM1 MADS-box protein can interact with transcriptional repressors to inhibit gene expression (Herschbach *et al.*, 1994; Smith and Johnson, 1992).

C. REGULATION OF *AP2* BY A MICRORNA

Genetic studies in *Arabidopsis* indicate that *AP2* is required for A-function and contributes to the specification of sepal and petal identities (Bowman *et al.*, 1989, 1991b; Haughn and Somerville, 1988; Kunst *et al.*, 1989). Although A-function is genetically restricted to the outer two whorls, molecular studies indicate that *AP2* mRNA is found throughout the floral primordium (Jofuku *et al.*, 1994). Therefore, there seems to be post-transcriptional regulation of *AP2*. It has been shown that microRNAs (miRNAs) are involved in the translational repression of *AP2* (Chen, 2004) (Fig. 2). An miRNA called miRNA172 was found to have a sequence that is complementary to the *AP2* mRNA (Chen, 2004). Elevated levels of miRNA172 cause *ap2* mutant-like phenotypes (Chen, 2004). In addition, the overexpression of mutant *AP2* mRNAs that have synonymous base pair mutations that are mismatched with miRNA172 lead to an increase in *AP2* function at the expense of *AG* function (Chen, 2004). As discussed later in the section on the regulation of C-function, several genes act to repress *AG* expression, such as *LEUNIG* (*LUG*), *SEUSS* (*SEU*), and *BELLRINGER* (*BLR*) (Bao *et al.*, 2004; Conner and Liu, 2000; Franks *et al.*, 2002; Liu and Meyerowitz, 1995). These genes interact positively with *AP2* to repress *AG* expression (see Section IV), providing a critical component of the antagonism between A- and C-functions.

III. REGULATORS OF B-FUNCTION

A. EXPRESSION PATTERNS OF B-FUNCTION GENES IN *ARABIDOPSIS*

In the wild-type flower of *Arabidopsis*, the expression patterns of *AP3* and *PI* are both temporally and spatially specific, as shown by RNA *in situ* hybridizations (Goto and Meyerowitz, 1994; Jack *et al.*, 1992). These experiments indicate that *AP3* expression is not detected in the inflorescence meristem or in stage-1 to -2 floral buds. From stage 3, when the sepal primordia are formed from the floral meristem, *AP3* expression becomes detectable between the center of the flower and the newly formed sepal primordia in a region that is destined to become petals and stamens in whorls two and three, respectively (Jack *et al.*, 1992). Later in development, *AP3* expression accumulates in this region and achieves a high level at stage 5 (Jack *et al.*, 1992). At this stage, both the stamen and petal primordia are formed (Smyth *et al.*, 1990). Following this, the stamens enlarge in size and the petals experience a rapid expansion that lasts until anthesis (Smyth *et al.*, 1990). During these stages, *AP3* expression is maintained at a high level in all the cells of the petals and stamens, and begins to drop as the flower reaches maturity (Jack *et al.*, 1992). Low levels of *AP3* expression are also detected in the ovules at late developmental stages (Jack *et al.*, 1992), but the functional significance of this is not known. The other B-function gene, *PI*, has an expression pattern slightly different from that of *AP3*. *PI* expression lags slightly behind *AP3* expression in timing and *PI* is expressed earlier in the ovules at low levels (Goto and Meyerowitz, 1994). However, the expression patterns of these two genes are quite similar both spatially and temporally. High levels of expression for both genes are mostly restricted to the petals and stamens (Goto and Meyerowitz, 1994; Jack *et al.*, 1992). The expression of both genes can also be separated into two stages: an initiation stage with a relatively low-expression level and a maintenance stage with a high-expression level.

B. ACTIVATION OF B-FUNCTION GENES BY *LFY* AND *AP1*

The genes that determine floral meristem identity, *LFY* and *AP1*, also regulate the expression of the B-function genes (Fig. 3). This was first suggested by the fact that the *lfy-6* and *ap1-1* mutant also have petal and/or stamen defects besides deficiencies in meristem identity (Weigel *et al.*, 1992). Furthermore, experimental evidence shows that *LFY* and *AP1* are likely the direct activators of B-function genes (Lamb *et al.*, 2002; Ng and Yanofsky, 2001). *LFY* is expressed in the region of the inflorescence meristem that will initiate the next floral meristem and in the newly emerged floral meristem (Weigel *et al.*, 1992). In *lfy* mutants, the identities of both the petals and the stamens are defective and the expression of *AP3* and *PI* are greatly reduced, indicating that LFY is a positive regulator of the B-function genes (Lamb *et al.*, 2002; Weigel and Meyerowitz, 1993). In addition, both the *AP3* and *PI* promoters have a *cis*-element that binds to LFY in a yeast-one hybrid assay and confers activation by *LFY* to a reporter gene (Lamb *et al.*, 2002). Besides *LFY*, *AP1* is also important for *AP3/PI* expression (Weigel and Meyerowitz, 1993). In *ap1* mutants, *LFY*-dependent expression of *AP3* is reduced (Ng and Yanofsky, 2001). Given the positive regulation of *AP1* by *LFY*, it was postulated that *LFY* promotes *AP3* expression partially through *AP1* (Ng and Yanofsky, 2001). *CAL* may also positively regulate *AP3* and *PI* expression besides having a role in floral meristem identity. This is suggested by the fact that in *ap1* background, *CAL* is required for petal development together with *SEP3*, which also seems to activate B-function genes (Castillejo *et al.*, 2005; see also later sections).

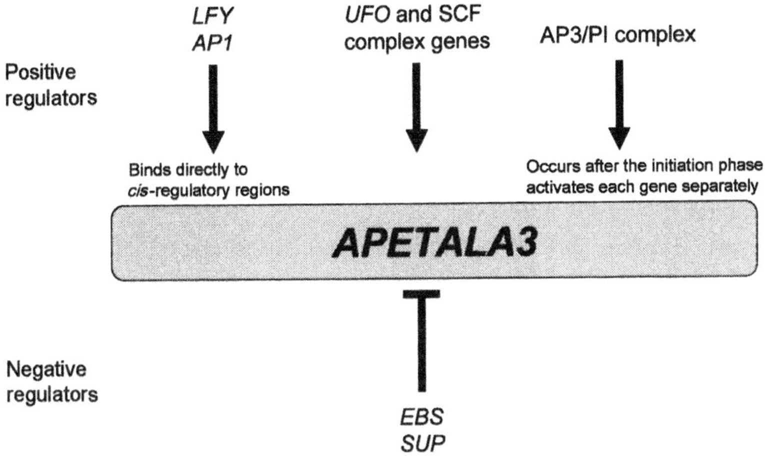

Fig. 3. Regulation of the B-function gene *AP3*.

C. SCF^{UFO} E3 UBIQUITIN LIGASES AS REGIONAL ACTIVATOR OF B-FUNCTION GENES

The accurate patterning of flower organs requires not only the activation but also the whorl-specific expression of floral homeotic genes, including *AP3* and *PI*. This is shown by the fact that ectopic expression of *AP3/PI* causes the homeotic transformation of the sepals and the carpels (Jack *et al.*, 1994). *LFY* is globally expressed in the floral meristem, but is only required for *AP3* and *PI* activation in whorls two and three. Therefore, it is not the only factor in the spatial regulation of *AP3/PI* expression and whorl-specific factors other than *LFY* are required to confine the expression of *AP3/PI* to specific whorls.

Several lines of evidence suggest that the activation of *AP3* by *LFY* also requires the function of the *UNUSUAL FLORAL ORGANS* (*UFO*) gene (Levin and Meyerowitz, 1995; Zhao *et al.*, 2001) (Fig. 3). The *ufo* mutants are defective in petal and stamen identity (Lee *et al.*, 1997; Levin and Meyerowitz, 1995; Wilkinson and Haughn, 1995). The strong *ufo-2* allele exhibits a reduced number and size of both petals and stamens, and has chimeric petaloid stamens and/or staminoid petals in whorls two and three, respectively (Levin and Meyerowitz, 1995). Also, in the *ufo* mutant, both *AP3* and *PI* expression are greatly reduced and *ufo* defects can be rescued by the overexpression of *AP3* (Lee *et al.*, 1997). A *35S:UFO* transgene is able to increase the numbers of petals and stamens and expand the region of *AP3/PI* expression (Lee *et al.*, 1997). This suggests that *UFO* is a regional activator of *AP3* and *PI*. LFY and UFO together are sufficient to activate *AP3*, as the leaves of transgenic plants carrying both *35S:LFY* and *35S:UFO* exhibit ectopic expression of *AP3* (Lee *et al.*, 1997). Therefore, the spatial expression of *AP3* and *PI* is regulated by the specific expression of *LFY* and *UFO* in the flower.

A comparison of UFO amino acid sequences with other proteins revealed that it contains an F-box domain (Samach *et al.*, 1999). An F-box protein is a subunit of the SCF (*S*kp1/*C*ullin/*F*-box) E3 ubiquitin ligases, which are multisubunit protein complexes consisting of SKP1 (S phase kinase-associated protein 1), Cullin, Rbx, and one F-box protein (Gagne *et al.*, 2002; Risseeuw *et al.*, 2003; Vierstra, 2003). In *Arabidopsis*, there are 21 *SKP1* homologs (named *ASK1–21*), 11 *Cullin* homologs (*AtCul1–11*), one functional *Rbx* gene (*AtRbx1a*), and about 700 F-box genes (Craig and Tyers, 1999; Gagne *et al.*, 2002; Kong *et al.*, 2004; Lechner *et al.*, 2002; Zhao *et al.*, 2003). The SCF complex brings together an E2 ubiquitin-conjugating enzyme with its target protein, facilitating the transfer of ubiquitin to the target (Vierstra, 2003). The ubiquitylated protein is then degraded by the 26S proteasome (Smalle and Vierstra, 2004; Vierstra, 2003). This process is necessary to control the turnover of a broad range of short-life regulators. It has been hypothesized that UFO is involved in SCF complex(es) that regulate B-function genes by ubiquitylating and degrading putative repressor(s) of these genes (Samach *et al.*, 1999).

Further evidence supporting the "SCF^{UFO}" hypothesis comes from the floral phenotype caused by either mutations in or the repression of genes encoding other putative subunits of the SCF^{UFO} complex (Ni *et al.*, 2004; Zhao *et al.*, 2001). In the *ask1-1* mutant, flowers have petals and stamens with reduced sizes, chimeric organs, and reduced *AP3*/*PI* expression (Zhao *et al.*, 2001). *AtRbx1a* and *AtCUL1* are indispensable in early development. Therefore, their functions in floral development were examined by a reduction in the functions of these genes. Plants carrying an inducible RNAi construct for *AtRbx1a* or a viable weak mutant allele of *AtCUL1* exhibited similar phenotypes to those observed in *ufo* and *ask1-1* mutant flowers (Ni *et al.*, 2004). Double-mutant analyses involving *ufo, ask1*, and other mutations yield stronger phenotypes than either single mutants, suggesting that other F-box and *ASK* genes might participate in the SCF complexes that regulate *AP3* and *PI* expression (Ni *et al.*, 2004). In particular, ASK2 is a likely subunit of such SCF complexes (Ni *et al.*, 2004). These observations are consistent with the data that support the physical interactions of those four SCF subunits from yeast-two hybrid and pull-down assays (Risseeuw *et al.*, 2003).

However, some components of the SCF^{UFO} pathway remain elusive as the identities of the targets of the SCF complexes important for floral development are still unknown. These targets are hypothesized to have the following three features: (1) they may be repressors of *AP3/PI*, if *UFO* directs their degradation (Zhao *et al.*, 2001); (2) they could be *AG*-dependent factors, since the petal defects of *ufo12* or *ask1* can be suppressed by *ag* (Durfee *et al.*, 2003; Ni *et al.*, 2004; see Section IV.B.5); and (3) they may only function for a short time during flower development, as transient expression of *UFO* can rescue the defects of the *ufo* mutant (Durfee *et al.*, 2003). Other unknowns include the upstream regulator of *UFO*. There is a CArG box, a putative binding site for a MADS-domain protein, at the distal region of the *UFO* promoter. In addition, at least one *cis*-element has been identified that regulates *UFO* expression in the petals. However, the *trans*-factors responsible for binding to these elements are not known. Uncovering the identity of both the regulators of UFO and the substrates of the SCF complexes will provide important insight into SCF function and may even possibly indicate potential interactions between B- and C-functions.

D. REGULATION OF B-FUNCTION BY *SEP3*

As discussed in the introduction, the *Arabidopsis SEP* genes function as important determinants of the floral state. *SEP3* is an E-function gene and a member of the *SEP* MADS-box subfamily (Ditta *et al.*, 2004; Pelaz *et al.*, 2000; Zahn *et al.*, 2005a). Although the function of *SEP* genes appears to be highly redundant (Ditta *et al.*, 2004; Pelaz *et al.*, 2000), *SEP3* seems to be more important in flower and ovule development than *SEP1,2,4* (Castillejo *et al.*, 2005; Ditta *et al.*, 2004; Favaro *et al.*, 2003; Pelaz *et al.*, 2001). This is supported by the fact that *SEP3* has a slightly different expression pattern from those of the *SEP1,2,4* genes (Ditta *et al.*, 2004; Flanagan and Ma, 1994; Ma *et al.*, 1991; Mandel and Yanofsky, 1998; Savidge *et al.*, 1995). Furthermore, *sep1, sep2*, and *sep4* single mutants have subtle or no floral phenotypes, whereas *sep3* plants have sepaloid petals (Castillejo *et al.*, 2005; Ditta *et al.*, 2004; Pelaz *et al.*, 2000, 2001). In addition, *sep1 sep2 sep3/+* plants produce nearly normal ovule and floral development and *sep1 sep2 sep3/+ sep4* plants exhibit floral organ formation (Ditta *et al.*, 2004; Favaro *et al.*, 2003), indicating that a single *SEP3* allele is able to provide much of the *SEP* function. This functional differentiation between *SEP3* and *SEP1,2,4* is in agreement with the phylogeny of the *SEP* subfamily (Zahn *et al.*, 2005a). *SEP3* is the only *Arabidopsis* gene found in the *AGL9* (*SEP3*) clade, whereas three genes, *SEP1,2,4,* are in the *AGL2/3/4* clade, and have highly redundant function (Castillejo *et al.*, 2005; Ditta *et al.*, 2004; Pelaz *et al.*, 2000; Zahn *et al.*, 2005a).

A recent study suggests that *SEP3* is an activator of both B- and C-functions (Castillejo *et al.*, 2005). The constitutive expression of *SEP3* in plants has been demonstrated to affect the floral meristem, resulting in early flowering and terminal flowers similar to the *35S::AP1* phenotype (Pelaz *et al.*, 2001). It was shown that *35S::SEP3* is able to rescue *ufo* plants, suggesting that *SEP3* regulates B-function downstream or independently of *UFO* (Castillejo *et al.*, 2005). This hypothesis is also supported by the fact that *35S::SEP3* plants exhibit ectopic expression of *AP3* in the leaves (Castillejo *et al.*, 2005). In plants transformed with both *35S::SEP3* and *35S::UFO* or both *35S::SEP3* and *35S::LFY* petals develop ectopically on nonfloral organs (Castillejo *et al.*, 2005), further supporting the idea that the SEP3 protein is able to activate B-function genes.

E. CONTROL OF THE INNER BOUNDARY OF B-FUNCTION GENE EXPRESSION BY *SUPERMAN*

In *Arabidopsis* flowers, the boundaries of B-function gene expression in petals and stamens with respect to adjacent whorls are precisely defined (Bowman *et al.*, 1992; Jack *et al.*, 1994; Sakai *et al.*, 1995; Schultz *et al.*, 1991). The *SUPERMAN* (*SUP*, also called *FLO10*) gene is important for controlling the inner boundary of *AP3*/*PI* expression between the third and fourth whorls (Bowman *et al.*, 1992; Schultz *et al.*, 1991) (Fig. 3). The *sup* mutants have extra stamens at the expense of the gynoecium (Bowman *et al.*, 1992; Sakai *et al.*, 1995). In *sup* mutants, the domain of *AP3*/*PI* expression is expanded into the fourth whorl, providing an explanation for the observed change of the inner boundary of stamens (Jacobsen and Meyerowitz, 1997; Sakai *et al.*, 1995). Conversely, the ectopic expression of *SUP* suppresses both petal and stamen development (Kater *et al.*, 2000; Nandi *et al.*, 2000; Yun *et al.*, 2002), indicating that *SUP* is a negative regulator of *AP3*/*PI* expression.

The expression of *AP3*/*PI* in the center of *sup* mutant flowers led to the proposal that SUP acts in the fourth whorl to repress *AP3* and *PI* expression (Goto and Meyerowitz, 1994; Sakai *et al.*, 1995). This hypothesis predicts that the *SUP* expression should be detected in the center of the floral meristem. However, *in situ* hybridization revealed that *SUP* is expressed in a small region near the junction between the third and fourth whorls (Sakai *et al.*, 1995). Therefore, *SUP* acts to mark the inner boundary of *AP3* and *PI* expression, rather than repressing *AP3* and *PI* expression at the center. The *SUP* gene encodes a zinc finger transcription factor that binds to DNA *in vitro* and negatively regulates cell proliferation (Dathan *et al.*, 2002). The ectopic expression of *SUP* suppresses cell proliferation in both *Arabidopsis* and other plants such as tobacco (Kater *et al.*, 2000; Yun *et al.*, 2002). These results suggest that the boundaries between different floral whorls are

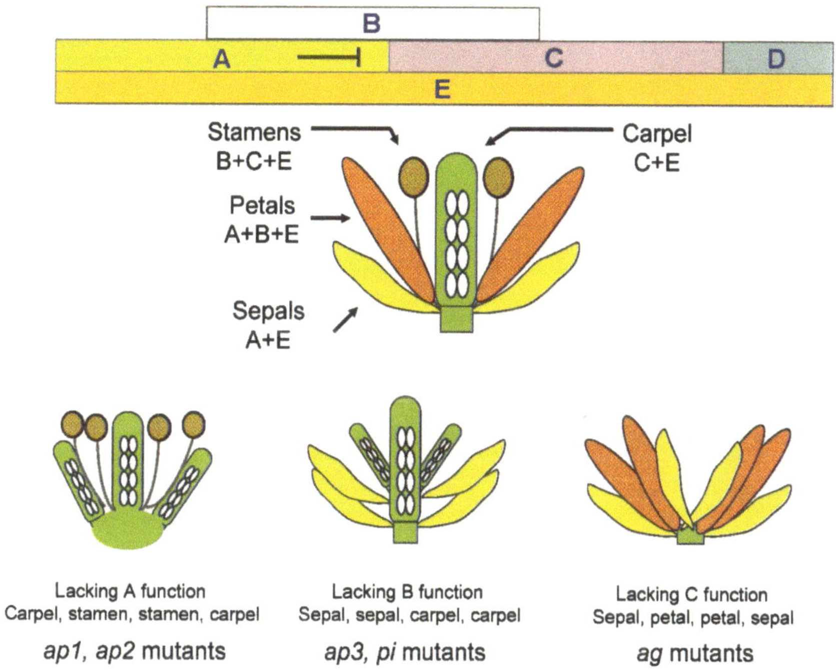
B
A
C
D
E
Stamens
B+C+E
Carpel
C+E
Petals
A+B+E
Sepals
A+E
Lacking A function
Carpel, stamen, stamen, carpel
ap1, ap2 mutants
Lacking B function
Sepal, sepal, carpel, carpel
ap3, pi mutants
Lacking C function
Sepal, petal, petal, sepal
ag mutants

Plate 1

affected by the genes controlling the balance of cell proliferation between the whorls. It is possible that *SUP* affects the domain of B-function gene expression by controlling the population of those cells that are competent for B-function gene expression.

Another level of regulation involving *AP3/PI* and *SUP* was uncovered by a careful comparison of *SUP* expression patterns in wild-type and *ap3* mutant flowers (Sakai *et al.*, 1995). *SUP* expression is first detected at stage 3, slightly later than the onset of *AP3* and *PI* expression (Jack *et al.*, 1992). However, in *ap3* and *pi* mutant flowers, *SUP* expression is reduced, indicating that *AP3* and *PI* are positive regulators of *SUP* (Ito *et al.*, 2003). This suggests that the normal number of whorls of stamens and carpels in *Arabidopsis* is controlled by a balance between *AP3/PI* and *SUP* acting in a negative feedback loop. Another aspect of *SUP* function is that its expression can be affected by epigenetic mechanisms (Jacobsen and Meyerowitz, 1997). The epigenetic alleles of *sup* (called the *clark kent* alleles) are hypermethylated at cytosine residues, have decreased levels of *SUP* RNA, and are phenotypically similar to loss-of-function *sup* alleles with changes in the DNA sequence (Jacobsen and Meyerowitz, 1997). Therefore, epigenetic regulation, such as DNA methylation, also plays an important role in regulating the activity of floral developmental genes.

F. THE AUTOREGULATORY CIRCUIT OF *AP3/PI*

As noted above, *AP3/PI* expression reaches and then is maintained at a high level at stage 5 (Jack *et al.*, 1992). Since *LFY* expression is not detectable after stage 5, and *AP1* expression is restricted to the outer two whorls starting at stage 3 (Huala and Sussex, 1992; Mandel *et al.*, 1992; Weigel *et al.*, 1992), the high level of *AP3/PI* expression must require an additional regulatory factor. It is unlikely that *UFO* is this factor, since a transient expression of *UFO* is sufficient to rescue *ufo* defects (Laufs *et al.*, 2003). This indicates that at late stages, even without *UFO*, *AP3/PI* expression is still maintained at a high level.

In the *ap3* mutant the expression of *PI* is greatly reduced, and *vice versa* (Goto and Meyerowitz, 1994; Jack *et al.*, 1994; McGonigle *et al.*, 1996). This suggests that both *AP3* and *PI* are required for each other's expression (Fig. 3). A series of deletions of the promoter region of *AP3* reveals that *AP3* expression is controlled by discrete elements in the *AP3* promoter region (Tilly *et al.*, 1998). Besides a distal element that is responsible for LFY binding, there are three CArG boxes, which are typical motifs for binding by MADS-box transcription factors (Tilly *et al.*, 1998). AP3 and PI are both MADS-box proteins and can bind to DNA as a heterodimer as well as interacting in yeast-two hybrid assays (Lamb *et al.*, 2002; Riechmann *et al.*, 1996a,b; Tilly *et al.*, 1998). The AP3/PI heterodimer is able to bind to CArG boxes to promote *AP3* expression (Tilly *et al.*, 1998) and the PI protein is also involved in such a positive regulation circuit (Honma and Goto, 2000). The ability of the AP3/PI heterodimer to bind to the *AP3* promoter and drive reporter gene expression has been verified using the yeast-one hybrid assay (Lamb *et al.*, 2002). Thus, the AP3/PI heterodimer and corresponding CArG boxes in the *AP3* promoter form a positive autoregulatory circuit, which greatly enhances *AP3* expression. Consistent with this, abolishing the CArG box greatly reduces the overall expression level of *AP3* (Tilly *et al.*, 1998). Similar to *AP3*, the *PI* gene is also autoregulated with crossregulation mediated by AP3/PI heterodimers (Honma and Goto, 2000). However, in the *PI* promoter region, there are no CArG boxes (Honma and Goto, 2000). Therefore, it has been hypothesized that another cofactor is needed in addition to the AP3/PI heterodimer to promote *PI* expression. This cofactor has not yet been identified.

Moreover, the dimerization of AP3 and PI is not only important for forming functional transcriptional complex binding to the *AP3* promoter but is also crucial for the nuclear co-localization of these proteins (McGonigle *et al.*, 1996). Simultaneous expression of both genes is required for their nuclear localization, while the transient expression of AP3 or PI separately results in cytoplasmic localization (McGonigle *et al.*, 1996). Therefore, it is likely that dimerization of these two proteins allows them to have the appropriate conformation for nuclear localization. Since the heterodimerization of both AP3 and PI is crucial for their high expression level, the domains that mediate the interaction of these proteins are of interest. It was found that the K-domain, an element of all described floral MADS-box proteins (Ma *et al.*, 1991), is crucial for their dimerization (Riechmann *et al.*, 1996a,b). Mutations in the K-domain of PI affect petal development, as shown by *pi-5* mutants (Yang *et al.*, 2003b).

G. GA SIGNALING AND EARLY BOLTING IN SHORTDAYS AFFECTS *AP3/PI* EXPRESSION

Gibberellin (GA) signaling has been suggested as a mechanism that promotes the establishment of the flower meristem identity (Cheng *et al.*, 2004; Yu *et al.*, 2004). It was reported that GA can promote *AP3* and *PI* expression, as suggested by the floral phenotypes of GA-deficient mutant *ga1-3* (Cheng *et al.*, 2004; Yu *et al.*, 2004). These mutants have defects in all floral organs including sterile stamens (Yu *et al.*, 2004). However, the defects in normal floral organ identity and floral homeotic gene expression can be rescued by GA application (Yu *et al.*, 2004). GA signaling requires the removal of DELLA proteins (RGA; GAI; RGL1, 2, 3), which are repressors of GA response (Sun and Gubler, 2004). It was found that *rga-t2* and *rgl2-1* suppress *ga1-3* floral defects, further indicating a need for inactivating DELLA proteins for normal GAs signaling in floral organ development (Cheng *et al.*, 2004; Yu *et al.*, 2004). This hypothesis was also supported by the results of examining the expression of floral homeotic genes in floral tissues of *ga1-3* mutants after GA treatment or in plants with inducible RGA activation (Yu *et al.*, 2004). Consistently, *AP3* and *PI* expression was increased in the floral tissues of *ga1-3* mutants after GA treatment and reduced after the activation of RGA (Yu *et al.*, 2004). Although it is clear that GA promotes *AP3* and *PI* expression by removing DELLA proteins, it is still not known whether the DELLA proteins have a direct effect on *AP3* and *PI* expression.

EARLY BOLTING IN SHORTDAYS (*EBS*) has been shown to regulate flowering in a GA-dependent fashion and also affects petal and stamen development (Gomez-Mena *et al.*, 2001) (Fig. 3). *ebs* mutants exhibit early flowering under short day conditions and this early flowering is suppressed by GA-deficient mutations (Gomez-Mena *et al.*, 2001). *ebs* can partially rescue the petal and stamen defects of *lfy-6*, and can increase both *AP3/PI* expression even in the *lfy-6* background (Gomez-Mena *et al.*, 2001). It is quite possible that *EBS* is a repressor that antagonizes GA response (Gomez-Mena *et al.*, 2001). The *EBS* gene has been identified and encodes a nuclear protein (Pineiro *et al.*, 2003). In this protein, there are two protein–protein interaction motifs: a bromoadjacent homology domain and a homeodomain zinc finger, which occurs in a chromatin-remodeling complex. Disruption of this complex by mutations in *EBS* prevents its repression of flowering-time genes, specifically *FLOWERING LOCUS T* (*FT*), and promotes early flowering (Pineiro *et al.*, 2003). The *ebs-1 lfy-6 ft* triple mutant produces flowers without petals and stamens, unlike the *ebs-1 lfy-6* double mutants (Gomez-Mena *et al.*, 2001; Pineiro *et al.*, 2003). The enhancement of the *ebs-1 lfy-6* phenotype by *ft* may be due to a reduction of *AP1* since *AP1* expression is reduced in *ft* mutants (Pineiro *et al.*, 2003). Altogether, the data suggest that the pathways promoting flowering transition may also directly regulate floral homeotic genes. However, evidence of direct transcriptional regulation is still lacking.

IV. REGULATORS OF C-FUNCTION

The first identified C-function gene, *AG*, is expressed starting at stage 3 (Smyth *et al.*, 1990), in the region of the floral meristem that will give rise to the third and forth whorls and in the stamens and carpel primordia of

Arabidopsis flowers (Drews *et al.*, 1991; Yanofsky *et al.*, 1990). At later floral stages, *AG* expression becomes limited to the wall and connective tissues of the anther, the stigma, and the endothelium of the ovules (Bowman *et al.*, 1991a). The *ag* mutants undergo a homeotic transformation of the stamens into petaloid organs and the carpels into sepaloid structures (Bowman *et al.*, 1989). In addition, floral meristem determinacy is lost such that the flower consists of repeating whorls of sepals, petals, and petals (Bowman *et al.*, 1989). Constitutive ectopic expression of *AG* results in a flower with carpelloid sepals and staminoid petals (Mizukami and Ma, 1992). *AG* expression is controlled by both positive and negative regulators (Busch *et al.*, 1999; Deyholos and Sieburth, 2000; Sieburth and Meyerowitz, 1997). Furthermore, *AG* expression is primarily mediated by *cis*-regulatory regions in the second intron (Busch *et al.*, 1999; Deyholos and Sieburth, 2000; Sieburth and Meyerowitz, 1997).

AG regulation is highly complicated and involves a number of genes, many of which have partially redundant or overlapping functions in positive and negative regulation (Ma, 2005). These regulatory genes often also have functions in other developmental processes (Aubert *et al.*, 2001; Byrne *et al.*, 2003; Chen *et al.*, 2002; Elliott *et al.*, 1996; Gallois *et al.*, 2004; Gross-Hardt *et al.*, 2002; Li and Chen, 2003; Mayer *et al.*, 1998; Smith and Hake, 2003). Regulation of *AG* expression involves many types of proteins belonging to several different protein families, some of which are well characterized and some of which represent the first-described members of a protein family (Aubert *et al.*, 2001; Bowman *et al.*, 1993; Byzova *et al.*, 1999; Conner and Liu, 2000; Elliott *et al.*, 1996; Franks *et al.*, 2002; Goodrich *et al.*, 1997; Jofuku *et al.*, 1994; Weigel *et al.*, 1992; Yoshida *et al.*, 2001). Positive regulators of *AG* that have been identified to date include *ENHANCER OF HUA1–4* (*HEN1, HEN2, HEN3, HEN4*), *HUA1, HUA2, LEAFY* (*LFY*), *PAUSED* (*PSD*, also known as *HEN5*), and *WUSCHEL* (*WUS*) (Chen and Meyerowitz, 1999; Chen *et al.*, 2002; Cheng *et al.*, 2003; Li and Chen, 2003; Mayer *et al.*, 1998; Wang and Chen, 2004; Weigel *et al.*, 1992; Western *et al.*, 2002) (Fig. 4).

A. POSITIVE REGULATORS OF *AG*

1. *Regulation of* AG *expression by* LFY *and* WUS

LFY was the first described positive regulator of *AG* (Weigel *et al.*, 1992). *AG* RNA levels are decreased in *lfy* mutants, and those flowers that arise late exhibit a restricted pattern of *AG* expression (Weigel and Meyerowitz, 1993). *AG* is ectopically expressed in the stem and bracts of *lfy* mutants, establishing that LFY also negatively regulates *AG* expression in these organs (Busch *et al.*, 1999; Parcy *et al.*, 1998). LFY controls *AG* expression in the floral meristem along with AP1. This is supported by the abnormal expression patterns of *AG* in *lfy ap1* double mutants (Weigel and Meyerowitz, 1993; Weigel *et al.*, 1992), although the biochemical role of AP1 in *AG* regulation has not yet been identified.

WUS is a novel homeodomain protein that is hypothesized to function as a transcriptional regulator (Mayer *et al.*, 1998) and acts together with LFY in regulating *AG* (Lohmann *et al.*, 2001). The *wus* mutant alleles have abnormal flowers with significantly reduced numbers of stamens, no carpels, and sometimes even shoot meristems, somewhat similar to those that occur in *ag* mutant flowers (Laux *et al.*, 1996). When *WUS* is ectopically expressed, plants have an increased number of stamens in additional whorls, a reduced number of sepals and petals, and staminoid carpels; also, new flowers are formed within the primary flowers and flowers have precocious carpel development (Lenhard *et al.*, 2001; Lohmann *et al.*, 2001). Most likely these observations are due to the activation of *AG* by the ectopic WUS. WUS directly binds to the second intron of *AG*, along with LFY (Lohmann *et al.*, 2001), and is in a feedback loop with *AG* such that WUS is an activator of *AG* whereas AG is a repressor of *WUS* (Lenhard *et al.*, 2001). However, WUS function in the flower is limited as it does not regulate the stamen and carpel identity function of *AG* (Laux *et al.*, 1996; Lenhard *et al.*, 2001).

In *ag* mutants, *WUS* is continuously expressed in the meristem, leading to the formation of many whorls of sepals and petals, unlike its wild-type transient expression in the center of the floral meristem (Lenhard *et al.*, 2001). On the other hand, *ag wus* double mutants produce determinate flowers (Laux *et al.*, 1996), contrasting to the flowers in *ag* single mutant, showing that *wus* is epistatic to *ag* in terms of floral meristem determinacy. These pieces of evidence suggest that *WUS* is responsible for the indeterminate meristem found in *ag* mutants.

2. *Regulation of* AG *transcript processing by the* HUA *and* HUA ENHANCER *proteins*

While LFY and WUS are required for *AG* expression in early floral development, they are not essential for late *AG* expression. This is evidenced by the fact that *AG* expression is initially reduced in *lfy* mutants but eventually *AG* accumulates to wild-type levels (Parcy *et al.*, 1998; Weigel and Meyerowitz, 1993). A set of genes that regulate *AG* expression at later stages has been identified genetically (Ma, 2005). These genes, including *HUA1, HUA2*, and *HUA ENHANCER1,2,3,4* (*HEN1,2,3,4*), encode proteins in diverse families and are hypothesized to regulate *AG* expression after floral stages 6 and 7 (Chen *et al.*, 2002).

Neither *hua1* nor *hua2* single mutants has an obvious phenotype but the *hua1 hua2* double mutant has minor defects in stamen and carpel development and *hua1 hua2* mutations enhance the phenotype of the weak *ag-4* allele (Chen and Meyerowitz, 1999), which has stamens but not carpels (Sieburth *et al.*, 1995). Both carpel identity and floral determinacy are lost in *ag-4 hua1 hua2* and *ag-1/+hua1 hua2* plants (Li *et al.*, 2001). The single *hua2* mutation alone does not affect *AG* transcript level (Cheng *et al.*, 2003). Moreover, *hua2* may directly affect floral timing as *hua2* mutants exhibit late flowering and have altered levels of the expression of genes regulating floral timing, including *FRIGIDA* (*FRI*), *FLOWERING LOCUS C* (*FLC*), *FLOWERING LOCUS M* (*FLM*)[also known as *MADS AFFECT FLOWERING 1* (*MAF1*)], *MAF2*, and *SHORT VEGETATIVE PHASE* (*SVP*) (Doyle *et al.*, 2005). Therefore, HUA2 may alter the mRNA levels of these genes possibly by enhancing the expression of genes that delay flowering (Doyle *et al.*, 2005).

HUA1 encodes a conserved eukaryotic nuclear protein with a CCCH-type zinc finger that binds to single-stranded RNA and DNA but not double-stranded DNA (Li *et al.*, 2001). HUA2 is a protein containing an RPR (arginine- and proline-rich) domain, a motif involved in RNA metabolism that may be involved in processing *AG* mRNA (Chen and Meyerowitz, 1999; Doyle *et al.*, 2005; Li *et al.*, 2001). *HUA1* and *HUA2* have high levels of expression in the inflorescence meristem, inflorescence stem, and flowers (Chen and Meyerowitz, 1999; Li *et al.*, 2001). HUA1 binds nonregionally to *AG* pre-mRNA *in vitro* (Cheng *et al.*, 2003) and has been hypothesized to be involved in posttranscriptional processing of *AG* mRNA (Li *et al.*, 2001).

Several other genes that positively regulate *AG* were identified by mutations that enhance the loss-of-C-function phenotypes of the *hua1 hua2* double mutant, and were named *HEN1–5* (Chen *et al.*, 2002; Cheng *et al.*, 2003; Ma, 2005; Wang and Chen, 2004; Western *et al.*, 2002). *hen1 hua1 hua2* mutants resemble *ag* mutants; this and other genetic studies support ectopic *AP1* and *AP2* activities as the reason for the mutant phenotypes, suggesting that these three genes function together to regulate *AG* expression through the negative regulation of *AP1* and *AP2* (Chen *et al.*, 2002; Park *et al.*, 2002). *hen2 hua1 hua2* triple mutants have flowers one-third to one-half the size of wild type and undergo a transformation of the stamens to organs that are mosaics of sepal-like and petal-like cells and the carpels to sepaloid organs (Western *et al.*, 2002). The phenotypes of A-, B-, and C-function mutants (*ap1, ap2, ap3, pi*, and *ag*) in a *hen2 hua1 hua2* background suggest that *HEN2* antagonizes A-function, and promotes stamen and carpel development most likely by promoting the maintenance of *AG* expression (Western *et al.*, 2002). *hen3 hua1 hua2* mutants have small leaves, which are most likely due to reduced cell size, and show a loss of both reproductive organ identity and floral determinacy (Wang and Chen, 2004). Specifically, *hen3 hua1 hua2* flowers have third floral whorl organs that are mostly petals and carpels with sepaloid cells (Wang and Chen, 2004). In addition, the fourth whorl of the triple mutant has an increased number of organs and surrounds internal flowers (Wang and Chen, 2004). Furthermore, *hen3 hua1 hua2* mutants are abnormal in *WUS* expression and have altered *AG* mRNA processing and increased levels of *AP1* and *AP2* transcripts (Wang and Chen, 2004). *hen4 hua1 hua2* triple mutants have decreased levels of *AG* transcripts, delayed *WUS*, and ectopic *AP1* expression that results in stamen to petal transformations, sepaloid carpels, and late flowers with enlarged gynoecia and additional floral organs (Cheng *et al.*, 2003). Quadruple mutants of *hen4hua1 hua2* with *ap1* or *ap2* display phenotypes similar to *ag ap2* double mutants, demonstrating that *HEN4* operates in the C-function pathway (Cheng *et al.*, 2003).

The *HEN* genes have dissimilar sequences. *HEN1* encodes a putative nuclear protein with homology to a protein found in animals and yeast that is required for miRNA activity (Cheng *et al.*, 2003). HEN1 was found to methylate miRNAs *in vitro* (Yu *et al.*, 2005), providing a mechanism for its function in miRNA biogenesis. Its role in promoting *AG* expression is thought to be mediated by a negative regulation of AP2 translation (see Section II) through its function in miRNA biogenesis. *HEN2* encodes a broadly expressed DEAD/xH-box RNA helicase, suggesting that it is involved in miRNA processing (Western *et al.*, 2002). *HEN3* encodes a cyclin-dependent kinase (CDKE) that is a homolog of a protein known to function in cell expansion in leaves and cell fate specification in floral meristems (Wang and Chen, 2004). *HEN4* is expressed in most parts of the plant and codes for a putative RNA-binding protein with a K homology (KH) domain (Cheng *et al.*, 2003). Given the sequence similarity to proteins involved in RNA processing it has been proposed that *HUA1, HUA2, HEN2*, and *HEN4* possibly function in a single pathway. This pathway has been hypothesized to promote *AG* expression by either splicing the *AG* RNA and/or processing *AG* pre-mRNA, or by preventing the termination of transcription within the large second intron of *AG* (Cheng *et al.*, 2003).

HEN5 [also known as PAUSED (PSD)] is another regulator of *AG*, possibly via LFY (Li and Chen, 2003). The PSD/HEN5 protein shows homology to human exportin-t and yeast Los1p, regulators of the nuclear export of tRNAs (Li and Chen, 2003). Phenotypes of *psd hua1* and *psd hua2* mutants include a transformation of the reproductive organs into perianth-like organs. This and the fact that *psd* single mutants have reduced *AG* levels (Telfer *et al.*, 1997) suggest that *PSD* regulates *AG* via a different pathway than that of *HUA1* and *HUA2* (Li and Chen, 2003). *PSD* might regulate *AG* either at the transcriptional or posttranscriptional level (Li and Chen, 2003). *psd* mutations enhance the phenotype of *lfy* and *hua1 hua2 psd ap1* mutants have a unique phenotype with shoots with sepalloid and leaf-like organs instead of flowers. The possible interaction between PSD and LFY and lack of floral organs in *hua1 hua2 psd ap1* plants suggest that besides a role in C function, *PSD* also functions in the floral meristem (Li and Chen, 2003).

3. *Regulation of C-function by SEP3*

As discussed in Section III.D, SEP3 may be a positive regulator of both B- and C-functions (Castillejo *et al.*, 2005). The positive effect of SEP3 on C-function is supported by the ectopic expression of *AG* in *35S::SEP3* plants (Castillejo *et al.*, 2005). SEP3 also interacts with AG, as suggested by the fact that ectopic expression of AG and SEP3 results in carpelloid structures instead of sepals and leaves (Goto *et al.*, 2001) and that these proteins, along with PI and AP3, result in staminoid structures instead of leaves (Goto *et al.*, 2001; Honma and Goto, 2001). Once *AG* expression is initiated it is hypothesized that SEP3 then acts together or in parallel with AG to further promote C-function. This is shown by the more complete transformation of sepals and leaves into carpelloid structures in *35S::SEP3 35S::AG* plants than in plants with only the *35S::AG* construct (Castillejo *et al.*, 2005). Moreover, the phenotype of doubly transgenic *35S::SEP3 35S::LFY* plants has ectopic carpelloid structures that replace the terminal floral structures seen in *35S::LFY* plants (Castillejo *et al.*, 2005). This interaction is also supported by yeast-two hybrid experiments that demonstrate an interaction between AG and SEP3 proteins (Fan *et al.*, 1997; Favaro *et al.*, 2003; Honma and Goto, 2001).

B. NEGATIVE REGULATORS OF *AG*

Like the positive regulators of *AG*, the negative regulators of *AG* work at several different spatial levels. The genes that have been identified to date include *AINTEGUMENTA* (*ANT*), *AP2, ASK1/ASK2, AtMSI1* [*Arabidopsis* gene similar to yeast *MSI1* (multicopy suppressor of *ira1*)], *BELL1* (*BEL1*), *BLR, CURLY LEAF* (*CLF*), *EMBRYONIC FLOWER1,2* (*EMF1,2*), *LUG, RABBIT EARS* (*RBE*), *SEU, STERILE APETALA* (*SAP*), and *UFO* (Fig. 4). Of these genes, *ANT, AtMSI1, BEL1, CLF*, and *EMF1,2* regulate *AG* in vegetative tissue or the ovules and may not be directly involved in the regulation of C-function, save that these genes affect *AG* expression.

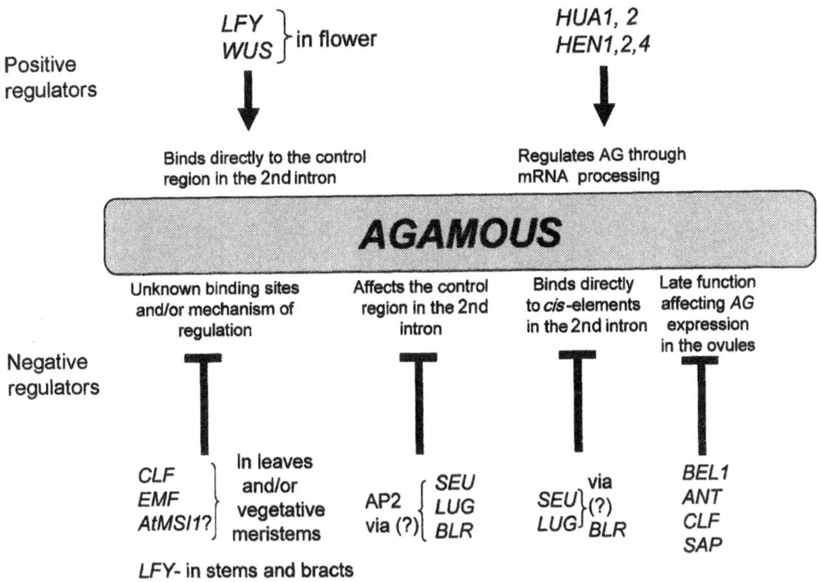

Fig. 4. Regulation of the C-function gene *AG*.

1. Negative regulators of AG *in vegetative tissues*

Negative regulation of *AG* occurs throughout the *Arabidopsis* shoot. Studies have shown that without the activity of specific repressors, such as *EMF1, EMF2*, and *CLF, AG* would be expressed throughout the developing shoot in most vegetative and floral organs. Strong *emf1* mutants produce flowers soon after germination, following the expansion of the cotyledons (Aubert *et al.*, 2001; Sung *et al.*, 1992). Weak *emf1* and *emf2* mutants have a greatly reduced vegetative phase and produce flowers much earlier than normal (Aubert *et al.*, 2001; Chen *et al.*, 1997; Moon *et al.*, 2003; Yoshida *et al.*, 2001). Mutant *clf* plants have short stems and curled leaves that have a similar appearance to that of plants with constitutive *AG* expression (Coupland *et al.*, 1993; Kim *et al.*, 1998; Mizukami and Ma, 1992). This phenotype appears to be due to ectopic expression of *AG* in the inflorescence stem, pedicels, and leaves. In addition, *clf* flowers after stage 9 are abnormal, have carpelloid sepals (Goodrich *et al.*, 1997), and flowering occurs earlier than in wild type (Coupland *et al.*, 1993). The negative regulation of *AG* apparently occurs by interactions between *CLF* and sequences within the second intron (Sieburth and Meyerowitz, 1997).

When *AtMSI1* was downregulated by cosuppression, it resulted in a phenotype similar to *clf* with both *AG* and *AP2* accumulating in the leaves (Hennig *et al.*, 2003). These data suggest that *AtMSI1* is a repressor of *AG* in the leaf, consistent with the observation that *AtMSI1* accumulates at the edges of the sepals. The EMF1 protein is a novel transcriptional regulator (Aubert *et al.*, 2001) and *CLF* and *EMF2* are members of the polycomb-group with homology to the *Drosophila* genes *Enhancer of zeste* (Goodrich *et al.*, 1997) and *Suppressor of zeste* (Yoshida *et al.*, 2001), respectively. *AtMSI1* encodes a WD-40 repeat protein similar to *MSI1*-like proteins found in animals and yeast and is hypothesized to be involved in chromatin structure and assembly (Ach *et al.*, 1997; Hennig *et al.*, 2003; Kenzior and Folk, 1998). The sequence similarity of the EMF1, EMF2, CLF, and AtMSI1 proteins to those of proteins known to regulate chromatin states suggests that these proteins also act to maintain stable chromatin structure and gene expression patterns (Goodrich *et al.*, 1997; Hennig *et al.*, 2003; Kim *et al.*, 1998; Moon *et al.*, 2003; Yoshida *et al.*, 2001).

2. AP2 as a repressor of AG *in the perianth*

The first described negative regulator of *AG* was AP2 (described in Sections I and II.C.). In *ap2* mutants *AG* expression is observed earlier in floral development and in all four floral whorls, showing that AP2 inhibits *AG* expression (Drews *et al.*, 1991). The *ap2-1* mutation primarily affects only sepal and petal identity but has only minor effects on the repression of *AG* (Bowman *et al.*, 1989; Drews *et al.*, 1991). The regulation of *AG* by AP2 is complicated as it seems to interact with proteins of several complexes that are transcriptional repressors of *AG* (Bomblies *et al.*, 1999). Several lines of evidence suggest that AP2 regulates *AG* indirectly. Most importantly the regulatory region of *AG* in the second intron known to be affected by AP2 (Sieburth and Meyerowitz, 1997) does not contain any known AP2-binding sites. Therefore, either AP2 regulates *AG* indirectly, as part of a protein complex, or else it can bind to a previously unidentified binding site (Deyholos and Sieburth, 2000). Moreover, *in situ* hybridization shows *AP2* expression in all floral organs, some of which overlap with *AG* expression (Jofuku *et al.*, 1994). Given that *AP2* translation is regulated by miRNAs (see earlier), it is necessary to look at protein expression in these organs to characterize the true function of AP2.

The interactions and possible function of negative regulators of *AG* have been carefully examined through the characterization of multiple mutants. One of the better defined means by which *AG* is repressed involves the genes that interact or regulate AP2 in the appropriately named AP2 pathway. Genes that function together with *AP2* to regulate *AG* at early stages of floral development include *LUG* (Liu and Meyerowitz, 1995) and *SEU* (Franks *et al.*, 2002). In addition, a second group of *AG*-regulators that function via AP2 include *ANT* (Krizek *et al.*, 2000), *CLF* (Goodrich *et al.*, 1997), and *SAP*

(Byzova *et al.*, 1999). These genes are discussed later. The regulation of *AG* may be affected by the age of the inflorescence meristem, as inflorescence age has been identified as a major factor in the effects of *ap2, lug, seu*, and other mutations (Bao *et al.*, 2004).

3. LUG and SEU form a corepressor complex

The initial *lug* mutations were isolated as enhancers of weak *ap2* mutants (Liu and Meyerowitz, 1995), suggesting that LUG acts in the AP2 pathway. In *lug* single mutants, *AG* is expressed earlier than in the wild type, suggesting that LUG negatively regulates *AG* (Conner and Liu, 2000). *lug* mutants have a split stigma, abnormal carpel and ovule development, reduced male and female fertility, and narrower leaves and floral organs (Conner and Liu, 2000). *lug* mutants have homeotic transformations in the first and second whorls of flowers resulting in carpelloid sepals and staminoid petals (Conner and Liu, 2000). In *seu* mutants, mild homeotic transformations are observed, including whorl one organs that are petaloid and carpelloid instead of sepals, whorl two organs that are narrow and can be staminoid, whorl three stamens with reduced size, and whorl four carpels often slightly split at the top with occasional protrusions of tissue (Franks *et al.*, 2002). In addition, the number of floral organs in whorls two and three is reduced; plants are shorter than wild type and demonstrate increased lateral branching (Franks *et al.*, 2002). In *seu* mutant flowers, *AG* mRNA is sometimes detected in all floral whorls (Franks *et al.*, 2002). The fact that both *lug* and *seu* mutants exhibit ectopic *AG* expression supports the hypothesis that they work together to negatively regulate *AG* (Conner and Liu, 2000; Franks *et al.*, 2002). Furthermore, in *seu lug* double-mutant plants, ectopic *AG* expression is higher than that in *seu* single mutants and is detected early in the floral meristem (Franks *et al.*, 2002). In *ag seu* and *seu lug ag* mutants, the floral phenotypes were like those of *ag* mutant flowers, indicating that ectopic *AG* expression was the cause of *lug* and *seu* floral phenotypes (Franks *et al.*, 2002).

LUG encodes a glutamine-rich protein with 7 WD repeats and is hypothesized to be a transcriptional repressor (Conner and Liu, 2000). *SEU* encodes a glutamine (Q)-rich protein with a putative dimerization domain and is related to the animal family of LIM (*L*in-II *I*sl-1 *M*ec-3)-domain-binding proteins (Franks *et al.*, 2002). The phenotypic similarity of *lug* and *seu* mutants raised the possibility that these genes encode the subunits of a complex. This hypothesis is supported by the evidence that SEU can only repress gene regulation when interacting with LUG (Sridhar *et al.*, 2004) and that *seu lug* double mutants result in plants with more complete homeotic transformations and other effects on floral development than observed in either single mutant

(Franks *et al.*, 2002). These results strongly support the hypothesis that LUG is a transcriptional repressor and SEU is an adaptor for LUG; together they form a plant corepressor complex that negatively regulates *AG* expression (Sridhar *et al.*, 2004).

4. BLR *mediates repression of* AG

Ethylmethane sulphonate (EMS)-induced point mutations in the *BLR* gene [also called *PENNYWISE* (*PNY*) or *REPLUMLESS* (*RPL*)] result in homeotic transformations of sepals to carpels in flowers from old terminating inflorescences and cause defects in phyllotaxy (Bao *et al.*, 2004). Null and nearly null T-DNA insertional *blr* mutants have less severe floral phenotypes than those with point mutations, suggesting that point mutations may cause abnormal proteins that cannot function in regulation. Plants with T-DNA insertions in *blr* also exhibit defects in phyllotaxy, similar to the EMS *blr* alleles (Smith and Hake, 2003), as well as abnormal replum development (Bao *et al.*, 2004; Byrne *et al.*, 2003). Molecular studies of *blr* mutants indicate that BLR represses *AG* expression in the outer region of the floral meristem (Bao *et al.*, 2004). In *blr* mutants, *AG* is expressed ectopically in the region of floral meristem corresponding to whorls one and two (Bao *et al.*, 2004).

BLR encodes a protein similar to the BEL1 homeodomain protein (Byrne *et al.*, 2003; Roeder *et al.*, 2003; Smith and Hake, 2003). *BLR/PNY* has a close homolog in *Arabidopsis, POUND-FOOLISH* (*PNF*), the null mutants of which show no phenotype. However, *pny pnf* double mutants are unable to flower, indicating that they fail to make the transition from vegetative to reproductive development (Smith *et al.*, 2004). Therefore, *PNY* and *PNF* together are required for the ability of the plant to respond to signals that normally induce flowering, indicating that the functions of these genes overlap.

Electrophoretic mobility shift assays demonstrate that BLR binds to sites within the second *AG* intron and that these regions are conserved among some, but not all, Brassicaceae and non-Brassicaceae species (Bao *et al.*, 2004). Because BLR can bind to *cis*-elements in the second *AG* intron, it is likely that BLR is directly involved in repressing *AG* transcription. In addition, *ag blr* double mutants do not exhibit terminating carpelloid mutants like *blr* single mutants and have normal sepals and petals (Bao *et al.*, 2004), indicating that the *blr* single-mutant phenotypes are due to ectopic *AG* expression. It was observed that *blr lug* and *blr seu* double mutants exhibit more severe phenotypes than the *blr* mutants (Bao *et al.*, 2004). However, the *blr ap2* phenotype is similar to that of *ap2* (Bao *et al.*, 2004). This is consistent with the idea that BLR, LUG, and SEU act together in promoting the negative regulation of *AG* in the AP2 pathway. Furthermore, in *blr* plants, *AG* is ectopically expressed in the SAM, similar to strong *ap2* mutants but unlike *lug* and *seu* mutants (Smith and Hake, 2003). This suggests that BLR is a general *AG* repressor, while LUG and SEU are floral-specific regulators (Bao *et al.*, 2004). BLR was also demonstrated to interact with the KNOX gene coding for the BREVIPEDICELLUS (BP) protein and is hypothesized to be part of a complex that regulates early patterning in the inflorescence meristem (Smith and Hake, 2003).

5. RBE *and* UFO *regulate the boundary of* AG *expression*

Plants defective in the *RABBIT EARS* (*RBE*) gene produce flowers with abnormal petals (Takeda *et al.*, 2003). It has been shown that in *rbe* mutant flowers, *AG* expression is ectopically expressed in the second whorl (Krizek *et al.*, 2006), providing an explanation for the abnormal petals in the *rbe* flower. Furthermore, the *rbe* mutation enhances the abnormalities in the second whorl of *ap2, lug*, and *clf* mutant flowers (Krizek *et al.*, 2006), which all exhibit *AG* misexpression. Therefore, *RBE* regulates the *AG* expression boundary by preventing *AG* from being expressed in the second whorl. *RBE* encodes a zinc finger protein similar to the SUP protein (Takeda *et al.*, 2003), which is important for controlling the boundary of B-function genes *AP3* and *PI* between the third and fourth whorls (Bowman *et al.*, 1992; Schultz *et al.*, 1991).

Several alleles of *ufo*, including *ufo-14*, have been reported to produce flowers that usually lack petals (Durfee *et al.*, 2003); in addition, the introduction of an *ag* mutation into these *ufo* mutants restored petal formation, indicating that ectopic *AG* function was responsible for the lack of petal development in these *ufo* mutants (Durfee *et al.*, 2003). It was found that *AG* is misexpressed in the few second-whorl organs produced in the *ufo-14* mutant flowers, supporting the idea that *UFO* normally plays a role in keeping *AG* expression in the third and fourth whorls (Krizek *et al.*, 2006). Furthermore, mutations in the *ASK1* and *ASK2* genes together cause the formation of carpel-like organs in the first whorl; in addition, this phenotype is suppressed by the *ag* mutation (Ni *et al.*, 2004). This suggests that *ASK1/ASK2* also negatively regulate *AG*, as supported by *in situ* hybridization experiments. It is possible that UFO- and ASK1/ASK2-containing SCF ubiquitin ligases are involved in restricting the *AG* expression domain. Alternatively, there might be other UFO- and ASK-dependent mechanisms for regulating *AG* expression and function (Ni *et al.*, 2004).

6. SAP *has both positive and negative functions*

Another gene that affects AG function is *SAP*, a putative novel transcriptional regulator (Byzova *et al.*, 1999). SAP both positively regulates *AG* in the early floral meristem (Byzova *et al.*, 1999) and negatively regulates *AG* subsequently by interacting with *AP2*. *sap* mutants have short inflorescence internodes, carpelloid sepals, absent or short and narrow petals, a reduced number of stamens with sterile anthers, abnormally long stigmatic papillae, and aberrant callose deposits on the ovules (Byzova *et al.*, 1999). Additionally, ovule development is arrested during megasporogenesis after the first meiotic division in *sap* mutants (Byzova *et al.*, 1999). Strong *sap ap2* double mutants have stronger phenotypes than either single-mutant phenotype (Byzova *et al.*, 1999), most likely due to an increase in ectopic *AG* expression. The double-mutant phenotypes include a complete transformation of carpels into sepals and the formation of a gynoecium-like structure that contains four carpels in the interior whorls (Byzova *et al.*, 1999). This enhanced phenotype in the double mutant suggests that AP2 and SAP interact genetically to regulate *AG* expression (Byzova *et al.*, 1999).

Similar to *ag* flowers, *sap ag* double-mutant flowers lack determinacy, have sepaloid first-whorl organs, and the overall organ numbers in the second and third whorls of the double mutant are similar to those of the *ag* mutant flower (Byzova *et al.*, 1999). However, *sap ag* flowers have shorter and narrower petals than those in *ag* mutants (Byzova *et al.*, 1999). In late-appearing *sap ag* flowers, a reiterative set of axillary flowers with elongated nodes form in the axils of the second-whorl organs, suggesting that in these plants there has been a partial reversion from floral to inflorescence identity (Byzova *et al.*, 1999). The phenotypes of *sap ag* double mutants, along with expanded *AG* expression in *sap* mutants suggest that SAP represses *AG* in the outer two floral whorls (Byzova *et al.*, 1999).

7. *AINTEGUMENTA*

ANT is a member of the *AP2* gene family that is expressed in both floral and vegetative tissues (Elliott *et al.*, 1996). *ant* mutants often have a reduced number of floral organs that may exhibit developmental defects such as serrated or fused sepals, narrow and short, and sometimes threadlike petals, reduced stamen numbers with thin anthers and filaments that may be petal/stamen mosaics and an unfused, narrow gynoecium (Elliott *et al.*, 1996). Mutants in *ant* also have ovules that fail to develop properly and lack integuments and have deformed nucellus development (Elliott *et al.*, 1996). ANT acts redundantly with AP2 as a negative regulator of *AG* in the development of the second-whorl petals as evidenced by the enhancement of weak *ap2* mutant phenotypes is *ant ap2* double mutants (Krizek *et al.*, 2000). However, the boundaries of *AG* expression do not expand in *ant* mutants as they do in mutants defective in one of several other genes that directly repress *AG* expression (Krizek *et al.*, 2000), suggesting that ANT has a different mode of regulation.

8. C-Function regulators that do not directly regulate AG

Other genes have been hypothesized to also negatively regulate *AG* at specific stages of floral development. Studies have uncovered potential regulators of WUS, which act upstream of *AG*. One characterized case involves the regulation of the *AG*/*WUS* pathway by a histone acetyltransferase, *AtGCN5*. When *AtGCN5* is mutated, both *WUS* and *AG* are ectopically expressed or overexpressed (Bertrand *et al.*, 2003), suggesting that AtGCN5 is a negative regulator of *WUS* and *AG*. Additionally, the *SHOOT MERISTEMLESS* (*STM*) gene acts upstream of *WUS* and *ZWILLE* (*ZLL*) in maintaining the undifferentiated state in the central meristem (Endrizzi *et al.*, 1996) and may be involved in this same pathway.

C. REGULATORS OF *AG* IN THE OVULE

AG also plays a role in ovule development, but it is not a critical gene (Favaro *et al.*, 2003; Pinyopich *et al.*, 2003). Besides roles in stamen and carpel development, regulators of *AG*, such as the positive regulator SEP3 (Castillejo *et al.*, 2005; Favaro *et al.*, 2003) and the negative regulators ANT (Klucher *et al.*, 1996), BEL1 (Modrusan *et al.*, 1994; Ray *et al.*, 1994; Western *et al.*, 2002), SAP (Byzova *et al.*, 1999), and possibly LUG (Liu *et al.*, 2000), may also function in ovule development. Genes that affect *AG* expression in the ovules include *BEL1, ANT*, and *LUG*. BEL1, a homeodomain DNA-binding transcription factor (Reiser *et al.*, 1995), appears to partially redundantly regulate ovule and embryo sac development (Western *et al.*, 2002) and may repress *AG* during ovule development in the integuments. This is supported by the fact that *bel1* mutants result in ectopic *AG* expression in late embryo formation (Modrusan *et al.*, 1994; Ray *et al.*, 1994; Western *et al.*, 2002). However, some studies have challenged the hypothesis that BEL1 is a regulator of *AG* based on the fact that their expression overlaps in the ovule in wild-type plants (Reiser *et al.*, 1995), suggesting that BEL1 has function in ovule development independent of *AG* regulation (Western *et al.*, 2002). Other positive and negative regulators of *AG* have expression in and/or defects in the development of the ovules as well. For example, *lug ant* double mutants lack ovules besides exhibiting severe defects in all floral organs, especially the carpel, style, and stigma (Liu *et al.*, 2000).

It is important to note that the function of these genes may not be *AG* specific. This is suggested by the fact that *BLR* (as *RPL*) was shown to prevent ectopic expression of the *AG* paralogs, *SHP1*, and *SHP2* in the fruit (Roeder *et al.*, 2003). Extending from this observation, we can postulate that genes identified as regulating *AG* in ovule development may also regulate the closely related genes *SHP1, SHP2*, and *STK*, all of which are members of the *AG* subfamily of MADS-box genes (Kramer *et al.*, 2004; Zahn *et al.*, 2006). For example, *lug* mutants have a split stigma, abnormal carpel and ovule development, reduced male and female fertility, and narrow leaves and floral organs (Conner and Liu, 2000). These phenotypes are consistent with the idea that *LUG* also negatively regulates *SHP*. Whether or not these genes affect *SHP1/2* and/or *STK* in ovule development remains to be determined. Regardless of the role in regulating *SHP* and *STK*, the ovule function of the genes that regulate *AG* expression, as indicated by mutant phenotypes, suggest that these genes may have evolved before the origin of angiosperms, or the carpel. If this is the case, *AG* may have started as an ovule development gene before it evolved its function in carpel development (Krizek *et al.*, 2000; Zahn *et al.*, 2006).

V. SUMMARY AND PERSPECTIVES

Forward genetic approaches have uncovered a large number of genes that regulate the ABC-genes. The ABC-genes all encode transcription factors, as do many of the ABC regulatory genes, further enhancing the view that transcriptional regulation is critical for developmental control. Among these, LFY is a central regulator that activates, along with others, genes in each of the ABC-functions. In addition, other ABC regulatory genes act at posttranscriptional levels, including mRNA processing, miRNA biogenesis, translational control, protein modification, and protein degradation. Therefore, multiple regulatory mechanisms are important for normal flower development. Among these, there are a greater number of genes that have been identified that affect the expression of *AG* or *AP3/PI* than those for *AP1* or *AP2*. This difference in number of genes may reflect the complexity of reproductive organs in comparison with the perianth organs. Another interesting aspect may be that A-function seems to be the least conserved function of the three and that the sepal may not represent an evolutionary conserved organ identity (Litt and Irish, 2003).

Furthermore, many of the genes that regulate B- and C-functions also play roles in other floral and nonfloral processes, indicating that flower development shares regulatory mechanisms with other developmental programs. It is also of interest to note that the pathways and mechanisms that regulate A-, B-, and C-functions are often different, suggesting that the control of organ identity has differentiated upstream of the organ identity genes. Given that B- and C-functions controlling stamen and carpel identities are highly conserved within the angiosperms and, even to possibly some extent, in gymnosperms (Becker and Theissen, 2003; Kramer *et al.*, 2004; Zahn *et al.*, 2005a,b, 2006), one interesting question for future study is whether the various programs of regulating B- and C-genes are conserved among angiosperms and gymnosperms.

An examination of those genes that have been identified to date as important regulators of A-, B-, and C-functions shows that several genes play multiple roles as both enhancers and repressors of floral organ development. The ABC-genes, along with their regulators, also act in feedback loops of positive and negative regulations that pattern the floral meristem. This is demonstrated by the antagonistic effects between A- and C-functions. Other regulatory loops include, most notably, the interaction between *AP1* and *TFL1*, between *AP3*/*PI* and *SUP*, and the positive regulation of *AG* by *WUS*, which in turn is negatively regulated by AG. It is of great interest to identify the regulatory networks and pathways that involve genes that have been previously described, although perhaps not as affecting floral development.

The majority of known ABC regulatory genes have been identified using forward genetic techniques, including enhancer screens with existing mutants. Although the ABC-gene functions have been well characterized, the function of the genes that regulate the ABC-genes are less well understood, and it is very likely that additional ones remain to be discovered. Further characterization of those genes that have already been identified as ABC regulatory genes will undoubtedly contribute to the understanding of flower development. At the same time, expression profiling and reverse genetics should play an increasing role in the identification and characterization of floral genes. Studies have incorporated microarray hybridization techniques to identify gene expression in a multitude of tissues, including early developing flowers (Hennig *et al.*, 2004; Schmid *et al.*, 2003, Wellmer *et al.*, 2004; Werner *et al.*, 2005; Zhang *et al.*, 2005). The functions of most of the florally expressed genes are yet to be determined.

Reverse genetic studies have been successful in demonstrating the function of redundant genes, such as *AP1*/*CAL*/*FUL*, *SHP1*/*2*, and *SEP1*/*2*/*3*/*4* (Ditta *et al.*, 2004; Ferrandiz *et al.*, 2000; Liljegren *et al.*, 2000; Pelaz *et al.*, 2000; Pinyopich *et al.*, 2003), and will be necessary for many plants, including *Arabidopsis*, due to genetic redundancy (Ditta *et al.*, 2004; Liljegren *et al.*, 2000; Moore *et al.*, 2005; Pelaz *et al.*, 2000; Pinyopich *et al.*, 2003; Zahn *et al.*, 2005a,b, 2006). This is especially important in plants like *Arabidopsis* which underwent a genome-wide duplication event ~30 MYA (Blanc and Wolfe, 2004; Blanc *et al.*, 2003; Moore *et al.*, 2005; Simillion *et al.*, 2002; Vision *et al.*, 2000) but still retains redundant function among genes with important function in floral development (Ditta *et al.*, 2004; Liljegren *et al.*, 2000; Moore *et al.*, 2005; Pelaz *et al.*, 2000; Pinyopich *et al.*, 2003). In addition, comparative genomic approaches are effective means to uncovering genes that are conserved among angiosperm floral transcriptomes (Albert *et al.*, 2005; Soltis *et al.*, 2002; see Leebens-Mack *et al.*, Chapter 14). The combination of genetic, genomic, and evolutionary approaches will reveal ever greater extent of the secrets behind the complexity and diversity of the flower.

ACKNOWLEDGMENTS

We thank D. Soltis for helpful comments on this chapter. Our research on plant reproductive development has been supported by grants from the National Science Foundation (IBN-0077832, MCB-9896340, MCB-0092075, DBI-0115684), the National Institutes of Health (R01 GM63871), the US Department of Agriculture (2001-35301-10570, 2003-35301-13313), and the Department of Energy (DE-FG02-02ER15332), and by funds from The Huck Institutes of the Life Sciences and the Biology Department, The Pennsylvania State University. H. M. gratefully acknowledges the generous support from the John Simon Guggenheim Memorial Foundation and the National Institutes of Health (F33 GM72245-1).

REFERENCES

Ach, R. A., Taranto, P. and Gruissem, W. (1997). A conserved family of WD-40 proteins binds to the retinoblastoma protein in both plants and animals. *Plant Cell* **9**, 1595–1606.

Albert, V. A., Soltis, D. E., Carlson, J. E., Farmerie, W. G., Wall, P. K., Ilut, D. C., Mueller, L. A., Landherr, L. L., Hu, Y., Buzgo, M., Kim, S., Yoo, M.-J., *et al.* (2005). Floral gene resources from basal angiosperms for comparative genomics research. *BMC Plant Biology*. http://www.biomedcentral.com/1471-2229/5/5.

Angenent, G. C., Franken, J., Busscher, M., van Dijken, A., van Went, J. L., Dons, H. J. and van Tunen, A. J. (1995). A novel class of MADS box genes is involved in ovule development in petunia. *Plant Cell* **7**, 1569–1582.

Aoki, S., Uehara, K., Imafuku, M., Hasebe, M. and Ito, M. (2004). Phylogeny and divergence of basal angiosperms inferred from *APETALA3-* and *PISTILLATA*-like MADS-box genes. *Journal of Plant Research* **117**, 229–244.

Aubert, D., Chen, L., Moon, Y. H., Martin, D., Castle, L. A., Yang, C. H. and Sung, Z. R. (2001). EMF1, a novel protein involved in the control of shoot architecture and flowering in *Arabidopsis*. *Plant Cell* **13**, 1865–1875.

Bao, X., Franks, R. G., Levin, J. Z. and Liu, Z. (2004). Repression of *AGAMOUS* by BELLRINGER in floral and inflorescence meristems. *Plant Cell* **16**, 1478–1489.

Becker, A. and Theissen, G. (2003). The major clades of MADS-box genes and their role in the development and evolution of flowering plants. *Molecular Phylogenetics Evolution* **29**, 464–489.

Bertrand, C., Bergounioux, C., Domenichini, S., Delarue, M. and Zhou, D. X. (2003). *Arabidopsis* histone acetyltransferase AtGCN5 regulates the floral meristem activity through the *WUSCHEL/AGAMOUS* pathway. *Journal of Biological Chemistry* **278**, 28246–28251.

Blanc, G. and Wolfe, K. H. (2004). Widespread paleopolyploidy in model plant species inferred from age distributions of duplicate genes. *Plant Cell* **16**, 1667–1678.

Blanc, G., Hokamp, K. and Wolfe, K. H. (2003). A recent polyploidy superimposed on older large-scale duplications in the *Arabidopsis* genome. *Genome Research* **13**, 137–144.

Bomblies, K., Dagenais, N. and Weigel, D. (1999). Redundant enhancers mediate transcriptional repression of *AGAMOUS* by APETALA2. *Developmental Biology* **216**, 260–264.

Bowman, J. L., Smyth, D. R. and Meyerowitz, E. M. (1989). Genes directing flower development in *Arabidopsis*. *Plant Cell* **1**, 37–52.

Bowman, J. L., Drews, G. N. and Meyerowitz, E. M. (1991a). Expression of the *Arabidopsis* floral homeotic gene *AGAMOUS* is restricted to specific cell types late in flower development. *Plant Cell* **3**, 749–758.

Bowman, J. L., Smyth, D. R. and Meyerowitz, E. M. (1991b). Genetic interactions among floral homeotic genes of *Arabidopsis*. *Development* **112**, 1–20.

Bowman, J. L., Sakai, H., Jack, T., Weigel, D., Mayer, U. and Meyerowitz, E. M. (1992). *SUPERMAN*, a regulator of floral homeotic genes in *Arabidopsis*. *Development* **114**, 599–615.

Bowman, J. L., Alvarez, J., Weigel, D., Meyerowitz, E. M. and Smyth, D. R. (1993). Control of flower development in *Arabidopsis thaliana* by *APETALA1* and interacting genes. *Development* **119**, 721–743.

Bradley, D., Ratcliffe, O., Vincent, C., Carpenter, R. and Coen, E. (1997). Inflorescence commitment and architecture in *Arabidopsis*. *Science* **275**, 80–83.

Busch, M. A., Bomblies, K. and Weigel, D. (1999). Activation of a floral homeotic gene in *Arabidopsis*. *Science* **285**, 585–587.

Buzgo, M., Soltis, D. E., Soltis, P. S. and Ma, H. (2004). Towards a comprehensive integration of morphological and genetic studies of floral development. *Trends in Plant Science* **9**, 164–173.

Byrne, M. E., Groover, A. T., Fontana, J. R. and Martienssen, R. A. (2003). Phyllotactic pattern and stem cell fate are determined by the *Arabidopsis* homeobox gene *BELLRINGER*. *Development* **130**, 3941–3950.

Byzova, M. V., Franken, J., Aarts, M. G., de Almeida-Engler, J., Engler, G., Mariani, C., Van Lookeren Campagne, M. M. and Angenent, G. C. (1999). *Arabidopsis STERILE APETALA*, a multifunctional gene-regulating inflorescence, flower, and ovule development. *Genes & Development* **13**, 1002–1014.

Castillejo, C., Romera-Branchat, M. and Pelaz, S. (2005). A new role of the *Arabidopsis SEPALLATA3* gene revealed by its constitutive expression. *Plant Journal* **43**, 586–596.

Chen, L., Cheng, J. C., Castle, L. and Sung, Z. R. (1997). *EMF* genes regulate *Arabidopsis* inflorescence development. *Plant Cell* **9**, 2011–2024.

Chen, X. (2004). A microRNA as a translational repressor of *APETALA2* in *Arabidopsis* flower development. *Science* **303**, 2022–2025.

Chen, X. and Meyerowitz, E. M. (1999). *HUA1* and *HUA2* are two members of the floral homeotic *AGAMOUS* pathway. *Molecular Cell* **3**, 349–360.

Chen, X., Liu, J., Cheng, Y. and Jia, D. (2002). *HEN1* functions pleiotropically in *Arabidopsis* development and acts in C function in the flower. *Development* **129**, 1085–1094.

Cheng, H., Qin, L., Lee, S., Fu, X., Richards, D. E., Cao, D., Luo, D., Harberd, N. P. and Peng, J. (2004). Gibberellin regulates *Arabidopsis* floral development via suppression of DELLA protein function. *Development* **131**, 1055–1064.

Cheng, Y., Kato, N., Wang, W., Li, J. and Chen, X. (2003). Two RNA-binding proteins, HEN4 and HUA1, act in the processing of *AGAMOUS* pre-mRNA in *Arabidopsis thaliana. Developmental Biology* **4**, 53–66.

Coen, E. S. and Meyerowitz, E. M. (1991). The war of the whorls: Genetic interactions controlling flower development. *Nature* **353**, 31–37.

Colombo, L., Franken, J., Koetje, E., van Went, J., Dons, H. J., Angenent, G. C. and van Tunen, A. J. (1995). The petunia MADS box gene *FBP11* determines ovule identity. *Plant Cell* **7**, 1859–1868.

Conner, J. and Liu, Z. (2000). LEUNIG, a putative transcriptional corepressor that regulates *AGAMOUS* expression during flower development. *Proceedings of the National Academy of Sciences of the United States of America* **97**, 12902–12907.

Coupland, G., Dash, S., Goodrich, J., Lee, K., Long, D., Martin, M., Puangsomlee, P., Putterill, J., Robson, F. and Wilson, K. (1993). Molecular and genetic analysis of the control of flowering time in response to daylength in *Arabidopsis thaliana. Flower Newsletter* **16**, 27–32.

Craig, K. L. and Tyers, M. (1999). The F-box: A new motif for ubiquitin dependent proteolysis in cell cycle regulation and signal transduction. *Progress in Biophysics and Molecular Biology* **72**, 299–328.

Crane, P. R., Friis, E. M. and Pedersen, K. R. (1995). The origin and early diversification of angiosperms. *Nature* **374**, 27–33.

Dathan, N., Zaccaro, L., Esposito, S., Isernia, C., Omichinski, J. G., Riccio, A., Pedone, C., Di Blasio, B., Fattorusso, R. and Pedone, P. V. (2002). The *Arabidopsis* SUPERMAN protein is able to specifically bind DNA through its single Cys2-His2 zinc finger motif. *Nucleic Acids Research* **30**, 4945–4951.

Davies, B., Egea-Cortines, M., de Andrade Silva, E., Saedler, H. and Sommer, H. (1996). Multiple interactions amongst floral homeotic MADS box proteins. *EMBO Journal* **15**, 4330–4343.

Davies, B., Motte, P., Keck, E., Saedler, H., Sommer, H. and Schwarz-Sommer, Z. (1999). PLENA and FARINELLI: Redundancy and regulatory interactions between two *Antirrhinum* MADS-box factors controlling flower development. *EMBO Journal* **18**, 4023–4034.

De Bodt, S., Maere, S. and Van de Peer, Y. (2005). Genome duplication and the origin of angiosperms. *Trends in Ecology and Evolution* **20**, 591–597.

Deyholos, M. K. and Sieburth, L. E. (2000). Separable whorl-specific expression and negative regulation by enhancer elements within the *AGAMOUS* second intron. *Plant Cell* **12**, 1799–1810.

Ditta, G., Pinyopich, A., Robles, P., Pelaz, S. and Yanofsky, M. F. (2004). The *SEP4* gene of *Arabidopsis thaliana* functions in floral organ and meristem identity. *Current Biology* **14**, 1935–1940.

Doyle, M. R., Bizzell, C. M., Keller, M. R., Michaels, S. D., Song, J., Noh, Y. S. and Amasino, R. M. (2005). HUA2 is required for the expression of floral repressors in *Arabidopsis thaliana. Plant Journal* **41**, 376–385.

Drews, G. N., Bowman, J. L. and Meyerowitz, E. M. (1991). Negative regulation of the *Arabidopsis* homeotic gene *AGAMOUS* by the *APETALA2* product. *Cell* **65**, 991–1002.

Durfee, T., Roe, J. L., Sessions, R. A., Inouye, C., Serikawa, K., Feldmann, K. A., Weigel, D. and Zambryski, P. C. (2003). The F-box-containing protein UFO and AGAMOUS participate in antagonistic pathways governing early petal development in *Arabidopsis*. *Proceedings of the National Academy of Sciences of the United States of America* **100**, 8571–8576.

Elliott, R. C., Betzner, A. S., Huttner, E., Oakes, M. P., Tucker, W. Q., Gerentes, D., Perez, P. and Smyth, D. R. (1996). *AINTEGUMENTA*, an *APETALA2*-like gene of *Arabidopsis* with pleiotropic roles in ovule development and floral organ growth. *Plant Cell* **8**, 155–168.

Endrizzi, K., Moussian, B., Haecker, A., Levin, J. Z. and Laux, T. (1996). The *SHOOT MERISTEMLESS* gene is required for maintenance of undifferentiated cells in *Arabidopsis* shoot and floral meristems and acts at a different regulatory level than the meristem genes *WUSCHEL* and *ZWILLE*. *Plant Journal* **10**, 967–979.

Fan, H. Y., Hu, Y., Tudor, M. and Ma, H. (1997). Specific interactions between the K domains of AG and AGLs, members of the MADS domain family of DNA-binding proteins. *Plant Journal* **12**, 999–1010.

Favaro, R., Pinyopich, A., Battaglia, R., Kooiker, M., Borghi, L., Ditta, G., Yanofsky, M. F., Kater, M. M. and Colombo, L. (2003). MADS-box protein complexes control carpel and ovule development in *Arabidopsis*. *Plant Cell* **15**, 2603–2611.

Ferrandiz, C., Gu, Q., Martienssen, R. and Yanofsky, M. F. (2000). Redundant regulation of meristem identity and plant architecture by *FRUITFULL, APETALA1* and *CAULIFLOWER*. *Development* **127**, 725–734.

Flanagan, C. A. and Ma, H. (1994). Spatially and temporally regulated expression of the MADS-box gene *AGL2* in wild-type and mutant *Arabidopsis* flowers. *Plant Molecular Biology* **26**, 581–595.

Franks, R. G., Wang, C., Levin, J. Z. and Liu, Z. (2002). SEUSS, a member of a novel family of plant regulatory proteins, represses floral homeotic gene expression with LEUNIG. *Development* **129**, 253–263.

Frohlich, M. W. and Meyerowitz, E. M. (1997). The search for flower homeotic gene homologs in basal angiosperms and gnetales: A potential new source of data on the evolutionary origin of flowers. *International Journal of Plant Sciences* **158**, S131–S142.

Gagne, J. M., Downes, B. P., Shiu, S. H., Durski, A. M. and Vierstra, R. D. (2002). The F-box subunit of the SCF E3 complex is encoded by a diverse superfamily of genes in *Arabidopsis*. *Proceedings of the National Academy of Sciences of the United States of America* **99**, 11519–11524.

Gallois, J. L., Nora, F. R., Mizukami, Y. and Sablowski, R. (2004). *WUSCHEL* induces shoot stem cell activity and developmental plasticity in the root meristem. *Genes & Development* **18**, 375–380.

Gomez-Mena, C., Pineiro, M., Franco-Zorrilla, J. M., Salinas, J., Coupland, G. and Martinez-Zapater, J. M. (2001). *Early bolting in short days*: An *Arabidopsis* mutation that causes early flowering and partially suppresses the floral phenotype of *leafy*. *Plant Cell* **13**, 1011–1024.

Goodrich, J., Puangsomlee, P., Martin, M., Long, D., Meyerowitz, E. M. and Coupland, G. (1997). A Polycomb-group gene regulates homeotic gene expression in *Arabidopsis*. *Nature* **386**, 44–51.

Goto, K. and Meyerowitz, E. M. (1994). Function and regulation of the *Arabidopsis* floral homeotic gene *PISTILLATA*. *Genes & Development* **8**, 1548–1560.

Goto, K., Kyozuka, J. and Bowman, J. L. (2001). Turning floral organs into leaves, leaves into floral organs. *Current Opinion in Genetics & Development* **11**, 449–456.

Gross-Hardt, R., Lenhard, M. and Laux, T. (2002). *WUSCHEL* signaling functions in interregional communication during *Arabidopsis* ovule development. *Genes & Development* **16**, 1129–1138.

Gu, Q., Ferrandiz, C., Yanofsky, M. F. and Martienssen, R. (1998). The *FRUITFULL* MADS-box gene mediates cell differentiation during *Arabidopsis* fruit development. *Development* **125**, 1509–1517.

Gustafson-Brown, C., Savidge, B. and Yanofsky, M. F. (1994). Regulation of the *Arabidopsis* floral homeotic gene *APETALA1*. *Cell* **76**, 131–143.

Haughn, G. W. and Somerville, C. R. (1988). Genetic control of morphogenesis in *Arabidopsis*. *Developmental Genetics* **9**, 73–89.

Hennig, L., Taranto, P., Walser, M., Schonrock, N. and Gruissem, W. (2003). *Arabidopsis* MSI1 is required for epigenetic maintenance of reproductive development. *Development* **130**, 2555–2565.

Hennig, L., Gruissem, W., Grossniklaus, U. and Kohler, C. (2004). Transcriptional programs of early reproductive stages in *Arabidopsis*. *Plant Physiology* **135**, 1765–1775.

Herschbach, B. M., Arnaud, M. B. and Johnson, A. D. (1994). Transcriptional repression directed by the yeast alpha 2 protein *in vitro*. *Nature* **370**, 309–311.

Honma, T. and Goto, K. (2000). The *Arabidopsis* floral homeotic gene *PISTILLATA* is regulated by discrete *cis*-elements responsive to induction and maintenance signals. *Development* **127**, 2021–2030.

Honma, T. and Goto, K. (2001). Complexes of MADS-box proteins are sufficient to convert leaves into floral organs. *Nature* **409**, 525–529.

Huala, E. and Sussex, I. M. (1992). *LEAFY* interacts with floral homeotic genes to regulate *Arabidopsis* floral development. *Plant Cell* **4**, 901–913.

Huijser, P., Klein, J., Lonnig, W. E., Meijer, H., Saedler, H. and Sommer, H. (1992). Bracteomania, an inflorescence anomaly, is caused by the loss of function of the MADS-box gene *squamosa* in *Antirrhinum majus*. *EMBO Journal* **11**, 1239–1249.

Immink, R. G. H., Ferrario, S., Busscher-Lange, J., Kooiker, M., Busscher, M. and Angenent, G. C. (2003). Analysis of the petunia MADS-box transcription factor family. *Molecular Genetics and Genomics* **268**, 598–606.

Irish, V. F. and Sussex, I. M. (1990). Function of the *apetala-1* gene during *Arabidopsis* floral development. *Plant Cell* **2**, 741–753.

Ito, T., Sakai, H. and Meyerowitz, E. M. (2003). Whorl-specific expression of the *SUPERMAN* gene of *Arabidopsis* is mediated by *cis* elements in the transcribed region. *Current Biology* **13**, 1524–1530.

Ito, T., Wellmer, F., Yu, H., Das, P., Ito, N., Alves-Ferreira, M., Riechmann, J. L. and Meyerowitz, E. M. (2004). The homeotic protein AGAMOUS controls microsporogenesis by regulation of *SPOROCYTELESS*. *Nature* **430**, 356–360.

Jack, T., Brockman, L. L. and Meyerowitz, E. M. (1992). The homeotic gene *APETALA3* of *Arabidopsis thaliana* encodes a MADS box and is expressed in petals and stamens. *Cell* **68**, 683–697.

Jack, T., Fox, G. L. and Meyerowitz, E. M. (1994). *Arabidopsis* homeotic gene *APETALA3* ectopic expression: Transcriptional and posttranscriptional regulation determine floral organ identity. *Cell* **76**, 703–716.

Jacobsen, S. E. and Meyerowitz, E. M. (1997). Hypermethylated *SUPERMAN* epigenetic alleles in *Arabidopsis*. *Science* **277**, 1100–1103.

Jofuku, K. D., den Boer, B. G., Van Montagu, M. and Okamuro, J. K. (1994). Control of *Arabidopsis* flower and seed development by the homeotic gene *APETALA2*. *Plant Cell* **6**, 1211–1225.

Kater, M. M., Franken, J., van Aelst, A. and Angenent, G. C. (2000). Suppression of cell expansion by ectopic expression of the *Arabidopsis SUPERMAN* gene in transgenic petunia and tobacco. *Plant Journal* **23**, 407–413.

Kaufmann, K., Melzer, R. and Theissen, G. (2005). MIKC-type MADS-domain proteins: Structural modularity, protein interactions and network evolution in land plants. *Gene* **347**, 183–198.

Keck, E., McSteen, P., Carpenter, R. and Coen, E. (2003). Separation of genetic functions controlling organ identity in flowers. *EMBO Journal* **22**, 1058–1066.

Keller, E. T., Fu, Z. and Brennan, M. (2005). The biology of a prostate cancer metastasis suppressor protein: Raf kinase inhibitor protein. *Journal of Cellular Biochemistry* **94**, 273–278.

Kempin, S. A., Savidge, B. and Yanofsky, M. F. (1995). Molecular basis of the cauliflower phenotype in *Arabidopsis*. *Science* **267**, 522–525.

Kenzior, A. L. and Folk, W. R. (1998). AtMSI4 and RbAp48 WD-40 repeat proteins bind metal ions. *FEBS Letters* **440**, 425–429.

Kim, G. T., Tsukaya, H. and Uchimiya, H. (1998). The *CURLY LEAF* gene controls both division and elongation of cells during the expansion of the leaf blade in *Arabidopsis thaliana*. *Planta* **206**, 175–183.

Kim, S., Yoo, M.-J., Albert, V. A., Farris, J. S., Soltis, P. S. and Soltis, D. E. (2004). Phylogeny and diversification of B-function MADS-box genes in angiosperms: Evolutionary and functional implications of a 260-million-year-old duplication. *American Journal of Botany* **91**, 2102–2118.

Kim, S., Koh, J., Ma, H., Hu, Y., Endress, P. K., Hauser, B. A., Buzgo, M., Soltis, P. S. and Soltis, D. E. (2005). Sequence and expression studies of A-, B-, and E-class MADS-box genes in *Eupomatia* (Eupomatiaceae): Support for the bracteate origin of the calyptra. *International Journal of Plant Science* **166**, 185–198.

Klucher, K. M., Chow, H., Reiser, L. and Fischer, R. L. (1996). The *AINTEGUMENTA* gene of *Arabidopsis* required for ovule and female gametophyte development is related to the floral homeotic gene *APETALA2*. *Plant Cell* **8**, 137–153.

Kong, H.-Z., Leebens-Mack, J., Ni, W., dePamphilis, C. W. and Ma, H. (2004). Highly heterogeneous rates of evolution in the *SKP1* gene family in animals and plants: Functional and evolutionary implications. *Molecular Biology and Evolution* **21**, 117–128.

Kramer, E. M., Jaramillo, M. A. and Di Stilio, V. S. (2004). Patterns of gene duplication and functional evolution during the diversification of the *AGAMOUS* subfamily of MADS box genes in angiosperms. *Genetics* **166**, 1011–1023.

Krizek, B. A., Prost, V. and Macias, A. (2000). AINTEGUMENTA promotes petal identity and acts as a negative regulator of *AGAMOUS*. *Plant Cell* **12**, 1357–1366.

Krizek, B. A., Lewis, M. W. and Fletcher, J. C. (2006). *RABBIT EARS* is a second-whorl repressor of *AGAMOUS* that maintains spatial boundaries in *Arabidopsis flowers*. *Plant Journal* **45**, 369–383.

Kunst, L., Klenz, J. E., Martinezzapater, J. and Haughn, G. W. (1989). *AP2* Gene determines the identity of perianth organs in flowers of *Arabidopsis thaliana*. *Plant Cell* **1**, 1195–1208.

Lamb, R. S., Hill, T. A., Tan, Q. K. and Irish, V. F. (2002). Regulation of *APETALA3* floral homeotic gene expression by meristem identity genes. *Development* **129**, 2079–2086.

Laufs, P., Coen, E., Kronenberger, J., Traas, J. and Doonan, J. (2003). Separable roles of *UFO* during floral development revealed by conditional restoration of gene function. *Development* **130**, 785–796.

Laux, T., Mayer, K. F., Berger, J. and Jurgens, G. (1996). The *WUSCHEL* gene is required for shoot and floral meristem integrity in *Arabidopsis*. *Development* **122**, 87–96.

Lechner, E., Xie, D., Grava, S., Pigaglio, E., Planchais, S., Murray, J. A., Parmentier, Y., Mutterer, J., Dubreucq, B., Shen, W. H. and Genschik, P. (2002). The AtRbx1 protein is part of plant SCF complexes, and its down-regulation causes severe growth and developmental defects. *Journal of Biological Chemistry* **277**, 50069–50080.

Lee, I., Wolfe, D. S., Nilsson, O. and Weigel, D. (1997). A LEAFY co-regulator encoded by *UNUSUAL FLORAL ORGANS*. *Current Biology* **7**, 95–104.

Lenhard, M., Bohnert, A., Jurgens, G. and Laux, T. (2001). Termination of stem cell maintenance in *Arabidopsis* floral meristems by interactions between *WUSCHEL* and *AGAMOUS*. *Cell* **105**, 805–814.

Levin, J. Z. and Meyerowitz, E. M. (1995). *UFO*: An Arabidopsis gene involved in both floral meristem and floral organ development. *Plant Cell* **7**, 529–548.

Li, J. and Chen, X. (2003). PAUSED, a putative exportin-t, acts pleiotropically in *Arabidopsis* development but is dispensable for viability. *Plant Physiology* **132**, 1913–1924.

Li, J., Jia, D. and Chen, X. (2001). HUA1, a regulator of stamen and carpel identities in *Arabidopsis*, codes for a nuclear RNA-binding protein. *Plant Cell* **13**, 2269–2281.

Liljegren, S. J., Gustafson-Brown, C., Pinyopich, A., Ditta, G. S. and Yanofsky, M. F. (1999). Interactions among *APETALA1, LEAFY,* and *TERMINAL FLOWER1* specify meristem fate. *Plant Cell* **11**, 1007–1018.

Liljegren, S. J., Ditta, G. S., Eshed, Y., Savidge, B., Bowman, J. L. and Yanofsky, M. F. (2000). *SHATTERPROOF* MADS-box genes control seed dispersal in *Arabidopsis*. *Nature* **404**, 766–770.

Litt, A. and Irish, V. F. (2003). Duplication and diversification in the *APETALA1/FRUITFULL* floral homeotic gene lineage: Implications for the evolution of floral development. *Genetics* **165**, 821–833.

Liu, Z. and Meyerowitz, E. M. (1995). LEUNIG regulates *AGAMOUS* expression in *Arabidopsis* flowers. *Development* **121**, 975–991.

Liu, Z., Franks, R. G. and Klink, V. P. (2000). Regulation of gynoecium marginal tissue formation by *LEUNIG* and *AINTEGUMENTA*. *Plant Cell* **12**, 1879–1892.

Lohmann, J. U., Hong, R. L., Hobe, M., Busch, M. A., Parcy, F., Simon, R. and Weigel, D. (2001). A molecular link between stem cell regulation and floral patterning in *Arabidopsis*. *Cell* **105**, 793–803.

Ma, H. (1994). The unfolding drama of flower development: Recent results from genetic and molecular analyses. *Genes & Development* **8**, 745–756.

Ma, H. (2005). Molecular genetic analyses of microsporogenesis and microgametogenesis in flowering plants. *Annual Review of Plant Biology* **56**, 393–434.

Ma, H. and dePamphilis, C. (2000). The ABCs of floral evolution. *Cell* **101**, 5–8.

Ma, H., Yanofsky, M. F. and Meyerowitz, E. M. (1991). *AGL1-AGL6*, an *Arabidopsis* gene family with similarity to floral homeotic and transcription factor genes. *Genes & Development* **5**, 484–495.

Malcomber, S. T. and Kellogg, E. A. (2005). *SEPALLATA* gene diversification: Brave new whorls. *Trends in Plant Science* **10**, 427–435.

Mandel, M. A. and Yanofsky, M. F. (1995). The *Arabidopsis AGL8* MADS box gene is expressed in inflorescence meristems and is negatively regulated by *APETALA1*. *Plant Cell* **7**, 1763–1771.

Mandel, M. A. and Yanofsky, M. F. (1998). The *Arabidopsis AGL9* MADS box gene is expressed in young flower primordia. *Sex Plant Reproduction* **11**, 22–28.

Mandel, M. A., Gustafson-Brown, C., Savidge, B. and Yanofsky, M. F. (1992). Molecular characterization of the *Arabidopsis* floral homeotic gene *APETALA1*. *Nature* **360**, 273–277.

Mayer, K. F., Schoof, H., Haecker, A., Lenhard, M., Jurgens, G. and Laux, T. (1998). Role of *WUSCHEL* in regulating stem cell fate in the *Arabidopsis* shoot meristem. *Cell* **95**, 805–815.

McGonigle, B., Bouhidel, K. and Irish, V. F. (1996). Nuclear localization of the *Arabidopsis APETALA3* and *PISTILLATA* homeotic gene products depends on their simultaneous expression. *Genes & Development* **10**, 1812–1821.

Mizukami, Y. and Ma, H. (1992). Ectopic expression of the floral homeotic gene *AGAMOUS* in transgenic *Arabidopsis* plants alters floral organ identity. *Cell* **71**, 119–131.

Modrusan, Z., Reiser, L., Feldmann, K. A., Fischer, R. L. and Haughn, G. W. (1994). Homeotic transformation of ovules into carpel-like structures in *Arabidopsis*. *Plant Cell* **6**, 333–349.

Moon, Y. H., Jung, J. Y., Kang, H. G. and An, G. (1999). Identification of a rice APETALA3 homologue by yeast two-hybrid screening. *Plant Molecular Biology* **40**, 167–177.

Moon, Y. H., Chen, L., Pan, R. L., Chang, H. S., Zhu, T., Maffeo, D. M. and Sung, Z. R. (2003). *EMF* genes maintain vegetative development by repressing the flower program in *Arabidopsis*. *Plant Cell* **15**, 681–693.

Moore, R. C., Grant, S. R. and Purugganan, M. D. (2005). Molecular population genetics of redundant floral-regulatory genes in *Arabidopsis thaliana*. *Molecular Biology Evolution* **22**, 91–103.

Nandi, A. K., Kushalappa, K., Prasad, K. and Vijayraghavan, U. (2000). A conserved function for *Arabidopsis* SUPERMAN in regulating floral-whorl cell proliferation in rice, a monocotyledonous plant. *Current Biology* **10**, 215–218.

Ng, M. and Yanofsky, M. F. (2001). Function and evolution of the plant MADS-box gene family. *Nature Reviews Genetics* **2**, 186–195.

Ni, W., Xie, D., Hobbie, L., Feng, B., Zhao, D., Akkara, J. and Ma, H. (2004). Regulation of flower development in Arabidopsis by SCF complexes. *Plant Physiology* **134**, 1574–1585.

Ohshima, S., Murata, M., Sakamoto, W., Ogura, Y. and Motoyoshi, F. (1997). Cloning and molecular analysis of the *Arabidopsis* gene *Terminal Flower 1*. *Molecular & General Genetics* **254**, 186–194.

Okamuro, J. K., den Boer, B. G., Lotys-Prass, C., Szeto, W. and Jofuku, K. D. (1996). Flowers into shoots: Photo and hormonal control of a meristem identity switch in *Arabidopsis*. *Proceedings of the National Academy of Sciences of the United States of America* **93**, 13831–13836.

Parcy, F., Nilsson, O., Busch, M. A., Lee, I. and Weigel, D. (1998). A genetic framework for floral patterning. *Nature* **395**, 561–566.

Park, W., Li, J., Song, R., Messing, J. and Chen, X. (2002). CARPEL FACTORY, a Dicer homolog, and HEN1, a novel protein, act in microRNA metabolism in *Arabidopsis thaliana*. *Current Biology* **12**, 1484–1495.

Pelaz, S., Ditta, G. S., Baumann, E., Wisman, E. and Yanofsky, M. F. (2000). B and C floral organ identity functions require *SEPALLATA* MADS-box genes. *Nature* **405**, 200–203.

Pelaz, S., Tapia-Lopez, R., Alvarez-Buylla, E. R. and Yanofsky, M. F. (2001). Conversion of leaves into petals in *Arabidopsis*. *Current Biology* **11**, 182–184.

Pineiro, M., Gomez-Mena, C., Schaffer, R., Martinez-Zapater, J. M. and Coupland, G. (2003). *EARLY BOLTING IN SHORT DAYS* is related to chromatin remodeling factors and regulates flowering in *Arabidopsis* by repressing. *FT Plant Cell* **15**, 1552–1562.

Pinyopich, A., Ditta, G. S., Savidge, B., Liljegren, S. J., Baumann, E., Wisman, E. and Yanofsky, M. F. (2003). Assessing the redundancy of MADS-box genes during carpel and ovule development. *Nature* **424**, 85–88.

Ratcliffe, O. J., Bradley, D. J. and Coen, E. S. (1999). Separation of shoot and floral identity in *Arabidopsis*. *Development* **126**, 1109–1120.

Ray, A., Robinson-Beers, K., Ray, S., Baker, S. C., Lang, J. D., Preuss, D., Milligan, S. B. and Gasser, C. S. (1994). *Arabidopsis* floral homeotic gene *BELL* (*BEL1*) controls ovule development through negative regulation of *AGAMOUS* gene (*AG*). *Proceedings of the National Academy of Sciences of the United States of America* **91**, 5761–5765.

Reiser, L., Modrusan, Z., Margossian, L., Samach, A., Ohad, N., Haughn, G. W. and Fischer, R. L. (1995). The *BELL1* gene encodes a homeodomain protein involved in pattern formation in the *Arabidopsis* ovule primordium. *Cell* **83**, 735–742.

Riechmann, J. L. and Meyerowitz, E. M. (1997). MADS domain proteins in plant development. *Journal of Biological Chemistry* **378**, 1079–1101.

Riechmann, J. L., Krizek, B. A. and Meyerowitz, E. M. (1996a). Dimerization specificity of *Arabidopsis* MADS domain homeotic proteins *APETALA1, APETALA3, PISTILLATA*, and *AGAMOUS*. *Proceedings of the National Academy of Sciences of the United States of America* **93**, 4793–4798.

Riechmann, J. L., Wang, M. and Meyerowitz, E. M. (1996b). DNA-binding properties of *Arabidopsis* MADS domain homeotic proteins *APETALA1, APETALA3, PISTILLATA* and *AGAMOUS*. *Nucleic Acids Research* **24**, 3134–3141.

Risseeuw, E. P., Daskalchuk, T. E., Banks, T. W., Liu, E., Cotelesage, J., Hellmann, H., Estelle, M., Somers, D. E. and Crosby, W. L. (2003). Protein interaction analysis of SCF ubiquitin E3 ligase subunits from *Arabidopsis*. *Plant Journal* **34**, 753–767.

Roeder, A. H., Ferrandiz, C. and Yanofsky, M. F. (2003). The role of the REPLUMLESS homeodomain protein in patterning the *Arabidopsis* fruit. *Current Biology* **13**, 1630–1635.

Sakai, H., Medrano, L. J. and Meyerowitz, E. M. (1995). Role of *SUPERMAN* in maintaining *Arabidopsis* floral whorl boundaries. *Nature* **378**, 199–203.

Samach, A., Kohalmi, S. E., Motte, P., Datla, R. and Haughn, G. W. (1997). Divergence of function and regulation of class B floral organ identity genes. *Plant Cell* **9**, 559–570.

Samach, A., Klenz, J. E., Kohalmi, S. E., Risseeuw, E., Haughn, G. W. and Crosby, W. L. (1999). The *UNUSUAL FLORAL ORGANS* gene of *Arabidopsis thaliana* is an F-box protein required for normal patterning and growth in the floral meristem. *Plant Journal* **20**, 433–445.

Savidge, B., Rounsley, S. D. and Yanofsky, M. F. (1995). Temporal relationship between the transcription of two *Arabidopsis* MADS box genes and the floral organ identity genes. *Plant Cell* **7**, 721–733.

Schmid, M., Uhlenhaut, N. H., Godard, F., Demar, M., Bressan, R., Weigel, D. and Lohmann, J. U. (2003). Dissection of floral induction pathways using global expression analysis. *Development* **130**, 6001–6012.

Schultz, E. A., Pickett, F. B. and Haughn, G. W. (1991). The *FLO10* gene product regulates the expression domain of homeotic genes *AP3* and *PI* in *Arabidopsis* flowers. *Plant Cell* **3**, 1221–1237.

Schwarz-Sommer, Z., Hue, I., Huijser, P., Flor, P. J., Hansen, R., Tetens, F., Lonnig, W. E., Saedler, H. and Sommer, H. (1992). Characterization of the *Antirrhinum* floral homeotic MADS-box gene *deficiens*: Evidence for DNA binding and autoregulation of its persistent expression throughout flower development. *EMBO Journal* **11**, 251–263.

Sieburth, L. E. and Meyerowitz, E. M. (1997). Molecular dissection of the *AGAMOUS* control region shows that *cis* elements for spatial regulation are located intragenically. *Plant Cell* **9**, 355–365.

Sieburth, L. E., Running, M. P. and Meyerowitz, E. M. (1995). Genetic separation of third and fourth whorl functions of *AGAMOUS*. *Plant Cell* **7**, 1249–1258.

Simillion, C., Vandepoele, K., Van Montagu, M. C. E., Zabeau, M. and Van de Peer, Y. (2002). The hidden duplication past of *Arabidopsis thaliana*. *Proceedings of the National Academy of Sciences of the United States of America* **99**, 13627–13632.

Smalle, J. and Vierstra, R. D. (2004). The ubiquitin 26S proteasome proteolytic pathway. *Annual Review of Plant Biology* **55**, 555–590.

Smith, D. L. and Johnson, A. D. (1992). A molecular mechanism for combinatorial control in yeast: MCM1 protein sets the spacing and orientation of the homeodomains of an alpha 2 dimer. *Cell* **68**, 133–142.

Smith, H. M. and Hake, S. (2003). The interaction of two homeobox genes, *BREVIPEDICELLUS* and *PENNYWISE*, regulates internode patterning in the *Arabidopsis* inflorescence. *Plant Cell* **15**, 1717–1727.

Smith, H. M., Campbell, B. C. and Hake, S. (2004). Competence to respond to floral inductive signals requires the homeobox genes *PENNYWISE* and *POUNDFOOLISH*. *Current Biology* **14**, 812–817.

Smyth, D. R., Bowman, J. L. and Meyerowitz, E. M. (1990). Early flower development in *Arabidopsis*. *Plant Cell* **2**, 755–767.

Soltis, P. S., Soltis, D. E., Savolainen, V., Crane, P. R. and Barraclough, T. G. (2002). Rate heterogeneity among lineages of tracheophytes: Integration of molecular and fossil data and evidence for molecular living fossils. *Proceedings of the National Academy of Sciences of the United States of America* **99**, 4430–4435.

Sommer, H., Beltran, J. P., Huijser, P., Pape, H., Lonnig, W. E., Saedler, H. and Schwarz-Sommer, Z. (1990). *Deficiens*, a homeotic gene involved in the control of flower morphogenesis in *Antirrhinum majus*: The protein shows homology to transcription factors. *EMBO Journal* **9**, 605–613.

Sridhar, V. V., Surendrarao, A., Gonzalez, D., Conlan, R. S. and Liu, Z. (2004). Transcriptional repression of target genes by *LEUNIG* and *SEUSS*, two interacting regulatory proteins for *Arabidopsis* flower development. *Proceedings of the National Academy of Sciences of the United States of America* **101**, 11494–11499.

Stellari, G. M., Jaramillo, M. A. and Kramer, E. M. (2004). Evolution of the *APETALA3* and *PISTILLATA* lineages of MADS-box containing genes in the basal Angiosperms. *Molecular Biology and Evolution* **21**, 506–519.

Sun, T. P. and Gubler, F. (2004). Molecular mechanism of gibberellin signaling in plants. *Annual Review of Plant Biology* **55**, 197–223.

Sung, Z. R., Belachew, A., Shunong, B. and Bertrandgarcia, R. (1992). *Emf*, an *Arabidopsis* gene required for vegetative shoot development. *Science* **258**, 1645–1647.

Takeda, S., Matsumoto, N. and Okada, K. (2003). *RABBIT EARS*, encoding a SUPERMAN-like zinc-finger protein, regulates petal development in *Arabidopsis thaliana*. *Development* **131**, 425–434.

Telfer, A., Bollman, K. M. and Poethig, R. S. (1997). Phase change and the regulation of trichome distribution in *Arabidopsis thaliana*. *Development* **124**, 645–654.

Theissen, G. (2001). Development of floral organ identity: Stories from the MADS house. *Current Opinion in Plant Biology* **4**, 75–85.

Theissen, G. and Saedler, H. (2001). Plant biology. Floral quartets. *Nature* **409**, 469–471.

Tilly, J. J., Allen, D. W. and Jack, T. (1998). The CArG boxes in the promoter of the *Arabidopsis* floral organ identity gene *APETALA3* mediate diverse regulatory effects. *Development* **125**, 1647–1657.

Trobner, W., Ramirez, L., Motte, P., Hue, I., Huijser, P., Lonnig, W. E., Saedler, H., Sommer, H. and Schwarz-Sommer, Z. (1992). *GLOBOSA*: A homeotic gene which interacts with *DEFICIENS* in the control of *Antirrhinum* floral organogenesis. *EMBO Journal* **11**, 4693–4704.

Vierstra, R. D. (2003). The ubiquitin/26S proteasome pathway, the complex last chapter in the life of many plant proteins. *Trends in Plant Science* **8**, 135–142.

Vision, T. J., Brown, D. G. and Tanksley, S. D. (2000). The origins of genomic duplications in *Arabidopsis. Science* **290**, 2114–2117.

Wagner, D., Sablowski, R. W. and Meyerowitz, E. M. (1999). Transcriptional activation of *APETALA1* by *LEAFY. Science* **285**, 582–584.

Wang, W. and Chen, X. (2004). HUA ENHANCER3 reveals a role for a cyclin-dependent protein kinase in the specification of floral organ identity in *Arabidopsis. Development* **131**, 3147–3156.

Weigel, D. and Meyerowitz, E. M. (1993). Activation of floral homeotic genes in *Arabidopsis. Science* **261**, 1723–1726.

Weigel, D. and Meyerowitz, E. M. (1994). The ABCs of floral homeotic genes. *Cell* **78**, 203–209.

Weigel, D. and Nilsson, O. (1995). A developmental switch sufficient for flower initiation in diverse plants. *Nature* **377**, 495–500.

Weigel, D., Alvarez, J., Smyth, D. R., Yanofsky, M. F. and Meyerowitz, E. M. (1992). *LEAFY* controls floral meristem identity in *Arabidopsis. Cell* **69**, 843–859.

Wellmer, F., Riechmann, J. L., Alves-Ferreira, M. and Meyerowitz, E. M. (2004). Genome-wide analysis of spatial gene expression in *Arabidopsis* flowers. *Plant Cell* **16**, 1314–1326.

Werner, J. D., Borevitz, J. O., Warthmann, N., Trainer, G. T., Ecker, J. R., Chory, J. and Weigel, D. (2005). Quantitative trait locus mapping and DNA array hybridization identify an FLM deletion as a cause for natural flowering-time variation. *Proceedings of the National Academy of Sciences of the United States of America* **102**, 2460–2465.

Western, T. L., Cheng, Y., Liu, J. and Chen, X. (2002). HUA ENHANCER2, a putative DExH-box RNA helicase, maintains homeotic B and C gene expression in *Arabidopsis. Development* **129**, 1569–1581.

Whipple, C. J., Ciceri, P., Padilla, C. M., Ambrose, B. A., Bandong, S. L. and Schmidt, R. J. (2004). Conservation of B-class floral homeotic gene function between maize and *Arabidopsis. Development* **131**, 6083–6091.

Wilkinson, M. D. and Haughn, G. W. (1995). *UNUSUAL FLORAL ORGANS* controls meristem identity and organ primordia fate in *Arabidopsis. Plant Cell* **7**, 1485–1499.

William, D. A., Su, Y., Smith, M. R., Lu, M., Baldwin, D. A. and Wagner, D. (2004). Genomic identification of direct target genes of *LEAFY. Proceedings of the National Academy of Sciences of the United States of America* **101**, 1775–1780.

Yang, Y., Fanning, L. and Jack, T. (2003a). The K domain mediates heterodimerization of the *Arabidopsis* floral organ identity proteins, APETALA3 and PISTILLATA. *Plant Journal* **33**, 47–59.

Yang, Y., Xiang, H. and Jack, T. (2003b). *pistillata-5*, an *Arabidopsis* B class mutant with strong defects in petal but not in stamen development. *Plant Journal* **33**, 177–188.
Yanofsky, M. F., Ma, H., Bowman, J. L., Drews, G. N., Feldmann, K. A. and Meyerowitz, E. M. (1990). The protein encoded by the *Arabidopsis* homeotic gene *agamous* resembles transcription factors. *Nature* **346**, 35–39.
Yoshida, N., Yanai, Y., Chen, L., Kato, Y., Hiratsuka, J., Miwa, T., Sung, Z. R. and Takahashi, S. (2001). EMBRYONIC FLOWER2, a novel polycomb group protein homolog, mediates shoot development and flowering in *Arabidopsis*. *Plant Cell* **13**, 2471–2481.
Yu, B., Yang, Z., Li, J., Minakhina, S., Yang, M., Padgett, R. W., Steward, R. and Chen, X. (2005). Methylation as a crucial step in plant microRNA biogenesis. *Science* **307**, 932–935.
Yu, H., Ito, T., Zhao, Y., Peng, J., Kumar, P. and Meyerowitz, E. M. (2004). Floral homeotic genes are targets of gibberellin signaling in flower development. *Proceedings of the National Academy of Sciences of the United States of America* **101**, 7827–7832.
Yun, J. Y., Weigel, D. and Lee, I. (2002). Ectopic expression of *SUPERMAN* suppresses development of petals and stamens. *Plant & Cell Physiology* **43**, 52–57.
Zahn, L. M., Kong, H., Leebens-Mack, J. H., Kim, S., Soltis, P. S., Landherr, L. L., Soltis, D. E., dePamphilis, C. W. and Ma, H. (2005a). The evolution of the *SEPALLATA* subfamily of MADS-box genes: A pre-angiosperm origin with multiple duplications throughout angiosperm history. *Genetics* **169**, 2209–2223.
Zahn, L. M., Leebens-Mack, J., dePamphilis, C. W., Ma, H. and Theissen, G. (2005b). To B or not to B a flower: The role of *DEFICIENS* and *GLOBOSA* orthologs in the evolution of the angiosperms. *The Journal of Heredity* **96**, 225–240.
Zahn, L. M., Leebens-Mack, J., Arringtona, J. M., Hu, Y., Landherr, L., dePamphilis, C. W., Becker, A., Theissen, G. and Ma, H. (2006). Conservation and divergence in the *AGAMOUS* subfamily of MADS-box genes: Evidence of independent sub- and neofunctionalization events. *Evolution & Development* **8**, 30–45.
Zhang, X., Feng, B., Zhang, Q., Zhang, D., Altman, N. and Ma, H. (2005). Genome-wide expression profiling and identification of gene activities during early flower development in *Arabidopsis*. *Plant Molecular Biology* **58**, 401–419.
Zhao, D., Yu, Q., Chen, M. and Ma, H. (2001). The *ASK1* gene regulates B function gene expression in cooperation with *UFO* and *LEAFY* in *Arabidopsis*. *Development* **128**, 2735–2746.
Zhao, D., Ni, W., Feng, B., Han, T., Petrasek, M. G. and Ma, H. (2003). Members of the *Arabidopsis-SKP1-like* gene family exhibit a variety of expression patterns and may play diverse roles in *Arabidopsis*. *Plant Physiology* **133**, 203–217.

Missing Links: DNA-Binding and Target Gene Specificity of Floral Homeotic Proteins

RAINER MELZER, KERSTIN KAUFMANN* AND GÜNTER THEIßEN

Friedrich-Schiller-Universität Jena, Lehrstuhl für Genetik, Philosophenweg 12, D-07743 Jena, Germany

*Present Address: Plant Research International, Business Unit Bioscience, Wageningen, 6700 AA, Netherlands.

Advances in Botanical Research, Vol. 44
Incorporating Advances in Plant Pathology

0065-2296/06 $35.00
DOI: 10.1016/S0065-2296(06)44005-2

ABSTRACT

Interactions of floral homeotic genes have so far mainly been defined genetically and in terms of protein complex formation. However, to better understand the molecular evolutionary dynamics of the network controlling floral organ identity, its complete "hard-wiring" has to be determined, that is, the physiochemical interactions on which floral homeotic gene function is based. Especially important are the protein–DNA interactions that lead to the specific regulation of target genes. Here we review what is already known about target gene specificity of floral homeotic MIKC-type MADS-domain proteins. Two extreme possibilities can be envisioned for how target gene specificity is achieved: (1) each MIKC-type protein complex binds to a completely different subset of target genes or (2) all MIKC-type proteins potentially regulate the same set of targets, but in different ways. Several aspects of MIKC-type protein function are of special interest in this respect: DNA bending, DNA-sequence specificity, protein–protein interactions, and protein concentration. We discuss how these features fit to either of the models and to our increasing knowledge about real target genes. Finally, the evolutionary capability of MIKC-type proteins to create floral diversity might in part be due to their evolutionary flexibility in regulating sets of potential target genes. In this respect, understanding target gene specificity may enlighten our understanding of molecular mechanisms on which the dramatic evolutionary diversification of flower form is based.

I. INTRODUCTION

Since the end of the 1980s, a rapidly increasing number of genes involved in inflorescence and flower development has been identified. Because of this a detailed picture is emerging as to how floral form is genetically regulated. The determination of floral organ identity is especially well understood. Based on the phenotypes of floral homeotic mutants, the ABC-model was developed, with an A-function specifying the identity of sepals, A and B together determining petals, B and C defining stamen identity, and C alone specifying carpels (Coen and Meyerowitz, 1991). An extended model added a D-function important for ovule development (Angenent and Colombo, 1996) and an E-function, which is important for the identity of the organs in all four floral whorls (Ditta *et al.*, 2004; Pelaz *et al.*, 2000, Theißen, 2001). The respective genes of *Arabidopsis* are *APETALA1* and *APETALA2* (*AP1* and *AP2*, A-function), *APETALA3* and *PISTILLATA* (*AP3* and *PI*, B-function), *AGAMOUS* (*AG*, C-function), *SHATTERPROOF1* (*SHP1*, formerly known as *AGL1*), *SHATTERPROOF2* (*SHP2*, formerly known as *AGL5*), *SEEDSTICK* (*STK*, formerly known as *AGL11*) and, again *AGAMOUS* (D function). The E-function is performed by *SEPALLATA1* to *SEPALLATA4* (*SEP1–SEP4*, formerly known as *AGL2, AGL4, AGL9*, and *AGL3*, respectively) (for a review, see Jack, 2004). With the exception of *AP2* the products of all of these genes belong to the MIKC-type transcription factor family, a special class of MADS-domain proteins that is only present in plants. They are composed of at least four

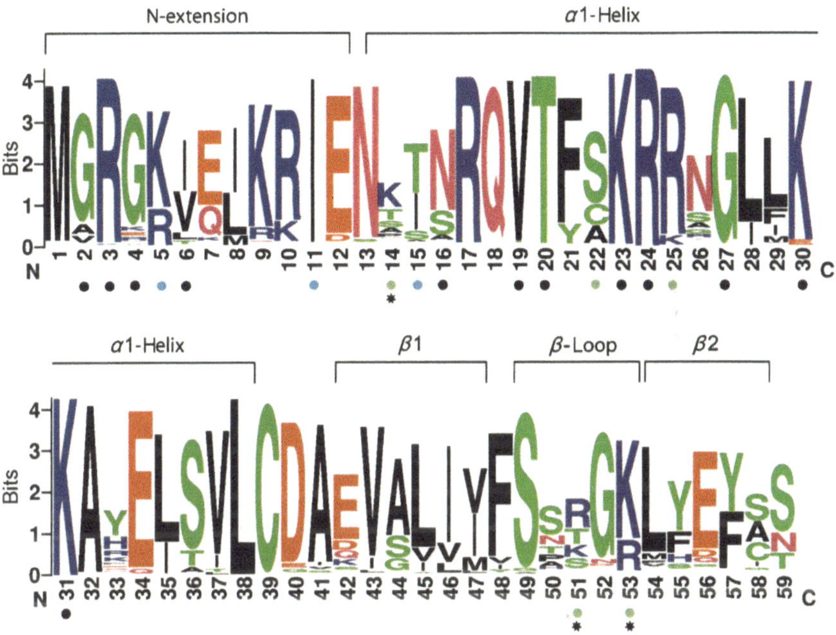

Plate 2

	−3	−2	−1	1	2	3	4	5	6	7	8	9	10	11	12	13
AG*	T	T	A/T	C	C	A/T	A/t	A	A/T	A	A/T	G/a	G	A/t	A	A
AG**	T	T	A/T	C	C	A/T	A/t	A/t	T/A	N	N	G	G	-G	A	A
SHP1	T	T	-C	C	C	A/T	A/T	A/t	T/A	N	N	G	G	T/A	A	A
SEP1	N	T/a	N	C	C	A	T/A	A/t	T/A	A/T	T	A/G	G	A/t	A	A
SEP4	T	T	N	C	C	A	T/A	A/T	T/A	A/t	T	A/g	G	-G	A/t	A

CArG-core

Plate 3

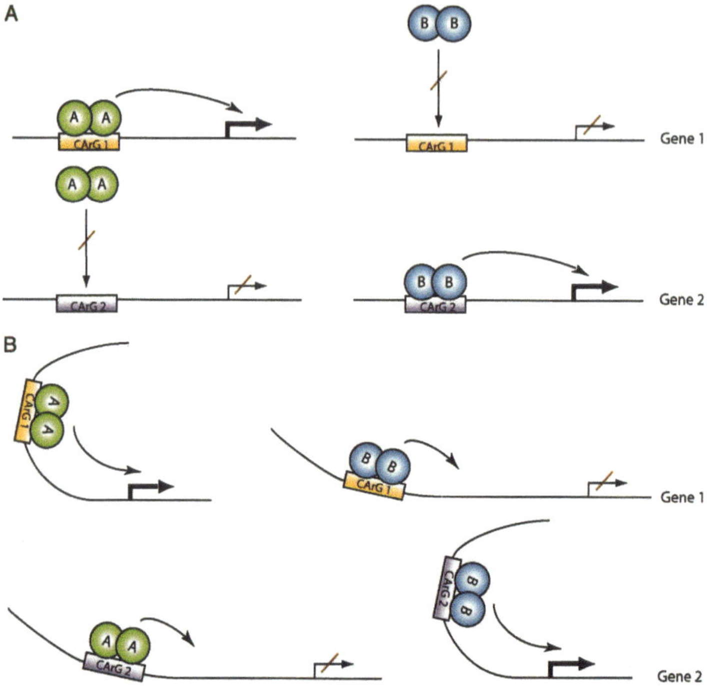

Plate 4

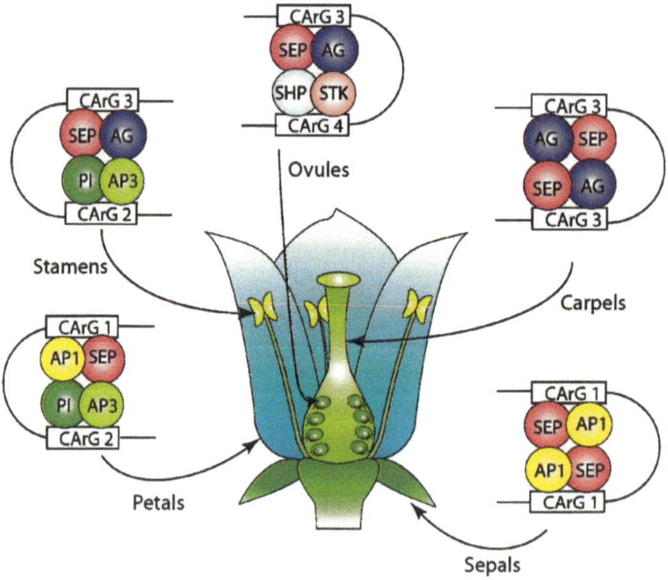

Plate 5

conserved domains, the *M*ADS-, *I*ntervening-, *K*eratin-like, and *C*-terminal (MIKC) domains (for a review, see Kaufmann *et al.*, 2005b).

It is reasonable to assume that MIKC-type proteins have to activate or repress well-defined sets of target genes in order to act as key regulators of flower development. In the case of floral homeotic proteins, for example, differential modulation of gene expression can be expected for each kind of floral organ. However, despite a wealth of available data on combinatorial interactions of organ identity genes and proteins, almost nothing is known about how target gene specificity is actually brought about. Thus, a central aspect of MIKC-type protein function, the physicochemical process of DNA binding and its role in target gene regulation, remained almost completely unknown. This lack of knowledge significantly hampers our understanding of plant development. Changes in the interactions of MIKC-type proteins with DNA may well have affected target gene selection and hence flower development during evolution at any stage from floral induction to the ontogeny of fruits and seeds. Coevolution of the protein–DNA and protein–protein interaction domains of MIKC-type proteins on the one hand and *cis*- regulatory elements of target genes on the other hand has thus very likely shaped the regulatory interactions in flower development. Since changes in these interactions underlie the evolution of floral form, they are of utmost importance for our understanding of the evolution of floral characters. In a similar vein the diversification of MIKC-type proteins from a common ancestral protein to dozens of players in angiosperms with diverse functions, such as organ development or the control of flowering time, cannot be understood without taking the evolutionary dynamics of protein–DNA interactions into account.

II. BINDING OF MADS-DOMAIN PROTEINS TO DNA

By definition, all MADS-domain transcription factors have the highly conserved MADS-domain in common, which is responsible for DNA binding (Fig. 1). Sequence-specific DNA-binding was first demonstrated for the MADS-domain proteins SRF and MCM1 from animals and yeasts, respectively. These transcription factors bind to DNA sequences with the core consensus 5′-CC(A/T)$_6$GG-3′, termed CArG boxes (Passmore *et al.*, 1989; Pollock and Treisman, 1990; Wynne and Treisman, 1992). As suspected from the palindromic or nearly palindromic CArG-box sequence, all MADS-domain proteins tested so far bind to DNA as dimers, either homo- or heterodimers (reviewed in Kaufmann *et al.*, 2005b).

The DNA-binding specificity of the *Arabidopsis* MIKC-type proteins SHP1, AG, SEP1, AP1, PI, and AP3 has been studied using gel retardation assays.

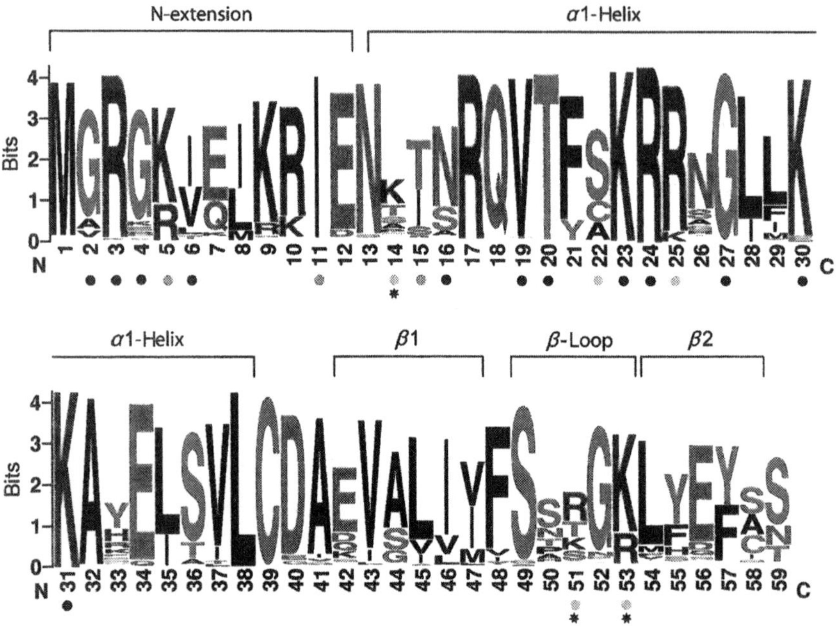

Fig. 1. Sequence logo of the MADS-domain. The logo was derived from more than 800 MADS-domain protein sequences obtained from GenBank. ClustalW with default parameters was used to align the sequences. The alignment was adjusted manually. The sequence logo was created with WebLogo [http://weblogo.berkeley.edu/; (Crooks *et al.*, 2004; Schneider and Stephens, 1990)]. At each position in the alignment, amino acid identity is represented by the single letter code. For positions in the alignment represented by more than one amino acid, the letters are stacked on top of one another. Letter height is proportional to the frequency of that amino acid at that position in the alignment. The *x*-axis is alignment position, and the *y*-axis is information content (in bits as described in Schneider and Stephens, 1990). Black circles indicate residues important for DNA-binding in both SRF and MEF2A. Green and blue circles indicate residues important for DNA-binding in SRF or MEF2A, respectively (Huang *et al.,* 2000; Pellegrini *et al.,* 1995). Amino acids of special importance for differential DNA bending are indicated by asterisks (West and Sharrocks, 1999; West *et al.,* 1997). (See color Plate 2.)

They all can bind to CArG-box sequences, although with somewhat different affinities. Moreover, the consensus sequences determined for binding of AG, SHP1, and SEP1 are similar, but not identical (Fig. 2) (Huang *et al.*, 1993, 1995, 1996; Riechmann *et al.*, 1996; Shiraishi *et al.*, 1993).

More data concerning DNA-binding specificity of MIKC-type proteins come from *Antirrhinum majus* (snapdragon). Homodimers of SQUA and PLE (which are closely related to AP1 and AG, respectively) as well as heterodimers of DEF/GLO (which are putative orthologs of AP3 and PI, respectively) have been shown to bind the "typical" CArG-box sequence 5′-CC(A/T)$_6$GG-3′

	−3	−2	−1	1	2	3	4	5	6	7	8	9	10	11	12	13
AG*			A/T	C	C	A/T	A/t	A	A/T	A	A/T	G/a	G	A/t	A	A
AG**			A/T	C	C	A/T	A/t	A/t	T/A	N	N	G	G	-G	A	A
SHP1			-C	C	C	A/T	A/T	A/t	T/A	N	N	G	G	T/A	A	A
SEP1	N	T/a	N	C	C	A	T/A	A/t	T/A	A/T		A/G	G	A/t	A	A
SEP4			N	C	C	A	T/A	A/T	T/A	A/t		A/g	G	-G	A/t	A

CArG-core

Fig. 2. Comparison of CArG-box consensus sequences bound by different *Arabidopsis* MIKC-type proteins (Huang *et al.*, 1993, 1995, 1996; Shiraishi *et al.*, 1993). Consensus rules according to Huang *et al.* (1996). The CArG-box core is indicated. The AG-consensus was determined two times independently by Shiraishi *et al.* (1993) (AG*) and Huang *et al.* (1993) (AG**). (See color Plate 3.)

but SQUA was also able to bind to the sequence 5′-C(A/T)$_8$G-3′, indicating a somewhat different binding specificity compared to PLE and DEF/GLO (West *et al.*, 1998).

In general, two different interpretations of the experiments described in the previous section have been proposed. Huang *et al.* (1996) argued that subtle differences in DNA-binding specificity are responsible for the differential activity of MIKC-type proteins. On the contrary, Riechmann *et al.* (1996) suggested that the DNA-binding specificity of MIKC-type protein dimers contributes only marginally to differences in target gene regulation.

In planta data obtained by Krizek and Meyerowitz (1996) and Riechmann and Meyerowitz (1997) support the latter hypothesis. Plants constitutively expressing chimeric genes composed of the MADS-box of *AGAMOUS* (C function) and the I-, K-, and C-boxes of *APETALA3* (B function) showed phenotypes typical for plants constitutively expressing *AP3*, that is, carpels in the fourth whorl were replaced by normal stamens or carpelloid stamens (Krizek and Meyerowitz, 1996). The same is true for chimeric constructs consisting of the MADS-box of *AGAMOUS* and the I-, K-, and C-boxes of *PISTILLATA* (also B-function). Riechmann and Meyerowitz (1997) consitutively expressed chimeric genes of *AP3, PI, AG*, or *AP1* in *Arabidopsis thaliana* wild type and mutant plants. In these chimeric genes, the 5′-terminal half of the MADS-domain coding sequence was substituted by the corresponding sequence of *SRF* or *MEF2A*. The resulting phenotypes were similar to the ones observed when the unmodified floral homeotic genes were constitutively expressed. Even rescue of the respective loss-of-function phenotypes was similarly achieved between lines ectopically expressing the chimeric constructs and the ones overexpressing the wild-type genes (Riechmann and Meyerowitz, 1997), suggesting that the substituted part of the MADS-domain influences functional specificity of floral homeotic proteins only to a minor extent. Even more puzzling, gel retardation analyses indicated that the DNA-binding specificity of the chimeric proteins differed from the ones of the wild-type proteins. For example, AP1 and AG bound to CArG-boxes resembling the consensus $CC(A/T)_6GG$, but failed to bind a CArG-box similar to the consensus sequence $CTA(A/T)_4TAG$, which is bound by MEF2A. In contrast, the chimeric constructs MEF2A-AG and MEF2A-AP1 did not bind to the $CC(A/T)_6GG$-CArG-box but to the one recognized by MEF2A (Riechmann and Meyerowitz, 1997). In other words, even though the proteins were still able to fulfill their natural function, their DNA-binding specificity appeared to be considerably changed.

Altogether, these results indicate that DNA-binding specificity of MIKC-type protein dimers has only limited influence on target gene specificity. At first sight, this is quite surprising, because binding to specific DNA sequences is thought to be the main determinant of functional specificity. However, this conundrum is not unique for MIKC-type transcription factors. In other transcription factor families, mechanisms have been suggested explaining how well-defined sets of target genes can be regulated even if this cannot be inferred from the DNA-binding specificity.

III. LEARNING FROM HOMEODOMAIN PROTEINS IN ANIMALS

Developmental processes in both plants and animals are mainly controlled by transcription factors. While in plants members of the MIKC-type MADS-domain protein family are encoded by the most prominent homeotic selector genes, this role is assigned to homeodomain transcription factors in animals. Besides the surprising parallels in how MIKC-type and homeodomain proteins control development (Meyerowitz, 2002), the problems in explaining target gene specificity are also similar. Much effort has been made to unravel this problem in the case of homeodomain proteins, and the progress in the animal field might give clues about the differential target gene regulation by MIKC-type proteins.

Particularly well studied is one special group of homeodomain proteins called Q50, named after a highly conserved glutamine residue at position 50 of the homeodomain. This group has members as famous as the *Drosophila* pair-rule proteins Even-skipped (Eve) and Fushi tarazu (Ftz), the segment polarity protein Engrailed (En) and the Hox proteins Abdominal-A (Abd-A), Deformed (Dfd), and Ultrabithorax (Ubx). The DNA-binding specificity of all of these transcription factors is quite similar (for a review, see Biggin and McGinnis, 1997). As in the case of MIKC-type proteins, this was taken as evidence that the DNA-binding specificity alone is not sufficient to confer target gene specificity (Biggin and McGinnis, 1997; Mann and Chan, 1996).

Moreover, a homeodomain-deleted version of Ftz is still able to regulate segmentation (Copeland *et al.*, 1996), implying that additional factors mediate functional specificity of homeodomain proteins. Cofactors have been shown to be important for homeodomain function (Mann and Chan, 1996; Mann and Morata, 2000; Nasiadka *et al.*, 2000 and references therein). However, the proposed mechanism by which target gene specificity is modulated by these cofactors remains controversial. Two major models have been proposed. The "*binding site selection*" model suggests that DNA-binding specificity of homeodomain proteins is modified by interaction with additional factors (Mann and Chan, 1996; Mann and Morata, 2000). In contrast, the "*widespread binding*" or "*activity regulating*" model assumes that all Q50 homeodomain proteins potentially bind a considerable number of shared targets, with cofactors modulating the ability to regulate these genes (Biggin and McGinnis, 1997). There are several experiments supporting either of the two models. On the one hand, *in vitro* studies suggest that Hox-cofactor complexes increase DNA-binding specificity (Chan and Mann, 1996). This is in agreement with *in vivo* data which show that changes in specific Hox-cofactor DNA-binding sites are sufficient to shift the Hox-dependent expression of a reporter gene (Chan *et al.*, 1997; Ryoo and Mann, 1999). On the other hand, *in vivo* cross-linking experiments suggest that the two homeodomain proteins Eve and Ftz bind to many genes and that the DNA-binding specificity of the two proteins is quite similar (Walter and Biggin, 1996; Walter *et al.*, 1994). Moreover, if the Hox-protein Ubx is fused to the strong transcription activation domain VP16, the chimeric protein can partially rescue an *antennapedia* (*antp*) loss-of-function mutant. This was taken as evidence that Hox proteins bind to similar targets but differ in their ability to activate or repress them (Li and McGinnis, 1999).

It is not clear which model is closer to reality. However, more recent data support the view that both—DNA-binding specificity and differential regulation of target genes—contribute to target gene specificity of homeodomain proteins (Nasiadka *et al.*, 2000).

With the data from homeodomain proteins in mind, two possibilities can be envisaged how target gene specificity is achieved in the case of MIKC-type proteins: (1) each MIKC-type protein complex binds to a completely different set of target genes or (2) all MIKC-type proteins regulate potentially the same target genes, but in different ways. It is important to note that in the case of model (2) target gene specificity does not depend on DNA-binding specificity. Rather, a post-DNA-binding mechanism is responsible for differential regulation of target genes (Fig. 3).

It appears likely that the truth lies somewhere between these two scenarios. However, discussing several aspects of MIKC-function with these alternatives in mind might not only give clues which hypothesis reflects reality better, but also provide hints to facilitate the design of future experiments.

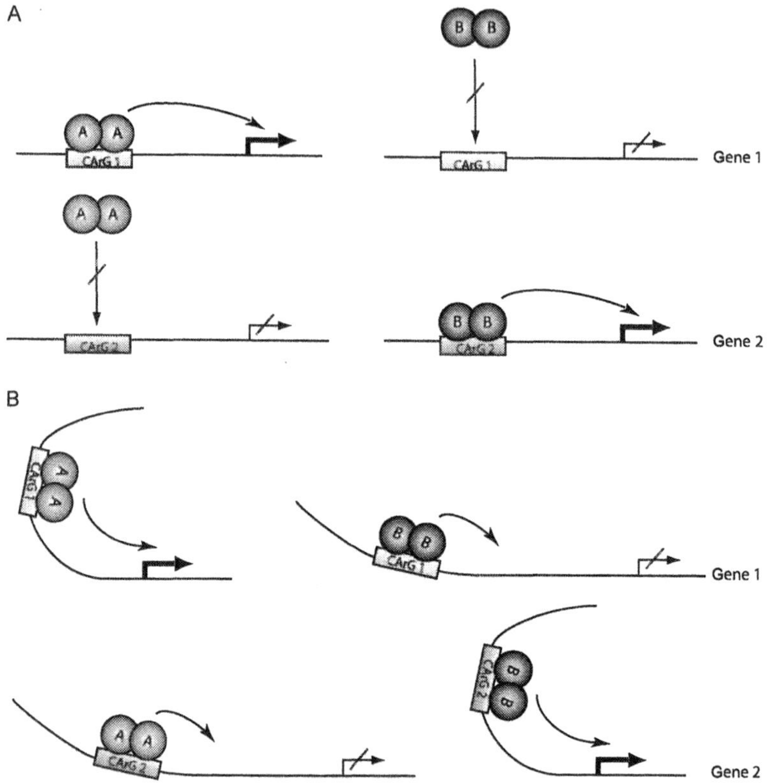

Fig. 3. Two extreme possibilities showing how target gene specificity can be achieved. (A) Target gene specificity is solely brought about by differences in DNA-binding specificity. (B) Target gene specificity is determined by a post-DNA-binding mechanism. In the example shown two CArG-boxes are bent to different extents and hence transcription is differentially regulated. The influence of other factors, like the formation of higher order complexes, the differential interaction with cofactors, the presence or absence of transcription activation domains, or the number and position of CArG-boxes are not shown for simplicity but might be of importance for either of the two models (for details, see text). CArG: CArG-box of gene 1 and 2, respectively; (A, B) MIKC-type proteins; an arrow crossed by a red line indicates that no binding or activation occurs. (See color Plate 4.)

IV. CARG GALORE: WHAT FOR?

The DNA sequence recognized by most transcription factors, including MIKC-type proteins, is 5–20 base pairs in length (Wray *et al.*, 2003). The core CArG-box consists of 10 nucleotides. As some of the surrounding bases are conserved as well, the whole consensus has a length of 16 nucleotides (Huang *et al.*, 1996). However, not every position is equally conserved, since at some positions, two, three, or even all four bases are possible. This reduces the information content significantly, thereby increasing the likelihood that a matching sequence appears just by chance. For example, about 49% of 27,186 upstream 3-kb sequences of *Arabidopsis* genes contain at least one CArG-box similar to the consensus sequence bound by AGAMOUS *in vitro* (Gomez-Mena *et al.*, 2005). If all of these CArG-boxes are of biological significance, about 13,000 genes would be controlled by floral homeotic proteins in *Arabidopsis*. This number exeeds by far what one would expect based on microarray data (Gomez-Mena *et al.* 2005; Wellmer *et al.*, 2004; Zik and Irish, 2003). The number of CArG-boxes is even higher when looking at regions downstream of the transcription units or within introns; it is therefore likely that not every CArG-box present in the vicinity of a gene is a target of MIKC-type proteins. Similar problems in determining target gene specificity are common to almost all transcription factor families. Powerful bioinformatics and experimental tools have been developed to overcome these difficulties. However, it is still poorly understood how functional and nonfunctional binding sites are distinguished at the molecular level. For example, whether a CArG-box is bound by an MIKC-type protein and hence transcription of the target is influenced might depend on (1) whether the chromatin is condensed or not, (2) the presence of cofactors, or (3) the presence of other proteins competing for the same or overlapping binding sites (Wray *et al.*, 2003).

It has been shown in yeast that the nucleosome density in promoter regions is significantly lower than in nonpromoter regions, therefore offering a mechanism by which nonfunctional binding sites are excluded (Sekinger *et al.*, 2005). In *Arabidopsis*, the polycomb-protein CURLY LEAF prevents ectopic expression of *AGAMOUS*, likely by chromatin modification (Goodrich *et al.*, 1997). DNA-methylation increases during *Arabidopsis* development, in conjunction with this a role of methylation in maintaining developmental states has been discussed (Ruiz-Garcia *et al.*, 2005). Moreover, it has been shown that chromatin modifications play a key role in the control of flowering time (reviewed in He and Amasino, 2005).

Generally, if only a subset of all CArG-boxes in the genome is accessible in a given cell at a given time, this could help to explain target gene specificity with only limited DNA-binding specificity.

V. DNA BENDING

DNA has a three-dimensional structure, and modulation of this structure might be as fundamental for gene regulation as the recognition of specific stretches of nucleotides. One possibility to modulate the structure of promoters is protein-induced DNA bending. It is quite well established that bending of the DNA is an important mechanism of gene regulation in bacteria. For example, protein-induced bending can enable or disable the interaction of two proteins binding to otherwise distant sites on the DNA (reviewed in PerezMartin and deLorenzo, 1997). Bending might be of similar importance in eukaryotic genomes, in which complex promoters necessitate architectural elements. Examples illustrate that bending: (1) facilitates the interaction of distantly spaced transcription factors (Kumari *et al.*, 2005), (2) reverses intrinsic DNA-curvatures and hence enables formation of the enhanceosome (Yie *et al.*, 1999), and (3) can regulate transcription in a fine-tuned way, depending on the degree of bending (Scaffidi and Bianchi, 2001).

Evidence that MADS-domain proteins bend DNA comes from a number of experiments. MADS proteins bind DNA as dimers. Two α-helices within the MADS-domain, one from each protein, are oriented in an antiparallel fashion and contact the DNA mainly in the major groove. In addition, an N-terminal extension ranging from the α-helices to the beginning of the MADS-domain contacts the DNA in the minor groove (Huang *et al.*, 2000; Pellegrini *et al.*, 1995). Residues of the α-helix and a β-loop located C-terminal to this α-helix confer the ability of MADS-domain proteins to bend DNA. This bending varies between MADS-domain proteins as exemplified by SRF and MEF2A, which is due to specific amino acid residues that differ in their ability to interact with DNA (Fig. 1) (Huang *et al.*, 2000; West *et al.*, 1997). It is not clear as to what extent DNA bending plays a role in transcriptional regulation by SRF and MEF2A, but it has been proposed that it influences the binding of additional *trans*-acting factors to their presumptive *cis*-regulatory sites (West *et al.*, 1997). In the case of the yeast MADS-domain transcription factor MCM1, strong experimental evidence points toward a role of DNA bending in transcriptional regulation, as MCM1 mutants defective in DNA bending also show defects in the formation of ternary transcription factor complexes (Lim *et al.*, 2003).

It has been shown that floral homeotic proteins are able to induce DNA bending as well (Riechmann *et al.*, 1996; West *et al.*, 1998). AG homodimers introduce a more pronounced bend compared to AP3/PI heterodimers or AP1 homodimers (Riechmann *et al.*, 1996). Moreover, in the case of SQUA, the magnitude of bending depends on the sequence of the binding site (West and Sharrocks, 1999; West *et al.*, 1998). Another difference was detected when comparing the direction of bending. While AP3/PI, AG, AP1, and SQUA all bend DNA toward the minor groove with respect to the center of the CArG-box, the direction of bending is different in the case of DEF/GLO heterodimers (Riechmann *et al.*, 1996; West *et al.*, 1998).

Altogether, these results suggest that DNA bending plays a role in the target gene specificity of MIKC-type proteins. For example, the set of target gene promoters potentially bound by two MIKC-type proteins might be similar, but depending on the degree or direction of bending, the expression of the respective target genes is on/off or even finely regulated (Fig. 3). In this respect, the multiple interactions of MIKC-type proteins with members of the same family are another potential source of variation. It could well be that different dimer combinations have distinct abilities to bend DNA. Similar observations have been made in the bZIP family of transcription factors (Kerppola and Curran, 1991). Finally, the fact that the magnitude of bending is influenced by the binding site sequence adds another level of complexity. It may then also depend on the promoter architecture of the target gene how transcription is regulated (West *et al.*, 1998).

A role of DNA bending in differential target gene regulation by MIKC-type proteins seems to be in conflict with the results of Riechmann and Meyerowitz (1997), who showed that chimeric proteins consisting of the MADS-domain of SRF or MEF2A and the I-, K-, and C-domains of floral homeotic proteins are able to rescue floral homeotic mutants. Important amino acid residues determining the degree of DNA bending are located within the MADS-domain, in particular in the DNA-binding α-helix but also within the β-loop (Huang *et al.*, 2000; West and Sharrocks, 1999 and references therein). The C-terminal half of the α-helix as well as the β-loop are not exchanged in the chimeric proteins used by Riechmann and Meyerowitz (1997). Therefore, it remains open as to what extent differential DNA bending was affected.

In summary, the ability of MIKC-type proteins to bend DNA seems to be one possible factor which might confer target gene specificity by a post-DNA-binding mechanism, although more experimental evidence is needed to clarify this issue.

VI. DOES THE PROTEIN CONCENTRATION MATTER?

The interaction of a transcription factor with *cis*-regulatory sites or other *trans*-acting factors critically depends on the concentration of the different components. Moreover, if two factors compete for the same binding site (protein or DNA), the ratio between them might be important for transcriptional regulation.

In the case of MIKC-type genes controlling flowering time, dosage dependency is a well-known phenomenon (Hartmann *et al.*, 2000; Scortecci *et al.*, 2001; Yu *et al.*, 2002). It has been proposed that different gene dosages of the floral repressor *FLC* contribute significantly to the natural variation of flowering time observed in the genus *Brassica* (Osborn, 2004). These dosage effects are most easily explained by variations in protein concentration. There are repressors as well as activators of flowering time among MIKC-type genes. One might imagine that they potentially all control the same set of target genes but in an opposite fashion, that is, not the binding site specificity, but the molecular action at the presumptive *cis*-sites (repression or activation) is critical in regulating target genes. Therefore, a high concentration of a floral activator would outcompete binding of the respective repressors and hence the plant flowers earlier and *vice versa*. This is consistent with models proposing that flowering time is controlled by numerous pathways which act to regulate the same set of target genes (Mouradov *et al.*, 2002).

Dosage effects may also be important for floral homeotic protein function. Jack *et al.* (1997) demonstrated a dosage dependency of *AGAMOUS* function. The phenotypic severity observed when *AGAMOUS* is ectopically expressed in whorl two of *Arabidopsis* flowers correlates with the number of transgenes present in the genome. Another example for dosage dependency is reported from the *SEPALLATA* genes. Whereas *SEP1 sep2 sep3* double mutants are phenotypically normal, *SEP1/sep1 sep2 sep3* plants fail to develop ovules (Favaro *et al.*, 2003). However, no haploinsufficiency has been observed in the case of the floral homeotic A-, B-, and C-function genes in normal (nontransgenic) *Arabidopsis*. It could simply be that 50% of the original protein concentration is still sufficient to mediate organ identity. This assumption is supported by data obtained from *Petunia*. *Petunia* carries two largely redundant *GLO*-like genes (*PhGLO1* and *PhGLO2*), enabling an experimental reduction of the gene dosage down to 25% (Vandenbussche *et al.*, 2004). Whereas *phglo1* as well as *phglo2*-mutants show no or only minor morphological defects, plant heterozygous for one and homozygous for the other mutant allele show full or partial conversion of stamens into carpels (Vandenbussche *et al.*, 2004).

Alternatively, the lack of ABC-mutants that show haploinsufficiency in *Arabidopsis* might be due to autoregulatory circuits that can compensate for the loss of one allele. Evidence for autoregulation is available for *APETALA3, PISTILLATA*, and *AGAMOUS* (Gomez-Mena *et al.*, 2005; Honma and Goto 2000; Jack *et al.*, 1994).

Protein concentration might also be crucial for the formation of the hybrid organs often observed in basal angiosperms. In *Amborella*, for example, the transition from outer tepals to inner tepals to stamens is gradual (Buzgo *et al.*, 2004). Inferring from this phenotype, Buzgo *et al.* (2004) proposed a "fading border" model, according to which variations in floral homeotic B- and C-protein concentration are responsible for the gradual transition between the floral organs. For example, low concentrations of B-function proteins in conjunction with A-function proteins would give rise to petaloid tepals. Gradual elevation of B-protein concentration in the succeeding inner whorls successively heightens the petaloid characteristics (Buzgo *et al.*, 2004). This would be in line with what is observed in *Arabidopsis* plants ectopically expressing floral homeotic proteins. Different levels of *AGAMOUS* expression, for example, cause varying severities of phenotypes, ranging from short petals to staminoid petals to complete absence of petals in the second whorl (Mizukami and Ma, 1992).

One possible line of reasoning explaining the important role of protein concentration is that different floral homeotic protein complexes compete in part for the same target genes. Elevating the concentration of one protein would then increase the fraction of target genes controlled by it. This mechanism is similar to what is proposed in the control of flowering time and would explain the occurrence of floral hybrid organs. However, one also might imagine that low protein levels are insufficient to activate all target genes appropriately. This could cause the dosage dependency of SEP-proteins in ovule development.

VII. PROTEIN–PROTEIN INTERACTIONS: ELECTIVE AFFINITIES

Target gene specificity can be dramatically enhanced by cooperative binding of several proteins. Furthermore, interactions with non-DNA binding cofactors or posttranslational modifications can influence the regulation of gene expression. In general, one can distinguish between interactions with other MIKC-type proteins and with non-MADS factors. These two possibilitics will be discussed separately here (for a detailed discussion on MIKC-type protein–protein interactions, see Kaufmann *et al.*, 2005b).

A. ROLE OF NON-MADS FACTORS

Animal and yeast MADS-domain transcription factors regulate a wide range of processes. For example, the yeast protein MCM1 is involved in cell cycle control, cell type specification, and arginine metabolism (reviewed in Messenguy and Dubois, 2003). The required specificity is brought about by the interaction with diverse unrelated proteins. Cooperative binding of several factors to one promoter requires several binding sites and hence enhances specificity.

Some cofactors mediate MADS-domain protein activity simply by binding to them, without requiring an additional DNA-binding domain. Evidence for this comes from interactions of MCM1 with homeodomain proteins (Pramila *et al.*, 2002). On the other hand, it is also possible that MADS-domain proteins regulate transcription without binding to DNA. MEF2A acts synergistically with bHLH proteins during myogenesis. However, at least in some cases, transcriptional regulation requires either a CArG-box (MEF2A binding) or an E-box (bHLH/E12 binding) (for reviews, see Messenguy and Dubois, 2003; Molkentin and Olson, 1996), suggesting the existence of MADS-domain protein target genes lacking a CArG-box in the promoter. Activity of MADS-domain transcription factors can also be regulated posttranscriptionally. For example, it has been suggested that phosphorylation of MEF2A by the p38 MAP kinase pathway enhances MEF2A-dependent gene expression (Zhao *et al.*, 1999).

There is only limited evidence about non-MADS cofactors of MIKC-type proteins. AGAMOUS interacts with a complex composed of FLOR1 (a novel protein containing a leucine rich repeat) and VSP1 (flower-specific vegetative storage protein) *in vitro* and in yeast two-hybrid assays. As AGAMOUS has a phosphatase recognition/phosphorylation site and VSP1 has phosphatase-activity it has been hypothesized that FLOR1 mediates the interaction between VSP1 and AGAMOUS (Gamboa *et al.*, 2001).

OsMADS18, a SEP-like protein from rice, has been shown to interact with NF-YB, which belongs to a group of CCAAT-box-binding transcription factors (Masiero *et al.*, 2002). Putative binding sites for CCAAT-box binding proteins are found in the regulatory region of *AG* genes from various eudicotyledonous plants. In addition, these sites seem to be important for correct *AG* expression *in vivo* (Hong *et al.*, 2003). As AG is target of direct autoregulation, the cooperative binding with CCAAT-box-binding proteins is an appealing possibility.

MADS-box INTERACTING PROTEIN1 (MIP1), the founding member of a protein family unique to plants, interacts with the *Antirrhinum* MADS-domain proteins PLENA, FARINELLI, DEFH72, and DEFH200 in yeast two-hybrid assays. MIP1 contains a leucine zipper motif and is able to activate yeast reporter genes. As MIP1 also contains a putative nuclear localization signal, it has been speculated that it confers transcriptional activity to MADS-domain proteins lacking a transcriptional activation domain (Causier *et al.*, 2003).

Honma and Goto (2001) screened a flower-specific cDNA library using PISTILLATA and APETALA3 as a bait. Besides several MIKC-type proteins, they also identified ATA20 as an interaction partner, providing the first evidence for a higher order complex consisting of MIKC-type proteins and non-MADS-domain proteins. However, ATA20 is an anther-specific protein that is thought to be secreted from tapetal cells, making it unlikely that it interacts with transcription factors *in planta* (Rubinelli *et al.*, 1998).

Although some non-MADS interaction partners of MIKC-type proteins have been identified, the functional significance of these factors is still unknown. This is in sharp contrast to animal and yeast MADS-domain proteins, where non-MADS cofactors have been shown to play an essential role in conferring functional specificity (for a review, see Messenguy and Dubois, 2003). Mutant analyses of the identified non-MADS interaction partners will reveal their biological significance in the context of MIKC-function. Remarkably, most putative cofactors show a flower-specific expression pattern. However, ectopic expression of the floral homeotic proteins AG, AP3/PI, and SEP in *Arabidopsis* results in the transformation of leaves into stamens, constitutive expression of AP3/PI and SEP converts leaves into petals (Honma and Goto, 2001). These data suggest that the floral homeotic proteins alone are sufficient to specify stamen and petal identity outside of the flower (Honma and Goto, 2001). Therefore, if any non-MADS cofactor is needed to specify petal or stamen identity, it has to be expressed throughout the plant or at least in the leaf primordia, making them susceptible for the ectopic expression of the floral homeotic proteins. There are several possible explanations why almost no such cofactor has been identified so far. For example, it is conceivable that binding to DNA is necessary for the cofactor interaction and hence the interaction has escaped detection with the yeast two-hybrid system. In addition, it is known that at least some MIKC-type proteins act in higher order complexes. Perhaps this complex formation is a prerequisite for the interaction with additional cofactors. In this case, screening with the "floral quartets" rather than with single proteins might be more appropriate. The possibility remains that the floral homeotic proteins specify floral organ identity without any cofactors.

B. HIGHER ORDER COMPLEXES AND THEIR INFLUENCE ON TARGET GENE SPECIFICITY

Egea-Cortines *et al.* (1999) demonstrated that the *Antirrhinum* MADS-domain proteins DEF, GLO, and SQUA interact with each other simultaneously, providing the first evidence for higher order complex formation among MIKC-type proteins. Shortly thereafter, Honma and Goto (2001) showed that the *Arabidopsis* floral homeotic proteins PI, AP3, and SEP (or AP1) as well as PI, AP3, SEP, and AG form higher order complexes in yeast *n*-hybrid assays. Moreover, constitutive expression of the respective genes leads to the conversion of leaves into petals and stamens, respectively (Honma and Goto, 2001). These findings led to the "floral quartet" model according to which the identity of the organs of each floral whorl is determined by a specific higher order complex composed of floral homeotic proteins. Sepals are specified by a complex of AP1 and SEP, an AP1/AP3/PI/SEP complex confers petal identity, stamens are determined by an AP3/PI/SEP/AG quartet and carpels by a complex of AG and SEP (Fig. 4) (Theißen, 2001; Theißen and Saedler, 2001). Meanwhile, similar experiments in *Petunia* strengthened the suggested fundamental role of higher order complex formation in flower development (Ferrario *et al.*, 2003). Moreover, a complex composed of SHP/STK/AG/SEP has been proposed to confer ovule identity in *Arabidopsis* (Fig. 4) (Favaro *et al.*, 2003; Pinyopich *et al.*, 2003). An ABS/STK/SEP complex has also been suggested to function in ovule development (Kaufmann *et al.*, 2005a).

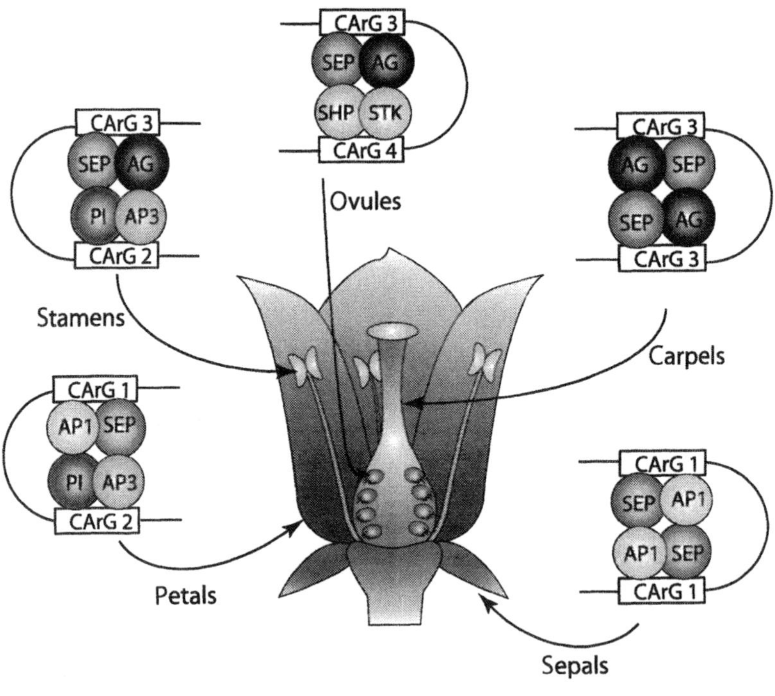

Fig. 4. The floral quartet model mapped on a generic flower of a core eudicot. Different higher order MIKC-type protein complexes determine the identity of each floral organ by binding to specific CArG-box sequences. The model is a refinement of previous ones (Kaufmann *et al.*, 2005b; Theißen, 2001; Theißen and Saedler, 2001). (See color Plate 5.)

Using FRET, circumstantial evidence was reported that higher order complexes exist *in planta* (Nougalli Tonaco *et al.*, 2006), supporting the floral quartet model and its way to explain target gene specificity. Although other scenarios are conceivable (Jack, 2001), a complex of four MADS-domain transcription factors might necessitate the presence of two CArG-boxes in target gene promoters. Multiple CArG-boxes have been found in close vicinity of genes potentially regulated by MIKC-type MADS-domain transcription factors (Table I). For some of them, *in vitro* and *in vivo* analyses indicate an important *in planta* function (Deyholos and Sieburth 2000; Gomez-Mena *et al.* 2005; Hill *et al.*, 1998; Ito *et al.*, 2004; Lee *et al.*, 2005; Tilly *et al.*, 1998). The presence of higher order complexes may also explain the results obtained by Riechmann and Meyerowitz (1997), that chimeric proteins in which the N-terminal half of the MADS-domain of a floral homeotic protein was replaced by the respective region of MEF2A are able to rescue the floral homeotic mutant phenotypes. One can imagine that even if one protein does not perfectly bind to the presumptive target sequence, it is recruited to this site within a higher order complex (Yang and Jack, 2004).

Apart from the results of Egea-Cortines *et al.* (1999), who showed that a multimeric complex composed of DEF/GLO and SQUA has a higher DNA-binding affinity than DEF/GLO heterodimers or SQUA homodimers, the DNA-binding specificity of higher order complexes is almost uncharacterized. However, complex formation can enhance specificity, as exemplified by Hox proteins. Whereas most Hox proteins alone bind to similar DNA-sequences, their binding specificity can be modulated by cofactors, leading to distinct binding sites for every Hox-cofactor complex (reviewed by Mann and Morata, 2000). Therefore, the formation of ternary or quarternary MIKC-type protein complexes might increase the DNA-binding specificity for distinct CArG-box sequences. But even if this is not the case, the simple need for two CArG-boxes to regulate one gene would narrow the number of potential targets and confer additional specificity. It is also possible that higher order complexes themselves do not confer target gene specificity but rather build a "platform" for the recruitment of additional factors (Kaufmann *et al.*, 2005b).

Besides this, ternary and quarternary complexes might also influence the promoter architecture more drastically than dimers. For example, differential DNA bending could be more significant in higher order complexes, providing another possibility for differential gene regulation.

It should also be noted that, even if the formation of higher order complexes is a plausible hypothesis for explaining organ identity, one can still imagine that MIKC-type protein dimers also play a role in target gene regulation (Kaufmann *et al.*, 2005b).

TABLE I

Summary of Identified Direct Target Genes of Floral Homeotic Proteins

Regulatory protein	Target protein	Proposed function	Number of CArG boxes identified	Reference
AP3/PI	NAC	NAC-domain containing transcription factor; involved in the transition from cell division to cell expansion in stamens and petals	1 (first intron)	(Sablowski and Meyerowitz, 1998)
AG, AP3/PI	AP3	MADS-domain transcription factor; confers petal and stamen identity	3 (upstream of the coding sequence)	(Gomez-Mena *et al.*, 2005; Hill *et al.*, 1998; Jack *et al.*, 1994; Tilly *et al.*, 1998)
AP1	AGL24	MADS-domain transcription factor; floral activator	Unknown	(Yu *et al.*, 2004)
AG	SPL	Transcriptional regulator; regulation of sporogenesis	2 (downstream of the coding sequence)	(Ito *et al.*, 2004)
AG	CRC	YABBY-transcription factor; involved in carpel and nectary development	2 (upstream of the coding sequence)	(Gomez-Mena *et al.*, 2005; Lee *et al.*, 2005)
AG	AG	MADS-domain transcription factor; determines carpel and stamen identity	2 (second intron)	(Gomez-Mena *et al.*, 2005; Hong *et al.*, 2003)
AG	At3g11000	Unknown	1 (upstream of the coding sequence)	(Gomez-Mena *et al.*, 2005)
AG	GA4	Enzyme; gibberellin biosynthesis	2 (upstream of the coding sequence)	(Gomez-Mena *et al.*, 2005)
AG	ATH1	Homeodomain transcription factor; regulation of gibberellin biosynthetic genes	1 (upstream of the coding sequence)	(Gomez-Mena *et al.*, 2005)
AG	At3g17010	B3 domain transcription factor; function unknown	1 (upstream of the coding sequence)	(Gomez-Mena *et al.*, 2005)
AG	SEP3	MADS-domain transcription factor; specifies petal, stamen, and carpel identity	1 (upstream of the coding sequence)	(Gomez-Mena *et al.*, 2005)
AG	At1g47610	WD-40 repeat family protein; function unknown	1 (upstream of the coding sequence)	(Gomez-Mena *et al.*, 2005)
AG	At1g13400	Zinc-finger transcription factor; function unknown	1 (upstream of the coding sequence)	(Gomez-Mena *et al.*, 2005)

VIII. TARGET GENES: HINTS FOR TARGET GENE SPECIFICITY?

Identifying direct targets of floral homeotic proteins is a difficult task, partly because it is hard to differentiate between direct and indirect targets. Sablowski and Meyerowitz (1998) solved this problem by fusing the coding region of *AP3* to the hormone-binding domain of the rat glucocorticoid receptor. The chimeric gene was expressed *in planta*. Normally, the glucocorticoid receptor is bound by HSP90 and resides in the cytosol, hindering transcriptional regulation by AP3. However, upon applying a steroid, HSP90 dissociates from the complex and the AP3-glucocorticoid receptor fusion can enter the nucleus and hence regulate transcription. By applying the translational inhibitor cycloheximide, only the mRNA levels of genes directly activated by AP3 are affected. Therefore, analyzing the expression patterns of genes in flowers where AP3 was induced compared to those where it was not is suitable for identifying direct target genes. Using this strategy, *NAP* was identified (Sablowski and Meyerowitz, 1998). Similar approaches led to the detection of *AGL24* as a target of AP1 and *SPL* as a target of AG (Ito *et al.*, 2004; Yu *et al.*, 2004). Employing a steroid inducible version of AG, Gomez-Mena *et al.* (2005) analyzed the expression pattern in early stamen and carpel organogenesis. As translation was not inhibited, *in vitro* binding of AG to CArG-boxes in the presumptive regulatory regions was used to confirm suspected direct target genes. For some of them also *in vivo* binding of AG was demonstrated (Gomez-Mena *et al.*, 2005). A summary of all identified direct target genes of floral homeotic proteins is given in Table I.

While most direct targets of floral homeotic proteins known are activated, *AGL24* is repressed by AP1. Moreover, there is also evidence that AP1 can act as a transcriptional activator (Honma and Goto, 2001; Ng and Yanofsky, 2001). This is important to note as it indicates that repression might be as crucial for proper function of the floral homeotic proteins as activation (Yu *et al.*, 2004). The ability to activate as well as to repress transcription is an essential prerequisite to explain target gene specificity with only limited DNA-binding specificity via a post-DNA-binding mechanism.

All direct target genes of MIKC-type proteins studied so far contain at least one CArG-box, indicating its importance for genes that are regulated by MIKC-type proteins. However, in the case of *NAP* and *AG*, the CArG-box is located in the first and second intron, respectively; in the case of *SPOROCYTELESS*, it is downstream of the transcription unit. This indicates that transcriptional regulation by MIKC-type proteins is not restricted to regions upstream of the coding sequence. The functional importance of the CArG-boxes in the second intron of *AGAMOUS* and downstream of *SPOROCYTELESS* was demonstrated using fusions of reporter genes with the sequences containing these CArG-boxes (Deyholos and Sieburth, 2000; Hong *et al.*, 2003; Ito *et al.*, 2004). However, deviations from the CArG-consensus are common; in the case of the *AP3*-promoter, even binding of AP3/PI to a promoter-fragment without a clear CArG-box was detected *in vitro* (Gomez-Mena *et al.*, 2005; Hill *et al.*, 1998). Moreover, changing the "imperfect" CArG-box in the 3′-region of *SPL* into a perfect one increased binding of AG *in vitro* (Ito *et al.*, 2004). This may suggest that binding with lower affinities is preferred over strong binding in some genes, allowing a differential regulation of target gene expression level. Alternatively, additional factors may enhance binding to "non-perfect" CArG-boxes *in vivo*.

More comprehensive target gene-studies compared microarray-based expression data of floral homeotic mutant and wild type plants (Wellmer *et al.*, 2004; Zik and Irish, 2003), of different floral organs (Bey *et al.*, 2004) or of different floral stages (Gomez-Mena *et al.*, 2005).

Wellmer *et al.* (2004) identified target genes of class A-, B-, and C-floral homeotic proteins. Surprisingly, only 13 and 18 genes specific for sepals and petals, respectively, were identified. In contrast, the number of stamen- and carpel-specific genes was assessed to be 1162 and 260, respectively. Importantly, no enrichment of CArG-boxes in the upstream regions of the organ-specific genes was detectable, indicating that most genes are not directly regulated by the floral homeotic proteins. Similar results were obtained by Gomez-Mena *et al.* (2005). However, almost all identified direct target genes are involved in transcriptional regulation (Table I). This supports the notion that floral homeotic genes activate a regulatory cascade rather than acting directly to regulate basic cellular functions that lead to organ differentiation (Gomez-Mena *et al.*, 2005). It will be interesting to see whether there is an enrichment of CArG-boxes in the regulatory regions of flower-specific transcription factors.

IX. EVOLUTIONARY IMPLICATIONS: NOTHING IN THE MADS WORLD MAKES SENSE EXCEPT IN THE LIGHT OF TARGET GENE SPECIFICITY

Changes in the regulatory network of floral homeotic genes underlie the morphological evolution of flowers and hence the generation of floral biodiversity (Kaufmann *et al.*, 2005b). However, network changes often have no obvious effect if target gene specificity remains unchanged. Knowledge about how target gene specificity is brought about might, therefore, dramatically increase our understanding of flower evolution. This is especially true for the evolutionary diversification of floral organs. In evolutionary terms, floral organs, such as sepals, petals, stamens, and carpels, are not strictly standardized structures but come in very different forms and functions, from very reduced or even absent to very large, conspicuous, and attractive for pollinators (or even human beings). It is well conceivable that changes in floral organ structure and function are brought about by changes in the interaction of floral organ identity proteins with their target genes. For any individual complex of floral homeotic proteins, target genes can be lost or gained during evolution, or activation and repression might be modified.

We suppose that at least three factors have contributed to differential target gene regulation: (1) *cis*-regulatory sequence mutations in target genes; (2) variations in expression pattern and protein concentration; and (3) changes in the protein coding sequence of the floral homeotic genes.

As some targets of floral homeotic proteins constitute transcription factors, changes in the respective *cis*-sites might be of great evolutionary significance. For example, it has been suggested that *CRC* plays an important role in carpel evolution (Fourquin *et al.*, 2005). Although highly speculative, one can imagine that the *de novo* origin of a CArG-box in the *CRC*-promoter made it susceptible for control by a preangiosperm AG-like protein, thereby recruiting it to the developmental pathway leading to carpel formation.

The most dramatic effect observed by changes in expression patterns of floral homeotic genes is homeosis, that is, one organ is completely transformed into another, the evolutionary significance of which has been emphasized by Ronse De Craene (2003). However, homeosis is only an extreme case on a continuous scale. Hybrid organs might be another result of changes in both pattern and level of floral homeotic gene expression (see Section VI). These hybrid organs often adopt new functions. For example, petaloid staminodes are recruited to prevent self-pollination or to attract pollinators in some Magnoliales (Walker-Larsen and Harder, 2000). If floral homeotic gene expression is changed in these organs, it would be plausible to assume that new "compositions" of target genes can lead to new floral forms. It is important to note that not only variations in the pattern but also in the level of expression might have an influence on target gene regulation.

Evidence that changes in the coding sequence of developmental key regulators are of evolutionary significance comes from homeodomain-proteins. It has been shown convincingly that the loss of phosphorylation sites, the gain of transcriptional repression motifs, changes in DNA-binding specificity and differences in the interaction with cofactors have shaped the evolutionary dynamics of differential target gene regulation and hence contributed to morphological diversity in the animal kingdom (reviewed by

Hsia and McGinnis, 2003). In MIKC-type proteins, small C-terminal motifs specific for the different functional classes of floral homeotic proteins have been detected (Kramer *et al.*, 1998, 2004; Litt and Irish, 2003; Zahn *et al.*, 2005). Moreover, due to frameshift mutations and subsequently fixation, some of these motifs changed dramatically during evolution (Litt and Irish, 2003; Vandenbussche *et al.*, 2003). As these frameshifts are found in two proteins involved in the specification of petal identity and coincide with the origin of core eudicots plants, they have been suggested to be of major importance for the origin of the petals in core eudicots (Litt and Irish, 2003; Vandenbussche *et al.*, 2003). The K-domain is another example thought to be responsible for differential complex formation and hence functional divergence of MIKC-type proteins (Kaufmann *et al.*, 2005b). However, it remains to be determined to what extent evolutionary changes in the K- and C-domains contributed to differential target gene regulation and hence the evolution of floral organs and flowers.

ACKNOWLEDGMENTS

Part of our work on the evolution of floral homeotic proteins is supported by a grant of the Deutsche Forschungsgemeinschaft (TH 417/5 to G. T.). R. M. receives a fellowship from the Studienstiftung des deutschen Volkes. We are grateful to Pam and Doug Soltis, Hong Ma, Claude dePamphilis, Jim Leebens-Mack, and other members of the Floral Genome Project for helpful discussions and the possibility to contribute to this volume. Special thanks to David Oppenheimer for valuable comments that helped to improve the manuscript.

REFERENCES

Angenent, G. C. and Colombo, L. (1996). Molecular control of ovule development. *Trends in Plant Science* **1**, 228–232.

Bey, M., Stüber, K., Fellenberg, K., Schwarz-Sommer, Z., Sommer, H., Saedler, H. and Zachgo, S. (2004). Characterization of *Antirrhinum* petal development and identification of target genes of the class B MADS box gene DEFICIENS. *Plant Cell* **16**, 3197–3215.

Biggin, M. D. and McGinnis, W. (1997). Regulation of segmentation and segmental identity by *Drosophila* homeoproteins: The role of DNA binding in functional activity and specificity. *Development* **124**, 4425–4433.

Buzgo, M., Soltis, P. S. and Soltis, D. E. (2004). Floral developmental morphology of *Amborella trichopoda* (Amborellaceae). *International Journal of Plant Sciences* **165**, 925–947.

Causier, B., Cook, H. and Davies, B. (2003). An *Antirrhinum* ternary complex factor specifically interacts with C-function and SEPALLATA-like MADS-box factors. *Plant Molecular Biology* **52**, 1051–1062.
Chan, S. K. and Mann, R. S. (1996). A structural model for a homeotic protein-extradenticle-DNA complex accounts for the choice of HOX protein in the heterodimer. *Proceedings of the National Academy of Sciences of the United States of America* **93**, 5223–5228.
Chan, S. K., Ryoo, H. D., Gould, A., Krumlauf, R. and Mann, R. S. (1997). Switching the *in vivo* specificity of a minimal Hox-responsive element. *Development* **124**, 2007–2014.
Coen, E. S. and Meyerowitz, E. M. (1991). The war of the whorls—genetic interactions controlling flower development. *Nature* **353**, 31–37.
Copeland, J. W., Nasiadka, A., Dietrich, B. H. and Krause, H. M. (1996). Patterning of the *Drosophila* embryo by a homeodomain-deleted Ftz polypeptide. *Nature* **379**, 162–165.
Crooks, G. E., Hon, G., Chandonia, J. M. and Brenner, S. E. (2004). WebLogo: A sequence logo generator. *Genome Research* **14**, 1188–1190.
Deyholos, M. K. and Sieburth, L. E. (2000). Separable whorl-specific expression and negative regulation by enhancer elements within the AGAMOUS second intron. *Plant Cell* **12**, 1799–1810.
Ditta, G., Pinyopich, A., Robles, P., Pelaz, S. and Yanofsky, M. F. (2004). The SEP4 gene of *Arabidopsis thaliana* functions in floral organ and meristem identity. *Current Biology* **14**, 1935–1940.
Egea-Cortines, M., Saedler, H. and Sommer, H. (1999). Ternary complex formation between the MADS-box proteins SQUAMOSA, DEFICIENS and GLOBOSA is involved in the control of floral architecture in *Antirrhinum majus*. *EMBO Journal* **18**, 5370–5379.
Favaro, R., Pinyopich, A., Battaglia, R., Kooiker, M., Borghi, L., Ditta, G., Yanofsky, M. F., Kater, M. M. and Colombo, L. (2003). MADS-box protein complexes control carpel and ovule development in *Arabidopsis*. *Plant Cell* **15**, 2603–2611.
Ferrario, S., Immink, R. G. H., Shchennikova, A., Busscher-Lange, J. and Angenent, G. C. (2003). The MADS box gene FBP2 is required for SEPALLATA function in petunia. *Plant Cell* **15**, 914–925.
Fourquin, C., Vinauger-Douard, M., Fogliani, B., Dumas, C. and Scutt, C. P. (2005). Evidence that CRABS CLAW and TOUSLED have conserved their roles in carpel development since the ancestor of the extant angiosperms. *Proceedings of the National Academy of Sciences of the United States of America* **102**, 4649–4654.
Gamboa, A., Paez-Valencia, J., Acevedo, G. F., Vazquez-Moreno, L. and Alvarez-Buylla, R. E. (2001). Floral transcription factor AGAMOUS interacts *in vitro* with a leucine-rich repeat and an acid phosphatase protein complex. *Biochemical and Biophysical Research Communications* **288**, 1018–1026.
Gomez-Mena, C., de Folter, S., Costa, M. M. R., Angenent, G. C. and Sablowski, R. (2005). Transcriptional program controlled by the floral homeotic gene AGAMOUS during early organogenesis. *Development* **132**, 429–438.
Goodrich, J., Puangsomlee, P., Martin, M., Long, D., Meyerowitz, E. M. and Coupland, G. (1997). A polycomb-group gene regulates homeotic gene expression in *Arabidopsis*. *Nature* **386**, 44–51.
Hartmann, U., Hohmann, S., Nettesheim, K., Wisman, E., Saedler, H. and Huijser, P. (2000). Molecular cloning of SVP: A negative regulator of the floral transition in *Arabidopsis*. *Plant Journal* **21**, 351–360.

He, Y. H. and Amasino, R. M. (2005). Role of chromatin modification in flowering-time control. *Trends in Plant Science* **10**, 30–35.

Hill, T. A., Day, C. D., Zondlo, S. C., Thackeray, A. G. and Irish, V. F. (1998). Discrete spatial and temporal cis-acting elements regulate transcription of the *Arabidopsis* floral homeotic gene APETALA3. *Development* **125**, 1711–1721.

Hong, R. L., Hamaguchi, L., Busch, M. A. and Weigel, D. (2003). Regulatory elements of the floral homeotic gene AGAMOUS identified by phylogenetic footprinting and shadowing. *Plant Cell* **15**, 1296–1309.

Honma, T. and Goto, K. (2000). The *Arabidopsis* floral homeotic gene PISTILLATA is regulated by discrete cis-elements responsive to induction and maintenance signals. *Development* **127**, 2021–2030.

Honma, T. and Goto, K. (2001). Complexes of MADS-box proteins are sufficient to convert leaves into floral organs. *Nature* **409**, 525–529.

Hsia, C. C. and McGinnis, W. (2003). Evolution of transcription factor function. *Current Opinion in Genetics & Development* **13**, 199–206.

Huang, H., Mizukami, Y., Hu, Y. and Ma, H. (1993). Isolation and characterization of the binding sequences for the product of the *Arabidopsis* floral homeotic gene AGAMOUS. *Nucleic Acids Research* **21**, 4769–4776.

Huang, H., Tudor, M., Weiss, C. A., Hu, Y. and Ma, H. (1995). The *Arabidopsis* Mads-box gene AGL3 is widely expressed and encodes a sequence-specific DNA-binding protein. *Plant Molecular Biology* **28**, 549–567.

Huang, H., Tudor, M., Su, T., Zhang, Y., Hu, Y. and Ma, H. (1996). DNA binding properties of two *Arabidopsis* MADS domain proteins: Binding consensus and dimer formation. *Plant Cell* **8**, 81–94.

Huang, K., Louis, J. M., Donaldson, L., Lim, F. L., Sharrocks, A. D. and Clore, G. M. (2000). Solution structure of the MEF2A-DNA complex: Structural basis for the modulation of DNA bending and specificity by MADS-box transcription factors. *EMBO Journal* **19**, 2615–2628.

Ito, T., Wellmer, F., Yu, H., Das, P., Ito, N., Alves-Ferreira, M., Riechmann, J. L. and Meyerowitz, E. M. (2004). The homeotic protein AGAMOUS controls microsporogenesis by regulation of SPOROCYTELESS. *Nature* **430**, 356–360.

Jack, T. (2001). Relearning our ABCs: New twists on an old model. *Trends in Plant Science* **6**, 310–316.

Jack, T. (2004). Molecular and genetic mechanisms of floral control. *Plant Cell* **16**, S1–S17.

Jack, T., Fox, G. L. and Meyerowitz, E. M. (1994). *Arabidopsis* homeotic gene APETALA3 ectopic expression: Transcriptional and posttranscriptional regulation determine floral organ identity. *Cell* **76**, 703–716.

Jack, T., Sieburth, L. and Meyerowitz, E. (1997). Targeted misexpression of AGAMOUS in whorl 2 of Arabidopsis flowers. *Plant Journal* **11**, 825–839.

Kaufmann, K., Anfang, N., Saedler, H. and Theissen, G. (2005a). Mutant analysis, protein-protein interactions and subcellular localization of the *Arabidopsis* B-sister (ABS) protein. *Molecular Genetics and Genomics* **274**, 103–118.

Kaufmann, K., Melzer, R. and Theissen, G. (2005b). MIKC-type MADS-domain proteins: Structural modularity, protein interactions and network evolution in land plants. *Gene* **347**, 183–198.

Kerppola, T. K. and Curran, T. (1991). Fos-Jun heterodimers and Jun homodimers bend DNA in opposite orientations—implications for transcription factor cooperativity. *Cell* **66**, 317–326.

Kramer, E. M., Dorit, R. L. and Irish, V. F. (1998). Molecular evolution of genes controlling petal and stamen development: Duplication and divergence within the APETALA3 and PISTILLATA MADS-box gene lineages. *Genetics* **149**, 765–783.

Kramer, E. M., Jaramillo, M. A. and Di Stilio, V. S. (2004). Patterns of gene duplication and functional evolution during the diversification of the AGAMOUS subfamily of MADS box genes in angiosperms. *Genetics* **166**, 1011–1023.

Krizek, B. A. and Meyerowitz, E. M. (1996). Mapping the protein regions responsible for the functional specificities of the *Arabidopsis* MADS domain organ-identity proteins. *Proceedings of the National Academy of Sciences of the United States of America* **93**, 4063–4070.

Kumari, D., Gabrielian, A., Wheeler, D. and Usdin, K. (2005). The roles of Sp1, Sp3, USF1/USF2 and NRF-1 in the regulation and three-dimensional structure of the Fragile X mental retardation gene promoter. *Biochemical Journal* **386**, 297–303.

Lee, J. Y., Baum, S. F., Alvarez, J., Patel, A., Chitwood, D. H. and Bowman, J. L. (2005). Activation of CRABS CLAW in the nectaries and carpels of *Arabidopsis*. *Plant Cell* **17**, 25–36.

Li, X. L. and McGinnis, W. (1999). Activity regulation of Hox proteins, a mechanism for altering functional specificity in development and evolution. *Proceedings of the National Academy of Sciences of the United States of America* **96**, 6802–6807.

Lim, F. L., Hayes, A., West, A. G., Pic-Taylor, A., Darieva, Z., Morgan, B. A., Oliver, S. G. and Sharrocks, A. D. (2003). Mcm1p-induced DNA bending regulates the formation of ternary transcription factor complexes. *Molecular and Cellular Biology* **23**, 450–461.

Litt, A. and Irish, V. F. (2003). Duplication and diversification in the APETALA1/FRUITFULL floral homeotic gene lineage: Implications for the evolution of floral development. *Genetics* **165**, 821–833.

Mann, R. S. and Chan, S. K. (1996). Extra specificity from extradenticle: The partnership between HOX and PBX/EXD homeodomain proteins. *Trends in Genetics* **12**, 258–262.

Mann, R. S. and Morata, G. (2000). The developmental and molecular biology of genes that subdivide the body of *Drosophila*. *Annual Review of Cell and Developmental Biology* **16**, 243–271.

Masiero, S., Imbriano, C., Ravasio, F., Favaro, R., Pelucchi, N., Gorla, M. S., Mantovani, R., Colombo, L. and Kater, M. M. (2002). Ternary complex formation between MADS-box transcription factors and the histone fold protein NF-YB. *Journal of Biological Chemistry* **277**, 26429–26435.

Messenguy, F. and Dubois, E. (2003). Role of MADS box proteins and their cofactors in combinatorial control of gene expression and cell development. *Gene* **316**, 1–21.

Meyerowitz, E. M. (2002). Plants compared to animals: The broadest comparative study of development. *Science* **295**, 1482–1485.

Mizukami, Y. and Ma, H. (1992). Ectopic expression of the floral homeotic gene AGAMOUS in transgenic *Arabidopsis* plants alters floral organ identity. *Cell* **71**, 119–131.

Molkentin, J. D. and Olson, E. N. (1996). Combinatorial control of muscle development by basic helix-loop-helix and MADS-box transcription factors. *Proceedings of the National Academy of Sciences of the United States of America* **93**, 9366–9373.

Mouradov, A., Cremer, F. and Coupland, G. (2002). Control of flowering time: Interacting pathways as a basis for diversity. *Plant Cell* **14**, S111–S130.

Nasiadka, A., Grill, A. and Krause, H. M. (2000). Mechanisms regulating target gene selection by the homeodomain-containing protein Fushi tarazu. *Development* **127**, 2965–2976.

Ng, M. and Yanofsky, M. F. (2001). Activation of the *Arabidopsis* B class homeotic genes by APETALA1. *Plant Cell* **13**, 739–753.

Nougalli Tonaco, I. A., Borst, J. W., de Vries, S. C., Angenent, G. C. and Immink, R. G. H. (2006). *In vivo* imaging of MADS-box transcription factor interactions. *Journal of Experimental Botany* **57**, 33–42.

Osborn, T. C. (2004). The contribution of polyploidy to variation in *Brassica* species. *Physiologia plantarum* **121**, 531–536.

Passmore, S., Elble, R. and Tye, B. K. (1989). A protein involved in minichromosome maintenance in yeast binds a transcriptional enhancer conserved in eukaryotes. *Genes & Development* **3**, 921–935.

Pelaz, S., Ditta, G. S., Baumann, E., Wisman, E. and Yanofsky, M. F. (2000). B and C floral organ identity functions require SEPALLATA MADS-box genes. *Nature* **405**, 200–203.

Pellegrini, L., Song, T. and Richmond, T. J. (1995). Structure of serum response factor core bound to DNA. *Nature* **376**, 490–498.

PerezMartin, J. and deLorenzo, V. (1997). Clues and consequences of DNA bending in transcription. *Annual Review of Microbiology* **51**, 593–628.

Pinyopich, A., Ditta, G. S., Savidge, B., Liljegren, S. J., Baumann, E., Wisman, E. and Yanofsky, M. F. (2003). Assessing the redundancy of MADS-box genes during carpel and ovule development. *Nature* **424**, 85–88.

Pollock, R. and Treisman, R. (1990). A sensitive method for the determination of protein-DNA binding specificities. *Nucleic Acids Research* **18**, 6197–6204.

Pramila, T., Miles, S., GuhaThakurta, D., Jemiolo, D. and Breeden, L. L. (2002). Conserved homeodomain proteins interact with MADS box protein Mcm1 to restrict ECB-dependent transcription to the M/G1 phase of the cell cycle. *Genes & Development* **16**, 3034–3045.

Riechmann, J. L. and Meyerowitz, E. M. (1997). Determination of floral organ identity by Arabidopsis MADS domain homeotic proteins AP1, AP3, PI, and AG is independent of their DNA-binding specificity. *Molecular Biology of the Cell* **8**, 1243–1259.

Riechmann, J. L., Wang, M. Q. and Meyerowitz, E. M. (1996). DNA-binding properties of Arabidopsis MADS domain homeotic proteins APETALA1, APETALA3, PISTILLATA and AGAMOUS. *Nucleic Acids Research* **24**, 3134–3141.

Ronse De Craene, L. P. (2003). The evolutionary significance of homeosis in flowers: A morphological perspective. *International Journal of Plant Sciences* **164**, S225–S235.

Rubinelli, P., Hu, Y. and Ma, H. (1998). Identification, sequence analysis and expression studies of novel anther-specific genes of *Arabidopsis thaliana*. *Plant Molecular Biology* **37**, 607–619.

Ruiz-Garcia, L., Cervera, M. T. and Martinez-Zapater, J. M. (2005). DNA methylation increases throughout *Arabidopsis* development. *Planta* **222**, 301–306.

Ryoo, H. D. and Mann, R. S. (1999). The control of trunk Hox specificity and activity by Extradenticle. *Genes & Development* **13**, 1704–1716.

Sablowski, R. W. M. and Meyerowitz, E. M. (1998). A homolog of NO APICAL MERISTEM is an immediate target of the floral homeotic genes APETALA3/PISTILLATA. *Cell* **92**, 93–103.

Scaffidi, P. and Bianchi, M. E. (2001). Spatially precise DNA bending is an essential activity of the Sox2 transcription factor. *Journal of Biological Chemistry* **276**, 47296–47302.

Schneider, T. D. and Stephens, R. M. (1990). Sequence Logos: A new way to display consensus sequences. *Nucleic Acids Research* **18**, 6097–6100.

Scortecci, K. C., Michaels, S. D. and Amasino, R. M. (2001). Identification of a MADS-box gene, FLOWERING LOCUS M, that represses flowering. *Plant Journal* **26**, 229–236.

Sekinger, E. A., Moqtaderi, Z. and Struhl, K. (2005). Intrinsic histone-DNA interactions and low nucleosome density are important for preferential accessibility of promoter regions in yeast. *Molecular Cell* **18**, 735–748.

Shiraishi, H., Okada, K. and Shimura, Y. (1993). Nucleotide sequences recognized by the AGAMOUS-MADS domain of *Arabidopsis thaliana in vitro*. *Plant Journal* **4**, 385–398.

Theißen, G. (2001). Development of floral organ identity: Stories from the MADS house. *Current Opinion in Plant Biology* **4**, 75–85.

Theißen, G. and Saedler, H. (2001). Plant biology. Floral quartets. *Nature* **409**, 469–471.

Tilly, J. J., Allen, D. W. and Jack, T. (1998). The CArG boxes in the promoter of the Arabidopsis floral organ identity gene APETALA3 mediate diverse regulatory effects. *Development* **125**, 1647–1657.

Vandenbussche, M., Theißen, G., Van de Peer, Y. and Gerats, T. (2003). Structural diversification and neo-functionalization during floral MADS-box gene evolution by C-terminal frameshift mutations. *Nucleic Acids Research* **31**, 4401–4409.

Vandenbussche, M., Zethof, J., Royaert, S., Weterings, K. and Gerats, T. (2004). The duplicated B-class heterodimer model: Whorl-specific effects and complex genetic interactions in *Petunia hybrida* flower development. *Plant Cell* **16**, 741–754.

Walker-Larsen, J. and Harder, L. D. (2000). The evolution of staminodes in angiosperms: Patterns of stamen reduction, loss, and functional re-invention. *American Journal of Botany* **87**, 1367–1384.

Walter, J. and Biggin, M. D. (1996). DNA binding specificity of two homeodomain proteins *in vitro* and in *Drosophila* embryos. *Proceedings of the National Academy of Sciences of the United States of America* **93**, 2680–2685.

Walter, J., Dever, C. A. and Biggin, M. D. (1994). Two homeo domain proteins bind with similar specificity to a wide-range of DNA sites in *Drosophila* embryos. *Genes & Development* **8**, 1678–1692.

Wellmer, F., Riechmann, J. L., Alves-Ferreira, M. and Meyerowitz, E. M. (2004). Genome-wide analysis of spatial gene expression in *Arabidopsis* flowers. *Plant Cell* **16**, 1314–1326.

West, A. G. and Sharrocks, A. D. (1999). MADS-box transcription factors adopt alternative mechanisms for bending DNA. *Journal of Molecular Biology* **286**, 1311–1323.

West, A. G., Shore, P. and Sharrocks, A. D. (1997). DNA binding by MADS-box transcription factors: A molecular mechanism for differential DNA bending. *Molecular and Cellular Biology* **17**, 2876–2887.

West, A. G., Causier, B. E., Davies, B. and Sharrocks, A. D. (1998). DNA binding and dimerisation determinants of *Antirrhinum majus* MADS-box transcription factors. *Nucleic Acids Research* **26**, 5277–5287.

Wray, G. A., Hahn, M. W., Abouheif, E., Balhoff, J. P., Pizer, M., Rockman, M. V. and Romano, L. A. (2003). The evolution of transcriptional regulation in eukaryotes. *Molecular Biology and Evolution* **20**, 1377–1419.

Wynne, J. and Treisman, R. (1992). SRF and MCM1 have related but distinct DNA-binding specificities. *Nucleic Acids Research* **20**, 3297–3303.

Yang, Y. Z. and Jack, T. (2004). Defining subdomains of the K domain important for protein-protein interactions of plant MADS proteins. *Plant Molecular Biology* **55**, 45–59.

Yie, J. M., Merika, M., Munshi, N., Chen, G. Y. and Thanos, D. (1999). The role of HMG I(Y) in the assembly and function of the IFN-beta enhanceosome. *EMBO Journal* **18**, 3074–3089.

Yu, H., Xu, Y. F., Tan, E. L. and Kumar, P. P. (2002). AGAMOUS-LIKE 24, a dosage-dependent mediator of the flowering signals. *Proceedings of the National Academy of Sciences of the United States of America* **99**, 16336–16341.

Yu, H., Ito, T., Wellmer, F. and Meyerowitz, E. M. (2004). Repression of AGAMOUS-LIKE 24 is a crucial step in promoting flower development. *Nature Genetics* **36**, 157–161.

Zahn, L. M., King, H. Z., Leebens-Mack, J. H., Kim, S., Soltis, P. S., Landherr, L. L., Soltis, D. E., dePamphilis, C. W. and Ma, H. (2005). The evolution of the SEPALLATA subfamily of MADS-Box genes: A preangiosperm origin with multiple duplications throughout angiosperm history. *Genetics* **169**, 2209–2223.

Zhao, M., New, L., Kravchenko, V. V., Kato, Y., Gram, H., di Padova, F., Olson, E. N., Ulevitch, R. J. and Han, J. H. (1999). Regulation of the MEF2 family of transcription factors by p38. *Molecular and Cellular Biology* **19**, 21–30.

Zik, M. and Irish, V. F. (2003). Global identification of target genes regulated by APETALA3 and PISTILLATA floral homeotic gene action. *Plant Cell* **15**, 207–222.

Genetics of Floral Development in *Petunia*

ANNEKE RIJPKEMA, TOM GERATS
AND MICHIEL VANDENBUSSCHE

*IWWR/Plant Genetics, Toernooiveld 1, 6525ED Nijmegen,
The Netherlands*

Advances in Botanical Research, Vol. 44
Incorporating Advances in Plant Pathology

0065-2296/06 $35.00
DOI: 10.1016/S0065-2296(06)44006-4

ABSTRACT

In the last two decades the genetic and molecular research on floral development has advanced tremendously. Initially the research focused mostly on the two species of which the homeotic floral development mutants formed the basis for the ABC-model: *Arabidopsis* and *Antirrhinum*. In recent years the importance of studying a wider range of species, especially in an "evo-devo" context, has become more and more evident. This review summarizes advances in the understanding of the genetic control of floral induction, inflorescence formation, and floral organ formation in *Petunia*. Moreover, we put the knowledge on *Petunia* floral development in the broader perspective of what is known on floral development in other species, thus pointing out differences and resemblances in the regulatory systems that ultimately form the basis for the enormous variation in flower shapes.

I. INTRODUCTION

A. SOLANACEAE

Petunia belongs to the Solanaceae, which is a plant family of great economic importance. Solanaceaous species are used for food (e.g., potato, tomato, pepper, eggplant), as drugs (e.g., tobacco, deadly nightshade, mandrake), and as ornamentals (e.g., petunia, velvet tongue, *Datura* spp., *Schizanthus* spp.) (Knapp *et al.*, 2004). Solanaceae can be found worldwide, from the driest deserts to tropical rainforests. The highest species diversity is found in the Neotropics. Estimates of species diversity in the family range from 3000 to 4000 species, almost half of which belonging to the large cosmopolitan genus *Solanum* (see Knapp, 2002b for a review of the genera in the family). The family is diverse, both in terms of life form, with species ranging from ephemeral herbs (*Leptoglossis* and *Schizanthus* of the Chilean deserts) to large forest trees (*Duckeodendron* of the Amazon), and in flower and fruit morphology (Knapp, 2002a,b). A literature and illustrations database on Solanaceae can be found at (http://www.bgard.science.ru.nl/solanaceae). Moreover, a huge international effort, SOL, aims at sequencing the tomato genome (Mueller *et al.*, 2005). Members of the Solanaceae family show a wide range of floral morphologies. Some species have zygomorphic or monosymmetric flowers, while others have actinomorphic or radially symmetric flowers (Knapp, 2002b).

B. THE GENUS *PETUNIA*

In 1803, Jussieu established the genus *Petunia* (Solanaceae), later referred to as *Petunia sensu* Jussieu. In the 1980–1990s *P. sensu* Jussieu was divided into two genera: *Petunia* and *Calibrachoa*. Ando *et al.* (2005) performed a detailed phylogenetic analysis of *P. sensu* Jussieu and demonstrated that the separation of *Petunia* and *Calibrachoa* into different genera is supported by chloroplast DNA RFLP data. Several clades in the *Petunia* phylogenetic trees were found to correspond with geographic distribution, suggesting that recent speciation occurred independently in different regions. To date, around 30 *Petunia* (sub)species have been described. The geographic origin of *Petunia* is the southern/central part of South America, and various species have been documented from collections made in Argentina, Brazil, Paraguay, and Uruguay (Ando *et al.*, 2005).

C. *PETUNIA* FLOWER CHARACTERISTICS

Petunia plants, as other Solanaceae species, exhibit determinate inflorescences composed of scorpioid or cincinnus cymes (Souer *et al.*, 1998; Weberling, 1989). At the base of each flower, two bracts are formed, each with a dormant (vegetative) meristem in its axil. A wild type flower contains five sepals, five petals, five stamens, and two carpels arranged in four concentric whorls. The five petals are fused. Stamen filaments are partly fused to the tube of the flower. The *Petunia* flower is zygomorphic in all floral whorls, which is partly due to the whorled or irregular arrangement of sepals and petals in the floral bud before it opens (Knapp, 2002b). Ovules have a single integument (Angenent *et al.*, 1995). The fruit is a capsule, containing variable amounts of seeds for different *Petunia* species (Gunn and Gaffney, 1974; Sink and Power, 1978). Seed capsules are conic, widest at the base and tapering to the apex. The mature fruit is surrounded by an enlarged, glandular-hairy calyx composed of five lobes (the sepals) which are equal to or longer than the capsule, depending on the species (Gunn and Gaffney, 1974) (Fig. 1).

D. FLORAL DIVERSITY IN THE *PETUNIA* CLADE: POLLINATION SYNDROMES

Hawkmoth (in *Petunia axillaris*) and bee (in *P. integrifolia*) pollination form typical examples of pollination syndromes in the genus *Petunia*. These two representatives of two groups of *Petunia* species have a complex set of morphological and physiological traits that are adapted to their respective pollinators. *P. axillaris* has white flowers, with long petal tubes that exactly fit the length of the tongue of the hawkmoths that pollinate them (*Manduca contracta* and *M. diffissa* ssp. *Petuniae*). Moreover, for the nocturnally active hawkmoths a colored flower is not as important as a strong scent, and in accordance with that *P. axillaris* has white, nocturnally scented flowers (Ando *et al.*, 2001; Stuurman *et al.*, 2004). *P. integrifolia* has unscented purple colored flowers, with a wide and short petal tube. Flowers of

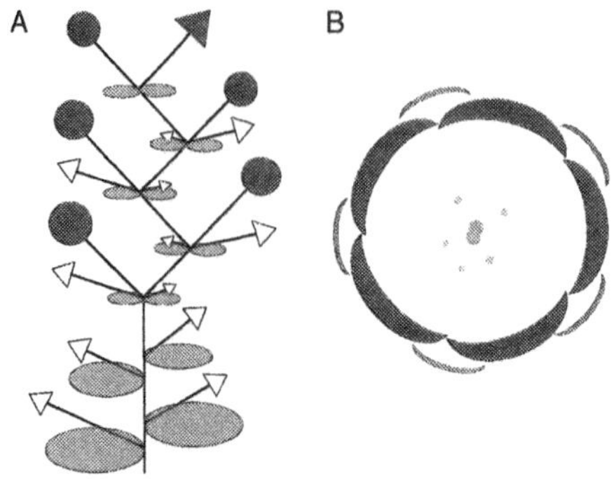

Fig. 1. (A) Schematic representation of *Petunia* branching in the reproductive phase (after Souer *et al.*, 1998). The position of the flowers is indicated by colored circles and the apical inflorescence meristem by a colored triangle. Leaves and bracts are shown by large and small green ovals, respectively. Vegetative axillary meristems are shown by open triangles. A smaller size of these triangles indicates a stronger dormancy. (B) Floral diagram of *Petunia* (after Knapp, 2002b). The floral organs, sepals (whorl 1), petals (whorl 2), stamens (whorl 3), and carpels (whorl 4) are depicted in green, red, yellow, and light green, respectively. The smaller stamen indicates the stamen in this position is reduced in some *Petunia* varieties. (See color Plate 6.)

P. integrifolia exhibit diurnal opening and closing movements synchronous with the activity period of the bee that pollinates them (*Hexantheda* sp.). Amounts of floral nectar in *P. axillaris* and *P. integrifolia* are within the range of hawkmoth- and bee-pollinated flowers, respectively (Ando *et al.*, 2001; Stuurman *et al.*, 2004). A thorough study in which the *Petunia* pollination syndromes have been dissected into their most important phenotypic and genetic components has been conducted by Stuurman *et al.* (2004) (Fig. 2).

Natural hybrids of *P. axillaris* and *P. integrifolia* have never been reported, even though artificial crosses can produce fertile hybrid offspring (garden petunias are known to be descendants of such a hybrid) and the two species grow together. Therefore, Ando *et al.* (2001) studied the reproductive isolation between the two species. Differential insect visitation of *P. integrifolia* and *P. axillaris* in sympatric populations was observed, suggesting an important biological meaning of the floral differences in color, scent, and amount of nectar. Insect visitation is not the only reproductive barrier among *Petunia* species, as genetic incompatibilities between ecotypes can also be considerable (Ando *et al.*, 2001). Still, the extensive divergence in the *Petunia* floral pollination syndromes indicates that insect visitation has certainly had a huge impact on the evolutionary history of the *Petunia* species (Stuurman *et al.*, 2004).

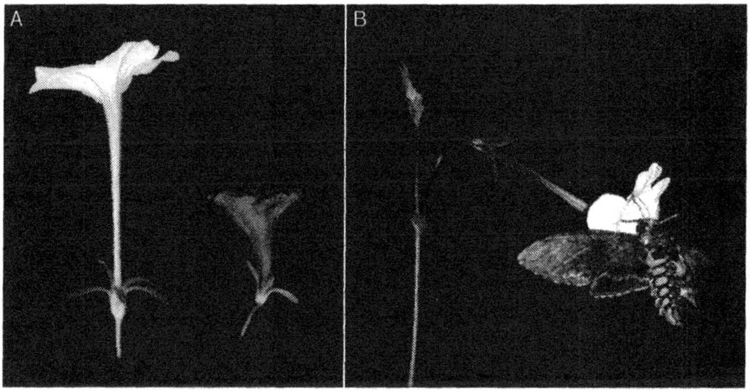

Fig. 2. (A) *P. axillaris* flower (left) and a *P. integrifolia* flower (right). Picture courtesy of Mary Hoballah and Jeroen Stuurman. (B) *Manduca sexta* hawkmoth on a *P. axillaris* flower. Picture courtesy of Mary Hoballah and Jeroen Stuurman. (See color Plate 7.)

Two important characteristics in pollination syndromes are scent production and floral color. In *Petunia hybrida*, the floral aroma is predominantly determined by volatile benzoids (Verdonk *et al.*, 2003). Verdonk *et al.* (2005) identified *ODORANT1* (*ODO1*), a member of the R2R3-type MYB family, as a candidate for the regulation of volatile benzoid production in *P. hybrida* "Mitchell" (W115) flowers. Underwood *et al.* (2005) demonstrated, using transgenic ethylene insensitive "Mitchell" lines, that the production of volatile organic compounds is regulated by ethylene. Once the flower has been pollinated and attraction of pollinators is no longer necessary, ethylene acts as a signal to downregulate the expression of scent biosynthetic genes (Negre *et al.*, 2003). Flower color is mainly determined by flavonoid components. The genetics, biochemistry, and molecular biology of flavonoid synthesis are fairly well understood in *Petunia* (Koes *et al.*, 2005; Martin and Gerats, 1993; Spelt *et al.*, 2002).

E. RESEARCH ON *PETUNIA* FLOWER DEVELOPMENT

Petunia has been studied since around 1830, and there are some early papers that mention specific flower developmental mutants. To quote Bailey (1896), who was referring to work by Harrison from around 1838: "Various curiously marked types of petunias have appeared and are lost. One of the early forms had a red body color with grass-green borders." Further, Bailey quotes Carman (Proc. Sixth Conv. Soc. Am. Flor., 1890) as obtaining plants with "rosettes of green leaves without the rudiments of calyx, corolla, stamens, or pistils." The last description is reminiscent of a full *sep* phenotype

(Ditta *et al.*, 2004). Levan (1937) describes a mutant in which ovules have been replaced by leaf-like structures, comparable to the phenotype of a double mutant for two MADS-box genes, *fbp2/fbp5* (Vandenbussche *et al.*, 2003b). More information on various aspects of the use of *Petunia* in research can be found in Gerats and Vandenbussche (2005). On the Petunia platform website (http://www.petuniaplatform.net) most groups working with *Petunia* as a main system are presented. A valuable resource for background information on culture and various research aspects of *Petunia* still is Sink's 1984 monograph "Petunia."

F. *PETUNIA* IN MOLECULAR STUDIES

A range of materials, techniques, and strategic approaches make *Petunia* a feasible system to work with. Besides easy culture conditions, an endogenous transposable element system is available, which can be used efficiently in both forward and reverse approaches. Two extensively used *Petunia* varieties in molecular research are the easy-to-transform "Mitchell diploid" and the high copy-number *dTph1* transposon line "W138" (for details see Gerats and Vandenbussche, 2005). Forward approaches are primarily performed by Transposon Display methods (De Keukeleire *et al.*, 2001; Van den Broeck *et al.*, 1998). Reverse approaches have been optimized over the years (Koes *et al.*, 1995; Vandenbussche *et al.*, 2003b). Many of the genes involved in floral development (floral transition, floral patterning) are MADS-box genes (see Irish, Chapter 3; Kramer and Zimmer, Chapter 9; and Soltis *et al.*, Chapter 12). Over time, a number of these have been studied by transgenic methods (e.g., Angenent *et al.*, 1994, 1995; Immink *et al.*, 1999) or by insertional mutagenesis as mentioned in an earlier section. For all major clades of MADS-box genes *Petunia* members are known (for an overview see Vandenbussche *et al.*, 2003b). In this review we only focus on the genes for which functional data are present.

G. FLORAL DEVELOPMENT

Flower development can be divided into several distinct phases: (1) transition to flowering, (2) inflorescence/flower meristem formation, and (3) floral organ patterning. It appears that, while in general molecular aspects of flower development are quite comparable for *Arabidopsis* and *Antirrhinum*, this can not always be generalized to fully encompass other species like *Petunia*. Thus, to discern ornamental differences from fundamental ones, it is important to develop insights in a range of systems.

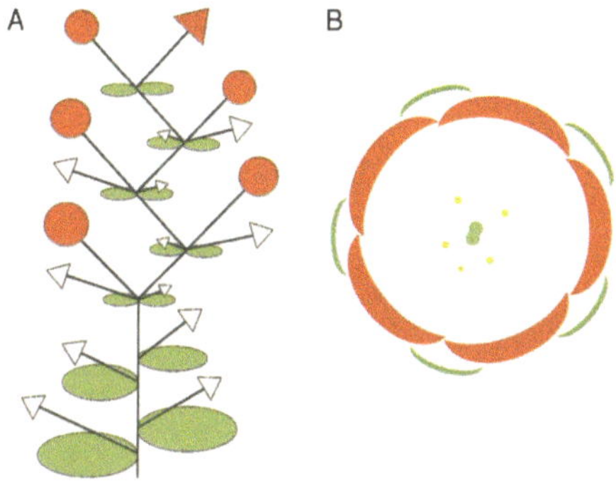

Plate 6

Plate 7

II. THE TRANSITION TO FLOWERING

The transition from the vegetative to the reproductive phase is an important developmental shift in the plant life cycle, and its timing is critical for reproductive success. This shift is characterized by the induction and development of an inflorescence meristem that generates floral meristems. This morphogenetic change is controlled by endogenous factors, where the program to flowering is turned on after a certain time of vegetative growth or when a defined number of leaves or biomass is produced, and by environmental conditions. In *Arabidopsis*, a number of genetic pathways controlling flowering time (see Engelmann and Purugganan, Chapter 13) have been identified, and a lot of genes involved in these pathways have been studied extensively. Models now extend beyond "primary" controlling factors and show an ever-increasing number of cross-talks between pathways triggered or influenced by various environmental factors and hormones (mainly gibberellins) (reviewed in Bernier and Perilleux, 2005; Boss *et al.*, 2004).

For *Petunia* there is less extensive knowledge on the regulatory mechanisms and genes involved in floral transition. We do know flowering in *Petunia* is photoperiodically controlled, and long day conditions or a night interruption with artificial light promote early flowering (Adams *et al.*, 1999). Moreover, quite some work on the participation of gibberellins and gibberellin-induced proteins in diverse developmental processes in *Petunia*, including flower induction, development, and pigmentation, has been done by the group of David Weiss (e.g. Ben-Nissan *et al.*, 2004; Izhaki *et al.*, 2001; Weiss, 2000).

When studying floral transition in *Petunia* and genes involved in the genetic pathways controlling flowering time, obvious candidates are genes homologous to *Arabidopsis* genes with a known function in floral transition. The key genes integrating multiple floral transition promoting pathways in *Arabidopsis* are *FLOWERING LOCUS T* (*FT*), *LEAFY* (*LFY*), and *SUPPRESSOR OF OVEREXPRESSION OF CONSTANS1* (*SOC1*; also known as *AGAMOUS-LIKE20*) (Blazquez and Weigel, 2000; Kardailsky *et al.*, 1999; Kobayashi *et al.*, 1999; Lee *et al.*, 2000; Nilsson *et al.*, 1998; Samach *et al.*, 2000). These three floral pathway integrators have both overlapping and independent functions in the determination of flowering time and floral initiation (Moon *et al.*, 2005).

SOC1 is a MADS-box gene that has a promotive effect on flowering. *SOC1* is activated during the transition to flowering; transgenic plants overexpressing *SOC1* flower early while *soc1* mutants are delayed in flowering (Borner *et al.*, 2000; Samach *et al.*, 2000). The *SOC1* gene integrates signals from the photoperiod, vernalization, and gibberellin pathways (Blazquez, 2000;

Borner *et al.*, 2000; Moon *et al.*, 2003; Samach *et al.*, 2000). As can be seen in the phylogenetic tree in Fig. 3 four genes have been identified in *Petunia* belonging to the *SOC1/TM3* clade (Immink *et al.*, 2003). *Arabidopsis* also has five additional *SOC1/TM3* clade members besides *SOC1* itself (*AGAMOUS LIKE14, AGL19, AGL42, AGL71, AGL72*). Therefore, without thorough functional analyses it is impossible to find out which *Petunia* gene, or which combination of genes, is functionally orthologous to the *Arabidopsis SOC1* gene. The *Arabidopsis SOC1* gene is expressed in most organs at variable levels, but upon floral induction its expression is rapidly upregulated in the apical meristems, whereas in vegetative plants only very little *SOC1* transcript can be detected in these meristems. Later, during floral development, *SOC1* is expressed in apical meristems and in procambial strands of developing inflorescences. Although *SOC1* is not expressed in emerging floral meristems, it was detectable in the center of floral meristems at a later stage (Borner *et al.*, 2000).

The *Petunia* members of the *SOC1/TM3* clade (Fig. 3), *FLORAL BINDING PROTEIN21* (*FBP21*), *FBP22, FBP28*, and *UNSHAVEN* (*UNS*; formerly called *FBP20*), have related expression patterns. All are mainly expressed in the vegetative tissues of the plant, however some differences in expression patterns have been observed (Immink *et al.*, 2003). It is not yet clear if an upregulation of expression of either of these *Petunia SOC1/TM3* clade genes upon floral transition takes place (as for *SOC1* in *Arabidopsis*).

Transposon insertion knockout mutants have so far only been identified for *UNS* and *FBP28*. The *uns* and *fbp28* single mutants are similar to

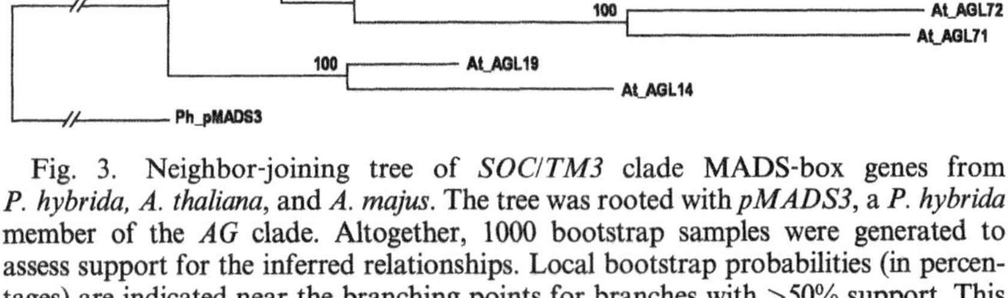

Fig. 3. Neighbor-joining tree of *SOC/TM3* clade MADS-box genes from *P. hybrida, A. thaliana*, and *A. majus*. The tree was rooted with *pMADS3*, a *P. hybrida* member of the *AG* clade. Altogether, 1000 bootstrap samples were generated to assess support for the inferred relationships. Local bootstrap probabilities (in percentages) are indicated near the branching points for branches with >50% support. This neighbor-joining tree, and the ones shown in Figs. 4–7, were obtained according to the methodology described previously (Vandenbussche *et al.*, 2003a).

wild-type plants, indicating that if the *Petunia SOC1/TM3* genes have a function in floral transition or development, they probably act redundantly.

Constitutive expression of *UNS* under control of the *Cauliflower mosaic virus* 35S (CaMV 35S) promoter indicated that *UNS* might nevertheless have a function in floral transition similar to that of *SOC1*. *UNS* overexpression leads to an acceleration of flowering, as also found in *SOC1* overexpressing plants. In addition, these *UNS* overexpressing transgenic plants exhibit ectopic trichome formation on floral organs and a conversion of petals into organs with leaf-like features, the so-called unshaven floral phenotype (Ferrario *et al.*, 2004). Ferrario *et al.* set up an experiment to find out if part of, or the whole phenotype could be due to a dominant-negative action of the protein, rather than showing the native protein function. A truncated version of *UNS*, lacking the MADS-box domain, was introduced. This truncated protein was shown not to be translocated to the nucleus, and any phenotype resulting from its introduction in the plant could, therefore, only be due to a dominant negative action of the protein. With overexpression of a truncated version of *UNS* the same floral phenotype, accompanied by a delay in flowering, was obtained. Thus, the conclusion was that the "unshaven" phenotype had nothing to do with the protein's function. However, the early flowering of the plants constitutively expressing *UNS* under control of the CaMV 35S promoter did represent the native function of the UNS protein (Ferrario *et al.*, 2004). As for its *Arabidopsis* homolog *SOC1*, overexpression of *UNS* has a promotive effect on flowering, which indicates that *UNS* is most likely also involved in the floral transition. The absence of a phenotype for the *uns* mutant leads to the conclusion that, contrary to *SOC1, UNS* must act in a redundant manner, probably with other *SOC1/TM3* genes.

The *Arabidopsis FRUITFULL* (*FUL*) gene belongs to the *APETALA1/SQUAMOSA* (AP1/SQUA) clade as do *AP1* and *CAULIFLOWER* (*CAL*). *FUL* plays a redundant role with *AP1* and *CAL* in *LFY* upregulation, thus promoting floral meristem specification. Moreover, *FUL* was found to have a floral meristem identity promoting activity independent of *LFY* (Ferrandiz *et al.*, 2000).

In *Petunia* four genes have so far been identified that belong to the *AP1/SQUA* clade: *PETUNIA FLOWERING GENE* (*PFG*), *FLORAL BINDING PROTEIN26* (*FBP26*), *FBP29* and *P. hybrida FRUITFULL*-like (*PhFL*) (Fig. 4). All of these harbor a paleo*AP1*/eu*FUL*-motif just like the *Arabidopsis FUL* gene (Ferrandiz *et al.*, 2000; Immink *et al.*, 1999, 2003; Litt and Irish, 2003; Vandenbussche *et al.*, 2003a). *FBP26, FBP29*, and *PFG* are expressed in most plant tissues, except stamens. Highest expression levels for *PFG* are found in vegetative and inflorescence meristems (Immink

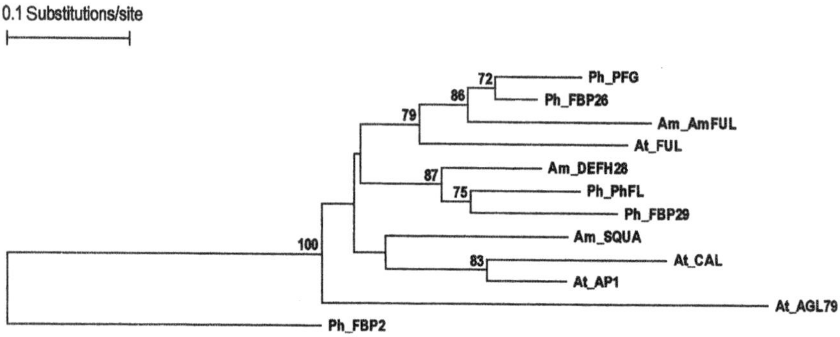

Fig. 4. Neighbor-joining tree of *SQUA/AP1* clade MADS-box genes from *P. hybrida, Arabidopsis, A. majus*, and a tomato *TM6* lineage gene. The tree was rooted with *FBP2*, a *P. hybrida* member of the *SEP* clade. See legend of Fig. 3 for technical details.

et al., 1999, 2003). No expression studies have been performed for *PhFL*. Vandenbussche *et al.* (2003b) isolated knockout alleles for both *FBP26* and *PFG* and demonstrated that single insertion mutants do not exhibit a phenotype when homozygous. No transposon insertion mutants have yet been found for *FBP29* or *PhFL*.

An indication for a role for *PFG* in the floral transition came from transgenic plants in which *PFG* expression was inhibited by cosuppression. In these plants the formation of inflorescences was completely blocked and vegetative growth was maintained, with the morphological characteristics typical of vegetative growth (Immink *et al.*, 1999). This nonflowering phenotype of *PFG* cosuppression plants is much more drastic than the slightly late flowering phenotype of *ful* single mutants. The flower-promoting activity of the *Arabidopsis FUL* gene is thought to be largely obscured by other highly redundant activities (Ferrandiz *et al.*, 2000). Not surprisingly, when the mutants were analyzed, not only the expression of *PFG* but also that of *FBP26* turned out to be downregulated (Immink *et al.*, 1999). This could well be expected as the putative protein sequences of *FBP26* and *PFG* are very similar. Vandenbussche *et al.* (2003b) showed that, in contrast to the *PFG* cosuppression line which gave a drastic nonflowering phenotype, homozygous *fbp26/pfg* double mutants only exhibit a subtle phenotype. Thus, to obtain the drastic nonflowering phenotype of the *PFG* cosuppression line, besides downregulation of *PFG* and *FBP26* at least a third gene needs to be knocked out. Looking at the sequences of *PFG* and *FBP26*, candidates to be knocked out by the *PFG* cosuppression construct (CaMV 35S promoter with the full-length *PFG* gene) are likely found in the *AP1/SQUA* clade of MADS-box genes.

III. MERISTEM IDENTITY GENES: INFLORESCENCE AND FLOWER ARCHITECTURE

Inflorescence architecture is highly variable in *Petunia*. In some species, the inflorescence consists of a single flower, whereas other species generate more complex inflorescences with multiple flowers arranged in various patterns. The diversity in inflorescence architecture is the result of a difference in action of meristematic cells, also called stem cells, in the inflorescence meristem of the different species. Development and maintenance of stem cells in general, both in inflorescence meristems as well as in the vegetative shoot apical meristem, is governed by regulatory circuits that integrate cues from different cellular origins, like the meristem itself or the young lateral organ primordia. Several genes have been identified that play an important role in these processes. *WUSCHEL* (*WUS*) expression is required for stem cell maintenance, while the *CLAVATA* (*CLV*) genes act antagonistically by inhibiting the proliferation of stem cells in a feedback loop with *WUS* (Brand *et al.*, 2000; Laux *et al.*, 1996; Schoof *et al.*, 2000). Sharing labor with *WUS* is *SHOOTMERISTEMLESS* (*STM*), which is required to suppress differentiation throughout the meristem dome, thus allowing stem cell division to occur, while the daughter cells differentiate into organs (Lenhard *et al.*, 2002).

Stuurman *et al.* (2002) identified the *Petunia WUS* homolog *TERMINATOR* (*TER*, also called *PhWUS*) and the *Petunia STM* homolog *PhSTM*, in a study on the *Petunia HAIRY MERISTEM* (*HAM*) gene. The *HAM* gene is essential for shoot apical meristem maintenance. *HAM* is a GRAS protein family member, like the *Arabidopsis SCARECROW* protein, which is required to prevent stem cells in the root meristem from adopting the fate of their differentiated neighbors (Bolle, 2004; Sabatini *et al.*, 2003). The *Petunia HAM* protein was shown to act in parallel with *TER/PhWUS*, and is required for the cellular response to *TER/PhWUS* and *PhSTM*. *HAM* mRNA is expressed in L3-derived cells of lateral organ primordia and stem provasculature. This expression pattern suggests that the *HAM* gene acts non-cell-autonomously in a signaling system through which the differentiating tissues play a role in maintaining the undifferentiated state of the shoot apical meristem (Stuurman *et al.*, 2002). Like *Arabidopsis wus* mutants, *Petunia ter* mutants stop shoot development after the first true leaves, continuously reiterating ectopic leaves and defective meristems from flat apices. This leads to very bushy plants that flower only occasionally. When flowers appear on these plants they have fewer organs per whorl, strongly resembling *wus* mutants (Laux *et al.*, 1996; Stuurman *et al.*, 2002).

A. *PETUNIA* INFLORESCENCE ARCHITECTURE

Members of the Solanaceae, such as *Petunia,* are considered to possess a cymose inflorescence that terminates in a flower. Growth continues from a sympodial meristem in the axis of this flower (Child, 1979). The formation of floral meristems in wild-type *Petunia* plants starts with the simultaneous generation of two bracts by the inflorescence meristem, before a bifurcation of the central dome yields two diversifying meristems (Souer *et al.*, 1998). One develops as a determinate floral meristem, that soon after the bifurcation starts to generate sepals, the first floral organs. The other remains meristematic and will continue with a new division, perpendicular to the last division, to form two new bracts and a new floral meristem. The same floral meristem initiation pattern is also found in tomato and pea, where flower formation also involves bifurcation of the inflorescence meristem (Souer *et al.*, 1998). *Petunia* inflorescence development thus is mainly directed by two processes: bifurcation at a predetermined position and induction of floral meristem identity. Once these processes have taken place and the floral meristem is established, meristem identity genes are necessary to determine the position of floral organ primordia. The last step than is the determination of organ identity for the primordia, which is regulated according to the ABCDE-model of flower development (see in a later section).

The *Petunia EXTRA PETALS* (*EXP*) gene is required for the split of the inflorescence apex into a floral and an inflorescence meristem (Souer *et al.*, 1998). *exp* mutant inflorescences consist of a single terminal flower that almost completely lacks the pedicel. Several lines of evidence indicate that the formation of a terminal flower in *exp* mutants is due to the complete transformation of the apical meristem into a floral meristem. First, no remains of the inflorescence meristem are detectable after this transformation. Moreover, the flower is located apically and once a terminal flower is generated the *exp* mutant loses its apical dominance (just like wild-type plants from which the inflorescence is manually removed). Consequently, the dormancy of the vegetative meristems in the axils of existing leaves is broken and a new stem with leaves will be generated from these axils, which will terminate again with the formation of a single flower (Souer *et al.*, 1998). The *exp* mutant is not the only *Petunia* mutant in which sympodial branching is lost and a single solitary flower is formed per infloresence; *hermit* and *sympodial* mutants also show this phenotype. These three mutants represent at least two different loci. *EXP* and *HERMIT* have been transposon tagged and cloned and are now being studied in more detail (Angenent *et al.*, 2005).

In *Arabidopsis*, meristem identity genes, such as *LFY*, *AP2*, and the *AP1/SQUA* clade genes *AP1, CAL*, and *FUL*, which are expressed in the newly formed floral primordia, are responsible for the fate of floral meristems, and thereby their determinacy (Bowman *et al.*, 1993; Ferrandiz *et al.*, 2000; Weigel *et al.*, 1992). *FUL* plays a redundant role with *AP1* and *CAL* in *LFY* upregulation, thus promoting floral meristem specification (Ferrandiz *et al.*, 2000). Ditta *et al.* (2004) demonstrated by mutant and overexpression analysis that the *SEPALLATA* clade gene *SEP4* also plays a role in promoting flower meristem identity. In the complete absence of meristem identity gene activity, the floral meristems remain fully or partially inflorescence meristems, the apparent default pathway.

ABERRANT LEAF AND FLOWER (*ALF*) is the *Petunia* ortholog of *LFY* from *Arabidopsis* (Gerats *et al.*, 1988; Souer *et al.*, 1998). A study showed that *LFY* and *ALF* are very similar both in structure and in function in specifically inducing floral fate during the reproductive phase, whereas *LFY* homologs from mosses have a truly different biochemical function (Maizel *et al.*, 2005). No differences between *alf* and wild-type plants can be detected during their vegetative phase. Only after transition of the vegetative shoot meristem to an inflorescence meristem the differences become evident. The *alf* mutant inflorescence is a continuously bifurcating structure bearing only bracts, but no flowers because floral meristems fail to adopt their identity and develop as inflorescence meristems instead. The expression of *ALF* marks the formation of the floral meristem in the inflorescence (as *LFY* expression does in *Arabidopsis*), before the bifurcation of the apex becomes visible (Souer *et al.*, 1998). As is clear from scanning electron microscope studies, the bifurcation of the inflorescence meristem takes place as normal in *alf* mutants, and it is only the subsequent transition from inflorescence meristem identity to floral meristem identity that is affected. *exp/alf* double-mutant plants, like *alf* mutants, have an indeterminate inflorescence that contains bracts and completely lacks flowers. In addition, sympodial branching is lost in the *exp/alf* double mutants due to the *exp* mutation. *EXP* and *ALF* thus function in two distinct processes (Souer *et al.*, 1998).

DOUBLE TOP (*DOT*) is the *Petunia* ortholog of *UNUSUAL FLORAL ORGANS* (*UFO*) of *Arabidopsis*, and is, together with *ALF*, required to specify floral meristem identity (Tobena-Santamaria *et al.*, 2002).

All four *AP1/SQUA* clade genes identified in *Petunia* so far harbor a paleo*AP1*/eu*FUL*-motif and therefore are most likely *FUL*-homologs (as noted), but there are undoubtedly more clade members (Litt and Irish, 2003; Vandenbussche *et al.*, 2003a). The available single mutants (*fbp26* and *pfg*) exhibit a wild-type phenotype. At this point, therefore, no conclusions can be drawn on the role of *FUL* or *AP1* homologs in *Petunia* meristem identity. To ascertain these roles, first the entire set of *Petunia AP1/SQUA* clade genes needs to be known and mutants need to be isolated for all of them.

Several *Petunia* genes are known to be required for marking the boundaries between different floral organ primordia, and thereby determining the position of the different primordia. The *Petunia NO APICAL MERISTEM* (*NAM*) gene and its identified putative orthologs from *Arabidopsis CUC1–CUC3* are involved in the formation of the shoot apical meristem during embryogenesis and are required for establishing the boundary between the cotyledons (Aida *et al.*, 1997; Souer *et al.*, 1996; Vroemen *et al.*, 2003). The *CUC* genes are thought to act upstream of *SHOOTMERISTEMLESS* (*STM*), as they are redundantly required for the expression of *STM* in the initiation pathway of the shoot apical meristem (Aida *et al.*, 1999). The expression pattern of the *Petunia NAM* gene in the inflorescence suggests that the *NAM* gene product acts very early in floral development, as the gene expression already marks the boundaries between different primordia before their separation becomes visible. The phenotype of occasional flowers on *nam* mutants (which in most cases even fail to produce the first leaves, let alone flowers) indicates that *NAM* is required in the cells around the stamen primordia in whorl three to prevent this region from developing into a primordium (Souer *et al.*, 1996). Even though *NAM* is also expressed at other sites in the inflorescence apex (e.g., at the boundaries of the site at which bract primordia will appear, and between developing carpel primordia), no corresponding phenotypic changes are observed in *nam* inflorescences. This is most likely due to redundancy in *NAM* function at these sites, as *NAM* is a member of a gene family; likewise, a high degree of redundancy was found for its *Arabidopsis* homologs, the *CUC*-genes (Souer *et al.*, 1996; Vroemen *et al.*, 2003). It will be interesting to analyze the effect of the *nam* mutation on the expression of *PhSTM*, to see if the relation between *CUC* and *STM* as found in *Arabidopsis* is conserved in *Petunia*.

In *Arabidopsis* the *SUPERMAN* (*SUP*) gene is involved in establishing a boundary between whorls three and four and in ovule development (Sakai *et al.*, 2000). *PhSUP1* from *Petunia* can partly complement the *Arabidopsis sup* mutant, indicating *PhSUP1* is an ortholog of *SUP* (Nakagawa *et al.*, 2004). *PhSUP1* plays a role in ovule development in *Petunia* as *SUP* does in *Arabidopsis*. Moreover, the gene may contribute to flower morphogenesis by preventing over-progression of intercalary growth. Presumably, this particular role of *PhSUP1* has co-evolved with the flower structure of *Petunia*. *PhSUP1* also seems to play a role in placenta and anther morphogenesis. In summary, the early floral meristem function and late function in ovule development of the *SUP* gene, originally discovered in *Arabidopsis*, are conserved in *Petunia*. Furthermore, *PhSUP1* has some additional functions in placenta and anther morphogenesis, which have not been described for the *Arabidopsis SUP* gene (Nakagawa *et al.*, 2004).

As in the process of floral transition, hormones are also involved in inflorescence and flower formation. Tobena-Santamaria *et al.* (2002) analyzed the *FLOOZY* (*FZY*) gene, which is involved in synthesizing a signalling compound, most likely auxin, required for floral organ initiation. In *fzy* mutants the formation of floral organ primordia in the outermost three floral whorls and one of the two bracts at the base of the flower is blocked at an early stage (Tobena-Santamaria *et al.*, 2002).

IV. FLORAL ORGAN IDENTITY DETERMINATION

During floral organogenesis, five different types of organ primordia emerge from the floral meristem and differentiate into the floral organs. These floral organs are organized in concentric whorls: sepals, petals, stamens, carpels, and in the center of the flower, the placenta bearing the ovules. What organ is formed where is specified by a combinatorial action of five functional classes of genes. This was first formulated in the famous ABC-model, which has formed the foundation for our understanding of floral development, and was later extended with two extra functional classes D and E (Angenent *et al.*, 1994, 1995; Coen and Meyerowitz, 1991; Honma and Goto, 2001; Pelaz *et al.*, 2001). Almost all of the identified players in this model belong to closely related paralogous lineages of the MADS-box gene family. The different lineages have arisen by duplication events, although the exact timing of the duplications and the exact relationship of the lineages to each other is not yet fully known (Becker *et al.*, 2000; Nam *et al.*, 2003; Purugganan *et al.*, 1995; Theissen *et al.*, 2000; see Irish, Chapter 3 and Kramer and Zimmer, Chapter 9). The B- and C-function lineages appear to be among the oldest lineages, as genes belonging to these clades are involved in the development of the female and male reproductive organs already in gymnosperms (Becker *et al.*, 2000; Tandre *et al.*, 1998). Younger lineages, which are angiosperm specific, are those of the *SQUA/AP1* clade and *SEP* clade (Litt and Irish, 2003). One of the most important characteristics of MADS-box proteins is that they can form (multimeric) protein complexes with each other and probably also with other partners (de Folter *et al.*, 2005; Gutierrez-Cortines and Davies, 2000; Immink *et al.*, 2003; Masiero *et al.*, 2002). This efficiently creates a large collection of different transcription activation complexes that can regulate different sets of target genes, thus resulting in the formation of specific organs at specific times and specific positions in the floral meristem (Theissen and Saedler, 2001).

A. THE A-FUNCTION GENES

In *Arabidopsis* two genes are generally considered to represent the A-function: the MADS-box gene *APETALA1* (*AP1*) and *AP2*, the only non MADS-box gene in the ABC-model so far. However, *AP1* may be interpreted as a meristem identity and flower induction pathway gene, as it does not truly belong to this functional class. *AP1* function is not essential to identify sepals and petals, as it is actually the overexpression of *AGL24* in *ap1* mutants that is responsible for many aspects of the *ap1* floral phenotype, including defects in the first and second whorl floral organ development. Some floral organ defects of *ap1–1* mutants, especially the absence of petals, can partly be rescued by the absence of *AGL24* in an *ap1/agl24* double mutant (Kramer and Hall, 2005; Yu *et al.*, 2004).

The *Arabidopsis AP2* gene fulfils two roles in the process of floral organ identity determination: a cadastral function consisting of repressing the C-function gene *AGAMOUS* (*AG*) and promoting an organ specification function in the perianth (sepals and petals) (Jofuku *et al.*, 1994). In *Antirrhinum* the *AP2*-like genes *LIPLESS1* (*LIP1*) and *LIP2* are together essential for sepal and petal specification, but unlike the *ap2* mutants in *Arabidopsis*, the *lip1/lip2* double mutants do not show any ectopic C-class gene expression (Keck *et al.*, 2003). In *Petunia* three *AP2*-like genes have been identified: *P. hybrida APETALA2A* (*PhAP2A*), *PhAP2B*, and *PhAP2C*. *PhAP2A* has a high overall sequence similarity with the *Arabidopsis AP2* gene and a similar expression pattern during flower development, suggesting that they are orthologs. *PhAp2B* and *PhAp2C* encode for *AP2*-like proteins that belong to a different subgroup of the *AP2* family of transcription factors and exhibit divergent, nearly complementary expression patterns during flower development compared to *PhAp2A*. The only clear overlap in expression between the three *PhAp2* genes is in the endosperm where all three are strongly expressed (Maes *et al.*, 2001). *PhAP2A* is the functional ortholog of the *Arabidopsis AP2* gene, as it can complement the *Arabidopsis ap2–1* mutant. Surprisingly, several *phap2a* transposon insertion mutants in which the *PhAP2A* gene was knocked out, did not exhibit a mutant phenotype in floral development. Thus, *PhAP2A* is not essential for normal perianth development (Maes *et al.*, 2001). Because the sequences of *PhAP2B*/*PhAP2C*, and their expression patterns during flower development, are very different from those of *PhAP2A*, it is very unlikely that they are functionally equivalent, and would act in a redundant way in the *phap2a* mutant. *Petunia* thus might differ from both *Antirrhinum* and *Arabidopsis* in this respect. The A-function, as encoded by *AP2* of *Arabidopsis* and *LIP1* and *LIP2* from *Antirrhinum*, does not seem to exist as such in *Petunia*. It seems as if these three species have each found a different way to encode the A-function. In *Arabidopsis* one gene both has a cadastral and an organ identity function. In *Antirrhinum* two homologous genes function in organ identity specification together, while other genes must be responsible for the cadastral function. In *Petunia* all the knowledge suggests that *AP2*-like genes are not involved in either organ identity specification, or setting boundaries for expression of C-function genes.

Nevertheless, a *Petunia* A-function mutant has been known for a long time, *blind* (*bl*) (Maes *et al.*, 2001; Vallade *et al.*, 1987). Unfortunately the identity of the *BLIND* (*BL*) gene is still unkown. *bl* mutant flowers display a homeotic conversion of the corolla limb into antheroid structures in the second whorl and, under certain conditions, homeotic conversion of the tips of the first whorl sepals into carpelloid tissue (Vallade *et al.*, 1987). The *bl* phenotype is quite variable, but the pistil tube is never affected and the mutant does not show the complete A-function conversion as observed in the *Arabidopsis ap2* mutant.

Tsuchimoto *et al.* (1993) and Kater *et al.* (1998) demonstrated that the *bl* phenotype is caused by ectopic expression of the C-function genes *pMADS3* and *FBP6* in the first two floral whorls of the *bl* mutant. In addition, ectopic expression of *pMADS3* and *FBP6* was also observed in leaves of the *bl* mutant, although the *FBP6* hybridization signal was only detectable after long exposure. These results indicate that the *BL* gene product is involved in the suppression of both petunia *AG* homologs in leaves and in the first two floral whorls (Kater *et al.*, 1998).

In search of the *BL* gene, Mayama *et al.* (2003) studied the *Petunia* orthologs of one of the *Arabidopsis* cadastral genes, *CURLY LEAF* (*CLF*), which is required to repress transcription of the class C gene *AG* in the first and second floral whorls and also in vegetative organs. *CLF* encodes for a protein with extensive similarity to the product of the *Drosophila* Polycomb-group gene *Enhancer of zeste* (*E*(*Z*)) (Goodrich *et al.*, 1997). *Petunia* harbors at least two *CLF* homologs (*PhCLF1* and *PhCLF2*). The two *PhCLF* proteins share two conserved domains with related proteins. Both *PhCLF1* and *PhCLF2* are expressed in all the floral organs, but the amounts of *PhCLF1* and *PhCLF2* transcripts differ. The *PhCLF1* transcript contains alternatively spliced RNA species encoding proteins truncated in the C-terminal region. Neither *PhCLF1* nor *PhCLF2* appears to coincide with the *BL* gene, but their expression is affected by homeotic transformations in the *bl* mutant flower (Mayama *et al.*, 2003).

An important step in understanding how the A-function is regulated in *Petunia* will be the discovery of the sequence underlying the mutation causing the *bl* phenotype. Currently, everything points in the direction that at least some aspects of A-function regulation in *Petunia* will be organized differently compared to *Arabidopsis*.

B. THE B-FUNCTION GENES

The most extensively studied B-function genes are from *Arabidopsis* and *Antirrhinum, AP3* and *PISTILLATA* (*PI*), and *DEFICIENS* (*DEF*) and *GLOBOSA* (*GLO*), respectively. They are mainly expressed in the second and third whorl, consistent with their function in petal and stamen identity specification. The *DEF/AP3* and *GLO/PI* lineage genes are thought to represent paralogous genes that arose from a duplication event that occurred before the origin of the angiosperms (Kramer *et al.*, 1998; Purugganan, 1997; Theissen *et al.*, 2000; Kim *et al.*, 2004). The AP3 and DEF proteins form heterodimers with respectively PI and GLO (Riechmann *et al.*, 1996; Schwarz-Sommer *et al.*, 1992). These heterodimers are important in the autoregulation of the expression of *DEF/AP3* and *GLO/PI*, as the heterodimer formation enhances the initially low-expression levels of the genes and maintains their expression (Honma and Goto, 2000; Saedler and Huijser, 1993; Schwarz-Sommer *et al.*, 1992; Zachgo *et al.*, 1995). This was long believed to be the general system of B-function, but more and more deviating systems are being discovered in different species. Often, for one or both of the gene lineages *DEF/AP3* and *GLO/PI* more representatives are found which can have (partly) redundant but also diverged functions. Not uncommon is a shift in the expression pattern of one or more B-function genes, often resulting in different floral morphologies (Kanno *et al.*, 2003; Kramer *et al.*, 2003; Nakamura *et al.*, 2005).

1. GLO/PI *lineage genes*

While *Arabidopsis* and *Antirrhinum* each have only one *GLO/PI* lineage gene, *Petunia* harbors two *GLO/PI* lineage genes: *P. hybrida GLOBOSA1* (*PhGLO1*; formerly called *FBP1*) and *PhGLO2* (formerly called *PMADS2* or *FBP3*) (Fig. 5). In wild-type *Petunia* flowers, the expression domain of *PhGLO1* and *PhGLO2* is mainly confined to the second and third whorl, and signals are slightly stronger in younger buds (Angenent *et al.*, 1992; van der Krol *et al.*, 1993; Vandenbussche *et al.*, 2004). The expression patterns of *PhGLO1* and *PhGLO2* are thus very similar to those of their *Arabidopsis* and *Antirrhinum* counterparts. *PhGLO1* and *PhGLO2* act largely redundant in petal and stamen formation. The differences between the function of the two genes become visible as unique phenotypical aspects of *phglo1* single mutants: petal midveins are greenish (sepaloid) and stamen filaments are not fused to the petal tube. This indicates that *PhGLO1*, and not *PhGLO2*,

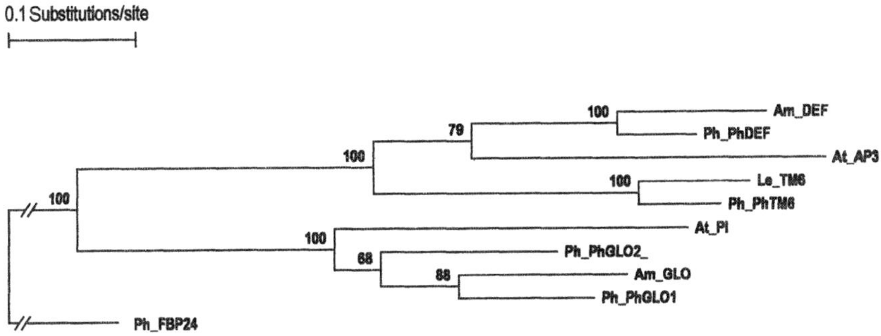

Fig. 5. Neighbor-joining tree of B-class MADS-box genes from *P. hybrida, Arabidopsis, A. majus*, and a tomato *TM6* lineage gene. The tree was rooted with *FBP24*, a *P. hybrida* member of the B_{sister} (B_s) MADS-box subfamily (Becker *et al.*, 2002). See legend of Fig. 3 for technical details.

controls the formation of the petal midvein and growth under the zone of petal and stamen initiation, which causes the corolla tube and stamen filaments to emerge as a congenitally fused structure (Vandenbussche *et al.*, 2004). Apart from these differences the two *Petunia GLO/PI* lineage genes act redundantly in petal and stamen formation and only the *phglo1/phglo2* double mutant shows a complete conversion from petals to sepals and stamens to carpels. The two *Petunia GLO/PI* genes together thus function in the same way as their *Arabidopsis* and *Antirrhinum* orthologs *PI* and *GLO*.

2. DEF/AP3 *lineage genes*

Within the *DEF/AP3* lineage, two clades can clearly be distinguished on the basis of their completely different C-terminal motifs (Kramer *et al.*, 1998). The first motif is referred to as the paleo*AP3* motif and is found in *DEF/AP3* proteins from basal eudicots, magnoliids monocots, and basal angiosperms, while a second type, named the eu*AP3* motif, is uniquely present in *DEF/AP3* proteins from core eudicots. A number of core eudicot species contain both the eu*AP3* and paleo*AP3* type of genes, termed eu*AP3* and *TOMATO MADS BOX GENE6* (*TM6*) lineages, respectively (Kramer and Irish, 2000).Vandenbussche *et al.* (2003a) have shown that the eu*AP3* motif most likely resulted from a simple frameshift mutation in one of the copies of the duplicated ancestral paleo*AP3*-type gene. Lamb and Irish (2003) published data indicating that paleo*AP3* and eu*AP3* motifs encode different functions: a chimeric construct made up of an *Arabidopsis AP3*-gene, containing a paleo*AP3*-motif from *Dicentra eximia* instead of its own eu*AP3*-motif, could not rescue petal formation in an *ap3* mutant. In contrast to these findings, Whipple *et al.* (2004) demonstrated that the full-length maize paleo*AP3* encoding gene *Silky* is capable of identifying and properly regulating the genes necessary for normal petal and stamen development in the *Arabidopsis* eudicot flower. Therefore, at the moment it is not clear what the overall functional significance is of the acquirement of the new eu*AP3* motif during evolution.

Both the *Antirrhinum DEF* gene and the *Arabidopsis AP3* gene belong to the "modern" clade of the *DEF/AP3* lineage and harbor a eu*AP3*-motif. These species do not have an ancestral *DEF/AP3*, with a paleo*AP3*-motif. *Petunia* however harbors both types present within the *DEF/AP3* lineage: *PhDEF* (formerly known as *GREEN PETAL* (*GP*) or *PMADS1*), and *P. hybrida TM6* (*PhTM6*) (Angenent *et al.*, 1992; Kramer and Irish, 2000; van der Krol *et al.*, 1993; Vandenbussche *et al.*, 2004). The *PhDEF* gene contains a eu*AP3* motif, while the *PhTM6* gene contains a paleo*AP3* motif. Thus, while several core eudicots apparently have lost the gene copy containing the paleo*AP3* motif, *Petunia,* as well as at least two other Solanaceous species, tomato and potato, harbors a paleo*AP3* as well as a eu*AP3* gene (Fig. 5) (Kramer *et al.*, 1998; Vandenbussche *et al.*, 2003a). For a recent and more comprehensive overview of B-class MADS-box gene phylogeny, we refer to Kim *et al.* (2004).

3. PhDEF

In wild-type *Petunia* flowers, the expression domain of the eu*AP3*-type gene *PhDEF* is mainly confined to the second and third whorl, with slightly stronger expression in younger buds (Angenent *et al.*, 1992; van der Krol *et al.*, 1993; Vandenbussche *et al.*, 2004). Low levels of *PhDEF* are detectable in the first and fourth whorls (Tsuchimoto *et al.*, 2000; Vandenbussche *et al.*, 2004), which has also been reported for *DEF* in *A. majus* (Schwarz-Sommer *et al.*, 1992). Surprisingly, mutations in *PhDEF* cause homeotic transformations only in one whorl: petals are converted to sepals, whereas stamens remain unaffected (de Vlaming *et al.*, 1984; van der Krol *et al.*, 1993). This indicates that *PhDEF* is essential for petal formation, but might act redundantly with other factors in stamen development (Vandenbussche *et al.*, 2004).

4. PhTM6: *an atypical and interesting B-function gene*

While the expression patterns of *PhGLO1, PhGLO2*, and *PhDEF* are very similar to those of their *Arabidopsis* and *Antirrhinum* counterparts, the expression of the paleo*AP3*-type gene *PhTM6* differs drastically (Vandenbussche *et al.*, 2004). In small buds the strongest signals for *PhTM6* transcripts are detected in carpels and stamens, while the expression level in sepals and petals is much lower. Later in development, the expression level for *PhTM6* remains high in the fourth whorl, while declining in the stamens at the time of maturation (Vandenbussche *et al.*, 2004). Remarkably, the expression pattern of *PhTM6* thus is much more C-class-like. Moreover, in the A-function *blind* (*bl*) mutant, *PhTM6* expression is extended from the third and fourth whorl to all floral whorls, which is exactly what happens with the expression pattern of the *Petunia* C-class MADS-box genes *pMADS3* and *FBP6* (see later section).

The *PhTM6* expression pattern offers a logical explanation for the phenotype of both *phdef* flowers and *phdef*/*bl* double mutant flowers (see later section). Since in wild type plants, *PhTM6* is mainly expressed in whorls three and four, *PhDEF* is the only *DEF*/*AP3* lineage member expressed at high levels in petals, while expression of both *PhDEF* and *PhTM6* in anthers suggests that they might act redundantly in stamen formation. Likewise, *phdef* mutants only display a homeotic conversion of petals to sepals, while anthers remain virtually unaffected (Vandenbussche *et al.*, 2004). The question remains whether it is the lack of expression of *PhTM6* in the second whorl, or the inability of the paleo*AP3*-clade protein PhTM6 itself, that blocks a function in the petal developmental program.

Although the full homeotic conversion of petals to sepals in *phdef* single mutants suggests full absence of B-function activity in the second whorl of *phdef* flowers, *phdef*/*bl* double mutants develop antheroid structures in the second whorl, as in *bl* single mutants, although one would rather expect carpels in the second whorl as would be predicted for an A/B double mutant. This indicates ectopic B-function activity in the second whorl of *phdef*/*bl* flowers, which is not present in *phdef* single mutants, suggesting that the ectopic *PhTM6* expression in the *bl* mutant background might account for this (Vandenbussche *et al.*, 2004).

5. *Interactions between the* Petunia *B-function proteins*

The *phdef*/*phglo2* double mutant shows a complete conversion of petals to sepals and stamens to carpels, which clearly demonstrates that the PhTM6-PhGLO1 heterodimer is either not formed or not sufficient to confer petal and stamen identity. Yeast two-hybrid studies suggest that this might be due to the specificity of the PhTM6 protein for PhGLO2, as PhTM6 only interacts with PhGLO2 and not with PhGLO1, while the PhDEF protein does interact with both PhGLO1 and PhGLO2 (Vandenbussche *et al.*, 2004). *PhTM6* together with *PhGLO2* expression on the other hand, is sufficient to induce stamen development. This is supported by the phenotype of the *phdef*/*phglo1* double mutant, which shows no additional phenotype compared with the *phdef* single mutant (Vandenbussche *et al.*, 2004).

In conclusion, the *PhGLO1* and *PhGLO2* genes act largely redundantly in petal and stamen formation, with the only apparent differences between the two being the function of *PhGLO1* in the formation of the petal midveins and the fusing process of stamen filaments and tube. More divergence is observed in the *DEF/AP3*-lineage. *PhTM6* apparently does not have a function in petal development, where *PhDEF* has. In addition, while *PhDEF* expression is sufficient for stamen formation together with either *PhGLO1* or *PhGLO2*, *PhTM6* interacts specifically with *PhGLO2* and not with *PhGLO1* in the induction of stamen development (Vandenbussche *et al.*, 2004). Analysis of *phtm6* single mutants and double and triple mutants of *phtm6* with the other *Petunia* B-function gene mutants *phdef*, *phglo1*, and *phglo2*, will certainly provide more clarity on the B-function as encoded in *Petunia*.

The C-class expression pattern of the B-function gene *PhTM6* allows for speculation on the origin of *PhTM6* and B-function genes in general. At this point it is impossible to decide whether the C-class expression pattern of *PhTM6* reflects the original function of *PhTM6* (and thus of B-function genes in general), or that these characteristics are the result of a divergence in function that is specific for *Petunia* (or maybe Solanaceaous species). The *PhTM6* homologs from tobacco and potato or other Solanaceaous species, have not yet been studied in enough detail to allow a final interpretation.

C. THE C-FUNCTION GENES

In *Arabidopsis* the gene responsible for the C-function is *AGAMOUS* (*AG*). Loss of *AG* function results in the conversion of stamens into petals and in the absence of the fourth whorl carpels, which are replaced by indeterminate perianth whorls (Yanofsky *et al.*, 1990). *AG* thus has two functions: establishing stamen and carpel organ identity and maintaining meristem determinacy. In *Petunia*, and other species like *Antirrhinum* and maize (Davies *et al.*, 1999; Mena *et al.*, 1996) the C-function is encoded by two or more genes in a redundant manner.

Two *Petunia* genes are known with sequences highly homologous to that of *AG*: *Petunia MADS3* (*pMADS3*) and *floral-binding protein 6* (*FBP6*) (Angenent *et al.*, 1993; Tsuchimoto *et al.*, 1993) (Fig. 6). At an early stage, when the sepal primordia become apparent on the flanks of the floral meristem, *pMADS3* and *FBP6* transcripts start to accumulate in cells that later give rise to the stamen and carpel primordia. When the stamen primordia are clearly visible and carpel primordia start to develop, *pMADS3* and *FBP6* are expressed throughout the central part of the floral apex that develops into the pistil. No expression can be detected in sepal or petal primordia. At later stages during flower development, *pMADS3* and *FBP6*

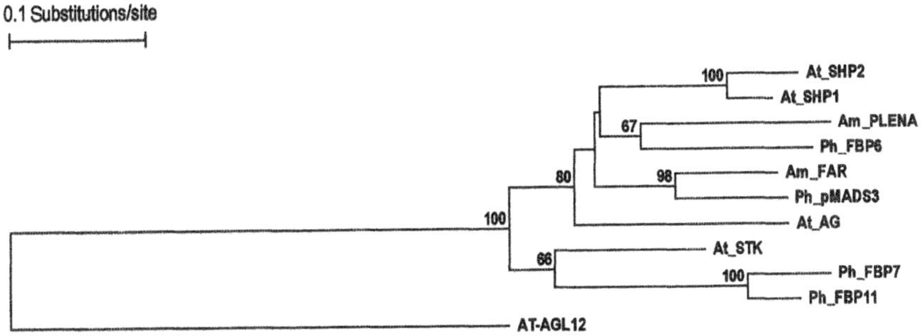

Fig. 6. Neighbor-joining tree of C- and D-class MADS-box genes from *P. hybrida*, *Arabidopsis*, and *A. majus*. The tree was rooted with *AGL12*, an *Arabidopsis* MADS-box gene. See legend of Fig. 3 for technical details.

become differentially expressed. *FBP6* then is highly expressed in the stigma and transmitting tissue of the style, while *pMADS3* is more abundant in the ovules, vascular tissue and the nectaries (Kater *et al.*, 1998). Kapoor *et al.* (2002) found that the gene structure of *pMADS3* is consistent with that of the other dicot C-function genes. In *AG, cis*-elements including the binding sites for regulatory proteins have been identified in the second intron. The same kind of *cis*-elements (*LEAFY*, homeodomain protein, and MADS-box protein consensus binding sites), and a conserved stretch of 70 bp, were found in the second intron of *pMADS3* (Kapoor *et al.*, 2002).

Due to a lack of transposon insertion mutants that knock out the expression of *pMADS3* or *FBP6*, the functional characterization of these two genes has so far only been carried out by the analysis of overexpression and cosuppression mutants (Kapoor *et al.*, 2002; Kater *et al.*, 1998; Tsuchimoto *et al.*, 1993). The conclusion can be drawn that *pMADS3* is the *Petunia* ortholog of *AG* and is required for stamen and carpel development (Kapoor *et al.*, 2002; Kater *et al.*, 1998; Tsuchimoto *et al.*, 1993). Several lines of evidence support this conclusion: first, the spatial and temporal expression pattern of *pMADS3*, and the overall sequence similarity with other C-function genes completely correspond with a C-function role. Second, in the *Petunia* A-function mutant *blind, pMADS3* is ectopically expressed in the first and second whorl, where the homeotic conversions take place: corolla limbs into antheroid tissue and small parts of sepals into carpelloid tissue (Tsuchimoto *et al.*, 1993). Thirdly, transgenic plants overexpressing *pMADS3* under control of the constitutive CaMV 35S promoter phenocopy the A-function mutant *blind.* These transgenics show petal limbs that are largely reduced in size and have antheroid tissue at the fusion site of the petals.

Their sepals are curled up at the tip and stylar and stigmatic tissues are sometimes present on these sepal tips (Kater *et al.*, 1998; Tsuchimoto *et al.*, 1993). A fourth indication for the function of *pMADS3* came from a transposon insertion mutant, in which the transposon was inserted in one of the *pMADS3* introns in such a way that it induced *pMADS3* overexpression. This was the third type of plant in which *pMADS3* was overexpressed, and again it showed the same *blind*-like mutant phenotype (Kater *et al.*, 1998).

In contrast to *AG* and *PLE* (Mizukami and Ma, 1992; Saedler and Huijser, 1993), ectopic expression of *pMADS3* is not able to induce a complete homeotic conversion of the sepals and petals into reproductive organs. The sepals are typically largely unchanged, and especially the petal tube is always completely unaffected. This suggests that C-activity repression in the outer two whorls might be difficult to override by ectopic C-function gene expression (also implying that repression of C-activity in the *blind* mutant is not completely abolished), or alternatively, that *pMADS3* requires additional factors to give a full spectrum of C-function activity. On the other hand, ectopic expression of cucumber *AG* homolog *CUM1* in *Petunia* did result in a much more complete conversion of petals to anthers and sepals to carpels (Kater *et al.*, 1998). At first sight, this might also point in the direction that the C-function in *Petunia* is controlled by two or more genes whose functions are combined in this single cucumber gene, *CUM1* (Kater *et al.*, 1998). Equally possible, the cucumber CUM1 protein is less prone to C-activity repression in the outer whorls of the *Petunia* flower due to the heterologous nature of this experiment.

An obvious candidate for defining the C-function together with *pMADS3* is *FBP6*. Yet, despite the similarities between *pMADS3* and *FBP6* with respect to sequence and expression pattern, overexpression of *FBP6* did not result in a homeotic conversion of sepals into carpels and petals into stamens (Kater *et al.*, 1998). In line with this, in the *blind* mutant in which both *pMADS3* and *FBP6* are overexpressed, there was no additional phenotype when compared to the *pMADS3* overexpressor (in which *FBP6* was normally expressed in whorls three and four). Only the *fbp6/pmads3* double mutant will give solid proof if it is really *pMADS3* together with *FBP6* that defines the C-function in *Petunia*, or whether additional genes are involved.

All analyzed *pMADS3* overexpressors only gave indications for a role of *pMADS3* in stamen and carpel development. However, the phenotype of the transgenic plants in which the *pMADS3* gene was silenced (Kapoor *et al.*, 2002), suggests an additional function for *pMADS3* in controlling determinacy in the flower as has been found for *AG* (Yanofsky *et al.*, 1990). Silencing of *pMADS3* resulted in homeotic conversion of stamens into petaloid structures, whereas the carpels were only weakly affected. But most remarkable were the emerging ectopic secondary inflorescences from the interstamenal region in the third whorl, while the fourth-whorl carpels were unaffected. Third-order inflorescences emerged at corresponding positions in the third whorl of inner flowers of secondary inflorescences, indicating reiterative conversion of parts of the floral meristem into an inflorescence meristem (Kapoor *et al.*, 2002). Noteworthy is that, whereas *ag* mutant flowers develop indeterminate floral organs in the fourth floral whorl, the *pMADS3* knock-out plants demonstrate indeterminate organ formation in the third floral whorl. It is interesting to speculate on the question whether this is a fundamental difference between *Petunia* and *Arabidopsis*, or whether it is simply the absence of the carpels in *Arabidopsis* that makes the difference, while the location of formation of indeterminate organs is in fact the same. In *Arabidopsis* a negative feedback loop in the floral meristem, involving *WUS*, the floral meristem identity gene *LEAFY* (*LFY*) and the C-function gene *AG*, is thought to be responsible for *WUS* suppression in the floral meristem. Suppression of *WUS* then leads to termination of the floral meristem (Lenhard *et al.*, 2001; Lohmann *et al.*, 2001). In *Petunia, pMADS3* together with the E-function protein FBP2 could be responsible for terminating meristematic activity in the third whorl region of the floral meristem, by suppressing *TER* in the center of the flower (see later section and Kapoor *et al.*, 2002).

D. THE D-FUNCTION GENES

1. *Pistil and ovule development in* Petunia

The *Petunia* pistil is composed of two completely fused carpels that arise separately from the floral apex in the center of the flower. Only immediately after the induction of sepal, petal, and stamen primordium formation are the two carpel primordia morphologically distinguishable. The two horseshoe-shaped primordia soon fuse to form a circular structure (Angenent *et al.*, 1995). This primordial cylinder extends, and before the gynoecium closes at the top, the placenta starts to develop in the center of the flower. As the gynoecium closes, style formation starts. The style elongates and transmitting tissue differentiates to form a tract through which pollen tubes can grow (Angenent *et al.*, 1995). At the same time ovule primordia arise from the placental tissue as a dense group of meristematic cells. Within each ovule primordium a single megasporocyte is formed from which eventually a seven-cell embryo sac develops. During this process, the ovule becomes stalked and an integument is initiated at the base of the nucellus. This integument elongates and grows over the nucellus and finally forms the micropyle. At the micropyle, a pollen tube penetrates the ovule to deliver the sperm cell into the embryo sac for the double-fertilization process (Angenent *et al.*, 1995).

In 1995, a novel functional class of MADS-box genes, highly homologous to C-class MADS-box genes (Fig. 6), was discovered in *Petunia*, involved in ovule development (Angenent *et al.*, 1995; Colombo *et al.*, 1995). More recently, the MADS-box genes involved in the process of ovule development in *Arabidopsis* were described (Favaro *et al.*, 2003; Pinyopich *et al.*, 2003). The genes in *Petunia* are *FLORAL BINDING PROTEIN7* (*FBP7*) and *FLORAL BINDING PROTEIN11* (*FBP11*). The putative protein products of these genes share ~90% of their overall amino acid sequence (Angenent *et al.*, 1995). At the sequence level the putative proteins of *FBP7* and *FBP11* are most similar to the *Arabidopsis SEEDSTICK* (*STK*, formerly called *AGAMOUS LIKE11*), which was shown to play a role in ovule development in *Arabidopsis* (Pinyopich *et al.*, 2003). The *STK* gene is also required for normal development of the funiculus, a stalk-like structure that connects the developing seed to the fruit, and for dispersal of the seeds when the fruit matures. In promoting ovule identity, *STK* acts redundantly with the C-class genes *SHATTERPROOF1* (*SHP1*), *SHP2*, and *AG* (Pinyopich *et al.*, 2003).

FBP7 and *FP11* are expressed in the center of the gynoecium before ovule primordia become visible. At a later stage they are restricted to the ovules, predominantly the endothelium, which is the innermost cell layer of the integument. The expression levels of both *FBP11* and *FBP7* increase immediately after pollination and decline in developing seeds (Colombo *et al.*, 1995, 1997).

The conclusion that *FBP11* and *FBP7* encode a new floral function that specifies ovule identity was based on the analysis of *FBP11* cosuppression plants. In the ovary of these transformants, at many of the positions normally taken up by ovules carpelloid spaghetti-shaped structures developed. These carpelloid structures originate directly from the placenta and consist of tissues characteristic of style and stigma. Although at early developmental stages, irregular structures with a chimeric identity were observed, the identity of these structures in mature ovaries was more uniform. Either these structures elongated and developed into carpelloids, or morphologically normal ovules were formed (Angenent *et al.*, 1995). All seeds produced by these ovules had developmental defects (Colombo *et al.*, 1997). The frequency of ovule conversion seemed to be related to the residual *FBP11* expression in the mutants. Low-residual gene expression in the primary transformants was sufficient to overcome a certain threshold, required for normal ovule development. As the overall sequence similarity between *FBP11* and *FBP7* is very high, it was not surprising to find that in the *fbp11* cosuppression plants, the expression of *FBP7* was also reduced to approximately the same extent as *FBP11* (Angenent *et al.*, 1995). The suppression of *FBP7* expression could also point to a regulatory role for *FBP11* determining *FBP7* expression levels. This is quite unlikely, however, since in the *FBP11* overexpression mutants, *FBP7* expression was not upregulated. Thus, although primordia are still formed from the placenta without *FBP7* and *FBP11*, the expression levels of *FBP11* and *FBP7* determine which type of development takes place after this primordium formation, that is, toward the formation of real ovules or carpelloid structures (Cheng *et al.*, 2000).

The phenotype of the *FBP11/FBP7* cosuppression mutant is reminiscent of that of the *stk/shp1/shp2* triple mutant. In the *stk/shp1/shp2* triple mutant, normal ovule and seed development is completely disrupted, with some ovules converted to leaf-like or carpel-like structures. In addition, the *Arabidopsis* C-class gene *AG* was also found to play a role in promoting ovule identity (Pinyopich *et al.*, 2003). If the redundancy between D- and C-class genes is conserved between *Arabidopsis* and *Petunia*, this would suggest that in *Petunia FBP11* and *FBP7* might act redundantly with C-function genes *pMADS3* and/or *FBP6*.

When ectopically expressed, *FBP11* can induce the formation of ovule-like structures on sepals, and, rarely, on petals (Colombo *et al.*, 1995). The presence of ovule-like structures on the adaxial side of the sepals is accompanied by a transformation of the sepal inner epidermis into placenta-like tissue. However, even though ovule-like structures are sometimes also found on the petals of these *FBP11* overexpressing plants, there their presence is not accompanied by the presence of placenta-like tissue (Colombo *et al.*, 1995). Ectopic expression of *FBP11* thus is sufficient to promote ovule development, as is ectopic expression of *STK* in *Arabidopsis* (Favaro *et al.*, 2003; Pinyopich *et al.*, 2003).

2. Interacting proteins

In yeast two-hybrid experiments FBP11 was shown to interact specifically with the three very closely related E-function (*SEPALLATA*) MADS-box proteins *FLORAL BINDING PROTEIN2* (*FBP2*), *FBP5*, and *FBP9* (Ferrario *et al.*, 2003; Immink *et al.*, 2002). Furthermore, Immink *et al.* (2002) demonstrated in a *FRET-FLIM* experiment that *FBP11* is only transported to the nucleus when a physical interaction takes place with the E-function protein *FBP2*. Expression analysis showed that *FBP2, FBP5*, and *FBP9* are expressed in ovules (Ferrario *et al.*, 2003). Further, *in situ* hybridization on sepals of the *FBP11* overexpression plants revealed the presence of *FBP2* mRNA in the ectopically formed ovules. This suggests that there might be a function for *FBP2-FBP11* and possibly *FBP5-FBP11* protein complexes in ovule development. Definite proof that the *SEP*-genes are involved in ovule development comes from the *fbp2/fbp5* double mutants (see later section), in which leaf-like organs emerge from the positions normally occupied by ovules in the wild type. Remarkably, hardly any ectopic ovules were found on floral organs other than the sepals in the *FBP11* overexpression plants, even though *FBP2* is also expressed in petals and stamens (Colombo *et al.*, 1995; Immink *et al.*, 2002). This suggests the presence of (an) other interaction partner(s), indispensable for ovule formation. An indication that these other interaction partners of *FBP11, FBP7*, and *SEPALLATA* proteins might be C-funtion proteins came from experiments on *Arabidopsis* proteins. Favaro *et al.* (2003) showed that the *Arabidopsis* counterparts of these *Petunia* proteins *STK, AG, SHP1*, and *SHP2* can form multimeric complexes and that these interactions require *SEP* proteins.

E. THE E-FUNCTION GENES

Indications for the existence of an E-function were presented in 1994 based on the phenotypes of *FLORAL BINDING PROTEIN2* (*FBP2*) and (*TOMATO MADS5*) *TM5* cosuppression lines in *Petunia* and tomato, respectively (Angenent *et al.*, 1994; Pnueli *et al.*, 1994). However, the E-functional class was generally accepted and understood only in 2000, when Pelaz *et al.* published a triple mutant of the *Arabidopsis* homologs of *FBP2/TM5*, the *SEPALLATA* genes *SEP1* (formerly called *AGAMOUS-LIKE2*), *SEP2* (*AGL4*), and *SEP3* (*AGL9*). From this *sep1/sep2/sep3* mutant it was evident that B and C floral organ identity functions require SEP1, SEP2, and SEP3 for the formation of petals, stamens, and carpels because in the triple mutant all these organs are converted into sepals. In addition, these three genes are required to prevent the indeterminate growth of the flower meristem (Pelaz *et al.*, 2000). Ditta *et al.* (2004) characterized another *SEPALLATA* gene *SEP4* (formerly called *AGL3*), which turned out to be involved in the flower meristem identity and organ identity together with the other three *SEP*s. Although the *sep4* single mutant appears wild type, the floral organs are converted into leaf-like organs in *sep1/sep2/sep3/sep4* quadruple mutants, indicating the involvement of all four *SEP* genes in the development of sepals. Moreover, *sep4* also contributes to the development of petals, stamens, and carpels, and plays an important role in meristem identity (Ditta *et al.*, 2004).

In *Petunia*, six genes have so far been identified that belong to the *SEP* clade: *FLORAL BINDING PROTEIN2* (*FBP2*), *FBP4, FBP23, FBP5, FBP9*, and *PETUNIA MADS BOX GENE12* (*pMADS12*) (Angenent *et al.*, 1992; Ferrario *et al.*, 2003; Immink *et al.*, 2003; Vandenbussche *et al.*, 2003b).

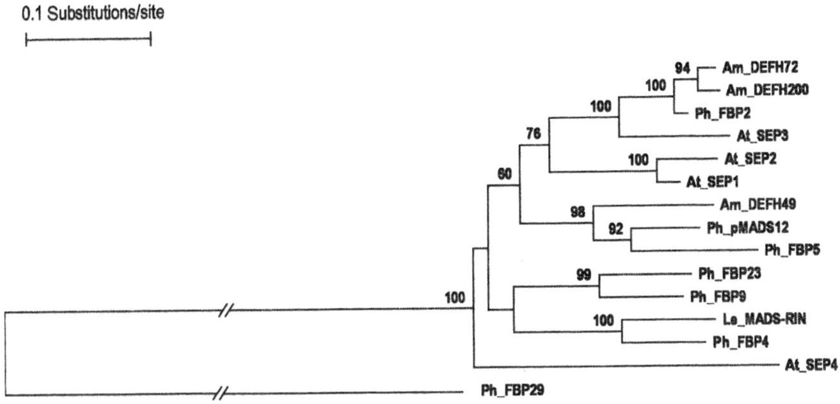

Fig. 7. Neighbor-joining tree of *SEP* clade MADS-box genes from *P. hybrida, Arabidopsis, A. majus*, and a tomato *SEP* clade gene. The tree was rooted with *FBP29*, a *P. hybrida* member of the *AP1/SQUA* clade. See legend of Fig. 3 for technical details.

See Fig. 7 for a simplified phylogeny of the *Petunia SEP* clade genes together with those of some other species. In an article by Zahn *et al.* (2005), a thorough phylogenetic analysis of the entire *SEP*-clade can be found. *FBP2, FBP5, FBP23*, and *pMADS12* are solely expressed in the floral domains, whereas *FBP4* and *FBP9* are also expressed outside of the floral organs (Ferrario *et al.*, 2003). *FBP5* and *pMADS12* transcripts can be detected already in the inflorescence meristem, while *FBP2* can only be detected later, in the central dome of the floral meristem, after it splits from the inflorescence meristem. In contrast to *FBP2*, the expression of both *FBP5* and *pMADS12* can be detected throughout the floral meristem, whereas at later stages when sepal primordia emerge, it becomes confined to the inner three floral whorls, like *FBP2* (Ferrario *et al.*, 2003). *FBP4, FBP9*, and *FBP23* are expressed in all floral whorls, except for the stamens. Furthermore, *FBP4* is also expressed in bracts, while *FBP9* accumulates in all green tissues of the plant; no transcript of *FBP23* can be detected in vegetative tissues. *FBP23* and *FBP4* are expressed in seed pods (Ferrario *et al.*, 2003). More details on expression patterns and protein–protein interactions for the *Petunia* E-function genes can be found in Ferrario *et al.* (2003).

In *Arabidopsis* three *SEP* genes need to be knocked out to obtain a full conversion of the second, third, and fourth whorl organs to sepals and meristem indeterminacy in the center of the flower (Pelaz *et al.*, 2000). The *Petunia FBP2* cosuppression mutant showed a quite similar phenotype.

An explanation for this might be that the E-function in *Petunia* is mainly encoded by a single gene, more particularly *FBP2*. But alternatively, cosuppression in these transgenics might have caused downregulation of multiple *SEP* genes simultaneously. Research suggests that the latter possibility most likely explains this seeming difference between *Petunia* and *Arabidopsis*. Ferrario *et al.* (2003) showed that another *SEP* homolog, *FBP5*, a gene unknown at the time of publication of the *FBP2* cosuppression experiments, was downregulated together with *FBP2*. Moreover, the phenotype of *fbp2/fbp5* double mutants in which the expression of *fbp2* and *fbp5* was specifically knocked out by transposon insertions (Vandenbussche *et al.*, 2003b, see later section) still was much less severe than that of the *FBP2* cosuppression mutants and the *Arabidopsis sep1/sep2/sep3* triple mutants. Therefore, in the *FBP2* cosuppression lines, at least three *SEP* genes must be downregulated. This indicates that the full E-function in *Petunia* is redundantly encoded by multiple *SEP* genes, as has been found in *Arabidopsis*. Nevertheless, the unique phenotype of *fbp2* insertion mutants and the *fbp2/fbp5* double mutant indicate differences in the degree of redundancy among the *SEP* genes between *Arabidopsis* and *Petunia*.

Two independent insertion alleles for *FBP2*, which both contained a *dTph1* insertion in the K-domain region, were identified. Plants homozygous for either insertion allele display an identical phenotype, in which the normal shaped petals exhibit an overall diffuse green hue, which is strongest in the areas surrounding the main veins and at the edges of the petals (Vandenbussche *et al.*, 2003b), indicating a partial conversion of petal to sepal identity in these regions. The most remarkable phenotype, however, is the presence of secondary inflorescences in the third whorl, positioned between the stamens near the nectaries at the base of the pistil. These secondary inflorescences are formed relatively late during development, when all organs of the primary flower have already been formed, and they rarely develop beyond a very young stage. The appearance of secondary inflorescences strongly suggests a loss of determinacy in the third whorl, and is exactly what Kapoor *et al.* found in flowers in which the C-function gene *pMADS3* is downregulated (see previous part and Kapoor *et al.*, 2002). In *Arabidopsis* loss of C-function is also associated with indeterminacy in the center of the flower, although the location of the formation of indeterminate floral organs seems different in *Arabidopsis* (as previously discussed and Yanofsky *et al.*, 1990). A yeast four-hybrid experiment revealed interactions between the FBP2 protein, a B-function heterodimer, and the C-function protein pMADS3 (Ferrario *et al.*, 2003). All together, these results strongly indicate that *FBP2*, together with *pMADS3*, is essential for meristem identity (Vandenbussche *et al.*, 2003b).

While loss of *FBP2* function by itself is sufficient to induce an E-function mutant phenotype, *fbp5* single mutants morphologically appear as wild-type (Vandenbussche *et al.*, 2003b) suggesting functional redundancy as observed in *Arabidopsis*. This was indeed confirmed by the phenotype of *fbp2/fbp5* double mutants. Flowers of *fbp2-2/fbp5-1* double mutants display an enhanced phenotype compared with *fbp2* mutant flowers. The petals of *fbp2-2/fbp5-1* plants show an increased petal-to-sepal conversion compared with *fbp2* petals, and sepal-like structures covered by trichomes develop on top of the anthers. In the fourth whorl a dramatic phenotypical change occurs in the *fbp2-2/fbp5-1* mutants: a huge pistil-like structure, without transmitting tissue, develops, covered with trichomes and often consisting of more than two carpels that never fuse at the top. Inside these pistils, leaf-like organs develop instead of ovules, supporting a function for *FBP2* and/or *FBP5* in directing ovule development, as discussed before. The development of secondary inflorescences in the third whorl of the double mutant is not enhanced significantly compared with the *fbp2* mutant (Vandenbussche *et al.*, 2003b).

The enhanced phenotype of *fbp2-2/fbp5-1* double mutants, compared with *fbp2* mutants, demonstrates that *FBP2* and *FBP5* act in a largely redundant manner, while *FBP2* has a unique function in the maintenance of deteminacy in the third whorl. Furthermore, the sepaloid characteristics of the petals, stamens, and pistil of the *fbp2-2/fbp5-1* double mutant indicate that *FBP2* and *FBP5* are required for B and C organ identity functions as the *Arabidopsis SEP* genes are. *FBP2* is essential for meristem determinacy, most likely together with the C-function gene *pMADS3*. Compared with the *Arabidopsis SEP* gene analyses (Ditta *et al.*, 2004; Goto *et al.*, 2001; Pelaz *et al.*, 2000), the research on these two *Petunia SEP* genes already shows that clear differences in redundancy, within the *SEP* clade, exist between the two species. These differences in redundancy between species can be very helpful in uncovering functions that would otherwise be missed. The role of the *Arabidopsis SEP* genes in ovule development could only be determined by indirect evidence (as discussed in an earlier section) because the phenotype of the *sep1/sep2/sep3* triple mutant is so strong that no ovary is formed at all in the fourth whorl. The *Petunia fbp2/fbp5* double mutant, however, does make ovaries, and clearly shows that these *SEP* genes are essential for ovule formation.

So far, for only two *Petunia SEP*-genes a detailed functional analysis has been performed. The expression patterns and different protein–protein interaction partners suggest different roles for the other *Petunia SEP*-genes. In order to fully analyze the functions and redundancy within this subfamily, transposon insertion mutants will have to be identified for all of the genes belonging to this subfamily.

V. CONCLUSIONS

So far, the research on the genetic regulation of floral transition in *Petunia* has focused mostly on genes from two MADS-box gene clades: representatives of the *TM3/SOC1* clade, and FUL-like genes from the *AP1/SQUA*-clade. Genes from these clades were shown to be important in floral transition in *Petunia* in a redundant manner, as is also the case for their *Arabidopsis* homologs. More extensive work has been done on meristem identity genes and their role in inflorescence architecture. With its cymose inflorescence *Petunia* clearly differs from racemose species like *Arabidopsis* and *Antirrhinum*, implying that the initiation of meristematic processes in inflorescence development will proceed differently.

Three main processes direct *Petunia* inflorescence and flower architecture: the bifurcation of the inflorescence meristem in two parts, the determination of floral meristem identity in one part, and the establishment of boundaries between the different floral organ primordia. The *EXP* gene is essential for the bifurcation process, while *ALF*, the *Petunia* ortholog of *LFY*, is indispensable in the establishment of floral meristem identity. After that, meristem identity genes like *PhSUP* and *NAM* are involved in determining the boundaries between different floral whorls and thus in positioning of the floral organ primordia. More genes important in determining inflorescence architecture in *Petunia* are known and are being studied, so considerable progress can be expected in this field in the years to come.

To date, the most intensively studied part of floral development in *Petunia* is the process of floral organ patterning. However, regarding the A-function numerous questions still remain (see Irish, Chapter 3; Kramer and Zimmer, Chapter 9; Soltis *et al.*, Chapter 12). The function of the *Petunia PhAP2* seems different from that of its *Arabidopsis* ortholog *AP2*; alternatively, the *PhAP2* gene may act in a redundant manner with other genes. And most important: which gene product is affected in the *blind* A-function mutant? It will be interesting to see if the A-function as encoded by the *BL* gene is conserved in other species.

Petunia harbors four B-function genes: two *GLO/PI* lineage representatives that are nearly completely complementary and two *DEF/AP3* lineage genes, of which *PhDEF* harbors a eu*AP3* motif, while *PhTM6* represents the ancestral gene with a paleo*AP3* motif. The *Petunia* B-function gene set clearly shows the result of divergence in function that has occurred after duplication of both the ancestral *AP3/DEF*-lineage gene (probably at the same time the core eudicots arose), and the *GLO/PI*-lineage gene (probably more recent). Divergence in function is most obvious in the *DEF/AP3* gene lineage. Whereas the eu*AP3* gene *PhDEF* contributes to petal and stamen formation, the paleo*AP3* gene *PhTM6* is involved only in stamen formation. Moreover, the *PhTM6* protein has evolved a dimerization preference for *PhGLO2*. These two genes offer a great opportunity to study the significance of gaining a novel C-terminal motif and gene expression pattern shifts in evolution. In addition, research on the *Petunia* B-function genes revealed a novel function specifically controlled by one of the possible petunia B-function heterodimers. In wild-type *Petunia* flowers, as in many other species of Solanaceae, the stamen filaments are partially fused to the petal tube, probably out of the necessity to support the long thin filaments in an upward position. In both *phglo1* and *phdef* mutants, the stamens emerge as free-standing structures, indicating that the *PHDEF/PHGLO1* heterodimer specifically controls this process. In *Arabidopsis* and *Antirrhinum*, such a function does not exist, since anthers emerge as free-standing structures in these species. This might be an example of a subtle difference in function that accounts for species-specific differences in floral architecture.

Two *AG*-homologs have been identified in *Petunia*: *pMADS3* and *FBP6*. *pMADS3* has been shown to be required for stamen and carpel development, while the role of *FBP6* is less clear; there might even be other *AG*-homologs and/or different genes involved in the C-function process. *pMADS3* is also thought to be involved in meristem determinacy in the third floral whorl, together with the E-function gene *FBP2*. The *Petunia* D-function genes *FBP7* and *FBP11* are involved in ovule formation and thereby also important for seed formation. In the ovule formation process, D-function proteins act together with E-function proteins, while C-function proteins might very well be involved, too.

Like *Arabidopsis*, *Petunia* has several different *SEP/AGL2* clade genes. So far only two of the six *Petunia SEP/AGL2* clade genes have been analyzed in detail. *FBP2* and *FBP5* were found to act in a redundant manner in the development of petals, anthers, carpels, and ovule formation. The *Petunia SEP/AGL2* genes vary in sequence, expression pattern, and protein–protein interaction partners. This, together with the mutant phenotypes of the *fbp2* single and the *fbp2/fbp5* double mutant, leads to the conclusion that the *Petunia SEP/AGL2* clade genes have diverged in function. A detailed study on the other clade members will have to show what functions have been acquired by its other representatives. Though we have focused on MADS-box genes for which functional data are available, genes belonging to other major MADS-box gene family clades have also been identified in *Petunia*; these can thus also become a subject of further research.

The analyses of the regulatory systems in *Petunia* floral development contribute to the elucidation of the mechanisms that have been at work in the evolutionary development of the flower as a sophisticated set of organs that ensure successful reproduction. Moreover, further comparative research will enable us to better understand the molecular basis for the enormous diversity in floral (organ) development and function. One of the main forces in this process undoubtedly has been the high rate of gene duplications, resulting initially in a release of selection pressure as long as the original function is maintained by both duplicates. Subsequent divergence in gene sequence in either of the copies may lead to a shift in gene expression or a change in protein structure, thereby enabling a divergence in function. Both of these two overall mechanisms are probably important in causing functional divergence. Kramer *et al.* (2003), Kanno *et al.* (2003), and Nakamura *et al.* (2005) present several examples of how variations in gene expression patterns result in variations in floral forms. However, the diverged coding sequences of the different subfamilies within the MADS-box gene family also indicate that differences in coding sequence have a huge impact on gene function. In fact, one may conclude that it is not a matter of either/or: Nature itself provides the biggest laboratory, where all options we can think of (and more) have been and still are being tested.

ACKNOWLEDGMENTS

We want to thank Jeroen Stuurman and Mary Hoballah for kindly providing the pictures of the *P. axillaris* and *P. integrifolia* flowers. The work of A.R. is funded by NWO grant 814.02.009.

REFERENCES

Adams, S. R., Pearson, S., Hadley, P. and Patefield, W. M. (1999). The effects of temperature and light integral on the phases of photoperiod sensitivity in *Petunia* × *hybrida*. *Annals of Botany* **83**, 263–269.

Aida, M., Ishida, T., Fukaki, H., Fujisawa, H. and Tasaka, M. (1997). Genes involved in organ separation in *Arabidopsis*: An analysis of the cup-shaped cotyledon mutant. *Plant Cell* **9**, 841–857.

Aida, M., Ishida, T. and Tasaka, M. (1999). Shoot apical meristem and cotyledon formation during *Arabidopsis* embryogenesis: Interaction among the CUP-SHAPED COTYLEDON and SHOOT MERISTEMLESS genes. *Development* **126**, 1563–1570.

Ando, T., Nomura, M., Tsukahara, J., Watanabe, H., Kokubun, H., Tsukamoto, T., Hashimoto, G., Marchesi, E. and Kitching, I. J. (2001). Reproductive isolation in a native population of *Petunia sensu* Jussieu (Solanaceae). *Annals of Botany* **88**, 403–413.

Ando, T., Kokubun, H., Wantabe, H., Tanaka, N., Yukawa, T., Hashimoto, G., Marchesi, E., Suarez, E. and Basualdo, I. L. (2005). Phylogenetic analysis of *Petunia sensu* Jussieu (Solanaceae) using Chloroplast DNA RFLP. *Annals of Botany* **96**, 289–297.

Angenent, G. C., Busscher, M., Franken, J., Mol, J. and van Tunen, A. J. (1992). Differential expression of two MADS box genes in wild-type and mutant Petunia flowers. *Plant Cell* **4**, 983–993.

Angenent, G. C., Franken, J., Busscher, M., Colombo, L. and van Tunen, A. J. (1993). Petal and stamen formation in *Petunia* is regulated by the homeotic gene *Fbp1*. *The Plant Journal* **4**, 101–112.

Angenent, G. C., Franken, J., Busscher, M., Weiss, D. and van Tunen, A. J. (1994). Co-suppression of the petunia homeotic gene *fbp2* affects the identity of the generative meristem. *The Plant Journal* **5**, 33–44.

Angenent, G. C., Franken, J., Busscher, M., van Dijken, A., van Went, J. L., Dons, H. and van Tunen, A. J. (1995). A novel class of MADS box genes is involved in ovule development in Petunia. *Plant Cell* **7**, 1569–1582.

Angenent, G. C., Stuurman, J., Snowden, K. C. and Koes, R. (2005). Use of Petunia to unravel plant meristem functioning. *Trends in Plant Science* **10**, 243–250.

Bailey, L. H. (1896). Evolution of the Petunia. *In* "The Survival of the Unlike." Chapter XXIV, pp. 465–472. MacMillan, New York.

Becker, A., Winter, K.-U., Meyer, B., Saedler, H. and Theissen, G. (2000). MADS-box gene diversity in seed plants 300 million years ago. *Molecular Biology amd Evolution* **17**, 1425–1434.

Becker, A., Kaufmann, K., Freialdenhoven, A., Vincent, C., Li, M.-A., Saedler, H. and Theissen, G. (2002). A novel MADS-box gene subfamily with a sister-group relationship to class B floral homeotic genes. *Molecular Genetics and Genomics* **266**, 942–950.

Ben-Nissan, G., Lee, J.-Y., Borohov, A. and Weiss, D. (2004). GIP, a *Petunia hybrida* GA-induced cysteine-rich protein: A possible role in shoot elongation and transition to flowering. *The Plant Journal* **37**, 229–238.

Bernier, G. and Perilleux, C. (2005). A physiological overview of the genetics of flowering time control. *Plant Biotechnology Journal* **3**, 3–16.

Blazquez, M. A. (2000). Flower development pathways. *Journal of Cell Science* **113**, 3547–3548.

Blazquez, M. A. and Weigel, D. (2000). Integration of floral inductive signals in *Arabidopsis*. *Nature* **404**, 889–892.

Bolle, C. (2004). The role of GRAS proteins in plant signal transduction and development. *Planta* **218**, 683–692.

Borner, R., Kampmann, G., Chandler, J., Gleißner, R., Wisman, E., Apel, K. and Melzer, S. (2000). A MADS domain gene involved in the transition to flowering in *Arabidopsis*. *The Plant Journal* **24**, 591–599.

Boss, P. K., Bastow, R. M., Mylne, J. S. and Dean, C. (2004). Multiple pathways in the decision to flower: Enabling, promoting, and resetting. *Plant Cell* **16**, S18–S31.

Bowman, J. L., Alvarez, J., Weigel, D., Meyerowitz, E. M. and Smyth, D. R. (1993). Control of flower development in *Arabidopsis thaliana* by APETALA1 and interacting genes. *Development* **119**, 721–743.

Brand, U., Fletcher, J. C., Hobe, M., Meyerowitz, E. M. and Simon, R. (2000). Dependence of stem cell fate in *Arabidopsis* on a feedback loop regulated by CLV3 Activity. *Science* **289**, 617–619.

Cheng, X.-F., Wittich, P. E., Kieft, H., Angenent, G., XuHan, X. and van Lammeren, A. A. M. (2000). Temporal and spatial Expression of MADS Box genes, FBP7 and FBP11, during initiation and early development of ovules in wild type and mutant *Petunia hybrida*. *Plant Biology* **2**, 693–702.

Child, A. (1979). A review of branching patterns in the Solanaceae. *In* "The Biology and Taxonomy of the Solanaceae" (J. G. Hawkes, R. N. Lester and A. D. Skelding, eds.), pp. 345–356. Academic Press, London.

Coen, E. S. and Meyerowitz, E. M. (1991). The war of the whorls: Genetic interactions controlling flower development. *Nature* **353**, 31–37.

Colombo, L., Franken, J., Koetje, E., van Went, J., Dons, H., Angenent, G. C. and van Tunen, A. J. (1995). The Petunia MADS box gene FBP11 determines ovule identity. *Plant Cell* **7**, 1859–1868.

Colombo, L., Franken, J., Van der Krol, A. R., Wittich, P. E., Dons, H. and Angenent, G. C. (1997). Downregulation of ovule-Specific MADS Box Genes from Petunia results in maternally controlled defects in seed development. *Plant Cell* **9**, 703–715.

Davies, B., Motte, P., Keck, E., Saedler, H., Sommer, H. and Schwarz-Sommer, Z. (1999). PLENA and FARINELLI: Redundancy and regulatory interactions between two Antirrhinum MADS-box factors controlling flower development. *The EMBO Journal* **18**, 4023–4034.

de Folter, S., Immink, R. G. H., Kieffer, M., Parenicova, L., Henz, S. R., Weigel, D., Busscher, M., Kooiker, M., Colombo, L., Kater, M. M., Davies, B. and Angenent, G. C. (2005). Comprehensive interaction map of the *Arabidopsis* MADS box transcription factors. *Plant Cell* **17**, 1424–1433.

De Keukeleire, P., Maes, T., Sauer, M., Zethof, J., Van Montagu, M. and Gerats, T. (2001). Analysis by transposon display of the behavior of the *dTph1* element family during ontogeny and inbreeding of *Petunia hybrida*. *Molecular Genetics and Genomics* **265**, 72–81.

de Vlaming, P., Cornu, A., Farcy, E., Gerats, A. G. M., Wiering, H. and Wijsman, H. J. W. (1984). *Petunia hybrida*: A short description of the action of 91 genes, their origin and their map locations. *Plant Molecular Biology Reports* **2**, 21–42.

Ditta, G., Pinyopich, A., Robles, P., Pelaz, S. and Yanofsky, M. F. (2004). The SEP4 gene of *Arabidopsis thaliana* functions in floral organ and meristem identity. *Current Biology* **14**, 1935–1940.

Favaro, R., Pinyopich, A., Battaglia, R., Kooiker, M., Borghi, L., Ditta, G., Yanofsky, M. F., Kater, M. M. and Colombo, L. (2003). MADS-Box protein complexes control carpel and ovule development in *Arabidopsis*. *Plant Cell* **15**, 2603–2611.

Ferrandiz, C., Gu, Q., Martienssen, R. and Yanofsky, M. (2000). Redundant regulation of meristem identity and plant architecture by FRUITFULL, APETALA1 and CAULIFLOWER. *Development* **127**, 725–734.

Ferrario, S., Immink, R. G. H., Shchennikova, A., Busscher-Lange, J. and Angenent, G. C. (2003). The MADS box gene FBP2 is required for SEPALLATA function in Petunia. *Plant Cell* **15**, 914–925.

Ferrario, S., Busscher, J., Franken, J., Gerats, T., Vandenbussche, M., Angenent, G. C. and Immink, R. G. H. (2004). Ectopic expression of the Petunia MADS box gene UNSHAVEN accelerates flowering and confers leaf-like characteristics to floral organs in a dominant-negative manner. *Plant Cell* **16**, 1490–1505.

Gerats, A. G. M., Kaye, C., Collins, C. and Malmberg, R. L. (1988). Polyamine levels in Petunia genotypes with normal and abnormal floral morphologies. *Plant Physiology* **86**, 390–393.

Gerats, T. and Vandenbussche, M. (2005). A model system for comparative research: Petunia. *Trends in Plant Science* **10**, 251–256.

Goodrich, J., Puangsomlee, P., Martin, M., Long, D., Meyerowitz, E. M. and Coupland, G. (1997). A Polycomb-group gene regulates homeotic gene expression in *Arabidopsis*. *Nature* **368**, 44–51.

Goto, K., Kyozuka, J. and Bowman, J. L. (2001). Turning floral organs into leaves, leaves into floral organs. *Current Opinion in Genetics & Development* **11**, 449–456.

Gunn, C. R., and Gaffney, F. B. (1974). Seed characteristics of 42 economically important species of Solanaceae in the United States. *United States Department of Agriculture Technical Bulletin* **1471**, 1–32.

Gutierrez-Cortines, M. E. and Davies, B. (2000). Beyond the ABCs: Ternary complex formation in the control of floral organ identity. *Trends in Plant Science* **5**, 471–476.

Honma, T. and Goto, K. (2000). The *Arabidopsis* floral homeotic gene PISTILLATA is regulated by discrete cis-elements responsive to induction and maintenance signals. *Development* **127**, 2021–2030.

Honma, T. and Goto, K. (2001). Complexes of MADS-box proteins are sufficient to convert leaves into floral organs. *Nature* **409**, 525–529.

Immink, R., Hannapel, D., Ferrario, S., Busscher, M., Franken, J., Lookeren Campagne, M. and Angenent, G. (1999). A petunia MADS box gene involved in the transition from vegetative to reproductive development. *Development* **126**, 5117–5126.

Immink, R. G. H., Gadella, T. W. J., Jr., Ferrario, S., Busscher, M. and Angenent, G. C. (2002). Analysis of MADS box protein-protein interactions in living plant cells. *Proceedings of the National Academy of Sciences of the United States of America* **99**, 2416–2421.

Immink, R. G. H., Ferrario, S., Busscher-Lange, J., Kooiker, M., Busscher, M. and Angenent, G. C. (2003). Analysis of the petunia MADS-box transcription factor family. *Molecular Genetics and Genomics* **268**, 598–606.

Izhaki, A., Swain, S. M., Tseng, T.-S., Borochov, A., Olszewski, N. E. and Weiss, D. (2001). The role of SPY and its TPR domain in the regulation of gibberellin action throughout the life cycle of *Petunia hybrida* plants. *Plant Journal* **28**, 181–190.

Jofuku, K. D., Boer, B., Montagu, M. V. and Okamuro, J. K. (1994). Control of *Arabidopsis* flower and seed development by the homeotic gene APETALA2. *Plant Cell* **6**, 1211–1225.

Kanno, A., Saeki, H., Kameya, T., Saedler, H. and Theissen, G. (2003). Heterotopic expression of class B floral homeotic genes supports a modified ABC model for tulip (*Tulipa gesneriana*). *Plant Molecular Biology* **52**, 831–841.

Kapoor, M., Tsuda, S., Tanaka, Y., Mayama, T., Okuyama, Y., Tsuchimoto, S. and Takatsuji, H. (2002). Role of petunia pMADS3 in determination of floral organ and meristem identity, as revealed by its loss of function. *The Plant Journal* **32**, 115–127.

Kardailsky, I., Shukla, V. K., Ahn, J. H., Dagenais, N., Christensen, S. K., Nguyen, J. T., Chory, J., Harrison, M. J. and Weigel, D. (1999). Activation tagging of the floral inducer FT. *Science* **286**, 1962–1965.

Kater, M. M., Colombo, L., Franken, J., Busscher, M., Masiero, S., Van Lookeren Campagne, M. M. and Angenent, G. C. (1998). Multiple AGAMOUS homologs from cucumber and Petunia differ in their ability to induce reproductive organ fate. *Plant Cell* **10**, 171–182.

Keck, E., McSteen, P., Carpenter, R. and Coen, E. (2003). Separation of genetic functions controlling organ identity in flowers. *The EMBO Journal* **22**, 1058–1066.

Kim, S., Yoo, M.-J., Albert, V. A., Farris, J. S., Soltis, P. S. and Soltis, D. E. (2004). Phylogeny and diversification of B-function MADS-box genes in angiosperms: Evolutionary and functional implications of a 260-million-year-old duplication. *American Journal of Botany* **91**, 2102–2118.

Knapp, S. (2002a). Tobacco to tomatoes: A phylogenetic perspective on fruit diversity in the Solanaceae. *Journal of Experimental Botany* **53**, 2001–2022.

Knapp, S. (2002b). Floral diversity and evolution in the Solanaceae. *In* "Developmental Genetics and Plant Evolution" (Q. C. B. Cronk, R. M. Bateman and J. A. Hawkins, eds.), pp. 267–297. Taylor & Francis, London.

Knapp, S., Bohs, L., Nee, M. and Spooner, D. M. (2004). Conference Review: Solanaceae—a model for linking genomics with biodiversity. *Comparative and Functional Genomics* **5**, 285–291.

Kobayashi, Y., Kaya, H., Goto, K., Iwabuchi, M. and Araki, T. (1999). A pair of related genes with antagonistic roles in mediating flowering signals. *Science* **286**, 1960–1962.

Koes, R., Souer, E., Van Houwelingen, A., Mur, L., Spelt, C., Quattrocchio, F., Wing, J., Oppedijk, B., Ahmed, S., Maes, T., Gerats, T. Hoogeveen, P., *et al.* (1995). Targeted gene inactivation in petunia by PCR-based selection of transposon insertion mutants. *Proceedings of the National Academy of Sciences of the United States of America* **92**, 8149–8153.

Koes, R., Verweij, W. and Quattrocchio, F. (2005). Flavonoids: A colorful model for the regulation and evolution of biochemical pathways. *Trends in Plant Science* **10**, 236–242.

Kramer, E. M. and Hall, J. C. (2005). Evolutionary dynamics of genes controlling floral development. *Current Opinion in Plant Biology* **8**, 13–18.

Kramer, E. M. and Irish, V. F. (2000). Evolution of petal and stamen developmental programs: Evidence from comparative studies of the lower eudicots and basal angiosperms. *International Journal of Plant Sciences* **161**, S29–S40.

Kramer, E. M., Dorit, R. L. and Irish, V. F. (1998). Molecular evolution of genes controlling petal and stamen development: Duplication and divergence within the APETALA3 and PISTILLATA MADS-Box gene lineages. *Genetics* **149**, 765–783.

Kramer, E. M., Di Stilio, V. S. and Schlüter, P. M. (2003). Complex patterns of gene duplication in the *APETALA3* and *PISTILLATA* lineages of the Ranunculaceae. *International Journal of Plant Sciences* **164**, 1–11.

Lamb, R. S. and Irish, V. F. (2003). Functional divergence within the APETALA3/PISTILLATA floral homeotic gene lineages. *Proceedings of the National Academy of Sciences of the United States of America* **100**, 6558–6563.

Laux, T., Mayer, K., Berger, J. and Jurgens, G. (1996). The WUSCHEL gene is required for shoot and floral meristem integrity in *Arabidopsis*. *Development* **122**, 87–96.

Lee, H., Suh, S.-S., Park, E., Cho, E., Ahn, J. H., Kim, S.-G., Lee, J. S., Kwon, Y. M. and Lee, I. (2000). The AGAMOUS-LIKE 20 MADS domain protein integrates floral inductive pathways in Arabidopsis. *Genes & Development* **14**, 2366–2376.

Lenhard, M., Bohnert, A., Jurgens, G. and Laux, T. (2001). Termination of stem cell maintenance in Arabidopsis floral meristems by interactions between WUSCHEL and AGAMOUS. *Cell* **105**, 805–814.

Lenhard, M., Jurgens, G. and Laux, T. (2002). The WUSCHEL and SHOOTMERISTEMLESS genes fulfil complementary roles in *Arabidopsis* shoot meristem regulation. *Development* **129**, 3195–3206.

Levan, A. (1937). Eine erbliche anomalie der samenanlage bei Petunia. *Botaniska Notiser* **1**, 35–55.

Litt, A. and Irish, V. F. (2003). Duplication and Diversification in the APETALA1/FRUITFULL floral homeotic gene lineage: Implications for the evolution of floral development. *Genetics* **165**, 821–833.

Lohmann, J. U., Hong, R. L., Hobe, M., Busch, M. A., Parcy, F., Simon, R. and Weigel, D. (2001). A molecular link between stem cell regulation and floral patterning in *Arabidopsis*. *Cell* **105**, 793–803.

Maes, T., Van de Steene, N., Zethof, J., Karimi, M., D'Hauw, M., Mares, G., Van Montagu, M. and Gerats, T. (2001). Petunia Ap2-like genes and their role in flower and seed development. *Plant Cell* **13**, 229–244.

Maizel, A., Busch, M. A., Tanahashi, T., Perkovic, J., Kato, M., Hasebe, M. and Weigel, D. (2005). The floral regulator LEAFY evolves by substitutions in the DNA binding domain. *Science* **308**, 260–263.

Martin, C. and Gerats, T. (1993). Control of pigment biosynthesis genes during petal development. *Plant Cell* **5**, 1253–1264.

Masiero, S., Imbriano, C., Ravasio, F., Favaro, R., Pelucchi, N., Gorla, M. S., Mantovani, R., Colombo, L. and Kater, M. M. (2002). Ternary complex formation between MADS-box transcription factors and the histone fold protein NF-YB. *The Journal of Biological Chemistry* **277**, 26429–26435.

Mayama, T., Ohtstubo, E. and Tsuchimoto, S. (2003). Isolation and expression analysis of Petunia *CURLY LEAF*-like genes. *Plant Cell Physiology* **44**, 811–819.

Mena, M., Ambrose, B. A., Meeley, R. B., Briggs, S. P., Yanofsky, M. F. and Schmidt, R. J. (1996). Diversification of C-function activity in maize flower development. *Science* **274**, 1537–1540.

Mizukami, Y. and Ma, H. (1992). Ectopic expression of the floral homeotic gene AGAMOUS in transgenic Arabidopsis plants alters floral organ identity. *Cell* **71**, 119–131.

Moon, J., Suh, S.-S., Lee, H., Choi, K.-R. H., Choo Bong Paek, N.-C., Kim, S.-G. and Lee, I. (2003). The SOC1 MADS-box gene integrates vernalization and gibberellin signals for flowering in *Arabidopsis*. *The Plant Journal* **35**, 613.

Moon, J., Lee, H., Kim, M. and Lee, I. (2005). Analysis of flowering pathway integrators in *Arabidopsis*. *Plant Cell Physiol.* **46**, 292–299.

Mueller, L. A., Solow, T. H., Taylor, N., Skwarecki, B., Buels, R., Binns, J., Lin, C., Wright, M. H., Ahrens, R., Wang, Y., Herbst, E. V. Keyder, E. R., *et al.* (2005). The SOL Genomics Network. A Comparative Resource for Solanaceae Biology and Beyond. *Plant Physiol.* **138**, 1310–1317.

Nakagawa, H., Ferrario, S., Angenent, G. C., Kobayashi, A. and Takatsuji, H. (2004). The Petunia ortholog of Arabidopsis SUPERMAN plays a distinct role in floral organ morphogenesis. *Plant Cell* **16**, 920–932.

Nakamura, T., Fukuda, T., Nakano, M., Hasebe, M., Kameya, T. and Kanno, A. (2005). The modified ABC model explains the development of the petaloid perianth of *Agapanthus praecox* ssp. *orientalis* (Agapanthaceae) flowers. *Plant Molecular Biology* **58**, 435–445.

Nam, J., dePamphilis, C. W., Ma, H. and Nei, M. (2003). Antiquity and evolution of the MADS-box gene family controlling flower development in plants. *Molecular Biology and Evolution* **20**, 1435–1447.

Negre, F., Kish, C. M., Boatright, J., Underwood, B., Shibuya, K., Wagner, C., Clark, D. G. and Dudareva, N. (2003). Regulation of methylbenzoate emission after pollination in snapdragon and Petunia flowers. *Plant Cell* **15**, 2992–3006.

Nilsson, O., Lee, I., Blazquez, M. A. and Weigel, D. (1998). Flowering-time genes modulate the response to LEAFY activity. *Genetics* **150**, 403–410.

Pelaz, S., Ditta, G. S., Baumann, E., Wisman, E. and Yanofsky, M. F. (2000). B and C floral organ identity functions require *SEPALLATA* MADS-box genes. *Nature* **405**, 200–203.

Pelaz, S., Tapia-López, R., Alvarez-Buylla, E. R. and Yanofsky, M. F. (2001). Conversion of leaves into petals in *Arabidopsis*. *Current Biology* **11**, 182–184.

Pinyopich, A., Ditta, G. S., Savidge, B., Liljegren, S. J., Baumann, E., Wisman, E. and Yanofsky, M. F. (2003). Assesing the redundancy of MADS-box genes during carpel and ovule development. *Nature* **424**, 85–88.

Pnueli, L., Hareven, D., Broday, L., Hurwitz, C. and Lifschitz, E. (1994). The TM5 MADS box gene mediates organ differentiation in the three inner whorls of tomato flowers. *Plant Cell* **6**, 175–186.

Purugganan, M. D. (1997). The MADS-box homeotic gene lineages predate the origin of seed plants: Phylogenetic and molecular clock estimates. *Journal of Molecular Evolution* **45**, 392–396.

Purugganan, M. D., Rounsley, S. D., Schmidt, R. J. and Yanofsky, M. F. (1995). Molecular evolution of flower development: Diversification of the plant MADS-box regulatory gene family. *Genetics* **140**, 345–356.

Riechmann, J. L., Krizek, B. A. and Meyerowitz, E. M. (1996). Dimerization specificity of Arabidopsis MADS domain homeotic proteins APETALA1, APETALA3, PISTILLATA, and AGAMOUS. *Proceedings of the National Academy of Sciences of the United States of America* **93**, 4793–4798.

Sabatini, S., Heidstra, R., Wildwater, M. and Scheres, B. (2003). SCARECROW is involved in positioning the stem cell niche in the *Arabidopsis* root meristem. *Genes & Development* **17**, 354–358.

Saedler, H. and Huijser, P. (1993). Molecular biology of flower development in *Antirrhinum majus* (snapdragon). *Gene* **135**, 239–243.

Sakai, H., Krizek, B. A., Jacobsen, S. E. and Meyerowitz, E. M. (2000). Regulation of SUP expression identifies multiple regulators involved in *Arabidopsis* floral meristem development. *Plant Cell* **12**, 1607–1618.

Samach, A., Onouchi, H., Gold, S. E., Ditta, G. S., Schwarz-Sommer, Z., Yanofsky, M. F. and Coupland, G. (2000). Distinct roles of CONSTANS target genes in reproductive development of *Arabidopsis*. *Science* **288**, 1613–1616.

Schoof, H., Lenhard, M., Haecker, A., Mayer, K. F. X., Jurgens, G. and Laux, T. (2000). The stem cell population of *Arabidopsis* shoot meristems is maintained by a regulatory loop between the CLAVATA and WUSCHEL genes. *Cell* **100**, 635–644.

Schwarz-Sommer, Z., Hue, I., Huijser, P., Flor, P., Hansen, R., Tetens, F., Lonnig, W., Saedler, H. and Sommer, H. (1992). Characterization of the Antirrhinum floral homeotic MADS-box gene deficiens: Evidence for DNA binding and autoregulation of its persistent expression throughout flower development. *The EMBO Journal* **11**, 251–263.

Sink, K. C. (1984). "Monographs on Theoretical and Applied Genetics 9: Petunia." Springer Verlag, Heidelberg.

Sink, K. C. and Power, J. B. (1978). Incongruity of interspecific and intergeneric crosses involving Nicotiana and Petunia species that exhibit potential for somatic hybridization. *Euphytica* **27**, 725–730.

Souer, E., van Houwelingen, A., Kloos, D., Mol, J. and Koes, R. (1996). The No apical meristem gene of Petunia is required for pattern formation in embryos and flowers and is expressed at meristem and primordia boundaries. *Cell* **85**, 159–170.

Souer, E., van der Krol, A., Kloos, D., Spelt, C., Bliek, M., Mol, J. and Koes, R. (1998). Genetic control of branching pattern and floral identity during Petunia inflorescence development. *Development* **125**, 733–742.

Spelt, C., Quattrocchio, F., Mol, J. and Koes, R. (2002). ANTHOCYANIN1 of Petunia controls pigment synthesis, vacuolar pH, and seed coat development by genetically distinct mechanisms. *Plant Cell* **14**, 2121–2135.

Stuurman, J., Jaggi, F. and Kuhlemeier, C. (2002). Shoot meristem maintenance is controlled by a GRAS-gene mediated signal from differentiating cells. *Genes & Development* **16**, 2213–2218.

Stuurman, J., Hoballah, M. E., Broger, L., Moore, J., Basten, C. and Kuhlemeier, C. (2004). Dissection of floral pollination syndromes in Petunia. *Genetics* **168**, 1585–1599.

Tandre, K., Svenson, M., Svensson, M. E. and Engstrom, P. (1998). Conservation of gene structure and activity in the regulation of reproductive organ development of conifers and angiosperms. *The Plant Journal* **15**, 615–623.

Theissen, G. and Saedler, H. (2001). Floral quartets. *Nature* **409**, 469–471.

Theissen, G., Becker, A., Di Rosa, A., Kanno, A., Kim, J. T., Münster, T., Winter, K.-U. and Saedler, H. (2000). A short history of MADS-box genes in plants. *Plant Molecular Biology* **42**, 115–149.

Tobena-Santamaria, R., Bliek, M., Ljung, K., Sandberg, G., Mol, J. N. M., Souer, E. and Koes, R. (2002). FLOOZY of petunia is a flavin mono-oxygenase-like protein required for the specification of leaf and flower architecture. *Genes & Development* **16**, 753–763.

Tsuchimoto, S., van der Krol, A. R. and Chua, N. H. (1993). Ectopic expression of pMADS3 in transgenic *Petunia* phenocopies the *Petunia* blind mutant. *Plant Cell* **5**, 843–853.

Tsuchimoto, S., Mayama, T., van der Krol, A. and Ohtsubo, E. (2000). The whorl-specific action of a petunia class B floral homeotic gene. *Genes Cells* **5**, 89–99.

Underwood, B. A., Tieman, D. M., Shibuya, K., Dexter, R. J., Loucas, H. M., Simkin, A. J., Sims, C. A., Schmelz, E. A., Klee, H. J. and Clark, D. G. (2005). Ethylene-regulated floral volatile synthesis in *Petunia* corollas. *Plant Physiology* 104.051144.

Vallade, J., Maizonnier, D. and Cornu, A. (1987). La morphogenese florale chez le petunia. Analyze d'un mutant à corolle staminée. *Canadian Journal of Botany* **65**, 761–764.

Van den Broeck, D., Maes, T., Sauer, M., Zethof, J., De Keukeleire, P., D'Hauw, M., Van Montagu, M. and Gerats, T. (1998). Transposon display identifies individual transposable elements in high copy number lines. *The Plant Journal* **13**, 121–129.

van der Krol, A., Brunelle, A., Tsuchimoto, S. and Chua, N. (1993). Functional analysis of petunia floral homeotic MADS box gene pMADS1. *Genes & Development* **7**, 1214–1228.

Vandenbussche, M., Theissen, G., Van de Peer, Y. and Gerats, T. (2003a). Structural diversification and neo-functionalization during floral MADS-box gene evolution by C-terminal frameshift mutations. *Nucleic Acids Research* **31**, 4401–4409.

Vandenbussche, M., Zethof, J., Souer, E., Koes, R., Tornielli, G. B., Pezzotti, M., Ferrario, S., Angenent, G. C. and Gerats, T. (2003b). Toward the analysis of the *Petunia* MADS box gene family by reverse and forward transposon insertion mutagenesis approaches: B, C, and D floral organ identity functions require SEPALLATA-like MADS box genes in Petunia. *Plant Cell* **15**, 2680–2693.

Vandenbussche, M., Zethof, J., Royaert, S., Weterings, K. and Gerats, T. (2004). The duplicated B-class heterodimer model: Whorl-specific effects and complex genetic interactions in *Petunia hybrida* flower development. *Plant Cell* **16**, 741–754.

Verdonk, J. C., de Vos, C. H. R., Verhoeven, H. A., Haring, M. A., van Tunen, A. J. and Schuurink, R. C. (2003). Regulation of floral scent production in petunia revealed by targeted metabolomics. *Phytochemistry* **62**, 997–1008.

Verdonk, J. C., Haring, M. A., van Tunen, A. J. and Schuurink, R. C. (2005). ODORANT1 regulates fragrance biosynthesis in Petunia flowers. *Plant Cell* **17**, 1612–1624.

Vroemen, C. W., Mordhorst, A. P., Albrecht, C., Kwaaitaal, M. A. C. J. and de Vries, S. C. (2003). The CUP-SHAPED COTYLEDON3 gene is required for boundary and shoot meristem formation in *Arabidopsis*. *Plant Cell* **15**, 1563–1577.

Weberling, F. (1989). Morphology of flowers and inflorescences Cambridge University Press, Cambridge.

Weigel, D., Alvarez, J., Smyth, D., Yanofsky, M. and Meyerowitz, E. (1992). LEAFY controls floral organ meristem identity in *Arabidopsis*. *Cell* **69**, 843–859.

Weiss, D. (2000). Regulation of flower pigmentation and growth: Multiple signaling pathways control anthocyanin synthesis in expanding petals. *Physiologia Plantarum* **110**, 152–157.

Whipple, C. J., Ciceri, P., Padilla, C. M., Ambrose, B. A., Bandong, S. L. and Schmidt, R. J. (2004). Conservation of B-class floral homeotic gene function between maize and Arabidopsis. *Development* **131**, 6083–6091.

Yanofsky, M. F., Ma, H., Bowman, J. L., Drews, G. N., Feldmann, K. A. and Meyerowitz, E. M. (1990). The protein encoded by the *Arabidopsis* homeotic gene agamous resembles transcription factors. *Nature* **346**, 35–39.

Yu, H., Ito, T., Wellmer, F. and Meyerowitz, E. M. (2004). Repression of AGAMOUS-LIKE 24 is a crucial step in promoting flower development. *Nature Genetics* **36**, 157–161.

Zachgo, S., Silva, E., Motte, P., Trobner, W., Saedler, H. and Schwarz-Sommer, Z. (1995). Functional analysis of the *Antirrhinum* floral homeotic DEFICIENS gene *in vivo* and *in vitro* by using a temperature-sensitive mutant. *Development* **121**, 2861–2875.

Zahn, L. M., Kong, H., Leebens-Mack, J. H., Kim, S., Soltis, P. S., Landherr, L. L., Soltis, D. E., dePamphilis, C. W. and Ma, H. (2005). The evolution of the SEPALLATA subfamily of MADS-box genes: A preangiosperm origin with multiple duplications throughout angiosperm history. *Genetics* **169**, 2209–2223.

Flower Development: The *Antirrhinum* Perspective

BRENDAN DAVIES,* MARIA CARTOLANO† AND ZSUZSANNA SCHWARZ-SOMMER†

**Centre for Plant Sciences, University of Leeds, Leeds LS2 9JT, United Kingdom*
†Max Planck Institut für Züchtungsforschung, 5089 Köln, Germany

Advances in Botanical Research, Vol. 44
Incorporating Advances in Plant Pathology

0065-2296/06 $35.00
DOI: 10.1016/S0065-2296(06)44007-6

ABSTRACT

Research with the snapdragon, *Antirrhinum majus*, has a long history with many highlights, making this species a significant model system for comparative genetic, molecular, and ecological studies. In this chapter, we focus interest on flower development, in particular the genetic control of floral organ identity, floral asymmetry, and petal cell-type specification, where results obtained with *Antirrhinum* provided the first insights into the underlying molecular mechanisms, leading to advances in the field. In addition to reviewing past and recent scientific achievements, we propose simple models to aid the understanding of complex genetic observations and also to resolve the inconsistencies and contradictions, which inhibit the general application of existing models to other species. In particular, we propose a revision of the ABC-model to reflect better the experimental results obtained in a variety of model species including *Arabidopsis*. For this we revive the (*A*)BC-scheme in which sepal identity follows from floral meristem identity and represents the ground state for floral organs. In addition to controlling sepal identity, the complex (*A*)-function performs several roles that are necessary for the initiation, maintenance, spatial restriction, and functionality of the B- and C-organ identity functions. By providing information on current resources for molecular research and newly arising research areas we intend to encourage the scientific community to utilize *Antirrhinum* for research in the future.

I. INTRODUCTION

A diversity of model organisms serves three distinct purposes. First, studying similar processes in diverse species allows us to identify the common threads which control life. Second, the very differences between mechanisms, whether these are different implementations of the same overall mechanism or the substitutions of entirely new ones, allow us to understand the variety of life that we observe around us. Finally, not all models are suitable or even competent for the study of all biological processes. All of these purposes are of value and lead directly to an enhanced understanding of biological processes and an increased ability to engineer those processes. This makes a strong argument for the use of a variety of different model organisms. In this chapter, we consider *Antirrhinum* as a model species, predominantly concentrating on the contribution it has made, and is continuing to make, to our understanding of flower development.

Antirrhinum has a long history of use as a model plant (Schwarz-Sommer *et al.*, 1990, 2003a) being used by both Darwin and Mendel, to different effect, for studies of inheritance. The systematic application of mutant studies, the advent of plant molecular biology, and the characterization of transposable elements, in that order, made it possible to identify genes responsible for morphological traits. These studies opened an era of dramatic success in our increased understanding of flower development leading to the isolation of such key genes (*Arabidopsis* orthologs shown in brackets) as *DEFICIENS* (*APETALA3*), *GLOBOSA* (*PISTILLATA*), *FLORICAULA* (*LEAFY*), *SQUAMOSA* (*APETALA1*), *FIMBRIATA* (*UNUSUAL FLORAL ORGANS*), and *CENTRORADIALIS* (*TERMINAL FLOWER 1*).

These genes, and many more involved in flower development, have subsequently been identified and characterized in the best studied model plant, *Arabidopsis thaliana*. It is certainly the case, therefore, that the processes controlling flower development are understood to a greater extent in *Arabidopsis* than in any other plant species. However, as we will show in this chapter, there is a great deal of variability in these processes when they are examined in other species. Sometimes the differences are related to individual peculiarities of the less commonly studied species, but this will not always be the case. It would be unfortunate if universal theories were extrapolated from exceptions to the rule. With this in mind, we will highlight some of the inconsistencies revealed from a comparison of flower development in *Antirrhinum* with flower development in *Arabidopsis* and suggest what these differences could tell us about the conservation of the core mechanisms. A model that satisfies the universality criteria for the genetic control of floral organ identity more adequately than the textbook ABC-model will also be discussed.

II. THE CONTROL OF FLORAL IDENTITY: WHERE THE DIFFERENCES BEGIN

Once the transition from the vegetative to reproductive phase of development is accomplished the shoot meristem, which previously contributed to leaf and axillary meristem formation, changes to an inflorescence meristem, initiating leaf-like bracts with flower primordia in their axils. Defects in the genetic control of flower initiation are revealed by partial or complete lack of flowering and the development of axillary inflorescences instead of flowers.

The *Antirrhinum* genes *FLORICAULA* (*FLO*) (Coen *et al.*, 1990) and *SQUAMOSA* (*SQUA*) (Huijser *et al.*, 1992), the orthologs of *LEAFY* (*LFY*) (Weigel *et al.*, 1992) and *APETALA1* (*AP1*) (Mandel *et al.*, 1992) in *Arabidopsis*, were the first of these genes to be isolated and studied in detail. The phenotypes of these meristem identity mutants differ, on close inspection, in that the *Antirrhinum* mutants display a more severe defect of flower to inflorescence conversion. For instance, *flo* mutants never flower, in contrast to *lfy*, and *squa* plants flower rarely whereas *ap1* mutants regularly flower (Bowman *et al.*, 1991). Thus, it appears that the conditions to determine the floral state are more stringent in *Antirrhinum* compared to *Arabidopsis*.

An intriguing difference between the properties of *Arabidopsis* and *Antirrhinum* floral meristem identity genes is revealed by their cell autonomy; that is their ability to influence functions in a neighboring cell layer. This aspect was first studied in *Antirrhinum*, where transposon-induced, genetically unstable *flo* mutants were used to generate chimeras, with active or inactive copies of the *FLO* gene in different cell layers (Carpenter and Coen, 1995; Hantke *et al.*, 1995). These studies showed that expression of the wild-type *FLO* gene in the outermost cell layer is sufficient to confer wild-type phenotype. Expression of the B- and C-organ identity controlling functions (see later section) was established in all layers suggesting that *FLO* can activate these genes nonautonomously. *FLO* activity in the inner two cell layers, however, results in aberrant phenotypes. In contrast, expression of *LFY* in any of the cell layers was sufficient to confer a wild-type phenotype to the plants in *LFY* chimeras generated by a transgenic approach in *Arabidopsis* (Sessions *et al.*, 2000). Although the *in vivo* significance of noncell autonomy, achieved by protein trafficking or cell-to-cell signaling, in controlling flowering is not clear, this difference suggests some deviation in cellular properties and the ability of cells to communicate in *Arabidopsis* and *Antirrhinum*. This is corroborated by the behavior of other floral transcription factors, showing that cell–cell communication between layers is facilitated in *Arabidopsis* and more restricted in *Antirrhinum* (Efremova *et al.*, 2001).

III. THE (*A*)BC-MODEL: PAST DISCOVERIES AND CURRENT PROBLEMS

The long tradition of observing and collecting "monstrous," homeotically altered flowers in various species has provided a valuable tool to study the genetic control of floral organ identity. In fact, the availability of mutants with homeotic defects and the apparent similarity of the features of such mutants in different species served as the basis for constructing the first generally applicable models of this control.

The textbook ABC-model (Fig. 1B, top) assumed three partly overlapping homeotic functions to control, either alone or in combination, the identity of floral organs (Coen and Meyerowitz, 1991; Weigel and Meyerowitz, 1994). It was based on three types of loss-of-function mutants in *Arabidopsis,* two of which (the B- and C-types) shared phenotypic characteristics with mutants in *Antirrhinum*, and described in a very simple way how loss or (ectopic) gain of these functions could explain virtually every homeotic mutant phenotype in many flowering species.

However, the first version of this model, which we will refer to as the (*A*)BC-model (Fig. 1B, middle), was based only on the *Antirrhinum* and *Arabidopsis* mutants displaying comparable homeotic features. This model contained two developmental pathways corresponding to B and C of the ABC-model, which conferred either petal or carpel identity when acting alone, stamen identity when acting in combination, and sepal identity when absent (Schwarz-Sommer *et al.*, 1990). An unknown and uncharacterized function determining sepal identity was supposed; in the currently used terminology it would be the control of floral meristem identity that acts throughout all floral whorls and all other functions necessary for the B- and C-functions to exert their effects (Fig. 1B, middle). In this interpretation, sepals represented the "ground state" of floral organs, the identity of which follows from floral meristem identity. In contrast, in the ABC-model the ground state was assumed to be leaves and the A-function was defined as an organ specification function controlling sepal identity and, together with B, petal identity. In this section, we address questions that necessitate rigorous modifications to the textbook ABC-model from the perspective of both *Antirrhinum* and *Arabidopsis* research, mainly by explaining and extending the complex roles of the A-function. For simplicity, genes whose role can be partly or fully assigned to the A-, B-, or C-functions and their mutants will be designated as A-, B-, or C-genes and A-, B-, or C-mutants.

A. THE A-FUNCTION AND ITS DUBIOUS ROLE IN THE CONTROL OF PERIANTH IDENTITY: THE RELATION BETWEEN FLORAL MERISTEM AND SEPAL IDENTITY

According to the textbook ABC-model, the A-function can be separated into two subfunctions (Jack, 2004; Weigel and Meyerowitz, 1994). The first subfunction involves the direct specification of organ identity and is revealed by leaf-like characters of sepals of the A mutants *ap1* and *ap2*, which also show defects in the control of floral meristem identity (Section II).

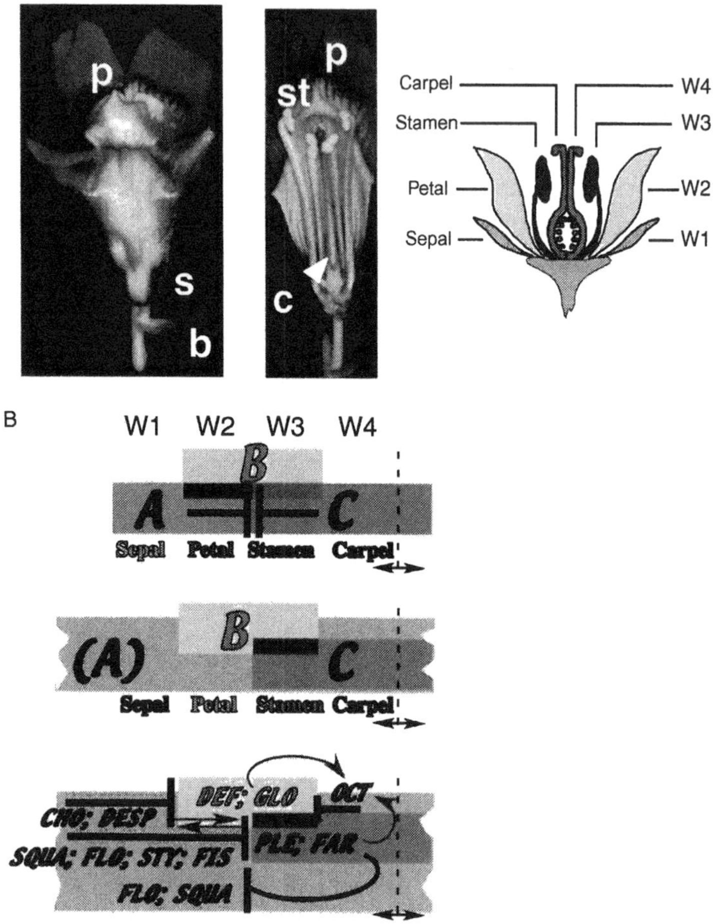

Fig. 1. The (*A*)-, B-, and C-control functions in the control of floral organ identity. (A) The photographs on the left show the mature *Antirrhinum* flower in front view and with the lower second-whorl petals removed to reveal the reproductive organs. The drawing on the right shows the flower in a longitudinal section with the names of the organs and the number of the respective whorls (W). In *Antirrhinum* five sepals (W1) form the calyx, five petals form the zygomorphic corolla (W2), and together with the sepals the perianth. Five stamens are initiated in W3, the dorsal (adaxial or upper) of which is retarded and is named the staminode. The flower terminates with two carpels in W4 forming the gynoecium in the center of the flower. (B) The block schemes illustrate the textbook ABC-model (top; Weigel and Meyerowitz, 1994) and the (*A*)BC-model [middle and bottom; modified from Schwarz-Sommer *et al.* (1990)]. Only the left half of the flower is shown and the center is indicated by a dotted vertical line. B and C are organ identity control functions, expressed in two adjacent whorls. Alone they control petal and carpel identity, respectively, and their combined expression results in stamen identity. The complementary colors yellow for B (petals) and blue for C (carpels) illustrate the B and C expression domains, their overlap (according to chromatics) results in

The second subfunction involves setting the boundaries of expression of other genes, referred to as a "cadastral" function, which is revealed by the expansion of the C-function into the perianth (discussed in Section III.B). Several observations indicate, however, that sepal identity is inherently linked to the establishment of floral identity rather than determined by a separable floral function (Litt and Irish, 2003; Motte *et al.*, 1998a,b).

The control of sepal identity by the A-genes *AP2* and *AP1* in *Arabidopsis* was first challenged by experiments showing that by changing the conditions in which A-mutants are grown, the inflorescence characters become dramatically enhanced (Okamuro *et al.*, 1997). The inflorescence-like traits, such as leaf-like first-whorl organs in the mutant flowers, thus do not result from a homeotic sepal-to-leaf conversion, but reflect the role of *AP2* (or, similarly, *AP1*) in controlling floral meristem identity. The more strictly nonfloral phenotype of the *Antirrhinum* meristem identity mutants compared to those in *Arabidopsis* (Section II) thus suggests a more stringent dependence of sepal initiation on floral determination and explains the lack of mutants with leaf-like floral organs identifying the A-function in *Antirrhinum* (Motte *et al.*, 1998a,b). The close association between sepal identity and floral meristem identity is supported by phylogenetic analyses of *AP1*-like genes in several angiosperms (Litt and Irish, 2003).

Observations on mutants of *AP2*-like genes in plants other than *Arabidopsis* are scant and suggest their species-specific involvement in developmental events. In *Petunia*, the *AP2* ortholog appears to mainly participate in determination of inflorescence meristem identity without affecting sepal

---

green (stamens). The individual color coding for A in the textbook ABC scheme indicates the distinct role of the A-function in sepal and petal organ identity in the two outer whorls. In the absence of this function leaf-like organs (the ground state in the ABC-model) develop instead of sepals. The complex (*A*)-function in the (*A*)BC-scheme controls floral (meristem) identity, all functions necessary for manifestation of the B- and C-functions along with the control of the boundaries of the B- and C-domains (Section III.G). In concomitant absence of B and C sepals develop, the ground state in the (*A*)BC-model, defined by the (*A*)-function (Section III.G), which is expressed in all floral whorls and is represented by gray color. For simplicity, the intermediate genes (*Im* genes) (corresponding to the *SEPALLATAs* or the E-function in *Arabidopsis*; see Section III.C) are not shown. The genes involved are shown in their respective (*A*), B, or C colors in the lower block scheme. Barred black lines indicate negative effects and red lines with arrows indicate positive influence in the boundary control. The two parallel arrows emphasize the interdependence between the control of the outer B- and C-boundaries as discussed in the text (Section III.B). For the sake of completeness the involvement of the B- and C-genes in the control of the inner B-boundary (see Section III.F and Fig. 4 for details) is also indicated. (See color Plate 8.)

identity in the rarely forming terminal flowers (Maes *et al.*, 1999). In *Antirrhinum*, double mutants of the (duplicated) *AP2* orthologs *LIPLESS1* and *LIPLESS2* (*LIP1/LIP2*) display enlarged sepals with slightly bract-like morphology (Keck *et al.*, 2003). This anomaly is possibly related to a weak defect in floral meristem identity, although this aspect has not been experimentally pursued.

SEPALLATA4 (*SEP4*) is another gene discovered to play a role in the maintenance of *Arabidopsis* floral meristem identity, indicated by enhanced meristem identity defects of the *ap1 sep4* double mutant (Ditta *et al.*, 2004). *sep4* mutants have no phenotype, but leaf-like characters appear in the sepals of *sep4* flowers provided that all of the functionally redundant *SEP* genes (Section III.D) are nonfunctional. This, again, corroborates the link between sepal identity and floral meristem identity in *Arabidopsis*.

A common role of *AP1* or *AP2* in the homeotic control of petal identity is difficult to assess. Clearly, *AP1* in collaboration with other meristem identity-controlling genes, such as *LFY* and several other factors (Jack, 2004), is necessary for early activation of the B-function, and contributes in this way to petal identity. *AP2* and *LIP1/LIP2*, on the other hand, participate in controlling petal (as well as stamen and carpel) growth, although in different ways (Keck *et al.*, 2003). In particular, the *LIP1/LIP2* function relates to developmental events during elaboration of the complex shape of *Antirrhinum* petals rather than to an early homeotic control of petal identity.

If *AP1* and *AP2,* the genes which define the A-function, do not act by specifying *Arabidopsis* perianth organ identity, but rather by controlling meristem identity and thus conditioning sepal initiation, the presence of whorls of leaves in the A, B, C triple mutants (Weigel and Meyerowitz, 1994) cannot be interpreted to indicate that leaves are the "ground state" of floral organs. That floral organs are modified leaves is a long-standing and most likely correct idea (Goethe, 1790). Its approval by the phenotype of the homeotic *Arabidopsis* triple mutant, however, is due to a lucky coincidence in that, unlike *Antirrhinum*, *Arabidopsis* flowers can form in the absence of strict floral determination (Okamuro *et al.*, 1997). It will be interesting to learn in the future, what distinguishes *Arabidopsis* and *Antirrhinum* in this respect and whether this distinction is inherent to close relatives of *Antirrhinum* (in the asterid clade) or *Arabidopsis* (in the rosid clade).

In summary, there is no control function during flower development to concomitantly confer sepal and petal identity to the first and second floral whorls, analogous to the B- and C-functions in the control of petal, stamen, and carpel identity in the other whorls. Instead, sepal identity, the "ground state" of floral organs, is conferred by the floral meristem identity genes in both *Arabidopsis* and *Antirrhinum*. Because this issue changes the meaning of the A-function, we will use (*A*)-function at places where this meaning becomes relevant.

B. THE A-FUNCTION IN THE SPATIAL CONTROL OF C-GENE EXPRESSION

In the *Arabidopsis* A-function mutants *ap1* and *ap2*, the expression domain of the C-function, which is restricted to the inner whorls in the wild type, expands to the perianth. This suggests a role for the A-function in the control of C-gene expression and represents what is most commonly understood as "A-function." Vice versa, the C-function in the inner whorls negatively affects the A-function—a mutual negative control that has been of mechanistic importance in the *Arabidopsis* ABC-model (Jack, 2004; Weigel and Meyerowitz, 1994). These restrictive mechanisms were previously termed "cadastral" and included other genes whose floral functions were limited to the negative control of the C-function (Liu and Meyerowitz, 1995; Weigel and Meyerowitz, 1994).

The control and maintenance of floral meristem identity [one of the (*A*)-subfunctions, as shown in Section III.A, which was excluded from the A-function in the textbook ABC-model] along with meristematic activities *per se* have to be circumvented in the center of the flower to block organ initiation and to accomplish floral determinacy (Section III.F). This is achieved, in part, through negative control by the C-function (the cadastral C-subfunction) in both *Arabidopsis* and *Antirrhinum* and is detectable by reduced transcription of the floral meristem identity-controlling genes (Fig. 1, barred line from C to *FLO* and *SQUA*; Huijser *et al.*, 1992; Motte *et al.*, 1998a).

The cadastral A-subfunction involved in the repression of the C-function in the outer whorls appears to be complex and the gene-to-gene comparison between *Arabidopsis* and *Antirrhinum* is difficult. The *Antirrhinum* meristem identity gene *SQUA* in contrast to *AP1* in *Arabidopsis* does not seem to perform the cadastral subfunction. Similarly, and unlike *AP2*, *LIP1/LIP2* in *Antirrhinum* do not seem to participate in the spatial control of the C-function, although to date only one of the C-genes has been tested (Keck *et al.*, 2003; Motte *et al.*, 1998a). However, in contrast to the meristem identity gene *LFY* in *Arabidopsis* (not considered as a class A-gene), its ortholog *FLO* in *Antirrhinum* is not only involved in floral meristem establishment (Section II) but also in the negative control of the C-function in the perianth (Fig. 1). This is clear from double-mutant analyses where weak alleles of *flo* enhance the defects in mutants with aberrant control of the C-function (McSteen *et al.*, 1998; Motte *et al.*, 1998a). *FLO* therefore shares functional similarity in this respect with the A-genes *AP1* and *AP2* in *Arabidopsis*. This again supports the view of meristem identity control as a subfunction of the (*A*)-function, which

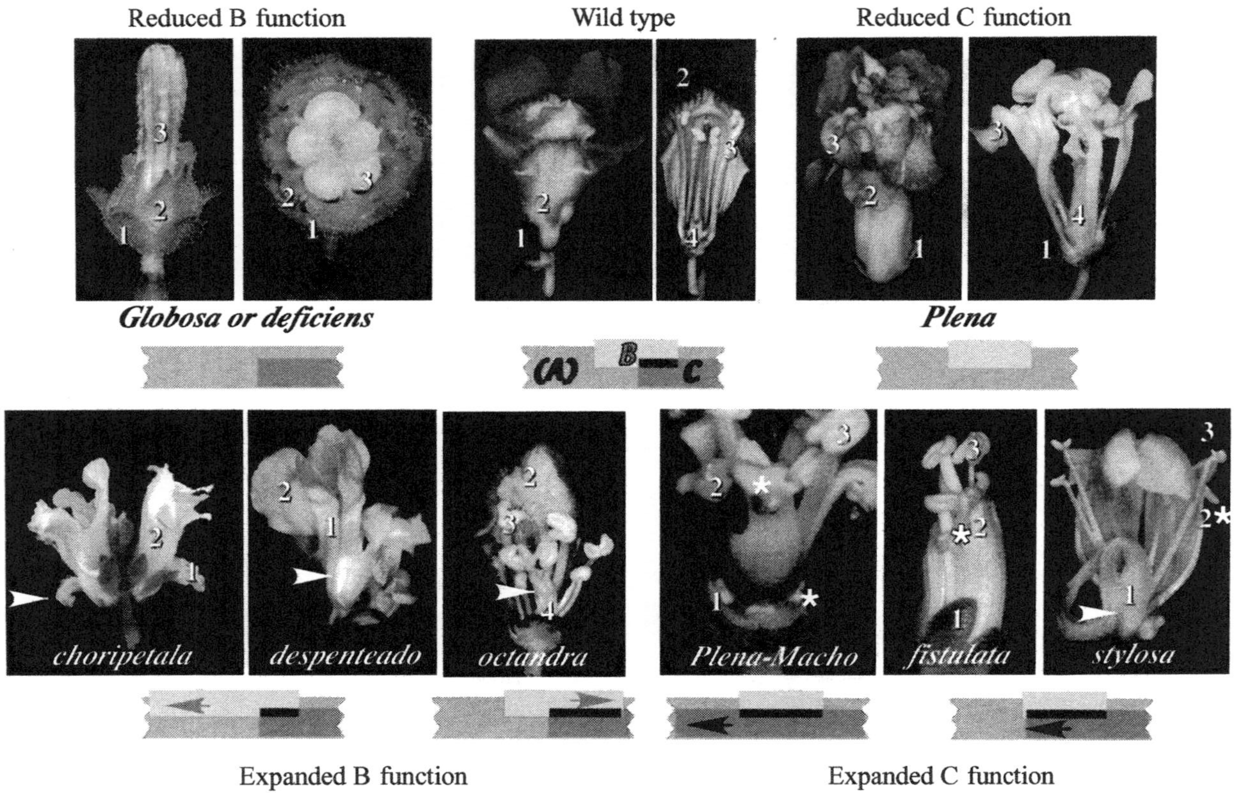
Reduced B function
Wild type
Reduced C function
Globosa or deficiens
(A)
B
C
Plena
choripetala
despenteado
octandra
Plena-Macho
fistulata
stylosa
Expanded B function
Expanded C function

consequently includes all meristem identity-controlling factors irrespective of their role in the cadastral control of C-genes.

Several mutants with homeotically transformed carpelloid sepals and/or stamenoid petals exist in *Arabidopsis* and *Antirrhinum*, indicative of alterations in the expression domain of the C-function (Fig. 2).

STYLOSA, *FISTULATA*, and *GRAMINIFOLIA* (*STY*, *FIS*, *GRAM*; Figs. 2 and 3) are such cadastral A-function genes in *Antirrhinum* which either alone and/or in combination control the expression level and the outer boundary of the C-domain (Motte *et al.*, 1998a; Navarro *et al.*, 2004; Wilkinson *et al.*, 2000). Concomitantly, the outer boundary of the B-domain is also affected (see, for instance, the petaloid sepals of *sty* in Fig. 2), suggesting interdependence of the two processes, or, perhaps, a common mechanism of boundary control. This idea is corroborated by the observation that mutants in genes primarily affecting the outer boundary of the B-domain, such as *CHORIPETALA* (*CHO*) or *DESPENTEADO* (*DES*) (Fig. 2), also reveal an influence on the establishment of the C-domain in certain double-mutant combinations [(Wilkinson *et al.*, 2000); Fig. 3], indicated by two red horizontal arrows running between the barred lines in the block scheme at the bottom in Fig. 1B.

The impaired identity of *fimbriata* (*fim*) mutant organs, affected by the reduced expression of the B- and C-genes (Simon *et al.*, 1994), improves toward wild-type-like in the background of mutations in most of these genes [(Wilkinson *et al.*, 2000) and unpublished observations summarized in Fig. 3]. It has been suggested that FIM, an F-box-containing protein, is involved in targeting repressors of the B- and C-gene products for early proteolytic degradation (Ingram *et al.*, 1997); in the absence of this process repressors of B and C are maintained, preventing manifestation of the organ

Fig. 2. *Antirrhinum* mutants with reduced and expanded domains of the B- or C-organ identity control functions. The photographs show mutants whose phenotypes reveal interference with the control of B- and C-gene expression boundaries. The names of the respective mutants are indicated under the photographs. Some or all petals were removed from the *plena*, *octandra*, and *plena*-macho flowers. *choripetala* and *fistulata* are shown from the side, *despenteado* from the back, and all other flowers from the front. In the bottom panel arrows point to organs whose identity is altered due to expansion of the B-function (petaloid sepals in *choripetala*, *despenteado*, *stylosa*, and stamenoid carpels in *octandra*). Asterisks show expansion of the C-function, which are aberrant or stamenoid petals in all respective mutants and ovules in the carpelloid first whorl of *plena*-macho flowers, carrying a semidominant *PLENA* allele (Lönnig and Saedler, 1994). Notice increased stamen number in the *octandra* mutant (Coen and Carpenter, unpublished data), which is a result of extended B-function (Section III.F and Fig. 4). The block schemes beneath the photographs show alterations in the organ identity functions in the mutants in the context of the model in Fig. 1. Defects in determinacy and the number of whorls (Fig. 4) are not considered. (See color Plate 9.)

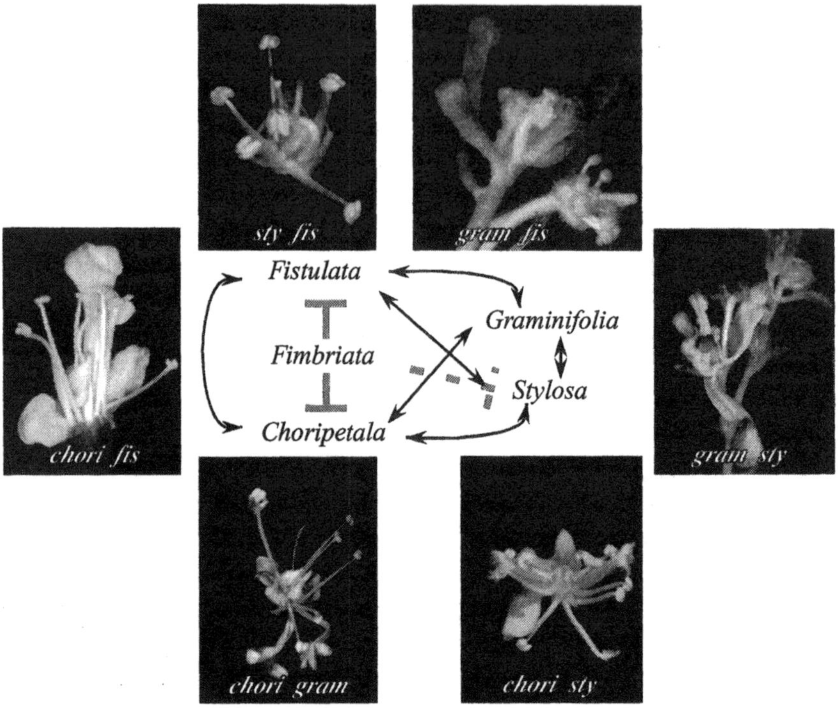

Fig. 3. Genetic interactions between genes controlling the boundaries of the B- and C-domains. The photographs show double mutants whose phenotype is more aberrant compared to the single-mutant phenotypes (Fig. 2). The genetic interactions (black lines with arrows) between the respective genes are schematically indicated in the graph in the middle. Red barred lines show negative interference between genes and are based on the epistatic relation of the respective genes to *FIMBRIATA* when combined in double mutants (details in Section III.B). Choripetala, fimbriata, fistulata, graminifolia, and stylosa are abbreviated as chori, fim, fis, gram, and sty respectively. (See color Plate 10.)

identity functions. Probably, FIS, CHO, and DESP (and to a lesser extent also STY) qualify as such repressors.

The precise molecular nature of these cadastral A-proteins is not clear except for STY, the ortholog of LEUNIG (LUG) in *Arabidopsis*. Based on their structural relation to GRO/TUP1-like proteins, LUG and STY most likely act as transcriptional corepressors (Conner and Liu, 2000; Navarro *et al.*, 2004). From studies of mRNA and protein expression patterns it seems that, unlike some of the meristem identity subclass of the (*A*)-function genes, the potential repressors of the cadastral subclass are not subject to negative control by the C-function (Navarro *et al.*, 2004).

STY interacts both genetically and at the protein level with GRAM [a YABBY protein and ortholog of FIL1 in *Arabidopsis* (Golz *et al.*, 2004)], revealed by synergistic floral homeotic defects in the *sty gram* double mutant (Fig. 3) and by yeast interaction assays (Navarro *et al.*, 2004). The influence of *GRAM* on the C-boundary is minor, but it becomes dramatically enhanced in double-mutant combinations with *sty*. Intriguingly, *gram* enhances the defects of several other mutants influencing B/C boundaries, such as *fis* and *cho* (Fig. 3). *fis* and *cho* also enhance each other's defects as well as that of *sty*. These multiple interactions suggest that the boundary control is quite complex and depends on the function of several factors. Efforts to clone the *Antirrhinum* genes are under way and will help to decipher this problem.

C. ADDITIONS TO (*A*): FLORAL FUNCTIONS NECESSARY FOR MANIFESTATION OF THE B- AND C-CONTROL

The B and C floral control genes turned out to encode MADS-box transcription factors (Schwarz-Sommer *et al.*, 1990; Sommer *et al.*, 1990; Yanofsky *et al.*, 1990) whose functional, genetic, and biochemical properties as well as evolutionary implications became widely studied in different species, as documented by various chapters in this book. One of the best studied molecular aspects of MADS-box proteins is their ability to form complexes, primarily heterodimers or homodimers (Tröbner *et al.*, 1992; Zachgo *et al.*, 1995). Studies suggest that the proteins can also form higher order complexes (Ferrario *et al.*, 2003; Honma and Goto, 2001), which most likely stabilize the proteins in the dimers and/or enhance their DNA-binding affinity (Egea-Cortines *et al.*, 1999).

One class of *Antirrhinum* MADS-box proteins named DEFH200, DEFH72, and DEFH84, previously detected as partners which interact with C-proteins (Davies *et al.*, 1996b), deserves special attention. These were termed the "intermediate" or Im-proteins (Egea Gutierrez-Cortines and Davies, 2000) because the respective genes are expressed after the meristem identity genes, but before initiation of transcription of the organ identity genes in all floral whorls. Based on their interactions and expression patterns it appeared that these Im MADS-box proteins are involved in the organ identity control. This assumption turned out to be correct since mutation of three of the (redundant) orthologs of these genes in *Arabidopsis* (the *SEP1–3* genes; Pelaz *et al.*, 2000) and two in *Petunia* (*FBP2*/*FBP5*; Ferrario *et al.*, 2003) results in formation of sepaloid organs in all whorls. Given that *in vitro* higher order protein complexes can form between the B-, C-, and SEP-proteins, it seems likely that such complexes perform the B- and C-organ identity controls *in vivo*. Nevertheless, the existence of these complexes in plants awaits experimental confirmation.

Mutation of all four *Arabidopsis* orthologs (*SEP1–4*), as mentioned before, leads to the production of leaf-like sepals (Ditta *et al.*, 2004) suggesting a role in the control of meristem identity (Section III.A). A function in meristem identity control can also be postulated for *SEP3* as *sep3* mutants display some inflorescence characters such as secondary flowers in the axils of sepals (Pelaz *et al.*, 2001a). Furthermore, like other meristem identity genes, *SEP3* is also involved in activation of the B- and C-genes (Castillejo *et al.*, 2005). The functions of the Im/SEP-proteins are now integrated into the ABC-model as the "E-function" (Theissen, 2001). In our view (see Section III.G), the Im/SEP E-genes belong to the (*A*)-function because they are essential prerequisites for the control of meristem identity, activation of the B- and C-genes, and manifestation of the B- and C-organ identity. This is also more sensible considering that the alphabetical order of ABC correctly reflects the timing of initiation and pattern of expression of the controlled processes.

D. THE B-FUNCTION PROTEINS AND THEIR PARTNERS IN THE MAKING OF A PETAL

Class B genes are the major determinants in the control of petal identity (and, together with the C-function, that of stamens). B-function mutants are characterized by sepaloid organs instead of petals and carpelloid organs instead of stamens in the second and third whorls, respectively (Fig. 2). *Arabidopsis* and *Antirrhinum* B-mutants are comparable in terms of these organ identity changes, and also in their concomitant defects in both whorls. In both species a pair of orthologous MADS-box proteins, DEFICIENS and GLOBOSA (DEF/GLO in *Antirrhinum*) (Sommer *et al.*, 1990; Tröbner *et al.*, 1992) and APETALA3 (AP3) and PISTILLATA (PI) in *Arabidopsis* (Goto and Meyerowitz, 1994; Jack *et al.*, 1992) are responsible for governing the B-function. This is not a general rule, however; in other species, such as *Petunia*, for instance, the number of B components can be higher and their interdependence more complex than in *Arabidopsis* or *Antirrhinum* (Vandenbussche *et al.*, 2004).

The overall similarity in the B-control function in *Arabidopsis* and *Antirrhinum* is achieved by a similar principle, which is based on protein interaction between the two B-proteins, leading to stabilization of the proteins in the complex, and an autoregulatory circuit that maintains their expression. As a result, the gene pairs are expressed mainly in the second and third whorls, although some details in the initial expression patterns in the two species differ (Davies *et al.*, 1999b). The control process with respect to downstream target genes is similar as well as revealed by efforts to identify class B petal target genes by a genome-wide expression analysis in both *Arabidopsis* (Zik and Irish, 2003) and *Antirrhinum* (Bey *et al.*, 2004). Common to these studies is the observation that only a small fraction of the controlled genes are transcription factors; the majority of the target genes belong to cellular processes, which are not necessarily petal specific, but are required for petal morphogenesis. This finding agrees well with previous studies showing that the B-function is required until very late stages of petal development, thus suggesting that the B-proteins do not delegate their control function to a downstream transcription factor network that eventually overtakes their function (Zachgo *et al.*, 1995).

Class-B genes can govern organ identity within a floral context, as shown by their ability to condition petaloid development when artificially expressed in the sepals of transgenic plants (Davies *et al.*, 1996a; Krizek and Meyerowitz, 1996). To produce petals outside of the flower, however, additional factors are needed. Findings show that the SEP-proteins, at least one of which is expressed in any of the floral whorls, qualify for this role. First, the SEP-proteins are required for the manifestation of the B-function (Section III.C), likely by forming higher order complexes with the B-proteins (Honma and Goto, 2001; Pelaz *et al.*, 2000). This association provides the protein complex with a domain for transcriptional activation, which is lacking in the B-proteins. Second, overexpression of *SEP3* not only induces early flowering (Pelaz *et al.*, 2001a) but, when combined with the two B-proteins, also converts *Arabidopsis* rosette and cauline leaves to petaloid organs (Honma and Goto, 2001). Similar to *SEP3*, overexpression of *AP1* also results in early flowering and, when coexpressed with the two B-proteins, cauline leaves become converted to petals (Honma and Goto, 2001; Pelaz *et al.*, 2001b).

That both SEP3 and AP1 can complement the B-proteins in the petal identity-controlling function is not surprising, given that they both can form higher order complexes with the B-proteins and both contain transcriptional activation domains (Honma and Goto, 2001). Thus, *in vitro* or when overexpressed in vegetative tissues, AP1 can associate with the B-proteins and perform the same function as SEP3. However, in the flower, AP1 in *Arabidopsis* or SQUA in *Antirrhinum* are not absolutely necessary for petal identity, as petals develop in the *agl24 ap1* (Yu *et al.*, 2004) or *inco squa* (Masiero *et al.*, 2004) double mutants. Furthermore, other experiments made clear that *AP1* prevents a C-function-dependent pathway from inhibiting organ primordia formation in the second (petal) whorl (Durfee *et al.*, 2003), rather than controlling, together with the B-function, petal identity. This then leaves the SEP- and B-proteins as the major determinants for petal identity.

In summary, petal development is controlled by the B-proteins in collaboration with the SEP-proteins which *in vitro* can form higher order protein complexes. This is the essence of the "floral quartet" model (Theissen and Saedler, 2001) which assumes direct interactions between four of the ABC-proteins (AP1 for A, SEP, and two Bs in petals; C, SEP, and two Bs in stamens) and the existence of the respective tetramers. This model, however, should not be interpreted literally, since it simply represents a guess at the stoichiometry and composition of complexes. This is particularly true for the hypothetical AP1/PI/AP3/SEP (or ABE) quartet in petal organ identity, where the role of one of the members of the corps, AP1, is highly questionable.

E. THE C-FUNCTION: PLE AND FAR TOGETHER CONTROL REPRODUCTIVE DEVELOPMENT

The C-function in *Antirrhinum* is governed by PLENA (PLE) and FARINELLI (FAR), two MADS-box proteins with extended amino acid sequence similarity and high structural and sequence resemblance of their transcription units (Davies *et al.*, 1999a). In spite of this relatedness, which suggests a duplication event of a previously unified function, the *in vivo* control governed by PLE and FAR differs considerably. This is revealed by the phenotypes of the *ple* and *far* mutants. In *ple* flowers carpels are in the "ground state," that is, sepaloid and, owing to the fact that only the B-genes are functional, stamens become petaloid (Fig. 2). *far* flowers in contrast are male sterile, but otherwise appear wild-type-like; the role of FAR in stamen development is only evident from the enhanced petaloidy of third-whorl organs in the *ple far* double mutant. Clearly, loss of FAR function can be largely compensated by PLE, but not vice versa. The reason for this is in part due to complex regulatory relations between the two genes, in that expression of *FAR* depends to some extent on *PLE*, but *FAR* negatively regulates *PLE* in regions where expression of the two genes does not overlap. When ectopically expressed in transgenic tobacco plants, *PLE* and *FAR* confer distinct homeotic conversions of sepals and petals (Davies *et al.*, 1996a, 1999a) suggesting that male characters (in the second whorl) depend more on *FAR* while female characters (in the first whorl) depend more on *PLE*. Similar results are obtained when these two genes are ectopically expressed in both *Antirrhinum* and *Arabidopsis* (Causier *et al.*, 2005). Together these observations suggest that the two C-function genes play distinct and separable roles in *Antirrhinum* reproductive development.

The *ple* mutant phenotype closely resembles that of the *Arabidopsis* C-mutant *ag*, although, intriguingly, the AG and FAR proteins are more closely related than PLE and AG (Davies *et al.*, 1999a), while PLE appears to be the ortholog of the SHP-proteins (Kramer *et al.*, 2004). This illustrates the random nature of evolution, since the nonorthologous pair that arose by duplication of a gene in a common ancestor retained the genuine developmental function (Causier *et al.*, 2005).

F. THE LINK BETWEEN THE B- AND C-FUNCTIONS IN CONTROLLING FLORAL DETERMINACY

The *(A)*BC-model is based on the phenotypes of B- and C-mutants in *Arabidopsis* and *Antirrhinum*, but not all aspects of the expected phenotypes fully correspond to reality. These usually concealed differences in the phenotypes of *Arabidopsis* and *Antirrhinum* B- and C-mutants and their deviation from the model can be explained as shown in Fig. 4 by differences in the link between the B- and C-functions in the control of termination of floral organ initiation (for references in the following section, see Davies *et al.*, 1999a; Schwarz-Sommer *et al.*, 2003a).

Flowers are determinate shoots as no further lateral organ primordia initiate after the central carpels, ignoring the ovules, which are sometimes considered as floral organs (Colombo *et al.*, 1995; Ferrario *et al.*, 2004). The control of floral determinacy is complex and both the B- and C-genes participate in it, albeit in contrasting ways; the B-function promotes organ initiation and the C-function counteracts it. Abolishment of the B-function inside whorl 3 is therefore a prerequisite for floral determinacy, which is usually completed by C-governed carpel development in whorl 4. When the B-function is artificially maintained, either by ectopic expression in transgenic plants, or in the *superman* (*sup*) and *octandra* (*oct*) mutants in *Arabidopsis* and *Antirrhinum*, supernumerary stamens develop inside whorl 3 (Fig. 2). Thus the inner boundary of the B-function is controlled by *SUP* in *Arabidopsis* and by its ortholog *OCT* (Sakai and Coen, personal communication) in *Antirrhinum*.

The logic of the effects of B, C, and SUP/OCT on determinacy enables predictions to be made for B- and C-mutant phenotypes. B-mutants should be composed of three whorls because there is no B-function to prevent C in the third whorl from terminating initiation of the next, fourth whorl. In C-mutants, SUP/OCT will prevent maintenance of B inside whorl 3 and in the absence of B and C a new internal flower can form because the remaining (*A*)-function defines a new floral meristem (which then initiates sepals in the first floral whorl). Obviously, as schematically depicted in Fig. 4A, B-mutants in *Antirrhinum* and C-mutants in *Arabidopsis* are in accord with these "predictions," but C-mutants in *Antirrhinum* form sepaloid carpels in

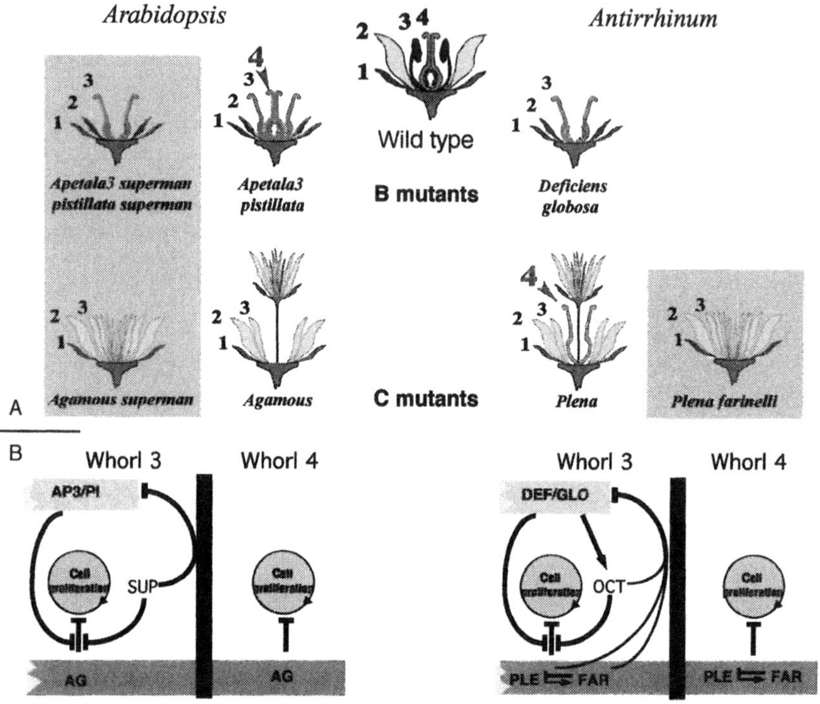

Fig. 4. Differences in the genetic control of determinacy in *Arabidopsis* (left) and *Antirrhinum* (right). The colored flower diagrams in panel A show the principal structure of the wild-type flower, common to both *Arabidopsis* and *Antirrhinum* and the phenotypes of class B- and C-homeotic mutants, which differ in the two species in terms of the number of whorls present. In panel B, a mechanistic explanation for these differences is given which is based on deviations in the regulatory relation between several genes as deduced from genetic analyses. The phenotypic differences among mutants of the same class in the two species become abolished in certain double-mutant combinations (shaded in pink). This adjustment is contributed by mutation in the *SUP* gene in *Arabidopsis* and in *FAR*, the second class C gene, in *Antirrhinum*. Because B mutants in *Antirrhinum* equal the *sup*–B mutant combination in *Arabidopsis*, it appears that *OCT*, the ortholog of *SUP*, is controlled by the B-function in *Antirrhinum*. This is shown by a black arrow pointing from B to *OCT* in panel B. Similarly, the differences in the C mutant phenotypes become abolished in double-mutant combination with *superman* in *Arabidopsis* and with *far* in *Antirrhinum*. This suggests that the C genes in *Antirrhinum* support *OCTANDRA* in negatively controlling the B-function at the boundary of the third and fourth whorls. This is indicated in panel B by joining the C-function to the barred line from *OCT* to the B-domain. In contrast, *SUPERMAN* in *Arabidopsis* prevents expansion of the B-domain without support from the B- or C-functions. Notice that the function of *SUPERMAN* or *OCTANDRA* is limited to the boundary region between the third and the fourth whorls and that the indicated mechanism is an oversimplification by not including many other genes involved. Furthermore, positive influence indicated by arrows and negative influence indicated by barred lines are not necessarily direct. Modified from Davies *et al.* (1999a). (See color Plate 11.)

their fourth whorl (followed by a new flower initiating with sepals) and B-mutants in *Arabidopsis* display fourth-whorl carpels (unexpected fourth whorls highlighted in red in Fig. 4A). The possible solution to this mind-boggling puzzle is presented in a model in Fig. 4B, where differences in the regulatory relation of the B- and C-genes to each other and to *SUP* or *OCT* in *Arabidopsis* and *Antirrhinum* are postulated to account for the differences in the B- and C-mutant phenotypes.

Regardless of such underlying differences in regulatory mechanisms, the overall structure of the wild-type flower in *Antirrhinum* and *Arabidopsis* is comparable. This demonstrates that a similar outcome in terms of terminal structures in development, such as the flower composed of four whorls of organs, does not necessarily reflect the complexity of the underlying control processes which achieve this structure. Comparative analyses thus provide insights into the variability and flexibility of genetic mechanisms controlling development.

G. A NOVEL DEFINITION OF THE (*A*)-FUNCTION IN THE (*A*)BC-MODEL TO RECOVER ITS UNIVERSALITY

In the earlier sections, we have provided several arguments against a role of the originally defined A-function in the control of floral organ identity, as proposed in the textbook ABC-model. We have shown that, *sensu stricto*, an additional function comparable to B and C does not exist, as only these functions fulfill the following requirements.

First, the B and C developmental functions are the only ones in which a few key genes clearly act, within a floral context, to determine organ identity. Second, the expression domains are limited to certain whorls, usually the second/third (B) and third/fourth (C). The complex case of the B-function in *Petunia* is then just a variation within this theme, as are the simple changes in expression domains necessary to explain petaloid first-whorl development for tepals in the tulip flower or the gradual "transformation" of petals to stamens in *Nymphaea*. Third, the absence of either or both functions has no consequence for floral identity, defining sepals in a whorled structure as the ground state of floral organ identity. Fourth, the functions have both individual (B for petal and C for carpel) and combined contributions (BC for stamen) to organ identity control.

Since B and C alone or together are not relevant outside of the flower, the floral environment in which they can operate has to be established. This is one of the subfunctions of the (*A*)-function, where the brackets signify its distinction from the B- and C-functions (Fig. 1). So what do we expect from the (*A*)-function? *First*, that it is expressed before the genes controlling

B- and C-functions are expressed. *Second*, establishment of sepal identity, the ground state of floral organs, performed by the meristem identity genes, as a consequence of the acquisition of floral identity. In some plants (e.g., *Antirrhinum*) this control is tighter and in others (e.g., *Arabidopsis*) more lax, perhaps resulting from differences in requirements for activation of the B- and C-functions. *Third*, activation of the B- and C-organ identity functions. *Fourth*, establishment and maintenance of the expression boundaries of B and C, performed by the cadastral control. Some genes share a role in meristem identity and cadastral control and some others just participate in one of them; orthologs in different species can behave differently in this respect. *Fifth*, concurrent support for B and C to exert their function. The Im/*SEP* (termed the E-function) genes (and perhaps several others) belong to this category. These genes, unlike the B- and C-genes themselves, contribute to both the B- and C-functions, as well as to the meristem identity control. Thus, the (*A*)-function stands for several more or less distinct subfunctions each performing one or more tasks to enable or complete floral organ identity control by the B- and C-functions.

This concept of the (*A*)-function enables the (*A*)BC-model to regain its universal applicability, a necessary requirement in view of the unsatisfactory interpretation of the A-function even in *Arabidopsis* in the ABC-model. Although the (*A*)-function is complex and diverse, the (*A*)BC-model remains simple and easily adaptable to describe the morphology of flowers, for instance for teaching, where its cadastral subfunction is the most easily and broadly recognized component (mistakenly understood as a genuine organ identity-controlling function). It is also flexible enough to allow the addition of newly discovered functions, which can be expected in the future, for instance those controlling the outer B-domain. *Finally*, the (*A*)BC-model is compatible with variations in the mechanistic details, for instance "floral quartets," that might be found in different species, as the key factors and complexes become better characterized.

IV. INFLORESCENCE AND FLORAL ARCHITECTURE

Inflorescences are found in a wide array of different shapes and sizes. However, from our simplistic viewpoint we can consider two main types of inflorescence organization—determinate and indeterminate. Determinate inflorescences eventually convert each inflorescence terminal meristem into a floral meristem, producing a terminal flower. Indeterminate inflorescences maintain the meristematic nature of their terminal meristem, producing lateral floral meristems only. Despite this simplification, it can be seen that there is still considerable variation within each class, since *Antirrhinum* and *Arabidopsis* are both members of the indeterminate class, yet their inflorescences are readily distinguishable. In this section, we will consider two genes which contribute to the organization of the inflorescence and to its distinctive nature in each species.

A. CEN: A MAJOR INFLUENCE ON INFLORESCENCE ARCHITECTURE

The apical meristems of both *Antirrhinum* and *Arabidopsis* inflorescences produce floral meristems on their flanks, but are not themselves converted into floral meristems, giving rise to an indeterminate inflorescence. However, mutants exist in both species in which the inflorescence meristem becomes inappropriately converted into a floral meristem. The result of this conversion is that the indeterminate inflorescence meristem is converted into a determinate meristem bearing a terminal flower. The genes affected in these mutants were cloned, first from *Antirrhinum* (*CENTRORADIALIS*, *CEN*) and subsequently from *Arabidopsis* (*TERMINAL FLOWER 1*, *TFL1*) (Bradley *et al.*, 1996, 1997). The genes were found to be homologous, if slightly enigmatic; they encode proteins which show homology to phosphatidylethanolamine-binding proteins (PEBPs). PEBPs are also found in animals, yeasts, and bacteria where they play a range of roles in diverse processes. *TFL1* belongs to a small gene family in *Arabidopsis* (Mimida *et al.*, 2001).

Both *CEN* and *TFL1* have been ectopically expressed in different species and the common feature to emerge is their effect to delay phase change. For example, overexpression of *CEN* in tobacco plants results in an extended vegetative phase (Amaya *et al.*, 1999). Overexpression of *TFL1* in *Arabidopsis* similarly results in lengthened vegetative, inflorescence, and floral stages. However, *cen* mutants only affect the inflorescence, producing a terminal flower but not causing early flowering, whereas *tfl1* mutants flower early and produce terminal flowers (Amaya *et al.*, 1999). This difference appears to result from regulatory differences between the *CEN* and *TFL1* genes. Whereas *TFL1* is expressed in both inflorescence and vegetative phases, expression of *CEN* is confined to the inflorescence phase. *TFL1* is therefore capable of exerting a greater degree of influence over the regulation of phase change in *Arabidopsis* than *CEN* can in *Antirrhinum*. It remains to be seen whether this variation plays a significant part in the differences that are observed in the induction of flowering in the two species.

Despite regulatory differences between *CEN* and *TFL1*, both genes act as a general brake on phase change in tissues in which they are expressed. It is thought that by this mechanism they indirectly act to delay activation of the floral meristem identity genes *FLO*/*LFY* and *SQUA*/*AP1*. The effect of *CEN* or *TFL1* expression in the inflorescence apex is to exclude the floral meristem identity genes and thereby preserve the undifferentiated apical meristem. In support of this, in plants ectopically expressing *LFY*, *TFL1* expression is not detected (Ratcliffe *et al.*, 1999). Such plants flower early and show conversion of inflorescence meristems to floral meristems, an effect that is partially alleviated in 35S::*LFY* 35S::*TFL1* inflorescences. In contrast to the slowing down of phase change which results from increased levels of *CEN* and *TFL1*, other related genes exist in these and other species, which have the opposite effect. An example of such a gene is the *Arabidopsis* flowering time gene *FLOWERING LOCUS T* (*FT*), another member of the *TFL*-like family, which, when overexpressed, results in increase in the rate of phase change and early flowering. How two closely related genes can influence the progression of developmental phases in opposite directions is unknown, but must be related to the mechanism through which this class of regulatory genes work. Some progress has been made in understanding this, by the demonstration that it is possible to change a single amino acid at the supposed ligand-binding pocket, and flip the effect of the gene from activation to repression or vice versa (Hanzawa *et al.*, 2005). This discovery has two immediate implications. First, the identification of a critical residue in a binding pocket should facilitate the identification of the other molecule(s) involved in the regulation of phase change and thereby open up a new signaling and control mechanism in plant development. Second, the ability of localized amino acid substitutions to entirely reverse the effect of a regulator raises the possibility that such changes could have been used in nature to cause dramatic changes in development. An increase in our understanding of these genes and the way they exert their effects will require further detailed investigation in *Arabidopsis* and other model plant species.

Such studies of *CEN*-like genes in plants with different inflorescence architectures have already begun. Tomato has a sympodial growth habit, where the reproductive and vegetative stages alternate. This is characterized by inflorescence termination and the initiation of a new shoot from a lower axillary bud. The new shoot produces an inflorescence which terminates again, and so on, giving rise to the typical tomato inflorescence structure. Loss-of-function of a *CEN*-like gene in tomato, *SELF PRUNING* (*SP*), alters this architecture by shortening the intervening vegetative phases leading to the production of two consecutive inflorescence phases and termination of development (Pnueli *et al.*, 1998). *CEN*-like genes exist as small families in plants (Carmel-Goren *et al.*, 2003). It is very likely that variation in their activity, either acting as activators or repressors of phase change, combined with subtle regulation of their spatial and temporal expression patterns, will be responsible for at least some of the natural variety in inflorescence architecture and growth habit.

B. *INCOMPOSITA*: ABERRANT PROPHYLL INITIATION IMPAIRS FLORAL ARCHITECTURE

Prophylls are the first leaves or leaf-pair that develops on a shoot. During flower development prophylls (also termed bracteoles) can protect the developing bud, but their presence is not obligatory. Even within genera species can exist with or without prophylls, or with initiated but then suppressed organ primordia at their position (Prenner, 2004b). Due to their morphogenetic potential with respect to initiation and position of first-whorl organs of the flower, the presence or absence of prophylls (together with the structure and developmental dynamics of the first whorl followed by scanning electron microscopy) is frequently applied to plant systematics (Prenner, 2004a).

Lack of prophylls can either be due to their incorporation into the sepal whorl during evolution or to suppression due to the lack of initiation (Weberling, 1989). The order of organ initiation in the first whorl is indicative for incorporation or loss of prophylls; in the former case the sepals initiate at the position of the prophylls and in the latter case sepals initiate as if prophylls were present. According to this model *Antirrhinum* belongs to the species where prophylls are suppressed, as sepal initiation starts at the abaxial (lower or ventral) side of the flower primordium (Fig. 5). Results suggest that the *INCO* gene (mentioned in Section II as a component in the control of floral meristem identity) controls suppression of prophyll formation because *inco* mutants form prophylls as two additional organs at lateral positions before sepals initiate (Fig. 5; Masiero *et al.*, 2004).

The presence of the mistakenly established prophyll primordia has severe consequences for the placing of sepal primordia and the timing of their initiation. The two lateral organs become displaced toward the petal whorl (Fig. 5) and can undergo homeotic conversions to petals. Petal development in turn is also impaired in that the second-whorl organs can fuse to the first whorl resulting in a misshapen corolla in the perianth. Thus, failure to suppress prophyll development impairs floral architecture, which is crucial for fertilization of the flower (Sections V and VI).

The evolutionary history of *INCO* is intriguing and deserves some attention in the future. The protein belongs to the StMADS11 subfamily of MADS-box proteins which contains a fairly diverse group of transcription factors involved in distinct developmental control events (Masiero *et al.*, 2004). This contrasts with the frequent overall functional similarity of MADS proteins belonging to a subfamily (Theissen *et al.*, 1996) and might indicate a recent acquisition of functions in the StMADS11 group. In fact, although the lack of suppression of prophyll initiation is the only visible consequence of impaired INCO function, the gene is also integrated in

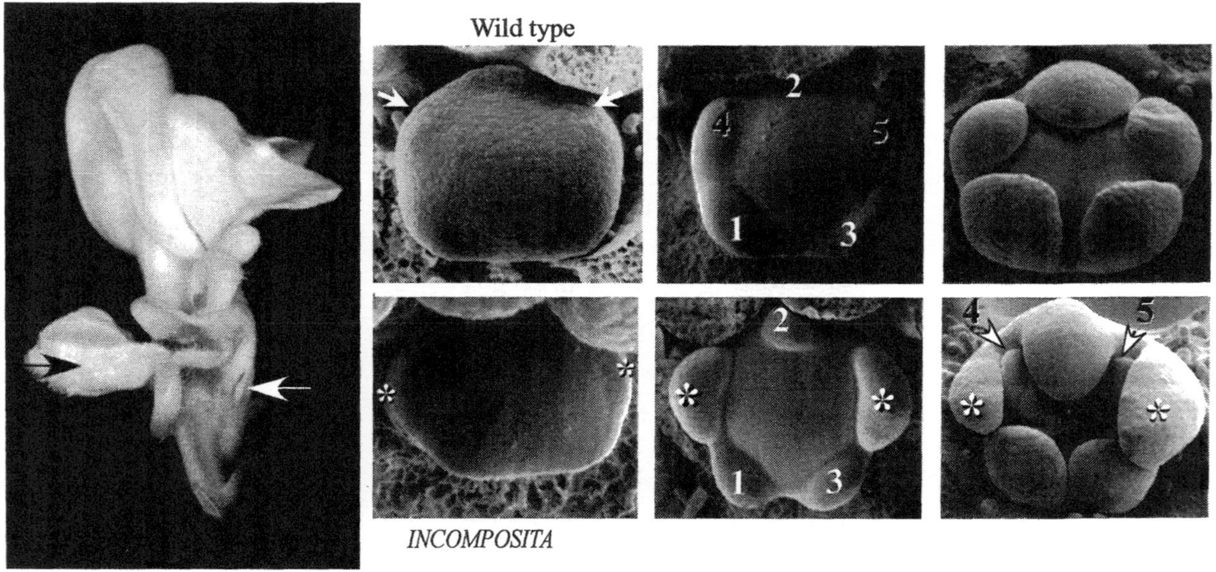

Fig. 5. *INCOMPOSITA* prevents prophyll initiation and thereby contributes to wild-type floral architecture. The photograph at the left shows an *inco* mutant flower, with petaloid sepals (black arrow), prophylls (white arrow), and misshapen petals. The scanning electron micrographs show early events in organ initiation in wild-type (upper panel) and *inco* mutant (lower panel) flower buds. Sepals are numbered in the order of their initiation and prophylls are highlighted with asterisks. The arrows in the youngest wild-type flower primordium indicate the initiating lateral sepals to be compared with the similar position of prophylls in the mutant. The black numbers 4 and 5 draw attention to the delayed development of lateral sepals in *inco* and the arrowheads indicate their displacement toward the petal whorl. As discussed in Section IV.B, impaired *inco* floral architecture is the consequence of the failure to suppress prophyll initiation and follow from delayed and displaced initiation of organ primordia. Modified from Masiero *et al.* (2004). (See color Plate 12.)

a complex way into the control of floral meristem identity (Masiero *et al.*, 2004). This function is only evident in double-mutant combinations between *inco* and the *squa* or *flo* mutants (see Section II) suggesting that in wild-type development the meristem identity role of *INCO* is fully or partially redundant. Since several other members of the StMADS11 group are involved in the control of meristem identity or flowering time, it is possible that a progenitor of INCO performed functions related to establishment of floral identity.

V. FLORAL SYMMETRY

The striking zygomorphic flower of *Antirrhinum* has proved to be an ideal model system to study the genetic control of floral asymmetry. The occurrence of bilaterally symmetrical flowers in a variety of different angiosperm clades has suggested that this trait evolved independently more than 25 times (Cubas, 2004). It is thought that the change from radial symmetry (actinomorphy) to bilateral symmetry (zygomorphy) facilitates specialized pollinator interactions. The availability of actinomorphic mutants of *Antirrhinum* has enabled researchers to begin to dissect the molecular nature of the signals that regulate shape and symmetry, and has also provided a framework against which to test symmetry mechanisms in other species.

Experiments studying *Antirrhinum* petal shape have also been at the forefront of attempts to apply quantitative methods to the analysis of the genetic control of organ shape (Coen *et al.*, 2004; Rolland-Lagan *et al.*, 2003). In these experiments organ growth was measured by clonal analysis and the results integrated into growth models. Using this approach petal asymmetries were seen to be generated by orientation of the direction of growth, rather than by differentially regulating growth in different zones of the petal. This type of modeling is providing the framework with which it will become possible to understand the genetic control of organ size and three-dimensional shape.

A. THE PLAYERS: MUTANTS WITH SYMMETRY DEFECTS AND THE GENES INVOLVED

Wild-type *Antirrhinum* flowers are zygomorphic (Figs. 1 and 6A). A single plane of symmetry runs from the top of the flower (dorsal) to the bottom of the flower (ventral). The zygomorphic nature of the flower is evident by differences between the petals and stamens, which occupy dorsal and ventral positions in the flower. Wild-type flowers have two large dorsal petals, two

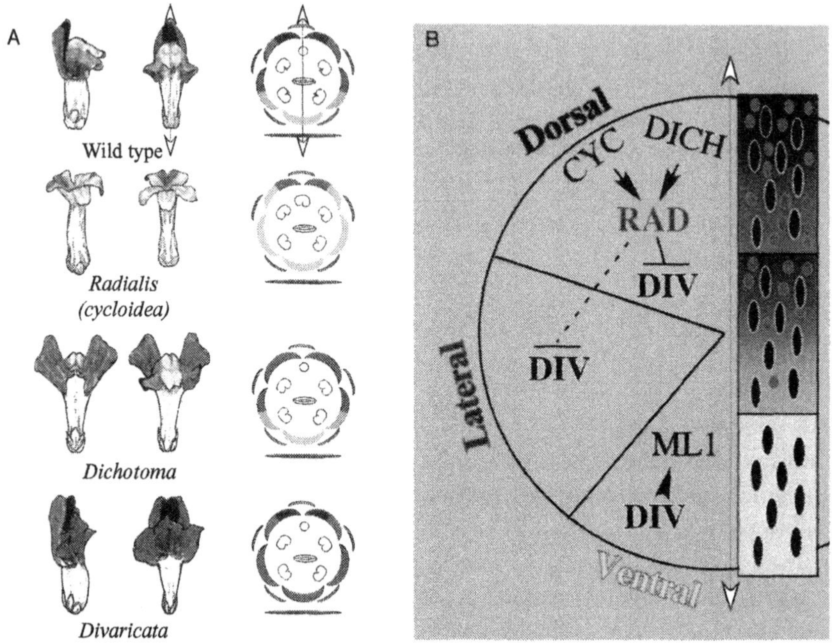

Fig. 6. The control of floral asymmetry. (A) The dorsoventral asymmetry, which is most evident in petals, results in wild-type flowers having a bilateral axis of symmetry, shown by arrows. The lateral and dorsal petals are also asymmetrical (as shown by the color scheme; dorsal, light blue/dark blue; lateral, orange/red). The ventral petal is symmetrical. The key mutants referred to in the text are illustrated schematically. *Radialis, dichotoma,* and *cycloidea* all show reduced dorsality, although this is mild in *dichotoma. Divaricata* (a semidominant mutant), in contrast, shows reduced ventrality. The flowers in the drawings at the left are shown in lateral view (except for *dichotoma,* which is in a dorsal view) and the flowers at the right are shown from the front. Their petal lobes are stained in the same color code as in the flower diagrams to clarify the symmetry relations. Modified from Schwarz-Sommer *et al.* (2003a). (B) A possible scheme for the interaction between these key factors is shown schematically in the half flower on the left. CYC and DICH, two TCP transcription factors which are expressed in the dorsal petals, promote expression of RAD, a truncated MYB transcription factor. DIV, a MYB transcription factor, is expressed in all petals, but its activity is inhibited by the presence of RAD. Either the RAD protein, or a downstream effector of the RAD signal, acts to decrease the effectiveness of DIV expression in the lateral petals. The ventral petal, defined by the expression of DIV alone, begins to express AmMYBML1 (ML1), which induces some of the specific changes which define the characteristics of that petal. The bar on the right shows one possible interpretation of this model in which the relative amounts of the two MYB factors (DIV, blue oval and RAD, green oval) play a major role in defining position on the dorsoventral axis. In this scheme, we make the assumption that movement of RAD into the lateral zone creates a RAD gradient. Alternative models are also possible. (See color Plate 13.)

smaller lateral petals, and a single small ventral petal. These petals can be distinguished by differences in size, shape, pigmentation, and in the production of certain specialized cell types (Section VI.A). The third-whorl stamens are offset with respect to the petals, so that a single stamen should form in the dorsal position and two lateral and two ventral stamens form. In fact, the single dorsal stamen is arrested during development to form the staminode. A series of mutants have been characterized which affect either the dorsality or ventrality of these organs to different degrees. The genes affected in these mutants—*CYCLOIDEA* (*CYC*), *DICHOTOMA* (*DICH*), *RADIALIS* (*RAD*), and *DIVARICATA* (*DIV*)—have all been isolated and together they form a regulatory network controlling floral asymmetry.

The *cyc* and *dich* mutants were the first to be studied in detail (Luo *et al.*, 1996, 1999; Fig. 6A). Double mutants between *cyc* and *dich* form radially symmetrical flowers in which all petals appear similar to the wild-type ventral petal. In *cyc* single mutants the dorsal petals show some lateral characteristics and the lateral petals adopt a ventral identity. The loss of dorsal identity in *cyc* mutants reveals a dorsalizing effect for the *CYC* gene in normal development. Similarly, the *DICH* gene has a dorsalizing effect on normal flowers and the *dich* single mutant has slightly altered dorsal petals. Both genes encode related members of the TCP family of transcription factors and both are expressed in the dorsal region of the developing flower from early stages. Other members of this family of transcription factors have been shown or suggested to play a role in the control of growth (Cubas, 2004). The expression pattern of *DICH* is more restricted than that of *CYC* within the dorsal domain, providing a potential explanation for the less severe phenotype observed in *dich* single mutants. Confirmation of the dorsalizing effect of these TCP genes was obtained in a semidominant allele of *CYC* (*Backpetals*) in which *CYC* expression becomes expanded into the ventral domain. In this mutant the ventral and lateral petals become dorsalized (Luo *et al.*, 1999).

In addition to the dorsalizing function of *CYC* and *DICH*, there is a ventralizing function provided by *DIV*. Mutants in *DIV* result in the ventral petals developing as lateral petals, indicating that *DIV* provides a ventralizing function in wild-type flowers. Unlike *CYC* and *DICH*, *DIV* encodes a member of the MYB family of transcription factors (Galego and Almeida, 2002). It can be shown that *DIV* promotes ventral identity because *cyc dich* double mutants form fully radially symmetrical flowers containing ventral petals at all positions, whereas all the petals of the radially symmetrical *cyc dich div* triple mutants have lateral identity. Although the effects of *DIV* expression are confined to the ventral petal and ventral parts of the lateral petals when *CYC* and *DICH* are active, *CYC* and *DICH* cannot simply act to repress dorsal expression of *DIV*. This is because *DIV* is expressed throughout the wild-type flower, including tissues which express *CYC* and *DICH*.

A potential explanation for this apparent contradiction is provided by the fourth gene in this regulatory network, *RAD* (Corley *et al.*, 2005). Mutants of *RAD*, like *cyc* mutants, cause a partial conversion from zygomorphic to actinomorphic flowers, with the flowers showing a ventralization of the dorsal petals. Unlike the other dorsalizing factors, CYC and DICH, RAD is not a member of the TCP family of transcription factors. In fact, *RAD* encodes a single-MYB domain transcription factor with similarity to the two-MYB-domain DIV protein. *RAD* expression comes on in the dorsal region of the flower slightly after *CYC* and *DICH* suggesting that *RAD* might be a downstream target of *CYC* and *DICH*. This is supported by the observation that CYC-binding sites are found within the promoter and intron of *RAD* (Costa *et al.*, 2005). Although CYC expression is unaffected in *rad* mutants, *RAD* expression is greatly reduced in *cyc* mutants and abolished in *cyc dich* double mutants. Furthermore, in *Backpetals CYC* is ectopically expressed in the ventral domain (see earlier) and, as a consequence, *RAD* also becomes ectopically expressed. In contrast, *RAD* expression is unaffected in *div* mutants.

B. A POSSIBLE MECHANISM TO CONTROL SYMMETRY

The discovery of *RAD* leads to a model for asymmetric flower development in which the two TCP transcription factors, *CYC* and *DICH*, act autonomously to activate expression of a truncated MYB transcription factor, RAD, which promotes dorsal identity (Fig. 6B). Unfortunately, *cyc rad* double mutants have more severely radialized flowers than *rad* single mutants, suggesting that *CYC* is able to influence asymmetry by other routes that do not involve *RAD*. Nevertheless, a major part of the dorsalizing role of *CYC* and *DICH* is exerted through *RAD*.

One clue as to how *RAD* could act is provided by the mechanisms for patterning root hairs and trichomes in *Arabidopsis* (Pesch and Hülskamp, 2004). In these cases the cells that will produce trichomes and those that will produce the nonroot hair cells (atrichoblasts) are defined by the expression of a MYB transcription factor, a basic helix-loop-helix (bHLH) factor and a WD40 protein, acting cell autonomously and a truncated MYB transcription factor (CAPRICE in atrichoblasts and TRYPTICHON in trichoblasts). It is possible that these truncated MYB factors could act non-cell autonomously to interfere with the activity of the full length MYB factors in neighboring root hair and non-trichome cells, respectively. If a similar mechanism is at work in the establishment of floral asymmetry, it could be expected that RAD interferes with the activity of DIV in the dorsal region of the

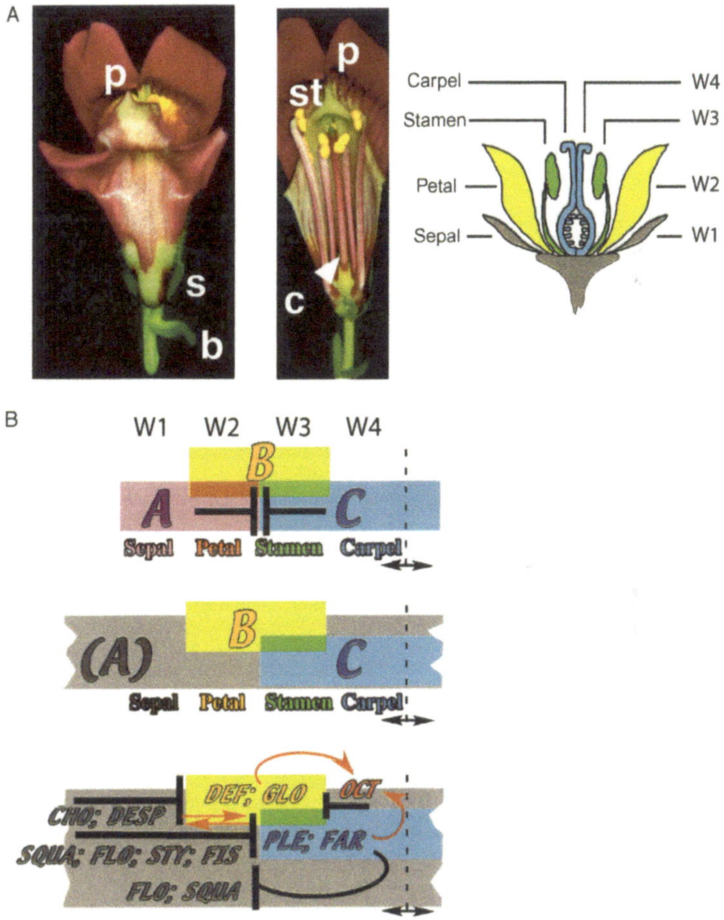
A
p
s
b
p
st
c
Carpel
Stamen
Petal
Sepal
W4
W3
W2
W1
B
W1
W2
W3
W4
B
A
C
Sepal
Petal
Stamen
Carpel
B
(A)
C
Sepal
Petal
Stamen
Carpel
DEF; GLO
OCT
CHO; DESP
PLE; FAR
SQUA; FLO; STY; FIS
FLO; SQUA

Plate 8

Reduced B function
Wild type
Globosa or deficiens
(A)
B
C
choripetala
despenteado
octandra
Plena-Macho
fistulata
stylosa
Expanded B function
Expanded C function

Plate 9

Plate 10

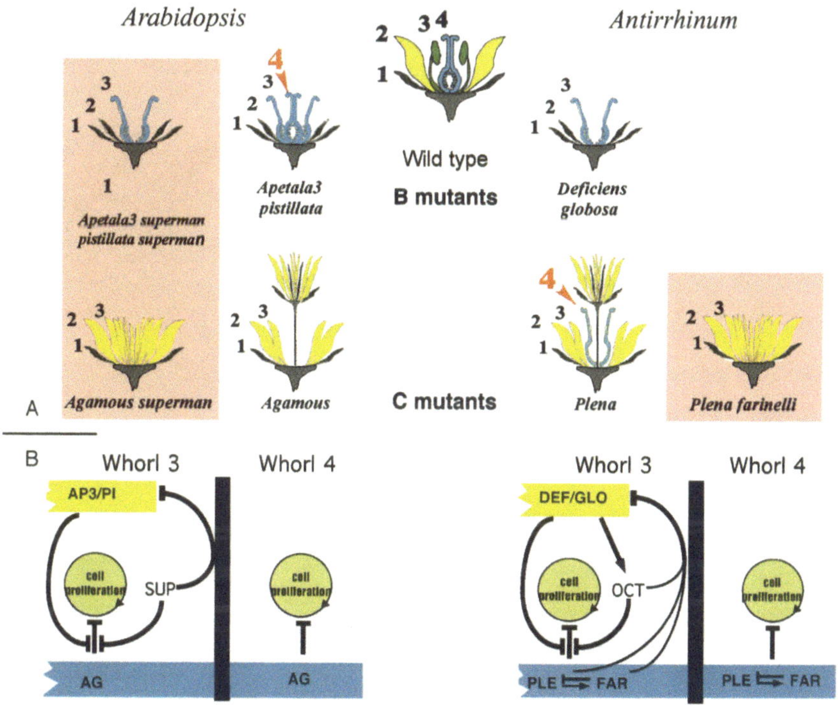

Plate 11

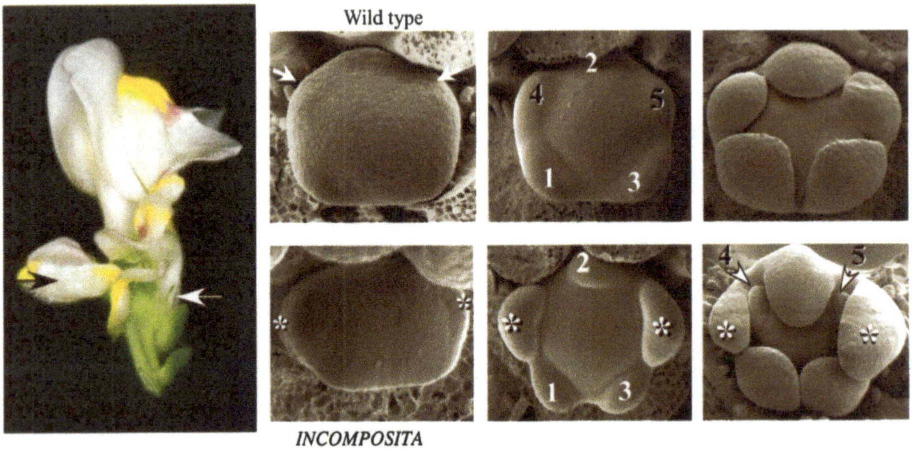

Plate 12

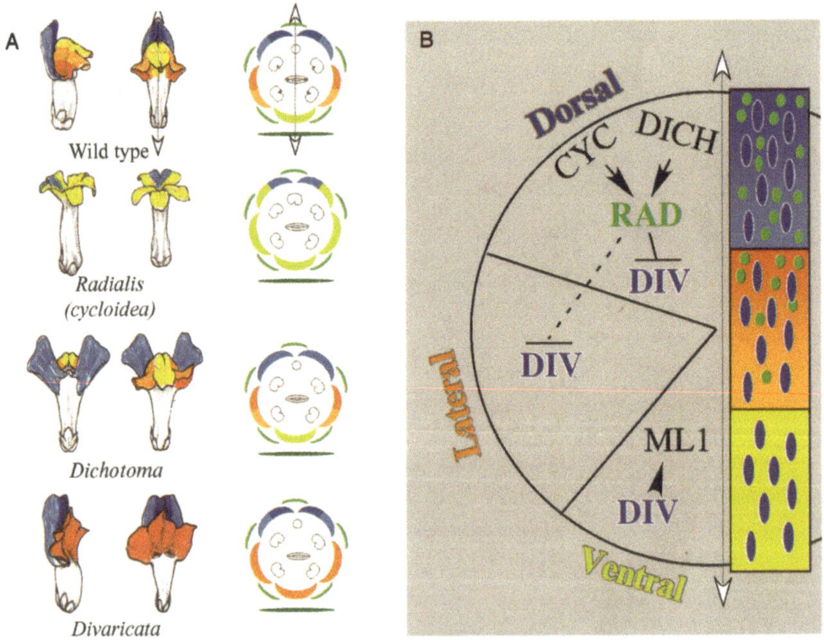

Plate 13

flower. RAD could act nonautonomously by being transported to lateral regions bordering the dorsal domains. In regions where both RAD and DIV proteins are present, RAD could compete with DIV for binding partners (protein or DNA) and thus ameliorate the effects of *DIV* expression in those tissues (Fig. 6B). This would explain the apparent insensitivity of *DIV* expression to the presence of RAD and also explain how dorsal expression of *DIV* has no ventralizing effect. It is also consistent with the fact that *DIV* activity is dosage dependent (Almeida *et al.*, 1997). However, *rad div* double mutants are distinct from *div* single mutants, indicating that RAD must be able to act in more ways than simply inhibiting DIV activity. It will be interesting to probe further into the similarities and differences between the apparently conserved truncated MYB interference system. If the system is conserved it should be possible to identify a bHLH protein that interacts with CYC and/or DICH.

C. FLORAL ASYMMETRY IN OTHER SPECIES

Now that a molecular mechanism has been established in *Antirrhinum* it becomes of great interest to analyze the control of zygomorphy in other species. An early investigation into this question looked at a peloric (radially symmetrical) mutant of *Linaria*, first described by Linneaus in the eighteenth century. In this case the *CYC* gene was again implicated in the establishment of asymmetry, but the mutation was discovered to be epigenetic (Cubas *et al.*, 1999). The gene duplication that gave rise to CYC and DICH was confined to the *Antirrhinum* lineage (including closely related species such as *Linaria*). However, the *Linaria cyc* mutant does not share all aspects of the *Antirrhinum* mutant, suggesting that CYC and DICH might have undergone subfunctionalization (Cubas, 2004).

A nice example of the way in which changes in the expression patterns of *CYC* and *DICH* can be correlated with both evolution of different floral structures and variation in the species ecology is provided by the study of the desert ghost flower, *Mohavea confertiflora* (Hileman *et al.*, 2003). Flowers of this species, a close relative of *Antirrhinum*, do not show the strong bilateral symmetry seen in *Antirrhinum* petals and carry just two stamens as opposed to the four found in *Antirrhinum*. The difference in stamen number can be accounted for by the fact that in *Mohavea* three stamens are aborted (two lateral and one adaxial), whereas only one *Antirrhinum* stamen is aborted. Expression of the *Mohavea CYC* and *DICH* genes was found to be expanded to the lateral stamens, providing an ideal explanation for the mechanism of stamen abortion in this species (Hileman *et al.*, 2003). It is also possible that the apparent radial symmetry of the *Mohavea* petals results from a reduction in expression of DICH in the second whorl, a change that could result from the selective advantage of presenting radially symmetrical flowers to pollen-collecting bees (Hileman *et al.*, 2003).

Antirrhinum and *Linaria* are both members of the Lamiales so it might not be surprising to find similar mechanisms using related genes to specify asymmetry in these two species. *CYC*-like genes have been identified from unrelated species, such as members of the legume family, which also show zygomorphy (Citerne *et al.*, 2003). However, it is still too early to say whether these genes function in the establishment of floral asymmetry. In the unrelated and mostly radially symmetrical flowered species *A. thaliana*, the homolog of *CYC* is *TCP1*. Although *tcp1* mutants have no visible phenotype (Cubas, 2004), the *TCP1* gene shows a *CYC*-like dorsal expression pattern in early flowers, which fails to be maintained at later stages and also in axillary shoot meristems. The *Arabidopsis RAD* homolog does not appear to have the CYC-binding sites which are found in the *Antirrhinum RAD* gene (Costa *et al.*, 2005). These results suggest two things. First, it could be the prolongation of *CYC* expression and/or CYC's ability to bind to *RAD* regulatory sequences in the developing *Antirrhinum* flower that directly led to zygomorphy. Second, asymmetric expression of CYC-like genes could represent an ancient condition that would be available for repeated use in the independent evolution of zygomorphy (Costa *et al.*, 2005; Hileman *et al.*, 2003). *Antirrhinum* can continue to contribute to studies on floral asymmetry by providing a model in which to further elaborate a basic mechanism from which comparisons can be made. If floral asymmetry has indeed evolved by duplication, subfunctionalization and neofunctionalization within the *CYC* subclade of the TCP transcription factor family and the phenotypic consequences are a result of changes in their expression, interaction, and targets, the details will only become apparent by careful further analysis in this species.

VI. FLORAL COLOR, SCENT, AND CELL SHAPE: THE SKILLS OF *ANTIRRHINUM* TO ATTRACT POLLINATORS

The *Antirrhinum* flower has several other distinctive characters that make it a useful and productive model to study biological questions. We have already discussed the zygomorphic nature of the flowers and this is just one of the adaptations to pollinator interaction that distinguishes the flower. *Antirrhinum* is pollinated by bees and the ventral petal is adapted in both shape and rigidity to form a landing platform on which bees can alight. The weight of the pollinating bee allows the petals to part at the hinge between the tube and lobes, to allow entry of the bee into the flower. Several other floral structures contribute to the attractiveness of the flower to bees. The petals are both brightly colored and the apparent intensity of the inherent color is intensified by modifications to specific epidermal cells. Other modified epidermal cells give rise to multicellular trichomes, a type not found in *Arabidopsis*, which enhance pollinator interactions by acting as nectar guides, facilitating pollen transfer from pollinator to plant and by producing scent to attract bees. Over the past few years progress has been made in understanding the genetic regulation of these specific epidermal structures in *Antirrhinum*.

A. FLORAL COLOR

The vivid colors and patterns observed in *Antirrhinum* flowers have certainly contributed to their history of cultivation and research into the genetic control of flower color has a long tradition in this species (Schwarz-Sommer *et al.*, 2003a). As the genes involved in the synthesis of the pigment anthocyanin were identified and isolated, attention turned to their regulation (Almeida *et al.*, 1989; Goodrich *et al.*, 1992; Luo *et al.*, 1991; Martin *et al.*, 1991). Many of the genes that regulate anthocyanin production are members of transcription factor families including MYBs and bHLHs. Several examples have now been obtained to show that mutations which place structural or regulatory genes under different spatial and temporal control can profoundly alter the pigmentation pattern of the flower. For example, imprecise excisions of a transposable element from the promoter of the structural gene *PALLIDA* (*PAL*) results in differential coloration in different parts of the petals. At least part of this effect results from changes in the ability of the bHLH transcription factor *DELILA* (*DEL*) to regulate *PAL* expression (Goodrich *et al.*, 1992).

Expression of the structural genes responsible for pigment production is regulated by similar genes to those that control pigment production in other plant species, but specialization and modification in the way these control mechanisms are used have allowed *Antirrhinum* to evolve flowers with specific patterns of coloration. The supposition is that specialization in pigmentation has evolved to enhance pollinator interactions and the availability of multiple gross and subtle pigmentation mutants will allow such theories to be tested (Niovi Jones and Reithel, 2001).

B. THE PRODUCTION OF SCENT

Scent represents another way in which pollinators can be both attracted to flowers and learn to distinguish flowers of one species from similarly shaped and colored flowers of another species. Scent production is, however, extremely complex, since it can consist of hundreds of different volatile compounds present in amounts which differ over an enormous range, giving rise to a wide range of specific scents that facilitate the identification of different flowers. The molecules involved in scent are usually low-molecular weight volatile products of secondary metabolism with phenylpropanoids, benzenoids, and terpenes being the most prevalent (Goodwin *et al.*, 2003). The major scent compound emitted by *Antirrhinum* flowers is methylbenzoate. As would be expected of a mechanism used to attract pollinators, scent production in *Antirrhinum* peaks around six days after anthesis (Dudareva *et al.*, 2000; Goodwin *et al.*, 2003) and bees have been shown to be able to discriminate even subtle changes in scent (Wright *et al.*, 2005). The production of methylbenzoate is regulated by the circadian rhythm in *Antirrhinum*, a mechanism which ensures that maximum scent production is coupled to the daytime periods of pollinator activity (Dudareva *et al.*, 2000). Methylbenzoate production is also halted following successful pollination once the pollen tubes reach the ovary, reflecting the lack of a further requirement to attract a pollinator at that stage of development (Negre *et al.*, 2002). Further research in the production of scent will provide insights into both the behavioral aspects of plant–pollinator interactions and also the intriguing secondary metabolism responsible for the production and regulation of scent.

C. MYB TRANSCRIPTION FACTORS CONTROL CELL SHAPE

The *mixta* mutant of *Antirrhinum* has led to an intriguing series of discoveries into the molecular control of cell shape. *mixta* is unaffected in pigment production, but the colors of wild-type petals appear deeper than those of the *mixta* mutant (Comba *et al.*, 2000; Glover and Martin, 1998). Scanning electron microscopy revealed that the specialized conical cells which normally form on the inner epidermis of the petals were flattened in the *mixta* mutants and that it was this change in cell shape that led to the apparent difference in color intensity (Glover and Martin, 1998). The *MIXTA* gene was cloned and found to encode a MYB transcription factor (Noda *et al.*, 1994). When the *Antirrhinum MIXTA* gene was ectopically expressed in transgenic tobacco, the leaves of the transgenic plants produced either conical cells, multicellular trichomes, or a mixture of both. The predominance of conical cells or trichomes on such leaves is determined by the stage of leaf development at which the *Antirrhinum MIXTA* function becomes established in those leaves. When *MIXTA* is expressed early, before the end of cell division in the developing leaves, trichomes are formed. However, if *MIXTA* is only expressed after the end of cell division, as is normally the case in the developing petals of *Antirrhinum*, conical cells are formed on the leaf epidermis instead. These results demonstrate that *MIXTA* can specify conical cell or trichome fate and implicate the cell's ability or inability to undergo further rounds of cell division in the decision as to which cell type is formed (Glover and Martin, 1998). The specification of conical cells and multicellular trichomes can therefore be seen to share a common mechanism, although the relationship, if any, of this mechanism to the control of unicellular trichomes, as found in *Arabidopsis*, is still unknown. It is also not known how a MYB transcription factor can act to determine cell shape and it will be extremely interesting to discover the target genes which are under the control of *MIXTA*.

MIXTA, however, is not the only MYB transcription factor to influence conical cell and trichome formation in *Antirrhinum* (Perez-Rodriguez *et al.*, 2005). Another *MIXTA*-like gene, *AmMYBML1*, was also discovered, presumably resulting from duplication of an ancestral MYB gene. Ectopic expression experiments in tobacco showed that, like *MIXTA*, *AmMYBML1* is capable of inducting the formation of both conical cells and multicellular trichomes. However, unlike *MIXTA*, these epidermal cell types could only be induced by *AmMYBML1* expression in floral tissues—*AmMYBML1* is unable to induce conical cells and trichomes on leaf epidermis. In wild-type *Antirrhinum* flowers *AmMYBML1* is expressed in the ventral petals, to which three distinct characteristics can be associated. First, the ventral petal has been shown to be predominantly responsible for the production of the large quantity of trichomes that arise in the throat of the wild-type flower. Second, the ventral petal is shaped and strengthened to form a landing surface of sufficient rigidity to discriminate bees (for which, as the natural pollinators of *Antirrhinum*, the hinge opens) from lighter insects. Third, in common with other petals, conical cells are produced on the epidermis of ventral petals, but unlike the conical cells found on other petals, reduced numbers of ventral petal conical cells still form in *mixta* mutants (Perez-Rodriguez *et al.*, 2005). This shows that the specification of conical cells on ventral petals is redundantly controlled by *MIXTA* and another factor(s). Evidence has indicated that *AmMYBML1* is the additional factor which has been co-opted to specify these distinctive aspects of ventral petal development and hence could have played a key role in the optimization of the pollination strategy of *Antirrhinum*.

The temporal and spatial expression pattern of *AmMYBML1* is consistent with its proposed role in the specification of trichomes and conical cells on the ventral petal epidermis, since it is expressed from early stages in the development of that petal (Perez-Rodriguez *et al.*, 2005). Unlike *MIXTA*, expression of *AtMYBML1* is also observed in adaxial mesophyll cells in the hinge of the ventral petal. Presumably to reinforce the rigidity of this crucial zone of the landing platform, this precise area of the mesophyll is composed of expanded and tightly packed cells. In support of the role of *AtMYBML1* in specifying this reinforcement of the mesophyll, ectopic expression of *AtMYBML1* in tobacco petals also led to thickening of the petals and increased expansion of the mesophyll cells, a morphological consequence which was not so apparent in *MIXTA* overexpression experiments. *MIXTA* and *AtMYBML1* therefore represent a very nice example of a gene pair which has been produced by gene duplication and subsequent subfunctionalization and neofunctionalization, the result of which has been the evolution of a distinctive pollinator interaction strategy. As is often the case in such examples, part of the subfunctionalization has clearly involved changes in patterns of gene expression. In this case, *MIXTA* remains expressed in the epidermal cells of all petals whereas *AtMYBML1* has become confined to the ventral petal and has expanded to include mesophyll expression there. However, neofunctionalization has also involved refinement of the capabilities of the proteins encoded by the duplicated genes, leading to the optimization of pollinator interaction. In this case it appears that modifications to the coding capacity and expression pattern of *AtMYBML1* has resulted in a ventral petal which enhances color to attract bees, is reinforced to limit reward access to appropriate pollinators, and has a mass of perfectly positioned trichomes to aid in the transfer of pollen from the bees to the stigma (Perez-Rodriguez *et al.*, 2005).

Of course, *AtMYBML1* can only aid in the specification of such specific pollinator interaction mechanisms if it is provided with spatial information at a higher level. We have already seen that specification of the ventral petal is accomplished by the *DIV* gene (Galego and Almeida, 2002). In *div* mutants, *MIXTA* expression is unaffected but *AtMYBML1* expression is significantly reduced (Perez-Rodriguez *et al.*, 2005). *div* mutants also lack throat trichomes and the petal in the ventral position is no longer reinforced to form a landing platform. This suggests that *AtMYBML1* is a prime candidate for a direct or indirect target of *DIV* (Fig. 6B).

VII. OUTLOOK

In the foregoing sections we have described the recent history of research into just one aspect of plant development, looking specifically at areas in which *Antirrhinum* has made a significant contribution. In the course of this we have highlighted several examples where a comparative approach is helping us to understand both the core-conserved mechanisms that control flower development and the distinctive modifications that play a role in generating the diversity of flower forms that we see in nature. In this final section, we would like to make a case for the wider adoption of the *Antirrhinum* model system in two ways. First, we will present the current resources available to the community to enable researchers to continue to address the type of questions that we have considered. Second, we will briefly introduce a particular aspect of *Antirrhinum* research that offers great potential for the future.

A. PRESENT AND FUTURE RESOURCES

One of the obvious advantages of *Antirrhinum* is that there are two large collections of mutants, totaling in excess of 700. These mutants can be accessed through the *Antirrhinum* database (www.antirrhinum.net/), by following the links for the IPK-Gatersleben and Enrico Coen's website, respectively. The vast majority of *Antirrhinum* mutants, which have been studied using molecular techniques, have resulted from the insertion of transposable elements (transposons), of which many have been characterized in this species. Historically, the most efficient method for going from mutant to gene in *Antirrhinum* has been simply to look for cosegregating polymorphisms resulting from transposon integration. Often the mutants show reversion, due to transposon excision, simplifying the task of assigning affected genes to mutants. However, the advent of a molecular linkage map (Schwarz-Sommer *et al.*, 2003b; Zhang *et al.*, 2004), together with the molecular resources described below, now means that it is possible to consider adopting a map-based cloning approach in *Antirrhinum* where the method of transposon tagging fails. The forward genetics resources are complemented by a reverse genetics system, based on transposons. Using this system it is possible to screen a series of pooled DNA samples by polymerase chain reaction (PCR), using gene-specific primers and primers specific to several of the more common *Antirrhinum* transposons, for insertion mutants in a gene of choice (Davies *et al.*, 1999a; Keck *et al.*, 2003).

Extensive publicly available molecular tools have also been developed (Schwarz-Sommer *et al.*, 2003a) including genomic and cDNA libraries, BAC and TAC libraries, yeast two-hybrid libraries, and *Agrobacterium*-mediated transformation of *Antirrhinum* by several methods (Cui *et al.*, 2003, 2004; Heidmann *et al.*, 1998). A large expressed sequence tag (EST) database has been generated containing ~12,000 unique sequences, facilitating expression analysis (Bey *et al.*, 2004), comparison of gene complements, and identification of candidate genes for reverse genetic analysis. A series of characterized fluorescence *in situ* hybridization (FISH) probes have been described (Zhang *et al.*, 2004). Despite the fact that *Antirrhinum* has only eight pairs of chromosomes, the similarity of several of them to each other makes conventional karyotyping based on length and arm ratios impossible. However, using a single centromere-specific probe it is possible to unambiguously distinguish all eight chromosomes. Additionally, a set of chromosome-specific probes has also been developed to allow the simple identification of any single chromosome of interest.

B. *ANTIRRHINUM* FOR STUDYING DIVERSITY AND ECOLOGY

Throughout this chapter we have used the term *Antirrhinum* to refer to a single species, *A. majus*. However *A. majus* is just 1 of about 20 species of the genus, which naturally occur in southwestern Europe and North Africa. Although these species are morphologically diverse, thriving in a range of habitats, they share the same chromosome complement ($2n = 16$) and form interfertile hybrids when artificially cross-pollinated. Together with the advanced molecular resources described above, the range and diversity of species available make it possible for future research to address biological questions at the interfaces between genetics and ecology. The various different *Antirrhinum* species are well adapted to suit their natural environment. The contrasts in morphology between *A. majus* (which has long been cultivated as an ornamental), *Antirrhinum charidemi* (naturally restricted to an area of Spain which is the driest in mainland Europe), and *Antirrhinum molle* (from Pyrenean cliffs and screes), for example, are immediately apparent (Schwarz-Sommer *et al.*, 2003a). Such variety provides an opportunity for introgression studies in which different species are crossed and the segregation of traits analyzed in subsequent generations and in recombinant inbred and near isogenic lines derived from them. In this way it becomes possible to address questions about the mechanisms of adaptation to the environment, the genetic control of habit and organ size and to study the evolution of a variety of morphological traits. In order to be able to address the genetics of the evolutionary changes in size and shape within the *Antirrhinum* species it is first necessary to apply quantitative approaches to define the parameters. The F2 progeny of a *majus* × *charidemi* cross have been used for just this purpose, to define the allometric space which encompasses the variety of leaf sizes and shapes found amongst *Antirrhinum* species (Langlade *et al.*, 2005). All the *Antirrhinum* taxa form interconnected clouds within the defined space, thus illustrating the various viable forms representative of the species diversity. QTL analysis could then be used to define at least 15 loci which contribute to the three principal components identified (Langlade *et al.*, 2005).

Antirrhinum also provides scope for ecological investigations looking at such questions as pollinator interactions and species isolation.

This combination of powerful and established molecular tools, together with a well-defined ecology and the ability to interbreed the different species, makes it likely that evolutionary and molecular ecological studies will flourish in the future, ensuring that *Antirrhinum* continues to make a significant contribution to our understanding of flowering plants.

ACKNOWLEDGMENTS

We would like to thank Drs. Enrico Coen and Hajime Sakai for providing information on the *OCT* gene and mutant prior to publication and Heinz Saedler as well as all members of the Schwarz-Sommer group for critical reading of this chapter.

REFERENCES

Almeida, J., Carpenter, R., Robbins, T. P., Martin, C. and Coen, E. S. (1989). Genetic interactions underlying flower color patterns in *Antirrhinum majus*. *Genes Development* **3**, 1758–1767.

Almeida, J., Rocheta, M. and Galego, L. (1997). Genetic control of flower shape in *Antirrhinum majus*. *Development* **124**, 1387–1392.

Amaya, I., Ratcliffe, O. J. and Bradley, D. J. (1999). Expression of *CENTRORADIALIS* (*CEN*) and CEN-like genes in tobacco reveals a conserved mechanism controlling phase change in diverse species. *Plant Cell* **11**, 1405–1418.

Bey, M., Stüber, K., Fellenberg, K., Schwarz-Sommer, Z., Sommer, H., Saedler, H. and Zachgo, S. (2004). Characterization of *Antirrhinum* petal development and identification of target Genes of the Class B MADS Box gene *DEFICIENS*. *Plant Cell* **16**, 3197–3215.

Bowman, J. L., Smyth, D. R. and Meyerowitz, E. M. (1991). Genetic interactions among floral homeotic genes of *Arabidopsis*. *Development* **112**, 1–20.

Bradley, D., Carpenter, R., Copsey, L., Vincent, C., Rothstein, S. and Coen, E. (1996). Control of inflorescence architecture in *Antirrhinum*. *Nature* **379**, 791–797.

Bradley, D., Ratcliffe, O., Vincent, C., Carpenter, R. and Coen, C. (1997). Inflorescence commitment and architecture in *Arabidopsis*. *Science* **275**, 80–83.

Carmel-Goren, L., Liu, Y. S., Lifschitz, E. and Zamir, D. (2003). The *SELF-PRUNING* gene family in tomato. *Plant Molecular Biology* **52**, 1215–1222.

Carpenter, R. and Coen, E. S. (1995). Transposon-induced chimeras show that *floricaula*, a meristem identity gene, acts non-autonomously between cell layers. *Development* **121**, 19–26.

Castillejo, C., Romera-Branchat, M. and Pelaz, S. (2005). A new role of the *Arabidopsis SEPALLATA3* gene revealed by it constitutive expression. *Plant Journal* **43**, 586–596.

Causier, B., Castillo, R., Zhou, J., Ingram, R., Xue, Y., Schwarz-Sommer, Zs. and Davies, B. (2005). Evolution in action: Following function in duplicated floral homeotic genes. *Current Biology* **15**, 1508–1512.

Citerne, H. L., Luo, D., Pennington, R. T., Coen, E. and Cronk, Q. C. (2003). A phylogenomic investigation of *CYCLOIDEA*-like TCP genes in the Leguminosae. *Plant Physiology* **131**, 1042–1053.

Coen, E. S. and Meyerowitz, E. M. (1991). The war of the whorls: Genetic interactions controlling flower development. *Nature* **353**, 31–37.

Coen, E. S., Romero, J. M., Doyle, S., Elliott, R., Murphy, G. and Carpenter, R. (1990). *Floricaula*: A homeotic gene required for flower development in *Antirrhinum majus*. *Cell* **63**, 1311–1322.

Coen, E., Rolland-Lagan, A. G., Matthews, M., Bangham, J. A. and Prusinkiewicz, P. (2004). The genetics of geometry. *Proceedings of the National Academy of Sciences of the United States of America* **101**, 4728–4735.

Colombo, L., Franken, J., Koetje, E., van Went, J., Dons, H., Angenent, G. C. and van Tunen, A. J. (1995). The Petunia MADS Box Gene *FBP11* determines ovule Identity. *Plant Cell* **7**, 1859–1868.

Comba, L., Corbet, S. A., Hunt, H., Outram, S., Parker, J. S. and Glover, B. (2000). The role of genes influencing the corolla in pollination of *Antirrhinum majus*. *Plant Cell Environment* **23**, 639–647.

Conner, J. and Liu, Z. C. (2000). LEUNIG, a putative transcriptional corepressor that regulates *AGAMOUS* expression during flower development. *Proceedings of the National Academy of Sciences of the United States of America* **97**, 12902–12907.

Corley, S. B., Carpenter, R., Copsey, L. and Coen, E. (2005). Floral asymmetry involves an interplay between TCP and MYB transcription factors in *Antirrhinum*. *Proceedings of the National Academy of Sciences of the United States of America* **102**, 5068–5073.

Costa, M. M., Fox, S., Hanna, A. I., Baxter, C. and Coen, E. (2005). Evolution of regulatory interactions controlling floral asymmetry. *Development* **132**, 5093–5101.

Cubas, P. (2004). Floral zygomorphy, the recurring evolution of a successful trait. *BioEssays* **26**, 1175–1184.

Cubas, P., Vincent, C. and Coen, E. (1999). An epigenetic mutation responsible for natural variation in floral symmetry. *Nature* **401**, 157–161.

Cui, M.-L., Handa, T. and Ezura, H. (2003). An improved protocol for *Agrobacterium*-mediated transformation of *Antirrhinum majus* L. *Molecular Genetics and Genomics* **270**, 296–302.

Cui, M.-L., Ezura, H., Nishimura, S., Kamada, H. and Handa, T. (2004). A rapid *Agrobacterium*-mediated transformation of *Antirrhinum majus* L. by using direct shoot regeneration from hypocotyl explants. *Plant Science* **166**, 873–879.

Davies, B., Di Rosa, A., Eneva, T., Saedler, H. and Sommer, H. (1996a). Alteration of tobacco floral organ identity by expression of combinations of *Antirrhinum* MADS-box genes. *Plant Journal* **10**, 663–677.

Davies, B., Egea-Cortines, M., de Andrade Silva, E., Saedler, H. and Sommer, H. (1996b). Multiple interactions amongst floral homeotic MADS box proteins. *EMBO Journal* **15**, 4330–4343.

Davies, B., Motte, P., Keck, E., Saedler, H., Sommer, H. and Schwarz-Sommer, Z. (1999a). *PLENA* and *FARINELLI*: Redundancy and regulatory interactions between two *Antirrhinum* MADS-box factors controlling flower development. *EMBO Journal* **18**, 4023–4034.

Davies, B., Sommer, H. and Schwarz-Sommer, Z. (1999b). Flower development: Genetic news and molecular views. *In* "Development: Genetics, Epigenetics and Environmental Regulation" (V. A. E. Russo, D. J. Cove, L. G. Edgar and F. Salamini, eds.), pp. 167–183. Springer, Berlin, Heidelberg, New York, Tokyo.

Ditta, G., Pinyopich, A., Robles, P., Pelaz, S. and Yanofsky, M. F. (2004). The *SEP4* gene of *Arabidopsis thaliana* functions in floral organ and meristem identity. *Current Biology* **14**, 1935–1940.

Dudareva, N., Murfitt, L. M., Mann, C. J., Gorenstein, N., Kolosova, N., Kish, C. M., Bonham, C. and Wood, K. (2000). Developmental regulation of methyl benzoate biosynthesis and emission in snapdragon flowers. *Plant Cell* **12**, 949–961.

Durfee, T., Roe, J. L., Sessions, R. A., Inouye, C., Serikawa, K., Feldmann, K. A., Weigel, D. and Zambryski, P. C. (2003). The F-box-containing protein UFO and AGAMOUS participate in antagonistic pathways governing early petal development in *Arabidopsis*. *Proceedings of the National Academy of Sciences of the United States of America* **100**, 8571–8576.

Efremova, N., Perbal, M. C., Yephremov, A., Hofmann, W. A., Saedler, H. and Schwarz-Sommer, Z. (2001). Epidermal control of floral organ identity by class B homeotic genes in *Antirrhinum* and *Arabidopsis*. *Development* **128**, 2661–2671.

Egea Gutierrez-Cortines, M. E. and Davies, B. (2000). Beyond the ABCs: Ternary complex formation in the control of floral organ identity. *Trends in Plant Sciences* **5**, 471–476.

Egea-Cortines, M., Saedler, H. and Sommer, H. (1999). Ternary complex formation between the MADS-box proteins SQUAMOSA, DEFICIENS and GLOBOSA is involved in the control of floral architecture in *Antirrhinum majus*. *EMBO Journal* **18**, 5370–5379.

Ferrario, S., Immink, R. G. H., Shchennikova, A., Busscher-Lange, J. and Angenent, G. C. (2003). The MADS Box gene FBP2 is required for SEPALLATA function in Petunia. *Plant Cell* **15**, 914–925.

Ferrario, S., Immink, R. G. and Angenent, G. C. (2004). Conservation and diversity in flower land. *Current Opinion in Plant Biology* **7**, 84–91.

Galego, L. and Almeida, J. (2002). Role of *DIVARICATA* in the control of dorsoventral asymmetry in *Antirrhinum* flowers. *Genes & Development* **16**, 880–891.

Glover, B. J. and Martin, C. (1998). The role of petal cell shape and pigmentation in pollination success in *Antirrhinum majus*. *Heredity* **80**, 778–784.

Goethe, J. W. V. (1790). "Versuch die Metamorphose der Pflanzen zu erklären," Gotha.

Golz, J. F., Roccaro, M., Kuzoff, R. and Hudson, A. (2004). GRAMINIFOLIA promotes growth and polarity of *Antirrhinum* leaves. *Development* **131**, 3661–3670.

Goodrich, J., Carpenter, R. and Coen, E. S. (1992). A common gene regulates pigmentation pattern in diverse plant species. *Cell* **68**, 955–964.

Goodwin, S. M., Kolosova, N., Kish, C. M., Wood, K. V., Dudareva, N. and Jenks, M. A. (2003). Cuticle characteristics and volatile emissions of petals in *Antirrhinum majus*. *Physiologia Plantarum* **117**, 435–443.

Goto, K. and Meyerowitz, E. M. (1994). Function and regulation of the *Arabidopsis* floral homeotic gene *PISTILLATA*. *Genes Development* **8**, 1548–1560.

Hantke, S. S., Carpenter, R. and Coen, E. S. (1995). Expression of *floricaula* in single cell layers of periclinal chimeras activates downstream homeotic genes in all layers of floral meristems. *Development* **121**, 27–35.

Hanzawa, Y., Money, T. and Bradley, D. (2005). A single amino acid converts a repressor to an activator of flowering. *Proceedings of the National Academy of Sciences of the United States of America* **102**, 7748–7753.

Heidmann, I., Efremova, N., Saedler, H. and Schwarz-Sommer, Z. (1998). A protocol for transformation and regeneration of *Antirrhinum majus*. *Plant Journal* **13**, 723–728.

Hileman, L. C., Kramer, E. M. and Baum, D. A. (2003). Differential regulation of symmetry genes and the evolution of floral morphologies. *Proceedings of the National Academy of Sciences of the United States of America* **100**, 12814–12819.

Honma, T. and Goto, K. (2001). Complexes of MADS-box proteins are sufficient to convert leaves into floral organs. *Nature* **409**, 525–529.

Huijser, P., Klein, J., Lönnig, W.-E., Meijer, H., Saedler, H. and Sommer, H. (1992). Bractomania, an inflorescence anomaly, is caused by the loss of function of the MADS-box gene *squamosa* in *Antirrhinum majus*. *EMBO Journal* **11**, 1239–1249.

Ingram, C. G., Doyle, S. D., Carpenter, R., Schultz, E. A., Simon, R. and Coen, E. S. (1997). Dual role for *fimbriata* in regulating floral homeotic genes and cell division in *Antirrhinum*. *EMBO Journal* **16**, 6521–6534.

Jack, T. (2004). Molecular and genetic mechanisms of floral control. *Plant Cell* **16**, S1–S17.

Jack, T., Brockman, L. L. and Meyerowitz, E. M. (1992). The homeotic gene *APETALA3* of *Arabidopsis thaliana* encodes a MADS box and is expressed in petals and stamens. *Cell* **68**, 683–697.

Keck, E., McSteen, P., Carpenter, R. and Coen, E. (2003). Separation of genetic functions controlling organ identity in flowers. *EMBO Journal* **22**, 1058–1066.

Kramer, E. M., Jaramillo, M. A. and Di Stilio, V. S. (2004). Patterns of gene duplication and functional evolution during the diversification of the AGAMOUS subfamily of MADS box genes in angiosperms. *Genetics* **166**, 1011–1023.

Krizek, B. A. and Meyerowitz, E. M. (1996). The *Arabidopsis* homeotic genes *APETALA3* and *PISTILLATA* are sufficient to provide the B class organ identity function. *Development* **122**, 11–22.

Langlade, N. B., Feng, X., Dransfield, T., Copsey, L., Hanna, A. I., Thebaud, C., Bangham, A., Hudson, A. and Coen, E. (2005). Evolution through genetically controlled allometry space. *Proceedings of the National Academy of Sciences of the United States of America* **102**, 10221–10226.

Litt, A. and Irish, V. F. (2003). Duplication and diversification in the *APETALA1/FRUITFULL* floral homeotic gene lineage: Implications for the evolution of floral development. *Genetics* **165**, 821–833.

Liu, Z. and Mexerowitz, E. (1995). *LEUNIG* regulates *AGAMOUS* expression in *Arabidopsis* flowers. *Development* **121**, 975–991.

Lönnig, W. E. and Saedler, H. (1994). The homeotic *Macho* mutant of *Antirrhinum majus* reverts to wild-type or mutates to the homeotic *plena* phenotype. *Molecular Genetics and Genomics* **245**, 636–643.

Luo, D., Coen, E. S., Doyle, S. and Carpenter, R. (1991). Pigmentation mutants produced by transposon mutagenesis in *Antirrhinum majus*. *Plant Journal* **1**, 59–69.

Luo, D., Carpenter, R., Vincent, C., Copsey, L. and Coen, E. (1996). Origin of floral asymmetry in *Antirrhinum*. *Nature* **383**, 794–799.

Luo, D., Carpenter, R., Copsey, L., Vincent, C., Clark, J. and Coen, E. (1999). Control of organ asymmetry in flowers of *Antirrhinum*. *Cell* **99**, 367–376.

Maes, T., Van Montagu, M. and Gerats, T. (1999). The inflorescence architecture of *Petunia hybrida* is modified by the *Arabidopsis thaliana Ap2* gene. *Developmental Genetics* **25**, 199–208.

Mandel, M. A., Gustafson-Brown, C., Savidge, B. and Yanofsky, M. F. (1992). Molecular characterization of the *Arabidopsis* floral homeotic gene *APETALA1*. *Nature* **360**, 273–277.

Martin, C., Prescott, A., Mackay, S., Bartlett, J. and Vrijlandt, E. (1991). Control of anthocyanin biosynthesis in flowers of *Antirrhinum majus*. *Plant Journal* **1**, 37–49.

Masiero, S., Li, M.-A., Will, I., Hartmann, U., Saedler, H., Huijser, P., Schwarz-Sommer, Z. and Sommer, H. (2004). *INCOMPOSITA*: A MADS-box gene controlling prophyll development and floral meristem identity in *Antirrhinum. Development* **131**, 5981–5990.

McSteen, P. C. M., Vincent, C. A., Doyle, S., Carpenter, R. and Coen, E. S. (1998). Control of floral homeotic gene-expression and organ morphogenesis in *Antirrhinum. Development* **125**, 2359–2369.

Mimida, N., Goto, K., Kobayashi, Y., Araki, T., Ahn, J. H., Weigel, D., Murata, M., Motoyoshi, F. and Sakamoto, W. (2001). Functional divergence of the *TFL1*-like gene family in *Arabidopsis* revealed by characterization of a novel homologue. *Genes Cells* **6**, 327–336.

Motte, P., Saedler, H. and Schwarz-Sommer, Z. (1998a). *STYLOSA* and *FISTULATA*: Regulatory components of the homeotic control of *Antirrhinum* floral organogenesis. *Development* **125**, 71–84.

Motte, P., Wilkinson, M. and Schwarz-Sommer, Z. (1998b). Floral meristem identity and the A function in *Antirrhinum. Flowering Newsletter* **25**, 41–43.

Navarro, C., Efremova, N., Golz, J. F., Rubiera, R., Kuckenberg, M., Castillo, R., Tietz, O., Saedler, H. and Schwarz-Sommer, Z. (2004). Molecular and genetic interactions between STYLOSA and GRAMINIFOLIA in the control of *Antirrhinum* vegetative and reproductive development. *Development* **131**, 3649–3659.

Negre, F., Kolosova, N., Knoll, J., Kish, C. M. and Dudareva, N. (2002). Novel S-adenosyl–methionine: Salicylic acid carboxyl methyltransferase, an enzyme responsible for biosynthesis of methyl salicylate and methyl benzoate, is not involved in floral scent production in snapdragon flowers. *Archives of Biochemistry and Biophysics* **406**, 261–270.

Niovi Jones, K. and Reithel, J. S. (2001). Pollinator-mediated selection on a flower color polymorphism in experimental populations of *Antirrhinum* (Scrophulariaceae). *American Journal of Botany*. **88**, 447–454.

Noda, K., Glover, B. J., Linstead, P. and Martin, C. (1994). Flower colour intensity depends on specialized cell shape controlled by a Myb-related transcription factor. *Nature* **369**, 661–664.

Okamuro, J. K., Szeto, W., Lotys-Prass, C. and Jofuku, K. D. (1997). Photo and hormonal-control of meristem identity in the *Arabidopsis* flower mutants *apetala2* and *apetala1*. *Plant Cell* **9**, 37–47.

Pelaz, S., Ditta, G. S., Baumann, E., Wisman, E. and Yanofsky, M. F. (2000). B and C floral organ identity functions require SEPALLATA MADS-box genes. *Nature* **405**, 200–203.

Pelaz, S., Gustafson-Brown, C., Kohalmi, S. E., Crosby, W. L. and Yanofsky, M. F. (2001a). APETALA1 and SEPALLATA3 interact to promote flower development. *Plant Journal* **26**, 385–394.

Pelaz, S., Tapia-Lopez, R., Alvarez-Buylla, E. R. and Yanofsky, M. F. (2001b). Conversion of leaves into petals in *Arabidopsis. Current Biology* **11**, 182–184.

Perez-Rodriguez, M., Jaffe, F. W., Butelli, E., Glover, B. J. and Martin, C. (2005). Development of three different cell types is associated with the activity of a specific MYB transcription factor in the ventral petal of *Antirrhinum majus* flowers. *Development* **132**, 359–370.

Pesch, M. and Hülskamp, M. (2004). Creating a two-dimensional pattern *de novo* during *Arabidopsis* trichome and root hair initiation. *Current Opinion in Genetics & Development* **14**, 422–427.

Pnueli, L., Carmel-Goren, L., Hareven, D., Gutfinger, T., Alvarez, J., Ganal, M., Zamir, D. and Lifschitz, E. (1998). The SELF-PRUNING gene of tomato regulates vegetative to reproductive switching of sympodial meristems and is the ortholog of CEN and TFL1. *Development* **125**, 1979–1989.

Prenner, G. (2004a). Floral development in *Polygala myrtifolia* (Polygalaceae) and its similarities with Leguminosae. *Plant Systematics and Evolution* **249**, 67–76.

Prenner, G. (2004b). New aspects in floral development of Papilionoideae: Initiated but suppressed bracteoles and variable initiation of sepals. *Annals of Botany* **93**, 537–545.

Ratcliffe, O. J., Bradley, D. J. and Coen, E. S. (1999). Separation of shoot and floral identity in *Arabidopsis*. *Development* **126**, 1109–1120.

Rolland-Lagan, A. G., Bangham, J. A. and Coen, E. (2003). Growth dynamics underlying petal shape and asymmetry. *Nature* **422**, 161–163.

Schwarz-Sommer, Z., Huijser, P., Nacken, W., Saedler, H. and Sommer, H. (1990). Genetic control of flower development by homeotic genes in *Antirrhinum majus*. *Science* **250**, 931–936.

Schwarz-Sommer, Z., Davies, B. and Hudson, A. (2003a). An everlasting pioneer: The story of *Antirrhinum* research. *Nature Reviews Genetics* **4**, 657–666.

Schwarz-Sommer, Z., de Andrade Silva, E., Berndtgen, R., Lönnig, W.-E., Müller, A., Nindl, I., Stüber, K., Wunder, J., Saedler, H., Gübitz, T., Borking, A., Golz, J. F., *et al.* (2003b). A linkage map of an F2 hybrid population of *Antirrhinum majus* and *A. molle*. *Genetics* **163**, 699–710.

Sessions, A., Yanofsky, M. F. and Weigel, D. (2000). Cell–cell signaling and movement by the floral transcription factors LEAFY and APETALA1. *Science* **289**, 779–781.

Simon, R., Carpenter, R., Doyle, S. and Coen, E. (1994). *Fimbriata* controls flower development by mediating between meristem and organ identity genes. *Cell* **78**, 99–107.

Sommer, H., Beltran, J. P., Huijser, P., Pape, H., Lönnig, W. E., Saedler, H. and Schwarz-Sommer, Z. (1990). *Deficiens*, a homeotic gene involved in the control of flower morphogenesis in *Antirrhinum majus*: The protein shows homology to transcription factors. *EMBO Journal* **9**, 605–613.

Theissen, G. (2001). Development of floral organ identity: Stories from the MADS house. *Current Opinion in Plant Biology* **4**, 75–85.

Theissen, G. and Saedler, H. (2001). Plant biology–Floral quartets. *Nature* **409**, 469–471.

Theissen, G., Kim, J. T. and Saedler, H. (1996). Classification and phylogeny of the MADS-box multigene family. *Journal of Molecular Evolution* **43**, 484–516.

Tröbner, W., Ramirez, L., Motte, P., Hue, I., Huijser, P., Lönnig, W.-E., Saedler, H., Sommer, H. and Schwarz-Sommer, Z. (1992). *GLOBOSA*: A homeotic gene which interacts with *DEFICIENS* in the control of *Antirrhinum* floral organogenesis. *EMBO Journal* **11**, 4693–4704.

Vandenbussche, M., Zethof, J., Royaert, S., Weterings, K. and Gerats, T. (2004). The duplicated B-Class heterodimer model: Whorl-specific effects and complex genetic interactions in *Petunia hybrida* flower development. *Plant Cell* **16**, 741–754.

Weberling, F. (1989). "Morphology of flowers and inflorescences." Cambridge University Press, Cambridge, UK.

Weigel, D. and Meyerowitz, E. M. (1994). The ABCs of floral homeotic genes. *Cell* **78**, 203–209.

Weigel, D., Alvarez, J., Smyth, D. R., Yanofsky, M. F. and Meyerowitz, E. M. (1992). *LEAFY* controls floral meristem identity in *Arabidopsis*. *Cell* **69**, 843–859.

Wilkinson, M., de Andrade Silva, E., Zachgo, S., Saedler, H. and Schwarz-Sommer, Z. (2000). *CHORIPETALA* and *DESPENTEADO*: General regulators during plant development and potential targets of FIMBRIATA-mediated degradation. *Development* **127**, 3725–3734.

Wright, G. A., Lutmerding, A., Dudareva, N. and Smith, B. H. (2005). Intensity and the ratios of compounds in the scent of snapdragon flowers affect scent discrimination by honeybees (*Apis mellifera*). *Journal of Comparative Physiology A: Sensory, Neural, and Behavioral Physiology* **191**, 105–114.

Yanofsky, M. F., Ma, H., Bowman, J. L., Drews, G. N., Feldmann, K. A. and Meyerowitz, E. M. (1990). The protein encoded by the *Arabidopsis* homeotic gene *agamous* resembles transcription factors. *Nature* **346**, 35–39.

Yu, H., Ito, T., Wellmer, F. and Meyerowitz, E. M. (2004). Repression of AGAMOUS-LIKE 24 is a crucial step in promoting flower development. *Nature Genetics* **36**, 157–161.

Zachgo, S., de Andrade Silva, E., Motte, P., Tröbner, W., Saedler, H. and Schwarz-Sommer, Z. (1995). Functional analysis of the *Antirrhinum* floral homeotic *DEFICIENS* gene *in vivo* and *in vitro* by using a temperature-sensitive mutant. *Development* **121**, 2861–2875.

Zhang, D., Yang, Q., Bao, W., Zhang, Y., Han, B., Xue, Y. and Cheng, Z. (2004). Molecular cytogenetic characterization of the *Antirrhinum majus* genome. *Genetics* **169**, 325–335.

Zik, M. and Irish, V. F. (2003). Global identification of target genes regulated by *APETALA3* and *PISTILLATA* floral homeotic gene action. *Plant Cell* **15**, 207–222.

Floral Developmental Genetics of *Gerbera* (Asteraceae)

TEEMU H. TEERI,*,† MIKA KOTILAINEN,* ANNE UIMARI,* SATU RUOKOLAINEN,* YAN PENG NG,* URSULA MALM,* EIJA PÖLLÄNEN,* SUVI BROHOLM,* ROOSA LAITINEN,* PAULA ELOMAA* AND VICTOR A. ALBERT‡

**Gerbera Laboratory, Department of Applied Biology, 00014 University of Helsinki, Finland*

†Department of Biology, University of Tromsø, NO-9037 Tromsø, Norway

‡Natural History Museum, University of Oslo, Blindern, NO-0318 Oslo, Norway

Advances in Botanical Research, Vol. 44
Incorporating Advances in Plant Pathology

0065-2296/06 $35.00
DOI: 10.1016/S0065-2296(06)44008-8

ABSTRACT

Inflorescence development in the angiosperm family Asteraceae has distinct features not found in the traditional model systems (e.g., *Arabidopsis*, *Petunia*, and *Zea*). In *Gerbera hybrida*, inflorescences are composed of morphologically different types of flowers tightly packed into a flower head (capitulum) that overtly resembles a single flower. Individual floral organs, such as pappus bristles (sepals), are developmentally specialized, petals and anthers form fused structures, stamens are aborted in marginal flowers, and ovaries are located inferior to other floral organs. These specific features have made *Gerbera* a rewarding target for comparative studies. We have shown that *Gerbera* MADS-box genes that group phylogenetically with B- and C-function genes from *Arabidopsis* and *Antirrhinum* function in organ determination in a very similar manner to their respective homologs from these model plants. However, MADS-box genes encoding proteins that interact with those of the ABC-genes behave differently. In *Arabidopsis*, three *SEPALLATA* (*SEP*) genes have redundant functions and are needed for development of petals, stamens, and carpels. Homologs of these *SEP* genes are found in *Gerbera* (*GRCD1*, *GRCD2*), but they show functional specialization. *GRCD1* is necessary for stamen development, but not for petal or carpel development. Similarly, *GRCD2* has a homeotic function restricted to carpel development. Remarkably, downregulation of the latter also results in floral reversion (which occurs in ovaries) and alters inflorescence architecture by switching off terminal, determinate growth. This integrated SEP-like control over reproductive meristem fate has not been detected in the well-known model systems, which have a different carpel design and normally bear indeterminate inflorescences. Moreover, the organization of flowers on the *Gerbera* capitulum reveals the presence of a radial morphogenetic gradient that appears to regulate ABC and other MADS-box genes differentially in a cell-nonautonomous manner. As such, there is some commonality in gene regulatory features between *Gerbera* flowers and inflorescences, which suggests that *Gerbera* capitula are more than simple analogs of the flowers they bear.

I. INTRODUCTION

Principles of floral developmental genetics have been largely established using *Arabidopsis thaliana* (Brassicaceae). Collaborative efforts by the plant community have raised this otherwise unimportant weed into one of the most important models for many aspects of plant biology (Meyerowitz and Somerville, 1994). The complete sequence of the *Arabidopsis* genome has been available for several years now (Arabidopsis Genome Initiative, 2000), and an immensely useful set of tools is now available for mutant isolation, characterization, and map-based cloning of the corresponding genes (Somerville and Koornneef, 2002). For understanding eudicot flower development, several other plant species have played a particularly important role, especially *Antirrhinum majus* (see Davies *et al.*, Chapter 7) and to a large extent also *Petunia hybrida* (see Rijpkema *et al.*, Chapter 6). Common to these two species is the possibility of effective transposon mutagenesis, again making mutant analysis and gene isolation possible (Koes *et al.*, 1995; Schwarz-Sommer *et al.*, 2003). Curiously, *Antirrhinum,* which has had the strongest influence of these two systems for understanding the developmental genetics of the flower, lacks one crucial tool of modern genetics: transformation. Although gene transfer has been described for *Antirrhinum* (Heidmann *et al.*, 1998), it has not yet grown into a high-throughput tool for reverse genetics.

However, the absence of transposon systems, transformation, and a complete genomic sequence is not the exception, but rather the rule now that diverse species of angiosperms have become the focus of developmental genetic investigation. For example, questions regarding the origin of flowers, and their amazingly well-conserved developmental genetic control in eudicots, have emerged as important subjects within the plant biological community. It is now widely recognized that genetic resources from a broad range of taxa are important, such as from nongrass monocots, basal angiosperms, and seed plants outside of the flowering plant lineage (Albert *et al.*, 2005; Soltis *et al.*, 2002).

Our goal in contributing to the genetics of flower development has been in dissecting conserved principles within the eudicots through use of the cultivated plant *Gerbera* (*Gerbera hybrida*; Asteraceae) as a model system. The sunflower family, Asteraceae, is one of the most successful families of flowering plants and is typified by compact inflorescences (heads or capitula) in which a higher-order organization has evolved into carrying out a role not only in bearing flowers, but also in attracting pollinators. The inflorescence of most species of Asteraceae resembles a solitary flower, and we have found that the organization of this inflorescence may require extra tiers of genetic control not exhibited in the more simple inflorescences of *Arabidopsis*, *Antirrhinum*, and *Petunia*.

Gerbera is a member of Mutisieae, a tribe placed in a basal phylogenetic position in Asteraceae (Bremer, 1994; Kim and Jansen, 1995). However, a basal position in phylogenies does not necessarily translate into "primitive" or "ancestral" developmental, morphological, or other characters. Nonetheless, because of its phylogenetic position, *Gerbera* occupies a distinct "outpost" in the large lineage of flowering plants known as the "asterids." There are two major subclades within the core or euasterids, campanulids, and lamiids (APGII, 2003), with *Gerbera* representing the former of these, and *Antirrhinum* and *Petunia* placed in the second. Several traits differentiate these two asterid lineages, including different secondary chemistries.

Gerbera is an important cut flower crop, making accessions of *Gerbera* an important source of genetic variation that breeders have selected for floral and inflorescence traits (e.g., http://www.terranigra.com/gerbera.html;

Fig. 1. Genetically modified (GM) and traditional varieties of *Gerbera.* (A) Variety Terra Marimo, which shows bract-like structures instead of floral organs in its capitulum. (B) Variety Terra Regina, which is the nontransformed reference for all GM lines derived from the variety. (C) GM Regina, transformed with antisense *GCHS1*, which encodes chalcone synthase. Flowers are white since anthocyanin biosynthesis is blocked (Elomaa *et al.*, 1993). (D) GM Regina, transformed with 35S-*GDEF2*, which encodes a B-function MADS-box regulatory gene. This individual shows cosuppression of *GDEF2* expression. The petals are abnormal (Eija Pöllänen, unpublished). (E) GM Regina, transformed with antisense *GAGA2*, which encodes a C-function MADS-box gene. Flowers bear extra petals in place of stamens and have indeterminate flower meristems (Yu *et al.*, 1999). (F) GM Regina, transformed with antisense *GRCD2*, which encodes an E-function MADS-box gene. Flowers show a homeotic change in whorl 4, reversion in the ovary, and loss of determinacy at the inflorescence level (Uimari *et al.*, 2004). (G) GM Regina, transformed with 35S-*GGLO1*, which encodes a B-function MADS-box transcription factor gene. This individual shows cosuppression of *GGLO1* expression. The petals develop into structures that partly resemble pappus hairs (Yu *et al.*, 1999). (H) GM Regina, transformed with 35S-*GAGA2*, which encodes a C-function MADS-box gene. Instead of petals, this transformant bears mosaic structures that resemble both petals and stamens (Yu *et al.*, 1999). (I) GM Regina, transformed with 35S-*AG*, the

http://www.lansbergen.com/; http://www.preesman.com/GB/gerbera.htm). These variations include patterns of anthocyanin pigmentation, morphological traits of the flowers and inflorescences, and most recently, "oddities" of flower development, an important but as yet largely unexplored resource for relating candidate genes with flower and inflorescence developmental processes (Fig. 1A). *Gerbera* is amenable to genetic transformation, making this tool one of the most efficient means for functional analysis of *Gerbera* genes (Elomaa and Teeri, 2001; Elomaa *et al.*, 1993).

A. DESIGN OF *GERBERA* FLOWERS

Typical of species in the family Asteraceae, *Gerbera* bears several types of flowers in its inflorescence (Harris, 1995). From the point of view of pollinating insects, the inflorescence is functionally similar to a single flower, with an attractive and colorful periphery and centrally located sex organs. However, the roles in this structure are not played by individual floral organs, but by differentially developed flowers on the tightly packed inflorescence.

The marginal ray flowers of *Gerbera* are strongly zygomorphic, with their corollas (fused petals) forming two to three cycles of showy petal-like structures in the pseudanthium (Yu *et al.*, 1999). The center of the inflorescence is occupied by disc flowers, in which the corolla whorl is much less conspicuous (Fig. 2). In addition to the size of the corolla, the ray and disc flowers also differ in sex, ray flowers being female and disc flowers hermaphrodite. Between the ray and disc flower cycles, *Gerbera* bears a third type of florets, the trans flowers. Trans flowers are similar to ray flowers in that they are female, but similar to disc flowers in that the corolla is not particularly elongated. Because the length of the trans flower corolla varies between cultivars, and because of the abrupt change in sex between trans and disc flowers, the marginal, female ray and trans flowers are sometimes classified together and distinguished from the central, hermaphroditic disc flowers.

Although the three types of flowers on the *Gerbera* inflorescence have different roles and different appearance, their structure and organization is similar and can be seen as a variation on the common structure shared by all eudicot angiosperm flowers. Typical of Asteraceae, the outermost whorl of floral organs, the calyx (sepals), is not green and leaf-like. For individually developing flowers of most eudicots, the leaf-like sepals form a protective covering until the flower opens, but in a dense capitulum, this role is not

Arabidopsis C-function MADS-box gene. Compared to wild-type Regina, the petals are smaller, and in the central flowers they develop stamen-like characteristics (Yu *et al.*, 1999). (See color Plate 14.)

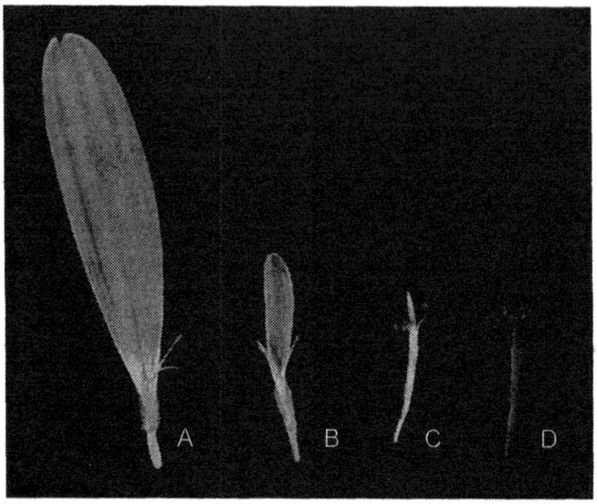

Fig. 2. *Gerbera* flower types. The marginal ray (A) and trans (B) flowers are female and strongly zygomorphic in *Gerbera*. In these flowers, stamen development starts normally, but then lags behind and only rudimentary staminodes are formed. Disc flowers (C and D) are hermaphrodite and develop both stamens and carpels. In all flower types, the corolla is composed of three fused and two rudimentary petals, the latter most clearly visible in C. The pronounced floral zygomorphy weakens toward the center, and the central-most disc flowers (D) are radially symmetrical. (See color Plate 15.)

required. The protective function is instead conferred by vegetative involucral bracts that surround the capitulum and cover the entire inflorescence during its early developmental stages. In *Gerbera*, the calyx has developed into a feathery structure, the pappus, which is attached to seeds and assists in their dispersal. Many species of the family utilize the calyx in this way to make a parachute for the seeds, but in some species the calyx is simply a scale-like structure, or is lacking altogether.

The second whorl of floral organs in *Gerbera*, the corolla, is composed of five petals. Petals first appear as separate bumps on the developing flower primordium, but they are already congenitally fused in their early development and form a characteristic ring primordium. Three of the petals remain fused during development and form the long ligule of the ray (and sometimes trans) flowers, whereas two (adaxial) petals remain small and rudimentary. In the inconspicuous disc flowers, the corollas are not much longer than the rudimentary petals, and the three abaxial petals remain fused together. Fusion of the abaxial petals in nearly all flowers of the *Gerbera* inflorescence makes the individual flowers bilaterally symmetrical, or zygomorphic. This is very obvious in the marginal flowers and less conspicuous in the disc flowers. Floral asymmetry gives way at the inflorescence center, where one can find perfectly symmetrical (actinomorphic) disc flowers with nonfused petals (Fig. 2).

The third floral whorl, occupied by stamens, shows a different developmental fate in marginal vs central flowers. In all flower types the stamen primordia begin development identically, but in the marginal flowers they soon lag behind and form only rudimentary filamentous structures called staminodes (Kotilainen *et al.*, 2000). In the disc flowers the stamens, which start developing separately, fuse postgenitally and cover the carpel. Therefore, whereas the marginal flowers are female by their structural design, the central flowers probably act in practice as male flowers, although being fully hermaphroditic.

Carpel development proceeds similarly in all flower types. The gynoecium is also a fused structure in *Gerbera*, composed of two fused carpels. A single ovary carries a single ovule, and the position of the ovary is inferior. This ovary position is different from the most common model species of flower development; *Arabidopsis*, *Antirrhinum*, and *Petunia* all have superior ovaries.

Flower initiation takes place centripetally in *Gerbera*, as in most species of Asteraceae (Harris, 1995). The marginal ray flowers are the first to appear, and as the inflorescence matures, trans and disc flowers appear successively. Eventually, the meristem in the center of the capitulum is fully consumed and occupied only by disc flower primordia. Thus, the number of flowers in the flower head is determinate (Uimari *et al.*, 2004). Flower primordia are all identical as they emerge, although they develop into distinct flower types as they mature. The position of the primordia on the inflorescence foretells their later fate, making it possible to microdissect flower primordia of different flower types at a very early stage.

Gerbera, as a member of Asteraceae, presents many themes of differential flower development that may be important factors for the evolutionary success of this group of flowering plants. In Asteraceae, marginal flowers are often feminized as in *Gerbera,* but some species, such as sunflower, carry hermaphrodite central flowers and sexually neutral ray flowers (Bremer, 1994). Many species of Asteraceae display a stronger difference in symmetry than *Gerbera* flowers, showing complete actinomorphy in all disc flowers, while still others have only a single type of flower that may be either zygomorphic or actinomorphic. Altogether, development of different types of flowers within inflorescences is a feature that the common model plants do not show but one which permits study of higher-order developmental regulation within single *Gerbera* genotypes.

B. REVERSE GENETICS TOOLS IN *GERBERA*

The cultivated *Gerbera* varieties are hybrids between the wild species *G. jamesonii* and *G. viridifolia* (Hansen, 1999). Like the cultivated *Petunia*, *G. hybrida* is a segmental diploid hybrid, that is, one with different chromosomal regions inherited from the two parental species. *Gerbera* is an outcrosser, and the varieties are bred by random or planned crossings and subsequent clonal propagation of the resulting lines. Micropropagation via tissue culture is the most important method of propagation for large scale production (Murashige *et al.*, 1974; Pierik *et al.*, 1973).

As an outcrossing plant, *Gerbera* shows strong inbreeding depression when selfed, making the use of homozygous inbred laboratory lines in research impractical. Still, methods for production of haploid and dihaploid lines from unfertilized ovules exist (Meynet and Sibi, 1984; Sitbon, 1981), although these have not been used extensively in either breeding or molecular research. Lack of pure lines, as well as the relatively large size of the plant, makes a forward genetics approach (mutation screening) a difficult task. However, *Gerbera* is amenable to reverse genetics, where gene transfer is used to modify the expression of previously isolated *Gerbera* genes. This is a great advantage of *Gerbera* as a model plant with complex inflorescences.

Gerbera transformation is based on the ability to regenerate plants from segments of leaf petioles. Transformation of a few varieties has been reported (Elomaa *et al.*, 1993; Hutchinson *et al.*, 1992; Orlikowska and Nowak, 1997), but the largest systematic reverse genetics approach has taken place in our laboratory with the variety Terra Regina (Eckermann *et al.*, 1998; Elomaa *et al.*, 1996; Kotilainen *et al.*, 1999, 2000; Uimari *et al.*, 2004; Yu *et al.*, 1999). This variety can be transformed with both *Agrobacterium* and particle-gun based methods (Elomaa and Teeri, 2001; Elomaa *et al.*, 1993; Paula Elomaa, unpublished). The timescale for transformation is about 12 months, from the start of the experiment to the first inflorescences in the greenhouse.

We have routinely used *Agrobacterium*-based transformation with cDNAs cloned in a sense or antisense orientation with respect to the promoter of the 35S transcript of Cauliflower mosaic virus. This typically results in three types of lines: ones that ectopically express the cDNA in all tissues, ones that show strong suppression of the corresponding endogenous gene, and ones that show partial suppression of the endogenous gene. The first two categories can be found among the sense lines, and the last category of plants in the antisense lines. We have also used constructs that lead to a double-stranded hairpin-like transcript when expressed (Smith *et al.*, 2000), from which we obtain strong suppression of the endogenous gene with high frequency.

The reverse genetics tools in *Gerbera* have been used for functional analysis of a large range of *Gerbera* genes, from genes coding for enzymes acting in secondary metabolism, structural proteins of the cell wall, MYB family transcription factors, and members of the MADS domain family of transcription factors.

Our reverse genetics tools meet hard challenges as the number of interesting genes that require functional testing has rapidly increased following analysis of our new 17,000-entry collection of *Gerbera* ESTs (Laitinen *et al.*, 2005). Even if functional assessments may be the most reliable using transgenic tools, a rapid functional screen would be a valuable additional resource. Generally applicable viral vectors that generate virus-induced gene silencing might become an important addition to the *Gerbera* toolbox in the future (Hileman *et al.*, 2005).

II. *GERBERA* MADS-BOX GENES

Mutation screens for identifying genes responsible for flower organ identity in *Arabidopsis* and *Antirrhinum* have remarkably coalesced on genes that encode members of a single family of transcription factors, the MADS-domain proteins. This family of transcription factors is shared by all kingdoms of eukaryotic species, with primitive forms identifiable even in prokaryotes (see http://www.ebi.ac.uk/interpro/DisplayIproEntry?ac=IPR002100). The acronym "MADS" derives from the initials of the first four sequenced MADS-box genes in yeast, plants, and man (Schwarz-Sommer *et al.*, 1990).

Fungal and animal genomes typically harbor only few MADS-box genes in their genomes. The human genome contains 16 MADS-box genes, whereas the genomes of *Drosophila melanogaster*, *Saccharomyces cerevisiae*, and *Caenorhabditis elegans* have 11, 4, and 3, respectively (see http://www.ebi.ac.uk/interpro/DisplayIproEntry?ac=IPR002100). In contrast, the number of MADS-box genes in flowering plants typically exceeds 100 (e.g., 107 in *Arabidopsis;* Parenicova *et al.*, 2003). Plant genomes contain a unique subgroup of MADS-box genes characterized by a second conserved domain, the K-box. These MIKC, or type II, MADS-box genes have been much more readily characterized functionally than the type I MADS-box genes of plants, which lack the K-box. While functions for a large number of type II MADS-box genes have been uncovered, principally but not exclusively relating to flower development, very little is known about the processes that type I MADS-box genes might be regulating (Kohler *et al.*, 2003).

The distinguishing feature of MADS-box genes, the nucleotides that encode the MADS domain with its DNA-binding properties, is a 168-base pair long sequence with absolute conservation concerning length and very high conservation concerning sequence (Alvarez-Buylla *et al.*, 2000). Isolation and identification of MADS-box genes in any given plant species is straightforward. Using a variety of techniques, from screening of *Gerbera* cDNA libraries with heterologous probes, RT-PCR with MADS-box specific primers, and mining of the *Gerbera* EST collection, we have isolated about 20 *Gerbera* cDNAs encoding MADS-domain proteins (Kotilainen *et al.*, 2000; Laitinen *et al.*, 2005; Yu *et al.*, 1999; Satu Ruokolainen, unpublished). All sequences represent type II MADS-domain proteins, and although the set is likely not complete, we argue that it represents *Gerbera*'s strongest expressed MADS-box genes.

As the number of plant MADS-box genes isolated from different species has grown (now exceeding 2000 sequence entries; http://www.ncbi.nlm.nih.gov/entrez/query.fcgi?db=nucleotide&cmd=search&term=mads+AND+viridiplantae), phylogenetic analysis has been carried out to deduce their evolutionary relationships. Immediately, it was realized that MADS-box genes with particular functions in determining floral organ identity according to the ABC-model (see later section), grouped together in phylogenetic analyses (Doyle, 1994; Purugganan *et al.*, 1995). Where functional analysis had been possible, A-, B-, and C-function MADS-box genes were each found in distinct clades. However, some genes without obvious functions grouped in otherwise well-delineated functional lineages. Nevertheless, these results have made it possible to use phylogenetic analysis as a first prediction of function for newly isolated MADS-box genes, thus putatively identifying *Gerbera* genes for characterization as encoding A-, B-, and C-function factors (Yu *et al.*, 1999). As a second step in functional analysis, we have used the pattern of expression in developing floral primordia. This is another characteristic of the ABC MADS-box genes—their expression is typically (but not always) limited to the tissues where they carry out their function (Johansen *et al.*, 2002).

The third level of evidence for gene function is generated using reverse genetics. Since we cannot yet routinely isolate mutations in the *Gerbera* genes we study, this is our final level of functional characterization. Generation of transgenic lines that express particular MADS-box genes at the wrong times and in the wrong places, or which have reduced levels of expression, have proven to be extremely useful in assigning functions for *Gerbera* MADS-box genes. Some genes we predicted to have the A-, B-, or C-function in *Gerbera* have been shown to function accordingly (Yu *et al.*, 1999). Intriguingly, others have failed the prediction, for example, *GSQUA1* and *GDEF1* (see later section) do not encode the A- and B-functions despite their close phylogenetic relationships with A- and B-function genes of *Arabidopsis* and *Antirrhinum* (Yu *et al.*, 1999).

A. ABC-MODEL OF FLOWER DEVELOPMENT IN *GERBERA*

Analysis of floral homeotic mutations in *Arabidopsis* and *Antirrhinum*—mutations that cause development of floral organs of the wrong identity in particular whorls of the flower—not only demonstrated that nearly all of the genes controlling determination of organ identity belonged to a plant-specific clade of the eukaryotic wide family of MADS-box regulatory factors, but also that a common model for the regulation of flower organ identity seemed applicable (Coen and Meyerowitz, 1991). The model is combinatorial, in remarkable resemblance to the combinatorial model of organ identity uncovered in vertebrate animals—although plants and animals evolved multicellularity independently (Meyerowitz, 1997). An important observation in floral homeotic mutations was that they affected successive pairs of the concentric whorls of flower organs. In the "typical" angiosperm flower, whorl 1 (the outermost whorl) is occupied by sepals, whorl 2 by petals, whorl 3 by stamens, and whorl 4 by carpels. In class A homeotic floral mutants, organs in whorls 1 and 2 develop into carpels and stamens, respectively. In class B-mutants, sepals and carpels develop instead of petals and stamens in whorls 2 and 3. Finally, the C-class mutants show petals in whorl 3, sepals in whorl 4, and a never-ending repetition of the structure sepals–petals–petals until the floral meristem stops proliferating. Thus, in C-class mutants, not only a homeotic conversion occurs, but also the determinacy of the flower is lost (Coen and Meyerowitz, 1991).

The A-, B-, and C-classes of homeotic mutations were observed in both *Arabidopsis* and *Antirrhinum*. Concluding that the domains (adjacent whorls) of mutational change corresponded to domains of regulatory function, the ABC-model was developed (Fig. 3). Thus, sepals develop when A-function is active alone, petals when A is active together with the B-function, stamens when B is active together with C, and finally, carpels develop when the C-function acts alone (Coen and Meyerowitz, 1991). This explains why loss of the B-function (in B-class mutants) converts petals to sepals and stamens to carpels. In order to explain the A- and C-class mutants, an additional postulation had to be made: the A- and C-functions had to be antagonistic, regulating each other negatively and expanding to the domain of the negative regulator when the latter was missing.

The A vs C antagonism, at least in *Arabidopsis*, functions through an intricate interplay between the A- and the C-function genes, including the non-MADS-box A-function gene *APETALA2* (*AP2*) (for a review, see Jack, 2004). Transgenic experiments show that not only mutations cause homeotic changes, but misexpression of the genes causes phenotypic conversions beyond the mutant range (Krizek and Meyerowitz, 1996). Double and triple mutants show that, in *Arabidopsis*, the "codeless" floral organ state is leaf (Bowman *et al.*, 1991a). However, in an ABC-triple mutant the leaf-like organs are still organized in a whorled phyllotaxis, demonstrating that these genes are responsible for organ identity in the flower, but not identity of the flower itself.

Major molecular components of the ABC-model are remarkably universal, but the A-function is an exception. Even between *Arabidopsis* and *Antirrhinum* there are important differences in the A-function. In *Arabidopsis*, the A-function is assigned to two genes, *APETALA1* (*AP1*) and *AP2*. Of these, only *AP1* is a MADS-box gene, expressed first throughout the flower primordium but later in whorls 1 and 2 (Mandel *et al.*, 1992). Plants with mutations in *AP1* have problems in flower formation, but when they finally emerge, the perianth whorls are abnormal and contain stamen and carpel-like organs, a phenotype otherwise associated with the observed ectopic expression of the *Arabidopsis* C-function MADS-box gene *AGAMOUS* (*AG*) in whorls 1 and 2 (Bowman *et al.*, 1991b). *Antirrhinum* also expresses a MADS-box gene during flower development that is orthologous to *AP1* and is expressed in whorls 1 and 2 (Huijser *et al.*, 1992). However, the phenotype of mutants in this gene, *SQUAMOSA* (*SQUA*), fails to initiate flower primordia on the inflorescence altogether. Still, *Antirrhinum* mutant collections do include an A-class mutation with carpels and stamens in whorls 1 and 2. These mutations map to the *Antirrhinum* C-class gene *PLENA* (*PLE*), which is ectopically expressed in the perianth whorls in the *ovulata* mutants (Bradley *et al.*, 1993). In fact, *Antirrhinum* flowers can be completely described by a model that lacks the A-function altogether (Schwarz-Sommer *et al.*, 1990). The *SQUA* gene matches the *Arabidopsis* A-function gene *AP1* in its expression pattern and phylogenetic position, but its role is solely in determination of flower meristem identity. *AP1* has also this function, in addition to the A-function in organ identity determination.

Phylogenetic analysis of *Gerbera* MADS-box genes has provided candidates representing all three functions of the ABC-model (Yu *et al.*, 1999). Our naming of the *Gerbera* ABC homologs has employed a strategy of placing the letter "G" (for *Gerbera*) before the name of the first characterized locus representing a particular gene lineage. Thus, the first *Gerbera* A-function gene candidate is *GSQUA1* after the corresponding *Antirrhinum* gene, and the *Gerbera* C function gene candidates are *GAGA1* and *GAGA2* after the corresponding *Arabidopsis* gene. The *Gerbera* B-function gene candidates are named after the *Antirrhinum* genes *DEFICIENS* (*DEF*) and *GLOBOSA* (*GLO*). Thus, the *Gerbera* B-function candidates are *GDEF1*, *GDEF2*, and *GGLO1*. The corresponding *Arabidopsis* genes are *APETALA3* (*AP3*) and *PISTILLATA* (*PI*), respectively.

The *Gerbera* C-function candidates *GAGA1* and *GAGA2* are expressed in the central part of each flower primordium, and later in whorls 3 and 4 from early stages of development onward (Yu *et al.*, 1999). This expression pattern matches well that of *AG* and *PLE*, the C-function genes of *Arabidopsis* and *Antirrhinum* (Davies *et al.*, 1999; Yanofsky *et al.*, 1990). *GAGA1* and *GAGA2* are 86% identical in their polypeptide-encoding nucleotide sequences, but we know from their inheritance in selfed progeny of the *Gerbera* cultivar Regina that they are not allelic. Still, so far as we have been able to proceed with functional analysis, they are operationally identical. Downregulation of either of these genes causes an eye-catching phenotype as the whorl 3 organs (including the rudimentary staminodes in marginal flowers) change into fully developed petals (Fig. 1E). Whorl 4 also changes; the inferior ovary appears normal from the outside, but it does not develop ovules, and the stigma and style are replaced by repeating structures with pappus-hair and petal-like character (Yu *et al.*, 1999). All of these changes, including loss of flower determinacy, closely match mutations in the C-class genes *AG* and *PLE*.

Ectopic expression of *GAGA1* or *GAGA2* throughout the developing flower primordia in transgenic *Gerbera* lines continues to follow the ABC-model in whorl 2, which normally develops petals. In 35S-*GAGA2* transgenic lines, or in heterologous 35S-*AG* lines, the whorl 2 organs in disc flowers do not develop anthocyanin pigmentation and all organs in the whorl fuse to cover the inner floral organs. In marginal ray flowers, the corresponding organs grow smaller in mild phenotypes still expressing anthocyanin (Fig. 1I), but in stronger phenotypes they convert into incompletely expanded and dry organs (Fig. 1H). Both of these characteristics are stamen-like, including the arrest of development in the marginal flowers. However, the whorl 1 organs, pappus hairs in *Gerbera*, do not react at all to the transgenic construct, although the *GAGA2* transcript is ectopically expressed in whorl 1. One interpretation of these results is that pappus hairs do not represent a calyx whorl at all, although they are still positionally homologous. Another, more likely explanation is that ectopic *GAGA2* expression is not sufficient to override the calyx program, a situation different from either *Arabidopsis* or *Antirrhinum*.

Of the three B-function gene candidates, *GDEF1*, *GDEF2*, and *GGLO1*, the first expresses at a very low level compared to the others (Yu *et al.*, 1999). The *GDEF1* transcript is flower specific, but the pattern of expression differs from the pattern known to be typical for B-function genes. Unlike the conventional whorls 2 and 3 specific B-function genes, in Northern blots, *GDEF1* shows strong expression only in whorl 3 (Suvi Broholm, unpublished). *GDEF1* also shows expression in whorls 1 and 4, although clearly weaker than in stamens, and no expression in whorl 2 is detectable. However, *in situ* analysis showed that *GDEF1* transcripts also occur in petals but only in a few spatially restricted cells (Suvi Broholm, unpublished). The lack of strong and continuous *GDEF1* expression in whorl 2 is in line with the expression pattern reported for other higher eudicot genes belonging to the same clade, designated as the *TM6* lineage based on the corresponding tomato gene (Kramer and Irish, 2000). Intriguingly, *Arabidopsis* lacks this lineage of MADS-box genes (Irish, 2003). We have generated a number of transgenic *Gerbera* lines that misexpress *GDEF1*, but they do not show consistently altered phenotypes.

Unlike *GDEF1*, both *GDEF2* and *GGLO1* are strongly expressed in flowers, and are very clearly whorl 2 and 3 specific, making them perfect candidates for the *Gerbera* pair of B-function MADS-box genes. Transgenic *Gerbera* lines provide full support for this hypothesis (Yu *et al.*, 1999). Downregulating either gene converts petals, in mild phenotypes, into greenish organs, and with more complete downregulation into hairy structures like the pappus hairs of whorl 1 (Fig. 1D and G). Accordingly, stamens lose their identity and gain carpel-like characteristics. In transgenic lines ectopically expressing *GGLO1*, carpels acquire stamen-like traits, the ovaries are disturbed, and pappus hairs convert into petals with full anthocyanin pigmentation (although narrower than whorl 2 petals; Yu *et al.*, 1999). Thus, in contrast to ectopic C-function gene expression, the calyx developmental program now converts into a corolla (petal whorl) program, suggesting that pappus bristles truly are modified sepals. At the molecular level, the floral B-function requires that both B-function MADS-box proteins occur in the same cells, and we do detect *GDEF2* transcripts in these altered whorl 1 organs (Eija Pöllänen, unpublished). We presume (although we have not determined) that *Gerbera* whorl 1 shows some early *GDEF2* expression, which is stabilized by the ectopic *GGLO1* expression through an autoregulatory loop shown to be operational in both *Arabidopsis* and *Antirrhinum*. In *Arabidopsis* it is ectopic *AP3* (the *DEF* ortholog) that can stabilize *PI* (the *GLO* ortholog) expression in whorl 1, whereas in *Gerbera* this is vice versa. Ectopic expression of *GDEF2* does not lead to phenotype changes in either whorl 1 or whorl 4, although downregulation of *GDEF2* causes similar homeotic changes to downregulation of *GGLO1* in whorls 2 and 3—but not in the ovaries, where *GDEF2* is strongly expressed (Eija Pöllänen, unpublished).

Finally, *GSQUA1* does not behave like an A-function gene, although it is clearly a member of the A-function clade in phylogenetic analyses. Unlike *AP1* and *SQUA* in *Arabidopsis* and *Antirrhinum*, respectively, *GSQUA1* is not expressed in whorls 1 and 2. *GSQUA1* is flower abundant, but expressed in the vascular tissue of the ovaries rather than in floral organ primordia (Yu *et al.*, 1999). Transgenic lines with *GSQUA1* would be interesting to analyze, but in spite of many attempts over more than 10 years, we have not been able to regenerate a single sense or antisense *GSQUA1* line. We have been able to identify at least four new *SQUA*-like genes in *Gerbera* with targeted PCR and EST mining. None of them seem to have an expression pattern typical of other A-function genes (Satu Ruokolainen, unpublished), and it remains to be determined whether *Gerbera* has an A-function in the same sense that *Arabidopsis* has (see Zahn *et al.*, Chapter 4). In any case, the unusually large collection of *SQUA* and *AP1*-like genes in *Gerbera* calls for further investigation.

In summary, the B- and C- functions (but not the A-function) of the "universal" ABC-model of flower development can be assigned to specific MADS-box genes in *Gerbera*. Application of the model toward understanding *Gerbera* (or Asteraceae)-specific features of flower development provides some important conclusions. Arrest of stamen development in marginal flowers is organ identity specific and not whorl specific. By changing identity of whorl 3 organs (with antisense *GAGA2*), the ray and trans flower whorl 3 organs develop fully, into petals instead of stamens. By changing identity of whorl 2 organs to stamen-like (with ectopic *GAGA2*), these organs face the similar fate of withering in marginal flowers. Sex identity does not seem to be controlled by ABC MADS-box genes in *Gerbera*. We have never seen altered sexuality among the transformants, and the B-function genes (*GDEF2* and *GGLO1*) that might be proposed to confer sex differences are fully expressed in the ray and trans flower stamens before their development arrests (Yu *et al.*, 1999).

Importantly, we conclude that pappus bristles are altered sepals. They change into petals as "normal" sepals do when exposed to the additional code of the B-function. But something makes the *Gerbera* calyx different from any other species investigated in this respect. Ectopic C-function gene expression does not alter the identity of organs in the *Gerbera* whorl 1. Something, we shall call it "A1," maintains whorl 1 identity in spite of the presence of a C-function gene. We have not investigated if the GAGA2 protein is unstable in whorl 1, or if coregulators present in whorl 4 are lacking in whorl 1. Still, an ectopic C-function works perfectly in whorl 2 according to the ABC-model. Thus, this whorl 2 specific "A2" function in *Gerbera* has different properties than the whorl 1 specific A1 function (Fig. 3).

B. BEYOND THE ABCS: SUBFUNCTIONALIZATION OF THE *SEPALLATA* GENES IN *GERBERA*

The universally applicable ABC-model of flower organ determination, including the subtle and possibly multiple solutions for setting up the A-function, clearly demonstrates that the ABC-function genes are necessary for floral organ determination, but not for floral organ identity per se. The latter is nicely shown in *Arabidopsis* triple mutants where none of the

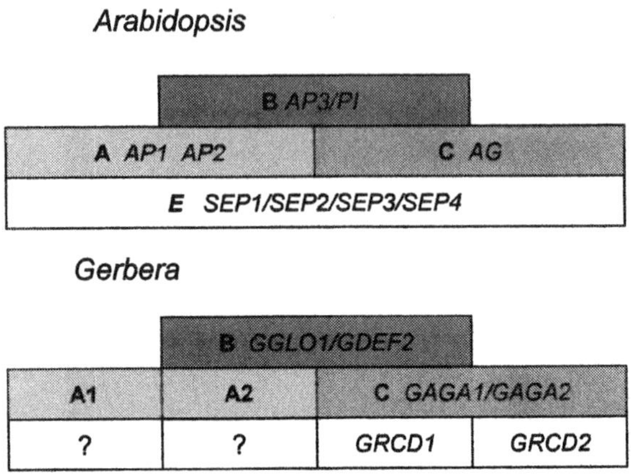

Fig. 3. Modified ABC(D)E model in *Gerbera.* Compared to *Arabidopsis*, the genes responsible for the homeotic functions that determine floral organ identity have undergone duplication and subfunctionalization in *Gerbera*. In *Arabidopsis*, four *SEP* genes encode the homeotic E-function in a redundant manner. In *Gerbera*, the genes *GRCD1* and *GRCD2* are whorl-specific counterparts of the *Arabidopsis SEP* genes. Our *Gerbera* model predicts that further, as yet uncharacterized, whorl-specific genes act in whorls 1 and 2, possibly though not necessarily redundantly as in *Arabidopsis*. Functional analysis of *Gerbera* MADS-box genes have shown that the putative A-function reacts differently to ectopically expressed B- and C-function genes. We propose that a further subdivision has occurred for the A-function in *Gerbera*, that is, A1 and A2. See text for more details. (See color Plate 16.)

three functions are effective: the whorled structure of the flowers remains, but all organs show leaf identity (Bowman *et al.*, 1991a). It has long been known that although necessary, the combinatory ABC-function is not sufficient for floral organ identity—ectopically expressed ABC-function genes do not convert leaves into floral organs (Krizek and Meyerowitz, 1996).

An impressive answer to the sufficiency question came through analysis of three *Arabidopsis* MADS-box genes, all grouping together phylogenetically, but outside of the A, B, and C clades. These genes were originally identified in general screens for MADS-box genes in *Arabidopsis*, and became known as *AGL2*, *AGL4*, and *AGL9* (*AGL* standing for *AGAMOUS-LIKE*). None of the floral mutations (or, in fact, any other mutation) analyzed mapped to these genes, so they were not readily recognized as determinants of floral organ identity. Generally, two simple reasons explain why mutations in particular genes are not found even when expected based on the breadth of the screen: the mutation may be lethal and therefore selected against, or the function may be redundantly encoded, that is, masked by other (related) genes that provide a similar or strongly overlapping function. For the three *AGL* genes, the latter proved to be the case. Due to the excellent genomic

Plate 14

Plate 15

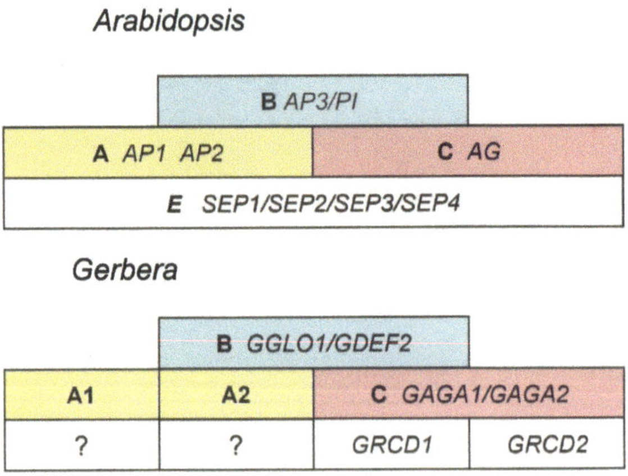

Plate 16

resources available for *Arabidopsis*, mutations in all three genes were available and known to have only subtle floral phenotypes. However, when combining all three mutations in the same genotype, the plants produced flowers composed of only sepals in each whorl (Pelaz *et al.*, 2000). Because of this phenotype, *AGL2*, *AGL4*, and *AGL9* were renamed to *SEPALLATA1* (*SEP1*), *SEP2*, and *SEP3*, respectively.

The *sep1/2/3* triple mutant resembles a double mutant where both B- and C-function MADS-box genes are mutated (e.g., *ap3/ag*). In such mutants, since the B- and C-codes are missing, all flower organs acquire the developmental fate of sepals. Furthermore, the sepal-containing flowers in the *sep1/2/3* triple mutant appear indeterminate, and extra whorls of flower organs emerge inside the fourth whorl (Pelaz *et al.*, 2000; but see later section for an alternative explanation for these organs). Therefore, it can be stated that both the B- and C-functions are dependent on the SEP function (encoded by any one member of the genes *SEP1–3*). Remarkably, the SEP function proved to be the missing link for sufficiency of flower organ identity. Expression of *SEP*, together with A-, B-, or C-function genes could convert vegetative leaves into floral organs, for example, into petals (with ectopic expression of *AP1*, *AP3*, *PI*, and *SEP3*), or into stamens (with *AP3*, *PI*, *AG*, and *SEP3*) (Honma and Goto, 2001; Pelaz *et al.*, 2001).

With SEP, the basic floral organ code was completed. It was also observed that sepal identity itself required a SEP function. For *AGL3*, then renamed to *SEP4*, the quadruple mutant *sep1/2/3/4* developed flowers made up of leaf-like organs only, phenocopying to a large degree an A/B/C-function triple mutant (Ditta *et al.*, 2004). Although the redundant SEP function is needed for development of all floral organs, including sepals themselves, it was reasonable to include it in the floral combinatorial code for two reasons. The first is that the combination between A, B, and C with SEP is sufficient for floral organ determination. The second reason is that the MADS-domain proteins encoded by the *SEP* genes together with the A-, B-, and C-function MADS-box genes appear to form transcriptional complexes. The first clues were demonstrated in the *Antirrhinum* system, where the SQUA protein was found to interact with DEF and GLO proteins (the B-function MADS-domain proteins of *Antirrhinum*) in a yeast two-hybrid assay (Egea-Cortines *et al.*, 1999). Although not directly verified *in planta*, the model of the day is that MADS-domain proteins act as pairs of dimers in higher-order complexes to carry out their function in gene regulation (Honma and Goto, 2001; Theißen, 2001; Theißen and Saedler, 2001). Accordingly, the ABC-model of flower organ identity determination has been expanded to the ABCDE-model, the SEP function named as the E-function (Fig. 3). The D-function had already been identified in *Petunia* as the

MADS-domain protein-mediated determinant of ovary identity, the proposed "fifth whorl" of the angiosperm flower (Angenent and Colombo, 1996; Colombo *et al.*, 1996).

The clade in which *SEP1–4* fall, originally referred to as the *AGL2* lineage but nowadays more often as the *SEP* gene clade, is a large group with many sequences from a great number of angiosperm species, including *Gerbera*, studied in detail. The *Gerbera SEP*-like genes are named as *GRCD* genes (for *GERBERA REGULATOR OF CAPITULUM DEVELOPMENT*), since the functional description of the *SEP* genes was not yet available when our analysis of *GRCD1* was published (Kotilainen *et al.*, 2000). Transgenic downregulation using a 35S-anti-*GRCD1* construct caused a homeotic alteration in whorl 3 of *Gerbera* flowers, where stamens (or staminodes in marginal flowers) normally develop. The phenotype was especially strong in the female marginal flowers, where the identity of the whorl 3 organs became strongly petal-like, with anthocyanin pigmentation typical of petals (in the variety Regina) and with petal-like epidermal sculpturing. Disc flowers also showed petal-like characteristics, but they were subtle. Anthers (unlike petals) never include stomata in wild-type *Gerbera*, but in the transgenic lines with downregulated *GRCD1* expression, the disc flower anthers displayed occasional stomata (Kotilainen *et al.*, 2000). These anthers were still functional and produced fertile pollen (although the pollen was not released). Thus, *GRCD1* has a function in proper stamen development in *Gerbera*, or, more strictly stated, in staminode development of the female flowers, since the disc flower stamens are nearly normal in antisense transformants. Furthermore, ectopic expression of *GRCD1* does not cause any detectable phenotype in *Gerbera* flowers.

Among the various *Gerbera* MADS-box genes isolated, the most similar gene to *GRCD1* is *GRCD2*, sharing 64% identity at nucleotide level (86% identity in the MADS-box). *GRCD2* has an expression pattern in developing *Gerbera* flowers similar to *GRCD1*, starting early in all cells of the primordia and concentrating later in development into whorls 3 and 4 (Uimari *et al.*, 2004). Both genes are expressed in the developing ovule, but *GRCD1* is only expressed in the outer integument while *GRCD2* is expressed throughout the ovule. In spite of these similarities, the phenotype of inflorescences and flowers in transgenic *Gerbera* lines in which *GRCD2* was downregulated differed dramatically from *GRCD1* downregulated lines. Whorl 3 (with stamens or staminodes) developed normally in anti-*GRCD2* plants, but whorl 4 was strongly altered. Weak phenotypes in which *GRCD2* was only moderately downregulated developed a green (leaf-like) structure in place of the stigma and style, but in lines with stronger suppression of expression, the stigma and style were replaced by an organ with petal-like traits including anthocyanin pigmentation and ridged epidermal sculpturing (Uimari *et al.*, 2004).

A conversion of staminodes into narrow petal-like organs, and the stigma and style into a green leaf-like organ, was also observed in plants with (moderate) downregulation of the *Gerbera* C-function, as encoded by the MADS-box genes *GAGA1* and *GAGA2* (Yu *et al.*, 1999). Thus, lack of full *GRCD1* or *GRCD2* expression separately phenocopied the whorls 3 and 4 phenotypes of downregulated C-function in *Gerbera*. As in *Arabidopsis*, the C-function is dependent on a SEP protein, but in *Gerbera* the SEP function is not encoded by the same genes in all floral whorls. Rather, the SEP function is provided by whorl-specific members of the *Gerbera SEP*-like MADS-box gene clade, demonstrating that the *SEP* (or *GRCD*) genes in *Gerbera* have undergone subfunctionalization after the events of gene duplication. In *Arabidopsis*, the SEP function is also needed for development of sepals in whorl 1 and petals in whorl 2, that is, the A- and B-function in *Arabidopsis* are also SEP dependent. In contrast, neither *GRCD1* nor *GRCD2* are needed for sepal (pappus hair) or petal development in *Gerbera*. A reasonable prediction is that further and as yet uncharacterized members of the *SEP* clade of MADS-box genes interact in the pappus and petal determining complexes of MADS-box genes (Fig. 3). In yeast two-hybrid analysis, the *Arabidopsis* SEP proteins have been shown to interact with B- and C-class proteins either directly or in ternary complexes (de Folter *et al.*, 2005; Honma and Goto, 2001). Similarly, in yeast, GRCD1 and GRCD2 both interact with GAGA1 and GAGA2. The former two genes do not interact with each other or with GDEF1 or GDEF2, but interestingly, and thus far without explanation, GRCD2 interacts with GGLO1 (Uimari *et al.*, 2004).

C. CONTROL OF FLORAL MERISTEM FATE BY *SEP* GENES

We noted in an earlier section that the *Arabidopsis* SEP function required for the determination of stamens and carpels has been divided in *Gerbera* between two members of the *SEP* clade of MADS-box genes. We observed a similar dependence on SEP proteins for the C-function in *Gerbera*, as has been observed in *Arabidopsis*. In *Arabidopsis* it has been interpreted that the second, nonhomeotic role of the C-function in controlling floral determinacy is also SEP dependent (Pelaz *et al.*, 2000). However, in *Gerbera*, although we can observe a repetition of floral structures inside whorl 4 in strong *GAGA1/2* downregulated transgenic lines (Yu *et al.*, 1999), the strong *GRCD2* transgenic lines show only homeotic conversion and not organ repetition in stigma and style (Uimari *et al.*, 2004). Instead, something dramatic takes place in the ovary. Outside, the ovary appears normal at first, but later in development it bursts open and a number of different *Gerbera* floral organs emerge. These include not only pappus hairs, petals, stamens, and carpels, but also leaf-like bracts and complete flower primordia—that is, a full *Gerbera* inflorescence seems to be crowded into the altered ovary (Uimari *et al.*, 2004).

The flower meristem in *GRCD2* downregulated lines is not consumed as in normal development, but instead reverts to an earlier meristem identity, that of the inflorescence meristem. In *Arabidopsis*, the *sep1/2/3* triple mutant flowers have been interpreted as indeterminate, that is, that extra whorls of floral organs (sepals in this case) grow continuously from the floral meristem. However, this "flower within a flower" includes an internode (Pelaz *et al.*, 2000), and has been interpreted as a new inflorescence by Ferrario *et al.* (2003). Therefore, the phenomenon of floral reversion in mutants or transgenic lines with reduced SEP activity may generally lead to floral reversion. As *Arabidopsis* has superior ovaries (positioned above the whorls of floral organs), reversion may misleadingly be confused with loss of flower meristem determinacy. The inferior positioning of the *Gerbera* ovary distinguishes these two phenotypes more clearly.

In addition to controlling the fate of the flower meristem, *GRCD2* also functions in the control of inflorescence meristem growth. In wild-type *Gerbera*, the number of flowers in the capitulum is fixed early in development, and the inflorescence meristem is consumed and replaced by floral meristems at the center of the capitulum. Although capitula have traditionally been interpreted as indeterminate (Harris, 1995; Stevens, 2001), the characteristics above are all signs of determinacy concerning the meristematic status of the inflorescence. In transgenic lines in which *GRCD2* is downregulated, the *Gerbera* inflorescence continues to produce flowers, and the center of the inflorescence remains undifferentiated until the inflorescence finally withers after prolonged flowering (Fig. 1F). More than double the normal number of flowers may develop in these inflorescences. As such, the evidently determinate state of these inflorescences has changed into indeterminate (Uimari *et al.*, 2004).

Most of the common model species of flower development (*Arabidopsis* and *Antirrhinum*, but not *Petunia*) bear indeterminate inflorescences that are not congested, so any involvement of the SEP function in controlling inflorescence determinacy is less obvious. However, in *Arabidopsis* and *Antirrhinum*, mutants are known that convert the indeterminate inflorescence into determinate. Interestingly, some of these (*terminal flower1* in *Arabidopsis* and *centroradialis* in *Antirrhinum*) are mutations in related genes of the phosphatidylethanolamine-binding protein family (see http://www.ebi.ac.uk/interpro/DisplayIproEntry?ac=IPR001858). The activity of these genes is unclear (they do not appear to be transcription factors), but it is intriguing to speculate that GRCD2 may be an upstream regulator of their inflorescence-meristem identity pathway. Interestingly, although loss of inflorescence determinacy cannot be directly investigated in *Arabidopsis*, ectopic expression of *SEP3* sometimes converts the indeterminate inflorescence into determinate (and thus phenocopies the *terminal flower1* mutant) (Honma and Goto, 2001; Pelaz *et al.*, 2001). As such, control of inflorescence determinacy state by the SEP function may be a general phenomenon among flowering plants, but one not easily detected in the common model systems.

Downregulation of *GRCD2* shows pleiotropic effects on at least three layers of flower development, from a homeotic function in organ identity determination, through control of flower meristem identity, to the control of determinacy of the inflorescence meristem. Although many of these effects may at first seem *Gerbera* or Asteraceae specific, instead we infer them to be general features of the SEP function. As such, meristem identity at the level of flowers and inflorescences appears to be interwoven by shared control mechanisms. In *Gerbera*, flowers are contained within inflorescences that appear to be modified versions of the same invention, like a babushka doll. These inflorescences can be interpreted as merely another fractal level of the floral theme. The resemblance of the capitulum to a single flower extends from the phenotypic to the genetic level, as will be discussed in greater detail in a later section.

III. *GERBERA* GENOMICS

The dissection of *Gerbera* floral development has strongly relied on reverse genetics approaches where individual genes have been isolated one by one using traditional methods of cDNA cloning. We initiated a genomic approach to analyze large collections of *Gerbera* genes by systematic sequencing of *Gerbera* cDNA libraries. Previously made inflorescence and petal libraries were replated, and complemented with libraries made from pappus hairs, stamens, leaves, and the floral stem (scape) to represent most of the aerial parts of the plant. About 17,000 sequences meeting quality criteria were entered in the *Gerbera* EST database for clustering and annotation. The collection represents about 8100 unigenes (clusters or individual sequences), and can be accessed through the EMBL Nucleotide Sequence Database (AJ750001–AJ766994) as well as the OpenSputnik Database (http://sputnik.btk.fi) and the Plant Genome Network (http://www.pgn.cornell.edu/). A number of new *Gerbera* MADS-box genes were identified among the ESTs (up to 10, since allelism of some closely related sequences has not been determined yet), and their phylogenetic positions, expression patterns, and transgenic phenotypes are under assessment.

All unique ESTs, along with all of the previously isolated individual *Gerbera* cDNAs, have been printed on a *Gerbera* cDNA microarray. The array was rigorously tested for its reliability using samples from *Gerbera* leaves and flowers (Laitinen *et al.*, 2005), and it will provide a valuable tool in the next phase of *Gerbera* floral developmental genetics. One of our best defined and long-term questions has been to understand how the developmental fate of the different flower types is determined on the capitulum. None of the floral regulators we have studied thus far has shown involvement in flower-type determination. To find lead performers in this drama, we have microdissected young and undifferentiated ray and disc flowers from developing capitula and analyzed their mRNA pools with the *Gerbera* cDNA microarray resource (Laitinen *et al.*, 2006). We have sliced the petals during their development into temporal and spatial sections to understand the dynamics of gene expression that lead to growth, its cessation, and pigment biosynthesis in the corolla (Kotilainen *et al.*, unpublished). The transgenic lines that have proved so valuable in the functional analysis of *Gerbera* MADS-box genes are also under transcriptional analysis using the *Gerbera* cDNA microarray to discover downstream targets for the MADS domain regulators, and to identify additional coexpressed regulators.

IV. THE *GERBERA* CAPITULUM

A. REGULATION BY A RADIAL MORPHOGENETIC GRADIENT?

The capitulum of Asteraceae has been used as a model system for studying inflorescence meristem development. Wounding experiments have been used to analyze inflorescence determination and phyllotaxis, and genetic studies of flower-type specific organ development have been accomplished. Both the wounding and genetic experiments have suggested that a radial morphogenetic gradient acts during capitulum development in the Asteraceae. Young sunflower capitula that are wounded cylindrically to produce isolated plugs of capitulum receptacle result in the formation of complete, miniature capitula surrounded by the "parent" capitulum (Hernandez and Palmer, 1998; Palmer and Marc, 1982). As such, the radially organized Asteraceae capitulum can be reset into further radially patterned zones upon disruption of cell–cell communication. Genetic studies in the genus *Microseris* also point to radial cell–cell communication. The hairy and yellow achene (fruit) characters in *Microseris* show radial phenotypes in *Microseris* capitula, although their regulation appears to be independent (Mauthe *et al.*, 1984). Hairy achenes (controlled by at least two genes) are always found peripherally to smooth achenes, and there is a region of overlap ("half-hairy" achenes) in which individual cells appear to respond specifically to a defining gradient. The yellow achenes trait also shows concentric zonation, the extent of which differs among *Microseris* strains segregating for two yellow-achene alleles. Bachmann and coworkers (Bachmann, 1983; Mauthe *et al.*, 1984) have hypothesized that the genes responsible for hairy achenes are participants in establishing a radial morphogenetic gradient, and that the alleles of yellow achenes show different response thresholds to this gradient.

Morphogenetic gradients with distinct patterning effects, such as production of ray and trans vs disc flowers in *Gerbera*, can be set up by simple threshold models of protein activation and inhibition, so long as the feedback loops are nonlinear (Meinhardt, 1978). An example of a nonlinear relationship is an inhibited activator protein that must function as a dimer. MADS-domain proteins form specific homo- and heterodimers and even higher-order complexes to conduct their function (Egea-Cortines *et al.*, 1999; Honma and Goto, 2001). Some MADS proteins are known to form obligate, autoregulatory heterodimers (e.g., DEF/GLO, AP3/PI) (Goto and Meyerowitz, 1994; Tröbner *et al.*, 1992). If we assume that the *Gerbera* B-function proteins GDEF2 and GGLO1 form a similar heterodimer in order to be active, then as argued in an earlier section, simple radial pattern-inducing regulatory models for flower development can be derived from this relationship. We provide possible models below, using known or speculated protein–protein interactions.

B. SOME SPECULATIVE MODELS

To avoid a circular argument, threshold models of pattern formation that rely on morphogenetic gradients require a mechanism for the initial establishment of pattern from shallower primary gradients (Meinhardt, 1978). It has been shown that this can be fulfilled by assuming a regulatory system that is based on short-range activation and long-range inhibition (Gierer and Meinhardt, 1972). Although the concept applies equally well to single cells, in multicellular tissues short-range translates easily to cell-autonomous and long range to cell-nonautonomous. Thus, diffusible regulatory molecules are key in pattern formation. A second requisite is nonlinearity of the activation response. In fact, the simplest models involve a bimolecular short range (cell-autonomous) activation and monomolecular long-range (cell-nonautonomous) inhibition (Gierer and Meinhardt, 1972).

Diffusible long-range effectors have in many cases been shown to be classic plant hormones. However, it is remarkable that within the family of MADS domain regulatory proteins, we can in the simplest case build models that rely entirely on these proteins. It has been demonstrated in *Antirrhinum* that DEF and GLO proteins can move through plasmodesmata (Perbal *et al.*, 1996). In a yeast two-hybrid assay, GDEF1 and GDEF2 both heterodimerize with GGLO1 (Anne Uimari and Satu Ruokolainen, unpublished). GDEF2 is a general B-function protein, whereas GDEF1 appears to be more stamen specific. Neither GDEF1 nor GDEF2 form heterodimers with GRCD2, but GRCD2 heterodimerizes with GGLO1 (Uimari *et al.*, 2004). To speculate, GDEF1, GDEF2, and GRCD2, therefore, probably compete for GGLO1 protein within cells. The easiest understood activation–inhibition example could involve only the B-class proteins. In cells that will form petals and stamens, GDEF1 can be envisioned to form the inhibitor to GDEF2/GGLO1, and vice versa. As DEF- and GLO-like proteins can move through plasmodesmata, the relative concentration of GDEF1 translation products in cells might set up a cell-nonautonomous gradient that will dictate the concentration of a GDEF2/GGLO1 autoregulatory heterodimer.

However, cell-autonomous MADS protein gradients might be set up equally well by simply altering their on/off expression, which relies on similar autoregulatory loops that have been demonstrated for DEF/GLO and AP3/PI. These expression patterns themselves could then be under the influence of another nonautonomous factor with its own regulatory feedback loop. For example, in *Arabidopsis*, LEAFY and TFL1 inhibit each other through a feedback loop, and it has been shown that LEAFY can move between cells (Sessions *et al.*, 2000). Moreover, the close TFL1 relative FT is also mobile (Huang *et al.*, 2005), but it is not known if TFL1 shares this trait. LEAFY in turn promotes the expression of B- and C-function MADS-box genes. As such, under our speculative model, differential concentrations of LEAFY across the floral meristem may alter the relative concentrations of GDEF1/GDEF2, GDEF1/GGLO1, and GDEF2/GGLO1 heterodimers in cells.

Following our model, GDEF1, which expresses at higher concentration in older as opposed to young disc flowers (Laitinen *et al.*, unpublished), could inhibit GDEF2/GGLO1 heterodimers in developing stamen cells. Further complexity could enter with differential concentration of the inhibitor GRCD2, which can also heterodimerize with GGLO1. Production of MADS multimers, as are thought to be the active state, could add significantly to the possible responses to a morphogenetic gradient as just described. Going further, regulation of LEAFY/TFL1 activity at the inflorescence level may be responsible for setting up the higher-order gradient effect seen in the Asteraceae capitulum. In summary, the MADS protein system provides a simple numerical scaffold (dimers, multimers) upon which to generate great diversity through differential regulation along a radial axis.

V. CONCLUSIONS

Gerbera has much to offer to plant developmental genetics. It represents a phylogenetic outpost in a different clade of asterid angiosperms than the model systems *Antirrhinum* and *Petunia*. Despite the agricultural importance of sunflower, *Gerbera* has eclipsed this species and another important crop, *Chrysanthemum*, as the model of choice for study of Asteraceae reproductive development. In turn, *Gerbera* presents plant molecular developmental research with a complex yet high-impact model system—in no other well-developed model species can an investigator simultaneously study the development of heteromorphic flowers borne in a compressed phyllotactic spiral that is amenable to fine-scale gene expression research. It may turn out that most if not all flowers in eudicots, including *Arabidopsis*, are cryptically heteromorphic, but their placement along extended rather than highly condensed inflorescences makes this possibility difficult to detect. Similarly, the indeterminate inflorescences of some common model systems have made studying phenotypes reflecting the regulatory aspects of their determinacy state more difficult than in *Gerbera*. If our speculations are correct, MADS-box genes may similarly control aspects of inflorescence development in the other model plants, but with effects that have been obscured experimentally by relatively mild gene expression patterns and phenotypes. Another possibility is that the babushka doll-like nature of the *Gerbera* capitulum is completely evolutionarily derived from the floral genetic program. Our preferred model is that *Gerbera*, through its unique inflorescence phenotype, displays well the general, fractal nature of plant reproductive development, from the ovule and its integuments to the carpels that bear them, through the flowers that contain carpels, the inflorescences upon which flowers are borne, and to the vegetative axes that interconnect reproductive zones.

REFERENCES

Albert, V. A., Soltis, D. E., Carlson, J. E., Farmerie, W. G., Wall, P. K., Ilut, D. C., Solow, T. M., Mueller, L. A., Landherr, L. L., Hu, Y., Buzgo, M. Kim, S., *et al.* (2005). Floral gene resources from basal angiosperms for comparative genomics research. *BMC Plant Biology* **5**, 5.

Alvarez-Buylla, E. R., Pelaz, S., Liljegren, S. J., Gold, S. E., Burgeff, C., Ditta, G. S., Ribas de Pouplana, L., Martinez-Castilla, L. and Yanofsky, M. F. (2000).

An ancestral MADS-box gene duplication occurred before the divergence of plants and animals. *Proceedings of the National Academy of Sciences of the United States of America* **97**, 5328–5333.

Angenent, G. C. and Colombo, L. (1996). Molecular control of ovule development. *Trends in Plant Sciences* **1**, 228–232.

Angiosperm Phylogeny Group II (Chase, M., Bremer, K., Bremer, B., Soltis, P. and Soltis, D.) (2003). An updated classification of the angiosperms. *Botanical Journal of the Linnaean Society* **141**, 399–436.

Arabidopsis Genome Initiative (2000). Analysis of the genome sequence of the flowering plant *Arabidopsis thaliana*. *Nature* **408**, 796–815.

Bachmann, K. (1983). Evolutionary genetics and the genetic-control of morphogenesis in flowering plants. *Evolutionary Biology* **16**, 157–208.

Bowman, J. L., Smyth, D. R. and Meyerowitz, E. M. (1991a). Genetic interactions among floral homeotic genes of *Arabidopsis*. *Development* **112**, 1–20.

Bowman, J. L., Drews, G. N. and Meyerowitz, E. M. (1991b). Expression of the *Arabidopsis* floral homeotic gene *AGAMOUS* is restricted to specific cell types late in flower development. *Plant Cell* **3**, 749–758.

Bradley, D., Carpenter, R., Sommer, H., Hartley, N. and Coen, E. (1993). Complementary floral homeotic phenotypes result from opposite orientations of a transposon at the *plena* locus of *Antirrhinum*. *Cell* **72**, 85–95.

Bremer, K. (1994). "Asteraceae: Cladistics and Classification." Timber Press, Portland.

Coen, E. S. and Meyerowitz, E. M. (1991). The war of the whorls: Genetic interactions controlling flower development. *Nature* **353**, 31–37.

Colombo, L., Franken, J., Koetje, E., van Went, J., Dons, H. J. M., Angenent, G. C. and van Tunen, A. J. (1996). The petunia MADS box gene *FBP11* determines ovule identity. *Plant Cell* **7**, 1859–1868.

Davies, B., Motte, P., Keck, E., Saedler, H., Sommer, H. and Schwarz-Sommer, Z. (1999). *PLENA* and *FARINELLI*: Redundancy and regulatory interactions between two *Antirrhinum* MADS-box factors controlling flower development. *EMBO Journal* **18**, 4023–4034.

de Folter, S., Immink, R. G., Kieffer, M., Parenicova, L., Henz, S. R., Weigel, D., Busscher, M., Kooiker, M., Colombo, L., Kater, M. M., Davies, B. and Angenent, G. C. (2005). Comprehensive interaction map of the *Arabidopsis* MADS Box transcription factors. *Plant Cell* **17**, 1424–1433.

Ditta, G., Pinyopich, A., Robles, P., Pelaz, S. and Yanofsky, M. F. (2004). The *SEP4* gene of *Arabidopsis thaliana* functions in floral organ and meristem identity. *Current Biology* **14**, 1935–1940.

Doyle, J. J. (1994). Evolution of a plant homeotic multigene family: Toward connecting molecular systematics and molecular developmental genetics. *Systematic Biology* **43**, 307–328.

Eckermann, S., Schröder, G., Schmidt, J., Strack, D., Edrada, R. A., Helariutta, Y., Elomaa, P., Kotilainen, M., Kilpeläinen, I., Proksch, P., Teeri, T. H. and Schröder, J. (1998). New pathway to polyketides in plants. *Nature* **396**, 387–390.

Egea-Cortines, M., Saedler, H. and Sommer, H. (1999). Ternary complex formation between the MADS-box proteins SQUAMOSA, DEFICIENS and GLOBOSA is involved in the control of floral architecture in *Antirrhinum majus*. *EMBO Journal* **18**, 5370–5379.

Elomaa, P. and Teeri, T. H. (2001). Transgenic *Gerbera*. *In* "Biotechnology in Agriculture and Forestry, Vol. 48, Transgenic Crops III" (Y. P. S. Bajaj, ed.), pp. 139–154. Springer-Verlag, Berlin.

Elomaa, P., Honkanen, J., Puska, R., Seppänen, P., Helariutta, Y., Mehto, M., Kotilainen, M., Nevalainen, L. and Teeri, T. H. (1993). *Agrobacterium*-mediated transfer of antisense chalcone synthase cDNA to *Gerbera hybrida* inhibits flower pigmentation. *Bio/Technology* **11**, 508–511.

Elomaa, P., Helariutta, Y., Kotilainen, M. and Teeri, T. H. (1996). Transformation of antisense constructs of the chalcone synthase gene superfamily into *Gerbera hybrida*: Differential effect on the expression of family members. *Molecular Breeding* **2**, 41–50.

Ferrario, S., Immink, R. G., Shchennikova, A., Busscher-Lange, J. and Angenent, G. C. (2003). The MADS box gene *FBP2* is required for *SEPALLATA* function in petunia. *Plant Cell* **15**, 914–925.

Gierer, A. G. and Meinhardt, H. (1972). A theory of biological pattern formation. *Kybernetik* **12**, 30–39.

Goto, K. and Meyerowitz, E. M. (1994). Function and regulation of the *Arabidopsis* floral homeotic gene *PISTILLATA*. *Genes & Development* **8**, 1548–1560.

Hansen, H. V. (1999). A story of the cultivated *Gerbera*. *New Plantsman* **6**, 85–95.

Harris, E. M. (1995). Inflorescence and floral ontogeny in Asteraceae: A synthesis of historical and current concepts. *Botanical Reviews* **61**, 93–278.

Heidmann, I., Efremova, N., Saedler, H. and Schwarz-Sommer, Z. (1998). A protocol for transformation and regeneration of *Antirrhinum majus*. *Plant Journal* **13**, 723–728.

Hernandez, L. F. and Palmer, J. H. (1998). Regeneration of the sunflower capitulum after cylindrical wounding of the receptacle. *American Journal of Botany* **75**, 1253–1261.

Hileman, L. C., Drea, S., Martino, G., Litt, A. and Irish, V. F. (2005). Virus-induced gene silencing is an effective tool for assaying gene function in the basal eudicot species *Papaver somniferum* (opium poppy). *Plant Journal* **44**, 334–341.

Honma, T. and Goto, K. (2001). Complexes of MADS-box proteins are sufficient to convert leaves into floral organs. *Nature* **409**, 525–529.

Huang, T., Bohlenius, H., Eriksson, S., Parcy, F. and Nilsson, O. (2005). The mRNA of the *Arabidopsis* gene *FT* moves from leaf to shoot apex and induces flowering. *Science* **309**, 1694–1696.

Huijser, P., Klein, J., Lönnig, W. E., Meijer, H., Saedler, H. and Sommer, H. (1992). Bracteomania, an inflorescence anomaly, is caused by the loss of function of the MADS-box gene *squamosa* in *Antirrhinum majus*. *EMBO Journal* **11**, 1239–1249.

Hutchinson, J. F., Kaul, V., Maheswaran, G., Moran, J. R., Graham, M. W. and Richards, D. (1992). Genetic improvement of floricultural crops using biotechnology. *Australian Journal of Botany* **40**, 765–787.

Irish, V. F. (2003). The evolution of floral homeotic gene function. *BioEssays* **25**, 637–646.

Jack, T. (2004). Molecular and genetic mechanisms of floral control. *Plant Cell* **16**, S1–S17.

Johansen, B., Pedersen, L. B., Skipper, M. and Frederiksen, S. (2002). MADS-box gene evolution-structure and transcription patterns. *Molecular Phylogenetics and Evolution* **23**, 458–480.

Kim, K.-J. and Jansen, R. K. (1995). *ndhF* sequence evolution and the major clades in the sunflower family. *Proceedings of the National Academy of Sciences of the United States of America* **92**, 10379–10383.

Koes, R., Souer, E., Van Houwelingen, A., Mur, L., Spelt, C., Quattrocchio, F., Wing, J., Oppedijk, B., Ahmed, S., Maes, T., Gerats, T. Hoogeveen, P., *et al.* (1995). Targeted gene inactivation in petunia by PCR-based selection of transposon insertion mutants. *Proceedings of the National Academy of Sciences of the United States of America* **92**, 8149–8153.

Kohler, C., Hennig, L., Spillane, C., Pien, S., Gruissem, W. and Grossniklaus, U. (2003). The Polycomb-group protein *MEDEA* regulates seed development by controlling expression of the MADS-box gene *PHERES1*. *Genes & Development* **17**, 1540–1553.

Kotilainen, M., Helariutta, Y., Mehto, M., Pöllänen, E., Albert, V. A., Elomaa, P. and Teeri, T. H. (1999). GEG participates in the regulation of cell and organ shape during corolla and carpel development in *Gerbera hybrida*. *Plant Cell* **11**, 1093–1104.

Kotilainen, M., Elomaa, P., Uimari, A., Albert, V. A., Yu, D. and Teeri, T. H. (2000). *GRCD1*, an *AGL2* like MADS box gene, participates in the C function during stamen development in *Gerbera hybrida*. *Plant Cell* **12**, 1893–1902.

Kramer, E. M. and Irish, V. F. (2000). Evolution of the petal and stamen developmental programs: Evidence from comparative studies of the lower eudicots and basal angiosperms. *International Journal of Plant Sciences* **161**, S29–S40.

Krizek, B. A. and Meyerowitz, E. M. (1996). The *Arabidopsis* homeotic genes *APETALA3* and *PISTILLATA* are sufficient to provide the B class organ identity function. *Development* **122**, 11–22.

Laitinen, R. A. E., Immanen, J., Auvinen, P., Rudd, S., Alatalo, E., Paulin, L., Ainasoja, M., Kotilainen, M., Koskela, S., Teeri, T. H. and Elomaa, P. (2005). Analysis of the floral transcriptome uncovers new regulators of organ determination and gene families related to flower organ differentiation in *Gerbera hybrida* (Asteraceae). *Genome Research* **15**, 475–486.

Laitinen, R. A. E., Broholm, S., Albert, V. A., Teeri, T., and Elomaa, P. (2006). Patterns of MADS-box gene expression mark flower-type development in *Gerbera hybrida* (Asteraceae). *BMC Plant Biology,* in press.

Mandel, M. A., Gustafson-Brown, C., Savidgem, B. and Yanofsky, M. F. (1992). Molecular characterization of the *Arabidopsis* floral homeotic gene *APETALA1*. *Nature* **360**, 273–277.

Mauthe, S., Bachmann, K., Chambers, K. L. and Price, H. J. (1984). Independent responses of two fruit characters to developmental regulation in *Microseris douglasii* (Asteraceae, Lactuceae). *Experientia* **40**, 1280–1281.

Meinhardt, H. (1978). Space-dependent cell determination under the control of a morphogen gradient. *Journal of Theoretical Biology* **74**, 307–321.

Meyerowitz, E. M. (1997). Plants and the logic of development. *Genetics* **145**, 5–9.

Meyerowitz, E. M. and Somerville, C. R. (1994). "*Arabidopsis*." Cold Spring Harbor Press, Cold Spring Harbor, NY.

Meynet, J. and Sibi, M. (1984). Haploid plants from *in vitro* culture of unfertilized ovules in *Gerbera jamesonii*. *Zeitschrift für Pflanzenzüchtung* **93**, 78–85.

Murashige, T., Serpa, M. and Jones, J. B. (1974). Clonal multiplication of gerbera through tissue culture. *HortScience* **9**, 175–180.

Orlikowska, T. and Nowak, E. (1997). Factors affecting transformation of gerbera. *Acta Horticulturae* **447**, 619–621.

Palmer, J. H. and Marc, J. (1982). Wound-induced initiation of involucral bracts and florets in the developing sunflower inflorescence. *Plant and Cell Physiology* **23**, 1401–1409.

Parenicova, L., de Folter, S., Kieffer, M., Horner, D. S., Favalli, C., Busscher, J., Cook, H. E., Ingram, R. M., Kater, M. M., Davies, B., Angenent, G. C. and Colombo, L. (2003). Molecular and phylogenetic analyses of the complete MADS-box transcription factor family in *Arabidopsis*: New openings to the MADS world. *Plant Cell* **15**, 1538–1551.

Pelaz, S., Ditta, G. S., Baumann, E., Wisman, E. and Yanofsky, M. F. (2000). B and C floral organ identity functions require *SEPALLATA* MADS-box genes. *Nature* **405**, 200–203.

Pelaz, S., Tapia-Lopez, R., Alvarez-Buylla, E. R. and Yanofsky, M. F. (2001). Conversion of leaves into petals in *Arabidopsis*. *Current Biology* **11**, 182–184.

Perbal, M.-C., Haughn, G., Saedler, H. and Schwarz-Sommer, Z. (1996). Non-cell-autonomous function of the *Antirrhinum* floral homeotic proteins DEFICIENS and GLOBOSA is exerted by their polar cell-to-cell trafficking. *Development* **122**, 3433–3441.

Pierik, R. L. M., Steegmans, H. H. M. and Marelis, J. J. (1973). *Gerbera* plantlets from *in vitro* cultivated capitulum explants. *Scientia Horticulturae* **1**, 117–119.

Purugganan, M. D., Rounsly, S. D., Schmidt, R. J. and Yanofsky, M. F. (1995). Molecular evolution of flower development: Diversification of plant MADS-box regulatory gene family. *Genetics* **140**, 345–356.

Schwarz-Sommer, Z., Huijser, P., Nacken, W., Saedler, H. and Sommer, H. (1990). Genetic control of flower development by homeotic genes in *Antirrhinum majus*. *Science* **250**, 931–936.

Schwarz-Sommer, Z., Davies, B. and Hudson, A. (2003). An everlasting pioneer: The story of *Antirrhinum* research. *Nature Reviews Genetics* **4**, 657–666.

Sessions, A., Yanofsky, M. F. and Weigel, D. (2000). Cell-cell signaling and movement by the floral transcription factors LEAFY and APETALA1. *Science* **289**, 779–782.

Sitbon, M. (1981). Production of haploid *Gerbera jamesonii* plants by *in vitro* culture of unfertilized ovules. *Agronomie* **1**, 807–812.

Smith, N. A., Singh, S. P., Wang, M. B., Stoutjesdijk, P. A., Green, A. G. and Waterhouse, P. M. (2000). Total silencing by intron-spliced hairpin RNAs. *Nature* **407**, 319–320.

Soltis, D. E., Soltis, P. S., Albert, V. A., dePamphilis, C. W., Frohlich, M. W., Ma., H., Oppenheimer, D. and Theißen, G. (2002). Missing links: The genetic architecture of flowers and floral diversification. *Trends in Plant Science* **7**, 22–30.

Somerville, C. and Koornneef, M. (2002). A fortunate choice: The history of *Arabidopsis* as a model plant. *Nature Reviews Genetics* **3**, 883–889.

Stevens, P. F. (2001) onwards. Angiosperm Phylogeny Website. Version 6, May 20, 2005, http://www.mobot.org/MOBOT/research/APweb/

Theißen, G. (2001). Development of floral organ identity: Stories from the MADS house. *Current Opinion in Plant Biology* **4**, 75–85.

Theißen, G. and Saedler, H. (2001). Floral quartets. *Nature* **409**, 469–471.

Tröbner, W., Ramirez, L., Motte, P., Hue, I., Huijser, P., Lönnig, W. E., Saedler, H., Sommer, H. and Schwarz-Sommer, Z. (1992). *GLOBOSA*: A homeotic gene which interacts with *DEFICIENS* in the control of *Antirrhinum* floral organogenesis. *EMBO Journal* **11**, 4693–4704.

Uimari, A., Kotilainen, M., Elomaa, P., Yu, D., Albert, V. A. and Teeri, T. H. (2004). Integration of reproductive meristem fates by a *SEPALLATA*-like MADS box gene. *Proceedings of the National Academy of Sciences of the United States of America* **101**, 15817–15822.

Yanofsky, M. F., Ma, H., Bowman, J. L., Drews, G. N., Feldmann, K. A. and Meyerowitz, E. M. (1990). The protein encoded by the *Arabidopsis* homeotic gene *agamous* resembles transcription factors. *Nature* **346**, 35–39.

Yu, D., Kotilainen, M., Pöllänen, E., Mehto, M., Elomaa, P., Helariutta, Y., Albert, V. A. and Teeri, T. H. (1999). Organ identity genes and modified patterns of flower development in *Gerbera hybrida* (Asteraceae). *Plant Journal* **17**, 51–62.

Gene Duplication and Floral Developmental Genetics of Basal Eudicots

ELENA M. KRAMER* AND ELIZABETH A. ZIMMER[†]

*Department of Organismic and Evolutionary Biology, Harvard University, Cambridge, Massachusetts 02138
[†]Department of Botany and Laboratories of Analytical Biology, National Museum of Natural History, Smithsonian Institution, Suitland, Maryland 20746

Advances in Botanical Research, Vol. 44
Incorporating Advances in Plant Pathology

0065-2296/06 $35.00
DOI: 10.1016/S0065-2296(06)44009-X

ABSTRACT

The diversification of floral morphology is likely to be due to multiple independent changes in many different components of the genetic pathways that control floral development. One of the best-understood participants in this developmental process is the floral organ identity program. Although many fundamental aspects of this program are conserved within angiosperms and even seed plants, significant developmental system drift has occurred, often in association with gene duplication events. In this review, we consider the many gene duplications that have been identified within the basal eudicot grade, a group of lineages which represent a critical stage of angiosperm evolution. These events are of particular interest since they appear to have laid the foundation for the canonical ABC-model that has been described in the core eudicots. Furthermore, the frequent occurrence of gene duplication in floral organ identity gene lineages offers an excellent opportunity to study the effects of processes, such as neo- and subfunctionalization, on a preexisting genetic program.

I. INTRODUCTION

The evolution of the flowering plants over the last ~150–170 My (Bell *et al.*, 2005; Sanderson *et al.*, 2004) has been marked by a series of dramatic radiations. The most successful of these is represented by the eudicots, a group that encompasses 75% of all angiosperm species (Magallon *et al.*, 1999). The rise to dominance by members of this clade is an intriguing subject that can be considered from many possible perspectives, including phylogenetic, morphological, developmental, genetic, and genomic. This chapter seeks to review our understanding of the eudicot radiation by keeping all of these viewpoints in mind but with special emphasis on genetic and genomic phenomena. Recent work has highlighted the frequent occurrence and apparent coincidence of gene duplication events during the early stages of eudicot diversification, particularly in genes that are critical to the control of floral organ identity (also reviewed by Irish and Litt, 2005; Kramer and Hall, 2005). We will, therefore, consider how significant these events may have been for the evolution of floral development and the angiosperms as a whole.

II. EVOLUTION AND MORPHOLOGY OF THE EUDICOTS

The recognition of the eudicot or tricolpate clade (Fig. 1) was initially based on morphological phylogenetic analyses (Donoghue and Doyle, 1989) but has subsequently been confirmed by a series of molecular studies (reviewed by Judd and Olmstead, 2004). The designation "tricolpates" refers to the main synapomorphy of the group, the possession of triaperturate pollen, but

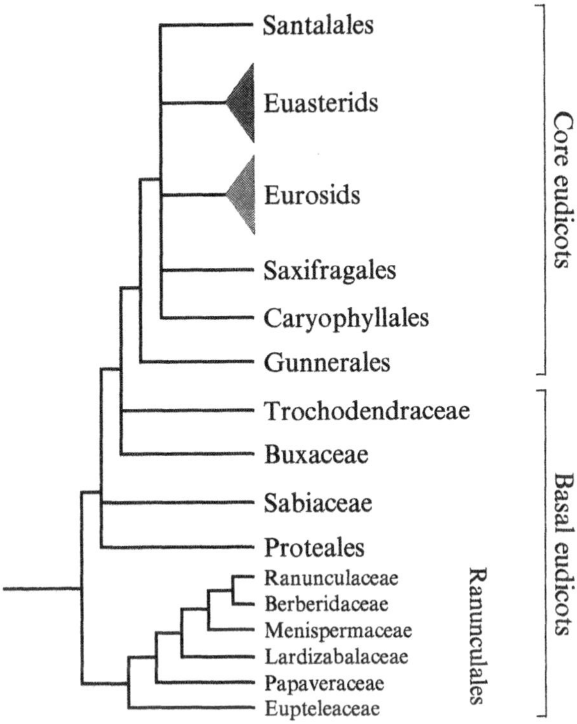

Fig. 1. Simplified eudicot phylogeny based on Soltis *et al.* (2003) and Kim *et al.* (2004a).

other shared traits include cyclic flowers with differentiated staminal filaments and S-type plastids in their sieve elements (Judd and Olmstead, 2004). In part, this somewhat weak set of characteristics reflects the fact that the early diverging lineages of the eudicot clade are quite variable, particularly in floral morphology. The basal grade includes a wide range of diversity in floral size, phyllotaxy, merosity, and organ fusion, as well as vegetative characters such as habit (Judd and Olmstead, 2004; Kim *et al.*, 2004a; Soltis *et al.*, 2003). Perhaps the best example of this phenomenon is the Proteales, a grouping of the families Platanceae, Nelumbonaceae, and Proteaceae that was previously unrecognized until the application of molecular techniques (Chase *et al.*, 1993). The morphology of the group ranges from large trees with highly reduced flowers to aquatic herbs with showy flowers to woody shrubs and trees with complex inflorescences and flowers (Kim *et al.*, 2004a; Magallon *et al.*, 1999). Members of the basal eudicot grade are also marked by their lability of perianth form with multiple instances of perianth loss, changes in merosity, variable degrees of petaloidy, and potential instances of independent petal derivation, most notably in the Ranunculales (Drinnan *et al.*, 1994). When taken on the whole, however, some general ancestral traits can be discerned for the eudicot flower: it was dimerous with whorled phyllotaxy and a differentiated bipartite perianth (Drinnan *et al.*, 1994; Soltis *et al.*, 2003; Zanis *et al.*, 2003). Within the basal eudicots there appear to have been several transitions to spiral phyllotaxy as well as multiple derivations of increased floral organ number. Overall, these lineages are still considered to have a more "open" floral plan that allowed floral morphology to remain flexible at the fundamental levels of phyllotaxy and merosity (Endress, 1990, 2001; see also Endress, Chapter 1; this volume).

This "open" condition underwent significant changes close to the base of the core eudicots. One major synapomorphy of this clade is syncarpy, a congenital fusion of the carpels to form a common transmitting tract (Endress, 1990). This character reflects a general commitment within the core eudicots to a fixed floral bauplan with the frequent occurrence of organ fusion both within and between whorls (Endress, 1987a,b, 1994). By abandoning evolutionary plasticity in basic aspects of floral architecture, such as merosity and phyllotaxy, the core eudicots achieved new flexibility at a "higher" level, what Endress (1990) refers to as synorganization. This increased reliance on organ fusion allowed greater elaboration of organ shape and floral symmetry, which facilitated adaptation to specific pollinators. Another component of this morphological canalization is seen at the node above Gunneraceae where there is a distinct shift to pentamery (Soltis *et al.*, 2003). Therefore, we can see a general trend over the course of eudicot evolution. This starts with the establishment of whorled phyllotaxy and a differentiated bipartite perianth in the basal eudicots and culminates with the evolution of complete floral synorganization in the core eudicots, potentially a major contributor to the dramatic radiation of this clade.

There are many genetic pathways that could have driven the evolutionary trends discussed in an earlier section. Given that organ position and number play a major role, obvious targets of study would be genes that control phyllotaxy and merosity. Unfortunately, these developmental phenomena are not well understood from the molecular standpoint (reviewed by Kramer, 2005). Mutations in *LEAFY* (*LFY*) transform meristem identity from floral to inflorescence, resulting in development of spiral rather than whorled phyllotaxy, as well as disruption of proper floral organ identity (Schultz and Haughn, 1991). The other *Arabidopsis* genes known to contribute to floral meristem identity, *APETALA1* (*AP1*) and *APETALA2* (*AP2*), exhibit whorled phyllotaxy as single mutants (Irish and Sussex, 1990; Kunst *et al.*, 1989) but enhance the development of spiral phyllotaxy when combined with alleles of *lfy* (Huala and Sussex, 1992). This indicates that *AP1* and *AP2* also promote the establishment of whorled floral phyllotaxy, although *LFY* is the major player. Despite numerous forward mutagenesis screens and the identification of *LFY* targets using microarray analyses (Schmid *et al.*, 2003; William *et al.*, 2004), nothing is known as to how floral meristem identity directly controls phyllotaxy. Similarly, the specific functions of other *Arabidopsis* loci affecting merosity, such as *PERIANTHIA* (*PAN*) (Running and Meyerowitz, 1996), *ETTIN* (*ETT*) (Sessions *et al.*, 1997), and *PETAL LOSS* (*PTL*) (Griffith *et al.*, 1999), remain under investigation. Of course, the best-understood genetic component of floral development is the floral organ identity pathway. Although this module functions independently from those controlling phyllotaxy and merosity, organ identity impacts several processes implicated in eudicot evolution, including organ fusion (Alvarez and Smyth, 1999; Mizukami and Ma, 1995; van der Krol and Chua, 1993). There may also be connections to the generally increased stability of organ identity that is seen in the core eudicots, with fewer instances of morphological gradients. Of particular interest is the fact that multiple duplication events have occurred in the participating gene lineages during the diversification of the eudicots, most notably within the basal eudicots close to the base of the core eudicots (reviewed by Kramer and Hall, 2005; Kramer, 2006).

III. GENETICS OF FLORAL ORGAN IDENTITY IN MODEL SPECIES

A. THE ABC-MODEL AND ITS ELABORATION

During the last 15 years significant genetic research using the model systems *Arabidopsis* and *Antirrhinum* has provided a detailed picture of how the identities of floral organs are established. The resultant model, known as the ABC program (Fig. 2A; Coen and Meyerowitz, 1991), outlines the existence of three classes of gene activity, termed A, B, and C, which function in overlapping domains to determine organ identity. Sepal identity is encoded by A-function alone; petal identity, by A + B; stamen identity, by B + C; and carpel identity, by C alone. Mutants in each class exhibit homeotic transformations of organ identity in two adjacent whorls; for instance, B-mutants have petals transformed into sepals and stamens into carpels (Bowman *et al.*, 1989). The genes corresponding to these classes have been well characterized and in *Arabidopsis* are represented by *APETALA1* (*AP1*) and *APETALA2* (*AP2*) in the A-class, *APETALA3* (*AP3*) and *PISTILLATA* (*PI*) in the B-class, and *AGAMOUS* (*AG*) in the C-class (Fig. 2B; Bowman *et al.*, 1989, 1991). All but one of these genes are members of the

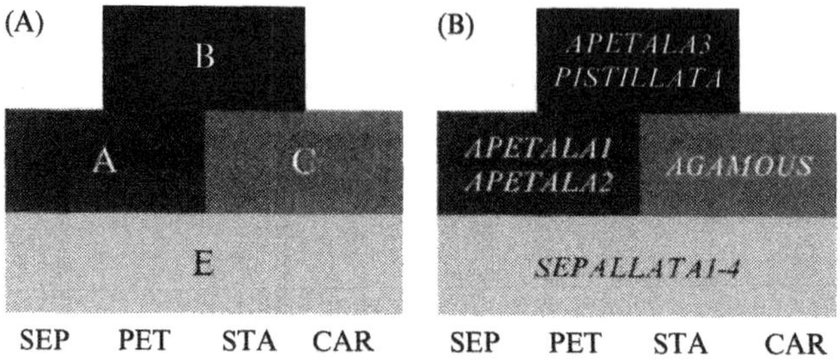

Fig. 2. (A) Schematic showing the basic ABC-model with addition of the E class domain. (B) Schematic showing the major *Arabidopsis* representatives of the A-, B-, C-, and E-class genes.

pan-eukaryotic MADS-box family of transcription factors (reviewed by Theissen *et al.*, 2000), the exception being *AP2*.

Recent work has shed potential light on the biochemical basis for the combinatorial nature of the ABC program. It has long been known that the products of the relevant MADS-box genes formed dimers and would bind DNA only in this state (Riechmann *et al.*, 1996a,b). Now, it has been suggested that these dimers interact with each other to form higher order protein complexes (Egea-Cortines *et al.*, 1999; Honma and Goto, 2001). Another class of MADS-box containing genes, the *SEPALLATA*s, also referred to as the E class (Theissen and Saedler, 2001), serves as critical components in these protein complexes (Honma and Goto, 2001; Pelaz *et al.*, 2000). For instance, in the second whorl, AP3/PI dimers would associate with AP1/SEP dimers to control petal identity. Presumably, differentiation of organ identities would result from the distinct DNA-binding specificities of each complex (Egea-Cortines *et al.*, 1999). It is important to note that although this "quartet" model is very attractive, it is currently supported by limited *in vivo* data (reviewed by Jack, 2004, but see also recent work in *Petunia* Tonaco *et al.*, 2006). The "D" class genes, which function in ovule development (reviewed by Kramer *et al.*, 2004), will not be considered here.

Analyses of homologs of these genes in *Antirrhinum*, as well as many other core eudicot taxa, indicate that their functions are generally conserved across the clade (reviewed by Becker and Theissen, 2003; see Irish, Chapter 3; Rijpkema *et al.*, Chapter 6; and Soltis *et al.*, Chapter 12). The major exception to this statement is the A-function genes, particularly orthologs of the *Arabidopsis* gene *AP1*, a contributor to both the early establishment of floral meristem identity and to perianth organ identity (Irish and Sussex, 1990; Mandel *et al.*, 1992). This dual function is not observed for the *Antirrhinum AP1*-like gene *SQUAMOSA* (*SQUA*), which clearly functions as a determinant of floral meristem identity but appears to be dispensable for organ identity (Huijser *et al.*, 1992). In keeping with this result, early formulations of a floral organ identity model in *Antirrhinum* were only based on two classes of gene function corresponding to the B- and C-classes of the ABC-model (Carpenter and Coen, 1990). Similarly, mutants of the *Pisum AP1* ortholog *PEAM4*, also known as *PROLIFERATING INFLORESCENCE MERISTEM* (*PIM*), are primarily defective in floral meristem identity and can produce almost normal perianth organs at later developmental stages (Taylor *et al.*, 2002). As a result of these findings, it has been suggested that the concept of A function may require revision (Gutierrez-Cortines and Davies, 2000; Kramer and Hall, 2005; Litt and Irish, 2003).

B. THE MIKC FAMILY OF MADS-BOX CONTAINING GENES

As mentioned in an earlier section, almost all of the organ identity genes identified to date are members of the pan-eukaryotic MADS transcription factor family (reviewed by Becker and Theissen, 2003; Messenguy and Dubois, 2003). More specifically, they are type II MADS proteins, which are characterized by a distinct "MIKC" domain structure (Alvarez-Buylla *et al.*, 2000). Our understanding of the biochemical nature of floral organ identity gene function is based on the distinct roles of each of these domains (Fig. 3A). DNA binding at sequence elements known as CArG boxes is controlled by the highly conserved N-terminal MADS domain (Riechmann *et al.*, 1996b). This only occurs, however, when the proteins are dimerized, which is primarily mediated by the adjacent I and K domains (Riechmann *et al.*, 1996a). Different dimerization preferences are observed between proteins and this appears to be important for determining functional specificity (Krizek and Meyerowitz, 1996; Riechmann *et al.*, 1996a). The AP3 and PI gene products are known to function as obligate heterodimers, while those of AP1, AG, and SEP1–4 have broader interaction potentials (Davies *et al.*, 1996; Honma and Goto, 2001; Riechmann *et al.*, 1996a). The current model is that AP1/SEP and AG/SEP are the critical heterodimers participating in higher order complexes, either with themselves or the AP3/PI heterodimer (Honma and Goto, 2001; Theissen and Saedler, 2001). This conclusion is based on several lines of evidence, including the fact that AG requires the presence of SEP in order to interact with an AP3/PI heterodimer (Honma and Goto, 2001) and that *SEP* gene function is required for the establishment of identity in all four floral whorls (Ditta *et al.*, 2004; Pelaz *et al.*, 2000). It is important to note that the delimitation of the K domain is somewhat variable. This region was originally defined based on two main criteria: first, sequence similarity among a group of initially identified MADS-box containing genes and second, predictions of secondary structure in this region that suggested similarity to the amphipathic helices of

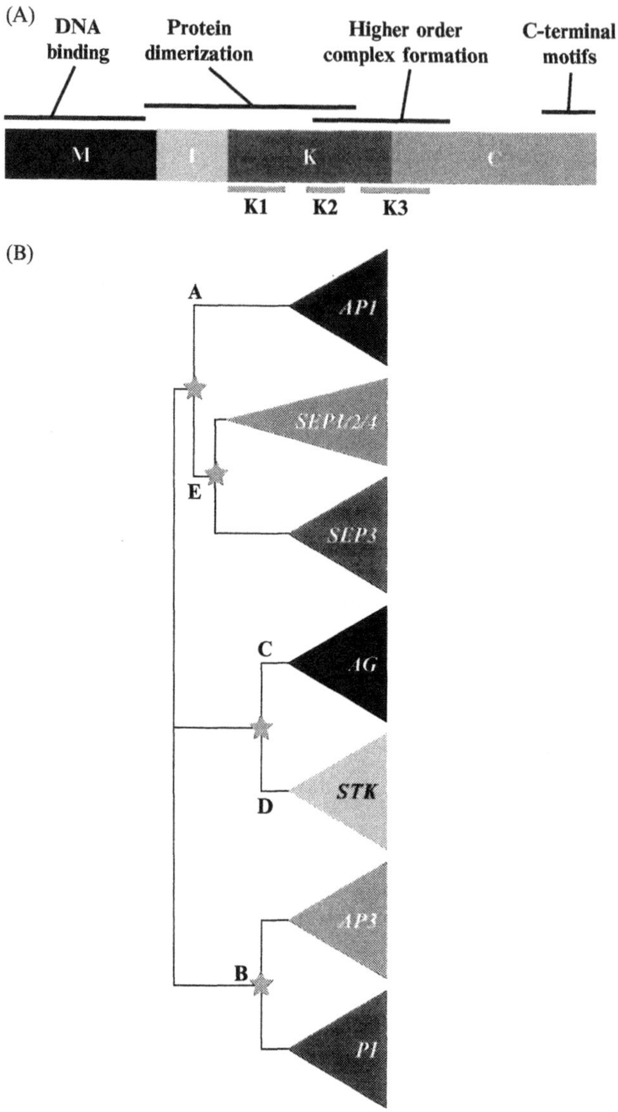

Fig. 3. (A) Schematic of a MIKC-type MADS box gene with the biochemical functions of the various domains indicated. See text for references. (B) Simplified phylogeny of MIKC-type MADS box genes showing only the major clades containing A-, B-, C-, and D-function genes from *Arabidopsis*. The correlations between *Arabidopsis* gene functions and each clade are indicated by the corresponding letters on each major node.

keratin (Ma *et al.*, 1991). Due to the fact that this initial group of "*AG*-like" genes actually represented a fairly diverse sampling of the MIKC family, the circumscription of the K domain was a conservative estimate. Although two α-helices (K1 and K2) are clearly present in the classically defined K domain, in many MIKC subfamilies sequence conservation extends beyond this region to include a third putative K3 helix (Fig. 3; Ma *et al.*, 1991; Trobner *et al.*, 1992; Yang and Jack, 2004). For AP3 and PI, it has been demonstrated that the K1 and K2 helices are primarily responsible for dimerization interactions at sites corresponding to the hydrophobic **a** and **d** positions in the $(\mathbf{abcdefg})_n$ heptad repeats (Yang *et al.*, 2003). Furthermore, the K2 appears to be involved in mediating interactions between PI and SEP, which *in vivo* most likely occur in the context of higher order protein interactions (Yang and Jack, 2004).

The final region, referred to simply as the C-terminal or C domain, exhibits much lower levels of overall sequence conservation (Purugganan *et al.*, 1995) but has previously been shown to be essential for proper *in vivo* gene function (Krizek and Meyerowitz, 1996). As discussed in an earlier section, the putative third α-helix overlaps the K/C domain boundary. Studies of MIKC proteins in *Antirrhinum* had implicated this region as important for protein complex formation (Egea-Cortines *et al.*, 1999), but these results could not be confirmed for *Arabidopsis* proteins (Yang and Jack, 2004). In addition, the C domain contains short, highly conserved motifs that are lineage specific (see later section) and are implicated in transcriptional activation in some cases (Honma and Goto, 2001; Moon *et al.*, 1999) and aspects of functional specificity in others (Lamb and Irish, 2003).

Each of the major floral organ identity genes identified in *Arabidopsis* (*AG, AP3, PI, AP1, SEP*) represents a distinct lineage within the larger MIKC family of MADS-box genes (Fig. 3B; reviewed Becker and Theissen, 2003). All of these were established before the diversification of the angiosperms and have experienced many subsequent duplication events (Aoki *et al.*, 2004; Kramer *et al.*, 2004; Kim *et al.*, 2004b; Litt and Stellari *et al.*, 2004; Zahn *et al.*, 2005a, see Irish, Chapter 3; Rijpkema *et al.*, Chapter 6; and Soltis *et al.*, Chapter 12). Deeper phylogenetic relationships among the lineages remain unresolved but three clades are well supported: one associating the B-gene lineages containing *AP3* and *PI*; a second that includes homologs of *AP1* and *SEP1–4*; and a third that comprises the *AG* homologs responsible for C-function and the *Arabidopsis SEEDSTICK* (*STK*) lineage, which includes genes related to the *Petunia* D-function genes *FBP7/11* (Becker and Theissen, 2003; Purugganan *et al.*, 1995). Identified gymnosperm homologs appear to be ancestral to each of these larger angiosperm clades, indicating that the gymnosperm loci predate the gene duplications that established the separate *AG*, *AP3*, *PI*, *AP1*, and *SEP* lineages. It has been suggested that this apparent series of angiosperm-specific gene duplications was significant to the evolution of flowers (Theissen *et al.*, 2002).

IV. GENE DUPLICATIONS WITHIN THE BASAL EUDICOT GRADE

As mentioned in an earlier section, a number of phylogenetic studies have shown that the *AG, AP3, AP1*, and *SEP* gene lineages all experienced duplication events within the basal eudicot grade (Kramer *et al.*, 1998, 2004; Litt and Irish, 2003; Zahn *et al.*, 2005a). We have already discussed the evolutionary significance of the core eudicots, but it is also important to consider the molecular evolutionary processes that shaped these gene lineages during this critical phase of angiosperm evolution. While many questions remain, our growing knowledge of gene function across the core eudicots has suggested that changes in both biochemical and developmental function correlate with these duplication events.

A. THE *APETALA3* LINEAGE

From the earliest phylogenetic analyses of MIKC-type MADS-box genes, it has been apparent that there were two distinct types of *AP3*-like genes in the core eudicots: one represented by *AP3* itself and the other by the tomato gene *TM6* (Purugganan *et al.*, 1995). Broader sampling of *AP3* homologs revealed that these two lineages, referred to as eu*AP3* and *TM6*, are specific to the core eudicots and are the products of a gene duplication that occurred in the basal eudicots (Kramer *et al.*, 1998). Subsequent studies have demonstrated that this event predates the entire core eudicot clade, including Gunneraceae (Kim *et al.*, 2004b), but postdates the divergence of the Proteales and Sabiaceae (Kramer *et al.*, 2003, 2006; Stellari *et al.*, 2004). *AP3* homologs from the Buxaceae and Trochodendraceae cannot clearly be resolved as pre- or postduplication (Kramer and Hu, unpublished data, Kramer *et al.*, 1998). Regardless of its exact timing, however, the formation of the eu*AP3* and *TM6* lineages is very significant from the standpoint of protein evolution. As it turns out, the ancestral *AP3* lineage, known as paleo*AP3* (Fig. 4A), is much more similar in sequence to the *TM6* lineage, suggesting that eu*AP3* represents a divergent paralogous lineage. This fact is most easily discerned by examining the diagnostic C-terminal motifs present in all predicted AP3-like proteins. The first, termed the PI-Motif Derived region, reflects the close relationship between *AP3* and *PI*. It shows relatively high conservation in paleo*AP3* and *TM6* representatives but decreased

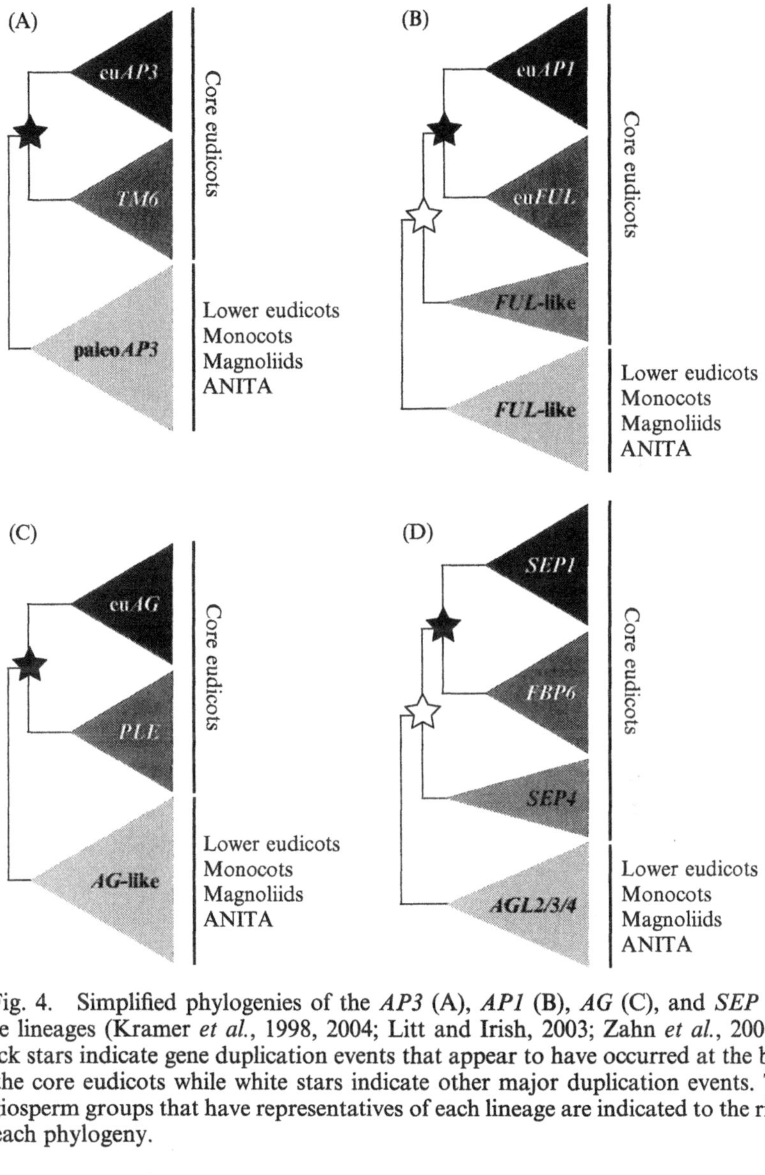

Fig. 4. Simplified phylogenies of the *AP3* (A), *AP1* (B), *AG* (C), and *SEP* (D) gene lineages (Kramer *et al.*, 1998, 2004; Litt and Irish, 2003; Zahn *et al.*, 2005a). Black stars indicate gene duplication events that appear to have occurred at the base of the core eudicots while white stars indicate other major duplication events. The angiosperm groups that have representatives of each lineage are indicated to the right of each phylogeny.

conservation in eu*AP3* members (Kramer *et al.*, 1998; Kim *et al.*, 2004b). More dramatically, the second C-terminal motif has two entirely different forms. The ancestral paleoAP3 motif is found throughout the paleo*AP3* and *TM6* lineages and has the consensus sequence YGxHDLRLA (where x represents a variable site) (Kramer *et al.*, 1998). This motif is absent from eu*AP3* representatives and is replaced by the euAP3 motif consensus,

SDLTTFALLE. It has recently been shown that this dramatic sequence remodeling was not the result of traditional diversifying selection but was due to a frameshift mutation that occurred at the 3′ end of the eu*AP3* coding region following the gene duplication (Vandenbussche *et al.*, 2003a).

In order to thoroughly consider the significance of the eu*AP3*/*TM6* duplication, it is constructive to briefly review what we know about the way *AP3* and *PI* work in *Arabidopsis*. The two gene products form obligate heterodimers that function in the establishment of petal and stamen identity (Bowman *et al.*, 1989; Riechmann *et al.*, 1996a,b). Although AP3 or PI homodimers are sometimes detected in yeast two-hybrid experiments (Sundstrom and Engstrom, 2002) or other contexts (Lamb and Irish, 2003), DNA binding has only been recovered for heterodimers (Hill *et al.*, 1998; Riechmann *et al.*, 1996b). This interaction is also required to promote the stability and nuclear localization of the proteins (Jack *et al.*, 1994; McGonigle *et al.*, 1996). Their expression is tightly integrated by cross- and autoregulation, which comes into play after the initiation phase of transcription (Goto and Meyerowitz, 1994; Jack *et al.*, 1992). This means that despite the fact that early *AP3* and *PI* expression patterns are not identical, at later stages their domains resolve to only the developing petals and stamens. Transcription in the stamens becomes progressively restricted to specific tissue types but in the petals the genes remain constitutively active until late in development. This broad and persistent expression in the petals is consistent with a continued requirement for AP3/PI function throughout the petal in order to maintain identity (Bowman *et al.*, 1989). While these functional details are best understood in *Arabidopsis*, they appear to be conserved in all of the core eudicots where they have been examined (reviewed Irish and Kramer, 1998; Zahn *et al.*, 2005b). It is important to note, however, that *Arabidopsis* has lost its *TM6* ortholog (Initiative, 2000) and no *TM6* ortholog has been identified in *Antirrhinum*, facts that complicate many of the experiments aimed at understanding the evolution of B-gene function.

A number of lines of evidence suggest that paleo*AP3* homologs and *PI*-like genes from outside the core eudicots may not function in the same way as the core eudicot loci, particularly in terms of petal identity. Expression studies in a wide array of taxa have shown that paleo*AP3* and *PI* are often expressed independently of one another, even at late developmental stages, demonstrating that their cross/autoregulation is not universal (Kim *et al.*, 2005; Kramer and Irish, 1999, 2000). Furthermore, although paleo*AP3*/*PI* expression in petaloid organs is commonly observed, the instances of spatial restriction and temporal variability differ distinctly from what is observed in core eudicots. This being said, stamen expression is very similar to core eudicot patterns and there is little doubt that the genes' role in stamen identity is conserved. In terms of protein interaction potentials, the ability to heterodimerize is detected in all taxa but in several monocots it appears that the PI and possibly the paleoAP3 homologs can also form stable homodimers (Kanno *et al.*, 2003; Tzeng *et al.*, 2004; Winter *et al.*, 2002). Given that obligate heterodimerization has been reported in more derived monocots (Whipple *et al.*, 2004), it remains unclear as to what the ancestral dimerization condition of the paleo*AP3*/*PI* lineages was or how flexible this trait may be across the angiosperms (Kim *et al.*, 2004b). Functional studies of noncore eudicot *AP3*/*PI* homologs have been limited to the grass model species *Oryza* and *Zea*. This work has shown that the grass paleo*AP3* and *PI* homologs are important for stamen identity as well as the development of the lodicules, grass-specific organs that seem to represent modified petals (Ambrose *et al.*, 2000; Kang *et al.*, 1998; Nagasawa *et al.*, 2003).

Overall, these findings indicate that many aspects of paleo*AP3*/*PI* function and regulation differ from what has been observed for eu*AP3*/*PI* in the core eudicots, although their role in stamen identity is clearly conserved. It is still difficult to determine whether we should favor a model where *AP3*/*PI* function in petal identity is also deeply conserved but has experienced developmental system drift or, rather, should conclude that *AP3*/*PI* homologs have been independently recruited many times to promote petal development (Kramer, 2005; Kramer and Irish, 2000; Kramer and Jaramillo, 2005 for further discussion; Kim *et al.*, 2005). Regardless, it does appear that many aspects of *AP3*/*PI* function and regulation underwent a transition in association with the eu*AP3*/*TM6* duplication event, which raises several questions. Primary among these is: What is the significance of the establishment of the euAP3 motif? Lamb and Irish addressed this issue in a study that tested the biochemical equivalence of the euAP3 and paleoAP3 motifs (Lamb and Irish, 2003). This was accomplished by swapping the euAP3 motif of *Arabidopsis* AP3 with the paleoAP3 motif of *Dicentra eximia* AP3 (*Dicentra* is in the Ranunculales). Transgenic analyses of this chimeric protein under the control of the endogenous *AP3* promoter demonstrated that it could not fully rescue *ap3* mutant phenotypes. Specifically, no complementation was observed for petal identity while weak rescue was recovered for the stamens. These results were interpreted to indicate that paleo*AP3* lineage members do not normally function in petal identity but may contribute to stamen identity. In this context, it is important to consider that *Arabidopsis* does not normally express any *AP3*-like protein with a paleoAP3 motif, so it is somewhat unclear as to whether this genomic architecture is a faithful representation of ancestral paleo*AP3* gene function. Other transgenic studies using entire heterologous paleo*AP3* genes have yielded inconsistent results. Overexpression of the *AP3* homolog from

Lilium causes a wide range of developmental defects, including some that could be interpreted as a dominant negative effect (Tzeng and Yang, 2001). In contrast, the *Zea* paleo*AP3* gene *SILKY1* (*Si1*) can rescue both petal and stamen development in *Arabidopsis ap3* mutants (Whipple *et al.*, 2004), but these lines carried high transgene copy numbers and expressed *Si1* at a level five times that of the endogenous *AP3*. Such high expression levels may have significantly skewed the dimerization kinetics of Si1 and PI, as well as the interactions of Si1/PI heterodimers with other proteins. On the whole, then, it appears that the most rigorous tests show that the paleoAP3 and euAP3 motifs are not biochemically equivalent, but it remains unclear as to whether paleo*AP3* genes function in a significantly different manner (biochemically or developmentally) from eu*AP3* lineage members.

Another pressing question is: What are the functions of the *TM6* lineage members? This issue is key to understanding core eudicot floral genetics as well as to providing insight into the possible ancestral functions of paleo*AP3*. As mentioned in an earlier section, the *TM6* ortholog has been lost from the *Arabidopsis* genome, but representatives of the *TM6* lineage are found in many other rosids as well as all of the other major core eudicot lineages (Kramer *et al.*, 2006; Kim *et al.*, 2004b). Luckily, both eu*AP3* and *TM6* orthologs have been characterized from the model species *Petunia hybrida*; these are termed *PhDEF* and *PhTM6*, respectively (Kramer *et al.*, 1998; Vandenbussche *et al.*, 2004; van der Krol *et al.*, 1993). Genetic studies indicate that while *PhDEF* contributes to petal and stamen identity, similar to other eu*AP3* orthologs, the role of *PhTM6* may be limited to stamen identity (Vandenbussche *et al.*, 2004). This intriguing finding seems to agree with the suggestion of Lamb and Irish that paleoAP3 containing loci are primarily stamen identity genes (Lamb and Irish, 2003). Further studies of *TM6* expression and function will be necessary to determine whether the role of *PhTM6* in *Petunia* is in fact representative of general *TM6* function and how it relates to ancestral paleo*AP3* function. Obviously, there are many critical questions yet to be answered in regards to the eu*AP3*/*TM6* duplication that will require more detailed analyses of both basal and core eudicots.

B. THE *APETALA1* LINEAGE

In a pattern strikingly similar to that observed for *AP3*, the *AP1* lineage experienced significant sequence changes following a gene duplication that occurred within the basal eudicot grade (Litt and Irish, 2003). This event appears to have postdated the divergence of the Buxaceae but predated the diversification of the main core eudicot clade (only Gunneraceae and Santalales remain unsampled from the core eudicots). The resultant paralogous lineages are termed eu*AP1* and eu*FUL* in reference to their *Arabidopsis* members, *AP1* and *FRUITFULL* (*FUL*), respectively (Fig. 4B). Mirroring the evolution of eu*AP3*/*TM6*, eu*AP1* members are divergent in sequence relative to the eu*FUL* and ancestral *FUL*-like lineages. Perhaps most surprisingly, the C-terminal *FUL*-like motif is conserved in eu*FUL* but was remodeled by a frameshift mutation in eu*AP1* to yield a conserved farnesylation motif (Litt and Irish, 2003; Vandenbussche *et al.*, 2003a). This so-called CaaX box can be targeted by farnesyltransferase both *in vitro* and *in vivo* but the functional significance of this posttranslation modification remains unclear (Yalovsky *et al.*, 2000). In addition, eu*AP1* members acquired a second motif immediately N-terminal to the CaaX box. This diagnostic euAP1 motif is rich in acidic residues and has been shown to act as a strong transcriptional activation domain (Cho *et al.*, 1999). The eu*FUL* and *FUL*-like loci possess glutamine and proline-rich regions that may similarly promote transcription but these sequences are not as highly conserved as the euAP1 motif and have not been functionally analyzed (Litt and Irish, 2003).

Two clear lines of inquiry regarding *AP1* are: (1) to determine what biochemical role the FUL-like/euFUL motif plays and (2) to establish how the function of *FUL*-like/eu*FUL* genes may differ from that of eu*AP1* homologs. Unlike the situation with eu*AP3*/*TM6*, *Arabidopsis* has retained eu*AP1* and eu*FUL* representatives, opening the door to more detailed biochemical studies of both motif types. This fact has already allowed genetic studies on the developmental functions of *Arabidopsis AP1* and *FUL*. As discussed in an earlier section, *AP1* plays an early role in establishing floral meristem identity and, later, promotes sepal and petal identity (Bowman *et al.*, 1993; Irish and Sussex, 1990). *FUL* also contributes to floral meristem identity, but this may not be due to simple redundancy given that *FUL*'s early expression in inflorescence meristems does not overlap with that of *AP1* (Ferrandiz *et al.*, 2000; Mandel and Yanofsky, 1995). Once floral development is underway, *FUL* acts to control aspects of carpel and fruit development (Gu *et al.*, 1998). Although nothing is known about the specific function of the euFUL motif, protein dimerization studies indicate that the interaction repertoire of FUL may be broader than that of AP1 (de Folter *et al.*, 2005). Additional investigation will be required in order to determine whether AP1 and FUL are biochemically equivalent *in vivo* and, if not, whether functional differences map to their distinct C-terminal motifs.

In terms of the developmental functions of these genes, we are faced with the question of how the ancestral *FUL*-like function relates to those of the paralogous eu*AP1* and eu*FUL*. Although functional data for members of the *AP1/FUL* lineage are very limited, comparative studies of gene expression are beginning to provide some information regarding the evolution of the lineage (reviewed Kim *et al.*, 2005). Across the core eudicots, all eu*AP1* orthologs exhibit very similar expression patterns, with transcript initially present throughout the early floral meristem and later expression restricted to the perianth whorls (Berbel *et al.*, 2001; Dong *et al.*, 2005; Huijser *et al.*, 1992; Taylor *et al.*, 2002). Although this is consistent with the *Arabidopsis AP1* expression pattern and canonical function, there are some caveats to consider. As discussed in an earlier section, clear evidence for sepal and petal identity function is lacking in other core eudicot models, and even to some degree in *Arabidopsis* (see Kramer and Hall, 2005 for discussion). It remains to be determined, therefore, how the conserved expression patterns of eu*AP1* orthologs relate to conservation of their function. What has become evident, however, is that the eu*AP1* expression pattern is not observed for the eu*FUL* or *FUL*-like homologs. First, it is important to note that phylogenetic analysis reveals that there is actually a third paralogous *FUL*-like lineage in the core eudicots, possibly due to an earlier duplication at the base of the eudicots (Fig. 4B; Litt and Irish, 2003). Although no ortholog of this lineage is present in *Arabidopsis*, one has been characterized in *Antirrhinum*, the *DEFH28* locus (Muller *et al.*, 2001). This gene is expressed in a manner very similar to *FUL*: in the inflorescence meristem and developing carpel. Muller *et al.* (2001) further demonstrated that the regulatory elements controlling carpel expression of *DEFH28* operate in a conserved manner in *Arabidopsis*. Further analysis of eu*FUL* and core eudicot *FUL*-like loci will be required in order to determine whether the similar expression patterns of the paralogous *FUL* and *DEFH28* loci reflect common inheritance of an ancestral function. Such a relationship would suggest that eu*AP1*'s expression pattern, and potentially its functional repertoire, were derived after the eu*AP1*/eu*FUL* duplication event. Along these lines, studies of *FUL*-like genes in the magnoliid dicots and ANITA grade have found that these loci display relatively variable expression patterns, which do not conform with the perianth-specific expression of eu*AP1* orthologs (Kim *et al.*, 2005).

In summary, the evolution of the *AP1* lineage along the backbone of the eudicots is somewhat more complex than that of *AP3*, having experienced two major duplications, but the second of these events resulted in a very similar pattern of sequence evolution in the divergent eu*AP1* lineage (Litt and Irish, 2003). Luckily, in the case of *AP1* we have some insight into the functions of the new motifs: in *Arabidopsis* the eu*AP1* motif acts as a transcriptional activation domain (Cho *et al.*, 1999) and the frameshift-acquired CaaX box is the target of posttranscriptional modification (Yalovsky *et al.*, 2000). There is also evidence that eu*AP1* differs markedly in expression and, possibly, in its functional repertoire relative to its *FUL*-like ancestors (reviewed by Kim *et al.*, 2005; Litt and Irish, 2003). Specifically, eu*AP1* may have lost a role in fruit development while possibly gaining one in perianth development. This being said, unlike the case with *AP3*, we have limited expression data and a complete absence of functional data for *FUL*-like loci, a situation that severely hampers our understanding of functional evolution within this lineage.

C. THE *AGAMOUS* LINEAGE

Unlike *AP3* and *AP1*, the basal eudicot duplication in the *AG* lineage was not followed by significant changes in protein sequence. The timing of this event has not been determined with certainty but seems to have occurred at some point between the divergence of the Sabiaceae and the major core eudicot radiation (Kramer *et al.*, 2004). Members of the resultant paralogous eu*AG* and *PLENA* (*PLE*) lineages (Fig. 4C) possess very similar C-terminal motifs that are conserved throughout the ancestral *AG*-like lineage. No function has been assigned to these regions, although their high degree of conservation suggests a critical role. The eu*AG*/*PLE* duplication is particularly notable for its potential as a model for studying independent patterns of subfunctionalization (Force *et al.*, 1999) following gene duplication. In *Arabidopsis, AG* (the eu*AG* representative) functions in stamen and carpel identity, floral meristem determinacy and ovule development (Bowman *et al.*, 1989; Mizukami and Ma, 1995; Pinyopich *et al.*, 2003). The *Arabidopsis PLE* lineage members *SHATTERPROOF1* and *2* (*SHP1/2*) contribute to carpel identity but primarily function in fruit and ovule development (Liljegren *et al.*, 2000; Pinyopich *et al.*, 2003). In contrast, the *Antirrhinum* eu*AG* ortholog *FARINELLI* (*FAR*) is specifically involved with stamen identity and development while *PLE* itself functions in a manner more similar to *AG* (Bradley *et al.*, 1993; Causier *et al.*, 2005; Davies *et al.*, 1999). Given that grass *AG*-like genes also function in floral meristem determinacy and reproductive organ identity, it is reasonable to conclude that the paralogs *AG* and *PLE* inherited their similar functional repertoires from their ancestor (reviewed Causier *et al.*, 2005; Kramer *et al.*, 2004). Following the eu*AG*/*PLE* duplication, subfunctionalization between the paralogs progressed independently in the asterids and rosids. Some neofunctionalization may also be at work given that the role of the *SHP* genes in fruit dehiscence is likely to be derived within the Brassicales (Hall *et al.*, 2002). Broader gene expression studies of eu*AG* and *PLE* members across the core eudicots indicate that these separate patterns of functional evolution have been the rule rather than the exception (reviewed Kramer *et al.*, 2004). In particular, there are a number of other examples of recruitment to carpel- or ovule-specific expression for either eu*AG* or *PLE* orthologs.

Thus, we see that the eu*AG*/*PLE* duplication does not conform to the model seen in *AP1* and *AP3*. There is no evidence of major sequence remodeling or dramatic changes in gene function. Possibly due to the critical importance of C-function genes in the development of reproductive organs, the evolutionary trajectory of these paralogs has primarily been one of subfunctionalization with lesser degrees of neofunctionalization (Causier *et al.*, 2005; Kramer *et al.*, 2004). The lineage does raise an important caution for the interpretation of functional evolution in other paralogous core eudicot lineages since it clearly demonstrates that the functions of genetic orthologs are not always conserved.

D. THE *SEPALLATA* SUBFAMILY

Functional and phylogenetic analyses of the *SEP* subfamily are much less advanced than those of the previously discussed lineages, largely due to the fact that their roles in floral development were only recently recognized (Ditta *et al.*, 2004; Pelaz *et al.*, 2000). It is important to note that the *SEP* subfamily is composed of two pan-angiosperm lineages, one represented by the *Arabidopsis* gene *SEP3* and the other, by the genes *SEP1, 2,* and *4* (Zahn *et al.*, 2005a). These were produced by a gene duplication event that occurred along the branch leading to flowering plants after its split from the gymnosperms. The *SEP3* lineage (also known as *AGL9*) appears to be relatively simple with no evidence of duplications within the basal eudicot grade. In contrast, Zahn *et al.* have shown that the *SEP1*/*2*/*4* genes form three distinct paralogous core eudicot lineages, much like *AP1* (Fig. 4D; Zahn *et al.*, 2005a). *Arabidopsis* possesses representatives of only two of these lineages, *SEP1* and *2*, which are very recent paralogs, and *SEP4*. Members of all three lineages have been identified in other rosids, such as *Malus*, and asterids including *Petunia*. The relationships of these clades to one another is not entirely clear but the presence of a basal eudicot sequence in the *SEP4* clade would suggest a preeudicot duplication event for that lineage. The *SEP1*/*2* and *FBP6* lineages only contain core eudicot sequences, perhaps indicating that their duplication occurred within the basal eudicot grade, similar to the eu*AP3*/*TM6*, eu*AP1*/*FUL*, and eu*AG*/*PLE* events. Regardless of the timing of their origin, all of these lineages retain similar C-terminal motifs that are, in turn, very much like the *FUL*-like and eu*FUL* motifs (Litt and Irish, 2003; Zahn *et al.*, 2005a). No detailed functional analysis has been conducted on these motifs, although the C-terminal domains of SEP3 and other SEP homologs have been shown to promote transcriptional activation (Davies *et al.*, 1996; Ferrario *et al.*, 2003; Honma and Goto, 2001).

In terms of functional evolution, it appears that members of the *SEP* subfamily are largely redundant with some aspects of distinct function. Their biochemical role is thought to be facilitating the function of other floral MIKC MADS-box proteins, particularly AG, SHP, STK, AP3, PI, and AP1, but possibly others as well (Castillejo *et al.*, 2005; Ditta *et al.*, 2004; Favaro *et al.*, 2003; Honma and Goto, 2001; Pelaz *et al.*, 2000). The current model is that the SEP proteins contribute transcriptional activation capacity to the larger quartet complex. *SEP1/2/3* are all expressed in the three inner whorls of the floral meristem while the *SEP4* expression domain expands to include the first whorl (Ditta *et al.*, 2004; Flanagan and Ma, 1994; Mandel and Yanofsky, 1998; Savidge *et al.*, 1995). Loss of function in any one locus has no significant effect on development but triple *sep1 sep2 sep3* mutants exhibit transformation of all internal floral organs into sepals with loss of floral meristem determinacy, similar to double B and C class mutants (Pelaz *et al.*, 2000). The flowers of quadruple *sep1–4* mutants have a complete loss of floral organ identity and produce only leaf-like organs (Ditta *et al.*, 2004). These findings demonstrate that the *SEP* loci contribute to the identity of all floral organs in a redundant and additive fashion. There is some evidence for functional distinctions insofar as *sep3* has a greater impact on the phenotype of *sep1 sep2* than does *sep4*. Various studies also suggest that *SEP3* and *SEP4* may contribute to floral meristem identity and the subsequent activation of both *AP3* and *AG* (Castillejo *et al.*, 2005; Ditta *et al.*, 2004; Pelaz *et al.*, 2001). Genetic analysis of *SEP* homologs in *Petunia* indicate that they are somewhat less redundant but function in a similar manner, aiding B-, C-, and D-gene function (Ferrario *et al.*, 2003; Vandenbussche *et al.*, 2003b). The *SEP*-like genes in *Antirrhinum* have not been analyzed functionally but can participate in similar protein complexes as those from other core eudicots (Davies *et al.*, 1996). Interestingly, in the composite *Gerbera* the *SEP*-like genes have acquired the additional function of patterning the inflorescence and controlling its determinacy (Uimari *et al.*, 2004). Also, two recent *Gerbera* paralogs of the *AGL2* lineage, *GRCD1* and GRCD2, appear to have become subfunctionalized such that they respectively contribute to stamen and carpel identity (Kotilainen *et al.*, 2000; Uimari *et al.*, 2004). Unfortunately, clear functional data for noneudicot *SEP* homologs has not been obtained (reviewed Malcomber and Kellogg, 2004, 2005). In the grass model species this is most likely because of extensive genetic redundancy due to independent rounds of gene duplications. Regardless, the high degree of functional similarity between the deeply diverging *SEP3* and *SEP1/2/4* clades argues for a broad conservation of function, possibly due to the highly pleiotropic nature of their functional repertoires. As with *AG*, the primary evolutionary trend in the *SEP* lineages appears to be variable degrees of redundancy combined with subfunctionalization. In some cases, such as *Gerbera*, the presence of multiple *SEP* gene copies may have provided raw material for subsequent neofunctionalization events.

E. BROADER IMPLICATIONS

There are a number of recurrent questions regarding the MADS-box gene duplications that occurred within the basal eudicot grade. The first is what are the functions of the C-terminal motifs? The eu*AP1* lineage is the only case where any specific information is available and, unfortunately, the other motifs do not have clear similarities to domains of known biochemical function. Several deletion studies have demonstrated that the C-terminal domain is essential to *in vivo* function (Krizek and Meyerowitz, 1996; Lamb and Irish, 2003) but the region has repeatedly been shown to be dispensable for protein dimerization (Davies *et al.*, 1996; Hill *et al.*, 1998; Riechmann *et al.*, 1996a,b; Yang *et al.*, 2003). Furthermore, removal of only the distal portion of the C domain, including the lineage specific motifs, has no effect on higher order complex formation in yeast (Egea-Cortines *et al.*, 1999; Honma and Goto, 2001). It remains possible that yeast systems are not sensitive enough to detect reductions in protein affinity that have significance *in vivo*. So the functions of the C-terminal motifs largely remain a mystery, making it that much more difficult to understand the evolutionary significance of changes in their sequence.

Another obvious issue requiring attention is whether the eu*AP3/TM6*, eu*AP1/FUL*, eu*AG/PLE*, and *SEP1/FBP6* duplications were truly coincident, quite possibly due to a genome duplication event (Irish, 2000; Zahn *et al.*, 2005a). Other lineages of MIKC-type MADS-box genes show similar patterns of evolution (Kramer *et al.*, unpublished data; Becker and Theissen, 2003) and several analyses of genomic evolution in the angiosperms point to a large-scale duplication event predating the core eudicots (Bowers *et al.*, 2003; Vision *et al.*, 2000). It is attractive to speculate that such an event may have contributed to the radiation of the core eudicots. Based on our sampling and the resolution of the *AP3, AP1/FUL, AG*, and *SEP* phylogenies, however, the exact timing of the various duplication events cannot be determined with certainty (Kramer *et al.*, 1998, 2004; Litt and Irish, 2003; Zahn *et al.*, 2005a). Even for the *AP3* lineage, which has been heavily sampled across the basal and core eudicots, the duplication cannot be fixed relative to the Buxales and Trochodendrales (Kramer *et al.*, 2006). In the case of the *AP1/FUL* and *SEP* lineages, the picture is further complicated by two rounds of gene duplication rather than just one (Litt and Irish, 2003; Zahn *et al.*, 2005a). Regardless, these duplication events cannot really be considered in isolation since the gene products are thought to function together in a protein complex, opening the possibility of coevolutionary processes. In particular, it has been suggested that there may be a connection between the acquisition of perianth function in eu*AP1* and the recruitment of eu*AP3*/*PI* to promote petal identity (Vandenbussche *et al.*, 2003). This hypothesis must be considered, but it is also important to note that evidence supporting a conserved function of eu*AP1* orthologs in petal identity is lacking. What can be concluded with certainty is that multiple participants in the floral organ identity program experienced gene duplication within the core eudicot grade and, in the case of eu*AP3* and eu*AP1*, this was followed by functionally significant changes in protein sequence. How these events are related to canalization of the ABCE program or the radiation of the core eudicots remains unclear.

V. GENE DUPLICATIONS IN THE EARLY BRANCHES OF THE EUDICOT RADIATION: THE RANUNCULALES

The Ranunculid clade at the base of the eudicots is of particular interest for a number of reasons. For one, they are easily the most species-rich group of the basal eudicots, with much of that diversity found in the family Ranunculaceae (Magallon *et al.*, 1999). Additionally, very specific hypotheses exist regarding patterns of petal evolution across the group (reviewed Drinnan *et al.*, 1994; Kramer *et al.*, 2003). In particular, the Ranunculaceae are frequently cited as one of the best cases for independently derived petaloid organs from stamens (Endress, 1995; Erbar *et al.*, 1998; Kosuge, 1994; Kosuge and Tamura, 1989). Given the MADS-box gene duplications discussed in an earlier section, the Ranunculales would seem like natural candidates for understanding the ancestral functions of the ABCE-genes. Unfortunately, the situation has not turned out to be straightforward as studies of the *AP3, PI, AG*, and *AP1* lineages have revealed multiple Ranunculid-specific duplication events (Kramer *et al.*, unpublished data; Kramer *et al.*, 2003, 2004; Litt and Irish, 2003). The best-understood gene lineages in this group are those of *AP3* and *PI*, which exhibit very different evolutionary patterns. In the case of *AP3*, two duplications gave rise to three lineages of paleoAP3 containing genes (Kramer *et al.*, 2003). The timing of these events can clearly be placed before the last common ancestor of Ranunculaceae and Berberidaceae but most likely occurred much earlier.

Representatives of all three lineages have been recovered across the Ranunculaceae and some evidence suggests that they have undergone both sub- and neofunctionalization (Kramer *et al.*, 2003; unpublished data). It remains to be determined how these specialized functions relate to the diverse perianth types found in the family or whether they are conserved within each paralogous lineage. In contrast to the pattern of early duplications, there is no concrete evidence for ancient duplications in *PI*. Rather, this lineage has retained paralogs from many independent duplications (over 10) that occurred during the diversification of the Ranunculaceae. These events do not appear to be due to genome-wide duplications because equivalent paralogs are not observed in any of the other MIKC-type MADS-box genes examined. Since there is no reason to believe that the different patterns of duplication in *AP3* and *PI* reflect changing rates of gene duplication, the conclusion would be that there have been different rates of paralog retention between the two lineages. Essentially, the retention of *PI* paralogs has increased in frequency within the Ranunculaceae as compared to other loci. One possible explanation for this trend is the heterodimerization of AP3 and PI. If the ancient *AP3* paralogs experienced sub- or neofunctionalization, it might become advantageous to retain *PI* paralogs that could specialize to function with specific *AP3* copies. This phenomenon has actually been observed in *Petunia* where the recent *PI* paralogs *PhGLO1* and *PhGLO2* each display biochemical and genetic preferences for the ancient *PhDEF* and *PhTM6* paralogs (Vandenbussche *et al.*, 2004).

For the *AG*-like, *FUL*-like, and *SEP* lineages, sampling has not been as thorough. Current evidence suggests that there was one duplication event in *AG* along the branch following the divergence of the Papaveraceae but before the radiation of the other families (Kramer *et al.*, 2004). Expression studies of the two *AG* paralogs in *Thalictrum* show that while one copy retains a typical C-gene expression pattern, the second has become ovule specific (Di Stilio *et al.*, 2004). Preliminary evidence from other Ranunculaceae indicates that this subfunctionalization may be a late event specific to the Isopyreae (Kramer *et al.*, unpublished data). *FUL*-like genes are present as two paralogous lineages in the Ranunculales, most likely due to a duplication predating the diversification of the order (Litt and Irish, 2003), but no Ranunculid-specific *SEP* duplications have been detected to date (Kramer *et al.*, unpublished data; Zahn *et al.*, 2005a). While these diverse *AP3, PI, AG,* and *AP1* duplications create difficulties for understanding the ancestral eudicot floral organ identity program, they also provide a distinct set of duplication events in which to study functional evolution. Moreover, the existence of genetic and genomic tools for several Ranunculid taxa offers the opportunity to directly analyze gene function, a critical need for the advancement of our understanding of eudicot floral evolution (Becker *et al.*, 2005; Hileman *et al.*, 2005; Kramer *et al.*, unpublished data; Park and Facchini, 2000).

VI. SUMMARY

Over the last 10 years, MIKC-type MADS box genes have arguably become one of the best-understood large eukaryotic gene families from an evolutionary perspective. Particularly within the angiosperms we have an unparalleled amount of information on the diversification of multiple gene lineages. Just for the *AP3* lineage, representatives have been identified from more than 100 diverse taxa. Few comparable datasets exist in animals that span such a relatively short phylogenetic timescale (exceptions include the globin gene family, for example, Prychitko *et al.*, 2005). Analyses of these genes have provided considerable insight into the evolutionary dynamics of gene lineages, revealing a surprising degree of lability. The constant process of gene birth-and-death coupled with shifting functional repertoires creates the impression of a highly mutable genetic program. Despite this, we know that the end points of these pathways are fairly constant—the establishment of predictable organ identities, several of which are thought to be deeply conserved. To a large degree, then, we are seeing evidence of widespread developmental system drift acting in a commonly inherited genetic module. One additional point to remember is that although the general floral organ identity programs are conserved, the elaboration of identity appears to be highly variable, suggesting further diversification of the downstream targets of the ABCE-class genes. Significant questions remain, especially regarding the evolution of *AP1* and *AP3*/*PI* function in the perianth. This case may represent a true instance of neofunctionalization, although it appears that any novel organ identity type was seamlessly integrated into the preexisting genetic program. Only through expanded functional genetic studies of critical taxa can we hope to determine how the biochemical and developmental functions of eu*AP3* and eu*AP1* compare to those of their ancestors. Establishing the biochemical function of the C-terminal motifs will play a critical part in addressing this issue and has the potential to greatly advance our understanding of MIKC MADS box-gene function in general. While it may remain unclear whether the clustered duplications at the base of the core eudicots were critical to the radiation of this important group, these lines of research will undoubtedly provide much more information regarding the process of functional evolution following gene duplication.

ACKNOWLEDGMENTS

We would like to thank Sarah Mathews, Amy Litt, Vivian Irish, and all of the members of the Kramer lab for many useful discussions and comments on the manuscript. The authors gratefully acknowledge financial support from the National Science Foundation and the Radcliffe Fellowship program at Harvard University.

REFERENCES

Alvarez, J. and Smyth, D. R. (1999). *CRABS CLAW* and *SPATULA*, two *Arabidopsis* genes that control carpel development in parallel with *AGAMOUS*. *Development* **126**, 2377–2386.

Alvarez-Buylla, E. R., Pelaz, S., Liljegren, S. J., Gold, S. E., Burgeff, C., Ditta, G. S., Ribas de Pouplana, L., Martinez-Castilla, L. and Yanofsky, M. F. (2000). An ancestral MADS-box gene duplication occurred before the divergence of plants and animals. *Proceedings of the National Academy of Sciences of the United States of America* **97**, 5328–5333.

Ambrose, B. A., Lerner, D. R., Ciceri, P., Padilla, C. M., Yanofsky, M. F. and Schmidt, R. J. (2000). Molecular and genetic analyses of the *Silky1* gene reveal conservation in floral organ specification between eudicots and monocots. *Molecules and Cell* **5**, 569–579.

Aoki, S., Uehara, K., Imafuku, M., Hasebe, M. and Ito, M. (2004). Phylogeny and divergence of basal angiosperms inferred from *APETALA3*- and *PISTILLATA*-like MADS-box genes. *Journal of Plant Research* **117**, 229–244.

Becker, A. and Theissen, G. (2003). The major clades of MADS-box genes and their role in the development and evolution of flowering plants. *Molecular Phylogenetics and Evolution* **29**, 464–489.

Becker, A., Gleissberg, S. and Smyth, D. R. (2005). Floral and vegetative morphogenesis in California poppy (*Eschscholzia californica* Cham.). *International Journal of Plant Sciences* **166**, 537–555.

Bell, C. D., Soltis, D. E. and Soltis, P. S. (2005). The age of the angiosperms: A molecular timescale without a clock. *Evolution* **59**, 1245–1258.

Berbel, A., Navarro, C., Ferrandiz, C., Canas, L. A., Madueno, F. and Beltran, J. P. (2001). Analysis of PEAM4, the pea AP1 functional homologue, supports a model for AP1-like genes controlling both floral meristem and floral organ identity in different plant species. *The Plant Journal* **25**, 441–451.

Bowers, J. E., Chapman, B. A., Rong, J. and Paterson, A. H. (2003). Unraveling angiosperm genome evolution by phylogenetic analysis of chromosomal duplication events. *Nature* **422**, 433–438.

Bowman, J. L., Smyth, D. R. and Meyerowitz, E. M. (1989). Genes directing flower development in *Arabidopsis*. *Plant Cell* **1**, 37–52.

Bowman, J. L., Smyth, D. R. and Meyerowitz, E. M. (1991). Genetic interactions among floral homeotic genes of *Arabidopsis*. *Development* **112**, 1–20.

Bowman, J. L., Alvarez, J., Weigel, D., Meyerowitz, E. M. and Smyth, D. R. (1993). Control of flower development in *Arabidopsis thaliana* by *APETALA1* and interacting genes. *Development* **119**, 721–743.

Bradley, D., Carpenter, R., Sommer, H., Hartley, N. and Coen, E. (1993). Complementary floral homeotic phenotypes result from opposite orientation of a transposon at the *plena* locus of Antirrhinum. *Cell* **72**, 85–95.

Carpenter, R. and Coen, E. (1990). Floral homeotic mutations produced by transposon-mutagenesis in *Antirrhinum majus*. *Genes Dev* **4**, 1483–1493.

Castillejo, C., Romera-Branchat, M. and Pelaz, S. (2005). A new role of the Arabidopsis SEPALLATA3 gene revealed by its constitutive expression. *The Plant Journal* **43**, 586–596.

Causier, B., Castillo, R., Zhou, J. L., Ingram, R., Xue, Y. B., Schwarz-Sommer, Z. and Davies, B. (2005). Evolution in action: Following function in duplicated floral homeotic genes. *Current Biology* **15**, 1508–1512.

Chase, M. W., Soltis, D. E., Olmstead, R. G., Morgan, D., Les, D. H., Mishler, B. D., Duvall, M. R., Price, R. A., Hills, H. G., Qiu, Y.-L., Kron, K. A. Rettig, J. H., *et al.* (1993). Phylogenetics of seed plants: An analysis of nucleotide sequences from the plastid gene rbcL. *Annals of the Missouri Botanical Garden* **80**, 528–580.

Cho, S., Jang, S., Chae, S., Chung, K. M., Moon, Y.-W., An, G. and Jang, S. K. (1999). Analysis of the C-terminal region of *Arabidopsis thaliana* APETALA1 as a transcription activation domain. *Plant Molecular Biology* **40**, 419–429.

Coen, E. S. and Meyerowitz, E. M. (1991). The war of the whorls: Genetic interactions controlling flower development. *Nature* **353**, 31–37.

Davies, B., Egea-Cortines, M., de Andrade Silva, E., Saedler, H. and Sommer, H. (1996). Multiple interactions amongst floral homeotic MADS box proteins. *EMBO* **15**, 4330–4343.

Davies, B., Motte, P., Keck, E., Saedler, H., Sommer, H. and Schwarz-Sommer, Z. (1999). PLENA and FARINELLI: Redundancy and regulatory interactions between two Antirrhinum MADS-box factors controlling flower development. *EMBO* **18**, 4023–4034.

Di Stilio, V. S., Kramer, E. M. and Baum, D. A. (2004). Floral MADS box genes and the evolution of homeotic gender dimorphism in meadow rues (Thalictrum, Ranunculaceae). *The Plant Journal* **41**, 755–766.

Ditta, G., Pinyopich, A., Robles, P., Pelaz, S. and Yanofsky, M. (2004). The SEP4 gene of *Arabidopsis thaliana* functions in floral organ and meristem identity. *Current Biology* **14**, 1935–1940.

Dong, Z. C., Zhao, Z., Liu, C. W., Luo, J. H., Yang, J., Huang, W. H., Hu, X. H., Wang, T. L. and Luo, D. (2005). Floral patterning in *Lotus japonicus*. *Plant Physiology* **137**, 1272–1282.

Donoghue, M. J. and Doyle, J. A. (1989). Phylogenetic analysis of angiosperms and the relationships of Hamamelidae. *In* "Evolution, Systematics, and Fossil History of the Hamamelidae" (P. R. Crane and S. Blackmore, eds.), Vol. 1. Introduction and "Lower" Hamamelidae." Clarendon Press, Oxford.

Drinnan, A. N., Crane, P. R. and Hoot, S. B. (1994). Patterns of floral evolution in the early diversification of non-magnoliid dicotyledons (eudicots). *In* "Early Evolution of Flowers" (P. K. Endress and E. M. Friis, eds.). Springer-Verlag, New York.

Egea-Cortines, M., Saedler, H. and Sommer, H. (1999). Ternary complex formation between the MADS-box proteins SQUAMOSA, DEFICIENS and GLOBOSA is involved in the control of floral architecture in Antirrhinum majus. *EMBO* **18**, 5370–5379.

Endress, P. K. (1987a). The early evolution of the angiosperm flower. *Trends in Ecology & Evolution* **2**, 300–304.

Endress, P. K. (1987b). Floral phyllotaxis and floral evolution. *Bot Jahr Syst* **08**, 417–438.

Endress, P. K. (1990). Patterns of floral construction in ontogeny and phylogeny. *Biological Journal of the Linnean Society* **39**, 153–175.

Endress, P. K. (1994). Floral structure and evolution of primitive angiosperms: Recent advances. *Plant Systematics and Evolution* **192**, 79–97.

Endress, P. K. (1995). Floral structure and evolution in Ranunculaceae. *Plant Systematics and Evolution* **9** (Suppl.), 47–61.

Endress, P. K. (2001). Origins of flower morphology. *The Journal of Experimental Zoology (Molecular and Developmental Evolution)* **291**, 105–115.

Erbar, C., Kusma, S. and Leins, P. (1998). Development and interpretation of nectary organs in Ranunculaceae. *Flora* **194**, 317–332.

Favaro, R., Pinyopich, A., Battaglia, R., Kooiker, M., Borghi, L., Ditta, G., Yanofsky, M. F., Kater, M. M. and Colombo, L. (2003). MADS-box protein complexes control carpel and ovule development in *Arabidopsis*. *Plant Cell* **15**, 2603–2611.

Ferrandiz, C., Gu, Q., Martienssen, R. and Yanofsky, M. F. (2000). Redundant regulation of meristem identity and plant architecture by FRUITFULL, APETALA1 and CAULIFLOWER. *Development* **127**, 725–734.

Ferrario, S., Immink, R. G. H., Shchennikova, A., Busscher-Lange, J. and Angenent, G. C. (2003). The MADS box gene *FBP2* is required for SEPALLATA function in Petunia. *Plant Cell* **15**, 914–925.

Flanagan, C. A. and Ma, H. (1994). Spatially and temporally regulated expression of the MADS-box gene AGL2 in wild type and mutant Arabidopsis flowers. *Plant Molecular Biology* **26**, 581–595.

de Folter, S., Immink, R. G. H., Kieffer, M., Parenicova, L., Henz, S. R., Weigel, D., Busscher, M., Kooiker, M., Colombo, L., Kater, M. M., Davies, B. and Angenent, G. C. (2005). Comprehensive interaction map of the *Arabidopsis* MADS box transcription factors. *Plant Cell* **17**, 1424–1433.

Force, A., Lynch, M., Pickett, F. B., Amores, A., Yan, Y.-L. and Postlethwait, J. (1999). Preservation of duplicate genes by complementary, degenerative mutations. *Genetics* **151**, 1531–1545.

Goto, K. and Meyerowitz, E. M. (1994). Function and regulation of the *Arabidopsis* floral homeotic gene PISTILLATA. *Genes Development* **8**, 1548–1560.

Griffith, M. E., da Silva Conceicao, A. and Smyth, D. R. (1999). PETAL LOSS gene regulates initiation and orientation of second whorl organs in the *Arabidopsis* flower. *Development* **126**, 5635–5644.

Gu, Q., Ferrandiz, C., Yanofsky, M. F. and Martienssen, R. (1998). The FRUITFULL MADS-box gene mediates cell differentiation during *Arabidopsis* fruit development. *Development* **125**, 1509–1517.

Gutierrez-Cortines, M. E. and Davies, B. (2000). Beyond the ABCs: Ternary complex formation in the control of floral organ identity. *Trends in Plant Science* **5**, 471–476.

Hall, J. C., Sytsma, K. J. and Iltis, H. H. (2002). Phylogeny of the Capparaceae and Brassicaceae based on Chloroplast sequences data. *American Journal of Botany* **89**, 1826–1842.

Hileman, L. C., Drea, S., de Martino, G., Litt, A. and Irish, V. (2005). Virus induced gene silencing is an effective tool to assay gene function in the basal eudicot species *Papaver somniferum* (opium poppy). *Plant Journal* **44**, 334–341.

Hill, T., Day, C. D., Zondlo, S. C., Thackeray, A. G. and Irish, V. F. (1998). Discrete spatial and temporal cis-acting elements regulate transcription of the *Arabidopsis* floral homeotic gene *APETALA3*. *Development* **125**, 1711–1721.

Honma, T. and Goto, K. (2001). Complexes of MADS-box proteins are sufficient to convert leaves into floral organs. *Nature* **409**, 525–529.

Huala, E. and Sussex, I. M. (1992). *LEAFY* interacts with floral homeotic genes to regulate *Arabidopsis* floral development. *Plant Cell* **4**, 901–913.

Huijser, P., Klein, J., Lonnig, W.-E., Meijer, H., Saedler, H. and Sommer, H. (1992). Bractomania, an inflorescence anomaly, is caused by the loss of function of the MADS-box gene *squamosa* in *Antirrhinum majus*. *EMBO Journal* **11**, 1239–1249.

Initiative, T. A. G. (2000). Analysis of the genome sequence of the flowering plant *Arabidopsis thaliana*. *Nature* **408**, 796–815.

Irish, V. F. (2000). Variations on a theme: Flower development and evolution. *Genome Biology* **1**, 1015.1–1015.4.

Irish, V. F. (2003). The evolution of floral homeotic gene function. *BioEssays* **25**, 637–646.

Irish, V. F. and Kramer, E. M. (1998). Genetic and molecular analysis of angiosperm flower development. *Advances in Botanical Research* **28**, 197–230.

Irish, V. F. and Litt, A. (2005). Floral development and evolution: Gene duplication, diversification and redeployment. *Current Opinion in Genetics & Development* **15**, 454–460.

Irish, V. F. and Sussex, I. M. (1990). Function of the *apetela-1* gene during *Arabidopsis* floral development. *Plant Cell* **2**, 741–753.

Jack, T. (2004). Molecular and genetic mechanisms of floral control. *Plant Cell* **16**, S1–S17.

Jack, T., Brockman, L. L. and Meyerowitz, E. M. (1992). The homeotic gene APETALA3 of *Arabidopsis thaliana* encodes a MADS box and is expressed in petals and stamens. *Cell* **68**, 683–697.

Jack, T., Fox, G. L. and Meyerowitz, E. M. (1994). Arabidopsis homeotic gene APETALA3 ectopic expression: Transcriptional and posttranscriptional regulation determine floral organ identity. *Cell* **76**, 703–716.

Judd, W. S. and Olmstead, R. G. (2004). A survey of tricolpate (eudicot) phylogenetic relationships. *American Journal of Botany* **91**, 1627–1644.

Kang, H.-G., Jeon, J.-S., Lee, S. and An, G. (1998). Identification of class B and class C floral organ identity genes from rice plants. *Plant Molecular Biology* **38**, 1021–1029.

Kanno, A., Saeki, H., Kameya, T., Saedler, H. and Theissen, G. (2003). Heterotopic expression of class B floral homeotic genes supports a modified ABC model for tulip (*Tulipa gesneriana*). *Plant Molecular Biology* **52**, 831–841.

Kim, S., Soltis, D. E., Soltis, P. S., Zanis, M. and Suh, Y. (2004a). Phylogenetic relationships among early-diverging eudicots based on four genes: Were the eudicots ancestrally woody? *Molecular Phylogenetics and Evolution* **31**, 16–30.

Kim, S., Yoo, M., Albert, V. A., Farris, J. S., Soltis, P. S. and Soltis, D. E. (2004b). Phylogeny and diversification of B-function genes in angiosperms: Evolutionary and functional implications of a 260-million year old duplication. *American Journal of Botany* **91**, 2102–2118.

Kim, S., Koh, J., Yoo, M. J., Konh, H., Hu, Y., Ma, H., Soltis, P. S. and Soltis, D. E. (2005). Expression of floral MADS-box genes in basal angiosperms: Implications for the evolution of floral regulators. *Plant Journal* **43**, 724–744.

Kosuge, K. (1994). Petal evolution in Ranunculaceae. *Plant Systematics and Evolution* **8** (Suppl.), 185–191.

Kosuge, K. and Tamura, M. (1989). Ontogenic studies on petals of the Ranunculaceae. *Journal of Japanese Botany* **64**, 65–74.

Kotilainen, M., Elomaa, P., Uimari, A., Albert, V. A., Yu, D. and Teeri, T. H. (2000). *GRCD1*, an *AGL2*-like MADS box gene, participates in the C function during stamen development in *Gerbera hybrida*. *Plant Cell* **12**, 1893–1902.

Kramer, E. M. (2005). Floral architecture: Regulation and diversity of floral shape and pattern. *In* "Plant Architecture and its Manipulation" (C. G. N. Turnbull, ed.), pp. 121–148. Blackwell Publishing, Oxford, UK.

Kramer, E. M. (in press). Floral patterning and control of floral organ formation. *In* "Flowering" (C. Ainsworth, ed.). Blackwell Publishing, Oxford, UK.

Kramer, E. M. and Hall, J. C. (2005). Evolutionary dynamics of genes controlling floral development. *Current Opinion in Plant Biology* **8**, 1–6.

Kramer, E. M. and Irish, V. F. (1999). Evolution of genetic mechanisms controlling petal development. *Nature* **399**, 144–148.

Kramer, E. M. and Irish, V. F. (2000). Evolution of the petal and stamen developmental programs: Evidence from comparative studies of the lower eudicots and basal angiosperms. *International Journal of Plant Science* **161**, S29–S40.

Kramer, E. M. and Jaramillo, M. A. (2005). The genetic basis for innovations in floral organ identity. *Journal of Experimental Zoology (Molecular and Developmental Evolution)* **304B**, 526–535.

Kramer, E. M., Dorit, R. L. and Irish, V. F. (1998). Molecular evolution of genes controlling petal and stamen development: Duplication and divergence within the *APETALA3* and *PISTILLATA* MADS-box gene lineages. *Genetics* **149**, 765–783.

Kramer, E. M., Di Stilio, V. S. and Schluter, P. (2003). Complex patterns of gene duplication in the *APETALA3* and *PISTILLATA* lineages of the Ranunculaceae. *International Journal of Plant Science* **164**, 1–11.

Kramer, E. M., Jaramillo, M. A. and Di Stilio, V. S. (2004). Patterns of gene duplication and functional evolution during the diversification of the AGAMOUS subfamily of MADS-box genes in angiosperms. *Genetics* **166**, 1011–1023.

Kramer, E. K., Su, H.-J. Wu, C.-C. and Hu, J.-M. (2006). A simplified explanation for the frameshift mutation that created a novel C-terminal motif in the *APETALA3* gene lineage. *BMC Evolutionary Biology* **6**, 30.

Krizek, B. A. and Meyerowitz, E. M. (1996). Mapping the protein regions responsible for the functional specificities of the *Arabidopsis* MADS domain organ-identity proteins. *Proceedings of the National Academy of Sciences of the United States of America* **93**, 4063–4070.

Kunst, L., Klenz, J. E., Martinez-Zapater, J. and Haughn, G. W. (1989). *AP2* gene determines the identity of perianth organs in flowers of Arabidopsis thaliana. *Plant Cell* **1**, 1195–1208.

Lamb, R. S. and Irish, V. F. (2003). Functional divergence within the APETALA3/PISTILLATA floral homeotic gene lineages. *Proceedings of the National Academy of Sciences of the United States of America* **100**, 6558–6563.

Liljegren, S. J., Ditta, G. S., Eshed, Y., Savidge, B., Bowman, J. L. and Yanofsky, M. F. (2000). *SHATTERPROOF* MADS-box genes control seed dispersal in *Arabidopsis*. *Nature* **404**, 766–770.

Litt, A. and Irish, V. F. (2003). Duplication and diversification in the *APETALA1/FRUITFULL* floral homeotic gene lineage: Implications for the evolution of floral development. *Genetics* **165**, 821–833.

Ma, H., Yanofsky, M. F. and Meyerowitz, E. M. (1991). *AGL1-AGL6*, an *Arabidopsis* gene family with similarity to floral homeotic and transcription factor genes. *Genes Development* **5**, 484–495.

Magallon, S., Crane, P. R. and Herendeen, P. S. (1999). Phylogenetic pattern, diversity, and diversification of eudicots. *Annals of the Missouri Botanical Garden* **86**, 297–372.

Malcomber, S. T. and Kellogg, E. A. (2004). Heterogeneous expression patterns and separate roles of the SEPALLATA gene LEAFY HULL STERILE1 in Grasses. *Plant Cell* **16**, 1692–1706.

Malcomber, S. T. and Kellogg, E. A. (2005). SEPALLATA gene diversification: Brave new whorls. *Trends in Plant Science* **10**, 427–435.

Mandel, M. A. and Yanofsky, M. F. (1995). The *Arabidopsis* AGL8 MADS box gene is expressed in inflorescence meristems and is negatively regulated by APETALA1. *Plant Cell* **7**, 1763–1771.

Mandel, M. A. and Yanofsky, M. F. (1998). The *Arabidopsis* AGL9 MADS box gene is expressed in young flower primordia. *Sex Plant Reproduction* **11**, 22–28.

Mandel, M. A., Gustafson-Brown, C., Savidge, B. and Yanofsky, M. F. (1992). Molecular characterization of the *Arabidopsis* floral homeotic gene *apetala1*. *Nature* **360**, 274–277.

McGonigle, B., Bouhidel, K. and Irish, V. F. (1996). Nuclear localization of the Arabidopsis APETALA3 and PISTILLATA homeotic gene products depends on their simultaneous expression. *Genes Development* **10**, 1812–1821.

Messenguy, F. and Dubois, E. (2003). Role of MADS box proteins and their cofactors in combinatorial control of gene expression and cell development. *Gene* **316**, 1–21.

Mizukami, Y. and Ma, H. (1995). Separation of AG function in floral meristem determinacy from that in reproductive organ identity by expressing antisense AG RNA. *Plant Molecular Biology* **28**, 767–784.

Moon, Y.-W., Kang, H.-G., Jung, J.-Y., Jeon, J.-S., Sung, S.-K. and An, G. (1999). Determination of the motif responsible for interaction between the rice APETALA1/AGAMOUS-LIKE9 family proteins using a yeast two-hybrid system. *Plant Physiology* **120**, 1193–1203.

Muller, B. M., Saedler, H. and Zachgo, S. (2001). The MADS-box gene *DEFH28* from *Antirrhinum* is involved in the regulation of floral meristem identity and fruit development. *Plant Journal* **28**, 169–179.

Nagasawa, N., Miyoshi, M., Sano, Y., Satoh, H., Hirano, H., Sakai, H. and Nagato, Y. (2003). *SUPERWOMAN1* and *DROOPING LEAF* genes control floral organ identity in rice. *Development* **130**, 705–718.

Park, S. U. and Facchini, P. J. (2000). Agrobacterium-mediated genetic transformation of California poppy, *Eschscholzia californica* Cham., via somatic embryogenesis. *Plant Cell Reports* **19**, 1006–1012.

Pelaz, S., Ditta, G. S., Baumann, E., Wisman, E. and Yanofsky, M. (2000). B and C floral organ identity functions require SEPALLATA MADS-box genes. *Nature* **405**, 200–203.

Pelaz, S., Gustafson-Brown, C., Kohalmi, S. E., Crosby, W. L. and Yanofsky, M. (2001). APETALA1 and SEPALLATA3 interact to promote flower development. *Plant Journal* **26**, 385–394.

Pinyopich, A., Ditta, G. S., Savidge, B., Liljegren, S. J., Baumann, E., Wisman, E. and Yanofsky, M. F. (2003). Assessing the redundancy of MADS-box genes during carpel and ovule development. *Nature* **424**, 85–88.

Prychitko, T., Johnson, R. M., Wildman, D. E., Gumucio, D. and Goodman, M. (2005). The phylogenetic history of New World monkey β globin reveals a platyrrhine β to δ gene conversion in the atelid ancestry. *Molecular Phylogenetics and Evolution* **35**, 225–234.

Purugganan, M. D., Rounsley, S. D., Schmidt, R. J. and Yanofsky, M. F. (1995). Molecular evolution of flower development: Diversification of the plant MADS-box regulatory gene family. *Genetics* **140**, 345–356.

Riechmann, J. L., Krizek, B. A. and Meyerowitz, E. M. (1996a). Dimerization specificity of *Arabidopsis* MADS domain homeotic proteins APETALA1, APETALA3, PISTILLATA, and AGAMOUS. *Proceedings of the National Academy of Sciences of the United States of America* **93**, 4793–4798.

Riechmann, J. L., Wang, M. and Meyerowitz, E. M. (1996b). DNA-binding properties of *Arabidopsis* MADS domain homeotic proteins APETALA1, APETALA3, PISTILLATA and AGAMOUS. *Nucleic Acids Research* **24**, 3134–3141.

Running, M. P. and Meyerowitz, E. M. (1996). Mutations in the PERIANTHIA gene of *Arabidopsis* specifically alter floral organ number and initiation pattern. *Development* **122**, 1261–1269.

Sanderson, M. J., Thorne, J. L., Wikström, N. and Bremer, K. (2004). Molecular evidence on plant divergence times. *American Journal of Botany* **91**, 1656–1665.

Savidge, B., Rounsley, S. D. and Yanofsky, M. F. (1995). Temporal relationship between the transcription of two *Arabidopsis* MADS box genes and the floral organ identity genes. *Plant Cell* **7**, 721–733.

Schmid, M., Uhlenhaut, N. H., Godard, F., Demar, M., Bressan, R., Weigel, D. and Lohmann, J. U. (2003). Dissection of floral induction pathways using global expression analysis. *Development* **130**, 6001–6012.

Schultz, E. A. and Haughn, G. W. (1991). LEAFY, a homeotic gene that regulates inflorescence development in *Arabidopsis*. *Plant Cell* **3**, 771–781.

Sessions, A., Nemhauser, J. L., McColl, A., Roe, J. L., Feldmann, K. A. and Zambryski, P. C. (1997). ETTIN patterns the *Arabidopsis* floral meristem and reproductive organs. *Development* **124**, 4481–4491.

Soltis, D. E., Senters, A. E., Zanis, M., Kim, S., Thompson, J. D., Soltis, P. S., Ronse De Craene, L. P., Endress, P. K. and Farris, J. S. (2003). Gunnerales are sister to other core eudicots: Implications for the evolution of pentamery. *American Journal of Botany* **90**, 461–470.

Stellari, G. M., Jaramillo, M. A. and Kramer, E. M. (2004). Evolution of the *APETALA3* and *PISTILLATA* lineages of MADS-box containing genes in basal angiosperms. *Molecular Biology Evolution* **21**, 506–519.

Sundstrom, J. and Engstrom, P. (2002). Conifer reproductive development involves B-type MADS-box genes with distinct and different activities in male organ primordia. *Plant Journal* **31**, 161–169.

Taylor, S. A., Hofer, J. M. I., Murfet, I. C., Sollinger, J. D., Singer, S. R., Knox, M. R. and Ellis, T. H. N. (2002). PROLIFERATING INFLORESCENCE MERISTEM, a MADS-box gene that regulates floral meristem identity in Pea. *Plant Physiology* **129**, 1150–1159.

Theissen, G., Becker, A., Di Rosa, A., Kanno, A., Kim, J. T., Munster, T., Winter, K.-U. and Saedler, H. (2000). A short history of MADS-box genes in plants. *Plant Molecular Biology* **42**, 115–149.

Theissen, G., Becker, A., Winter, K. U., Munster, T., Kirchner, C. and Saedler, H. (2002). How the land plants learned their floral ABCs: The role of MADS-box genes in the evolutionary origin of flowers. *In* "Developmental Genetics and Plant Evolution" (Q. C. B. Cronk, R. M. Bateman and J. A. Hawkins, eds.). Taylor and Francis, London.

Theissen, G. and Saedler, H. (2001). Floral quartets. *Nature* **409**, 469–471.

Tonaco, I. A. N., Borst, J. W., de Vries, S. C., Angenent, G. C. and Immink, R. G. H. (2006). *In vivo* imaging of MADS-box transcription factor interactions. *Journal of Experimental Botany* **57**, 33–42.

Trobner, W., Ramirez, L., Motte, P., Hue, I., Huijser, P., Lonnig, W. E., Saedler, H., Sommer, H. and Schwarz-Sommer, Z. (1992). Globosa—a homeotic gene which interacts with deficiens in the control of antirrhinum floral organogenesis. *EMBO Journal* **11**, 4693–4704.

Tzeng, T. Y., Liu, H.-C. and Yang, C. H. (2004). The C-terminal sequence of LMADS1 is essential for the formation of homodimers for B function proteins. *Journal of Biological Chemistry* **279**, 10747–10755.

Tzeng, T. Y. and Yang, C. H. (2001). A MADS box gene from lily (*Lilium longiflorum*) is sufficient to generate dominant negative mutation by interacting with PISTILLATA (PI) in *Arabidopsis thaliana*. *Plant Cell Physiology* **42**, 1156–1168.

Uimari, A., Kotilainen, M., Elomaa, P., Yu, D., Albert, V. A. and Teeri, T. H. (2004). Integration of reproductive meristem fates by a SEPALLATA-like MADS-box gene. *Proceedings of the National Academy of Sciences of the United States of America* **101**, 15817–15822.

van der Krol, A. R., Brunelle, A., Tsuchimoto, S. and Chua, N. H. (1993). Functional analysis of Petunia floral homeotic MADS box gene PMADS1. *Genes Development* **7**, 1214–1228.

van der Krol, A. R. and Chua, N.-H. (1993). Flower development in petunia. *Plant Cell* **5**, 1195–1203.

Vandenbussche, M., Theissen, G., Van de Peer, Y. and Gerats, T. (2003a). Structural diversification and neo-functionalization during floral MADS-box gene evolution by C-terminal frameshift mutations. *Nucleic Acids Research* **31**, 4401–4409.

Vandenbussche, M., Zethof, J., Souer, E., Koes, R., Torinelli, G. B., Pezzotti, M., Ferrario, S., Angenent, G. C. and Gerats, T. (2003b). Toward the analysis of the Petunia MADS box gene family by reverse and forward transposon insertion mutagenesis approaches: B, C, and D function organ identity functions require SEPALLATA-like MADS box genes in Petunia. *Plant Cell* **15**, 2680–2693.

Vandenbussche, M., Zethof, J., Royaert, S., Weterings, K. and Gerats, T. (2004). The duplicated B-class heterodimer model: Whorl-specific effects and complex genetic interactions in *Petunia hybrida* flower development. *Plant Cell* **16**, 741–754.

Vision, T. J., Brown, D. G. and Tanksley, S. D. (2000). The origins of genomic duplications in *Arabidopsis*. *Science* **290**, 2114–2117.

Whipple, C. J., Ciceri, P., Padilla, C. M., Ambrose, B. A., Bandong, S. L. and Schmidt, R. J. (2004). Conservation of B-class floral homeotic gene function between maize and Arabidopsis. *Development* **131**, 6083–6091.

William, D. A., Su, Y., Smith, M. R., Lu, M., Baldwin, D. A. and Wagner, D. (2004). Genomic identification of direct target genes of *LEAFY*. *Proceedings of the National Academy of Sciences of the United States of America* **101**, 1775–1780.

Winter, K. U., Weiser, C., Kaufmann, K., Bohne, A., Kirchner, C., Kanno, A., Saedler, H. and Theissen, G. (2002). Evolution of class B floral homeotic proteins: Obligate heterodimerization originated from homodimerization. *Molecular Biology Evolution* **19**, 587–596.

Yalovsky, S., Rodriguez-Concepcion, M., Bracha, K., Toledo-Ortiz, G. and Gruissem, W. (2000). Prenylation of the floral transcription factor APETALA1 modulates its function. *Plant Cell* **12**, 1257–1266.

Yang, Y., Fanning, L. and Jack, T. (2003). The K domain mediates heterodimerization of the *Arabidopsis* floral organ identity proteins, APETALA3 and PISTILLATA. *Plant Journal* **33**, 47–59.

Yang, Y. and Jack, T. (2004). Defining subdomains of the K domain important for protein-protein interactions of plant MADS proteins. *Plant Molecular Biology* **55**, 45–59.

Zahn, L. M., Kong, H., Leebens-Mack, J. H., Kim, S., Soltis, P. S., Landherr, L. L., Soltis, D. E., Depamphilis, C. W. and Ma, H. (2005a). The Evolution of the SEPALLATA Subfamily of MADS-Box Genes: A Preangiosperm Origin With Multiple Duplications Throughout Angiosperm History. *Genetics* **169**, 2209–2223.

Zahn, L. M., Leebens-Mack, J., dePamphilis, C. W., Ma, H. and Theissen, G. (2005b). To B or not to B a flower: The role of DEFICIENS and GLOBOSA orthologs in the evolution of the angiosperms. *Journal of Heredity* **96**, 225–240.

Zanis, M., Soltis, P. S., Qiu, Y.-L., Zimmer, E. A. and Soltis, D. E. (2003). Phylogenetic analyses and perianth evolution in basal angiosperms. *Annals of the Missouri Botanical Garden* **90**, 129–150.

Genetics of Grass Flower Development

CLINTON J. WHIPPLE AND ROBERT J. SCHMIDT

Section of Cell and Developmental Biology, Division of Biological Sciences, University of California at San Diego, La Jolla, California 92093

ABSTRACT

The developmental genetic analyses of floral organ specification that led to the well-known ABC-model of flower development were primarily performed in eudicot model species. To better understand how pathways controlling flower development have either been conserved or modified more broadly in the angiosperms, it is necessary to examine the genetic basis of flowering in plant groups more distantly related to *Arabidopsis* and *Antirrhinum*. Maize and rice are grass species with genomics and genetic resources that

Advances in Botanical Research, Vol. 44
Incorporating Advances in Plant Pathology

0065-2296/06 $35.00
DOI: 10.1016/S0065-2296(06)44010-6

make them amenable to both forward and reverse genetics. A combination of these two strategies is beginning to elucidate how the ABC-model is conserved, as well as ways in which grass flower development differs from eudicots. The ability to investigate the degree of conservation in developmental pathways, the evolution of derived morphologies, and the consequences of gene duplication events make the grass family an excellent model for studies on the evolution of flower development.

I. INTRODUCTION

The most striking characteristic of angiosperms is the flower, which, although it has been extremely modified in multiple angiosperm lineages, is basically composed of four organ types: (1) sepals, bract-like protective leaves; (2) petals, modified leaves serving as pollinator attractors; (3) the stamen, or male reproductive organ, and (4) the carpel, or female reproductive organ. These organs are found almost always (with the only known exception being *Lacandonia schismatica* (Ambrose *et al.*, 2006)) in a stereotypical order of concentric whorls starting from the outside and moving inwards: sepal, petal, stamen, and carpels. There are, of course, many species with missing whorls, or highly modified organs. Nevertheless, conservation of these organ types and this basic body plan allows us to recognize these shared features in even very modified and bizarre flowers (see Endress, Chapter 1).

Early developmental genetic research in model eudicot species *Arabidopsis thaliana* and *Antirrhinum majus* led to the establishment of a simple and powerful model for the genetic establishment of floral organ identity (Coen and Meyerowitz, 1991) (Fig. 1). The original ABC-model held that three classes of genes worked alone or in combination with another class to specify the identity of each of the four organ whorls. A-class alone provides sepal identity. A- and B-classes combine to establish petal identity. B- and C-classes together promote stamen identity, and C-class alone is responsible for carpel identity. As the genes underlying these activities were cloned, nearly all turned out to belong to a large family encoding transcription factors containing the conserved MADS domain, a protein domain that binds to the consensus CArG-box sequence in promoters of prospective target genes (Huang *et al.*, 1993; Mueller and Nordheim, 1991; Riechmann *et al.*, 1996). The only exception is *APETELA2* (*AP2*), a classically defined A-class homeotic gene, that belongs to a different family of putative transcription factors. As more was learned about the genetic control of flowering in *Arabidopsis*, it was clear that the ABC-model would need to be modified to include the activity of another set of MADS-box genes with redundant roles in regulating the activity of B- and C-class MADS proteins (Pelaz *et al.*, 2001). Other studies revealed further roles of additional MADS-box genes in specifying the identity of ovules, which together led to the proposal for the

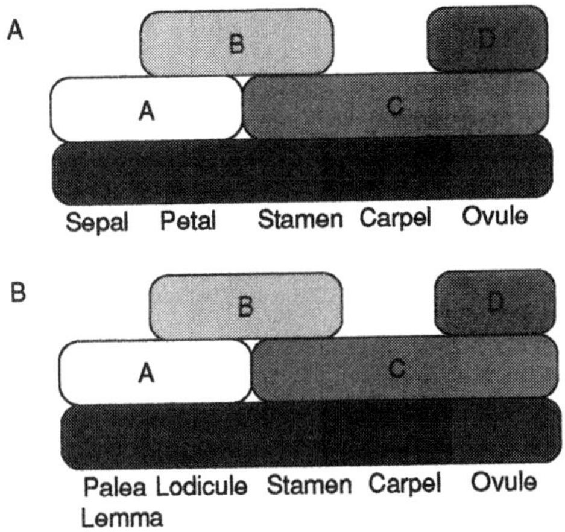

Fig. 1. The ABCDE-model illustrating the genetic interactions necessary for floral organ specification. (A) The ABCDE-model as it applies to eudicots, from Theissen (2001) with changes incorporating data from Ditta *et al.* (2004). (B) A modification of the ABCDE-model to explain the patterning of a grass floret, modified from Ambrose *et al.* (2000).

expansion of the ABC-model to include D- and E-classes, with D-activity specifying ovule identity and E-activity being required for the activity of B- and C-classes (Theissen, 2001). E-class activity was later shown to be necessary for A-class gene function in addition to B- and C-classes (Ditta *et al.*, 2004). Although it is not clear that ovule development should be included in a model that was designed to describe floral organ specification, it serves as a useful categorization of MADS-box gene members for the purpose of this chapter. These studies in model eudicot species have established the prominent role of so many different MADS-box genes in eudicot flower and ovule development, resulting in a large amount of interest and research on this family of genes. This is especially true for researchers interested in the evolution of developmental differences (evo-devo) in the origin and diversification of flowers (see Irish, Chapter 3; Kramer and Zimmer, Chapter 9; Rijpkema *et al.*, Chapter 6; Soltis *et al.*, Chapter 12; Zahn *et al.*, Chapter 4).

It was recognized early that modifications in the expression of the ABC-genes could be very useful in describing some of the floral variation seen in other angiosperm groups, such as the petaloid outer tepals seen in some monocots (van Tunen *et al.*, 1993). Modifications of the ABC-model have been invoked to describe morphological diversity in some basal eudicot flowers (Kim *et al.*, 2005; Kramer *et al.*, 2003; see Soltis *et al.*, Chapter 12). However, there is still a paucity of genetic or functional evidence to understand the degree to which the ABC-model is conserved outside the core eudicots.

There are at least two questions of broad general interest for understanding the evolution of flowers. First, what is conserved? That we can recognize a diversity of morphologies as flowers seems to indicate that there is a conserved genetic mechanism establishing the floral ground plan and organ identities. At first glance, it would seem that the ABC-model genes make good candidates for such a mechanism. The second question is: What is different? In other words, what kinds of changes are responsible for the often-striking differences in floral morphologies that occur in the angiosperms? Do modifications in ABC-genes or their expression domains correlate with changes in morphology, or is the ABC-model a basic ground plan, which most angiosperms maintain, and the morphological diversity of flowers is controlled by other pathways? To make progress on these two questions it is important to have a model group or species that is outside the core eudicots, one which has available forward and reverse genetic resources to rigorously test gene function. Additionally, to address the second question a group is needed that has some derived floral features. We would argue that the grasses, and maize (*Zea mays*) and rice (*Oryza sativa*) in particular, make an excellent choice for just such an inquiry. As monocots, grasses are distantly related to the species serving as plant models in the core eudicots. They have a derived floral morphology, especially in the two sterile outer whorls. Furthermore, an increasing array of genetic and genomic resources is being developed for these species. It is thus possible to begin examining grass ABC-model genes for conservation of function. It is also possible, through forward screens, to identify grass-specific factors involved in the derived morphology of this group.

In this chapter, we will discuss the genetics of grass flowering. Other reviewers have dealt broadly with advances in understanding grass inflorescence development (Bommert *et al.*, 2005; McSteen *et al.*, 2000; see Malcomber *et al.*, Chapter 11). For this reason, we will focus specifically on the genetics of flower development. As these studies lead naturally to a discussion of comparative floral development, we will finish by presenting the case for using grasses, with their emerging genomic resources, as a model group for evo-devo studies.

II. GRASS FLORAL MORPHOLOGY

Mature floral morphology in the grasses differs significantly from the typical monocot flower, so much so that traditional assignments of homology have been problematic (Clifford, 1987) (Fig. 2). It is clear, however, from phylogenetic analyses of the Poales [Grass Phylogeny Working Group (GPWG, 2001); Michelangeli *et al.*, 2003] that the closest extant outgroups to the grasses, *Joinvillea* and *Ecdeiocolea*, have a typical monocot floral plan with

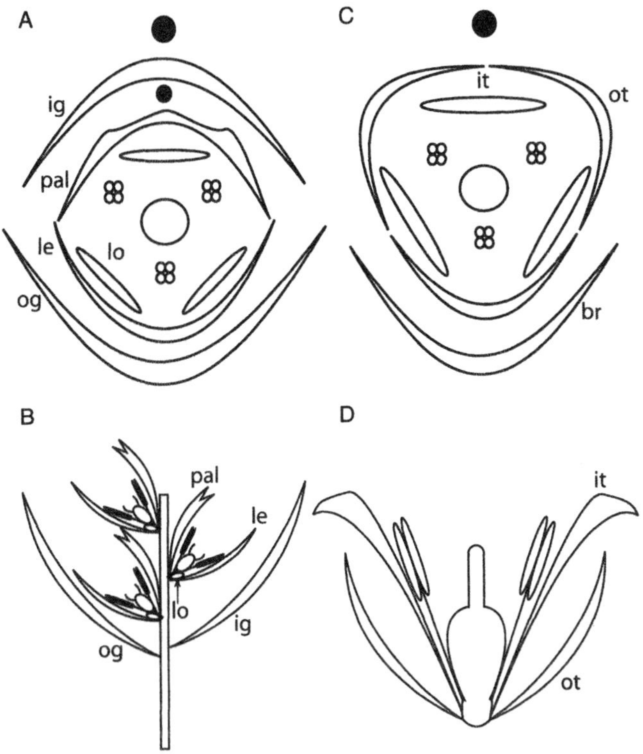

Fig. 2. Schematics for the grass spikelet and typical monocot flower. (A, B) Grass spikelet diagram. (C, D) Monocot flower diagram. Ig, inner glume; og, outer glume; pal, palea; le, lemma; lo, lodicule; br, bract; ot, outer tepal; it, inner tepal; black dots represent the relative position of the stem.

two outer whorls of tepals (more or less differentiated between the inner and outer tepal whorl), a whorl of stamens, and a central whorl of carpels. As typical in the monocots, the floral organs in each whorl occur in multiples of three. In the grasses, the basic unit of the inflorescence is the spikelet, or a small spike composed of one to many florets and subtended by two sterile bracts called glumes. The florets themselves consist of:

1. A lemma, often considered to be a bract in the axil of which the flower arises.
2. A palea, a generally two-veined bract-like organ, occasionally interpreted as the prophyll of the floral branch subtended by the lemma.
3. Lodicules, small scale-like or fleshy organs that swell at anthesis to open the floret and exsert the anthers. In some grasses (mostly bamboos), there are three lodicules that alternate with the anthers. However, in most grasses the medial (adaxial) lodicule aborts leaving only the two abaxial lodicules adjacent to the lemma.

4. Stamens, the male reproductive organs, generally occur in one or two whorls of three.
5. The female reproductive unit is composed of three fused carpels, generally with two stigmas and a single ovary.

The homology of the reproductive organs in grass flowers with those of eudicot flowers is clear and not in question. However, the outer sterile organs are derived in the grasses leading to their grass-specific nomenclature of lemma, palea, and lodicule. By making assumptions about the homology of the sterile floret organs, it is possible to adapt the ABC-model to the grass floret and propose a hypothesis of how ABC-genes may act in patterning a grass flower (Fig. 1) (Ambrose *et al.*, 2000). We will come back to a possible scenario for the evolution of the grass floret, and ways to test it in Section V.B.

III. GENETICS OF GRASS FLOWER DEVELOPMENT

Both forward and reverse genetic approaches have been taken to understand the genetic control of flowering in grasses. Forward genetics starts with a mutant known to cause a defect in a particular trait, but without any knowledge of the gene that has been mutated. This powerful approach was used in eudicot species to establish the ABC-model. While forward genetics is theoretically possible in any plant, practical difficulties in cloning genes underlying grass mutations initially limited its utility. However, candidate genes revealed from *Arabidopsis* and *Antirrhinum* could be examined in grasses. As these genes were cloned, their expression patterns were observed by a combination of *in situ* hybridization and Northern blot analyses, with the thought that conserved patterns of expression would indicate conserved function. Eventually more rigorous tests of function for some genes involved isolating transposon insertion mutants and specific knockdown of transcription by transgenes. Fortunately, increased genomic resources have made forward genetics more practical in grasses. This combination of forward and reverse genetics is beginning to unravel the developmental mechanisms of flowering in this interesting and economically important plant family.

A. MADS-BOX GENES AND GRASS FLOWER DEVELOPMENT: ROUNDING UP THE USUAL SUSPECTS

With studies in maize and rice, grasses emerged early as a model group to test the function of homeotic MADS-box genes outside the eudicots. While members from all classes (ABCD and E) have been identified, little is known about the function of these genes beyond the B- and C-classes. We will briefly review what is known about these groups in the grasses classifying the genes according to published phylogenetic work and focusing on studies that elucidate individual gene function. Although the ABCDE-model in Fig. 1 suggests that genes in these five classes have clear roles in organ identity, this is not always the case as will be discussed. However, genes in each class tend to be related phylogenetically, and thus the division into classes A–E is a convenient classification.

1. A-class

According to the ABC-model, A-class genes in *Arabidopsis* are necessary for sepal and petal identity. *Arabidopsis* A-class includes the MADS-box gene *AP1* and the non-MADS-box gene *AP2*. While mutations in *AP1* cause difficulties in sepal and petal development, it is not clear that *AP1* contributes directly to a discrete A-function for organ identity. These phenotypes may in fact be the result of incomplete loss of floral meristem identity as indicated by the phenotype of the triple-mutant combination of *ap1* with two closely related genes in *Arabidopsis, cauliflower* (*cal*) and *fruitful* (*ful*) (reviewed in Litt and Irish, 2003). Also, a true A-class was never defined in *Antirrhinum*, even though the *squamosa* mutant was shown to encode the *AP1* ortholog (Huijser *et al.*, 1992). Furthermore, *LIPLESS* (*LIP1*) and *LIP2*, the redundant *Antirrhinum* orthologs of the non-MADS-box A-class gene *AP2* neither have a completely conserved function with the *Arabidopsis* gene (Keck *et al.*, 2003) nor apparently the *Petunia hybrida AP2* homolog *PhAP2A* (Maes *et al.*, 2001). Thus, while the literature commonly refers to A-class specification of floral organ identity, the situation may be more subtle and complicated. Whatever the contribution of *AP1* to A-class gene function, there are clearly grass genes that belong to the *AP1/FRUITFUL* (*FUL*) group, and these appear to fall into three distinct lineages (Fig. 3A) (Münster *et al.*, 2002; Schmitz *et al.*, 2000). One lineage includes *Zea AP1a* (*Zap1a*) from maize and its duplicate *ZmMADS3* (*Zap1b*), *Barley MADS8* (*BM8*) from barley (*Hordeum vulgare*), and *O. sativa MADS15* (*OsMADS15*) from rice (Heuer *et al.*, 2001; Mena *et al.*, 1995; Moon *et al.*, 1999; Schmitz *et al.*, 2000). A second lineage includes *Z. mays MADS4* (*Zmm4*) and *-15* from maize, *OsMADS14* from rice, and *BM5* from barley (Fischer *et al.*, 1995; Moon *et al.*, 1999; Münster *et al.*, 2002; Schmitz *et al.*, 2000). A third lineage includes *Zmm28* from maize, *OsMADS18* from rice, and *BM3* from barley (Moon *et al.*, 1999; Münster *et al.*, 2002; Schmitz *et al.*, 2000).

Based on RNA blotting, expression of the first member of this group to be cloned, *Zap1a*, suggested a conserved pattern of expression with that of *AP1*, namely that transcripts were restricted to the nonreproductive organs of the maize floret (Mena *et al.*, 1995). This indicated a conserved role for *Zap1a* in specifying the outer whorls of the grass floret but not the inner

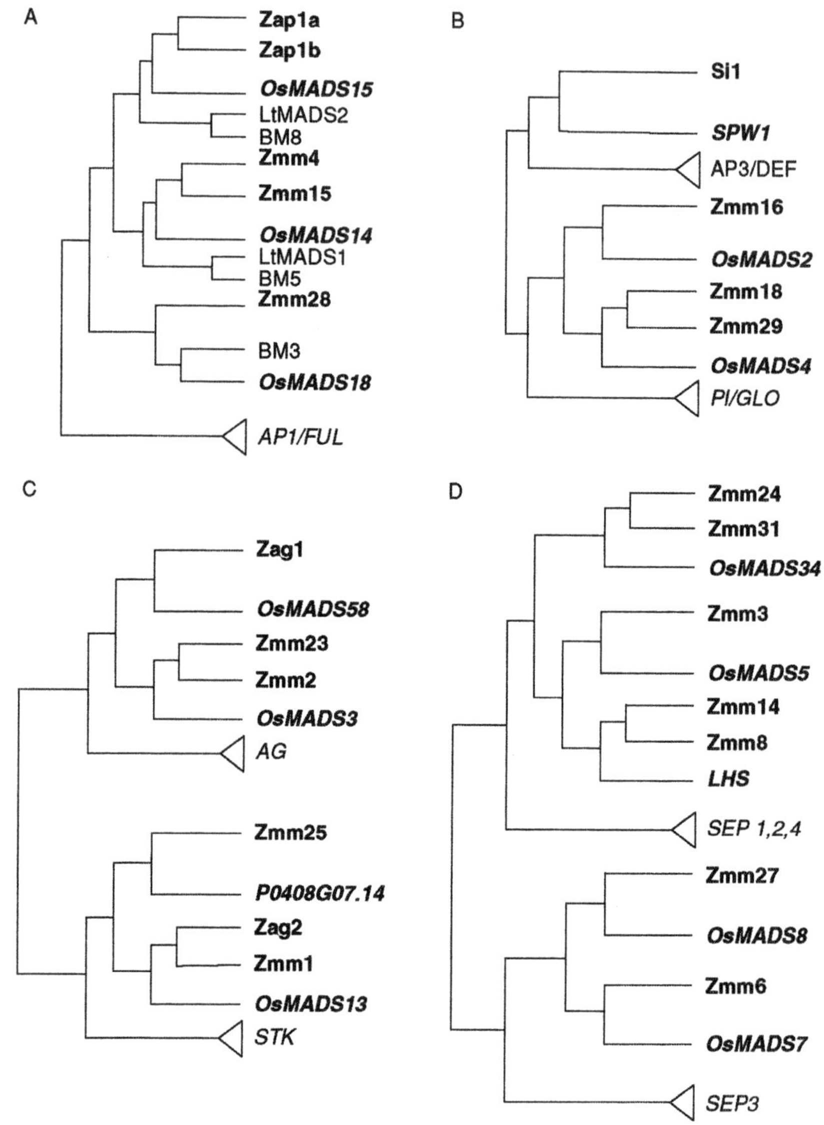

Fig. 3. Summary of phylogenetic relationships among grass ABCD- and E-classes of MADS-box genes. (A) A-class genes, (B) B-class genes, (C) C- and D-class genes, and (D) E-class genes. Maize genes in bolds, rice genes in bold italics, barley and *Lolium* genes in normal typefaces, and closest eudicot lineage (represented by *Arabidopsis*) in italics. Relationships in (A) and (B) are adapted from Münster *et al.* (2002), in (C) adapted from Kramer *et al.* (2004), and in (D) adapted from Zahn *et al.* (2005).

whorls. More revealing expression studies based on *in situ* hybridizations have been documented for genes from these lineages in rice, *Lolium temulentum*, and barley (*H. vulgare*). In agreement with the observations in maize, expression of the rice *RAP1* (*OsMADS14*) was not detected in developing stamen and carpel primordia, but was detected in developing palea, lemma, and lodicule (Kyozuka *et al.*, 2000). The situation in barley appears more complex, where all three genes (*BM8, BM5*, and *BM3*) appeared to be expressed in all developing floral organ primordia (Schmitz *et al.*, 2000). Unfortunately, only late stage inflorescences were examined, so the earliest onset of expression was not determined. Expression of the *Lolium* genes *LtMADS1* and *LtMADS2* (orthologs of *OsMADS14* and *OsMADS15*, respectively) was seen in the apical inflorescence meristem as well as the spikelet and floret meristems (Gocal *et al.*, 2001). Because the *in situ* hybridization studies in these different grass species did not always include equivalent stages of inflorescence development, it is difficult to extrapolate a consensus pattern of expression, if one exists. How any of these genes function in specifying grass floral organ or meristem identity is unknown, although the expression patterns are consistent with a role in meristem identity as seen for eudicot *AP1* genes, and for at least some of the species analyzed, the expression is consistent with a possible role in specifying or promoting the sterile outer whorls of the grass flower.

A possible role in promoting the transition to flowering for the *OsMADS14/BM5/Zmm4,15* clade is suggested by a study in which the diploid wheat (*Triticum monococum*) ortholog of this lineage mapped directly to the vernalization *Vrn1* locus which controls flowering time in response to cold treatment (Yan *et al.*, 2003). Spring wheat varieties that require no vernalization contain the dominant allele *Vrn1*, while winter wheat varieties that do require vernalization have the recessive allele *vrn1*. Three different spring wheat *Vrn1* alleles carry different size deletions that disrupt a putative CArG box in the 5′ proximal promoter of this *Triticum AP1* ortholog. Expression of this wheat *AP1* gene was shown to be induced by vernalization only in winter wheat varieties. These expression studies were confirmed by a study of the same gene in triploid bread wheat *Triticum aestivum* in which cosuppression of this gene leads to delayed flowering (Murai *et al.*, 2003).

Function for the *Zmm28/OsMADS18* clade has been investigated in only a single study. Overexpression of *OsMADS18* in rice leads to early flowering, however a transgeneic knockdown by RNAi had no obvious phenotype (Fornara *et al.*, 2004). As noted by the authors in this study, early flowering caused by overexpression is consistent with other studies in which *AP1*-like genes were overexpressed in rice and *Arabidopsis*, and thus may indicate a role in promoting flowering for this clade as well.

Although there is still limited understanding of the roles that A-class genes may play in grass flower development, it is interesting to note that the genes are commonly expressed in meristems other than the floral meristem, most notably in meristems that precede the floral meristem such as the spikelet meristem. In *Arabidopsis, AP1* and another non-MADS-box gene *LEAFY* (*LFY*) are both necessary and sufficient for establishing a floral meristem identity (Mandel and Yanofsky, 1995; Mandel *et al.*, 1992; Weigel and Nilsson, 1995; Weigel *et al.*, 1992). Grasses are different from *Arabidopsis* in that the first meristem that branches from the inflorescence meristem has a distinct identity that is not a floral meristem (McSteen *et al.*, 2000), namely the spikelet or spikelet pair meristem (Fig. 4). However, at least in some of the grasses, the *AP1* orthologs and the *LFY* orthologs are expressed in these early meristems without conferring a floral meristem identity (Bomblies *et al.*, 2003; Gocal *et al.*, 2001). This suggests that either they have been recruited to new roles in these earlier grass meristems, or that their role in promoting a floral meristem identity is somehow repressed until the later arising floral meristem is produced. The loss-of-function phenotype for the redundant maize *LFY* orthologs *Zfla* and *Zflb* supports the latter hypothesis. While the *zfla zflb* double mutant has floral defects similar to the *Arabidopsis lfy* mutant, indicating a conservation between monocots and eudicots in this pathway of promoting a floral identity, there are no apparent defects in the spikelet or other meristem identities (Bomblies *et al.*, 2003).

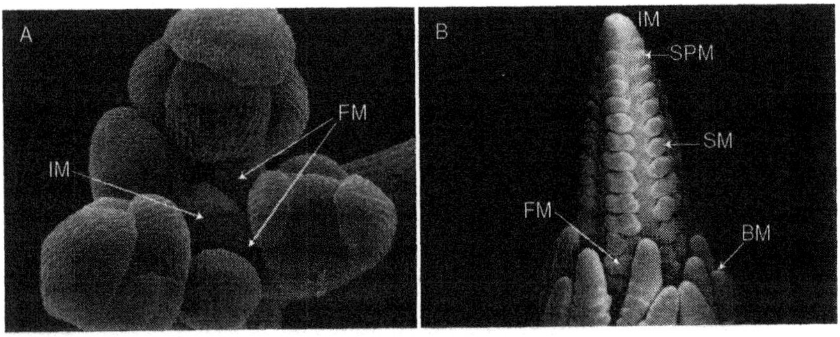

Fig. 4. Reproductive meristems in *Arabidopsis* vs the grasses. (A) The first meristem to form on the flanks of the *Arabidopsis* inflorescence meristem (IM) is the floral meristem (FM), which expresses *LFY* and *AP1*, thus conferring an FM identity. (B) With the exception of branch meristems (BM), the first meristems to form from the grass IM are either a spikelet pair meristem (SPM) or a spikelet meristem (SM). Shown here is maize, which produces an SPM that divides, forming two spikelet meristems. The spikelet meristems initiate two glumes and finally produce the floral meristems. Grass *AP1*-like genes and *Zfl* are both expressed in the SM and SPM without conferring a floral identity.

It remains to be seen if grass A-class mutants will also have defects outside of the floral meristem, or if, like *Zfla* and *Zflb*, they are expressed in these earlier meristems without effecting their identities. If it is true that their floral-promoting function is repressed during the early phases of grass inflorescence development, then there must be an unknown factor or factors responsible for this repression that are yet to be discovered in the grasses.

2. B-class

Angiosperm B-class genes belong to one of two groups that are the result of a duplication early in the evolution of angiosperms: *AP3/DEFICIENS* (*DEF*) and *PISTILATA* (*PI*)/*GLOBOSA* (*GLO*) (Stellari *et al.*, 2004). In the core eudicots, it is clear that these genes play a conserved role in conferring stamen and petal identity. As originally demonstrated in *Arabidopsis* (Goto and Meyerowitz, 1994; Jack *et al.*, 1992) and *Antirrhinum* (Sommer *et al.*, 1990; Trobner *et al.*, 1992) and in *Petunia* (Vandenbussche *et al.*, 2004), loss of B-class function leads to homeotic transformation of stamens into carpels and petals into sepals. Furthermore, AP3/DEF proteins and PI/GLO proteins are known to interact as obligate heterodimers to bind CArG-box elements in the promoters of target genes (Goto and Meyerowitz, 1994; Jack *et al.*, 1994; Riechmann *et al.*, 1996; Schwarz-Sommer *et al.*, 1992).

A duplication in the *AP3/DEF* lineage that occurred at the base of the core eudicots resulted in the paleo*AP3* and eu*AP3* lineages (see Kramer and Zimmer, Chapter 9). These two lineages are distinguished by a frame shift in the C-terminus that gave rise to the conserved euAP3 motif, distinct from the paleoAP3 motif found in basal eudicot, monocot, basal angiosperm, and core eudicot paleo*AP3* genes (Kramer *et al.*, 1998; Vandenbussche *et al.*, 2003). The expression of paleo*AP3* genes in the basal eudicots is consistently strong in stamens but often weak or patchy in petals (Kramer and Irish, 1999). This expression in petals is distinct from the strong stamen and petal expression seen for eu*AP3* genes in core eudicots. These observations have led to the proposal that B-class genes evolved a novel role in specifying petal identity in core eudicots coincident with the duplication creating the eu*AP3* lineage (Kramer and Irish, 1999, 2000). This hypothesis also appears consistent with classical morphological investigations of petal evolution, where petals are thought to have evolved independently multiple times in the angiosperms from either stamens or subtending sepals/bracts (Takhtajan, 1991). In this view, core eudicot petals are all interpreted as homologous organs derived a single time from stamens. A key prediction of the hypothesis proposed by Kramer and Irish is that paleo*AP3* genes would play no major role in specifying petal identity.

The only data for paleo*AP3* function comes from the grasses where knockouts have been characterized in both maize and rice. In both cases there is a single *AP3/DEF* ortholog, *Silky1* (*Si1*) in maize and *SUPERWOMAN1* (*SPW1*) in rice. In both *si1* and *spw1*, the same mutant phenotypes are seen, with homeotic conversion of stamens into carpels and lodicules into bract-like lemma/palea-type organs (Ambrose *et al.*, 2000; Nagasawa *et al.*, 2003). This has led to an interpretation of lodicules as second-whorl organs homologous to petals, and more loosely of palea/lemma-like bracts as first-whorl organs equivalent to sepals. The latter was further supported by the spikelet phenotype of the *si1 zag1* double mutant where these bract-like organs with palea/lemma characteristics proliferated within an otherwise normal pair of glumes (Ambrose *et al.*, 2000). This double-mutant phenotype was reminiscent of the mutant floral phenotypes observed in *ap3 ag* double mutants of *Arabidopsis* where only sepals are produced (Bowman *et al.*, 1989, 1991). Together, these observations led to the conclusion that B-class function is largely conserved across monocots and eudicots. However, another interpretation holds that lodicules may not be homologous to petals, and that B-class function was independently recruited to specify a lodicule identity in grasses (Irish, 2000). Whether B-class function in grasses should be interpreted as conserved or derived critically depends on the relationship of lodicules to second-whorl organs (petals or tepals) in other monocots, a subject addressed later (see Section V.B).

Unlike the single grass *AP3/DEF* lineage, there are two paralogous *PI/GLO* gene lineages in the grasses: *OsMADS2/Zmm16* and *OsMADS4/Zmm18/29* (Fig. 3B) (Chung *et al.*, 1995; Münster *et al.*, 2001). Expression of these *PI/GLO*-like genes in maize and rice is consistent with a role in B-class function, as their transcripts are present in developing stamen and lodicules (Chung *et al.*, 1995; Kyozuka *et al.*, 2000; Münster *et al.*, 2001; Whipple *et al.*, 2004). The only functional studies of these genes come from rice where an antisense suppression of *OsMADS4* resulted in homeotic conversion of lodicules and stamens similar to those observed in *spw1* and *si1* mutants (Kang *et al.*, 1998). However, the reported phenotype may not reflect a reduction in *OsMADS4* alone, as the authors did not rule out the possibility that the antisense construct used to silence *OsMADS4* may well have silenced *OsMADS2* as well. A more focused study by Prasad and Vijayraghavan (2003) used RNAi to specifically reduce *OsMADS2* levels, resulting in a loss of lodicule identity but no effect on stamens. This second study raises the interesting possibility that *OsMADS2* and *OsMADS4* have diverged in function, with *OsMADS2* playing a more key role in lodicule specification than stamen. The possibility of subfunctionalization is supported by the observation of Kyozuka *et al.* (2000) that *OsMADS2* expression is maintained in the lodicule, but quickly downregulated in the developing stamen primordia of the rice floret. A further understanding of the grass *PI/GLO*-like gene functions will require a more careful analysis of *OsMADS4* function as well as of the role of the corresponding maize genes.

The homeotic phenotype of the *si1* and *spw1* mutants discussed earlier make a strong case for conservation of B-class gene function in specifying organ identity. If it is accepted that lodicules are modified petals, then it appears that B-class function was conserved in the common ancestor of monocots and eudicots. A study of the biochemical function of maize B-class genes is consistent with this view (Whipple *et al.*, 2004). In this study, it was shown that the maize B-class genes *Si1* and *Zmm16* were capable of rescuing both petal and stamen identity in *Arabidopsis*. Furthermore, it was shown that the maize SILKY1 and ZMM16 proteins interact as an obligate heterodimer pair to bind DNA, as is the case with core eudicot AP3/DEF and PI/GLO proteins. Together, these findings appear consistent with a conserved B-class biochemical function between monocots and eudicots. However, the observation of obligate heterodimerization in maize is not so unambiguously interpreted as a conserved biochemical interaction as will be discussed in Section IV.

3. C-class

C-class genes in the eudicots control stamen and carpel organ identities, as well as conferring determinacy upon the floral meristem (Bowman *et al.*, 1989). In *Arabidopsis*, C-class is controlled by the action of a single gene, *AGAMOUS* (*AG*) (Yanofsky *et al.*, 1990), and loss-of-function mutants result in stamens converted to petals and a new flower arising in the position of the carpel reiterating the pattern: sepal, petal, petal, new flower.... In maize, two C-class genes were initially identified: *Zag1* and *Zmm2* (Mena *et al.*, 1996; Schmidt *et al.*, 1993; Theissen *et al.*, 1995). A knockout in *Zag1* shows that it plays a role in floral meristem determinacy (Mena *et al.*, 1996). However, the *zag1* mutant has little if any effect on carpel identity, and no discernable defect in stamen identity. This does not necessarily mean that maize C-class genes have no role in organ identity since *Zmm2* may be acting redundantly with *Zag1* to specify stamens and carpels. Expression domains of *Zag1* and *Zmm2* are overlapping but not identical. Both genes are expressed in stamens and carpels as would be expected for C-class genes, but *Zag1* is expressed more strongly in carpels and *Zmm2* is expressed more strongly in stamens (Mena *et al.*, 1996). This has led to the speculation that C-class function has been partitioned in maize, as would be expected from subfunctionalization following a gene duplication event (Force *et al.*, 1999). Overexpression and complementation studies in *Arabidopsis* indicate that such subfunctionalization is present at the level of protein function in addition to expression since *Zag1* is more capable of promoting carpel identity and *Zmm2* is better able to promote stamen identity (Whipple, C., Ambrose, B., Mena, M., and Schmidt, R., unpublished observation). *Zmm23*, a duplicate of *Zmm2* was also isolated from maize (Münster *et al.*, 2002). Further analysis of the maize C-class will require isolation of loss-of-function mutations for *Zmm2* and *Zmm23* to understand how they may contribute to organ identity, and if the apparent subfunctionalization at the level of gene expression is consistent with their mutant phenotypes, alone and in combination with *zag1*.

Studies of C-class function in rice focused initially on the *Zmm2* ortholog *OsMADS3* (Kang *et al.*, 1995; Kyozuka *et al.*, 2000). Overexpression of *OsMADS3* in tobacco leads to organ transformations consistent with C-class function (Kang *et al.*, 1995). A study in which *OsMADS3* was constitutively expressed in rice showed that it could convert lodicules into stamens, as would be predicted by the ABC-model, but had no effect on lemma or palea identity (Kyozuka and Shimamoto, 2002). Kang *et al.* (1998) expressed a transgenic *OsMADS3* antisense construct in an attempt to reduce *OsMADS3* expression in rice, with results largely consistent with a C-class function: stamens partially converted to lodicules, and carpels replaced by multiple flowers of undifferentiated carpels and stamens.

Expression and a functional characterization of both *OsMADS3* (the rice *Zmm2* ortholog) and *OsMADS58* (the rice *Zag1* ortholog) was recently reported (Yamaguchi *et al.*, 2006). Both genes are expressed in the floral meristem before the initiation of floral organs. Later on *OsMADS58* is maintained in both the developing stamen and carpel primordia, while *OsMADS3* expression is rapidly restricted to the ovule primordia. Two insertion alleles of *OsMADS3* had a phenotype, the strong T-DNA insertion line *osmads3-3* showed a near complete conversion of stamens into lodicules and a partial loss of meristem determinacy, while the weaker *Tos17* insertion line *osmads3-2* had a partial transformation of stamens and only an occasional loss of determinacy. The *OsMADS58* RNAi line *osmads58-s1* showed a complete loss of floral meristem determinacy. Additionally, carpel development was significantly affected, but only an occasional loss of stamen identity was reported. The phenotypes of *osmads58-s1* plants are similar to *zag1* mutants, although *OsMADS58* seems to have a more prominent role in carpel development.

The mutant phenotypes of *OsMADS3* and *OsMADS58* suggest that the rice C-class has been partitioned such that *OsMADS3* plays a more crucial role in the third whorl, and *OsMADS58* plays a more crucial role in the fourth whorl. While these results can largely be viewed as consistent with subfunctionalization of the two rice C-class genes as has been hypothesized for the maize C-class genes, some questions remain. For instance, it is still not clear what role C-class genes have in establishing carpel identity. The authors do not interpret the carpel defects seen in *osmads58* RNAi lines as loss of carpel identity, but rather as an indication that *OsMADS58* plays a later role in carpel development (it should be noted, however, that the carpels in *osmads58-s1* plants produce trichomes which are not formed on carpels of wild type plants and may indicate a partial loss of carpel identity). It is possible that *OsMADS3* is redundant with *OsMADS58* in establishing carpel identity. The authors report that silencing *OsMADS58* in the hypomorphic *osmads3-2* does not enhance the carpel defects seen in *osmads58-s1* plants. Since the *osmads3-2* allele retains some partial function, one cannot yet rule out a redundant role in establishing carpel identity. Another possibility, in light of an apparent role for *DROOPING LEAF* (*DL*) in establishing carpel identity (Section III.B), is that the rice C-class genes are redundant with *DL*.

Another interesting observation of Yamaguchi *et al.* (2006) was that *osmads3* mutants appear to develop extra lodicules in the second whorl, in the position where the medial lodicule aborts in wild type. Additionally, *OsMADS3* expression is detected in the position where the medial lodicule should develop. This may suggest that the rice C-class genes play a role in the abortion of the medial lodicule (abortion of the medial lodicule is common in many grasses although some bamboos have three). A role for C-class genes in establishing the lodicule asymmetry of grass flowers would represent an as yet undescribed function for this class of genes. There are some problems with this interpretation, however. If true, one would expect ectopic expression of *OsMADS3* in the second whorl to cause abortion of lodicules. However, Kyozuka and Shimamoto (2002) report that constitutive *OsMADS3* expression leads to conversion of lodicules to stamens. Furthermore, more than three lodicules develop in the second whorl of *osmad3-3* plants indicating that *OsMADS3* does not simply repress the medial lodicule. The indeterminacy of *osmads3-3* mutant flowers may obscure the whorl boundaries, such that these ectopic lodicules are produced by a larger, more indeterminate third whorl.

4. D-class

D-class genes were proposed as ovule identity genes when the *Petunia* MADS-box genes *FBP7* and *FBP11* were shown to be necessary for ovule specification (Angenent *et al.*, 1995; Colombo *et al.*, 1995; see Rijpkema *et al.*, Chapter 6). *FBP11* was also capable of forming ectopic ovules when constitutively expressed (Colombo *et al.*, 1995). *SEEDSTICK*, the

Arabidopsis homolog of this conserved clade (closely related to C-class genes), was shown to play a role in ovule development (Pinyopich *et al.*, 2003). A phylogenetic analysis of C- and D-class genes indicates that there are two clades of grass D-class genes (Fig. 3C) (Kramer *et al.*, 2004). The first clade includes *Zag2* and *Zmm1* from maize and *OsMADS13* from rice (Lopez-Dee *et al.*, 1999; Schmidt *et al.*, 1993; Theissen *et al.*, 1995). The second clade includes *Zmm25* from maize and the rice predicted gene P0408G07.14 (Kramer *et al.*, 2004; Münster *et al.*, 2002). There are, as yet, no informative functional studies for the grass D-class genes, but the maize *Zag2* and the rice *OsMADS13* expression patterns are consistent with other known D-class gene patterns, with expression early in the carpel primordia and subsequent restriction to the ovule (Lopez-Dee *et al.*, 1999; Schmidt *et al.*, 1993). Functional studies of these genes in grasses would provide insight into any conserved role they may play in ovule identity.

5. *E-class*

The *Arabidopsis SEPALLATA1, -2, -3,* and *-4* (*SEP1–4*) genes have been shown in a series of elegant genetic studies to be required for the activity of A-, B-, and C-class genes in specifying all the floral organs (Ditta *et al.*, 2004; Pelaz *et al.*, 2000). Without *SEP* function, the flower is replaced by an indeterminate shoot of lateral organs with a leaf-like identity (Ditta *et al.*, 2004). It is thought that the SEP proteins function as transcriptional activators in complexes of MADS-box proteins to mediate the activity of the ABC-genes (Honma and Goto, 2001; Pelaz *et al.*, 2001; Theissen and Saedler, 2001). The *SEP* clade is large and the result of both ancient and recent duplication events in the angiosperms (Zahn *et al.*, 2005). These complex duplication patterns are also evident in the grass representatives of the *SEP* clade (Fig. 3D) (Malcomber and Kellogg, 2004; Zahn *et al.*, 2005).

According to a phylogenetic analysis of *SEP* genes including basal angiosperms (Zahn *et al.*, 2005), there appear to be two large clades in the angiosperms: the *SEP3*(*AGL9*) clade and the *SEP1, -2, -4*(*AGL2, -4, -3*) clade. In the *SEP3* clade, a duplication has led to two grass lineages. The first clade, *OsMADS7*, includes maize *Zmm6* and rice *OsMADS7*, while the second, *OsMADS8*, clade includes maize *Zmm27* and rice *OsMADS8* (Fischer *et al.*, 1995; Kang *et al.*, 1997). In the *SEP1, -2, -4* clade there are three grass lineages: (1) the *OsMADS34* clade, including maize *Zmm24* and *Zmm31* and rice *OsMADS34* (also *OsMADS19*) (Münster *et al.*, 2002; Pelucchi *et al.*, 2002; Shinozuka *et al.*, 1999); (2) the *LHS1* clade, with maize *ZmLHS1a* and *ZmLHS1b* (aka *Zmm8* and *Zmm14*) and rice *LEAFY HULL STERILE1* (*LHS1* aka *OsMADS1*) (Cacharrón *et al.*, 1999; Chung *et al.*, 1994; Jeon *et al.*, 2000; Malcomber and Kellogg, 2004);

(3) the *OsMADS5* clade includes the maize *Zmm3* and rice *OsMADS5* (Fischer *et al.*, 1995; Kang and An, 1997).

It is clear that duplications have led to complexity in grass *SEP* lineages. Studies of grass *SEP* gene expression in both maize and rice indicate that their expression patterns are as complex as their lineages (Malcomber and Kellogg, 2004). The rice *lhs1* is the only described mutant in a grass *SEP*, and this lineage has been the most studied. The *lhs1* mutant results in flowers with lemma, palea, and lodicule transformed into leaf-like organs (Jeon *et al.*, 2000). However, *lhs1* is a semidominant mutation. For this reason, it is not clear what the true *lhs1* loss-of-function phenotype is. Nevertheless, expression in floral organs, and the semidominant mutant phenotype both indicate that this gene has a role in floral organ identity. Reduction of *OsMADS1* by RNAi led to a phenotype similar to *lhs1*, however some lines also had a conversion of stamens and carpels to leaf/glume-like organs indicating that *OsMADS1* is involved in establishing organ identity in all four whorls (Prasad *et al.*, 2005). A careful study of *LHS1* orthologs in multiple grass species has shown that expression patterns of this *SEP* lineage are highly variable (Malcomber and Kellogg, 2004). Such complexity in expression and lineage duplications, in combination with the relative lack of functional studies, make it difficult to understand how these genes may function in grasses. However, such complexity and lability in expression patterns also make this lineage an exciting group for continued study.

Analyzing the grass ABCDE MADS-box genes summarized in Table I, it becomes apparent that there is a lot of work yet to be accomplished to understand the function of these genes. True loss-of-function mutants are only available for two maize genes (*si1* and *zag1*) and two rice genes (*spw1* and *osmads3*). While patterns of gene expression and the few available mutant phenotypes appear consistent with what is predicted about their functions based on studies in eudicots, this appears mostly true for those genes controlling B- and C-organ identity functions. Perhaps this is not surprising; since the reproductive organs are thought to be homologous among all angiosperms it would seem logical that a single ancestral genetic mechanism for their specification would be conserved in divergent angiosperm groups. Little is known about how the nonreproductive organs of the grass flower are specified, and what role if any these MADS-box genes play. The major exception is the role of B-class genes in lodicule identity, although the interpretation of this remains controversial. Consequently, it will be interesting to see how genes predicted to have a role in first- and second-whorl organ identity (specifically A- and E-class genes) actually affect grass flower development.

TABLE I
Summary of Grass A B C D and E Class MADS-Box Genes Classified According to Their Phylogenetic Relationships[a]

	Eudicot lineaege	Maize	Rice	Other
A class	*AP1/CAL/FUL*	*Zap1a, Zap1b (ZmMADS3)**	*OsMDAS15*	*BM8, LtMADS2*
			OsMADS14	*BM5, LtMADS1, Vrn1**
		Zmm4, Zmm15	*OsMADS18*	
		Zmm28		*BM3*
B class	*AP3/DEF*	***Si1***	***SPW***	
	PI/GLO	*Zmm16*	*OsMADS2**	
		Zmm18, Zmm29	*OsMADS4**	
C class	*AG*	***Zag1***	*OsMADS58**	
		Zmm2, Zmm23	***OsMADS3****	
D class	*FBP11/STK*	*Zag2, Zmm1*	*OsMADS13*	
		Zmm25	P0408G07.14	
E class	*SEP3*	*Zmm6*	*OsMADS7*	
		Zmm27	*OsMADS8*	
	SEP1, 2, 4	*Zmm24, Zmm31*	*OsMADS34*	
		ZmLHS1a, ZmLHS1b	***LHS1****	
		Zmm3	*OsMADS5*	

[a]The first column indicates the closest eudicot lineage to each of the grass genes in that row. Bold genes have characterized mutants, and genes with a * indicate co-suppression or other studies of gene function.

B. THINKING OUTSIDE OF THE MADS-BOX: HOW FORWARD SCREENS REVEAL NON-MADS-BOX GENES IMPORTANT FOR GRASS FLORAL PATTERNING

It is clear from the work accomplished so far in grasses that MADS-box genes are playing a role in grass floral organ identity and patterning. It is not clear, however, that all aspects of the model are rigidly conserved. Testing the conservation of the ABC-model in grasses relies on the assumption that genes important in eudicots are good candidates for reverse genetic studies in maize and rice. While this strategy has been fruitful with B- and C-class genes, there is no reason *a priori* to believe that only MADS-box genes will be the major homeotic genes in grasses. In fact, work has shown that non-MADS-box genes may be playing crucial roles in grass floral patterning. These mutants demonstrate the importance of blind forward genetic strategies to understand the genes controlling grass floral development.

In rice, the *DL* mutation causes a homeotic conversion of carpels into stamens (Nagasawa *et al.*, 2003). In addition, *dl* mutants have a loss of floral meristem determinacy such that up to seven ectopic stamens are produced in severe alleles. *DL* was cloned and shown to be the rice ortholog of *CRABS CLAW* (*CRC*), a gene of the *YABBY* family that is important for proper development of the *Arabidopsis* carpel (Yamaguchi *et al.*, 2004). Unlike *CRC,* which is expressed primarily in abaxial tissues of the developing carpel, *DL* is expressed from the earliest stages of carpel initiation and maintained throughout all tissues of the carpel except the developing ovary, consistent with a direct role in carpel identity. Thus it appears that a non-MADS-box gene plays a crucial role in carpel specification in the grasses and that, unlike *Arabidopsis*, the *AG* pathway may not be the primary regulator of carpel identity. It is interesting that *Arabidopsis* also has an *AG*-independent pathway to specify carpels, as can be seen by the ectopic carpel growth in an *ag ap2* double mutant (Bowman *et al.*, 1991). This ectopic carpel identity is lost by removing *CRC* function, indicating that *CRC* and *AG* act in parallel pathways to specify *Arabidopsis* carpels, but that the *AG* pathway is more important (Alvarez and Smyth, 1999). In grasses, *DL* may have a more important role in establishing carpel identity than does its ortholog in *Arabidopsis*. It is also noteworthy that expression of a *CRC* ortholog in the basal angiosperm *Amborella trichopoda* is more similar to the abaxial pattern of expression observed in *Arabidopsis*, indicating that the carpel specification role of *DL* may be a derived function in the lineage leading to the grasses (Fourquin *et al.*, 2005).

While the data appear consistent with an important role for *DL* in establishing grass carpel identity, a more trivial explanation may be that *DL* is necessary for proper carpel *development* (not identity) as in *Arabidopsis*, and the homeotic transformation is an indirect result of *DL*'s role in keeping B-class expression out of the fourth whorl. A key prediction of this hypothesis is that removing B-class function from the *dl* mutant would result in a flower with malformed carpels in the central whorl. This experiment was done by Nagasawa *et al.* (2003), and they reported the growth of floral organs of "unknown identity" in the fourth whorl of a *spw1 dl* double mutant. It is possible that these organs are, in fact, the malformed carpels that would be predicted if *DL* function in rice was conserved with *CRC* function in *Arabidopsis*. Regardless of which explanation is more accurate, it is clear from the *dl* phenotype that the rice *CRC* ortholog has functions not seen in *Arabidopsis* that include control of the B-class expression domain, conferring determinacy upon the floral meristem, as well as a role in midrib development. These clear differences in *CRC* function between *Arabidopsis* and rice are interesting and demonstrate how the function of important developmental genes can change in the course of evolution.

The *palealess* (*pal*) mutant has been described in rice, which has defects in palea identity, but the other floral organs (lemma, lodicule, stamen, and carpel) are unaffected (Luo *et al.*, 2005). Most grasses have a palea with two prominent vascular bundles, but in the *pal* mutant two distinct leaf-like organs develop in the place of the palea. The authors suggest that the gene responsible for the *pal* phenotype could be a palea identity gene and thus may represent a grass A-class member. The *pal* mutation was mapped to a single BAC that has been sequenced, but contains neither a MADS-box gene nor an *AP2*-like gene, although there is another predicted transcription factor that may be *PAL*. Given that the mutant phenotype is only present in the palea without affecting the lodicules, and that it is not a MADS-box gene, the interpretation of *PAL* as an A-class gene seems premature. It will be interesting to learn the identity of *PAL*, and if further characterization is consistent with it being a palea identity gene. If true, this would suggest that grasses either have evolved a distinct mechanism to specify an organ unique to grasses, or that the grass outer "sepal" whorl is specified differently than what has been described in eudicots. Investigation of this gene in other grasses and outgroups may shed light on the evolution of the grass floret (Section V.B).

These two examples of non-MADS-box gene mutants affecting grass floral organ identity illustrate the importance of taking a forward, in addition to reverse, genetic approach to understand grass flowering. Unfortunately, grass floral organs are tightly enclosed in the developing spikelet, making large-scale mutant screens difficult. It is likely that many interesting grass mutants with defects in florets have been missed in the genetic screens performed. Our lab has undertaken a careful screen of spikelets and florets from ~1000 segregating EMS-mutagenized M2 maize families (for a searchable database of whole-plant phenotypes see http://www.maizegdb.org/ems-phenotype.php). Tassel branches were collected, dried, and stored until they could be carefully screened with the aid of a dissecting microscope for spikelet and floral phenotypes. Our initial results indicate that even in maize, which contains a significant number of duplicate genes, novel mutant phenotypes are to be found. Of the families screened, 12 mutants affecting spikelet and/or floret development were identified by this approach that were missed by field-based screens of the same families for inflorescence defects. Of these, at least two have floret phenotypes that have not been described previously. Such careful screens in a species with fewer gene duplications, such as rice, are likely to be even more productive.

IV. PHYSICAL INTERACTIONS AMONG MADS-BOX PROTEINS: FUNCTION AND EVOLUTION

MADS-box proteins are known to interact to form dimers and higher order complexes. The quartet model has been proposed to describe how tetramers of MADS-box proteins could interact to form a transcriptional activation complex sufficient to establish the identity of each of the four floral whorls

(Theissen and Saedler, 2001; see Melzer *et al.*, Chapter 5). Support for the quartet model is drawn in part from simultaneous overexpression of *Arabidopsis* MADS-box proteins known to interact *in vitro*, such as AP1/AP3/PI/SEP (a combination of A-, B-, and E-class proteins respectively), which results in leaves transformed into a petal identity (Honma and Goto, 2001; Pelaz *et al.*, 2001). These studies make it clear that the apparent protein–protein interactions of MADS-box gene products are critical to their function.

Protein–protein interaction among MADS-box gene products have been described in rice. Favaro *et al.* (2002) showed that the rice D-class protein OsMADS13 interacts with the E-class SEP homologs OsMADS8 and OsMADS7 in a yeast two-hybrid assay. An interaction of the rice A-class protein OsMADS18 with these same two SEPs was shown by yeast two-hybrid (Fornara *et al.*, 2004), and further verified by coimmunoprecipitation. Yeast two-hybrid interaction for the B-class protein SPW1 with OsMADS8 has also been reported (Lee *et al.*, 2003). These studies indicate an interaction between the diverse A-, B-, and D-class rice proteins and rice SEP orthologs. Such an interaction would be predicted from a conserved quartet model, where SEP proteins are necessary for the function of florally expressed MADS-box proteins. It is also interesting to note that each of these studies further indicated an interaction with the rice protein OsMADS6, an ortholog of the *Arabidopsis* AGL6. Although this gene has not been functionally characterized in *Arabidopsis*, it is closely related to the *SEP* subfamily (Zahn *et al.*, 2005), suggesting the *OsMADS6* lineage may also have an important role in grass floral development.

Unfortunately, while the above-mentioned studies suggest conservation of protein–protein interactions of grass SEP proteins with other ABCD-proteins, little is known about the *in vivo* relevance of such interactions in the grasses. Among MADS-box proteins analyzed, most form homodimers capable of binding DNA in EMSAs (Riechmann *et al.*, 1996). One exception to this ability to dimerize is the eudicot B-class proteins AP3 and PI which are known to interact as obligate heterodimers. This obligate heterodimerization appears to have direct functional consequences. Unlike other duplicate MADS-box genes such as *AP1, CALIFLOWER* (*CAL*), and *FUL* or *SEP1*, *-2*, *-3*, and *-4*, which are known to act redundantly in *Arabidopsis*, the duplicate B-class genes show no apparent functional redundancy, and mutations in either gene give nearly identical phenotypes, with no more severe phenotype in the double mutant. This is likely a result of obligate heterodimerization since any protein complex containing a single B-class group member would be incapable of binding DNA and thus incapable of promoting any aspect of B-class function. There has been interest in how B-class proteins interact, and how obligate heterodimerization evolved.

Winter *et al.* (2002) used a phylogenetic approach to examine the evolution of B-class protein interactions. A combination of EMSA and yeast two-hybrid assays indicated that a *Gnetum gnemon* (a gymnosperm) B-class protein and a *Lilium regale* (a monocot) PI/GLO ortholog were both able to form homodimers capable of binding DNA. The *Lilium* AP3/DEF ortholog required its PI/GLO partner to bind DNA. The most parsimonious interpretation of these findings is that the ancestral state for B-class proteins is to form homodimers, and that this was successively lost, first in the AP3/DEF lineage, and eventually in the PI/GLO lineage resulting in the obligate heterodimerization system described in eudicots. This study, while useful as an initial broad comparison, was unfortunately limited by a small sampling of angiosperm B-class proteins, and a robust parsimony reconstruction of this ancestral state would require more sampling. Another potential drawback is that it did not include any basal angiosperm members of the AP3 and PI lineages. The *Gnetum* B-class protein descended from a lineage that diverged prior to the duplication event that created the AP3 and PI lineages. Consequently, the ancestral states in the angiosperm AP3 and PI lineages are still ambiguous. It is entirely possible that obligate heterodimerization evolved early in the history of the angiosperms, and was subsequently lost in the *Lilium* PI/GLO lineage. Another possibility is that dimerization specificity is labile in noneudicot B-class proteins, resulting in multiple gains and losses, which would further complicate a limited survey of angiosperm B-class proteins. A larger analysis of basal monocot and basal angiosperm B-class proteins will help to clarify these issues.

That there could be lability in the dimerization properties of B-class genes is suggested by a study of the maize B-class proteins SILKY1 and ZMM16. Whipple *et al.* (2004) show that these proteins form an obligate heterodimer pair similar to *Arabidopsis* B-class proteins. Furthermore, no maize B-class homodimer is likely to be functional in the traditional B-class roles of specifying second- and third-whorl organ identities because the *silky1* mutant shows complete loss of organ identity and the expected homoeotic conversions in these whorls, indicating no functional redundancy and consistent with obligate heterodimerization. That two monocot PI/GLO proteins have different dimerization properties further complicates the parsimony reconstruction of Winter *et al.* (2002). Did obligate heterodimerization evolve independently in the grasses, or was it simply lost in the *Lilium* PI/GLO lineage? In order to address these questions our lab has begun examining the DNA-binding properties of B-class proteins in *Joinvillea ascendens*, a close outgroup to the grasses. Our results indicate that the *Joinvillea* PI protein (JaPI), unlike ZMM16, is capable of binding DNA as a homodimer, while the *Joinvillea* AP3 protein (JaAP3) is not (Fig. 5).

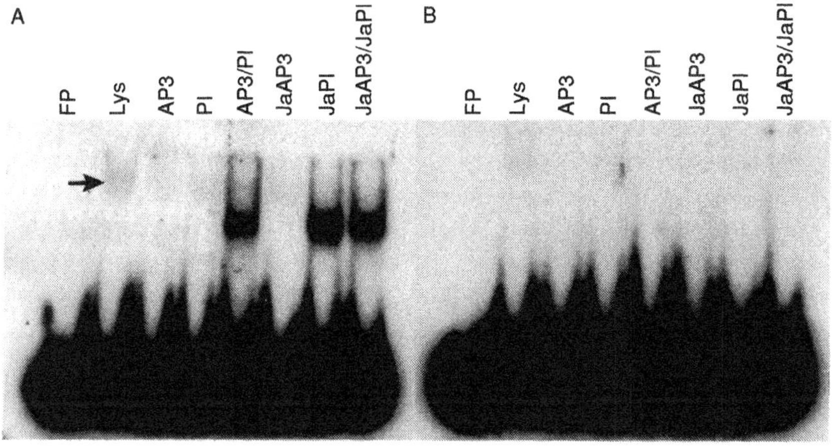

Fig. 5. Electromobility shift assay (EMSA) of *Arabidopsis* (AP3 and PI) and *Joinvillea ascendens* (JaAP3 and JaPI) B-class proteins. EMSA was performed as previously described (Whipple *et al.*, 2004). Binding specificity was tested using radiolabled probe containing a CArG box (5′-TTAGGCAATACTTTCCATTTTT-GGTAACTC-3′, CArG box underlined) derived from the *Arabidopsis AP3* promoter (A), or a mutated version (B). (A) Neither *Arabidopsis* protein binds DNA alone, but do together (AP3/PI lane) as indicated by upward shift in labeled probe. JaAP3 does not bind DNA alone, but JaPI does. Shift in JaAP3/JaPI lane could be due to JaPI homodimer or JaAP3-JaPI heterodimer, or a combination of the two. (B) A mutant *AP3* CArG-box (5′-TTAGGCAATACTTT**GG**ATTTTT**CC**TAACTC-3′, *mutations in bold*) abolishes all binding by both *Arabidopsis* and *Joinvillea* B-class proteins, indicating that the binding seen in A is specific. *Joinvillea* B-class genes (*JaAP3* and *JaPI*) were amplified by polymerase chain reaction (PCR) on cDNA from immature inflorescence tissue using a degenerate B-class-specific MADS-box primer and a poly-T primer. Phylogenetic analysis indicated that the B-class genes were sister to the grass genes (Whipple, C., Zanis, M., Kellogg, E., and Schmidt, R.; unpublished results). All B-class cDNAs were subcloned into the pSPUTK vector (Stratagene), then transcribed and translated using the TNT Quick Coupled Transcription Translation System (Promega). Arrow in A indicates a shift due to background proteins present in the lysate. FP, free probe; Lys, lysate control without plasmid added.

The ability of JaPI to bind DNA as a homodimer is similar to the other monocot PI proteins that have been analyzed. Together these results suggest that the ancestral state in the monocots is for PI proteins to homodimerize, and that the maize PI protein ZMM16 (and by extension other grass proteins of this lineage) lost this capacity. It is still unclear whether the other maize PI proteins, ZMM18 and -29, function as homodimers or as obligate heterodimers like ZMM16.

That JaPI can homodimerize, but the closely related maize ZMM16 cannot, suggests that this protein–protein interaction can be quickly lost. Such plasticity makes it problematic to assign an ancestral state to the monocot PI lineage. The possibility that this protein–protein interaction is labile can only be discarded after a broad and thorough analysis of monocot PI proteins. If it is true that homodimerization is easily gained and/or lost, perhaps it is a selectively neutral character and of little functional importance. If, however, it is found that the grasses have independently evolved another case of obligate heterodimerization, then it seems more likely that this change seen in the grass PI-like protein (compared to JaPI) is not neutral and has a functional relevance. What the function of a PI homodimer could be in the monocots is unclear. It is possible that the grass ZMM18, -29/OsMADS4 PI-like clade maintains the ability to homodimerize, in which case it would be possible to begin assessing any differential function by knocking out the genes in both *PI*-like clades in rice or maize.

Another avenue of research that could prove enlightening is to examine the domains and amino acids involved in this evolutionary change from JaPI to ZMM16. Domain swaps could be done between JaPI and ZMM16 to identify the region of the protein that is responsible. B-class MADS-box proteins have the stereotypical MIKC structure (MADS DNA-binding domain, Intervening region, Keratin-like coiled-coil domain, and C-terminal domain) (Riechmann and Meyerowitz, 1997). The K domain is thought to be important for heterodimerization (Yang and Jack, 2004; Yang *et al.*, 2003), although other domains may also play a role in protein–protein interactions. It will be interesting to see if the K domain is responsible for the difference between JaPI and ZMM16. After determining the domain, the high amino acid conservation between these closely related proteins could even allow identification of the critical amino acid(s), which could be tested by site-directed mutagenesis providing valuable structure-function insight into this evolutionary change of protein interactions.

V. THE GRASS FAMILY AS A MODEL SYSTEM FOR EVO-DEVO

So far this chapter has discussed primarily what is known about MADS-box gene function in the grasses, with emphasis on genetic studies from rice and maize. Unfortunately, there is as yet no other angiosperm genetic model species outside of the eudicots. Most floral evo-devo work has relied on extrapolations from gene expression patterns, inferring gene function based on what is known about the orthologous gene activity in *Arabidopsis* or other model species, and ectopic overexpression studies in *Arabidopsis* to indicate gene activity from gain of function phenotypes—all approaches that are prone to artifacts and ambiguous interpretations. Without some rigorous test of gene function, evo-devo hypotheses will continue to be hopeful "just-so stories" at best. The ability in maize and rice to obtain and analyze loss-of-function phenotypes has made the grasses an important group for understanding the conservation of genetic mechanisms elucidated primarily in eudicot model species. Maize and rice have established communities with a number of genetic resources that make them excellent for use in evo-devo studies, including approaches for performing forward and reverse genetics, and robust transformation technologies. We have discussed some ways these genetic strategies have already proven useful in understanding flower development in an angiosperm group distantly related to the core eudicots. Now we would like to consider some characteristics of the grass family itself that make it a model group for evo-devo, including some characteristics that at first glance might be considered drawbacks.

A. GENE DUPLICATION AND SUBFUNCTIONALIZATION

Phylogenetic studies of diverse MADS-box gene families in the angiosperms give evidence of gene duplication events occurring at multiple taxonomic levels (Kramer *et al.*, 1998, 2004; Litt and Irish, 2003; Stellari *et al.*, 2004; Zahn *et al.*, 2005). It is intriguing to speculate as to why so many duplicate copies have been maintained resulting in a radiation in MADS-box genes. An initial examination of MADS-box gene phylogenies shows that the gene radiations often correlate with radiations of successful plant groups, including the duplications at the base of the angiosperms and core eudicots. It is possible that duplications increase the molecular complexity necessary for morphological evolution, and thus mediate adaptive radiations. In other words, groups with more regulators of flower development can more easily evolve new morphologies and thus adapt more easily to diverse niches leading to increased speciation. Testing such speculative hypotheses will first require a careful examination of the roles of duplicate genes.

Ohno (1970) followed by Force *et al.* (1999) provided the theoretical background to the possible consequences of a gene duplication event. They define three processes that should occur after duplication. The first possibility is nonfunctionalization in which one of the duplicate copies begins to accumulate deleterious mutations until it becomes a nonfunctional pseudogene, and eventually disappears completely. Nonfunctionalization is thought to be common since duplications create completely redundant copies and there is no reason for selection to maintain both genes. Genome sequences show that such pseudogenes exist. Furthermore, that paralogs are often lost entirely is suggested by the existence of single-copy genes even though multiple other gene lineages are duplicated following a genome-wide duplication event as happened in an ancestor of maize. The second process, neofunctionalization, results in one of the gene copies gaining a new function that is selectively advantageous, and explains why some paralogs could be maintained. A neofunctionalization event is likely to be rare since most mutational events are deleterious. In the final process, subfunctionalization, both genes are maintained because they accumulate complementary mutations that make it necessary to have both paralogs present to maintain the function of the preduplication gene. As most genes are expressed in multiple tissues, mutations could occur in enhancer elements of the duplicates with the result that each gene is now expressed in different, but complementary, subdomains of the ancestral gene. Thus only by keeping both duplicates is the entire ancestral expression pattern maintained. Finally, mutations could also occur in the coding sequence of duplicate genes resulting in subfunctionalization at the level of protein function.

An interesting case of subfunctionalization appears to have occurred in the eudicot C-class lineage. In *Arabidopsis*, the C-class functions of organ identity and floral meristem determinacy are performed by *AG*, while in *Antirrhinum* the same activities are preformed by *PLENA* (*PLE*) (Coen and Meyerowitz, 1991), which was the purported ortholog based on its close sequence identity and function. However, evidence from more roboust gene sequence comparisons (Kramer *et al.*, 2004) and a direct comparison of syntenic regions from *Arabidopsis* and *Antirrhinum* (Causier *et al.*, 2005) have demonstrated that *AG* and *PLE* are not orthologs, as previously believed, but paralogs. In *Arabidopsis*, the *AG* duplicate *SHATTERPROOF* (actually encoded by two more recent, redundant duplicates *SHP1* and *-2*) has an essential role in establishing valve margin identity in the *Arabidopsis* fruit (Ferrandiz *et al.*, 2000; Liljegren *et al.*, 2000). In *Antirrhinum* the *AG* ortholog is *FARINELLI* (*FAR*), while the *SHP* ortholog is actually *PLE.* The organ identity and meristem determinacy functions in *Antirrhinum* are not controlled by the *AG* ortholog *FAR*, but by the *SHP* ortholog *PLE. far* mutants only have defects in some aspects of stamen development (Davies *et al.*, 1999), and so the function of *FAR*, like *SHP*, is more limited in scope. It is interesting that orthologous genes do not have the same function in *Arabidopsis* and *Antirrhinum*, but closely related paralogs do. This indicates that following the duplication event that created *AG/FAR* and *SHP/PLE* lineages, the paralogs subfunctionalized differently in each lineage. It remains an open question if such differential subfunctionalization is common, or if this represents and unusual occurrence. Either way it is clear that analyzing duplicate genes in divergent species that contain both genes can yield surprising insights into the evolution of developmental mechanisms. It also raises a cautionary note concerning assignment of functional orthology between closely related genes from different species even when evidence of similar function can be obtained.

As can be seen from Table I, multiple MADS-box lineages have been duplicated in the grass family. Duplications present in both rice and maize appear to be the result of a genome-wide polyploidization event that occurred before the radiation of the grasses (Paterson *et al.*, 2004). An allotetraploidy event in a common ancestor of *Tripsacum* and maize (Bomblies and Doebley, 2005; Gaut and Doebley, 1997) has led to additional duplications in maize gene lineages. Such duplications can confound initial reverse genetic attempts to assign function to the grass genes due to redundancy in function among closely related paralogs. However, such duplications are also an excellent chance to study the fate of duplicate genes involved in flower development, and test the proposals of Force and Lynch about subfunctionalization and neofunctionalization that should be responsible for maintaining these genes. If there is an obvious subfunctionalization of expression patterns for a duplicate pair in either maize or rice, other grasses can be examined to see if the pattern of subfunctionalization is conserved for orthologs, or if there is variability in the manner in which gene duplicates are expressed during the course of evolution. With a nearly complete rice genome and the maize genome project just beginning, it may be possible in the near future to use bioinformatics methods to quickly identify potential enhancers that are responsible for the expression differences using methods such as those described to identify conserved noncoding sequences (CNSs) (Guo and Moose, 2003; Inada *et al.*, 2003; Kaplinsky *et al.*, 2002). A complementary method to investigate subfunctionalization would be to compare the phenotypes from loss-of-function mutations for both duplicates in both maize and rice, which phylogenetically span the majority of grass diversity. Reverse genetic resources including transposon insertions (http://tos.nias.affrc.go.jp/~miyao/pub/tos17/, http://mtm.cshl.edu/) and TILLING (Till *et al.*, 2004) have been established, allowing such investigations of grass duplicate genes. Such reverse genetic methods have confirmed subfunctionalization of the duplicate rice C-class genes (Section III.A.3), although a full comparison with the maize C-class must await characterization of *Zmm2* and *Zmm23*.

These reverse genetic studies will take time, especially for maize where many of the genes have undergone an additional duplication. However, two established genetic model species make the grasses ideal to study family-level duplications, and will likely lead to insights into what happens when transcription factors controlling important developmental processes radiate.

B. THE EVOLUTION OF DERIVED MORPHOLOGIES

As discussed in Section II, grasses have highly modified flowers relative to other monocots. This is especially true for the nonreproductive organs where homology assignments are difficult. This can make interpreting grass floral mutants and expression patterns in relation to other monocots or eudicots difficult. However, these difficulties also present an opportunity to explore how derived morphologies evolve.

A prerequisite for any such study is a robust phylogeny, which has been worked out for the grasses by the GPWG (2001) (Fig. 6; see also Fig. 2, Malcomber *et al.*, Chapter 11). This phylogeny indicates that there are two major clades, which contain the majority of grass species. Members of these two clades have the standard grass spikelet as described in Section II. There

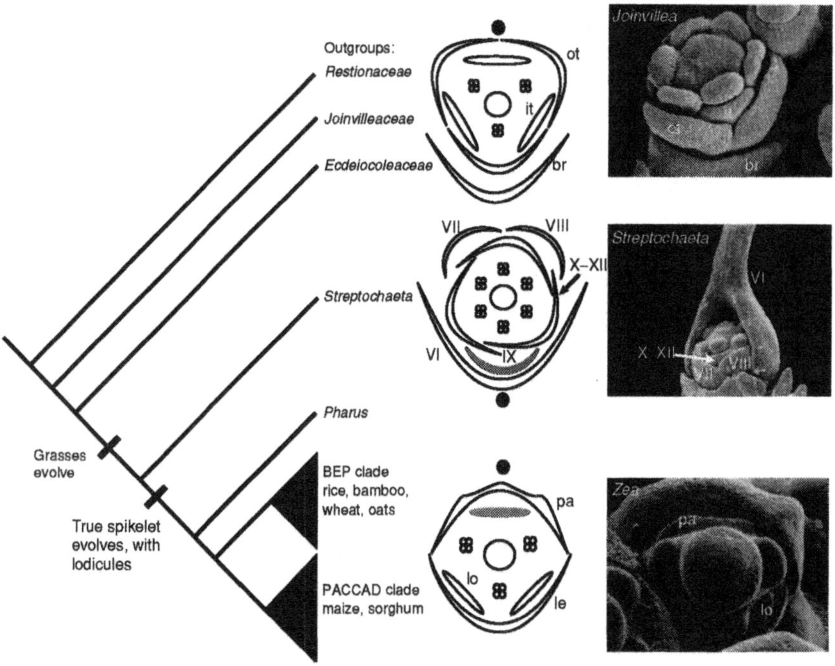

Fig. 6. Phylogeny of grasses and outgroups, with generalized floral diagrams for a grass spikelet, an outgroup monocot flower, and the intermediate morphology of the *Streptochaeta* spikelet equivalent. Bracts X–XII are in the relative position where lodicules (lo) are normally found in the standard grass floret, and also in the position where inner tepals (it) of outgroup monocot flowers are typically located. Bracts VII–IX of *Streptochaeta* are in the relative position of outer tepals (ot) of a standard monocot flower. Fusion of bracts VII and VIII, and abortion of bract IX would lead to the palea (pa) of the grass floret. The large bract VI could then correspond to a floral bract (br) in outgroups and the lemma (le) of the grass spikelet. Gray shaded organs in *Streptochaeta* and the grass floret indicate common abortion of bract IX and the medial lodicule. Black dots represent the relative position of the stem. BEP clade includes the grass subfamilies Bambusoideae, Ehrhartoideae, and Pooideae, while the PACCAD clade includes Panicoideae, Arundinoideae, Centothecoideae, Chloridoideae, Aristidoideae, and Danthonioideae.

are also some basal clades of herbaceous bamboos represented by *Pharus*, which also have a spikelet. However, the most basal extant grasses do not have a traditional spikelet and include the genera *Streptochaeta* and *Anomochloa* (Judziewicz and Soderstrom, 1989). *Anomochloa* has one species and is only known from a single population in Brazil. The *Anomochloa* spikelet equivalent consists of two successive bracts that initiate opposite each other, with the second bract enclosing the flower. There appear to be no sterile outer whorls as in other grasses, but a ciliated fringe does surround the stamen whorl. Unfortunately, *Anomochloa* is endangered and is not readily available for comparative studies. *Streptochaeta*, on the other hand, is more common and grows readily in greenhouse conditions, making it possible to examine expression of candidate genes. The *Streptochaeta* spikelet equivalent is composed of 12 bracts (generally designated I–XII). The first five basal bracts (I–V) are spirally arranged, and can occasionally contain axillary meristems that themselves grow into spikelet equivalents. Bract VI is large and encloses the rest of the floral organs, and appears to be modified for animal dispersal of the mature seed. Generally two shorter, pointed bracts (VII–VIII) develop opposite the large bract VI, although occasionally the remnant of a third bract (IX) can be detected in this whorl, indicating that it usually aborts. There is a whorl of three overlapping bracts (X–XII) that surround the stamens and enclose them until they emerge for pollination. Two whorls of three stamens each, and a gynoecium with three stigmas develop inside bracts X–XII. The closest outgroups to the grasses include the families Joinvilleaceae, Ecdeiocoleaceae, Restionaceae, and Flagellariaceae. These outgroups have a standard monocot floral plan, with two alternating whorls of outer tepals.

Knowing the phylogenetic position of *Streptochaeta* helps to form a hypothesis about how the derived grass floret evolved. The *Streptochaeta* spikelet equivalent is more likely to represent an ancestral grass floral morphology since it diverged before the evolution of true spikelets and the stereotypical grass floral organs. Even a cursory look at *Streptochaeta* in relation to true spikelets and outgroup flowers suggests that *Streptochaeta* can be interpreted as intermediate in morphology. Although it has no lodicules, it does have three laminar bracts with a distinct morphology in the position of lodicules, which also correspond to the inner tepal whorl of outgroups. Bracts VII and VIII are then in the position of an outer tepal whorl, with abortion of the medial bract IX. By fusion of bracts VII and VIII one can produce a palea of a true grass floret. This suggests that the grass palea represents a modified outer tepal whorl. Consistent with this view is the observation (discussed in an earlier section) that the *spw1* mutant of rice and the *si1* mutant of maize transforms lodicules into organs with a palea-like identity. It is also interesting that the rice *pal* mutant results in two unfused bracts growing in the place of a palea, indicating that the palea is in fact a fusion of two distinct primordia. This would then suggest that the large bract VI of *Streptochaeta* is an intermediate between the lemma and subtending floral bract of outgroups. Although there was no understanding of the current grass phylogeny or closest outgroups, this scheme for the homology of the derived grass floral organs was suggested nearly a century ago (Schuster, 1910). Similar interpretations of *Streptochaeta* have been suggested previously, although there is little consensus in the literature about the proper interpretation of the *Streptochaeta* spikelet equivalent (Judziewicz and Soderstrom, 1989; Page, 1951; Soderstrom, 1981).

This hypothesis of floret evolution can be tested using grass genes that are consistent markers of organ identity. For example, B-class genes are known to mark stamen and lodicule identity in grasses with a true spikelet. If bracts X–XII of *Streptochaeta* are, in fact, intermediates between the inner tepals of outgroups and the lodicules of other grasses, then one would expect to see B-class gene expression in stamens and bracts X–XII of *Streptochaeta*. Additionally, there should be expression in stamens and the inner tepal whorl of monocot outgroups. Our lab has isolated B-class genes from *Streptochaeta* as well as outgroups, and *in situ* expression patterns of these genes are entirely consistent with this hypothesis (Whipple, C. J., Zanis, M., Kellogg, E. A., and Schmidt, R. J., unpublished data). Unfortunately there are no robust markers of palea identity yet. However, as we learn more about genes controlling grass flowering other aspects of this hypothesis can be tested.

C. SYNTENY AND THE GRASS FAMILY AS AN INTEGRATED GENETIC SYSTEM

Comparative mapping studies in diverse grasses have indicated that grass genomes have a high degree of colinearity, or synteny, meaning broad regions of chromosomes in different species share the same genes in roughly the same order (Moore *et al.*, 1995). This colinearity among grass genomes has important implications for comparative studies of evolution and development in the grass family as has been noted by others (Freeling, 2001). With a complete draft sequence of the rice genome, and beginning sequencing of the maize and sorghum genomes, it is possible to use these species as "reference" genomes for mapping studies in nonmodel grasses. If it is possible to form fertile F1 hybrids from closely related species or subspecies that differ in morphological traits of interest, then it is possible to use the powerful methods of quantitative trait loci (QTL) mapping to define the approximate number and location of genes affecting the trait. An excellent example of this work is that of Doebley and colleagues in identifying the QTL that differentiate maize from its wild ancestor teosinte (Doebley and Stec, 1991). In theory it should be possible to map the loci differentiating any two grasses that form a fertile F1. With more sophisticated mapping techniques it is becoming reasonable to clone the genes underlying individual QTL (Salvi and Tuberosa, 2005).

Just such a QTL study with the nonmodel domesticated grass *Setaria italica* and its wild progenitor *Setaria viridis* was begun (Doust *et al.*, 2004, 2005). These two species differ significantly in vegetative and inflorescence branching habits. Maize RFLP probes were used to create an initial map of the *Setaria* genome, and a QTL analysis was performed using the same markers. Candidate genes such as *Tb1* from maize, a major determinant of vegetative branching differences between maize and teosinte, did not explain as much of the branching differences as did two other QTL to which no obvious candidate genes in maize mapped. This suggests that a QTL approach has identified genes for branching in *Setaria* that have yet to be identified in maize (perhaps due to maize gene redundancy, as discussed earlier), or that these mapping studies have identified grass genes that evolved more significant roles in this developmental pathway in the *Setaria* lineage than they did in maize. Markers flanking QTL of interest were located to syntenous regions on the rice genome and used to identify potential candidate genes that underlie important QTL. As these candidates are evaluated, new genes important for morphological changes selected during domestication should come to light.

The domestication studies in maize and *Setaria* demonstrate the power of QTL analysis to reveal the genetic basis of morphological variation among closely related species. That these techniques were applied successfully to a nonmodel grass is hopeful. The majority of QTL studies performed to date has been on agricultural species, and thus are biased toward traits selected by humans. The critical next step is to understand the genetic basis of morphological differences in wild, undomesticated species. There is impressive morphological variation among grass species, especially in inflorescence architecture. All that is needed is a fertile F1 from two wild species that differ for traits of interest. The work of Doust *et al.* (2004, 2005) demonstrates that it is not unreasonable to use grass synteny to move from an initial QTL study to a list of candidate genes to evaluate, even with little sequence information available for the grass species being studied.

VI. CONCLUSIONS

Because of its history as a great genetic system and due to its development as one of the first plant systems for reverse genetics, maize emerged early as a model for testing the applicability of the ABC-model of flower development. Rice, with its comparative ease of transformation, provided another grass species that was amenable to dissecting gene function through antisense and later, RNAi approaches. These two grasses afforded functional insights beyond gene sequence comparisons and comparative expression analyses of floral organ identify genes in other grass species. From these studies it appears that B-class gene activities are largely conserved, with these genes in the grasses specifying lodicule and stamen development, as compared to petal and stamen development in the core eudicots. C-class gene activity appears also conserved in terms of pattern of gene expression, having a role in stamen and possibly carpel identity and floral meristem determinacy, although as discussed, there may be elaborations on this program of development that are unique to the grasses. Gene redundancy in maize has complicated a thorough functional analysis to date. This is especially true regarding the role of genes in the A- and E-class gene lineages, where the lineage of *AP1/FUL*-like and *SEP*-like genes in the grasses appear to have radiated early in their evolutionary history. However, a combination of reverse genetics to understand the role of grass ABCDE-class genes, with forward genetics to uncover unknown genes important to grasses, promises to further advance our understanding of the genetics of grass flower development. Finally, the combined use of rice and maize genomic resources has facilitated map-based cloning of maize genes (Bortiri *et al.*, 2006; Wang *et al.*, 2005) portending a time when walking to a gene in these species will be as commonplace as in *Arabidopsis*.

On the completion of the closely related sorghum and maize genome sequences, the combined genomic resources for maize, sorghum, and rice will provide a strong platform from which to explore the evolution of floral development not only in these species but also across all the grasses. With rice and maize representing the majority of species diversity in the grass family, the analysis of the gene function and corresponding gene sequence changes that have occurred during their 50–60 million years of evolution will provide important insights into developmental genetics and comparative evolution of all the grasses. Approaches including comparative functional analyses of duplicated genes important in development, as well as QTL studies in morphologically distinct species, will provide a wealth of information on the processes by which evolution modifies morphology and developmental pathways. As derived members of the monocot lineage, the analysis of not only floral development, but many aspects of angiosperm development, in these groups will continue to provide important comparisons with the model eudicots.

ACKNOWLEDGMENTS

We wish to thank José Dinneny for the *Arabidopsis* SEM used in Fig. 4A, and Chris Padilla for the maize inflorescence SEM in Fig. 4B. This work was supported by funding from the National Science Foundation, and a fellowship from the ARCS Foundation to CJW.

REFERENCES

Alvarez, J. and Smyth, D. R. (1999). CRABS CLAW and SPATULA, two *Arabidopsis* genes that control carpel development in parallel with *AGAMOUS*. *Development* **126**, 2377–2386.

Ambrose, B. A., Lerner, D. R., Ciceri, P., Padilla, C. M., Yanofsky, M. F. and Schmidt, R. J. (2000). Molecular and genetic analyses of the silky1 gene reveal conservation in floral organ specification between eudicots and monocots. *Molecular Cell* **5**, 569–579.

Ambrose, B. A., Espinosa-Matias, S., Vazquez-Santana, S., Vergara-Silva, F., Martinez, E., Marquez-Guzman, J. and Alvarez-Buylla, E. R. (2006). Comparative developmental series of the Mexican triurids support a euanthial interpretation for the unusual reproductive axes of *Lacandonia schismatica* (Triuridaceae). *American Journal of Botany* **93**, 15–35.

Angenent, G. C., Franken, J., Busscher, M., van Dijken, A., van Went, J. L., Dons, H. J. and van Tunen, A. J. (1995). A novel class of MADS-box genes is involved in ovule development in petunia. *Plant Cell* **7**, 1569–1582.

Bomblies, K. and Doebley, J. F. (2005). Molecular evolution of *FLORICAULA/LEAFY* orthologs in the Andropogoneae (Poaceae). *Molecular Biology and Evolution* **22**, 1082–1094.

Bomblies, K., Wang, R. L., Ambrose, B. A., Schmidt, R. J., Meeley, R. B. and Doebley, J. (2003). Duplicate *FLORICAULA/LEAFY* homologs zfl1 and zfl2 control inflorescence architecture and flower patterning in maize. *Development* **130**, 2385–2395.

Bommert, P., Satoh-Nagasawa, N., Jackson, D. and Hirano, H. Y. (2005). Genetics and evolution of inflorescence and flower development in grasses. *Plant and Cell Physiology* **46**, 69–78.

Bortiri, E., Chuck, G., Vollbrecht, E., Rocheford, T., Martienssen, R. and Hake, S. (2006). ramosa2 Encodes a LATERAL ORGAN BOUNDARY domain protein that determines the fate of stem cells in branch meristems of maize. *Plant Cell* **18**, 574–585.

Bowman, J. L., Smyth, D. R. and Meyerowitz, E. M. (1989). Genes directing flower development in *Arabidopsis*. *Plant Cell* **1**, 37–52.

Bowman, J. L., Smyth, D. R. and Meyerowitz, E. M. (1991). Genetic interactions among floral homeotic genes of *Arabidopsis*. *Development* **112**, 1–20.

Cacharrón, J., Saedler, H. and Theissen, G. (1999). Expression of MADS-box genes ZMM8 and ZMM14 during inflorescence development of *Zea mays* discriminates between the upper and the lower floret of each spikelet. *Development Genes and Evolution* **209**, 411–420.

Causier, B., Castillo, R., Zhou, J., Ingram, R., Xue, Y., Schwarz-Sommer, Z. and Davies, B. (2005). Evolution in action: Following function in duplicated floral homeotic genes. *Current Biology* **15**, 1508–1512.

Chung, Y.-Y., Kim, S. R., Finkel, D., Yanofsky, M. F. and An, G. (1994). Early flowering and reduced apical dominance result from ectopic expression of a rice MADS-box gene. *Plant Molecular Biology* **26**, 657–665.

Chung, Y.-Y., Kim, S.-R., Kang, H.-G., Noh, Y.-S., Chul, P. M., Finkela, D. and An, G. (1995). Characterization of two rice MADS-box genes homologous to *GLOBOSA*. *Plant Science* **109**, 45–56.

Clifford, H. T. (1987). Spikelet and floral morphology. *In* "Grass Systematics and Evolution" (T. R. Soderstrom, K. W. Hilu, C. S. Campbell and M. E. Barkworth, eds.), pp. 21–30. Smithsonian Institution Press, Washington DC.

Coen, E. S. and Meyerowitz, E. M. (1991). The war of the whorls: Genetic interactions controlling flower development. *Nature* **353**, 31–37.

Colombo, L., Franken, J., Koetje, E., van Went, J., Dons, H. J., Angenent, G. C. and van Tunen, A. J. (1995). The petunia MADS-box gene *FBP11* determines ovule identity. *Plant Cell* **7**, 1859–1868.

Davies, B., Motte, P., Keck, E., Saedler, H., Sommer, H. and Schwarz-Sommer, Z. (1999). PLENA and FARINELLI: Redundancy and regulatory interactions between two *Antirrhinum* MADS-box factors controlling flower development. *EMBO Journal* **18**, 4023–4034.

Ditta, G., Pinyopich, A., Robles, P., Pelaz, S. and Yanofsky, M. F. (2004). The SEP4 gene of *Arabidopsis thaliana* functions in floral organ and meristem identity. *Current Biology* **14**, 1935–1940.

Doebley, J. and Stec, A. (1991). Genetic analysis of the morphological differences between maize and teosinte. *Genetics* **129**, 285–295.

Doust, A. N., Devos, K. M., Gadberry, M. D., Gale, M. D. and Kellogg, E. A. (2004). Genetic control of branching in foxtail millet. *Proceedings of the National Academy of Sciences of the United States of America* **101**, 9045–9050.

Doust, A. N., Devos, K. M., Gadberry, M. D., Gale, M. D. and Kellogg, E. A. (2005). The genetic basis for inflorescence variation between foxtail and green millet (poaceae). *Genetics* **169**, 1659–1672.

Favaro, R., Immink, R. G., Ferioli, V., Bernasconi, B., Byzova, M., Angenent, G. C., Kater, M. and Colombo, L. (2002). Ovule-specific MADS-box proteins have conserved protein–protein interactions in monocot and dicot plants. *Molecular Genetics and Genomics* **268**, 152–159.

Ferrandiz, C., Liljegren, S. J. and Yanofsky, M. F. (2000). Negative regulation of the *SHATTERPROOF* genes by FRUITFULL during *Arabidopsis* fruit development. *Science* **289**, 436–438.

Fischer, A., Baum, N., Saedler, H. and Theissen, G. (1995). Chromosomal mapping of the MADS-box multigene family in *Zea mays* reveals dispersed distribution of allelic genes as well as transposed copies. *Nucleic Acids Research* **23**, 1901–1911.

Force, A., Lynch, M., Pickett, F. B., Amores, A., Yan, Y. L. and Postlethwait, J. (1999). Preservation of duplicate genes by complementary, degenerative mutations. *Genetics* **151**, 1531–1545.

Fornara, F., Parenicova, L., Falasca, G., Pelucchi, N., Masiero, S., Ciannamea, S., Lopez-Dee, Z., Altamura, M. M., Colombo, L. and Kater, M. M. (2004). Functional characterization of OsMADS18, a member of the AP1/SQUA subfamily of MADS-box genes. *Plant Physiology* **135**, 2207–2219.

Fourquin, C., Vinauger-Douard, M., Fogliani, B., Dumas, C. and Scutt, C. P. (2005). Evidence that CRABS CLAW and TOUSLED have conserved their roles in carpel development since the ancestor of the extant angiosperms.

Proceedings of the National Academy of Sciences of the United States of America **102**, 4649–4654.

Freeling, M. (2001). Grasses as a single genetic system: Reassessment 2001. *Plant Physiology* **125**, 1191–1197.

Gaut, B. S. and Doebley, J. F. (1997). DNA sequence evidence for the segmental allotetraploid origin of maize. *Proceedings of the National Academy of Sciences of the United States of America* **94**, 6809–6814.

Gocal, G. F., King, R. W., Blundell, C. A., Schwartz, O. M., Andersen, C. H. and Weigel, D. (2001). Evolution of floral meristem identity genes. Analysis of *Lolium temulentum* genes related to *APETALA1* and *LEAFY* of *Arabidopsis*. *Plant Physiology* **125**, 1788–1801.

Goto, K. and Meyerowitz, E. M. (1994). Function and regulation of the *Arabidopsis* floral homeotic gene *PISTILLATA*. *Genes and Development* **8**, 1548–1560.

Grass Phylogeny Working Group (2001). Phylogeny and subfamilial classification of the grasses (Poaceae). *Annals of the Missouri Botanical Garden* **88**, 373–457.

Guo, H. and Moose, S. P. (2003). Conserved noncoding sequences among cultivated cereal genomes identify candidate regulatory sequence elements and patterns of promoter evolution. *Plant Cell* **15**, 1143–1158.

Heuer, S., Hansen, S., Bantin, J., Brettschneider, R., Kranz, E., Lorz, H. and Dresselhaus, T. (2001). The maize MADS-box gene *ZmMADS3* affects node number and spikelet development and is co-expressed with *ZmMADS1* during flower development, in egg cells, and early embryogenesis. *Plant Physiology* **127**, 33–45.

Honma, T. and Goto, K. (2001). Complexes of MADS-box proteins are sufficient to convert leaves into floral organs. *Nature* **409**, 525–529.

Huang, H., Mizukami, Y., Hu, Y. and Ma, H. (1993). Isolation and characterization of the binding sequences for the product of the *Arabidopsis* floral homeotic gene *AGAMOUS*. *Nucleic Acids Research* **21**, 4769–4776.

Huijser, P., Klein, J., Lonnig, W. E., Meijer, H., Saedler, H. and Sommer, H. (1992). Bracteomania, an inflorescence anomaly, is caused by the loss of function of the MADS-box gene squamosa in *Antirrhinum majus*. *EMBO Journal* **11**, 1239–1249.

Inada, D. C., Bashir, A., Lee, C., Thomas, B. C., Ko, C., Goff, S. A. and Freeling, M. (2003). Conserved noncoding sequences in the grasses. *Genome Research* **13**, 2030–2041.

Irish, V. F. (2000). Variations on a theme: Flower development and evolution. *Genome Biology* **1**, reviews 1015.1–1015.4.

Jack, T., Brockman, L. L. and Meyerowitz, E. M. (1992). The homeotic gene *APETALA3* of *Arabidopsis thaliana* encodes a MADS-box and is expressed in petals and stamens. *Cell* **68**, 683–697.

Jack, T., Fox, G. L. and Meyerowitz, E. M. (1994). *Arabidopsis* homeotic gene *APETALA3* ectopic expression: Transcriptional and posttranscriptional regulation determine floral organ identity. *Cell* **76**, 703–716.

Jeon, J. S., Jang, S., Lee, S., Nam, J., Kim, C., Lee, S. H., Chung, Y. Y., Kim, S. R., Lee, Y. H., Cho, Y. G. and An, G. (2000). Leafy hull sterile1 is a homeotic mutation in a rice MADS-box gene affecting rice flower development. *Plant Cell* **12**, 871–884.

Judziewicz, E. J. and Soderstrom, T. R. (1989). Morphological, anatomical, and taxonomic studies in *Anomochloa* and *Streptochaeta* (Poaceae: Bambusoideae). *Smithsonian Contributions to Botany* **68**, 1–52.

Kang, H. G. and An, G. (1997). Isolation and characterization of a rice MADS-box gene belonging to the AGL2 gene family. *Molecular Cells* **7**, 45–51.

Kang, H. G., Noh, Y. S., Chung, Y. Y., Costa, M. A., An, K. and An, G. (1995). Phenotypic alterations of petal and sepal by ectopic expression of a rice MADS-box gene in tobacco. *Plant Molecular Biology* **29**, 1–10.

Kang, H. G., Jang, S., Chung, J. E., Cho, Y. G. and An, G. (1997). Characterization of two rice MADS-box genes that control flowering time. *Molecular Cells* **7**, 559–566.

Kang, H. G., Jeon, J. S., Lee, S. and An, G. (1998). Identification of class B and class C floral organ identity genes from rice plants. *Plant Molecular Biology* **38**, 1021–1029.

Kaplinsky, N. J., Braun, D. M., Penterman, J., Goff, S. A. and Freeling, M. (2002). Utility and distribution of conserved noncoding sequences in the grasses. *Proceedings of the National Academy of Sciences of the United States of America* **99**, 6147–6151.

Keck, E., McSteen, P., Carpenter, R. and Coen, E. (2003). Separation of genetic functions controlling organ identity in flowers. *EMBO Journal* **22**, 1058–1066.

Kim, S., Koh, J., Yoo, M. J., Kong, H., Hu, Y., Ma, H., Soltis, P. S. and Soltis, D. E. (2005). Expression of floral MADS-box genes in basal angiosperms: Implications for the evolution of floral regulators. *Plant Journal* **43**, 724–744.

Kramer, E. M. and Irish, V. F. (1999). Evolution of genetic mechanisms controlling petal development. *Nature* **399**, 144–148.

Kramer, E. M. and Irish, V. F. (2000). Evolution of the petal and stamen developmental programs: Evidence from comparative studies of the lower eudicots and basal angiosperms. *International Journal of Plant Sciences* **161**, S29–S40.

Kramer, E. M., Dorit, R. L. and Irish, V. F. (1998). Molecular evolution of genes controlling petal and stamen development: Duplication and divergence within the *APETALA3* and *PISTILLATA* MADS-box gene lineages. *Genetics* **149**, 765–783.

Kramer, E. M., Di Stilio, V. S. and Schlüter, P. M. (2003). Complex patterns of gene duplication in the *APETALA3* and *PISTILLATA* lineages of the ranunculaceae. *International Journal of Plant Sciences* **164**, 1–11.

Kramer, E. M., Jaramillo, M. A. and Di Stilio, V. S. (2004). Patterns of gene duplication and functional evolution during the diversification of the *AGAMOUS* subfamily of MADS-box genes in angiosperms. *Genetics* **166**, 1011–1023.

Kyozuka, J. and Shimamoto, K. (2002). Ectopic expression of *OsMADS3*, a rice ortholog of *AGAMOUS*, caused a homeotic transformation of lodicules to stamens in transgenic rice plants. *Plant and Cell Physiology* **43**, 130–135.

Kyozuka, J., Kobayashi, T., Morita, M. and Shimamoto, K. (2000). Spatially and temporally regulated expression of rice MADS-box genes with similarity to *Arabidopsis* class A, B and C genes. *Plant Cell Physiology* **41**, 710–718.

Lee, S., Jeon, J. S., An, K., Moon, Y. H., Chung, Y. Y. and An, G. (2003). Alteration of floral organ identity in rice through ectopic expression of *OsMADS16*. *Planta* **217**, 904–911.

Liljegren, S. J., Ditta, G. S., Eshed, Y., Savidge, B., Bowman, J. L. and Yanofsky, M. F. (2000). *SHATTERPROOF* MADS-box genes control seed dispersal in *Arabidopsis*. *Nature* **404**, 766–770.

Litt, A. and Irish, V. F. (2003). Duplication and diversification in the *APETALA1/FRUITFULL* floral homeotic gene lineage: Implications for the evolution of floral development. *Genetics* **165**, 821–833.

Lopez-Dee, Z. P., Wittich, P., Enrico Pe, M., Rigola, D., Del Buono, I., Gorla, M. S., Kater, M. M. and Colombo, L. (1999). *OsMADS13*, a novel rice

MADS-box gene expressed during ovule development. *Developmental Genetics* **25**, 237–244.
Luo, Q., Zhou, K., Zhao, X., Zeng, Q., Xia, H., Zhai, W., Xu, J., Wu, X., Yang, H. and Zhu, L. (2005). Identification and fine mapping of a mutant gene for palealess spikelet in rice. *Planta* **221**, 222–230.
Maes, T., Van de Steene, N., Zethof, J., Karimi, M., D'Hauw, M., Mares, G., Van Montagu, M. and Gerats, T. (2001). Petunia *Ap2*-like genes and their role in flower and seed development. *Plant Cell* **13**, 229–244.
Malcomber, S. T. and Kellogg, E. A. (2004). Heterogeneous expression patterns and separate roles of the *SEPALLATA* gene *LEAFY HULL STERILE1* in grasses. *Plant Cell* **16**, 1692–1706.
Mandel, M. A. and Yanofsky, M. F. (1995). A gene triggering flower formation in *Arabidopsis*. *Nature* **377**, 522–524.
Mandel, M. A., Gustafson-Brown, C., Savidge, B. and Yanofsky, M. F. (1992). Molecular characterization of the *Arabidopsis* floral homeotic gene *APETALA1*. *Nature* **360**, 273–277.
McSteen, P., Laudencia-Chingcuanco, D. and Colasanti, J. (2000). A floret by any other name: Control of meristem identity in maize. *Trends in Plant Science* **5**, 61–66.
Mena, M., Mandel, M. A., Lerner, D. R., Yanofsky, M. F. and Schmidt, R. J. (1995). A characterization of the MADS-box gene family in maize. *Plant Journal* **8**, 845–854.
Mena, M., Ambrose, B. A., Meeley, R. B., Briggs, S. P., Yanofsky, M. F. and Schmidt, R. J. (1996). Diversification of C-function activity in maize flower development. *Science* **274**, 1537–1540.
Michelangeli, F. A., Davis, J. I. and Stevenson, D. W. (2003). Phylogenetic relationships among Poaceae and related families as inferred from morphology, inversions in the plastid genome, and sequence data from the mitochondrial and plastid genomes. *American Journal of Botany* **90**, 93–106.
Moon, Y. H., Kang, H. G., Jung, J. Y., Jeon, J. S., Sung, S. K. and An, G. (1999). Determination of the motif responsible for interaction between the rice *APETALA1/AGAMOUS*-LIKE9 family proteins using a yeast two-hybrid system. *Plant Physiology* **120**, 1193–1204.
Moore, G., Devos, K. M., Wang, Z. and Gale, M. D. (1995). Cereal genome evolution. Grasses, line up and form a circle. *Current Biology* **5**, 737–739.
Mueller, C. G. and Nordheim, A. (1991). A protein domain conserved between yeast MCM1 and human SRF directs ternary complex formation. *EMBO Journal* **10**, 4219–4229.
Münster, T., Wingen, L. U., Faigl, W., Werth, S., Saedler, H. and Theissen, G. (2001). Characterization of three *GLOBOSA*-like MADS-box genes from maize: Evidence for ancient paralogy in one class of floral homeotic B-function genes of grasses. *Gene* **262**, 1–13.
Münster, T., Deleu, W., Wingen, L. U., Ouzunova, M., Cacharrón, J., Faigl, W., Werth, S., Kim, J. T. T., Saedler, H. and Theissen, G. (2002). Maize MADS-box genes galore. *Maydica* **47**, 287–301.
Murai, K., Miyamae, M., Kato, H., Takumi, S. and Ogihara, Y. (2003). WAP1, a wheat *APETALA1* homolog, plays a central role in the phase transition from vegetative to reproductive growth. *Plant and Cell Physiology* **44**, 1255–1265.
Nagasawa, N., Miyoshi, M., Sano, Y., Satoh, H., Hirano, H., Sakai, H. and Nagato, Y. (2003). *SUPERWOMAN1* and *DROOPING LEAF* genes control floral organ identity in rice. *Development* **130**, 705–718.
Ohno, S. (1970). "Evolution by Gene Duplication." Springer-Verlag, New York.

Page, V. M. (1951). Morphology of the Spikelet of Streptochaeta. *Bulletin of the Torrey Botanical Club* **78**, 22–37.

Paterson, A. H., Bowers, J. E. and Chapman, B. A. (2004). Ancient polyploidization predating divergence of the cereals, and its consequences for comparative genomics. *Proceedings of the National Academy of Sciences of the United States of America* **101**, 9903–9908.

Pelaz, S., Ditta, G. S., Baumann, E., Wisman, E. and Yanofsky, M. F. (2000). B and C floral organ identity functions require *SEPALLATA* MADS-box genes. *Nature* **405**, 200–203.

Pelaz, S., Tapia-Lopez, R., Alvarez-Buylla, E. R. and Yanofsky, M. F. (2001). Conversion of leaves into petals in *Arabidopsis*. *Current Biology* **11**, 182–184.

Pelucchi, N., Fornara, F., Favalli, C., Masiero, S., Lago, C., Pè, M. E., Colombo, L. and Kater, M. (2002). Comparative analysis of rice MADS-box genes expressed during flower development. *Sexual Plant Reproduction* **15**, 113–122.

Pinyopich, A., Ditta, G. S., Savidge, B., Liljegren, S. J., Baumann, E., Wisman, E. and Yanofsky, M. F. (2003). Assessing the redundancy of MADS-box genes during carpel and ovule development. *Nature* **424**, 85–88.

Prasad, K. and Vijayraghavan, U. (2003). Double-stranded RNA interference of a rice *PI/GLO* paralog, *OsMADS2*, uncovers its second-whorl-specific function in floral organ patterning. *Genetics* **165**, 2301–2305.

Prasad, K., Parameswaran, S. and Vijayraghavan, U. (2005). *OsMADS1*, a rice MADS-box factor, controls differentiation of specific cell types in the lemma and palea and is an early-acting regulator of inner floral organs. *Plant Journal* **43**, 915–928.

Riechmann, J. L. and Meyerowitz, E. M. (1997). MADS domain proteins in plant development. *Biological Chemistry* **378**, 1079–1101.

Riechmann, J. L., Krizek, B. A. and Meyerowitz, E. M. (1996). Dimerization specificity of *Arabidopsis* MADS domain homeotic proteins *APETALA1, APETALA3, PISTILLATA*, and *AGAMOUS*. *Proceedings of the National Academy of Sciences of the United States of America* **93**, 4793–4798.

Salvi, S. and Tuberosa, R. (2005). To clone or not to clone plant QTLs: Present and future challenges. *Trends in Plant Science* **10**, 297–304.

Schmidt, R. J., Veit, B., Mandel, M. A., Mena, M., Hake, S. and Yanofsky, M. F. (1993). Identification and molecular characterization of *ZAG1*, the maize homolog of the *Arabidopsis* floral homeotic gene *AGAMOUS*. *Plant Cell* **5**, 729–737.

Schmitz, J., Franzen, R., Ngyuen, T. H., Garcia-Maroto, F., Pozzi, C., Salamini, F. and Rohde, W. (2000). Cloning, mapping and expression analysis of barley MADS-box genes. *Plant Molecular Biology* **42**, 899–913.

Schuster, J. (1910). Über die morphologie der Grasblüte. *Flora* **100**, 213–266.

Schwarz-Sommer, Z., Hue, I., Huijser, P., Flor, P. J., Hansen, R., Tetens, F., Lonnig, W. E., Saedler, H. and Sommer, H. (1992). Characterization of the *Antirrhinum* floral homeotic MADS-box gene deficiens: Evidence for DNA binding and autoregulation of its persistent expression throughout flower development. *EMBO Journal* **11**, 251–263.

Shinozuka, Y., Kojima, S., Shomura, A., Ichimura, H., Yano, M., Yamamoto, K. and Sasaki, T. (1999). Isolation and characterization of rice MADS-box gene homologues and their RFLP mapping. *DNA Research* **6**, 123–129.

Soderstrom, T. R. (1981). Some evolutionary trends in the Bambusoideae (Poaceae). *Annals of the Missouri Botanical Garden* **68**, 15–47.

Sommer, H., Beltran, J. P., Huijser, P., Pape, H., Lonnig, W. E., Saedler, H. and Schwarz-Sommer, Z. (1990). *Deficiens*, a homeotic gene involved in the control of flower morphogenesis in *Antirrhinum majus*: The protein shows homology to transcription factors. *EMBO Journal* **9**, 605–613.

Stellari, G. M., Jaramillo, M. A. and Kramer, E. M. (2004). Evolution of the *APETALA3* and *PISTILLATA* lineages of MADS-box-containing genes in the basal angiosperms. *Molecular Biology and Evolution* **21**, 506–519.

Takhtajan, A. (1991). "Evolutionary trends in flowering plants." Columbia University Press, New York.

Theissen, G. (2001). Development of floral organ identity: Stories from the MADS house. *Current Opinion in Plant Biology* **4**, 75–85.

Theissen, G. and Saedler, H. (2001). Plant biology. Floral quartets. *Nature* **409**, 469–471.

Theissen, G., Strater, T., Fischer, A. and Saedler, H. (1995). Structural characterization, chromosomal localization and phylogenetic evaluation of two pairs of *AGAMOUS*-like MADS-box genes from maize. *Gene* **156**, 155–166.

Till, B. J., Reynolds, S. H., Weil, C., Springer, N., Burtner, C., Young, K., Bowers, E., Codomo, C. A., Enns, L. C., Odden, A. R., Greene, E. A. Comai, L., *et al.* (2004). Discovery of induced point mutations in maize genes by TILLING. *BMC Plant Biology* **4**, 12.

Trobner, W., Ramirez, L., Motte, P., Hue, I., Huijser, P., Lonnig, W. E., Saedler, H., Sommer, H. and Schwarz-Sommer, Z. (1992). *GLOBOSA*: A homeotic gene which interacts with *DEFICIENS* in the control of *Antirrhinum* floral organogenesis. *EMBO Journal* **11**, 4693–4704.

van Tunen, A. J., Eikelboom, W. and Angenent, G. C. (1993). Floral organogenesis in *Tulipa*. *Flowering Newsletter* **16**, 33–37.

Vandenbussche, M., Theissen, G., Van de Peer, Y. and Gerats, T. (2003). Structural diversification and neo-functionalization during floral MADS-box gene evolution by C-terminal frameshift mutations. *Nucleic Acids Research* **31**, 4401–4409.

Vandenbussche, M., Zethof, J., Royaert, S., Weterings, K. and Gerats, T. (2004). The duplicated B-class heterodimer model: Whorl-specific effects and complex genetic interactions in *Petunia hybrida* flower development. *Plant Cell* **16**, 741–754.

Wang, H., Nussbaum-Wagler, T., Li, B., Zhao, Q., Vigouroux, Y., Faller, M., Bomblies, K., Lukens, L. and Doebley, J. F. (2005). The origin of the naked grains of maize. *Nature* **436**, 714–719.

Weigel, D. and Nilsson, O. (1995). A developmental switch sufficient for flower initiation in diverse plants. *Nature* **377**, 495–500.

Weigel, D., Alvarez, J., Smyth, D. R., Yanofsky, M. F. and Meyerowitz, E. M. (1992). *LEAFY* controls floral meristem identity in *Arabidopsis*. *Cell* **69**, 843–859.

Whipple, C. J., Ciceri, P., Padilla, C. M., Ambrose, B. A., Bandong, S. L. and Schmidt, R. J. (2004). Conservation of B-class floral homeotic gene function between maize and *Arabidopsis*. *Development* **131**, 6083–6091.

Winter, K. U., Weiser, C., Kaufmann, K., Bohne, A., Kirchner, C., Kanno, A., Saedler, H. and Theissen, G. (2002). Evolution of class B floral homeotic proteins: Obligate heterodimerization originated from homodimerization. *Molecular Biology and Evolution* **19**, 587–596.

Yamaguchi, T., Nagasawa, N., Kawasaki, S., Matsuoka, M., Nagato, Y. and Hirano, H. Y. (2004). The YABBY gene *DROOPING LEAF* regulates carpel specification and midrib development in *Oryza sativa*. *Plant Cell* **16**, 500–509.

Yamaguchi, T., Lee, D. Y., Miyao, A., Hirochika, H., An, G. and Hirano, H. Y. (2006). Functional diversification of the two C-class MADS-box genes OSMADS3 and OSMADS58 in *Oryza sativa*. *Plant Cell* **18**, 15–28.

Yan, L., Loukoianov, A., Tranquilli, G., Helguera, M., Fahima, T. and Dubcovsky, J. (2003). Positional cloning of the wheat vernalization gene VRN1. *Proceedings of the National Academy of Sciences of the United States of America* **100**, 6263–6268.

Yang, Y. and Jack, T. (2004). Defining subdomains of the K domain important for protein–protein interactions of plant MADS proteins. *Plant Molecular Biology* **55**, 45–59.

Yang, Y., Fanning, L. and Jack, T. (2003). The K domain mediates heterodimerization of the *Arabidopsis* floral organ identity proteins, *APETALA3* and *PISTILLATA*. *Plant Journal* **33**, 47–59.

Yanofsky, M. F., Ma, H., Bowman, J. L., Drews, G. N., Feldmann, K. A. and Meyerowitz, E. M. (1990). The protein encoded by the *Arabidopsis* homeotic gene *AGAMOUS* resembles transcription factors. *Nature* **346**, 35–39.

Zahn, L. M., Kong, H., Leebens-Mack, J. H., Kim, S., Soltis, P. S., Landherr, L. L., Soltis, D. E., Depamphilis, C. W. and Ma, H. (2005). The evolution of the *SEPALLATA* subfamily of MADS-box genes: A preangiosperm origin with multiple duplications throughout angiosperm history. *Genetics* **169**, 2209–2223.

Developmental Gene Evolution and the Origin of Grass Inflorescence Diversity

SIMON T. MALCOMBER,* JILL C. PRESTON,*
RENATA REINHEIMER,† JESSIE KOSSUTH*
AND ELIZABETH A. KELLOGG*

**Department of Biology, University of Missouri-St. Louis, One University Boulevard, Saint Louis, Missouri 63121*
†Vegetal Morphology, Facultad de Ciencias Agrarias (UNL), Kreder 2805, 3080 Esperanza, Santa Fe, Argentina

Advances in Botanical Research, Vol. 44
Incorporating Advances in Plant Pathology

0065-2296/06 $35.00
DOI: 10.1016/S0065-2296(06)44011-8

ABSTRACT

Grass inflorescences are diverse, developmentally complex, and provide many of the taxonomic characters used to differentiate the estimated 10,000 grass species. Here we review grass inflorescence development in detail and discuss which genes are involved at each developmental stage. We demonstrate that grass inflorescence development is complex, with multiple structures that are not present in *Arabidopsis*. New and published phylogenetic analyses of genes involved at each developmental stage indicate that the maize *FASCIATED EAR2* (*FEA2*) and *Arabidopsis CLAVATA2* (*CLV2*) genes are the sole remaining co-orthologs following multiple rounds of whole-genome duplication. Analyses of *BARREN STALK1/LAX PANICLE* (*BA1/LAX1*), *FRUITFULL* (*FUL*), *INDETERMINATE SPIKELET1* (*IDS1*), *KNOTTED1* (*KN1*), *LEAFY HULL STERILE1* (*LHS1*), and *RICE CENTRORADIALIS1/2* (*RCN1/2*) indicate that these genes are members of grass- or monocot-specific small gene families. The complex pattern of gene relationships mirrors a complex pattern of functional evolution. Maize FEA2 and *Arabidopsis* CLV2 have nonidentical roles, whereas distantly related grass KN1-like and RCN1/2 proteins show functional convergence and conservation, respectively. Duplications near the base of grasses in *BA1/LAX, FUL, IDS1* and *LHS1* have led to diverse roles in grass inflorescence development. We conclude that developmental gene duplication followed by functional diversification appears to have played a major role in the evolution of novel morphological structures within grass inflorescences.

I. INTRODUCTION

A. TOOLS FOR EVOLUTIONARY DEVELOPMENTAL GENETICS

The study of evolutionary developmental genetics requires a good phylogeny, precise description of developmental morphology, and a set of genes connected to the morphology (Baum, 2002; Kellogg, 1996). Over the last 15 years, increasingly robust phylogenies have become available for increasing numbers of taxa (Soltis *et al.*, 2005; Stevens, 2001 onwards). Description of developmental morphology, once pursued by only a handful of talented morphologists, is now undergoing a renaissance as more and more scientists see accurate description of the phenotype as key to understanding its evolution. The *Arabidopsis thaliana* genome sequence (Arabidopsis Genome Initiative, 2000), and several rice genome sequences (Goff *et al.*, 2002; Yu *et al.*, 2002, 2005) provide a list of the genes, and have vastly accelerated the effort to connect genotype with phenotype.

The major model systems provide the tools for studying evolution of development and are often the only species in which gene function can be assessed rigorously. But the model systems are just that—models. To understand diversification, one needs to look at diversity (see also comments by Gewin, 2005). The model systems are thus tools for understanding evolution, just as phylogenies and morphological descriptions are (Kellogg and Shaffer, 1993).

B. THE GRASS FAMILY

The grass family (Poaceae or Gramineae) is an ideal system for the study of evolution of development. The family is large (ca. 10,000 species) and morphologically diverse (Clayton and Renvoize, 1986; Watson and Dallwitz, 1992 onwards), and the developmental morphology of many species is well known. The family includes rice, with complete genome sequences from both *indica* and *japonica* varieties (Goff *et al.*, 2002; Yu *et al.*, 2002). It also includes maize, which has an unparalleled set of well-characterized morphological mutants, plus an ongoing effort to sequence the entire genome. Sorghum, wheat, barley, ryegrass (*Lolium*), meadow fescue (*Festuca pratensis*), sugarcane, tef, foxtail millet, and pearl millet also have varying amounts of genetic resources (genome sequencing projects, EST collections, genetic maps, well-organized stock centers). Informatic tools are available for comparing genetic maps (Gramene, Ware *et al.*, 2002) and for searching mutants and genes (Maize GDB, Lawrence *et al.*, 2005).

The grass family is monophyletic (Grass Phylogeny Working Group, 2001). All members of the family share a uniquely derived embryo structure. The embryo is highly differentiated, with clearly organized shoot and root meristems and two or more leaves initiated. This represents a heterochronic change relative to the grass sister groups and presumed ancestors, in which embryo development is accelerated relative to seed maturation (Kellogg, 2000a). In the sister taxa of the grasses (Joinvilleaceae and Ecdeiocoleaceae), the embryo reaches only an undifferentiated globular stage before the seed matures (Campbell and Kellogg, 1987; Rudall *et al.*, 2005). The grass embryo also has a unique haustorial organ, the scutellum, which could be derived from a cotyledon although it is so highly modified that direct comparisons are impossible.

A robust phylogeny is available for the family and provides the basis for the classification (Grass Phylogeny Working Group, 2001). Sister to the rest of the family is the tiny subfamily Anomochlooideae; the four species in this subfamily have the characteristic grass embryo, but lack spikelets, the flowering units of most grasses, composed of glumes, lemmas, paleas, lodicules, stamens, and pistil. Instead of spikelets, members of Anomochlooideae have "spikelet equivalents," which lack structures that are clearly homologous to glumes, lemmas, and paleas (Clark and Judziewicz, 1996). Subfamily Pharoideae is the earliest lineage with a true spikelet and is sister to all remaining spikelet-bearing grasses. Some time after divergence of Pharoideae, multiflowered spikelets arose. The earliest-diverging lineage of this group is Puelioideae. The major radiation of the grasses occurred much later, after the grasses had been around for several tens of millions of years. Bambusoideae, Pooideae, Ehrhartoideae, and a lineage leading to the PACCAD clade originated at this point, perhaps 40–50 million years ago, but the order of events is uncertain. The Grass Phylogeny Working Group (2001) phylogeny suggests that Bambusoideae, Pooideae, and Ehrhartoideae shared a common ancestor (and thus form the BEP clade), but support for this is weak. Nonetheless, each of the three subfamilies is strongly supported as monophyletic by both morphological and molecular data. The PACCAD clade includes the remaining subfamilies (Panicoideae, Arundinoideae, Centothecoideae, Chloridoideae, Aristidoideae, and Danthonioideae), relationships among which are not well supported.

Many morphological distinctions among grass taxa are based on characters of the inflorescence. In this chapter, we discuss this variation, and the current stage of knowledge about its development and genetics.

C. GRASS INFLORESCENCES VS *ARABIDOPSIS*

The grass inflorescence is a novel structure that is developmentally intricate, evolutionarily intriguing, and agronomically important. Its architecture controls both pollination and seed set, and is thus a target of natural and human selection. The major eudicot model, *Arabidopsis thaliana*, provides some useful information on the genes affecting inflorescence morphology but is not a substitute for looking directly at the grasses themselves.

The inflorescence of *Arabidopsis* (and most Brassicaceae) is simple (Fig. 1, Whipple and Schmidt, Chapter 10; this volume, Figs. 2 and 4). After receiving the appropriate environmental or endogenous signal, the shoot apical meristem produces a few leaves with inflorescence meristems in their axils, and then ceases to produce leaves entirely and begins to produce lateral floral meristems, in a phyllotactic spiral (Shannon and Meeks-Wagner, 1991; Smyth *et al.*, 1990). Unlike most angiosperms, the floral meristems are not subtended by bracts or bracteoles, which appear to be suppressed (Long and Barton, 2000) as in most Brassicaceae (Stevens, 2001 onwards). The shoot apical meristem itself (now the inflorescence meristem) eventually ceases to produce floral meristems but remains itself undifferentiated (Shannon and Meeks-Wagner, 1991). Secondary

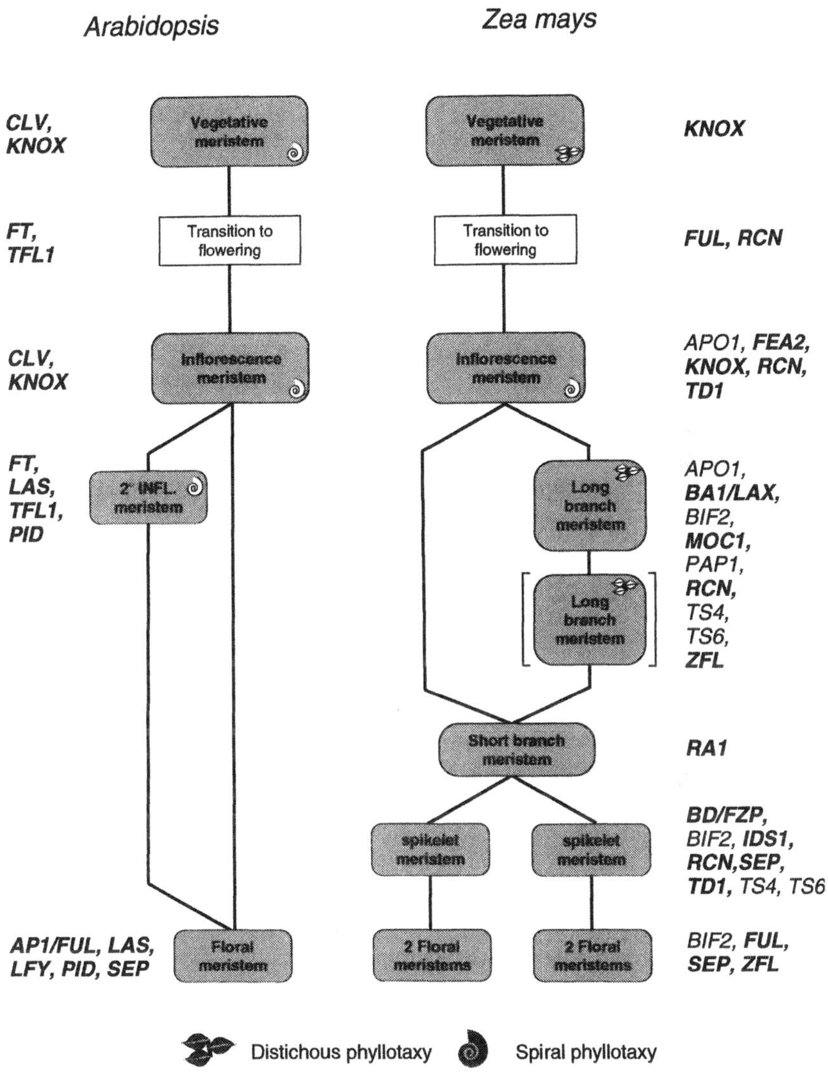

Fig. 1. Diagram comparing inflorescence development in *Arabidopsis* and maize, starting with the vegetative meristem and ending with the floral meristem, and the genes discussed in the text that are known to affect particular developmental stages. Short branch meristem is also referred to as the spikelet pair meristem in the maize literature. Boldface, cloned genes. *AP1, APETALA1; APO1, ABERRANT PANICLE ORGANIZATION1; BA1, BARREN STALK1; BD, BRANCHED SILKLESS1; BIF2, BARREN INFLORESCENCE2; CLV, CLAVATA; FEA2, FASCIATED EAR2; FON1, FLORAL ORGAN NUMBER1; FT, FLOWERING LOCUS T; FUL, FRUITFULL; FZP, FRIZZY PANICLE1; IDS1, INDETERMINATE SPIKELET1; KNOX, KNOTTED1-LIKE HOMEOBOX; LAS, LATERAL SUPPRESSOR; LAX, LAX PANICLE1; LFY, LEAFY; MOC1, MONOCULM1; PAP1, PANICLE PHYTOMER1; PID, PINOID; RA1, RAMOSA1; RCN, RICE CENTRORADIALIS; TD1, THICK TASSEL DWARF1; TFL, TERMINAL FLOWER1; TS4, TASSELSEED4; TS6, TASSELSEED6; SEP, SEPALLATA; ZFL, ZEA LEAFY.*

Fig. 2. Phylogenetic relationships of sampled plant taxa, as presented by the Angiosperm Phylogeny Web (Stevens, 2001 onwards). Grasses and *Arabidopsis* in boldface. Corn ear, grass clade; flower, *Arabidopsis*.

inflorescences in the leaf axils reiterate the developmental pattern of the main axis. The *Arabidopsis* inflorescence is thus an ideal system in which to study the transition from vegetative to inflorescence meristem, and the production of floral meristems from the inflorescence meristem. However, *Arabidopsis* provides only limited insight into the various mechanisms underlying the formation of inflorescence branches, changes in phyllotaxis, and termination of the inflorescence axis (or axes). Processes that do not occur in *Arabidopsis* need to be studied in plants in which they occur.

The grass inflorescence is, by contrast with that of Brassicaceae, baroque (Fig. 1, Whipple and Schmidt, Chapter 10; this volume, Figs. 2 and 4). The transition to flowering is often accompanied by a change in phyllotaxis, the inflorescence meristem generally produces branches which themselves may produce branches, and the branches may simply cease to develop (as in *Arabidopsis*) or may terminate in a unique structure, a spikelet, which is morphologically a contracted, flower-bearing branch, but also itself has some similarities to a flower. In classic taxonomic literature, agrostologists (grass taxonomists) commonly make an analogy between a spikelet and a flower, as the terminal structure on a branch, and then attempt to classify grass inflorescences as: (1) spikes, with the spikelets attached directly to the inflorescence axis, as in wheat, barley, and rye; (2) racemes, with the spikelets on pedicels, as in *Brachypodium* or *Brachyelytrum*; or (3) panicles, with the inflorescence branches themselves branched, as in most familiar grasses such as oats, fescue, sorghum, or proso millet (Clark and Pohl, 1996). This set of descriptors is misleading, and implies that grass inflorescence development is quite uniform, whereas grass inflorescences vary in multiple characteristics, some of which are described in more detail in a later section. In addition, the classic terminology is of limited use for describing the inflorescences of some species, such as big bluestem (*Andropogon gerardii*) or bamboo (e.g., *Phyllostachys*), which have intricate combinations of bracts, spathes, and inflorescence branches with both sessile and pedicellate spikelets. It is common to find descriptions such as the following for the inflorescence of *Andropogon*: "Inflorescence of spicate main branches, or paniculate (usually with paired or digitate "racemes," these often spatheate and aggregated into false panicles)... Inflorescence... a complex of "partial inflorescences" and intervening foliar organs..." (Watson and Dallwitz, 1992 onwards). The work of Vegetti and Weberling (1996) provides a better description of adult inflorescences but still attempts to create a typology for the entire flower-bearing portion of the plant, which makes it difficult to hypothesize underlying developmental processes.

Grass inflorescences may be better described by components of the phenotype that can vary in a combinatorial manner (Kellogg, 2000b). Specifically, any meristem can have one of three fates: it can terminate in a spikelet, it can continue producing lateral meristems, or it can cease developing. If it produces lateral meristems, then each lateral has the same three possible fates, and similarly with the lateral meristems on the lateral meristems. Importantly, the fate of each order of branching is potentially independent of the fate of each other order, so that over evolutionary time, virtually all conceivable combinations occur.

In this chapter, we describe the development of grass inflorescences in more detail and review current knowledge of the various genes involved at each stage of development. We show that grass inflorescences are far more complex than those of *Arabidopsis*. One could imagine multiple ways that such complexity might have evolved, but one possibility would be the origin of novel genes and gene functions. We show that gene duplication has been common among genes controlling morphogenesis. The sheer number of available genes and presumed gene functions may help explain the morphological variety that exists among grasses.

D. DUPLICATE GENES, THE RAW MATERIAL FOR EVOLUTION OF NOVEL FUNCTION

Haldane (1932, 1933), and later Ohno (1970) suggested that duplicate genes should accumulate mutations and become pseudogenes because there would be no selective pressure on them to retain function. This should happen quite rapidly, although the exact speed depends on population size (Walsh, 1995). Only rarely would a duplicate gene acquire a new function (neofunctionalization), and thus contribute to the diversity of the genome.

As gene and genomic sequences have accumulated, it has become abundantly clear that duplicated genes are common, contra the predictions of Haldane and Ohno. Most genes, in fact, belong to multigene families, which are generated by a variety of gene and genome-duplication processes (Zhang, 2003). Although duplicated genes are lost quite frequently, particularly after polyploidization events (Chantret *et al.*, 2005), an appreciable number of them persist for millions of years. The discrepancy between prediction and observation has led to a new round of theories to explain the long persistence of some duplicates.

One suggestion is that duplicate genes might divide up the ancestral function, so that, rather than acquire new functions, they simply became subfunctionalized (Force *et al.*, 1999; Hughes, 1999; Lynch and Force, 2000). Mechanisms have been proposed whereby this could occur via neutral processes, without having to invoke adaptation (positive selection) at all (Force *et al.*, 2005). Consistent with this theory, duplicate genes can be found that have apparently partitioned the ancestral expression domain or function (reviewed by Kellogg, 2003; Moore and Purugganan, 2005); much of the evidence for subfunctionalization comes from studies of gene expression.

But subfunctionalization cannot be the entire story, with new duplicates dividing up ancestral functions ever more finely. New functions must arise from time to time. Clearly, over time, a combination of subfunctionalization and neofunctionalization must occur. Models have been proposed by which this might happen (He and Zhang, 2005). Neofunctionalization may lead to a pattern of selective sweeps such as found by Moore and Purugganan (2003) in the history of multiple genes in *Arabidopsis*.

In summary, gene duplication can lead to diversification of function in many ways. Ancestral functions can be retained by one copy, or partitioned between two copies, and new functions can evolve in one or both copies. Acquisition of new functions is one way to generate novel morphology.

In the remainder of this chapter, we review knowledge of cloned genes affecting inflorescence architecture in the grasses, and present gene trees for some of them. The pattern of duplication is such that no genes in the grasses will have an ortholog in *Arabidopsis*. The *Arabidopsis* genome bears traces of a whole-genome duplication event that occurred at or near the base of the Brassicaceae (Bowers *et al.*, 2003). Likewise, whole-genome duplications are documented near the base of eudicots (Bowers *et al.*, 2003) and at or near the base of the grasses (Paterson *et al.*, 2004; Yu *et al.*, 2005). Thus, without gene loss, no one-to-one correspondence is expected between an *Arabidopsis* gene and a related gene in grasses because of these three rounds of duplication alone. Additional duplications within particular gene families lead to even less correspondence.

The pattern of extensive gene duplication greatly limits the utility of the traditional terminology used to describe gene relationships. "Orthologs" were originally defined as two genes in two different species that derive from a speciation event, and "paralogs" as genes that derive from a single gene that was duplicated within the genome (Fitch, 1970). Three whole-genome duplication events, such as occurred near the base of the Brassicaceae, eudicots, and grasses have produced three sets of gene paralogs, none of which is truly orthologous to any individual gene in the other lineages. This is even the case when only a single (paralogous) gene remains. To overcome this problem, Sonnhammer (2002) coined the term co-orthologs. Co-orthologs are defined as paralogs that are produced by duplications of orthologs subsequent to a given speciation event (see Sonnhammer, 2002 for additional details). Using this terminology, if a clade of paralogous grass genes is sister to a clade of paralogous *Arabidopsis* genes, then the two sets of genes are considered to be co-orthologous.

In species separated by multiple rounds of whole-genome duplication, such as *Arabidopsis* and grasses, the function of co-orthologs is unlikely to be fully conserved. If gene duplication is generally followed by divergence in function, then multiple duplications should lead to multiple variants in gene-expression pattern, biochemical function, and/or developmental role. Because of this, we expect that conservation of function between *Arabidopsis* and grasses should be the exception rather than the rule.

We show in a later section that multiple rounds of duplication have led to novel genes with novel functions in some cases, whereas duplicate genes appear to have been lost and functions retained in others.

II. METHODS

A. CHOICE OF GENES AND TAXA

Genes were chosen because: (a) they represent different stages of inflorescence and floral development and (b) some information is available on function and sequence diversity (Fig. 1). Other genes have been analyzed in more detail elsewhere in the literature, and we review these only briefly here, referring interested readers to the original publications for more detail. The genes included in this chapter are known to play roles in meristem identity and maintenance, inflorescence meristem identity and size, inflorescence branching, inflorescence determinacy, and spikelet development.

Taxon sampling outside the grasses reflects largely data available in GenBank (http://www.ncbi.nlm.nih.gov/); new sequences for some genes were generated from additional grasses and near grass relatives (Fig. 2). For some gene families, such as *KNOTTED1*-like genes, sequences were available from throughout land plants. In other less-studied gene families sampling was more biased toward model species such as *Arabidopsis*, maize, rice, and tomato. In all cases, we sampled broadly within grasses to explore patterns of gene duplication within the major diversification of the family [the BEP and PACCAD clades (Grass Phylogeny Working Group, 2001)]. In several cases, we were able to include early diverging grasses (*Streptochaeta* and *Pharus*) and near grass relatives (such as *Joinvillea*, Joinvilleaceae) to explore patterns of gene duplication near the origin of grasses. *Streptochaeta* and *Anomochloa* are the sole members of the earliest diverging grass lineage (subfamily Anomochlooideae) and lack true spikelets whereas *Pharus* is a member of the earliest diverging lineage with a true spikelet (Grass Phylogeny Working Group, 2001). Inclusion of these taxa is thus essential to understand how particular genes, gene duplications, and functional diversification might have been involved in the origin of the spikelet.

B. DATA, ALIGNMENT, AND POLYMERASE CHAIN REACTION PRIMER DESIGN

Gene sequences of *A. thaliana TERMINAL FLOWER1* (*AtTFL1*), *Oryza sativa LEAFY HULL STERILE1* (*OsLHS1*), and *Zea mays BARREN STALK1* (*ZmBA1*), *FASCIATED EAR2* (*ZmFEA2*), *INDETERMINATE*

SPIKELET1 (*ZmIDS1*), and *KNOTTED1* (*ZmKN1*) were retrieved from GenBank. Similar sequences were identified using BLAST, BLASTX, BLASTP, and tBLASTX (Altschul *et al.*, 1997) searches at NCBI (http://www.ncbi.nlm.nih.gov/BLAST/) and PlantGDB (http://www.plantgdb.org/PlantGDB-cgi/blast/PlantGDBblast). Nucleotide sequences were aligned, based on the conceptual amino acid translation, using CLUSTALX (Jeanmougin *et al.*, 1998) and MacClade 4 (Maddison and Maddison, 2003). Only sequences or regions within sequences that could be aligned reliably as determined by visual inspection of the aligned matrix of protein sequences were included in subsequent analyses. All aligned matrices were submitted to TREEBASE (www.treebase.org). To identify relationships among grass genes for designing polymerase chain reaction (PCR) primers, preliminary maximum parsimony trees were generated using heuristic searches with TBR and MULPARS on and 100 random addition sequence additions within PAUP*4.0 (Swofford, 2000). Gene-specific PCR primers (Table I) were designed based on available grass sequences (usually only rice and maize) using either OLIGO 4.0 (Molecular Biology Insights, Inc., Cascade, USA) or Primaclade (Gadberry *et al.*, 2005).

C. DNA ISOLATION, PCR, SUBCLONING, AND SEQUENCING

Plants were grown under standard greenhouse conditions at Missouri Botanical Garden or at the University of Missouri-Saint Louis from USDA seed stocks. Total DNA was isolated using either CTAB (Hillis *et al.*, 1996) or SDS-based (Dellaporta, 1994) extraction protocols. Total RNA was extracted from young inflorescences using RNAwiz solution (Ambion, Austin, TX), according to the manufacturer's instructions. cDNA was generated from the extracted total RNA using Superscript III (Invitrogen, Carlsbad, CA) following the manufacturer's instructions using a polyT with adaptor primer (5′-CCGGATCCTCTAGAGCGGCCGCTTTTTTTTTTTTTTTTT-3′).

Double-stranded PCR products were amplified in 50-μl reactions containing 2 units of *Taq* polymerase (Promega Corp., Madison, WI), 5 μl of 10× reaction buffer, 5 μl of 25 mM $MgCl_2$, 2 μl dNTP (2.5 mM stock solution), 20 μM each of the gene-specific forward and reverse primers (Table I), 5% DMSO (by volume), and approximately 200 ng of genomic DNA or 4 μl of cDNA. PCR reactions were performed using a hot-start PCR profile with a gene-specific annealing temperature (Table I). PCR fragments were purified and subcloned as described by Malcomber and Kellogg (2004). Dideoxy sequencing was conducted using the Big Dye 3.1 terminator cycle sequencing protocol (Applied Biosystems, Foster City, CA) using the plasmid primers T7 and SP6. Sequencing reactions were analyzed on either an ABI 377 or ABI 3100 automated DNA sequencer (Applied Biosystems). Base calling

TABLE I

Polymerase Chain Reaction (PCR) Primer Sequences for New Gene Sequences, Amplification Conditions and Proportion of the Coding Region Amplified

Gene	PCR primer name	Primer sequence	Annealing temperature	Amplicon size (cDNA or gDNA)	Percent coverage of rice (*Os*) or maize (*Zm*) gene coding region
FASCIATED EAR2 (*FEA2*)	FEA2-228F	5′-CTC CCC GCG TCT CTC CTT-3′	55 °C	~500 bp, cDNA	27% (*ZmFEA2*)
	FEA2-579R	5′-GGT TAC CAG CCA GTG ATA-3′			
INDETERMINATE SPIKELET 1 (*IDS1*)	IDS1-338F	5′-ATG GTG CTG GAT CTC AAT GT-3′	57 °C	~ 650 bp (gDNA)	50% (*ZmIDS1*)
	IDS1-AP2R	5″-GGT SAC GCC CCT GTA CTG CGA GCT-3″			
KNOTTED1 (*KN1*)	KN1-1F	5″-ATG GAG GAG ATC DSC CAM CAC TT-3″	57 °C	~1000 bp (cDNA)	96% (*ZmKN1*)
	KN1-1041R	5″-AGC CCG CYG TCG TTG AYG AAG TG-3″			
RICE CENTRORADIALIS (*RCN*)	RCN1-16F	5′-GAG CCT CTT RTT GTD GGK CGB GTS ATY GG-3′	59 °C	~1000 bp (gDNA)	71% (*OsRCN1*)
	RCN1-393R	5′-GCG YCT CCT GGC WGC AGT CTC YCT CTG-3′			
SEPALLATA-LEAFY HULL STERILE1 (*LHS1*)	MADS-1F	5″-ATG GGT MGS GGS AAG GTG GAG CTG AAG CGG-3′	57 °C	705 bp (cDNA)	100% (*OsLHS1*)
	LHS1-633R	5″-TAT CCA KCC RGA TSG RMY RTG YTC ATT SG-3″			
SEPALLATA3 (*SEP3*)	MADS-1F	See above	57 °C	600 bp, cDNA	75% (*OsMADS8*)
	SEP3-658R	5′- GAT CTG CAR GGT KGG CTC KGC KGC-3′			

within the chromatograms was checked and confidence scores assigned using PHRED (Ewing *et al.*, 1998). Only nucleotide sequences with PHRED scores >20 were used in subsequent analyses. Contiguous alignments were edited using Seqman II (DNASTAR Inc., Madison, WI), and all sequences were submitted to GenBank (DQ317417–DQ317439).

D. ALIGNMENT AND PHYLOGENETIC ANALYSIS

Nucleotide sequences were aligned, based on the conceptual amino acid translation, using RevTrans (Wernersson and Pedersen, 2003) or a combination of CLUSTALX and MacClade 4 (Maddison and Maddison, 2003), before being adjusted manually using MacClade 4. Optimal parameters for the Bayesian analyses were determined using MrModeltest2.0 (Nylander, 2004). Bayesian phylogenetic estimates were produced using MrBayes 3.1 (Huelsenbeck and Ronquist, 2001) on the parallel-processing cluster at the University of Missouri-St. Louis using 10 million generations and default flat priors. Trees were sampled every 500 generations and burn-in was determined empirically by plotting likelihood score against generation number. After burn-in trees had been removed, clade credibility (Bayesian posterior probability) was estimated using MrBayes (Huelsenbeck and Ronquist, 2001). The best topology visited by the Markov chain was retrieved by sorting the MrBayes probability (.p) output file. This topology is expected to be qualitatively identical to the maximum likelihood topology because flat priors were specified prior to conducting the analysis (Larget and Simon, 1999). Only branches supported by 0.95 or greater posterior probability were considered well supported, based on published simulation studies (Alfaro *et al.*, 2003), although other studies have noted that Bayesian phylogenetic analyses can overestimate clade credibility values (Simmons *et al.*, 2004).

III. MORPHOLOGICAL VARIATION AND MOLECULAR EVOLUTION OF GENES IN GRASS INFLORESCENCES

A. FORMATION OF LATERAL STRUCTURES AND NONCORRELATION OF MERISTEM FATES

1. *Morphological variation*

An inflorescence meristem forms lateral structures that themselves become meristems, either branches or spikelets. Although a few grasses have a single terminal spikelet (e.g., *Lygeum*), in the overwhelming majority the inflorescence meristem produces lateral branches that either end immediately in spikelets (e.g., *Lolium, Brachyelytrum*), or that go on to branch again. In addition, the fate of the inflorescence meristem is often different from that of lateral meristems (see examples in a later section).

Inflorescences develop both from terminal meristems (terminal inflorescences) and from meristems in the axils of vegetative leaves (axillary inflorescences). In most grasses, the morphology of terminal and axillary inflorescences is the same. On the other hand, in the grass tribe Andropogoneae (which includes maize, sorghum, and sugarcane), axillary inflorescences often have fewer branches than the terminal one (Kellogg, E. A., unpublished observations). This is most obvious in *Z. mays* ssp. *mays*, in which axillary branches terminate in an inflorescence without long branches (the ear), whereas the terminal inflorescence (the tassel) bears multiple long branches. In *Z. mays*, the position of the inflorescence also correlates with its sex expression, but this is unusual among grasses. *Tasselseed6* (*ts6*) and *tasselseed4* mutants of maize have more branches and more orders of branching than wild type (Irish, 1997), but the phenotype is much less severe in the ear (axillary inflorescence) than in the tassel (terminal inflorescence), also suggesting that an unknown factor suppresses branching in axillary inflorescences.

2. KNOTTED1-*like genes*

The ability to form and maintain meristems is affected by a number of genes, the best studied of which is *Z. mays KNOTTED1* (*ZmKN1*). *ZmKN1* is best known for its effects when over expressed; leaf morphology is altered, and ectopic meristems often form. However, the function of *ZmKN1* is apparently meristem formation and/or maintenance. Importantly for this review, *Zmkn1* mutants in permissive inbred backgrounds may become reproductive, but show defects in production of lateral inflorescence branches; if ears form they only produce a few kernels and tassels have fewer branches (Hake *et al.*, 2004).

ZmKN1 was the first plant protein found to contain a homeodomain (Vollbrecht *et al.*, 1991) and is named after the dominant leaf phenotype that has modified "knotted" leaf veins (Hake *et al.*, 2004). The *KNOTTED1*-like homeobox (*KNOX*) genes form a small gene family that has diversified extensively in flowering plants and falls into two monophyletic classes—class I and class II. Members of each class share intron positions, expression patterns, and characteristic residues within the homeodomain (Kerstetter *et al.*, 1994). Class I *KNOX* genes are usually expressed in the shoot apical meristem (SAM) whereas expression of Class II genes is more widespread (Hake *et al.*, 2004). Both classes contain representatives from flowering plants, gymnosperms, ferns, and mosses, suggesting the two clades probably resulted from a gene duplication event near the origin of land plants.

ZmKN1 is a class I *KNOX* gene. Our phylogenetic analysis of 65 class I *KNOX* genes (Fig. 3) is largely congruent with analyses by Harrison *et al.* (2005). Differences between ours and other analyses, such as Hake *et al.* (2004) and Reiser *et al.* (2000), probably reflect differences in sampling. Our phylogeny finds two major clades, *BP/KN1* and *KNAT2/LG3*, each of which must have originated early in angiosperm evolution. Within each of these clades are multiple duplications within monocots, leading to at least six different class I *KNOX* gene lineages. Of these, *KN1* forms a well-supported clade with *GN1/RS1, KNOX3*, and *KNOX8* sister to a clade of eudicots containing *Arabidopsis BREVIPEDICELLUS* (*AtBP*). The *KN1* clade contains diverse grass sequences including *Hordeum vulgare KNOX3* (*HvKNOX3*), *O. sativa HOMEOBOX1* (*OsH1*), and *ZmKN1*. The *KNOX3* clade comprises *Z. mays KNOX3* (*ZmKNOX3*) and *O. sativa HOMEOBOX3* (*OsH3*), the *KNOX8* clade, *Z. mays KNOX8* (*ZmKNOX8*) and *O. sativa HOMEOBOX43* (*OsH43*), and the *GN1/RS1* clade, *Z. mays GNARLY1* (*ZmGN1*), *Z. mays ROUGHSHEATH1* (*ZmRS1*), *O. sativa HOMEOBOX15* (*OsH15*), and *H. vulgare* "*HvKn1*." Sister to the *BP/KN1* clade is the *KNAT2/LG3* clade. *Z. mays LIGULELESS4* (*ZmLG4*) and *O. sativa HOMEOBOX71* (*OsH71*) are sisters as are *Z. mays LIGULELESS3* (*ZmLG3*) and *O. sativa HOMEOBOX6* (*OsH6*). The *LG3* and *LG4* clades themselves are well supported as sisters and together are sister to a clade of asterids containing the tomato *SeTKN4* gene.

Limited sampling within most of the grass clades makes the exact placement of gene duplication events unclear. However, all clades contain orthologous rice and maize sequences, and the *KN1* clade contains representatives of most grass subfamilies including the early diverging *Pharus* (Pharoideae), indicating that all lineages are at least as old as the grass family, and may be older. The *Arabidopsis* class I *KNOX* gene *BREVIPEDICELLUS* (*BP*) is the closest eudicot relative of *ZmKN1* and is co-orthologous to the monocot *KN1, KNOX3, KNOX8*, and *GN1/RS1* clade.

The *KN1, KNOX3, KNOX8*, and *GN1/RS1* clades have different expression patterns within vegetative tissues but overlapping patterns of expression in the inflorescence and floral meristems (Foster *et al.*, 1999; Hake *et al.*, 2004; Jackson *et al.*, 1994; Sato *et al.*, 1998; Sentoku *et al.*, 1999). *KN1* and *KNOX8* (which are not sister genes) are both expressed throughout maize and rice shoot meristems, but are downregulated as leaves are initiated. Likewise, maize and rice *GN1/RS1* and *KNOX3* gene expression is similar, even though the genes do not form a clade; expression is restricted to meristem and stem regions between successive leaf primordia (Foster *et al.*, 1999; Jackson *et al.*, 1994; Sato *et al.*, 1998; Sentoku *et al.*, 1999).

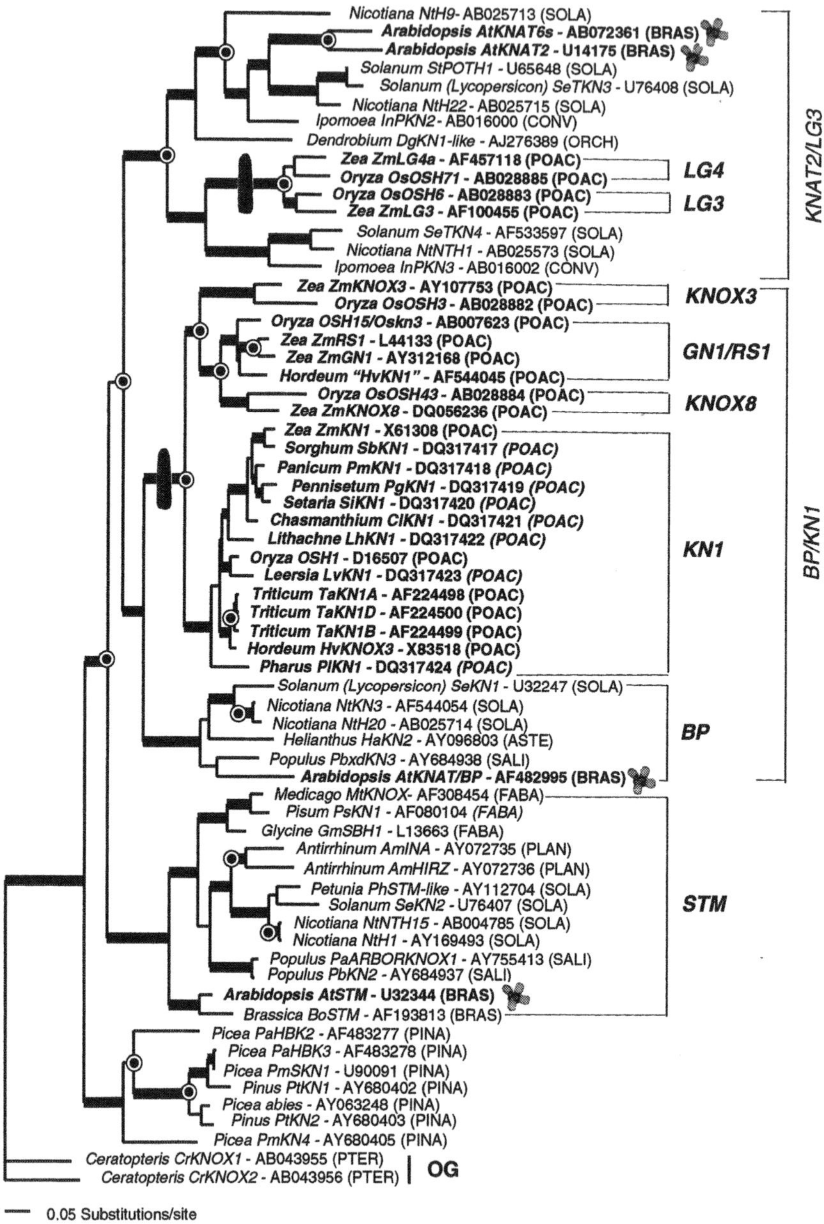

Fig. 3. Maximum likelihood phylogram of the 65 KNOTTED1-LIKE HOMEOBOX (*KNOX*) gene data set, comprising 555 bp, visited by Bayesian Markov Chain Monte Carlo search using the General Time Reversible model with some invariant sites and gamma distributed rates (GTR + I + G). –ln = 13505.68. Bold branches are supported by posterior probabilities >0.95. Boldface, grass and *Arabidopsis* sequences. ◉, Inferred duplication event; *Arabidopsis* flower, *Arabidopsis* sequence; corn

Like the expression studies, genetic studies find similar *KNOX* gene function among distantly related genes. Strong loss-of-function *Zmkn1* mutants have similar phenotypes to *Arabidopsis shoot meristemless* (*stm*) mutants, which produce cotyledons but no further elements of the shoot system (Kerstetter *et al.*, 1997; Long *et al.*, 1996). These phenotypes suggest a role in shoot apical meristem maintenance and/or initiation (Hake *et al.*, 2004). Distantly related maize and *Arabidopsis* genes also co-ordinate cell differentiation and internode growth (Hake *et al.*, 2004). Knockout mutants of the rice *GN1/RS1* gene *Ososh15* are shorter than wild-type plants due to reduced lower internode length (Sato *et al.*, 1998). This dwarf phenotype is similar to *Arabidopsis brevipedicellus* (*bp*) mutants that have shortened internodes and shorter down-pointing pedicels (Venglat *et al.*, 2002).

Similar loss-of-function phenotypes in the distantly related pairs *Zmkn1/Atstm* and *Atbp/Ososh15* indicate convergence in gene function in at least one of the pairs. If the *KN1/STM*1 function is ancestral, then the *GN1/RS1* and *BP* functions are convergent. Alternatively, if the *OSH15/BP* role is ancestral then the *KN1/STM* function originated independently. It is also possible, although considerably less parsimonious, that the ancestral protein was multifunctional, and that particular subfunctions have been lost repeatedly in evolutionary time. Functional information on other Class I *KNOX* genes would help to assess the ancestral pattern.

In summary, class I *KNOX* genes are generally involved in meristem maintenance and internode growth in grasses and eudicots. However, duplicate genes have diverged in function, and distantly related genes have converged.

3. LAX PANICLE1/BARREN STALK1

LAX PANICLE1 (*LAX*) in rice and *BARREN STALK1* (*BA1*) in maize affect lateral branching in the inflorescence, although neither affects growth of the apical meristem (Gallavotti *et al.*, 2004; Komatsu *et al.*, 2003a). *BA1* is required for initiation of axillary meristems throughout the aerial parts of the plant, both vegetative and reproductive. *LAX*, on the other hand, affects only the inflorescence. Plants carrying strong mutant *lax* alleles produce no spikelets and few panicle branches; those with weaker alleles produce primary branches, but these are almost devoid of spikelets except at their termini. The gene *SMALL PANICLE* (*SPA*) has a similar developmental role as *LAX*,

---

ear, grass clade. Plant families abbreviated to first four letters of name. *BP, BREVIPEDICELLUS; GN1, GNARLY1; KN1, KNOTTED1; KNAT2, KNOTTED1-LIKE HOMEOBOX PROTEIN FROM Arabidopsis thaliana 2; KNOX3, KNOTTED1-LIKE HOMEOBOX 3; KNOX8, KNOTTED1-LIKE HOMEOBOX 8; LG3, LIGULELESS3; LG4, LIGULELESS4*; OG, Outgroup; *RS1, ROUGHSHEATH1; STM, SHOOT MERISTEMLESS.*

and double mutants have more severe phenotypes than either alone; in the double mutants, all axillary meristems—both vegetative and reproductive—are affected. *SPA* is not yet cloned; Komatsu *et al.* (2003a) speculate that it might be another bHLH protein that dimerizes with *LAX*, or might belong to another class of transcription factor that binds to *LAX*. The developmental role of *BA1* is thus broader than that of *LAX*, and may imply that maize lacks a functional equivalent of *SPA*. Unlike *KN1, BA1* has no effect on the apical meristem during either the vegetative or reproductive phases, or plant growth, but strong mutant alleles have no axillary meristems.

Both *LAX* and *BA1* encode atypical bHLH transcription factors that map to syntenic positions in the rice and maize genomes, respectively. Both are expressed in a slender line of cells just above the point of attachment of lateral meristems (i.e., in the axil of the branches). In normal maize tassel development, *BA1* is expressed just above (adaxial to) initiating long- and short-branch meristems, a pattern analogous to that of *LAX* in rice. Gallavotti *et al.* (2004) hypothesize that *BA1* may be necessary to specify a set of cells that can become an axillary meristem.

Plants contain a very large number of distinct bHLH proteins, with 118 identified from *Arabidopsis* and at least 131 in rice (Buck and Atchley, 2003). The majority of the plant proteins, including 104 of those from *Arabidopsis*, are group B bHLH proteins (Groups A, C, D, and E are not found in plants.). Fourteen of the *Arabidopsis* proteins and 6 from *Oryza* fall into 2 distinct classes, which Buck and Atchley (2003) call PbHLH5 and 6. These proteins have distinctive and uncharacterized DNA-binding domains.

In our phylogenetic analyses including a sample of 43 plant bHLH genes, *LAX, BA1*, and a sequence from *Sorghum* are nested within a well-supported clade of PbHLH6 sequences (Fig. 4), and thus provide the first functional data for members of this clade. Komatsu *et al.* (2003) infer that *LAX* (and by inference, *BA1*, which was not cloned at the time of their paper) represents a type of bHLH protein unique to the grasses because all related proteins in *Arabidopsis* are quite dissimilar outside of the conserved DNA-binding domain. The phylogenetic analysis supports this hypothesis, although sampling of other angiosperm families would provide a more stringent test.

4. MONOCULM1

In contrast to *KN1* and *BA1*, the effect of *MOC1* on the inflorescence has not been extensively characterized. The gene was cloned from rice, in which mutants produce no tillers and very few primary branches in the inflorescence. The gene is expressed in axillary buds throughout their development. *OsOSH1* and *O. sativa TEOSINTE BRANCHED1* (also involved in tiller

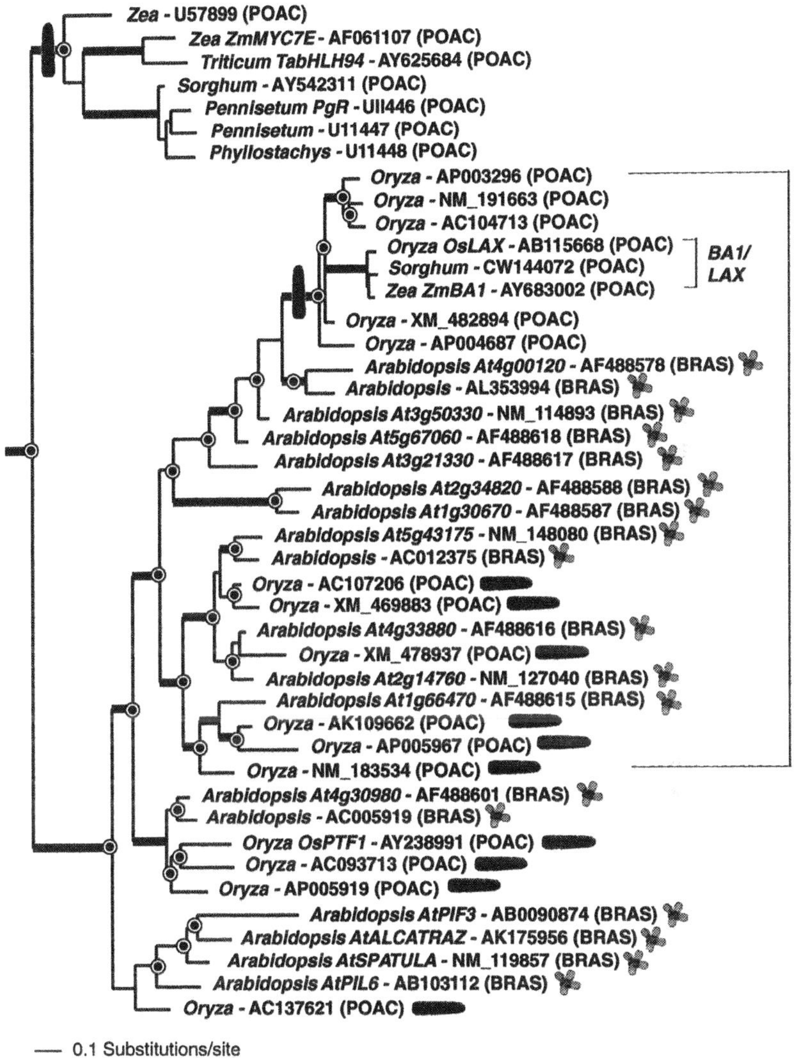

Fig. 4. Maximum likelihood phylogram of the 45 *BARREN STALK1/LAX PANICLE1*-like basic helix-loop-helix (bHLH) gene data set, comprising 156 bp, visited by Bayesian Markov Chain Monte Carlo search using the General Time Reversible model with some invariant sites and gamma distributed rates (GTR + I + G). Tree rooted using cat (NM_001009866) and rat (NM_022210) MAX sequences (not shown). –ln = 4803.03. Bold branches are supported by posterior probabilities >0.95. Boldface, grass and *Arabidopsis* sequences. ◉, Inferred duplication event; *Arabidopsis* flower, *Arabidopsis* sequence; corn ear, grass clade. Plant families abbreviated to first four letters of name. *PbHLH6*, plant basic helix-loop-helix family 6, following the classification of Buck and Atchley (2003).

elongation) were not expressed in axillary buds of *moc1* mutants, indicating that *MOC1* lies upstream in the developmental pathway of bud formation.

MOC1 encodes a GRAS family protein with some similarity to the *LATERAL SUPPRESSOR* proteins of tomato and *Arabidopsis* (*LS* and *LAS*, respectively). A phylogenetic analysis of GRAS proteins (Tian *et al.*, 2004), identified a second rice gene, which they called *OsGRAS-7*, that is sister to *MOC1*; these two are sister to the clade of *LS* and *LAS*. Except for OsGRAS-7, whose function is unknown, members of this clade appear to be similar in regulating formation of axillary meristems throughout the plant.

5. BARREN INFLORESCENCE2

Maize *barren inflorescence2* (*bif2*) mutants are characterized by fewer lateral inflorescence branches, spikelets and ear shoots than wild type plants suggesting that the *ZmBIF2* gene functions in inflorescence axillary meristem initiation and maintenance (McSteen and Hake, 2001). Although *Zmbif2* mutants sometimes have fasciated apical meristems (McSteen and Hake, 2001), it is notable that *ZmBIF2* affects apical and lateral meristems differently. *Zmbif2* mutants appear superficially similar to *pinoid* (*pid*) mutants in *Arabidopsis*. *AtPID* is a regulator of polar auxin transport (Benjamins *et al.*, 2001; Christensen *et al.*, 2000), and it will be of interest to know whether *ZmBIF2* has a similar function.

To summarize this section, three of the four proteins described here affect lateral meristems differently from the apical inflorescence meristem. This sets up a prepattern that allows the fate of lateral meristems to differ from that of the apical one. Such a prepattern permits (but probably does not cause) diversification of inflorescence form, and fits with the combinatorial model of inflorescence development (Kellogg, 2000b).

B. NUMBER OF ORDERS OF BRANCHING

Grass inflorescences vary in the number of times each branch itself branches (orders of branching). The literature is inconsistent in how these orders of branching are numbered, whether the inflorescence meristem is a first-order branch and it produces second-order branches (Doust and Kellogg, 2002; Vollbrecht *et al.*, 2005), or if the numbering should begin by calling the branches produced by the inflorescence meristem primary branches, implicitly calling the inflorescence meristem branch 0 (e.g., most of the rice literature). We will use the latter convention in this chapter and count the spikelets themselves as an order of branching.

In some grasses, such as *Lolium* or *Brachyelytrum*, the inflorescence meristem produces spikelets immediately (i.e., terminating the primary branches);

in some species with this architecture a pedicel elongates beneath the spikelet late in development. In other genera, the primary branches themselves branch, producing two orders of branching beyond the inflorescence axis itself. In *Eleusine* (finger millet), the primary branches each produce multiple short secondary axes, and the secondaries terminate in spikelets. *Oryza* has three orders of branching beyond the main inflorescence axis; the lower lateral meristems (primary branches) each produce several more lateral meristems (secondary branches) before ending in a spikelet; and the secondaries produce spikelets (third-order branches) laterally before themselves terminating in spikelets (Ikeda *et al.*, 2004). Four or more orders of branching are common in many grasses, including maize; and such densely branched species as *Pennisetum* (pearl millet) and *Setaria* (foxtail millet) may have six orders of branching or more (Doust and Kellogg, 2002).

The mechanisms controlling the number of orders of branching are largely unknown. The maize mutants *tasselseed4* and *tasselseed6* (Irish, 1997) both increase the numbers of orders of branching, with the strongest effect produced by *ts4*. Unfortunately, neither is cloned yet, so cross-species comparisons are not possible. Doust *et al.* (2005) identified two major QTL for this character in a mapping population of *Setaria*, which together explained nearly half of the variance in the character. The genes underlying the QTL are unknown.

C. NUMBER OF BRANCHES AT EACH ORDER

1. *Morphological variation*

The number of branches at each order of branching is often fixed or varies within a narrow range for a given species. For example, *Hordeum* always produces a single primary branch at a node, and the primary branch always produces two secondary branches before itself ending in a spikelet. The secondaries also terminate in spikelets, giving the characteristic of three spikelets per node. Similarly, in the tribe Andropogoneae, the final round of branching produces a pair of spikelets, one sessile and one pedicellate, from a common meristem (the spikelet pair meristem), which can be interpreted as a short branch meristem that itself produces a single branch. The two spikelets thus each terminate branches that represent different orders of branching, evidence for which comes from developmental studies and gene expression data. In *Tripsacum*, as in *Zea*, the sessile spikelet appears to develop slightly later and lateral to the pedicellate one (Orr *et al.*, 2001), as would be expected if it were a higher order branch. In maize, *BA1* is expressed in a narrow zone between the two spikelet meristems (Section III. A.2); because *BA1* is required for lateral meristem initiation, the expression pattern suggests that one of the spikelets is lateral to the other.

The number of primary branches produced by the inflorescence meristem varies independently of the number of secondary branches produced by the primaries. For example, the inflorescence meristems of *Digitaria sanguinalis* (crab grass) and *Cynodon dactylon* (Bermuda grass) each produce fewer than 15 primary branches, but each primary branch produces numerous secondaries; the secondaries may (*Digitaria*) or may not (*Cynodon*) produce a single tertiary and then terminate in a spikelet (Barkworth *et al.*, 2003). The inflorescences of these two species thus appear sparse and spreading (digitate, or more or less antenna-like). In contrast, the inflorescence meristem of *Setaria italica* (foxtail millet) produces numerous primary branches, but few secondaries per primary, and one at each of seven or more orders of branching (Doust and Kellogg, 2002). The result is an inflorescence that is tall, narrow, and very dense.

2. RICE CENTRORADIALIS

Overexpression of *RICE CENTRORADIALIS* (*RCN*) in rice resulted in more secondary branches per primary and more tertiary branches (spikelets, in rice) per secondary (Nakagawa *et al.*, 2002). Although it is difficult to infer wild-type role based on overexpression studies, *RCN1* might regulate the number of branches produced by each higher order meristem by delaying conversion of the branch meristem to a spikelet.

RCN1 and *RCN2* in rice are distantly related to *TERMINAL FLOWER1* (*TFL1*) in *Arabidopsis*, and its ancient paralog *FLOWERING LOCUS T* (*FT*) (Fig. 5), which are key regulators of flowering time and plant architecture in *Arabidopsis*. The two genes have opposite effects, with *TFL1* repressing and *FT* activating flowering. The difference in biochemical function is due to a single amino acid residue; replacing HIS88 in *TFL1* with TYR converts *TFL1* into an activator, and replacing TYR85 in *FT* with HIS converts *FT* into a repressor (Hanzawa *et al.*, 2005).

Our phylogenetic analysis of 43 *TFL1/FT* genes estimates a complex pattern of gene duplication (Fig. 5). Given the distribution of taxa within the respective clades, the *TFL1/FT* duplication occurred before separation of the eudicot and monocot lineages (Hanzawa *et al.*, 2005). Multiple duplication events occurred within the *TFL* lineage itself. Duplications within *Populus* and *Brassica* are likely associated with polyploidization events in Salicaceae and Brassicaceae, respectively. The placement of the *Joinvillea* sequence (*JaRCN1*) as sister to grass *RCN1* genes indicates that the *Oryza RCN1/RCN2* duplication occurred within monocots and before the origin of grasses. However, determining the exact timing of the event requires sequences from other Poales and more distantly related monocots. Yet another duplication occurred during the evolution of the grasses, apparently near the

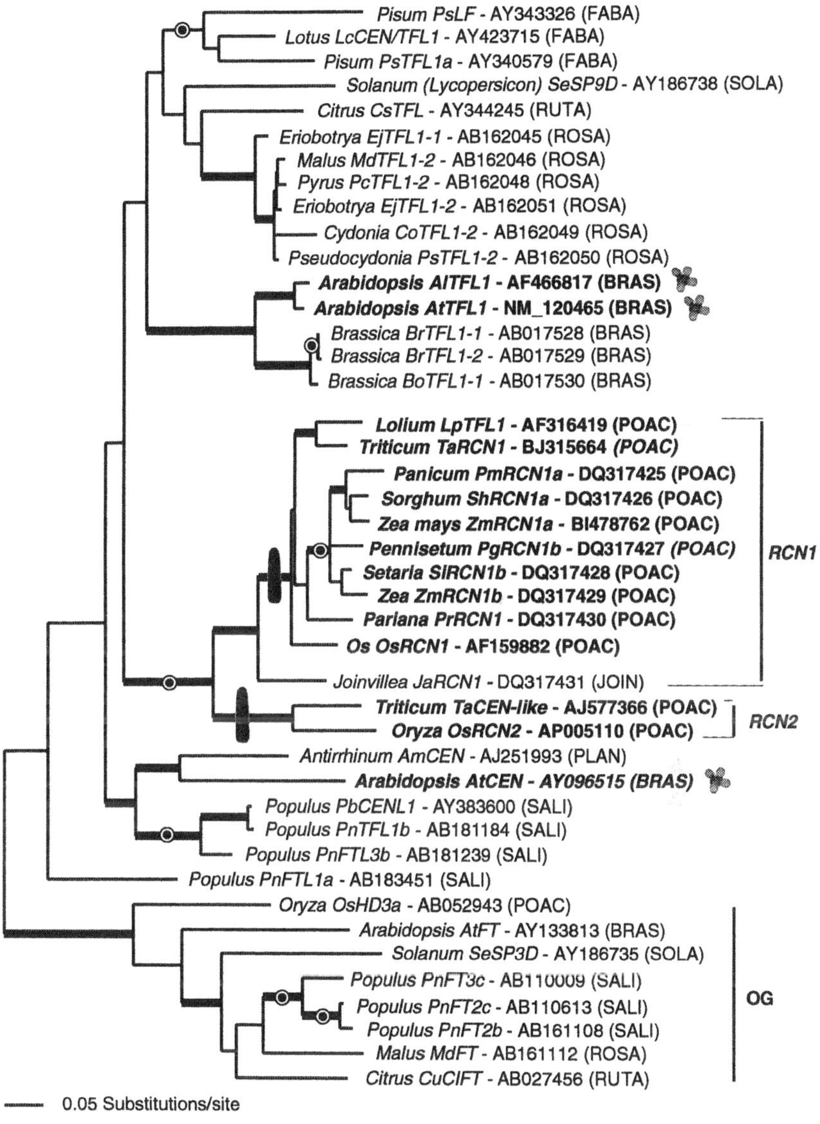

Fig. 5. Maximum likelihood phylogram of the 43 *CENTRORADIALIS, FLOWERING LOCUS T, RICE CENTRORADIALIS*, and *TERMINAL FLOWER1*-like gene data set, comprising 537 bp, visited by Bayesian Markov Chain Monte Carlo search using the General Time Reversible model with some invariant sites and gamma distributed rates (GTR + I + G). Tree rooted using *FLOWERING LOCUS T* sequences. $-\ln = 9556.78$. Bold branches are supported by posterior probabilities >0.95. Boldface = grass and *Arabidopsis* sequences. ◉, Inferred duplication event; *Arabidopsis* flower, *Arabidopsis* sequence; corn ear, grass clade. Plant families abbreviated to first four letters of name. OG, Outgroup; *RCN1, RICE CENTRORADIALIS1; RCN2, RICE CENTRORADIALIS2.*

base of panicoids. Because of the duplications, none of the grass *TFL*-like genes can be considered directly equivalent to *Arabidopsis TFL1*.

All known *TFL1/CEN/RCN* proteins have HIS rather than TYR at a position equivalent to amino acid 88 in the *Arabidopsis* TFL1 sequence, consistent with the functional analyses of Hanzawa *et al.* (2005), with the exception of the three *Brassica* TFL1-like proteins. In *Bo*TFL1 and *Br*TFL1, HIS is replaced by ARG, resulting from a G to A substitution at the second codon position. Both ARG and HIS are positively charged, with ARG being more strongly charged than HIS, whereas TYR is polar but neutral. Overexpression or knockouts of *BrTFL1* and *BoTFL1* genes will determine whether this amino acid substitution has any functional significance.

Functional analyses of *TFL*-like genes in grasses are available for rice (*O. sativa*) and ryegrass (*Lolium perenne*) (Jensen *et al.*, 2001; Nakagawa *et al.*, 2002). Overexpression of *OsRCN1, OsRCN2*, and *LpTFL1* in *Arabidopsis* caused delayed flowering and extensive branching (Jensen *et al.*, 2001; Nakagawa *et al.*, 2002), similar to overexpressed *AtTFL1* phenotypes (Ratcliffe *et al.*, 1998). Thus, the rice and ryegrass proteins might maintain biochemical functions similar to those of their eudicot co-ortholog, *Arabidopsis* TFL1.

3. RAMOSA1

In the maize tassel, branching is somewhat more complex than in rice. The first primary branches produced by the inflorescence meristem, long branches, themselves produce a large number of secondaries; in the maize literature the long branches are called indeterminate because the precise number of secondary branches is not fixed (Fig. 1). Later, the branches produced by the inflorescence meristem produce only one secondary branch, which produces one tertiary, together forming a spikelet pair (called determinate in the maize literature). The long branches also go on to produce spikelet pairs, which represent tertiary and quaternary branches. The developmental determination of long branches with many secondaries, vs short branches with only one, is controlled by *RAMOSA1* (Vollbrecht *et al.*, 2005). When *RA1* is mutated, more branches in the tassel develop as long branches, with large numbers of secondaries. Above these are long branches that produce a mix of branched and unbranched secondaries, such that some spikelets are paired and others are unpaired. These mixed branches ("spikelet multimers") are shorter toward the apex of the tassel, until conventional short branches are produced at the tip, giving a distinctive Christmas tree architecture. In maize, short branch (spikelet pair) production on the main inflorescence axis and on the long branches coincides with the onset of *RA1* expression, and in *ra1* mutants, more long branches are produced at the expense of short branches.

RA1 appears to have a similar function in other Andropogoneae. The genus *Miscanthus* produces long primary branches, each of which produces many short branches roughly simultaneously. Consistent with this morphology, *RA1* expression appears somewhat later in development than in maize, being delayed until all primary branches have initiated. The gene is then expressed for a very short time, corresponding to near-simultaneous short branch initiation. Unlike *Zea* and *Miscanthus, Sorghum* produces tertiary and quaternary branches before producing the final short branch meristem. Short branch production occurs on distal branches while proximal branches are themselves producing higher order branches. Consistent with this extended period of short-branch production, *RA1* in sorghum is expressed over a longer portion of inflorescence development than in the other species.

RA1 encodes an EPF-class zinc finger protein, with a DNA-binding domain that differs at one residue from the many other EPF class proteins in rice and *Arabidopsis*. The phylogeny of *RA1* and other EPF-class genes has yet to be explored. The short branches that produce spikelet pairs are only known in the panicoid grasses and are most common in Andropogoneae (Kellogg, 2000c). This observation, and the lack of an obvious *RA1*-like gene in rice, suggests that *RA1* could have originated from a duplication near the base of the Panicoideae.

4. ABERRANT PANICLE ORGANIZATION

ABERRANT PANICLE ORGANIZATION (*APO1*) in rice (Ikeda *et al.*, 2005) controls the conversion of inflorescence and branch meristems to spikelets. In *apo1* mutants, the number of primary branches is reduced, as are numbers of lateral structures (secondary branches or spikelets) on primary branches. The inflorescence meristem, which normally aborts in rice, is converted to a spikelet in many mutant plants. *APO1* has other effects on phyllotaxis (see Section III.D.3). The gene is not yet cloned, so comparative data are unavailable.

5. GRAIN NUMBER 1A/CKX2

A QTL for yield, called *grain number 1A* (*Gn1A*), has been cloned in rice (Ashikari *et al.*, 2005). Differing alleles at this locus affect the numbers of grains produced by controlling numbers of spikelets. One *indica* variety, Habataki, produces more spikelets (and hence has higher yield), but is shorter than a japonica variety Koshihikari; a mapping population derived from a cross between the two identified several loci controlling this phenotype, including *Gn1A*. *Gn1A* was identified as a cytokinin oxidase/dehydrogenase (*OsCKX2*) that is expressed preferentially in inflorescences. Higher levels of gene expression were observed in Koshihikari, and correlated with lower levels of cytokinin, whereas the opposite was true in Habataki.

The authors thus hypothesize that, by regulating the level of cytokinin in the inflorescence meristem, OsCKX2 controls the number of spikelets and hence the number of grains.

In summary, the number of branches produced at each order is fixed or varies within a narrow range for many species of grasses. The plant controls the number of branches in part by mechanisms that determine if and when the apical meristem of the inflorescence or inflorescence branch is converted to a spikelet; the timing of this conversion is affected by proteins such as RCN1, RCN2, APO1, and CKX2. The sharp distinction between long and short branches (spikelet pairs) may be synapomorphic (uniquely derived) for all or part of the Panicoideae; RA1 appears to be a candidate for control of this aspect of morphology.

D. PHYLLOTAXIS

1. Morphological variation

As the shoot apical meristem converts from a vegetative to an inflorescence meristem it elongates and begins producing primary lateral branches. In many grasses, the apical meristem also changes from distichous vegetative phyllotaxis to spiral inflorescence phyllotaxis. The shift from distichous to spiral is known to occur in Ehrhartoideae [*Oryza* (Ikeda *et al.*, 2005; Itoh *et al.*, 2005), *Zizania* (Weir and Dale, 1960)], Panicoideae [*Zea* (Bonnett, 1948; Sundberg and Orr, 1996), *Sorghum* (L. G. Le Roux and E. A. Kellogg, unpublished), *Ixophorus* (Kellogg *et al.*, 2004), *Panicum* (Bess *et al.*, 2005; Reinheimer *et al.*, 2005), *Pennisetum, Setaria* (Doust and Kellogg, 2002)], and Chloridoideae [*Eragrostis, Eleusine* (Moncur, 1981)]. Spiral inflorescence phyllotaxis is also clear in *Streptochaeta* (Anomochlooideae), which is sister to all other grasses, and in the grass outgroups *Joinvillea* (Joinvilleaceae; Malcomber and Kellogg, unpublished) and *Ecdeiocolea* (Ecdeiocoleaceae; Rudall *et al.*, 2005). Portions of the inflorescence of Restionaceae are also spirally arranged (Ronse Decraene *et al.*, 2002), but the early development of the inflorescence as a whole has not been studied in detail. Spiral inflorescence phyllotaxis is thus very probably ancestral in the grasses.

Some grasses do not undergo phyllotactic change. In all Pooideae [e.g., *Hordeum* (Klaus, 1966; Bossinger, 1990), *Avena* (Moncur, 1981), *Stipa* (Kellogg, E. A., unpublished), *Phaenosperma* (Kellogg, E. A., unpublished), and multiple other species (Evans, 1940)], the inflorescence is distichous, like the leaves. The failure to shift phyllotaxis also occurs in some Panicoideae [e.g. *Urochloa* (Reinheimer *et al.*, 2005); *Heteropogon* (Le Roux and Kellogg, 1999)], and in the woody bamboo *Fargesia* (Kellogg, E. A., unpublished). In all cases, producing distichous primary branches appears to be evolutionarily derived.

2. CLAVATA-*like*

The *CLAVATA*-like proteins [FASCIATED EAR2 and THICK TASSEL DWARF1 in maize (Bommert *et al.*, 2005a; Taguchi-Shiobara *et al.*, 2001) and FLORAL ORGAN NUMBER1 in rice (Suzaki *et al.*, 2004)] affect phyllotaxis by regulating the size of the inflorescence meristem. Maize *fasciated ear2* (*Zmfea2*) mutants are characterized by enlarged inflorescence and floral meristems. *ZmFEA2* encodes a membrane-localized leucine-rich repeat (LRR) receptor-like protein that is very similar to *Arabidopsis* CLAVATA2 (AtCLV2) (Taguchi-Shiobara *et al.*, 2001). Our phylogenetic analysis of 28 *FEA2*-like LRR genes indicates that *ZmFEA2* and *AtCLV2* are the sole remaining co-orthologs (Fig. 6). The *FEA2/CLV2* duplicates produced by the grass (Paterson *et al.*, 2004; Yu *et al.*, 2005), eudicot, and Brassicaceae (Bowers *et al.*, 2003) duplication events therefore must have been lost during evolutionary time. Although *FEA2/CLV2* genes have not been isolated from other eudicots, the clade appears to date back to at least the separation of the eudicot and monocot lineages.

ZmFEA2 and *AtCLV2* have similar broad-level expression patterns but nonidentical functional roles. Both genes are expressed in developing leaves and inflorescence tissues, but are downregulated in mature leaves (Jeong *et al.*, 1999; Taguchi-Shiobara *et al.*, 2001). *ZmFEA2* is also not expressed in roots (Taguchi-Shiobara *et al.*, 2001). Although both *Zmfea2* and *Atclv2* mutants have enlarged inflorescence and floral meristems, and in at least some flowers, longer pedicels and more stamens, only *Atclv2* mutants have enlarged vegetative meristems (Jeong *et al.*, 1999; Kayes and Clark, 1998; Taguchi-Shiobara *et al.*, 2001). Thus, both genes appear to limit meristem size, although the role of *ZmFEA2* seems to be restricted to inflorescence tissues.

The three *Arabidopsis* genes *AtCLV1, AtCLV2*, and *AtCLV3* form a single receptor–ligand complex with *AtCLV2* interacting with *AtCLV1*, and the *AtCLV1* + *AtCLV2* heterodimer acting as a receptor for the secreted *AtCLV3* signaling protein (Sharma *et al.*, 2003). This *CLV* signaling pathway functions interdependently with *WÜSCHEL* (*WUS*), with *WUS* promoting meristem fate and *CLV* restricting meristem size (Sharma *et al.*, 2003).

The maize *THICK TASSEL DWARF1* (*TD1*) and rice *FLORAL ORGAN NUMBER1* (*FON1*) genes form a clade that is sister to a eudicot clade containing *AtCLV1* (Bommert *et al.*, 2005a). All three mutants have enlarged meristems, although which meristems are enlarged depends upon the species. As in *Atclv2, Atclv1* mutants have enlarged vegetative, inflorescence, and floral meristems (Clark *et al.*, 1997). In contrast, only the inflorescence, spikelet and floral meristems are enlarged in *Zmtd1* mutants, and only floral meristems are affected in *Osfon1* mutants (Bommert *et al.*, 2005; Suzaki *et al.*, 2004). Expression patterns also vary among the three species.

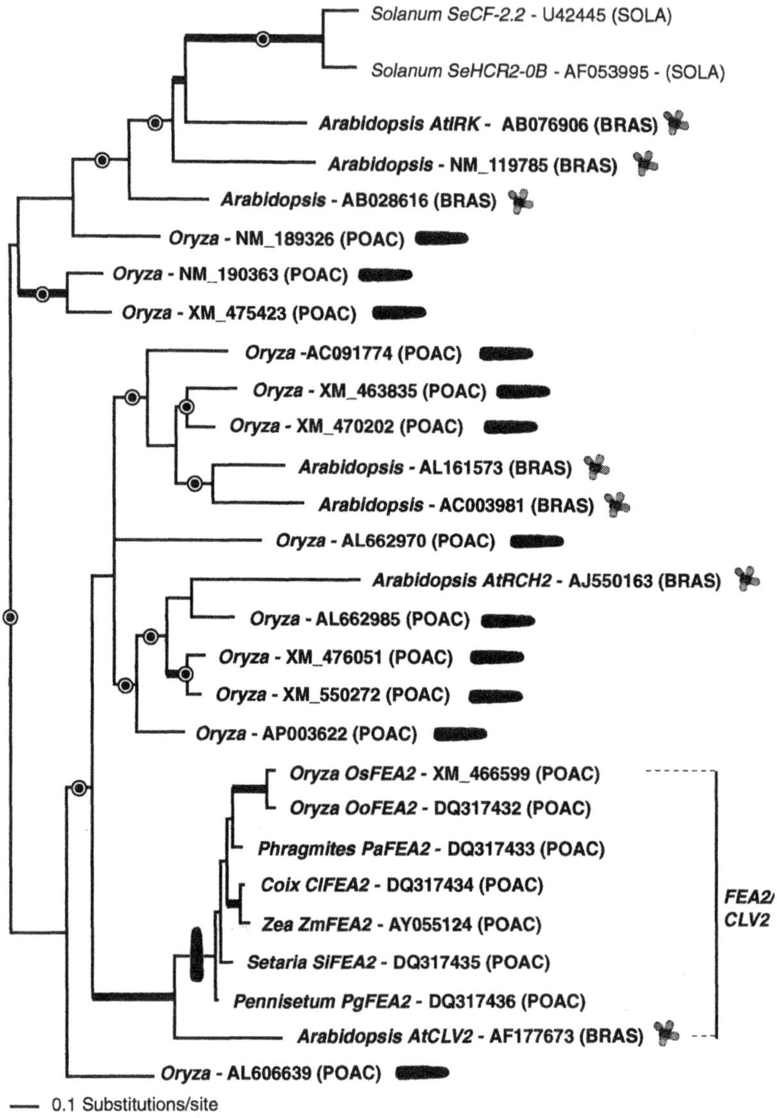

Fig. 6. Maximum likelihood phylogram of the 28 *FASCIATED EAR2* and *CLAVATA2*-like gene data set, comprising 375 bp, visited by Bayesian Markov Chain Monte Carlo search using the General Time Reversible model with some invariant sites and gamma distributed rates (GTR + I + G). Tree rooted at midpoint. $-\ln = 7072.01$. Bold branches are supported by posterior probabilities >0.95. Boldface = grass and *Arabidopsis* sequences. ◉, Inferred duplication event; *Arabidopsis* flower, *Arabidopsis* sequence; corn ear, grass clade. Plant families abbreviated to first four letters of name. *CLV2, CLAVATA2; FEA2, FASCIATED EAR2.*

AtCLV1 is restricted to inflorescence and floral meristems (Clark *et al.*, 1997), whereas *ZmTD1* and *OsFON1* are expressed broadly in all above ground tissues, including floral organs such as glumes, lemmas, and stamens (Bommert *et al.*, 2005a; Suzaki *et al.*, 2004). This shows that the different types of meristems in grasses are distinct in their development and regulation, a necessary condition for diversification of form.

The phenotypic similarity of *Zmtdl* and *Zmfea2* mutants suggests that the genes might belong to a single signaling pathway, like *AtCLV1* and *AtCLV2* (Bommert *et al.*, 2005). However, *Zmtdl/Zmfea2* double mutants exhibit more severe aberrations than either of the single mutants, with twice as many kernel rows and fewer leaves. If the two were simply components of a single signaling pathway, then the double mutant should be similar to either single mutant. Thus, unlike *Arabidopsis*, additional factors must be involved in the *TD1/FEA2* pathway. Despite broad similarity in *CLV*-like regulation of meristem size in grasses and *Arabidopsis*, there are significant differences in how such regulation is accomplished in the different species (see also Lunde and Hake, 2006).

3. ABPHYL1

Mutations in *ABPHYL1* in maize create plants with decussate phyllotaxis throughout the plant, including the first long branches of the tassel (Jackson and Hake, 1999). *Abphyl1* mutants have larger meristems than normal plants, which provide sufficient space for formation of two opposite leaves, rather than one, as is normal. The gene is a cytokinin-inducible response regulator related to the family of *Arabidopsis* response regulators (ARR, To *et al.*, 2004). A phylogenetic analysis of *ARR*-like genes in *Arabidopsis*, maize and rice indicates that *ABPHYL1* (labeled as *ZmRR3*) is sister to the rice gene BAC15873 within the type-A response regulator clade (To *et al.*, 2004). The grass *ABPHYL1* clade, in turn, is sister to a clade containing the maize genes *ZmRR1* and *ZmRR2*, indicating that the lineage is likely restricted to monocots and has duplicated at least once since the monocot/eudicot divergence. Distichous vegetative phyllotaxis is a synapomorphy for all graminoid Poales (Stevens, 2001 onwards), and conceivably *ABPHYL1* was involved in imposing this pattern, which then became fixed throughout the clade.

No grasses are known to have opposite leaves or inflorescence branches, although in some species with very dense inflorescences decussate phyllotaxis might be difficult to detect if it were present. No grasses have a phenotype similar to the *abphyl1* mutant inflorescence suggesting that this gene has a conserved function throughout the family. However, it remains possible that subtle changes in gene structure or regulation could have modest effects on meristem size and inflorescence phyllotaxis without creating truly opposite leaves.

4. ABERRANT PANICLE ORGANIZATION

Mutants of *APO1* in rice (Ikeda *et al.*, 2005; see also Section III.C) exhibit unusual phyllotaxis of the inflorescence axis. While this axis should be spiral in rice, in *apo1* mutants the phyllotaxis is distichous or biased distichous (two-ranked, but with a divergence angle other than 180°). Aberrant phyllotaxis correlates with modification of meristem shape, which is taller and narrower in the mutants. Some mutant alleles have a similar arrangement of branches as some pooids, raising the possibility that variation in the structure or regulation of this gene correlates with diversification in inflorescence form in the grasses. It will be of considerable interest to investigate this gene in other species once it is cloned.

Whatever the phyllotaxis of the inflorescence meristem, most higher order meristems in most species produce lateral structures (whether branch or spikelet meristems) distichously. Some exceptions to this occur, however. The phyllotaxis of primary branches in rice is biased distichous, with a divergence angle of 144° (Ikeda *et al.*, 2004). In some species such as *Urochloa* (Reinheimer *et al.*, 2005) or *Paspalum* (Kellogg, E. A. and LeRoux, L. G., unpublished), two ranks of secondary branches are produced on the abaxial side of the primary branch. And in genera such as *Zea* or *Tripsacum*, secondary branches are produced distichously, but the tertiaries are produced on the same side of the primary branch, creating a clear dorsiventral structure (e.g., Orr and Sundberg, 2004; Orr *et al.*, 2001). The genetic basis of these patterns is unknown.

E. BRACTS AND LEAVES SUBTENDING BRANCHES

1. Morphological variation

In the grasses as in most other angiosperms, inflorescence branches are always subtended by bracts, which are more or less prominent early in development as ridges forming just before and just below the inflorescence branches. In some species (e.g., *Stipa*), the bracts persist as small flaps, but more often are undetectable in mature inflorescences.

A number of species, particularly in Bambuseae (Bambusoideae) and in Andropogoneae (Panicoideae), also develop spathes subtending parts of the inflorescence (e.g., *Fargesia* in Bambuseae, *Heteropogon, Coelorachis, Hyparrhenia* in Andropogoneae). It is not clear whether these should be interpreted as cauline leaves with secondary inflorescences (paracladia) in their axils, or as subtending bracts that have not been developmentally suppressed, or if the distinction between the two is even meaningful.

2. LEAFY

The control of bract development is not well understood in any angiosperm. The best-studied gene is the *Arabidopsis* gene *LEAFY* (*LFY*) (Weigel *et al.*, 1992).

In *lfy* mutants of *Arabidopsis*, a floral bract develops beneath each flower. Maize has two *LFY*-like genes, *ZFL1* and *ZFL2*, which are largely redundant; single mutants have no obvious phenotype (Bomblies *et al.*, 2003). The double mutant, however, disrupts inflorescence development. Tassel architecture and sex determination are greatly altered; internodes subtending the terminal inflorescence are shortened and deformed, the pattern of branching is complex, and reduction in *ZFL* copy number correlates with reduction in number of kernel rows in the ear.

The lower branches of the tassel in *zfl1 zfl2* double mutants are replaced by peculiar "tassel ears" that are subtended by leaves. The total number of leaves per plant is higher in the mutants than in wild-type, which suggests that the leaves could correspond to derepressed inflorescence bracts. The tassel ears themselves terminate leafy branches whose leaves may correspond to additional bracts, although the structure is so abnormal that direct comparison to a normal inflorescence is difficult.

LFY-like genes have also been cloned from rice (*RFL*; Kyozuka *et al.*, 1998) and from *Lolium* (*LtLFY*; Pooideae; Gocal *et al.*, 2001), and are apparently single-copy in both species. Comparison with *ZFL* is unfortunately not possible because expression has not been studied at comparable stages in the different species and *rfl* and *Ltlfy* mutants have not been reported, so gene function is unknown. *RFL* is expressed very early in the development of young inflorescence axes, but is excluded from the primordia of primary and secondary branches. The comparable stage in maize would be during formation of the long branches of the tassel, but *ZFL* expression is not reported for that stage, so it is unclear if the *RFL* pattern is novel. Because *ZFL* mutations do affect long-branch morphology, it is conceivable that the mutations in maize correlate with the expression pattern in rice, and the role of the gene in early inflorescence development is conserved. The inflorescence of *Lolium* is unbranched, so is completely uninformative about branch production in general.

The two *ZFL* loci are the product of a gene duplication event that precedes the divergence of *Tripsacum* and maize, but follows the diversification of most Andropogoneae (Bomblies and Doebley, 2005). The genes are under purifying selection throughout the Andropogoneae, and within *Z. mays* itself. Thus sequence analysis suggests that gene function is conserved throughout the tribe.

In the grasses, *LFY* has no apparent function in either short branch (spikelet pair) formation or spikelet formation. Mutations of *ZFL1* and *ZFL2* in maize do not affect formation of short branches, and glumes, lemmas, and paleas are likewise normal. *ZFL* is expressed only weakly, if at all, in glumes, and *RFL* (rice) is not expressed in spikelets. Thus *LFY*-like genes are not involved in specification of developed bracts in spikelets.

The *Arabidopsis* gene *JAGGED* is also involved in development of inflorescence bracts, and may ultimately provide some insight (Dinneny *et al.*, 2004; Ohno *et al.*, 2004). *JAGGED*-like genes have not yet been cloned from grasses, however.

F. PRESENCE OF A TERMINAL SPIKELET

Grass species vary considerably in whether the inflorescence meristem terminates in a spikelet or not. For example, in rice (Ikeda *et al.*, 2004), maize (Bonnett, 1948), foxtail millet (Doust and Kellogg, 2002), and *Urochloa* (Reinheimer *et al.*, 2005), the apical meristem of the inflorescence never produces a spikelet and may remain clearly visible until quite late in development. Conversely, in ryegrass [*Lolium*; (Gocal *et al.*, 2001)], oats (*Avena*; Hiser, K. M., and Kellogg, E. A., unpublished), and proso millet [*Panicum miliaceum*, (Bess *et al.*, 2005)], among many others, the inflorescence meristem apparently is converted to a spikelet, although it is often hard to rule out completely the possibility of a tiny residual inflorescence meristem; in this case the "terminal" spikelet would actually be lateral.

Arabidopsis LEAFY is sufficient to convey floral identity (Weigel and Nilsson, 1995), and might be involved in formation of terminal flowers in grasses; however, expression of *LFY*-like genes in grasses does not correlate with presence of a terminal flower. *LFY*-like genes have been cloned from *Oryza* (Kyozuka *et al.*, 1998), *Lolium* (Gocal *et al.*, 2001), and maize (Bomblies *et al.*, 2003). *LtLFY, ZFL1*, and *ZFL2* are not expressed in the inflorescence meristems of *Lolium* or maize, respectively, even though the former has a terminal spikelet and the latter does not. *RFL* is expressed in the early inflorescence meristem of rice, but expression disappears by the time the secondary branches start to form on the primaries.

Mutants of *TFL1* in *Arabidopsis* convert the inflorescence meristem to a flower (Shannon and Meeks-Wagner, 1991), and when overexpressed delay the transition to flowering (Ratcliffe *et al.*, 1998). As noted in Section III.C, *RCN1* and *RCN2* in rice, which are related to *TFL1*, exhibit similar overexpression phenotypes and are thus good candidates for involvement in this aspect of diversification. Loss-of-function mutations in the rice genes are not known, however.

Grasses also vary in whether lateral branches end in spikelets or not. Most species do have terminal spikelets on secondary and higher order branches, but in genera such as *Paspalidium* (Panicoideae) and *Dactyloctenium* (Chloridoideae), secondary branches end in a stiff point. Candidate genes controlling this aspect of plant architecture are unknown.

G. ELONGATION OF INFLORESCENCE INTERNODES

Much of the obvious variation in grass inflorescences is created by differential elongation of internodes. For example, the spreading inflorescence of *Sorghum halepense* looks quite different from the contracted one of *Sorghum bicolor*. As with patterns of branching, elongation of the main inflorescence axis is independent of that of primary branches, and secondary branches may elongate independently from primaries. Thus, inflorescences that look very similar early in development can look very different later. In addition, elongation may not occur equally in all internodes even along the same axis. Reinheimer *et al.* (2005) have shown for *Panicum maximum* (= *Megathyrsus maximus*) that the most basal internodes on the main inflorescence axis fail to elongate at maturity, producing a pseudowhorl of primary branches. Such pseudowhorls are evident in other species as well (e.g., *Panicum mertensii, Setaria verticillata*), although their development has not been studied.

In most species, elongation occurs very late in development, after all branching has occurred and spikelets are largely formed (Bess *et al.*, 2005; Doust and Kellogg, 2002; Ikeda *et al.*, 2004). It is thus developmentally separable from specification of branch and spikelet identity.

Perhaps because it is a quantitative rather than qualitative character, few genes are reported to modify internode length. *PANICLE PHYTOMER1* in rice is one such gene, although it also increases the numbers of branches (Takahashi *et al.*, 1998). Several QTL have been identified that affect primary branch density in *Setaria* (Doust *et al.*, 2005). Also, treatment of plants with gibberellins and with GA inhibitors affects the length of panicle branches in *Setaria* (Bess, Doust, Preston, and Kellogg, unpublished). We thus expect that considerable information on internode length lies buried in the literature. Of most interest for comparative morphology would be loci that affect length of internodes in inflorescences but not in vegetative parts of the plant.

H. GLUMES AND SPIKELETS

1. Morphological variation

The ultimate unit of the grass inflorescence is the spikelet, literally a little spike. Although not actually synapomorphic for the family, the spikelet characterizes all grasses except the earliest diverging lineage, Anomochlooideae, that includes *Anomochloa* and *Streptochaeta* (Grass Phylogeny Working Group, 2001). The spikelet meristem produces lateral organs in a distichous arrangement, beginning with two bracts known as glumes. Above the glumes are one or more flowers, produced laterally on a short axis, the rachilla.

Floral development is discussed in detail by Whipple and Schmidt, Chapter 10 (this volume) so we will consider here only formation of glumes and the architecture of the spikelet itself.

The spikelet first appears as a lateral or terminal meristem on an inflorescence or inflorescence branch. The first product of the spikelet meristem is a leaf-like structure, the first glume. This is formed on the abaxial side of the spikelet meristem (relative to the axis on which the spikelet is borne) in many panicoid grasses but is lateral in most other species. The second glume forms opposite and slightly above the first, in a distichous arrangement.

Glumes are generally interpreted as bracts (Clifford, 1987), which are extrafloral structures. In a conventional monocot, flowers and inflorescence branches are subtended by bracts. Each inflorescence axis, and in a few taxa each floral axis, then bears an adaxial prophyll. To extend this conventional interpretation to the glumes, then, would require interpreting the first glume as an inflorescence bract and the second as a prophyll. Arguing against this interpretation is the position of the glumes and their morphology. Both glumes are produced by the spikelet meristem and are borne on the spikelet axis, suggesting that neither is an inflorescence bract and the first one (rather than the second) might be considered a prophyll. However, monocot prophylls are usually two-keeled structures and are adaxial, whereas the first glume is rarely adaxial, and even more rarely two keeled. The morphology and position of the first glume thus suggests that it may not be homologous to the prophyll in other monocots.

2. FRUITFULL 1/2/3

The MIKC MADS-box *FRUITFULL*-like (*FUL*-like) genes (*FRUITFULL* [*FUL*], *APETALA1* [*AP1*], and *CAULIFLOWER* [*CAL*]) play important roles in *Arabidopsis* floral development (Ferrandiz *et al.*, 2000) but act only after the genes determining the transition to flowering time such as *TFL1, FT, SUPPRESSOR OF OVEREXPRESSION OF CONSTANS1* (*SOC1*) and *CONSTANS* (*CO*), have initiated the floral pathway (Blazquez, 2000). The three genes have redundant roles in specifying floral meristems (Ferrandiz *et al.*, 2000), but *AtAP1* has a nonredundant role in specifying the sepal and petal whorls of the flower (Pelaz *et al.*, 2001), and *AtFUL* has a nonoverlapping role during fruit development, preventing ectopic activity of *INDEHISCENT* to maintain proper elongation of the silique (Liljegren *et al.*, 2004).

The three *Arabidopsis FRUITFULL*-like genes fall into two clades: the core eudicot *AP1*-like clade (*AtAP1* and *AtCAL*), and the core eudicot *FUL* clade (*AtFUL*) (Litt and Irish, 2003). Litt and Irish (2003) also showed that monocot genes of the *AP1-FUL* group are much more similar to *FUL* than to *AP1*. Their extensive *AP1-FUL* phylogeny demonstrated that orthologs of *Arabidopsis AP1* (*AtAP1*) and *CAULIFLOWER* (*AtCAL*) genes are restricted to Brassicaceae, and *AP1/CAL* genes are restricted to core eudicots. *FUL* genes form a grade, but magnoliid, monocot, noncore eudicot, and two core eudicot clades of *FUL* genes were recognized (Litt and Irish, 2003). Following their results, and contrary to much recent literature on grasses, we refer to the grass genes as *FUL*-like, rather than *AP1*-like.

Sampling within our phylogenetic analysis of 35 *FUL*-like genes concentrates on the grass *FUL* genes (Fig. 7). These fall into three clades: *FUL1* (containing *Triticum monococcum AP1, H. vulgare M5, L. temulentum MADS1, O. sativa MADS14, Z. mays MADS4, ZmMADS15*), *FUL2* (*H. vulgare M8, L. perenne MADS2, O. sativa MADS15, Z. mays AP1* and *ZmMADS3*), and *FUL3* (containing *H. vulgare M3, L. perenne MADS3, O. sativa MADS18* and *Z. mays MADS28*). *FUL1* and *FUL2* are sisters, and the *FUL1/FUL2* clade is sister to *FUL3*. All three clades result from gene duplications apparently near the base of grasses, although the exact placement of the duplication events requires additional sequences from related members of Poales.

All grass *FUL*-like genes appear to be expressed broadly throughout the plant, although most information is available for the *FUL1* and *FUL2* clades. Within spikelets, barley, ryegrass, rice, and maize *FUL1* and *2* genes are expressed in glumes, palea, and lemma of all species, but expression in other organs varies from species to species (J. Preston, unpublished data). *FUL1* and *2* thus may be involved in specifying the glumes, lemmas, and paleas, but orthologous genes in different species may have different roles in the inner three floral whorls (Pelucchi *et al.*, 2002; Schmitz *et al.*, 2000). All three *FUL* genes are expressed in floral meristems, consistent with a plesiomorphic (ancestral) role of the gene family in specifying floral meristem identity throughout angiosperms (Pelucchi *et al.*, 2002; Schmitz *et al.*, 2000). RNAi silencing of *OsMADS18* (a *FUL3* gene) did not produce an obvious phenotype (Fornara *et al.*, 2004), suggesting that *FUL3* may be at least partially redundant with *FUL1* and *FUL2* in rice.

Grass FUL proteins play a role in determining flowering time, unlike the *Arabidopsis* FUL-like proteins. Grass FUL1 proteins in barley, ryegrass, oats, and wheat have all been implicated in vernalization-induced competence to flower (Danyluk *et al.*, 2003; Fu *et al.*, 2005; Jensen *et al.*, 2005; Trevaskis *et al.*, 2003; Yan *et al.*, 2003). *H. vulgare M5* (*HvM5*) and *T. monococcum AP1* (*TmAP1*) show limited expression in winter cultivars that have not been vernalized but are strongly expressed following vernalization treatment (Trevaskis *et al.*, 2003; Yan *et al.*, 2003). Upregulation of *LpMADS2* (*FUL2*) and *LpMADS3* (*FUL3*) during floral transition in perennial ryegrass also points to a general role for the other grass *FUL*

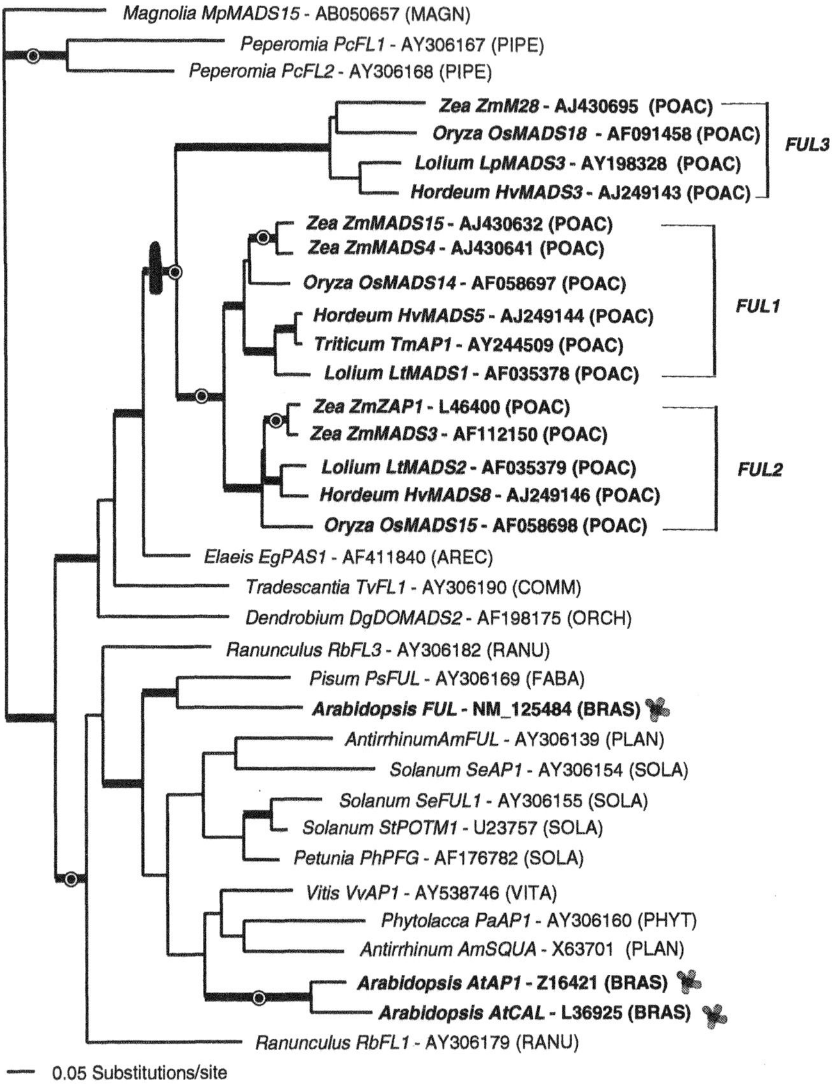

Fig. 7. Maximum likelihood phylogram of the 35 *FRUITFULL/APETALA1*-like gene data set, comprising 657 bp, visited by Bayesian Markov Chain Monte Carlo search using the General Time Reversible model with some invariant sites and gamma distributed rates (GTR + I + G). Tree rooted using *Magnolia MpMADS15* sequence. −ln = 11057.87. Bold branches are supported by posterior probabilities >0.95. Boldface, grass and *Arabidopsis* sequences. ◉, Inferred duplication event; *Arabidopsis* flower, *Arabidopsis* sequence; corn ear, grass clade. Plant families abbreviated to first four letters of name. *FUL1, FRUITFULL1; FUL2, FRUITFULL2; FUL3, FRUITFULL3.*

genes in flowering time (Andersen *et al.*, 2004). All species investigated to date are in subfamily Pooideae, so it will be interesting to know if the same proteins are involved in vernalization of species from other subfamilies.

These data suggest general roles for at least *FUL1* and *FUL2* genes, and conceivably *FUL3*, in specifying glumes, palea, lemma, and floral meristems, and also novel roles in the flowering time pathway in certain species following gene duplications near the origin of the grass family.

AP1-like genes in eudicots appear to be regulated by *LFY*, but this regulatory interaction is either reversed or lost in the grasses (see also Section III.E.1). The rice gene *RFL* is expressed in very young spikelets, but this expression is soon lost (Kyozuka *et al.*, 1998). *LtLFY* from *Lolium* and both *ZFL* genes are expressed in spikelet meristems, but expression in glumes is transitory (Bomblies *et al.*, 2003; Gocal *et al.*, 2001). *LtLFY* (in *Lolium*) is expressed notably later than *FUL*-like genes. In *zfl1 zfl2* double mutants, glumes, lemmas, and paleas—organs with conserved *FUL* expression—are unaffected. Together, these data imply that grass *LFY* orthologs are not involved in specifying spikelet meristem, glume, lemma, or palea identity. In addition, they do not regulate the *FUL*-like genes, because they are expressed after the *FUL*-like genes and are not expressed in the same organs; the lack of mutant phenotype in maize glumes also indicates that *LFY* does not function there.

3. *BRANCHED SILKLESS1/FRIZZY PANICLE1*

In most grasses, no meristem forms in the axil of either glume. Extensively branched bracteate structures do occur, however, in some woody bamboos (Judziewicz *et al.*, 1999). Because it is difficult to determine if highly branched units are at all homologous to conventional spikelets, they are generally called pseudospikelets; it is not known whether the bracts in the pseudospikelet are homologous to glumes. Occasional terata are also reported in which an extra spikelet forms in the axil of the glume (Sharman, 1947).

The BRANCHED SILKLESS1 (BD1) protein in maize, and its ortholog FRIZZY PANICLE1 (FZP) in rice control the outgrowth of the rachilla and production of flowers (Chuck *et al.*, 2002; Komatsu *et al.*, 2003). When the gene is mutated in maize or rice, glume-like structures are produced, but the "spikelet" meristem continues to produce bracts and to branch.

Both genes belong to the ethylene response element-binding factor (ERF) class of transcription factors that are involved in diverse developmental processes including ethylene-mediated responses to pathogens, cold and abiotic factors (Chuck *et al.*, 2002). Our phylogenetic analysis of 36 *BD1/FZP* genes using only the conserved ERF domain did not have enough nucleotide variation to provide a well-supported estimate of relationships (not shown). However, preliminary results from these analyses indicate that the *Arabidopsis LEAFY PETIOLE* gene is likely the closest relative of *BD1/FZP*.

BD1/FZP genes from seven diverse grasses have conserved amino acid sequences in the ERF domain and are 45–75% identical in other regions of the protein (Chuck *et al.*, 2002). Based on this sequence similarity and the similar mutant phenotypes in rice and maize, *BD1/FZP* genes are hypothesized to have similar roles in most BEP and PACCAD grasses. The exception appears to be the maize *BD1* duplicate that is conserved in the ERF domain, but divergent elsewhere, pointing to either a different or loss of gene function (Chuck *et al.*, 2002). Given the similarity between maize *bd1* mutant spikelets and bamboo pseudospikelets, it would be of considerable interest to isolate *BD1/FZP* genes from a bamboo.

I. GLUME VS LEMMA IDENTITY

1. Morphological variation

After it produces two glumes, the spikelet meristem continues to produce lateral structures. These may be morphologically similar to or distinct from the glumes and may bear an axillary floral meristem or not. Conventionally, any lateral bract-like structure above the second glume is considered to be a lemma, which is generally interpreted as a floral bract. The lemma is then described as sterile (if it has no axillary meristem) or fertile (if it does have an axillary meristem). For many grasses (e.g., *Ehrharta, Chasmanthium*), the sterile lemma is morphologically similar to the glumes, implying that there is a developmental gradient from the glumes through the sterile lemmas to the fertile ones, whereby floret identity is acquired gradually during development, rather than abruptly.

Evidence of such a gradient comes from overexpression studies of the *SEPALLATA* gene *LEAFY HULL STERILE1* in rice (Prasad *et al.*, 2001). In rice, the glumes are reduced to tiny flaps, called rudimentary glumes in the literature (Bommert *et al.*, 2005b). Above them are two larger subulate structures in the position of sterile lemmas, and morphologically distinct from either the true glumes or the fertile lemma; these are generally called "empty glumes." When *LHS1* is overexpressed, the sterile lemmas enlarge and become morphologically similar to the fertile lemma, whereas the true glumes are unaffected (Prasad *et al.*, 2001). These results, along with supporting data from *OsFZP* (Komatsu *et al.*, 2003b), support the hypothesis first proposed by Stapf (1917) that rice has a three-flowered spikelet with one fertile upper floret and two lower florets that are reduced to sterile lemmas.

2. SEPALLATA *genes*

SEPALLATA (*SEP*) genes play fundamental roles in the development of all floral whorls in *Arabidopsis* where they are hypothesized to act as cofactors with A-, B-, C-, and D-class MADS-box floral homeotic genes (Ditta *et al.*, 2004; Pelaz *et al.*, 2000). *Arabidopsis* has four *SEP* genes, *AtSEP1, AtSEP2, AtSEP3*, and *AtSEP4*. Expression of *AtSEP1, AtSEP2*, and *AtSEP3* is restricted to the flower, whereas *AtSEP4* is expressed in all above-ground organs (Huang *et al.*, 1995; Ma *et al.*, 1991; Savidge *et al.*, 1995). The four *Arabidopsis SEP* genes are developmentally redundant, with discernible phenotypes only when several genes have been removed. For example, *Arabidopsis sep1 sep2 sep3* (*sepallata*) triple mutant flowers are composed entirely of sepal-like structures, and *sep1 sep2 sep3 sep4* quadruple mutants have flowers composed entirely of leaf-like structures (Ditta *et al.*, 2004; Pelaz *et al.*, 2000).

The phylogenetic analysis of *SEPALLATA* genes by Zahn *et al.* (2005) includes 113 sequences that span flowering plant diversity. An early gene duplication event occurred before the origin of extant angiosperms and produced the *SEP3* and *LOFSEP* clades, containing *Arabidopsis AtSEP3*, and rice *LEAFY HULL STERILE1* (*LHS1*), *O. sativa MADS5* (*OsMADS5*) and *OsMADS34* genes, petunia *FLORAL BINDING PROTEIN9* (*PhFBP9*) and *PhFBP23* genes, and *Arabidopsis AtSEP1, AtSEP2*, and *AtSEP4* genes, respectively (Malcomber and Kellogg, 2005). Within the *LOFSEP* clade, additional duplications near the base of core eudicots produced the *SEP1/2, FBP9/23*, and *SEP4* clades, and conceivably the Brassicaceae duplication (Bowers *et al.*, 2003) produced the *AtSEP1* and *AtSEP2* genes. Additional duplications occurred near the origin of Solanaceae within the *SEP1/2* and *FBP9/23* clades. Two duplications near the base of grasses within the *LOFSEP* clade produced the *LHS1, OsMADS5*, and *OsMADS34* subclades. Polyploidy has also produced additional duplicates in monocots and eudicots throughout the phylogeny. These duplications imply that genes similar to *AtSEP3* are found throughout flowering plants, whereas *AtSEP1* and *AtSEP2* orthologs are restricted to Brassicaceae and *AtSEP4* orthologs to core eudicots (Malcomber and Kellogg, 2005).

Our phylogenetic analysis of 119 *SEP* genes comprises the Zahn *et al.* (2005) 113 *SEP* gene data set plus six additional sequences from early diverging grasses (*Streptochaeta* and *Pharus*) and additional monocots to localize putative grass-specific duplication events within the *LOFSEP* and *SEP3* clades (Fig. 8). This analysis is congruent with Zahn *et al.*'s (2005) analysis, with the exception of the placement of *Helianthus HaM137, Dendrathema DgCDM77*, and *Gerbera GhGRCD1* (ASTERACEAE SEP3, Malcomber and Kellogg, 2005). In our analysis, this clade appears to have

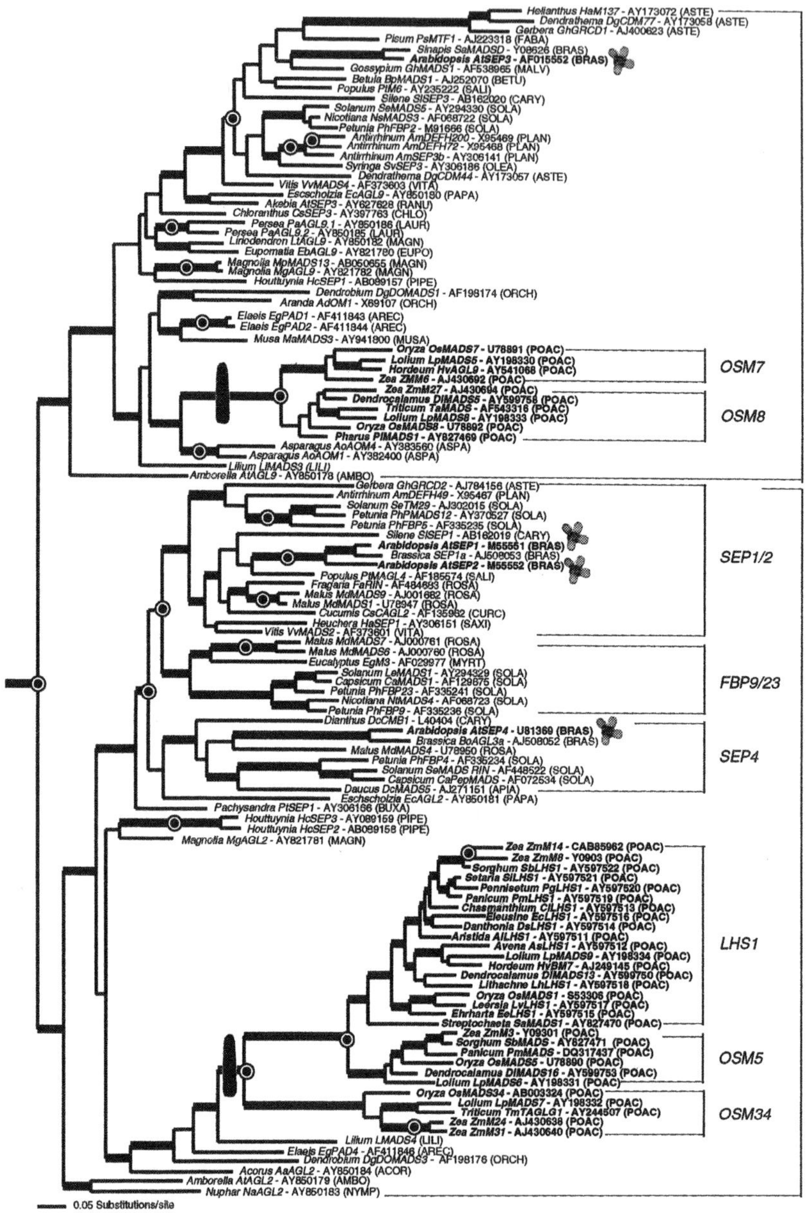

Fig. 8. Maximum likelihood phylogram of the 119 *SEPALLATA* gene data set, comprising 1200 bp, visited by Bayesian Markov Chain Monte Carlo search using the General Time Reversible model with some invariant sites and gamma distributed rates (GTR + I + G). Tree rooted using four *AP1/FUL*-like gene sequences (not shown). Analysis also included 26 *AGL6*-like gene sequences (not shown). −ln = 56818.93. Bold branches are supported by posterior probabilities >0.95. Boldface, grass and *Arabidopsis* sequences. ◉, Inferred duplication event; *Arabidopsis* flower, *Arabidopsis* sequence; corn ear, grass clade. Plant families abbreviated to first four letters of

been produced from a duplication near the base of eudicots, whereas in Zahn *et al.* (2005) the clade diverges within basal angiosperms.

Streptochaeta SaMADS1 falls within the *LHS1* clade of grass sequences, and shares C-terminal motifs with all other *LHS1* genes (Malcomber and Kellogg, 2005; Vandenbussche *et al.*, 2003). Placement of *SaMADS1* within the *LHS1* clade suggests the *LHS1/OsMADS5* duplication occurred prior to the origin of extant grasses, although the exact placement of this duplication requires additional sampling from related members of the Poales. The *Pharus PlMADS1* gene is sister to other *OsMADS8* genes, although its placement is not well supported. This position, or a placement sister to other *OsMADS7* and *OsMADS8*, is consistent with the grass *SEP3* duplication occurring near the origin of grasses. As in the *LHS1/OsMADS5* duplication, the exact placement of this duplication will require additional sequences from related members of the Poales.

SEP mRNA expression patterns are heterogeneous within flowering plants, and *SEP* gene function varies from redundant, as in *Arabidopsis*, to nonredundant with roles in fruit maturation, floral whorl identity and plant architecture (Malcomber and Kellogg, 2005). Expression patterns within the grass *LHS1* clade are particularly diverse (Malcomber and Kellogg, 2004). *LHS1* orthologs are always expressed in the palea and lemma, but expression in other organs of the floret and regions of the spikelet varies among species (Malcomber and Kellogg, 2004).

The semidominant negative rice *lhs1* mutant has leafy palea, lemma, and lodicules, fewer stamens, and occasionally an extra pistil or floret (Jeon *et al.*, 2000). Ectopic expression of this gene in rice produces plants with short panicles and irregularly positioned branches; the sterile lemmas of the two lower flowers are similar to the palea and lemma of the fertile upper flower (Prasad *et al.*, 2001). As discussed in an earlier section, these functional analyses indicate that *OsLHS1* influences palea and lemma morphology, but not glumes.

J. FLORET NUMBER

1. Morphological variation

Grasses vary considerably in the number of florets produced by the spikelet meristem. Multiflowered spikelets originated relatively late in grass evolution, just before the divergence of the Puelioideae, and reversals to single

---

name. *FBP9/23, FLORAL BINDING PROTEIN9/23; LOFSEP: LEAFY HULL STERILE1; OsMADS5, OsMADS34; FLORAL BINDING PROTEIN9/23 SEPALLATA1/2, SEPALLATA4* clade; *LHS1, LEAFY HULL STERILE1; OSM5, OsMADS5; OSM7, OsMADS7; OSM8, OsMADS8; OSM34, OsMADS34; SEP1/2, SEPALLATA1/2; SEP3, SEPALLATA3; SEP4, SEPALLATA4.*

flowered spikelets have occurred frequently (Grass Phylogeny Working Group, 2001). In some species, including all the panicoid grasses, the number of florets is fixed both within and between species, whereas in other species or clades the number varies within a range.

Spikelets may mature from the top-down (basipetal maturation) or the bottom-up (acropetal maturation). Spikelets with a fixed number of florets exhibit basipetal maturation, whereas in those with a variable number maturation is acropetal. The difference in maturation patterns also correlates with differences in expression domain of *LHS1*; in basipetal species, *LHS1* is expressed only in the uppermost floret, whereas it is expressed in multiple florets in acropetal species (Malcomber and Kellogg, 2004). The pattern in basipetal grass species supports the hypothesis that *LHS1* orthologs specify the terminal flower of the spikelet (the "selector gene" hypothesis; Cacharrón *et al.*, 1999), whereas the gene may have different or additional developmental roles in species with acropetal maturation of florets (Malcomber and Kellogg, 2004).

2. INDETERMINATE SPIKELET1

Mutations in the maize gene *INDETERMINATE SPIKELET1* (*IDS1*) lead to production of more than two flowers in the spikelet. It thus seems possible that modulation of *ids1* might result in some of the variation in floret number observed in the grasses.

IDS1 is an *APETALA2* (*AP2*)-like gene cloned and characterized in maize by Chuck *et al.* (1998). *ZmIDS1* is expressed broadly in both vegetative and floral tissues, but only appears to function in inflorescence development where it regulates the number of florets a spikelet produces (Chuck *et al.*, 1998).

Our phylogenetic analysis of 21 *IDS1*-like AP2 genes estimates a well-supported grass clade containing the *ZmIDS1* clade and two rice genes (Fig. 9). The position of these two rice genes suggests that *IDS1* and an *IDS1*-like rice sequence result from a duplication near the base of grasses, although additional *IDS1*-like sequences would help test this hypothesis. The wheat *IDS1* gene (*T. monococcum IDS1*) is the "*Q* gene" that confers the square-headed phenotype and free-threshing character of domesticated bread and durum wheat, and also affects the presence of keels on glumes, rachis toughness, spike length, spike type, and culm height (Faris *et al.*, 2003; Simons *et al.*, 2006). Molecular analyses reveal that the domesticated (*Q*) and wild type (*q*) gene products differ by a single amino acid at position 329; the *Q* gene has an isoleucine whereas *q* has a valine (Simons *et al.*, 2006). This sequence change increases the stability of *Q* gene homodimers relative

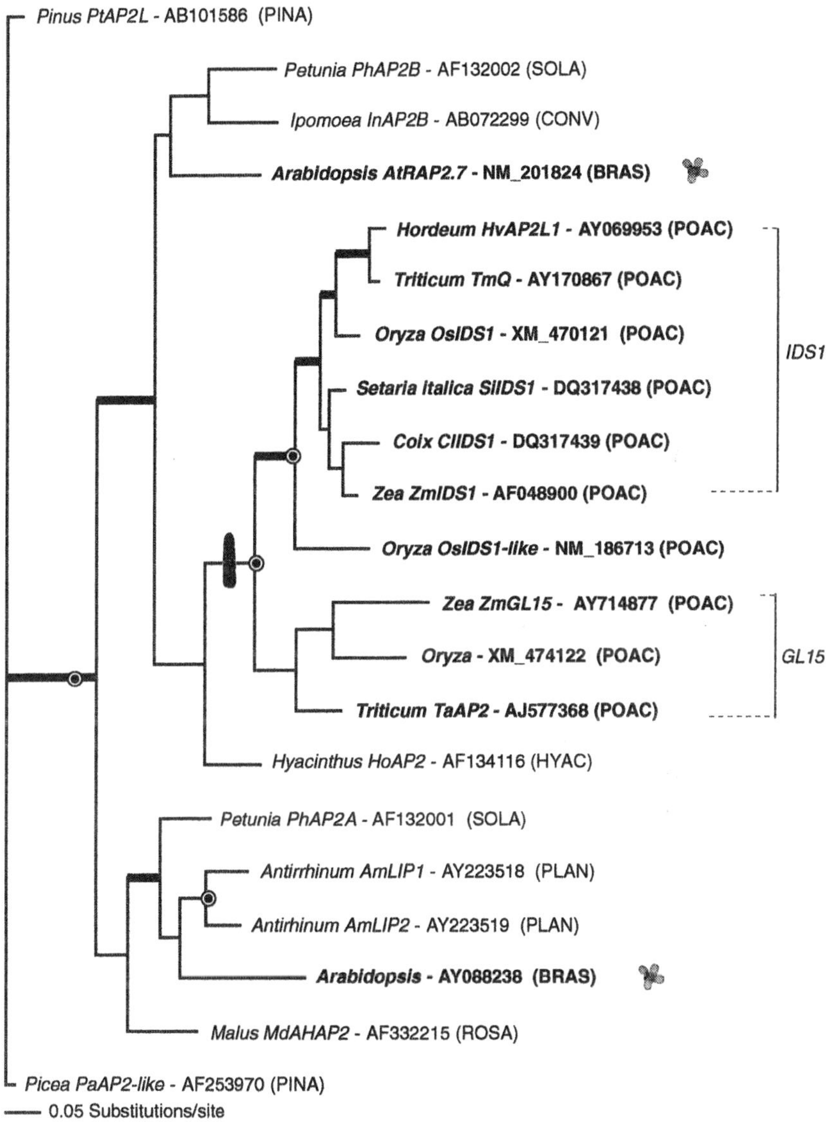

Fig. 9. Maximum likelihood phylogram of the 21 *INDETERMINATE SPIKELET1*-like gene data set, comprising 486 bp, visited by Bayesian Markov Chain Monte Carlo search using the General Time Reversible model with some invariant sites and gamma distributed rates (GTR + I + G). Tree rooted using *Pinus PtAP2L* sequence. −ln = 4601.91. Bold branches are supported by posterior probabilities >0.95. Boldface, grass and *Arabidopsis* sequences. ◉, Inferred duplication event; *Arabidopsis* flower, *Arabidopsis* sequence; corn ear, grass clade. Plant families abbreviated to first four letters of name. *GL15, GLOSSY15; IDS1, INDETERMINATE SPIKELET1.*

to the wild type gene, suggesting that the domesticated inflorescence phenotype is dosage related and caused by increased amounts of the *Q* gene transcript (Simons *et al.*, 2006). Sister to the *IDS1* + *IDS1*-like clade is a clade of rice, wheat, and maize sequences containing the maize *GLOSSY15* (*ZmGL15*) gene, which regulates leaf epidermal development (Lauter *et al.*, 2005; Moose and Sisco, 1996).

The closest eudicot relatives of *ZmIDS1* are *Arabidopsis AtRAP2.7, Petunia PhAP2B*, and *Ipomoea ImAP2B*. Information in addition to the nucleotide sequence is only available for *PhAP2B*, which is expressed strongly in the outer cells of young inflorescence bracts, the epidermis of sepals and the ovary, and seed endosperm (Maes *et al.*, 2001).

Available data for *IDS1* and related sequences suggest the *IDS1* clade was produced from one of several duplications within monocots. The *ZmGL15* clade seems to have maintained the expression and possible function of related eudicot genes, whereas all available information on the *IDS1* clade suggests a novel role restricted to spikelet development resulting from a duplication near the origin of grasses.

Transcription factors containing *AP2* domains play roles in regulating root, leaf, flower, seed, and ovule development, often via miRNA posttranscriptional regulation (Lauter *et al.*, 2005; Riechmann and Meyerowitz, 1998). Regulation by miRNAs has been demonstrated for *GLOSSY15* (Lauter *et al.*, 2005). We expect that this mode of regulation should apply to *IDS1* as well.

IV. CONCLUSIONS

A goal of evolutionary developmental genetics is to explain the diversity of life in terms of modification of underlying genes. This requires linking phenotype to genotype not just in a couple of model organisms, but in entire groups of species. In this chapter, we have described some of the phenotypic diversity among grass inflorescences and shown that multiple genes have been cloned that are good candidates for regulating that diversity. In drawing connections between genes and phenotypes, it has been essential to describe the inflorescence in terms of meristems and primordia and to focus on events in inflorescence development. The classic typology for describing inflorescences—spike, raceme, panicle—is not precise enough for this purpose.

The majority of the phenotypes described here do not occur in *Arabidopsis*. In addition, many of the genes we have analyzed have duplicated extensively in grasses or monocots. We suggest that the two observations are related—grass-specific genes have acquired new functions to create novel grass-specific morphology. Multiple model systems are going to be necessary to understand the genetic basis of evolutionary diversification, and this study is just one of many illustrations of that point (Gewin, 2005). Valuable insights can come from comparison to eudicots, and in a few cases close comparisons can be made (e.g., *FEA2* with *CLV2*) but more often the grasses appear to be a unique system, which can only be studied by direct analysis of genes in the grass models, rice and maize.

Despite tremendous progress in cloning relevant genes, only a few have so far been studied broadly enough to document diversification within the grasses themselves. *LHS1, FUL1/2/3*, and *RA1* all exhibit variation suggesting that they may be involved in diversification within the grasses. Most of the other genes described above are good candidates, however. Once orthologs have been cloned from a variety of species, the next step is clearly to investigate differences in expression pattern, which is the first step toward determining function.

Gene duplication is one source of novel genes. Extensive large-scale, possibly whole-genome, duplications have occurred frequently throughout the tree of life. Flowering plants are no exception, with large-scale duplications estimated near the base of Brassicaceae, grasses, core eudicots, and papilionoid legumes (Bowers *et al.*, 2003; Paterson *et al.*, 2004). The genes described here vary widely in how much evidence they retain of these duplication events. At one extreme, phylogenetic analyses including *ZmFEA2* (Fig. 6) and *ZmTD1* (Bommert *et al.*, 2005a) do not identify any duplication events in grasses, Brassicaceae, or elsewhere; thus any duplicate genes have suffered the fate of most gene copies and have been removed from the genome, likely via selection (Lynch and Force, 2000). Phylogenetic analyses of other developmental genes identify duplication events near the base of grasses and/or Brassicaceae, but the number of estimated events varies, as does the timing of the duplication. In the *KN1/BP* clade of *KNOX* genes, three duplications occurred before divergence of the grasses, but no duplicates are retained in Brassicaceae (Fig. 3). In the *LG3/KNAT2* clade, both the grass duplication and the Brassicaceae duplication are evident. Among *TFL*-like genes, the *RCN1/RCN2* duplication appears to have preceded the grass duplication (based on the position of *Joinvillea*), but the Brassicaceae duplication left only one descendant gene in *Arabidopsis* (Fig. 5). The *AP1/CAL/FUL* phylogeny reveals descendants of three monocot duplications, at least one of which may correspond to the grass duplication, as well as one at the base of eudicots and one in Brassicaceae (Fig. 7). One duplication event is also estimated to have occurred within the *SEP3* lineage, although only a single *Arabidopsis SEP3* gene has been identified (Fig. 8). As in the grass *FUL*-like clade, two duplication events are estimated near the base of grasses in the *SEPALLATA LOFSEP* clade, and in the *IDS1*-like gene family. The persistence of gene products following these gene duplication events has resulted in most of these grass genes being "novel."

Interpreting how and whether inflorescence genes have diversified functionally is limited by available data. However, the sparse data suggest a complex pattern. *ZmFEA2* and *AtCLV2*, along with *ZmTD1, OsFON1*, and the related *AtCLV1* have overlapping roles in defining meristem size, but their expression patterns and the pathways by which the genes regulate meristem size differ between grasses and *Arabidopsis*. In *Arabidopsis, clv1* and *clv2* mutants and *clv1 clv2*-double mutants have identical phenotypes, suggesting a simple pathway, whereas *Zmtd1 Zmfea2* double mutants have more pronounced phenotypes than either single mutant, indicating a more elaborate pathway in maize. Thus protein function has apparently diverged despite a lack of retained duplicates.

In contrast, the *TFL1* clade appears to be a small gene family in which successive gene duplications have produced little or no diversification in gene function. Despite different expression patterns and multiple duplications within flowering plants, the *Arabidopsis AtTFL1* and *AtCEN*, and grass *OsRCN1* and *OsRCN2* genes all regulate inflorescence branching and flowering time. Loss-of-function mutants in grasses, however, may uncover more and different gene functions.

Among class I *KNOX* genes, functional convergence has followed gene duplication. *ZmKN1* and *AtSTM* genes both play roles in meristem maintenance and *OsOSH15* and *AtBP* both regulate internodes (Long *et al.*, 1996; Sato *et al.*, 1998; Venglat *et al.*, 2002; Vollbrecht *et al.*, 2000). The phylogeny shows, however, that *ZmKN1* is actually more closely related to *AtBP*, whereas *AtSTM* is more closely related to *OsOSH15*. Thus, at least one pair of genes must have arrived at their similarity independently.

Gene duplications have also produced novel roles. Examples of this can be seen in the *SEPALLATA, FRUITFULL, IDS1*, and presumably *RA1* clades. The SEPALLATA protein LHS1 determines palea and lemma size and texture, whereas all *Arabidopsis* SEP proteins have redundant developmental roles (Ditta *et al.*, 2004; Pelaz *et al.*, 2000). In the case of FRUITFULL proteins, some aspects of function are conserved among family members (e.g., specifying meristem identity) whereas others appear to be novel roles that have evolved within grasses (vernalization-induced flowering). *Arabidopsis AP1/FUL* genes are restricted to floral tissue whereas grass *FUL*-like genes are expressed broadly throughout the plant. In the case of IDS1-like proteins, duplication has resulted in a diversification of functional roles, with *GLOSSY15* maintaining the apparent plesiomorphic condition of specifying epidermal morphology, whereas the role of IDS1 proteins is restricted to the spikelet. The function of the *IDS1*-like gene from rice that is sister to *IDS1* in our analysis might expand the functional domain of this small gene family even further. *RA1* also appears to be a novel gene, and it will be of interest to reconstruct its evolutionary history.

Morphological novelty might arise in many ways. Here we have focused on gene duplication as one important phenomenon that provides the raw material on which selection can act. The phenotypic result of gene duplication depends, however, on changes within the genes themselves or the sequences that regulate them. Changes in gene expression are likely to underlie much evolutionary change, and these changes may arise easily through modification of *cis*-regulatory sequences (Doebley and Lukens, 1998; Moore and Purugganan, 2005). Partial loss of function can change developmental pathways by leading to accumulation of intermediate gene products, and then selecting for enhanced processing of the intermediates (Grotewold, 2005).

Gene level changes can then lead to novel morphological outcomes by changing the timing (heterochrony) or location (heterotopy) of genetic programs (Kellogg, 2000). Vollbrecht *et al.* (2005) have suggested that the proteins that control branching in the grasses, such as RA1, BA1, and others, may modulate the timing of development. Thus by extending or narrowing the developmental window for primary branch production, more or fewer are produced, leading to a marked change in morphology. They suggest that this may occur by genetic regulation of a mobile signal, or possibly the gene products themselves are the signal.

In conclusion, the rich diversity of the grasses, coupled with the availability of several model systems, makes the family an ideal group for the study of morphological evolution. Much of the structure of the grass inflorescence has no homolog in *Arabidopsis* or Brassicaceae, and the terminal inflorescence unit, the spikelet, is unique in the angiosperms. The grass family thus has provided and will continue to provide new information on the origins of novel structures and mechanisms of morphological diversification.

ACKNOWLEDGMENTS

We thank Doug Soltis, Pam Soltis, and Jim Leebens-Mack for the opportunity to contribute to this volume and Volker Brendel, George Chuck, Sarah Hake, Dave Jackson, Rob Martienssen, Torbert Rocheford, Bob Schmidt, and Erik Vollbrecht for discussions, information, and for commenting on the ideas presented here. We also thank two anonymous reviewers for helpful suggestions, Lilla Bartkó for providing some of the grass *KNOTTED1* sequences, and Missouri Botanical Garden and the USDA for access to plant materials. This work was supported by National Science Foundation Grant DBI-0110189 (to E. A. K.).

REFERENCES

Alfaro, M. E., Zoller, S. and Lutzoni, F. (2003). Bayes or bootstrap? A simulation study comparing the performance of Bayesian Markov chain Monte Carlo sampling and bootstrapping in assessing phylogenetic confidence. *Molecular Biology and Evolution* **20**, 255–266.

Altschul, S. F., Madden, T. L., Schaffer, A. A., Zhang, J., Zhang, Z., Miller, W. and Lipman, D. J. (1997). Gapped BLAST and PSI-BLAST: A new generation of protein database search programs. *Nucleic Acids Research* **25**, 3389–3402.

Andersen, C. H., Jensen, C. S. and Petersen, K. (2004). Similar genetic switch systems might integrate the floral inductive pathways in dicots and monocots. *Trends in Plant Science* **9**, 105–107.

Arabidopsis Genome Initiative. (2000). Analysis of the genome sequence of the flowering plant *Arabidopsis thaliana*. *Nature* **408**, 796–815.

Ashikari, M., Sakakibara, H., Lin, S., Yamamoto, T., Takashi, T., Nishimura, A., Angeles, E. R., Qian, Q., Kitano, H. and Matsuoka, M. (2005). Cytokinin oxidase regulates rice grain production. *Science* **309**, 741–745.

Barkworth, M. E., Capels, M., Long, S. and Piep, M. B. (eds.) (2003). "Flora of North America: North of Mexico Volume 25, Magnoliophyta, Commelinidae (in part): Poaceae, part 2." Flora of North America. Oxford University Press, Oxford.

Baum, D. A. (2002). Identifying the genetic causes of phenotypic evolution: A review of experimental strategies. *In* "Developmental Genetics and Plant Evolution." (Q. E. B. Cronk, R. M. Bateman and J. Hawkins, eds.), pp. 493–507. Taylor and Francis, London.

Benjamins, R., Quint, A., Weijers, D., Hooykaas, P. and Offringa, R. (2001). The *PINOID* protein kinase regulates organ development in *Arabidopsis* by enhancing polar auxin transport. *Development* **128**, 4057–4067.

Bess, E. C., Doust, A. N. and Kellogg, E. A. (2005). A naked grass in the bristle clade: A phylogenetic and developmental study of *Panicum* section *Bulbosa* (Paniceae: Poaceae). *International Journal of Plant Sciences* **166**, 371–381.

Blazquez, M. (2000). Flower development pathways. *Journal of Cell Science* **113**, 3547–3548.

Bomblies, K. and Doebley, J. F. (2005). Molecular evolution of *FLORICAULA/LEAFY* orthologs in the Andropogoneae (Poaceae). *Molecular Biology and Evolution* **22**, 1082–1094.

Bomblies, K., Wang, R. L., Ambrose, B. A., Schmidt, R. J., Meeley, R. B. and Doebley, J. (2003). Duplicate *FLORICAULA/LEAFY* homologs *zfl1* and *zfl2* control inflorescence architecture and flower patterning in maize. *Development* **130**, 2385–2395.

Bommert, P., Lunde, C., Nardmann, J., Vollbrecht, E., Running, M., Jackson, D., Hake, S. and Werr, W. (2005a). *thick tassel dwarf1* encodes a putative maize ortholog of the *Arabidopsis CLAVATA1* leucine-rich repeat receptor-like kinase. *Development* **132**, 1235–1245.

Bommert, P., Satoh-Nagasawa, N., Jackson, D. and Hirano, H. Y. (2005b). Genetics and evolution of inflorescence and flower development in grasses. *Plant Cell and Physiology* **46**, 69–78.

Bonnett, O. T. (1948). Ear and tassel development in maize. *Annals of the Missouri Botanical Garden* **35**, 269–287.

Bossinger, G. (1990). *Klassifizierung von entwicklungsmutanten der gerste anhand einer interpretation des pflanzanaufbaus de Poaceae aus phytomeren*. Dissertation, Rheinischen Friedrich-Wilhelms-Universität zu Bonn, Bonn, Germany.

Bowers, J. E., Chapman, B. A., Rong, J. and Paterson, A. H. (2003). Unravelling angiosperm genome evolution by phylogenetic analysis of chromosomal duplication events. *Nature* **422**, 433–438.

Buck, M. and Atchley, W. (2003). Phylogenetic analysis of plant basic helix-loop-helix proteins. *Journal of Molecular Evolution* **56**, 742–750.

Cacharrón, J., Saedler, H. and Theissen, G. (1999). Expression of MADS box genes *ZMM8* and *ZMM14* during inflorescence development of *Zea mays* discriminates between the upper and the lower floret of each spikelet. *Developmental Genes and Evolution* **209**, 411–420.

Campbell, C. S. and Kellogg, E. A. (1987). Sister group relationships of the Poaceae. *In* "Grass Systematics and Evolution" (T. R. Soderstrom, K. W. Hilu, C. S. Campbell and M. E. Barkworth, eds.), pp. 217–224. Smithsonian Press, Washington, DC.

Chantret, N., Salse, J., Sabot, F., Rahman, S., Bellec, A., Laubin, B., Dubois, I., Dossat, C., Sourdille, P., Joudrier, P., Gautier, M. F., Cattolico, L., *et al.* (2005). Molecular basis of evolutionary events that shaped the hardness locus in diploid and polyploid wheat species (*Triticum* and *Aegilops*). *Plant Cell* **17**, 1033–1045.

Christensen, S. K., Dagenais, N., Chory, J. and Weigel, D. (2000). Regulation of auxin response by the protein kinase PINOID. *Cell* **100**, 469–478.

Chuck, G., Meeley, R. B. and Hake, S. (1998). The control of maize spikelet meristem fate by the *APETALA2-like* gene *indeterminate spikelet1*. *Genes and Development* **12**, 1145–1154.

Chuck, G., Muszynski, M., Kellogg, E., Hake, S. and Schmidt, R. J. (2002). The control of spikelet meristem identity by the *branched silkless1* gene in maize. *Science* **298**, 1238–1241.

Clark, L. G. and Judziewicz, E. J. (1996). The grass subfamilies Anomochlooideae and Pharoideae (Poaceae). *Taxon* **45**, 641–645.

Clark, L. G. and Pohl, R. W. (1996). "Agnes Chase's First Book of Grasses." Smithsonian Institution Press, Washington, DC.

Clark, S. E., Williams, R. W. and Meyerowitz, E. M. (1997). The *CLAVATA1* gene encodes a putative receptor kinase that controls shoot and floral meristem size in *Arabidopsis*. *Cell* **89**, 575–585.

Clayton, D. K. and Renvoize, S. A. (1986). "Genera Graminum: Grasses of the World." Her Majesty's Stationery Office, London.

Clifford, H. T. (1987). Spikelet and floral morphology. *In* "Grass Systematics and Evolution" (T. R. Soderstrom, K. W. Hilu, C. S. Campbell and M. E. Barkworth, eds.), pp. 21–30. Smithsonian Institute, Washington, DC.

Danyluk, J., Kane, N. A., Breton, G., Limin, A. E., Fowler, D. B. and Sarhan, F. (2003). *TaVRT-1*, a putative transcription factor associated with vegetative to reproductive transition in cereals. *Plant Physiology* **132**, 1849–1860.

Dellaporta, S. L. (1994). Plant DNA miniprep and microprep: Versions 2.1–2.3. *In* "The Maize Handbook" (M. Freeling and V. Walbot, eds.), pp. 522–525. Springer-Verlag, New York.

Dinneny, J. R., Yadegari, R., Fischer, R. L., Yanofsky, M. F. and Weigel, D. (2004). The role of *JAGGED* in shaping lateral organs. *Development* **131**, 1101–1110.

Ditta, G., Pinyopich, A., Robles, P., Pelaz, S. and Yanofsky, M. F. (2004). The *SEP4* gene of *Arabidopsis thaliana* functions in floral organ and meristem identity. *Current Biology* **14**, 1935–1940.

Doebley, J. and Lukens, L. (1998). Transcriptional regulators and the evolution of plant form. *Plant Cell* **10**, 1075–1082.

Doust, A. N. and Kellogg, E. A. (2002). Inflorescence diversification in the panicoid bristle grass clade (Paniceae, Poaceae): Evidence from molecular phylogenies and developmental morphology. *American Journal of Botany* **89**, 1203–1222.

Doust, A. N., Devos, K. M., Gadberry, M. D., Gale, M. D. and Kellogg, E. A. (2005). The genetic basis for inflorescence variation between foxtail and green millet (Poaceae). *Genetics* **169**, 1659–1672.

Evans, M. W. (1940). Developmental morphology of the growing point of the shoot and the inflorescence in grasses. *Journal of Agricultural Research* **61**, 481–520.

Ewing, B., Hillier, L., Wendl, M. C. and Green, P. (1998). Base-calling of automated sequencer traces using phred. I. Accuracy assessment. *Genome Research* **8**, 175–185.

Faris, J. D., Fellers, J. P., Brooks, S. A. and Gill, B. S. (2003). A bacterial artificial chromosome contig spanning the major domestication locus Q in wheat and identification of a candidate gene. *Genetics* **164**, 311–321.

Ferrandiz, C., Gu, Q., Martienssen, R. and Yanofsky, M. F. (2000). Redundant regulation of meristem identity and plant architecture by *FRUITFULL, APETALA1* and *CAULIFLOWER*. *Development* **127**, 725–734.

Fitch, W. M. (1970). Distinguishing homologous from analogous proteins. *Systematic Zoology* **19**, 99–113.

Force, A., Lynch, M., Pickett, F. B., Amores, A., Yan, Y. L. and Postlethwait, J. (1999). Preservation of duplicate genes by complementary, degenerative mutations. *Genetics* **151**, 1531–1545.

Force, A., Cresko, W. A., Pickett, F. B., Proulx, S. R., Amemiya, C. and Lynch, M. (2005). The origin of subfunctions and modular gene regulation. *Genetics* **170**, 433–446.

Fornara, F., Parenicova, L., Falasca, G., Pelucchi, N., Masiero, S., Ciannamea, S., Lopez-Dee, Z., Altamura, M. M., Colombo, L. and Kater, M. M. (2004). Functional characterization of *OsMADS18*, a member of the *AP1/SQUA* subfamily of MADS box genes. *Plant Physiology* **135**, 2207–2219.

Foster, T., Yamaguchi, J., Wong, B. C., Veit, B. and Hake, S. (1999). *Gnarley1* is a dominant mutation in the *knox4* homeobox gene affecting cell shape and identity. *Plant Cell* **11**, 1239–1252.

Fu, D., Szucs, P., Yan, L., Helguera, M., Skinner, J. S., von Zitzewitz, J., Hayes, P. M. and Dubcovsky, J. (2005). Large deletions within the first intron in *VRN-1* are associated with spring growth habit in barley and wheat. *Molecular Genetics and Genomics* **273**, 54–65.

Gadberry, M. D., Malcomber, S. T., Doust, A. N. and Kellogg, E. A. (2005). Primaclade—a flexible tool to find conserved PCR primers across multiple species. *Bioinformatics* **21**, 1263–1264.

Gallavotti, A., Zhao, Q., Kyozuka, J., Meeley, R. B., Ritter, M. K., Doebley, J. F., Pe, M. E. and Schmidt, R. J. (2004). The role of *barren stalk1* in the architecture of maize. *Nature* **432**, 630–635.

Gewin, V. (2005). Functional genomics thickens the biological plot. *PLoS Biology* **3**, e219.

Gocal, G. F., King, R. W., Blundell, C. A., Schwartz, O. M., Andersen, C. H. and Weigel, D. (2001). Evolution of floral meristem identity genes. Analysis of *Lolium temulentum* genes related to *APETALA1* and *LEAFY* of *Arabidopsis*. *Plant Physiology* **125**, 1788–1801.

Goff, S. A., Ricke, D., Lan, T. H., Presting, G., Wang, R., Dunn, M., Glazebrook, J., Sessions, A., Oeller, P., Varma, H., Hadley, D. Hutchison, D., *et al.* (2002). A draft sequence of the rice genome (*Oryza sativa* L. ssp. *japonica*). *Science* **296**, 92–100.

Grass Phylogeny Working Group. (2001). Phylogeny and subfamilial classification of the grasses (Poaceae). *Annals of the Missouri Botanical Garden* **88**, 373–457.

Grotewold, E. (2005). Plant metabolic diversity: A regulatory perspective. *Trends in Plant Science* **10**, 57–62.

Hake, S., Smith, H. M., Holtan, H., Magnani, E., Mele, G. and Ramirez, J. (2004). The role of *knox* genes in plant development. *Annual Review of Cell and Developmental Biology* **20**, 125–151.

Haldane, J. B. S. (1932). "The Causes of Evolution." Longmans, Green and Co., London.

Haldane, J. B. S. (1933). The part played by recurrent mutation in evolution. *American Naturalist* **67**, 5–19.

Hanzawa, Y., Money, T. and Bradley, D. (2005). A single amino acid converts a repressor to an activator of flowering. *Proceedings of the National Academy of Science of the United States of America* **102**, 7748–7753.

Harrison, C. J., Corley, S. B., Moylan, E. C., Alexander, D. L., Scotland, R. W. and Langdale, J. A. (2005). Independent recruitment of a conserved developmental mechanism during leaf evolution. *Nature* **434**, 509–514.

He, X. and Zhang, J. (2005). Rapid subfunctionalization accompanied by prolonged and substantial neofunctionalization in duplicate gene evolution. *Genetics* **169**, 1157–1164.

Hillis, D. M., Moritz, C. and Mable, B. K. (1996). "Molecular Systematics." Sinauer Associates, Sunderland, MA.

Huang, H., Tudor, M., Weiss, C. A., Hu, Y. and Ma, H. (1995). The *Arabidopsis* MADS-box gene *AGL3* is widely expressed and encodes a sequence-specific DNA-binding protein. *Plant Molecular Biology* **28**, 549–567.

Huelsenbeck, J. P. and Ronquist, F. (2001). MrBayes: Bayesian inference of phylogenetic trees. *Bioinformatics* **17**, 754–755.

Hughes, A. L. (1999). "Adaptive Evolution of Genes and Genomes." Oxford University Press, New York.

Ikeda, K., Sunohara, H. and Nagato, Y. (2004). Developmental course of inflorescence and spikelet in rice. *Breeding Science* **54**, 147–156.

Ikeda, K., Nagasawa, N. and Nagato, Y. (2005). *ABERRANT PANICLE ORGANIZATION 1* temporally regulates meristem identity in rice. *Developmental Biology* **282**, 349–360.

Irish, E. (1997). Class II tassel seed mutations provide evidence for multiple types of inflorescence meristems in maize (Poaceae). *American Journal of Botany* **84**, 1502–1515.

Itoh, J., Nonomura, K., Ikeda, K., Yamaki, S., Inukai, Y., Yamagishi, H., Kitano, H. and Nagato, Y. (2005). Rice plant development: From zygote to spikelet. *Plant Cell and Physiology* **46**, 23–47.

Jackson, D. and Hake, S. (1999). Control of phyllotaxy in maize by the *abphyl1* gene. *Development* **126**, 315–323.

Jackson, D., Veit, B. and Hake, S. (1994). Expression of maize *KNOTTED1* related homeobox genes in the shoot apical meristem predicts patterns of morphogenesis in the vegetative shoot. *Development* **120**, 405–413.

Jeanmougin, F., Thompson, J. D., Gouy, M., Higgins, D. G. and Gibson, T. J. (1998). Multiple sequence alignment with Clustal X. *Trends in Biochemical Sciences* **23**, 403–405.

Jensen, C. S., Salchert, K. and Nielsen, K. K. (2001). A *TERMINAL FLOWER1-Like* gene from perennial ryegrass involved in floral transition and axillary meristem identity. *Plant Physiology* **125**, 1517–1528.

Jensen, L. B., Andersen, J. R., Frei, U., Xing, Y., Taylor, C., Holm, P. B. and Lubberstedt, T. (2005). QTL mapping of vernalization response in perennial ryegrass (*Lolium perenne* L.) reveals co-location with an orthologue of wheat *VRN1*. *Theoretical and Applied Genetics* **110**, 527–536.

Jeon, J. S., Jang, S., Lee, S., Nam, J., Kim, C., Lee, S. H., Chung, Y. Y., Kim, S. R., Lee, Y. H., Cho, Y. G. and An, G. (2000). *leafy hull sterile1* is a homeotic mutation in a rice MADS box gene affecting rice flower development. *Plant Cell* **12**, 871–884.

Jeong, S., Trotochaud, A. E. and Clark, S. E. (1999). The *Arabidopsis CLAVATA2* gene encodes a receptor-like protein required for the stability of the *CLAVATA1* receptor-like kinase. *Plant Cell* **11**, 1925–1934.

Judziewicz, E. J., Clark, L. G., Londoño, X. and Stern, M. J. (1999). "American Bamboos." Smithsonian Institution Press, Washington, DC.

Kayes, J. M. and Clark, S. E. (1998). *CLAVATA2*, a regulator of meristem and organ development in *Arabidopsis*. *Development* **125**, 3843–3851.

Kellogg, E. A. (1996). Integrating genetics, phylogenetics, and developmental biology. *In* "The Impact of Plant Molecular Genetics." (B. W. S. Sobral, ed.), pp. 159–172. Birkhäuser, Cambridge, MA.

Kellogg, E. A. (2000a). The grasses: A case study in macroevolution. *Annual Review of Ecology and Systematics* **31**, 217–238.

Kellogg, E. A. (2000b). A model of inflorescence development. *In* "Monocots: Systematics and Evolution" (K. L. Wilson and D. A. Morrison, eds.), pp. 84–88. CSIRO, Melbourne.

Kellogg, E. A. (2000c). Molecular and morphological evolution in Andropogoneae. *In* "Grasses: Systematics and Evolution" (S. W. L. Jacobs and J. E. Everett, eds.). CSIRO, Melbourne, Australia.

Kellogg, E. A. (2003). Genome evolution: It's all relative. *Nature* **422**, 383–384.

Kellogg, E. A., Hiser, K. M. and Doust, A. N. (2004). Taxonomy, phylogeny, and inflorescence development of the genus *Ixophorus* (Panicoideae: Poaceae). *International Journal of Plant Sciences* **165**, 1089–1105.

Kellogg, E. A. and Shaffer, H. B. (1993). Model organisms in evolutionary studies. *Systematic Biology* **42**, 409–414.

Kerstetter, R., Vollbrecht, E., Lowe, B., Veit, B., Yamaguchi, J. and Hake, S. (1994). Sequence analysis and expression patterns divide the maize *knotted1-like* homeobox genes into two classes. *Plant Cell* **6**, 1877–1887.

Kerstetter, R. A., Laudencia-Chingcuanco, D., Smith, L. G. and Hake, S. (1997). Loss-of-function mutations in the maize homeobox gene, *knotted1*, are defective in shoot meristem maintenance. *Development* **124**, 3045–3054.

Klaus, H. (1966). Ontogenetische und histogentische Ungersuchungen an der Gerste (*Hordeum distichon* L.). *Botanisches Jahrbucher* **85**, 45–79.

Komatsu, K., Maekawa, M., Ujiie, S., Satake, Y., Furutani, I., Okamoto, H., Shimamoto, K. and Kyozuka, J. (2003a). *LAX* and *SPA*: Major regulators of shoot branching in rice. *Proceedings of the National Academy of Science of the United States of America* **100**, 11765–11770.

Komatsu, M., Chujo, A., Nagato, Y., Shimamoto, K. and Kyozuka, J. (2003b). *FRIZZY PANICLE* is required to prevent the formation of axillary meristems and to establish floral meristem identity in rice spikelets. *Development* **130**, 3841–3850.

Kyozuka, J., Konishi, S., Nemoto, K., Izawa, T. and Shimamoto, K. (1998). Down-regulation of *RFL*, the *FLO/LFY* homolog of rice, accompanied with panicle branch initiation. *Proceedings of the National Academy of Science of the United States of America* **95**, 1979–1982.

Larget, B. and Simon, D. L. (1999). Markov Chain Monte Carlo algorithms for the Bayesian analysis of phylogenetic trees. *Molecular Biology and Evolution* **16**, 750–759.

Lauter, N., Kampani, A., Carlson, S., Goebel, M. and Moose, S. P. (2005). *microRNA172* down-regulates *glossy15* to promote vegetative phase change in maize. *Proceedings of the National Academy of Science of the United States of America* **102**, 9412–9417.

Lawrence, C. J., Seigfried, T. E. and Brendel, V. (2005). The maize genetics and genomics database. The community resource for access to diverse maize data. *Plant Physiology* **138**, 55–58.

Le Roux, L. G. and Kellogg, E. A. (1999). Floral development and the formation of unisexual spikelets in the Andropogoneae (Poaceae). *American Journal of Botany* **86**, 354–366.

Liljegren, S. J., Roeder, A. H., Kempin, S. A., Gremski, K., Ostergaard, L., Guimil, S., Reyes, D. K. and Yanofsky, M. F. (2004). Control of fruit patterning in *Arabidopsis* by *INDEHISCENT*. *Cell* **116**, 843–853.

Litt, A. and Irish, V. F. (2003). Duplication and diversification in the *APETALA1/FRUITFULL* floral homeotic gene lineage: Implications for the evolution of floral development. *Genetics* **165**, 821–833.

Long, J. and Barton, M. K. (2000). Initiation of axillary and floral meristems in *Arabidopsis*. *Developmental Biology* **218**, 341–353.

Long, J. A., Moan, E. I., Medford, J. I. and Barton, M. K. (1996). A member of the *KNOTTED* class of homeodomain proteins encoded by the *STM* gene of *Arabidopsis*. *Nature* **379**, 66–69.

Lunde, C. and Hake, S. (2006). Florets and rosettes: Meristem genes in maize and *Arabidopsis*. *Maydica* **50**, 451–458.

Lynch, M. and Force, A. (2000). The probability of duplicate gene preservation by subfunctionalization. *Genetics* **154**, 459–473.

Ma, H., Yanofsky, M. F. and Meyerowitz, E. M. (1991). *AGL1-AGL6*, an *Arabidopsis* gene family with similarity to floral homeotic and transcription factor genes. *Genes and Development* **5**, 484–495.

Maddison, D. R. and Maddison, W. P. (2003). "MacClade: Analysis of Phylogeny and Character Evolution." Sinauer Associates, Sunderland, MA.

Maes, T., Van de Steene, N., Zethof, J., Karimi, M., D'Hauw, M., Mares, G., Van Montagu, M. and Gerats, T. (2001). *Petunia Ap2-like* genes and their role in flower and seed development. *Plant Cell* **13**, 229–244.

Malcomber, S. T. and Kellogg, E. A. (2004). Heterogeneous expression patterns and separate roles of the *SEPALLATA* gene *LEAFY HULL STERILE1* in grasses. *Plant Cell* **16**, 1692–1706.

Malcomber, S. T. and Kellogg, E. A. (2005). *SEPALLATA* gene diversification: Brave new whorls. *Trends in Plant Science* **10**, 427–435.

McSteen, P. and Hake, S. (2001). *barren inflorescence2* regulates axillary meristem development in the maize inflorescence. *Development* **128**, 2881–2891.

Moncur, M. W. (1981). "Floral Intiation in Field Crops." CSIRO, Melbourne.

Moore, R. C. and Purugganan, M. D. (2003). The early stages of duplicate gene evolution. *Proceedings of the National Academy of Sciences of the United States of America* **100**, 15682–15687.

Moore, R. C. and Purugganan, M. D. (2005). The evolutionary dynamics of plant duplicate genes. *Current Opinion in Plant Biology* **8**, 122–128.

Moose, S. P. and Sisco, P. H. (1996). *Glossy15*, an *APETALA2-like* gene from maize that regulates leaf epidermal cell identity. *Genes and Development* **10**, 3018–3027.

Nakagawa, M., Shimamoto, K. and Kyozuka, J. (2002). Overexpression of *RCN1* and *RCN2*, rice *TERMINAL FLOWER 1/CENTRORADIALIS* homologs, confers delay of phase transition and altered panicle morphology in rice. *Plant Journal* **29**, 743–750.

Nylander, J. A. A. (2004). *MrModeltest, Version 2*, Program distributed by the author. Evolutionary Biology Centre, Uppsala University.

Ohno, S. (1970). "Evolution by Gene Duplication." Springer-Verlag, London.

Orr, A. R. and Sundberg, M. D. (2004). Inflorescence development in a new teosinte: *Zea nicaraguensis* (Poaceae). *American Journal of Botany* **91**, 165–173.

Orr, A. R., Kaparthi, R., Dewald, C. L. and Sundberg, M. D. (2001). Analysis of inflorescence organogenesis in Eastern gamagrass, *Tripsacum dactyloides* (Poaceae): The wild type and the gynomonoecious *GSF1* mutant. *American Journal of Botany* **88**, 363–381.

Ohno, C. K., Reddy, G. V., Heisler, M. G. and Meyerowitz, E. M. (2004). The *Arabidopsis JAGGED* gene encodes a zinc finger protein that promotes leaf tissue development. *Development* **131**, 1111–1122.

Paterson, A. H., Bowers, J. E. and Chapman, B. A. (2004). Ancient polyploidization predating divergence of the cereals, and its consequences for comparative genomics. *Proceeding of the National Academy of Sciences of the United States of America* **101**, 9903–9908.

Pelaz, S., Ditta, G. S., Baumann, E., Wisman, E. and Yanofsky, M. F. (2000). B and C floral organ identity functions require *SEPALLATA* MADS-box genes. *Nature* **405**, 200–203.

Pelaz, S., Gustafson-Brown, C., Kohalmi, S. E., Crosby, W. L. and Yanofsky, M. F. (2001). *APETALA1* and *SEPALLATA3* interact to promote flower development. *Plant Journal* **26**, 385–394.

Pelucchi, N., Fornara, F., Favalli, C., Masiero, S., Lago, C., Pe, M. E., Colombo, L. and Kater, M. M. (2002). Comparative analysis of rice MADS-box genes expressed during flower development. *Sexual Plant Reproduction* **15**, 113–122.

Prasad, K., Sriram, P., Kumar, C. S., Kushalappa, K. and Vijayraghavan, U. (2001). Ectopic expression of rice *OsMADS1* reveals a role in specifying the lemma and palea, grass floral organs analogous to sepals. *Developmental Genes and Evolution* **211**, 281–290.

Ratcliffe, O. J., Amaya, I., Vincent, C. A., Rothstein, S., Carpenter, R., Coen, E. S. and Bradley, D. J. (1998). A common mechanism controls the life cycle and architecture of plants. *Development* **125**, 1609–1615.

Reinheimer, R., Pozner, R. and Vegetti, A. C. (2005). Inflorescence, spikelet, and floral development in *Panicum maximum* and *Urochloa plantaginea* (Poaceae). *American Journal of Botany* **92**, 565–575.

Reiser, L., Sanchez-Baracaldo, P. and Hake, S. (2000). Knots in the family tree: Evolutionary relationships and functions of *knox* homeobox genes. *Plant Molecular Biology* **42**, 151–166.

Riechmann, J. L. and Meyerowitz, E. M. (1998). The *AP2/EREBP* family of plant transcription factors. *Journal of Biological Chemistry* **379**, 633–646.

Ronse Decraene, L. P., Linder, H. P. and Smets, E. F. (2002). Ontogeny and evolution of the flowers of South African Restionaceae with special emphasis on the gynoecium. *Plant Systematics and Evolution* **231**, 225–258.

Rudall, P. J., Stuppy, W., Cunniff, J., Kellogg, E. A. and Briggs, B. G. (2005). Evolution of reproductive structures in grasses inferred by sister-group comparisons with their putative closest living relatives, Ecdeiocoleaceae. *American Journal of Botany* **92**, 1432–1443.

Sato, Y., Sentoku, N., Nagato, Y. and Matsuoka, M. (1998). Isolation and characterization of a rice homebox gene, *OSH15*. *Plant Molecular Biology* **38**, 983–997.

Savidge, B., Rounsley, S. D. and Yanofsky, M. F. (1995). Temporal relationship between the transcription of two *Arabidopsis* MADS box genes and the floral organ identity genes. *Plant Cell* **7**, 721–733.

Schmitz, J., Franzen, R., Ngyuen, T. H., Garcia-Maroto, F., Pozzi, C., Salamini, F. and Rohde, W. (2000). Cloning, mapping and expression analysis of barley MADS-box genes. *Plant Molecular Biology* **42**, 899–913.

Sentoku, N., Sato, Y., Kurata, N., Ito, Y., Kitano, H. and Matsuoka, M. (1999). Regional expression of the rice *KN1*-type homeobox gene family during embryo, shoot, and flower development. *Plant Cell* **11**, 1651–1664.

Shannon, S. and Meeks-Wagner, D. R. (1991). A mutation in the *Arabidopsis TFL1* gene affects inflorescence meristem development. *Plant Cell* **3**, 877–892.

Sharma, V. K., Carles, C. and Fletcher, J. C. (2003). Maintenance of stem cell populations in plants. *Proceedings of the National Academy of Science of the United States of America* **100** (Suppl. 1), 11823–11829.

Sharman, B. C. (1947). The biology and developmental morphology of the shoot apex in the Gramineae. *New Phytologist* **46**, 20–34.

Simmons, M. P., Pickett, K. M. and Miya, M. (2004). How meaningful are Bayesian support values? *Molecular Biology and Evolution* **21**, 188–199.

Simons, K. J., Fellers, J. P., Trick, H. N., Zhang, Z., Tai, Y.-S., Gill, B. S. and Faris, J. D. (2006). Molecular characterization of the major wheat domestication gene *Q*. *Genetics* **172**, 547–555.

Smyth, D. R., Bowman, J. L. and Meyerowitz, E. M. (1990). Early flower development in *Arabidopsis*. *Plant Cell* **2**, 755–767.

Soltis, D. E., Soltis, P. S., Endress, P. K. and Chase, M. W. (2005). "Phylogeny and Evolution of Angiosperms." Sinauer Associates, Sunderland, MA.

Sonnhammer, E. L. L. (2002). Orthology, paralogy and proposed classification for paralog subtypes. *Trends in Genetics* **18**, 619–620.

Stapf, O. (ed.) (1917). "Gramineae. Flora of Tropical Africa."London, Lowell Reeve & Co.

Stevens, P. F. (2001 onwards). Angiosperm Phylogeny Website, Version 6, http:/www.mobot.org/MOBOT/research/APweb/.

Sundberg, M. D. and Orr, A. R. (1996). Early inflorescence and floral development in *Zea mays* land race Chapalote (Poaceae). *American Journal of Botany* **83**, 1255–1265.

Suzaki, T., Sato, M., Ashikari, M., Miyoshi, M., Nagato, Y. and Hirano, H. Y. (2004). The gene *FLORAL ORGAN NUMBER1* regulates floral meristem size in rice and encodes a leucine-rich repeat receptor kinase orthologous to *Arabidopsis CLAVATA1*. *Development* **131**, 5649–5657.

Swofford, D. L. (2000). "PAUP*: Phylogenetic Analysis Using Parsimony." Sinauer Associates, Sunderland, MA.

Taguchi-Shiobara, F., Yuan, Z., Hake, S. and Jackson, D. (2001). The *fasciated ear2* gene encodes a leucine-rich repeat receptor-like protein that regulates shoot meristem proliferation in maize. *Genes and Development* **15**, 2755–2766.

Takahashi, M., Nagasawa, N., Kitano, H. and Nagato, Y. (1998). *panicle phytomer1* mutations affect panicle architecture of rice. *Theoretical and Applied Genetics* **96**, 1050–1056.

Tian, C., Wan, P., Sun, S., Li, J. and Chen, M. (2004). Genome-wide analysis of the GRAS gene family in rice and *Arabidopsis*. *Plant Molecular Biology* **54**, 519–532.

To, J. P. C., Haberer, G., Ferreira, F. J., Deruere, J., Mason, M. G., Schaller, G. E., Alonso, J. M., Ecker, J. R. and Kieber, J. J. (2004). Type-A *Arabidopsis* response regulators are partially redundant negative regulators of cytokinin signaling. *Plant Cell* **16**, 658–671.

Trevaskis, B., Bagnall, D. J., Ellis, M. H., Peacock, W. J. and Dennis, E. S. (2003). MADS box genes control vernalization-induced flowering in cereals. *Proceedings of the National Academy of Science of the United States of America* **100**, 13099–13104.

Vandenbussche, M., Theissen, G., Van de Peer, Y. and Gerats, T. (2003). Structural diversification and neo-functionalization during floral MADS-box gene evolution by C-terminal frameshift mutations. *Nucleic Acids Research* **31**, 4401–4409.

Vegetti, A. C. and Weberling, F. (1996). The structure of the paracladial zone in Poaceae. *Taxon* **45**, 453–460.

Venglat, S. P., Dumonceaux, T., Rozwadowski, K., Parnell, L., Babic, V., Keller, W., Martienssen, R., Selvaraj, G. and Datla, R. (2002). The homeobox gene *BREVIPEDICELLUS* is a key regulator of inflorescence architecture in *Arabidopsis*. *Proceedings of the National Academy of Science of the United States of America* **99**, 4730–4735.

Vollbrecht, E., Veit, B., Sinha, N. and Hake, S. (1991). The developmental gene *Knotted-1* is a member of a maize homeobox gene family. *Nature* **350**, 241–243.

Vollbrecht, E., Reiser, L. and Hake, S. (2000). Shoot meristem size is dependent on inbred background and presence of the maize homeobox gene, *knotted1*. *Development* **127**, 3161–3172.

Vollbrecht, E., Springer, P. S., Goh, L., Buckler, E. S., IV and Martienssen, R. (2005). Architecture of floral branch systems in maize and related grasses. *Nature* **436**, 1119–1126.

Walsh, J. B. (1995). How often do duplicated genes evolve new functions? *Genetics* **139**, 421–428.

Ware, D., Jaiswal, P., Ni, J., Pan, X., Chang, K., Clark, K., Teytelman, L., Schmidt, S., Zhao, W., Cartinhour, S., McCouch, S. and Stein, L. (2002). Gramene: A resource for comparative grass genomics. *Nucleic Acids Research* **30**, 103–105.

Watson, L. and Dallwitz, M. J. (1992 onwards). Grass Genera of the World: Descriptions, Illustrations, Identification, and Information Retrieval; including Synonyms, Morphology, Anatomy, Physiology, Phytochemistry, Cytology, Classification, Pathogens, World and Local Distribution, and References. http://biodiversity.uno.edu/delta/.

Weigel, D., Alvarez, J., Smyth, D. R., Yanofsky, M. F. and Meyerowitz, E. M. (1992). *LEAFY* controls floral meristem identity in *Arabidopsis*. *Cell* **69**, 843–859.

Weigel, D. and Nilsson, O. (1995). A developmental switch sufficient for flower initiation in diverse plants. *Nature* **377**, 495–500.

Weir, C. E. and Dale, H. M. (1960). A developmental study of wild rice, *Zizania aquatica* L. *Canadian Journal of Botany* **38**, 719–739.

Wernersson, R. and Pedersen, A. G. (2003). RevTrans: Multiple alignment of coding DNA from aligned amino acid sequences. *Nucleic Acids Research* **31**, 3537–3539.

Yan, L., Loukoianov, A., Tranquilli, G., Helguera, M., Fahima, T. and Dubcovsky, J. (2003). Positional cloning of the wheat vernalization gene VRN1. *Proceedings of the National Academy of Science of the United States of America* **100**, 6263–6268.

Yu, J., Hu, S., Wang, J., Wong, G. K., Li, S., Liu, B., Deng, Y., Dai, L., Zhou, Y., Zhang, X., Cao, M., Liu, J., *et al.* (2002). A draft sequence of the rice genome (*Oryza sativa* L. ssp. *indica*). *Science* **296**, 79–92.

Yu, J., Wang, J., Lin, W., Li, S., Li, H., Zhou, J., Ni, P., Dong, W., Hu, S., Zeng, C., Zhang, J., Zhang, Y., *et al.* (2005). The genomes of *Oryza sativa*: A history of duplications. *PLoS Biology* **3**, e38.

Zahn, L. M., Kong, H., Leebens-Mack, J. H., Kim, S., Soltis, P. S., Landherr, L. L., Soltis, D. E., Depamphilis, C. W. and Ma, H. (2005). The evolution of the *SEPALLATA* subfamily of MADS-box genes: A pre-angiosperm origin with multiple duplications throughout angiosperm history. *Genetics* **169**, 2225–2239.

Zhang, J. Z. (2003). Evolution by gene duplication: An update. *Trends in Ecology & Evolution* **18**, 292–298.

Expression of Floral Regulators in Basal Angiosperms and the Origin and Evolution of ABC-Function

PAMELA S. SOLTIS,* DOUGLAS E. SOLTIS,† SANGTAE KIM,*,† ANDRE CHANDERBALI*,† AND MATYAS BUZGO*,†

**Florida Museum of Natural History, University of Florida, Gainesville, Florida 32611*
†Department of Botany, University of Florida, Gainesville, Florida 32611

Advances in Botanical Research, Vol. 44
Incorporating Advances in Plant Pathology

0065-2296/06 $35.00
DOI: 10.1016/S0065-2296(06)44012-X

ABSTRACT

The ABC-model of floral organ identity explains the regular, sequential development of sepals, petals, stamens, and carpels in eudicot flowers. This general model, based on studies of the derived eudicots *Arabidopsis* and *Antirrhinum*, may apply to nearly all eudicots, most of which are characterized by discrete whorls of floral organs. However, floral morphology of basal angiosperms is typically characterized by variable numbers of floral parts and gradual transitions among floral organs, and it is unclear that the ABC-model applies to such flowers. Here we explore the origin and evolution of ABC-function through consideration of expression data for homologs of ABC-genes for basal angiosperms and conclude that the ABC-model represents an evolutionarily derived regulatory network that arose through spatial restriction of regulatory gene expression.

I. INTRODUCTION

The ABC-model of floral organ identity (Coen and Meyerowitz, 1991) is a unifying paradigm for much of the ongoing research in floral development and evolution. Its formulation based on mutant studies in *Arabidopsis* and *Antirrhinum*, representing two major clades of angiosperms, rosids and asterids, respectively, followed by its subsequent general agreement in the monocot *Zea* (Ambrose *et al.*, 2000), suggested that this model may in fact apply to all angiosperms (but see Davies *et al.*, Chapter 7). Furthermore, the development of the quartet model (Theissen, 2001) to explain the combinatorial action of the homeotic genes with A-, B-, and C-functions provided a mechanism that could conceivably apply to all flowering plants.

But what are the evolutionary origins of ABC-function? Does it characterize all flowering plants, having been inherited from their common ancestor, or did it evolve later, during the diversification of angiosperms? Studies of developmental morphology and gene expression in basal angiosperms are providing answers to these questions. In this chapter, we (1) summarize the ABC-model, along with alternative models that have been formulated on the basis of floral development and/or gene expression; (2) review the expression data available for basal angiosperms; (3) examine the correspondence between expression patterns and floral morphology, including possible origins of novel structures; (4) consider the evolutionary origin of the regulatory network described in the ABC-model; and (5) provide testable hypotheses for the origin and evolution of the underlying mechanisms of the ABC-model.

II. SUMMARY OF THE ABC-MODEL

A. THE ABC-MODEL

The ABC-model (Coen and Meyerowitz, 1991) posits that floral organ identity is controlled by the combinatorial action of three genetically independent functions. The A-function alone specifies sepal identity, A- and B-functions together specify petal identity, B- and C-functions together control stamen identity, and C-function alone specifies carpel identity (Fig. 1). A- and C-functions are typically antagonistic. In *Arabidopsis*, the A-function genes are *APETALA1* (*AP1*) and *AP2*, the B-function genes are *AP3* and *PISTILLATA* (*PI*), and the C-function gene is *AGAMOUS* (*AG*). In *Antirrhinum*, the ortholog of *AP1* is *SQUAMOSA* (*SQUA*), but it lacks A-function, although the homologs of *AP2, LIPLESS1* (*LIP1*) and *LIP2*, appear to have partial A-function by contributing to perianth identity (Keck *et al.*, 2003); the B-function genes are *DEFICIENS* (*DEF*) and *GLOBOSA* (*GLO*), which are orthologous to *AP3* and *PI*, respectively; and the C-function gene is *PLENA* (*PLE*). All of these genes except *AP2* are MADS-box genes.

Modification of the ABC-model includes additional MADS-box genes that specify ovule identity (D-function, Colombo *et al.*, 1995; e.g., *SEEDSTICK* (*STK*) [=*AGAMOUS-LIKE11* (*AGL11*), Pinyopich *et al.*, 2003;

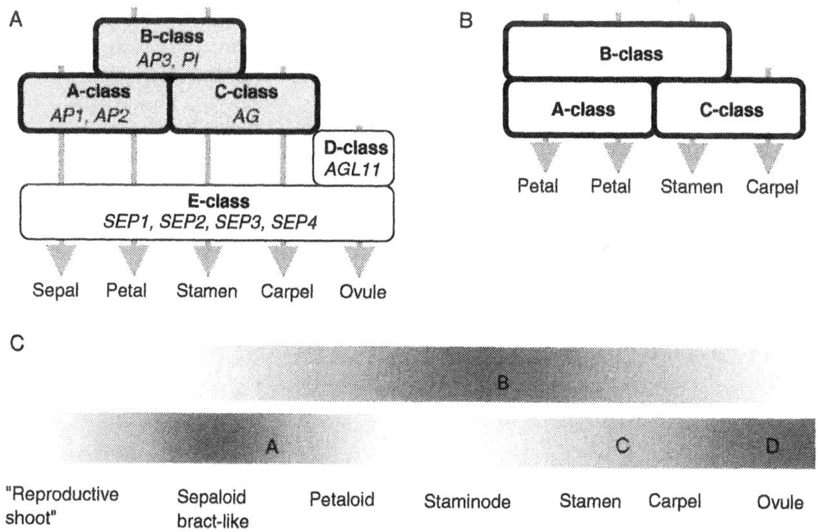

Fig. 1. Comparison of "classic" ABCDE-model of floral organ identity (A) with sliding boundaries (B) and fading borders (C) models.

Rounsley *et al.*, 1995], in *Arabidopsis*) and contribute to the specification of sepal, petal, stamen, and carpel identity (E-function; *AGL2, -4, -9*, and *-3*, renamed as *SEPALLATA1* (*SEP1*), *-2*, *-3*, and *-4*, in *Arabidopsis*; Ditta *et al.*, 2004; Pelaz *et al.*, 2000; Theissen, 2001). Although the *SEP* genes are expressed at different stages and in different primordia during floral development, they are nonetheless functionally redundant in the control of all floral organ identities, as all floral organs are replaced by leaf-like organs in the triple *SEP1,-2,-3* mutant; the SEP genes also play redundant roles in specifying floral meristem determinacy (Ditta *et al.*, 2004; Pelaz *et al.*, 2000). A more comprehensive shorthand for the combined effects of these functions is ABCDE. However, we will not consider D-function further in this chapter and will restrict our discussions to the ABC and ABCE components of the more complete ABCDE-model.

B. THE QUARTET MODEL

In *SEP* loss-of-function mutants in *Arabidopsis*, all floral organs resemble sepals, suggesting that (A + B + E)-function is needed for petals, (B + C + E)-function is needed for stamens, and (C + E)-function is required for carpels (Pelaz *et al.*, 2000). But how are the *SEP* genes involved in floral organ identity? The quartet model (Theissen, 2001) proposes that SEP proteins form heterotetramers with either A- and B-proteins (for petals), B- and C-proteins (for stamens), and C-protein (for carpels), and evidence from *Arabidopsis* supports this model (Honma and Goto, 2001). Does the quartet model apply to the ancestral flower and all extant flowering plants, or was it derived in a more recent ancestor of *Arabidopsis* and perhaps some of its closer relatives? Do the components of the quartet coevolve, or is their evolution independent? Understanding the evolutionary constraints of the quartet model may provide clues to possible mechanisms underlying the ABCDE-model.

III. ALTERNATIVES TO THE ABC-MODEL

Early angiosperms exhibited tremendous diversity in floral structure and organization (Friis *et al.*, 2000), and extant lineages of basal angiosperms have retained much of this diversity. For example, *Amborella, Cabomba*, and *Illicium* have perianths of largely undifferentiated tepals while some species of *Asimina* have morphologically distinct sepals and petals and species of *Ceratophyllum* and *Sarcandra* lack a perianth altogether. Other aspects of floral morphology also differ among basal lineages (e.g., Endress, 2001) and between basal angiosperms and the predictable, highly synorganized flowers of core eudicots. The fundamental differences in perianth morphology and organization among core eudicots, monocots, and many basal angiosperms suggest similar fundamental differences in the regulation of floral organ identity among these lineages. And, in fact, alternatives to the ABC-model have been proposed based on morphology and/or patterns of gene expression (Bowman, 1997; Buzgo *et al.*, 2004; Kanno *et al.*, 2003; Kim *et al.*, 2005a; Kramer *et al.*, 2003; van Tunen *et al.*, 1993). Evaluation of the ABC-model in basal angiosperms involves two basic questions: (1) do homologs of A-, B-, and C-function genes from model organisms exhibit expression patterns consistent with these functions in basal angiosperms? and (2) are the A-, B-, and C-functions specified by Coen and Meyerowitz (1991) in fact applicable to floral development in basal angiosperms, perhaps by different genes than in the models? Both questions will be addressed in the following paragraphs.

A. THE TULIP/LILY MODEL

Many monocots have flowers with morphologically identical, petaloid inner and outer perianth whorls. Van Tunen *et al.* (1993) suggested that the two petaloid whorls of *Lilium* were controlled by the same organ identity genes, and they proposed a modified ABC-model in which expression of B-function genes was extended outward into the "first" whorl, thus converting what would be the sepals, based on position, into petals (Fig. 1). The flowers of most tulips (barring some horticultural varieties) are similar to those of *Lilium* in having two morphologically identical and petaloid perianth whorls. In support of van Tunen *et al.*'s (1993) hypothesis, B-function homologs are expressed in the outer whorl, likely accounting for their petaloid appearance (Kanno *et al.*, 2003). This model, however, seems not to apply to all monocots, based on studies of *Asparagus* (Park *et al.*, 2004).

B. SHIFTING/SLIDING BOUNDARIES

The apparent mechanism responsible for multiple petaloid perianth whorls in *Lilium* and *Tulipa* has been described more generally as the "shifting boundary" (Bowman, 1997) and "sliding boundary" (Kramer *et al.*, 2003) models. The shifting/sliding boundary model allows the boundary of the B-function to "slide" across the floral meristem from its restricted location in *Arabidopsis* to include the outer perianth whorl (as in *Lilium, Tulipa*, and the basal eudicots *Ranunculus* and *Aquilegia*; see also Kramer and Zimmer, Chapter 9). Likewise, an inward shift of A-function, superimposed on B- and

C-functions, might lead to petaloid stamens and/or staminodes. This combined A- and C-function is not possible under the ABC-model because of their antagonistic effects; however, weak expression of either A- or C-function (or both) could allow for cofunction. This shifting/sliding in either direction might characterize all copies of A- or B-function genes or perhaps only some duplicate copies, if duplicates are present. The predicted result of shifting/sliding boundaries is that petaloid features would be extended beyond the typical (second) whorl; however, organ boundaries would remain discrete if A- and/or B-functions operated in specified regions.

C. FADING BORDERS

In *Amborella*, the perianth is considered "undifferentiated" (Endress and Igersheim, 2000; Posluszny and Tomlinson, 2003)—that is, there are no true sepals and petals. However, developmental studies suggest that in fact the perianth is differentiated into outer and inner tepals (Buzgo *et al.*, 2004). The margins of the outer tepals are smooth like the margins of the lateral receptacular bracts, whereas those of the inner tepals are papillose, more similar to those of the stamens. Furthermore, the tips of the inner tepals resemble the connective tips of the stamens. Therefore, in *Amborella*, there is a gradual transition from lateral receptacular bracts to tepals, from outer tepals to inner tepals, and from inner tepals to stamens in male flowers (and to carpels in female flowers, given the developmental similarities between stamens and carpels)—in other words, a gradual transition from outer to inner floral organs.

This view of floral morphology is not consistent with the discrete whorls of floral organs present in *Arabidopsis* (and typical of eudicots, see Endress, Chapter 1) and implied by the ABC-model. The sliding boundaries model applies to flowers with an undifferentiated perianth, such as lily and tulip and some Ranunculales, but this model fundamentally implies clear differentiation among floral organs—that is, no gradual transitions. However, gradual transitions among floral organs characterize two of the three basal lineages of angiosperms—*Amborella* and Austrobaileyales—and may therefore have been the ancestral condition (Ronse De Craene *et al.*, 2003; Soltis *et al.*, 2005a). Neither the ABC-model nor its sliding/shifting boundary modification can account for such morphological intergradations.

The fading borders model (Buzgo *et al.*, 2004) suggests that gradual transitions in organ morphology across the floral meristem result from a gradient in levels of expression of floral regulators across the meristem (Fig. 1). Weak expression at the edge of each gene's range of activity overlaps with expression of another regulator; strong expression of each regulator occurs distant from the adjacent gene's center of strong expression. This pattern of gene expression would impose some features of adjacent organs onto each other and thus produce morphologically intergrading rather than distinct floral organs. Expression of B-function homologs in *Amborella* supports this model (Kim *et al.*, 2005a).

IV. EXPRESSION OF FLORAL REGULATORS IN BASAL ANGIOSPERMS

To evaluate the applicability of the ABC-model in basal angiosperms, we will first consider two questions: (1) are homologs of *Arabidopsis* A-, B-, and C-function genes present in basal angiosperms? and (2) where and when are these homologs expressed? These experiments focus on *Amborella* (Amborellaceae), *Nuphar* (Nymphaeales; water lilies), *Illicium* (Austrobaileyales), *Eupomatia* (Eupomatiaceae: Magnoliales), *Magnolia* (Magnoliaceae: Magnoliales), *Asimina* (Annonaceae: Magnoliales), and *Persea* (Lauraceae: Laurales). *Amborella, Nuphar*, and *Illicium* represent the three basalmost lineages of angiosperms, and Magnoliales and Laurales are sister groups in the magnoliid clade (Barkman *et al.*, 2000; Qiu *et al.*, 1999; Soltis *et al.*, 1999, 2000; Zanis *et al.*, 2002); with this sampling, most of the major lineages of basal angiosperms are represented. Additional work on Piperales (Jaramillo and Kramer, 2004) and Canellales of the magnoliid clade and on Chloranthaceae (Li *et al.*, 2005) provides a broader phylogenetic perspective on basal angiosperm clades.

A. ARE HOMOLOGS OF *ARABIDOPSIS* A-, B-, AND C-FUNCTION GENES PRESENT IN BASAL ANGIOSPERMS?

Floral expressed sequence tags (ESTs) generated by the Floral Genome Project (FGP; Albert *et al.*, 2005) were searched for possible homologs to *Arabidopsis* ABC-genes, and reverse transcription polymerase chain reaction (RT-PCR) was used to target specific genes that were not initially present in the FGP database or to obtain genes from species that are not included in the FGP (Kim *et al.*, 2005a,b). The matrix of MADS-box genes from Becker and Theissen (2003) served as a framework for placing the MADS genes identified from basal angiosperms. Nearly all subfamilies of MADS-box genes have homologs in basal angiosperms, and in most cases, the sequences from the basal angiosperms are sister(s) to the rest of each gene subfamily (Fig. 2; Kim *et al.*, 2004, 2005a; Zahn *et al.*, 2005). It therefore appears that orthologs of the floral regulatory genes of *Arabidopsis* are pervasive among

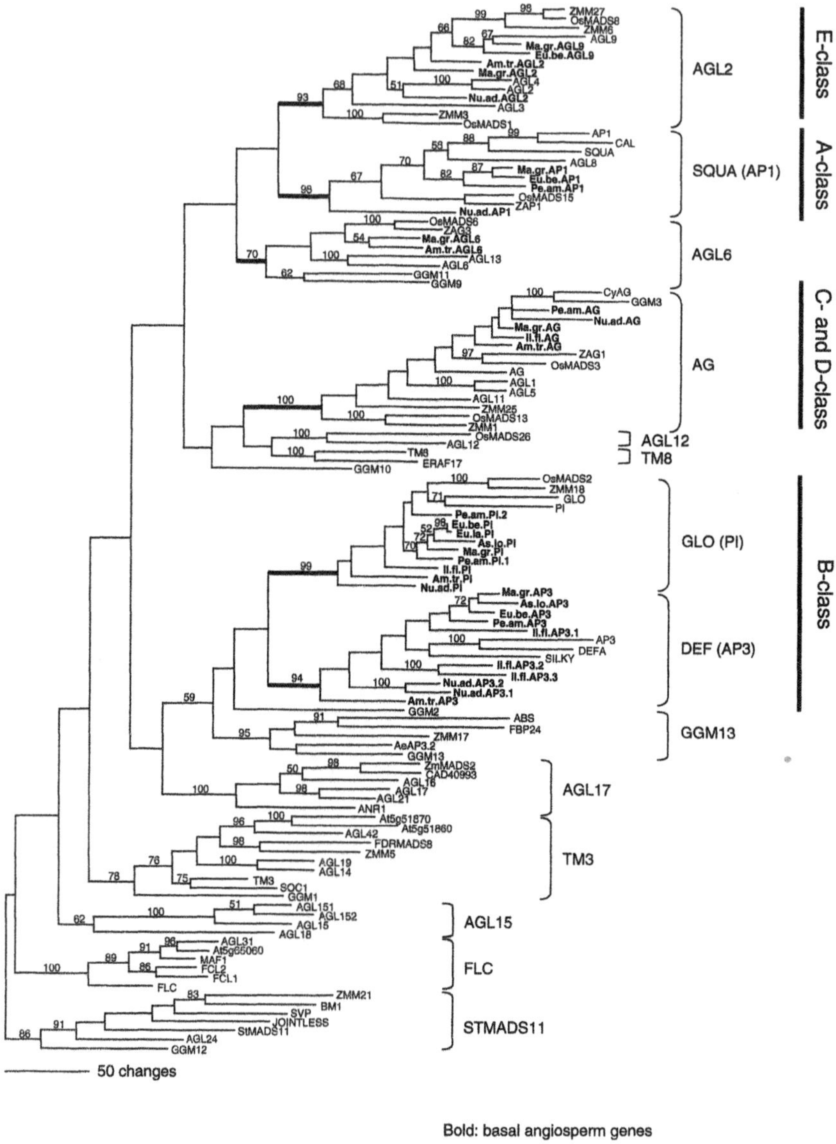

Fig. 2. Phylogenetic tree of MADS-box genes showing clades that correspond to the A-, B-, C-, D-, and E-function homologs of *Arabidopsis*. Sequences of basal angiosperms obtained through the Floral Genome Project and reported by Kim *et al.* (2005a) are indicated in bold.

early lineages of angiosperms and share a common ancestor that predates the angiosperms themselves. The question then becomes: what are these genes doing in basal lineages, most of which bear strikingly different floral morphologies from that of *Arabidopsis* (see Endress, Chapter 1)?

B. WHERE AND WHEN ARE HOMOLOGS OF THE A-, B-, AND C-FUNCTION GENES OF *ARABIDOPSIS* EXPRESSED?

Three types of data were used to assess patterns of expression of ABC-homologs in basal angiosperms: relative quantitative RT-PCR, real-time PCR, and *in situ* hybridization of young floral buds. The RT-PCR methods quantitatively examined gene expression in specific floral organs from buds of specified stages, and the *in situ* approach allowed for comparisons of expression at earlier stages of development than is possible via RT-PCR. Data were collected on A-, B-, C-, and E-function homologs, and patterns of expression were similar among methods, even though the methods differ in their sensitivities and the buds and organs used in the experiments were from different developmental stages.

Expression of MADS homologs in basal angiosperm flowers is generally consistent with the ABC-model based on functional studies in *Arabidopsis* and *Antirrhinum*, but with important differences: expression patterns in the basal angiosperms are often broader than specified by the ABC-model. This pattern is particularly evident for the B-function homologs: *AP3* and *PI* homologs are expressed in perianth parts and stamens in the basal angiosperm exemplars examined, just as they are in the petals and stamens of *Arabidopsis* and *Antirrhinum*. However, *AP3* and *PI* homologs are expressed in tepals, stamens, staminodes of female flowers, and carpels of *Amborella*, and in all floral organs of *Nuphar*, in both small and larger buds (Kim *et al.*, 2005a). Whereas weak expression of *AP3* and *PI* has also been observed in sepals and ovules of *Arabidopsis* (Irish, 2003), the expression patterns for basal angiosperms across the floral meristem were much stronger.

The A-function homolog *AP1* deserves special note. A gene duplication in the core eudicots generated two lineages: eu*AP1* and eu*FUL* (named for the *Arabidopsis FRUITFULL* (*FUL*) gene; Litt and Irish, 2003). *AP1* reportedly confers A-function in *Arabidopsis* (Bowman *et al.*, 1993). The *AP1* paralog *FUL* (in *Arabidopsis*) is important for fruit development (Gu *et al.*, 1998) and acts redundantly with *AP1* in regulating meristem identity (Ferrandiz *et al.*, 2000). Expression of *AP1* homologs in *Nuphar* and *Magnolia* was higher in leaves and carpels than in perianth or stamens and appears to be more similar to expression patterns for the eu*FUL* lineage than the eu*AP1* lineage (Irish, 2003; Kim *et al.*, 2005a). It may be that the A-function of the

ABC-model attributed to *AP1* is derived in *Arabidopsis* and perhaps relatives in the rosid clade; this function has not yet been demonstrated in other clades of angiosperms. However, the original inference of A-function for *AP1* in *Arabidopsis* may have been premature. *AP1* was reported to contribute to A-function because its expression is restricted to the outer two whorls of the *Arabidopsis* flower through repression by *AG* (Bowman *et al.*, 1993). However, *ap1* mutants do not display the floral morphology predicted by the ABC-model (Irish and Sussex, 1990), raising the possibility that *AP1* does not in fact confer A-function (Kramer and Hall, 2005). Furthermore, study suggests that *AP1* function is not essential for the identity of sepals and petals, even in *Arabidopsis* (Yu *et al.*, 2004). Finally, the *AP1* ortholog from *Antirrhinum, SQUA*, does not have A-function (Huijser *et al.*, 1992; Taylor *et al.*, 2002). Of course, *AP2* confers A-function in *Arabidopsis*, and its homologs *LIP1* and *LIP2* confer at least partial A-function in *Antirrhinum* (Keck *et al.*, 2003). Davies and colleagues (Davies *et al.*, Chapter 7) question whether A-function as specified in the ABC-model (Coen and Meyerowitz, 1991) is applicable to *Antirrhinum* and all flowering plants.

C. *AGL6*: A CANDIDATE A-FUNCTION GENE IN BASAL ANGIOSPERMS

Both B- and C-functions appear to operate in basal angiosperms, based on inferences from expression patterns. However, application of the *Arabidopsis* ABC-model in basal angiosperms requires A-function as well. To date, this A-function has not been identified in basal angiosperms or in any species other than *Arabidopsis*. Although other species clearly have orthologs of *AP1* (i.e., *SQUA* in *Antirrhinum*), the A-function suggested for *AP1* has not been recovered, and, as noted, it is even uncertain if *AP1* actually confers the so-called A-function in *Arabidopsis* (Yu *et al.*, 2004). In basal angiosperms, the A-class homologs clearly do not share expression patterns with *AP1*. However, strong expression of *AGL6* genes in basal angiosperms is found in the perianth: *Ma.gr.AGL6* (*Magnolia*) is expressed only in tepals, and *Am.tr. AGL6* (*Amborella*) is strongly expressed in tepals, with weak expression in other floral organs (Kim *et al.*, 2005a). If function can be inferred from expression, *AGL6* is a reasonable candidate for an A-function gene in basal angiosperms. This hypothesis should be tested using direct and indirect methods such as (1) protein–protein interactions of *AGL6* and B- and C-function homologs in basal angiosperms, (2) transformation of basal angiosperm *AGL6* genes into an *Arabidopsis AP1* mutant, and (3) *AGL6* knockouts in basal angiosperms. Of course, heterologous expression studies must be interpreted with caution particularly in this case when the function of *AGL6* in *Arabidopsis* is still unknown.

V. CORRESPONDENCE BETWEEN EXPRESSION PATTERNS AND FLORAL MORPHOLOGY

Basal angiosperms, with only a few exceptions, are not amenable to true functional studies, so hypotheses of gene function are currently limited to inferences based on expression. Molecular studies in *Arabidopsis* and *Antirrhinum* demonstrate that function of floral MADS-box genes is highly correlated with expression patterns (Ma and dePamphilis, 2000); therefore, expression patterns in other species may be good predictors of gene function. (Of course, use of antibodies or other means to detect protein, rather than simply transcript, would provide stronger evidence for possible function of ABC-homologs and other genes, but these approaches are still in their infancy in nonmodel species.) Given the premise that expression implies function in the case of floral MADS-box genes, we can formulate hypotheses of gene function in basal angiosperms and evaluate alternative models of floral organ specification.

The expression patterns obtained for basal angiosperms are generally consistent with the morphology predicted by the ABC-model. For example, expression of B-function homologs in the perianth is consistent with the petaloid nature of these organs in *Amborella, Nuphar*, and *Illicium*. Furthermore, broadly overlapping expression patterns of floral regulators in *Amborella* in particular may be responsible for the gradual intergradation of floral organs. In contrast, the floral morphology of *Asimina longifolia* (Fig. 3), which comprises discrete whorls of differentiated organs similar to those of *Arabidopsis*, appears to be controlled by restricted expression of B- and C-function homologs (Kim

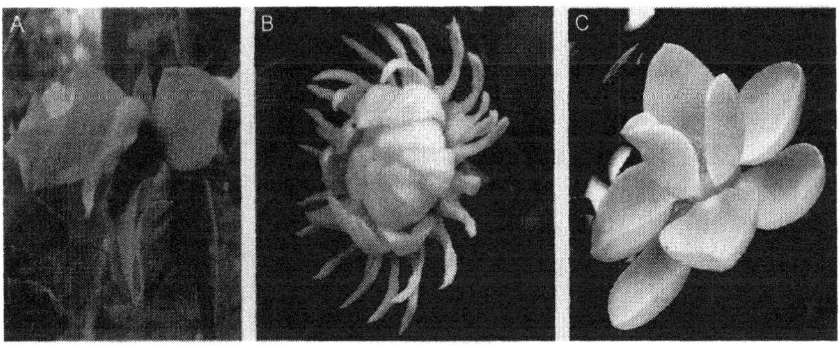

Fig. 3. Floral diversity in Magnoliales. (A) *A. longifolia* (Annonaceae), with discrete and differentiated perianth whorls (photo by S. Kim). (B) *Eupomatia bennettii* (Eupomatiaceae); earlier in development, the flower is enclosed by a calyptra, which has fallen away to reveal showy stamens and staminodes (photo courtesy of H. Teppner). (C) *Magnolia grandiflora* (Magnoliaceae), with showy tepals (photo courtesy of D. Callaway). (See color Plate 17.)

et al., 2005a): the patterns of gene expression are exactly as predicted by the ABC-model.

VI. NEW INSIGHTS INTO THE EVOLUTION OF NOVEL FLORAL STRUCTURES

The growing body of developmental genetic data for basal angiosperms provides a framework that can be used to extend to other basal angiosperms exhibiting highly unusual floral features to help elucidate the origin and evolution of novelty. The background provided earlier reveals that the ABC-model as deduced for *Arabidopsis* does not directly apply to basal angiosperms. Hence, understanding the evolution of floral novelty in basal angiosperms requires comparison to other basal angiosperms, rather than direct comparison to the known floral developmental genetics of *Arabidopsis*.

Although basal angiosperms represent only a small proportion of extant angiosperm diversity, they nonetheless exhibit considerable floral diversity, including the evolution of highly unusual or novel structures. Particularly noteworthy is *Eupomatia* (Eupomatiaceae), a genus of two species closely related to the *Magnolia* family (Magnoliaceae) of Magnoliales; flowers of *Eupomatia* are characterized by a highly unusual structure—termed a calyptra—that encloses and presumably protects the flower in bud (Fig. 3). A calyptra is known elsewhere in the angiosperms only in Himantandraceae, another basal angiosperm family that is also placed in Magnoliales. The origin of the calyptra has long been controversial; whereas some, on the basis of gross morphology (Cronquist, 1988), have considered it to represent a modified perianth, others, based on developmental data (Endress, 2003), have proposed that it represents a modified bract. Both genera of the related family Magnoliaceae (*Magnolia* and *Liriodendron*) have a well-developed perianth of showy tepals, as well as bracts that enclose the flower (Fig. 3), and Magnoliaceae represent a critical phylogenetic link for comparison with *Eupomatia*.

To provide insights into the origin of the calyptra, Kim *et al.* (2005b) examined the expression of A- and B-function homologs in this unusual structure. B-function homologs (*AP3* and *PI*) isolated from staminodes of species of *Eupomatia* were strongly expressed in developing stamens, staminodes, and carpels, but were either not expressed in the calyptra, or only weakly expressed at a level consistent with the expression observed in leaves of these plants. In basal angiosperms, including *Magnolia* (Magnoliaceae), a close relative of Eupomatiaceae, B-function homologs are expressed throughout the perianth. The pattern of expression of floral genes in the calyptra of *Eupomatia* generally matches the expectations for a vegetative organ, such as a leaf or bract, rather than the predictions for perianth. Hence, the evo-devo results of Kim *et al.*

(2005b) are consistent with the interpretation of the origin of the calyptra (Endress, 2003) based on developmental morphology.

This example illustrates that in some cases, such as in a clade of related taxa (in this case, multiple members of Magnoliales), comparative studies of floral gene expression may be useful in addressing the origins of enigmatic structures. However, in a basal angiosperm such as *Eupomatia*, comparisons are better made in light of the expectations based on close relatives rather than those derived from the classic ABC-model of *Arabidopsis* and other core eudicots.

VII. DERIVATION OF THE ABC-MODEL

In general, the B- and C-components of the ABC-model appear to apply to all basal angiosperms studied to date and therefore must have been part of the ancestral floral developmental program. For example, B-function homologs are apparently necessary for development of petaloid organs and stamens, and C-function homologs are expressed in stamens and carpels of all species examined. The A-function component of the ABC-model, however, has not been identified in basal angiosperms, or, in fact, in any species other than *Arabidopsis thaliana*. *AP1* homologs are not expressed in the perianth of basal angiosperms, although *AP2* is active in at least some basal taxa (Kim *et al.*, 2006). It is possible that *AGL6* supplies some or all of the A-function in at least some basal angiosperms. C-function is evidently repressed in the perianth of basal angiosperms, but the genetic basis of this repression is unknown. Repression of C-function in the perianth may have evolved in as many varied ways as the origin of the perianth organs themselves, which have arisen multiple times in angiosperm evolution (Albert *et al.*, 1998; Ronse De Craene *et al.*, 2003; Soltis *et al.*, 2005a).

The expression patterns of B-, C-, and E-homologs in most basal angiosperms are often broader than those specified by the ABCE-model. Therefore, these genes may have a conserved component that corresponds to their known function in genetic models as well as a broader component. We therefore hypothesize that the ABCE-model of *Arabidopsis* and *Antirrhinum* was derived from an ancestral genetic program that expressed floral regulators broadly across the floral meristem (Kim *et al.*, 2005a). Restrictions in the extent and pattern of gene expression ultimately resulted in the specification of discrete floral whorls controlled by the combinatorial actions of these floral regulators: the ABCE-model of *Arabidopsis*.

From this view of the ancestral floral genetic program, it is possible to envision the subsequent modifications that led to the evolution of diverse floral morphologies. For example, in the magnoliid *Asimina* and in the core eudicots, restriction of gene expression to localized areas appears to result in flowers with distinct and differentiated sepals, petals, stamens, and carpels. This innovation obviously represents an independent derivation of a similar floral structure in these distant lineages. Likewise, similar expression patterns in monocots (*Tulipa* and *Lilium*) and basal eudicots (*Ranunculus* and *Aquilegia*) support independent derivations of the sliding boundary model of gene expression in which B-function has been broadly retained across the meristem, perhaps while other floral regulators have become restricted, rather than outward movement of the B-function boundary previously envisioned (Kanno *et al.*, 2003; Kramer *et al.*, 2003; van Tunen *et al.*, 1993). This view assumes that broad B-expression in monocots and eudicots is homologous and is retained from their common ancestor (Kim *et al.*, 2005a). Gradual intergradation of floral organs, as in *Amborella*, can best be explained by broad, overlapping patterns of gene expression, as predicted by the fading borders model (Buzgo *et al.*, 2004; Kim *et al.*, 2005a).

Although fundamental components of the floral genetic/developmental program have been conserved across the angiosperms, evolutionary diversification of the hypothesized ancestral program may have occurred through localized expression of different regulators to different regions of the meristem. We should therefore view the ABC-model of *Arabidopsis*, with its discrete A-, B-, C-, and E-functions, as a derived condition. Likewise, we should expect that divergent floral morphologies may be controlled by underlying shifts in the timing and/or location of expression of floral regulators. In fact, no single model should be expected to explain all of floral diversity; it is likely that fine-tuning of the ancestral program has yielded a multitude of divergent forms. Perhaps it is time to turn our attention to those factors that control the location and timing of organ identity genes, that is, the regulators of the regulators (see Zahn *et al.*, Chapter 4).

VIII. FUTURE DIRECTIONS

A. THE ROLE OF GENE DUPLICATIONS

Many crucial genes that control floral initiation and development appear to have been duplicated either just prior to, or very early in, angiosperm evolution (see Irish, Chapter 3, and Kramer and Zimmer, Chapter 9). The B-function paralogs *AP3* and *PI* originated by duplication prior to the origin of the angiosperms, perhaps as much as 260 million years before present (Kim *et al.*, 2004). Likewise, *SEP* genes were duplicated to form the *AGL2/3/4* (*SEP1/2/4*) and *AGL9* (*SEP3*) lineages in the common ancestor of the angiosperms (Zahn *et al.*, 2005). The corresponding duplications of these key floral regulators prior to the origin of the angiosperms may have allowed diversification of the plant reproductive program, resulting in the early origin of the flower (Zahn *et al.*, 2005). Genomics data suggest that a genome-wide duplication may have preceded the origin of the angiosperms (Bowers *et al.*, 2003), and the duplicate floral regulators may therefore have arisen via polyploidy. It is possible that this genome duplication was a catalyst for the origin and early diversification of the angiosperms (Buzgo *et al.*, 2005; DeBodt *et al.*, 2005; Zahn *et al.*, 2005). Many other floral regulators have also undergone duplication (e.g., *AP2*; Kim *et al.*, 2006).

Additional floral regulators experienced duplication at a later time (*AP1*, Litt and Irish, 2003; *AP3*, Kramer *et al.*, 1998; *AG*, Kramer *et al.*, 2004), near the origin of the eudicots, a major clade that comprises 75% of all flowering plants (Drinnan *et al.*, 1994). These duplicatons may have resulted from another genome duplication (Bowers *et al.*, 2003). Changes in floral structure and rapid diversification seem to have followed closely on the heels of this duplication event, suggesting yet another example of floral diversification associated with gene duplications. Given the importance of floral morphology in angiosperm diversification, understanding the role of gene duplication in floral diversification will be key to understanding angiosperm evolution.

B. TRANSCRIPTION FACTOR COMPLEXES AND THE ORIGIN OF THE ABC-MODEL: TESTABLE HYPOTHESES

The quartet model (Theissen, 2001) proposes that SEP proteins form heterotetramers with either A- and B-proteins (for petals), B- and C-proteins (for stamens), and C-protein (for carpels), and evidence from *Arabidopsis* supports this model (Honma and Goto, 2001). In *Arabidopsis*, the heterotetramers involving B-proteins include both AP3 and PI, plus SEP and either an A- or C-protein, depending on its role in petal or stamen identity. Furthermore, the B-proteins obligately heterodimerize—that is, the tetramer will not contain two AP3 or two PI proteins; the K domain of these MADS-box genes contains signals that prevent homodimerization in *Arabidopsis*. In *Amborella*, however, the K-domain heterodimerization signals may not be as strong as in *Arabidopsis*, based on sequence similarities between *AP3* and *PI* amino acids, and sequence similarities in both the K and C domains have prompted the suggestion (Soltis *et al.*, 2005b) that *Amborella* B-function proteins may have different dimerization dynamics than core eudicots, such as *Arabidopsis*, and monocots, at least some of which can form PI/PI homodimers (Kanno *et al.*, 2003; Winter *et al.*, 2002). Both AP3 and PI may form homodimers in *Amborella*—AP3-1/AP3-1, AP3-2/AP3-2, and PI/PI.

If *AP3* and *PI* sequences are too similar for discrimination, then perhaps all possible combinations of AP3 and PI proteins are possible: AP3-1/AP3-1, AP3-2/AP3-2, and PI/PI homodimers and AP3-1/AP3-2, AP3-1/PI, and AP3-2/PI heterodimers. This array is much larger than the number of possible combinations in the obligately dimerizing core eudicots (one possible heterodimer, AP3/PI, in *Arabidopsis*) and even the monocots (AP3/PI heterodimer and PI/PI homodimer). Because *Nuphar AP3* and *PI* sequences contain similar signals to those of *Amborella*, it appears that the early angiosperms may have had greater flexibility in B-function protein complexes than are present in the core eudicots (Soltis *et al.*, 2005b). The derivation of obligate heterodimerization from homodimerizing ancestors is consistent with evidence from conifers (Sundström *et al.*, 1999) and Gnetales (Winter *et al.*, 2002). In gymnosperms, the single B-function homolog homodimerizes; heterodimerization may have been derived following the duplication that gave rise to the *AP3* and *PI* gene lineages (Winter *et al.*, 2002).

The possible multimeric combinations in *Amborella* may be even more extensive than previously suggested (Soltis *et al.*, 2005b), when the quartet structure is considered. If the X and Y subunits (SEP and A- or C-proteins) of the functioning quartet (Theissen, 2001) do not discriminate between B-function proteins in *Amborella*, then additional quartet types may be possible (Fig. 4)—21 in *Amborella* compared with 1 in *Arabidopsis* and 4 in monocots. In fact, the composition of possible active multimers may only be limited by: (1) available MADS translation products in cells and (2) the capacity of different AP3 and PI dimers to recognize these translation products for effective multimerization. Furthermore, the stoichiometry of AP3 and PI subunits may vary across developing floral organs. These proposed interactions in *Amborella*, and perhaps *Nuphar*, between pairs of B-function proteins and between B-proteins and SEP and other subunits of the quartet model are testable, for example using yeast two- and three-hybrid assays and/or immunolocalization.

IX. CONCLUSIONS

The evolutionary history of the floral developmental program is beginning to emerge, through the combined efforts of paleobotany, developmental morphology, phylogenetics, functional genetics, and comparative genomics. As information from these fields has accumulated, integration has focused on the *Arabidopsis*-based ABC-model of floral organ identity as a point of reference for inferring ancestral and derived aspects of the floral genetic

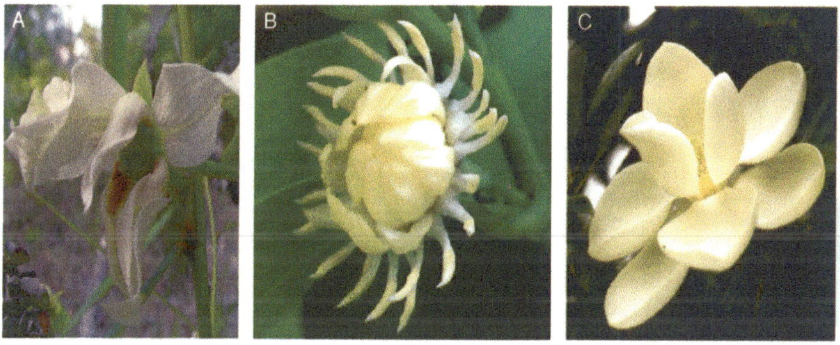

Plate 17

A

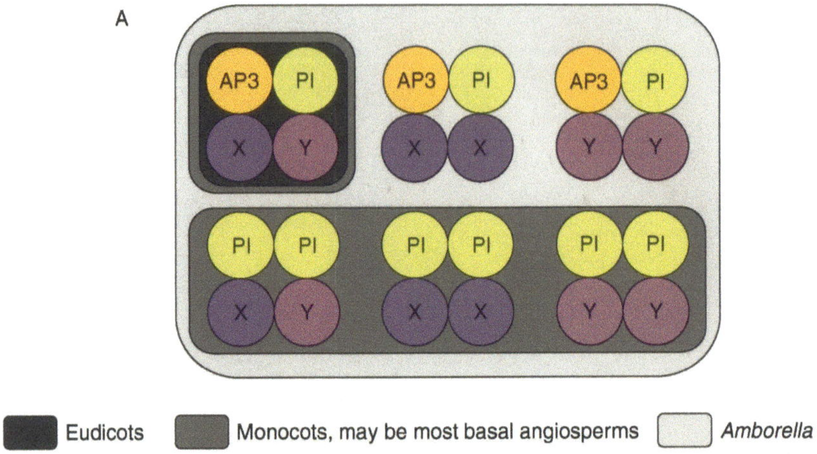

B

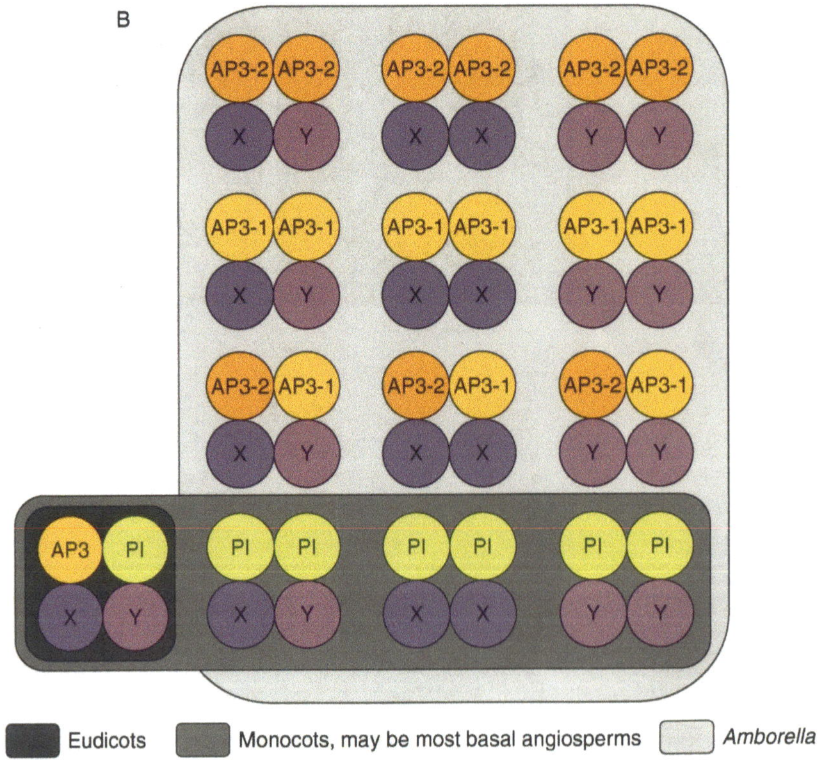

Plate 18 (*Continued*)

C

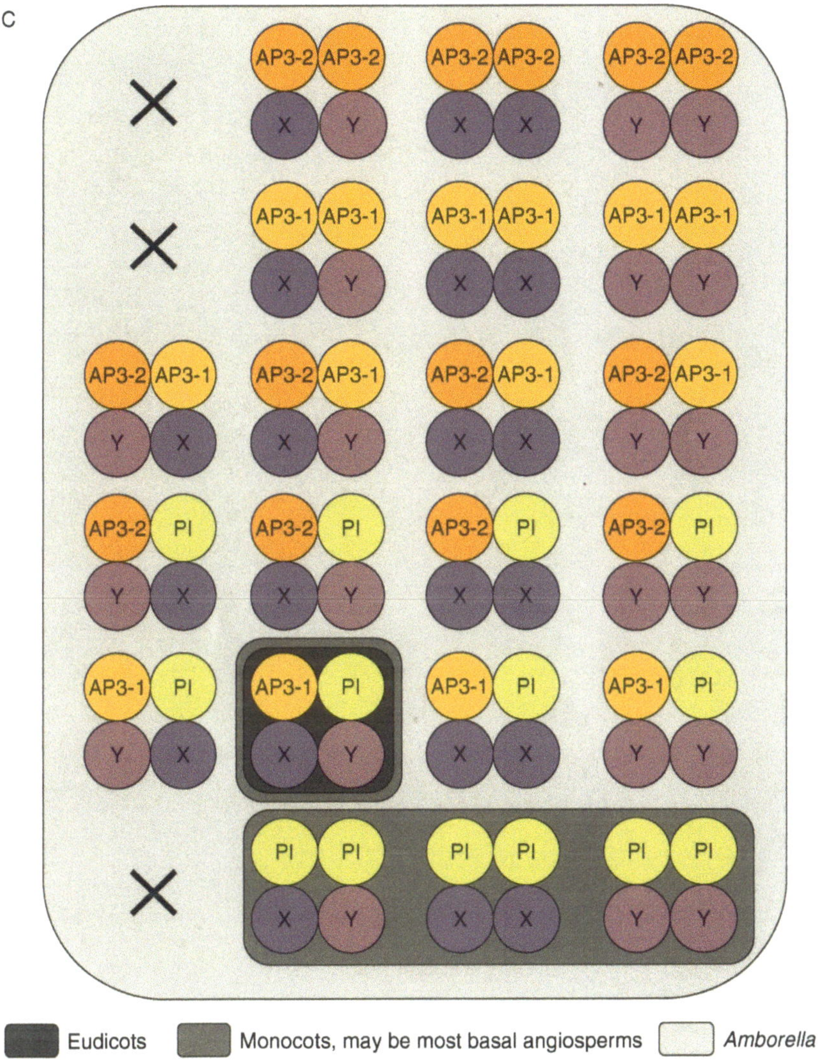
AP3-2 AP3-2 X Y
AP3-2 AP3-2 X X
AP3-2 AP3-2 Y Y
AP3-1 AP3-1 X Y
AP3-1 AP3-1 X X
AP3-1 AP3-1 Y Y
AP3-2 AP3-1 Y X
AP3-2 AP3-1 X Y
AP3-2 AP3-1 X X
AP3-2 AP3-1 Y Y
AP3-2 PI Y X
AP3-2 PI X Y
AP3-2 PI X X
AP3-2 PI Y Y
AP3-1 PI Y X
AP3-1 PI X Y
AP3-1 PI X X
AP3-1 PI Y Y
PI PI X Y
PI PI X X
PI PI Y Y
Eudicots
Monocots, may be most basal angiosperms
Amborella

Plate 18

A

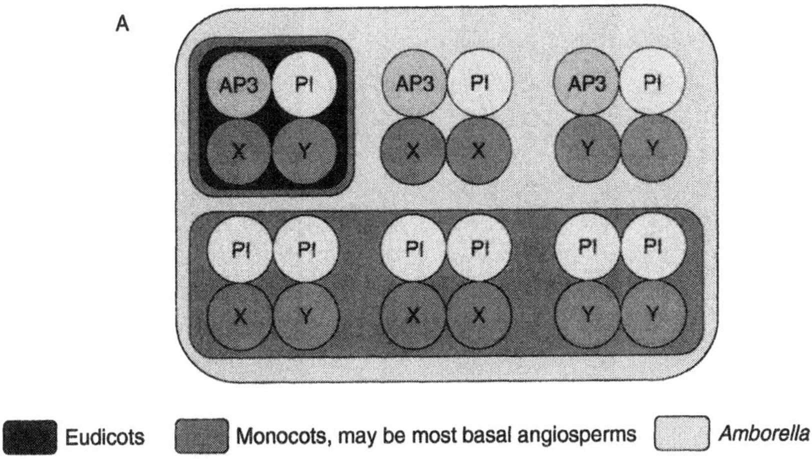

B

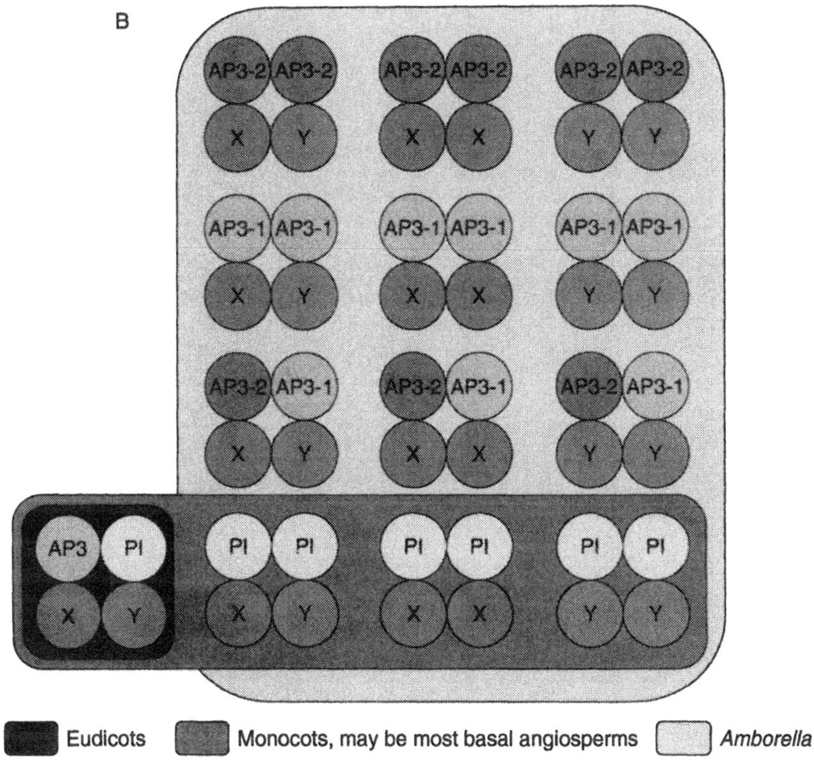

Fig. 4. (*Continued*)

C

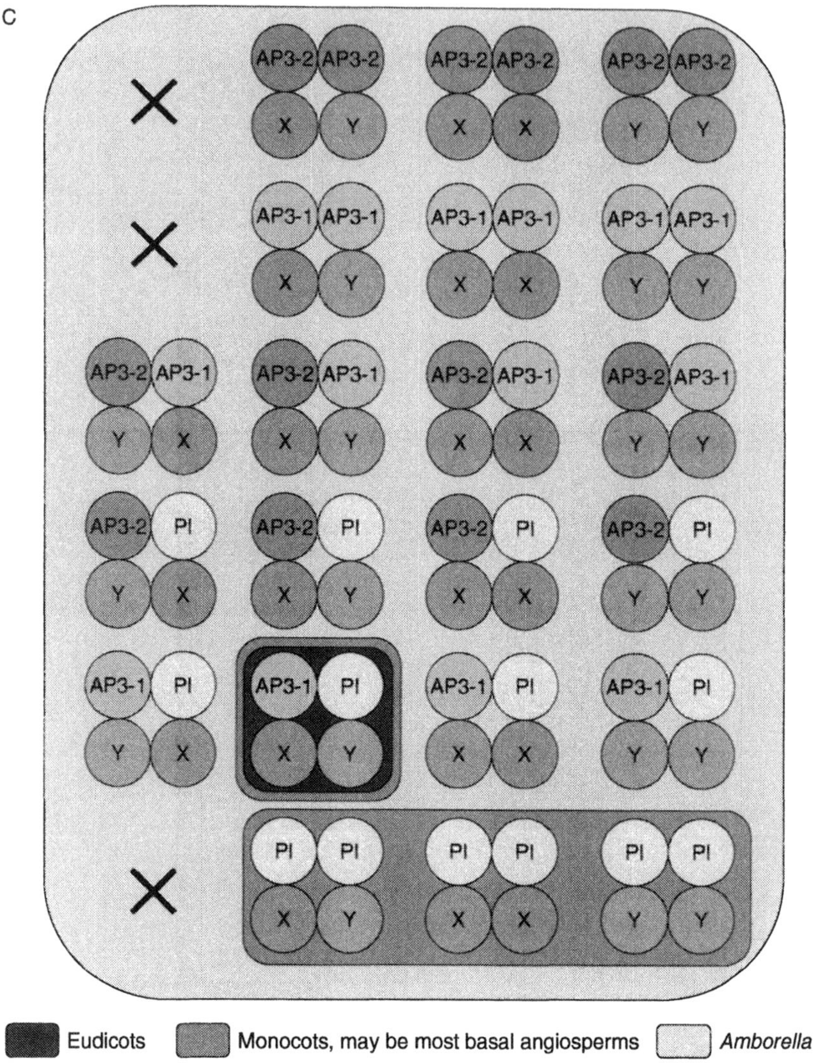

Fig. 4. Transcription factor complexes. (A) Obligate heterodimerization in eudicots, PI–PI homodimerization in monocots and possibly other basal angiosperms, and additional complexes in *Amborella* formed through XX and YY combinations with AP3 + PI, resulting in a ratio of 6 possibilities in *Amborella* to 4 in monocots (and other basal angiosperms?) to 1 in eudicots. (B) Similar complex formation except that duplicate AP3 proteins are incorporated, resulting in a ratio of 13:4:1. (C) If all components can combine in *Amborella* (and possibly *Nuphar* and other water lilies), even more complexes are possible, for a ratio of 21:4:1. (See color Plate 18.)

program. Phylogenetic studies of floral gene families suggest that most, if not all, of the fundamental genetic components of the ABC-model were present in the ancestral angiosperm genome. Significant differences between the flowers of basal angiosperms and core eudicots are therefore not due to underlying differences in their floral genomes but rather to differences in gene expression and perhaps interaction of gene products. For example, the flowers of many basal angiosperms exhibit gradual transitions among floral organs, with intergradations between adjacent organs rather than distinct organ types arranged in distinct whorls as is typical of *Arabidopsis* and other core eudicots. This fundamental difference in the identity and arrangement of floral parts suggests an underlying difference in the genetic processes that control specification of floral organs, and alternatives to the ABC-model have been proposed to account for these morphological differences. Expression of B- and C-function homologs in basal angiosperms is consistent with predictions of the ABC-model (i.e., B-function genes are expressed in petaloid structures and stamens), but expression is typically broader than observed in *Arabidopsis* and other eudicot models, extending into adjacent organs. This pattern of expression supports the "fading borders" model of floral organ identity (Buzgo *et al.*, 2004), one of the alternatives to the strict ABC-model, especially for flowers with gradual transitions among organs.

Given the phylogenetic distribution of gene expression patterns, the expression of B-function homologs across the floral meristem seems to represent the ancestral situation. The restriction of an initially broad deployment of floral organ identity genes across the developing flower to discrete domains has resulted in the distinct floral organs and whorled organization typical of core eudicot flowers.

Restriction in dimerization potential among floral regulators appears to have preceded restriction of gene expression patterns to specific locations on the meristem. Divergence in the C-terminal domains of *AP3* and *PI* appears to have resulted in obligate heterodimerization in core eudicots (cf. Tzeng *et al.*, 2004). Alternatively, obligate heterodimerization, determined by some other mechanism, may have allowed for divergence in the C-termini of these genes. Regardless of the cause and effect, the following questions remain: (1) is the potential extent of homodimerization/heterodimerization combinations in *Amborella* realized? and (2) at what phylogenetic point did the switch to obligate heterodimerization occur? Observations to date raise testable hypotheses about the relationship between sequence divergence and loss of homodimerization potential. However, several studies have not observed a role for the C-terminus in specifying dimerization (Egea-Cortines *et al.*, 1999; Riechmann *et al.*, 1996a,b; Yang *et al.*, 2003). Likewise, the role of the K domain in preventing homodimerization of B-proteins in

Arabidopsis (Egea-Cortines *et al.*, 1999; Honma and Goto, 2000) has been questioned (Yang and Jack, 2004), raising uncertainty about the roles of sequence and/or motif similarity in both the C and K domains.

Emerging data indicate a strong correspondence between expression patterns of floral regulators and morphology, at multiple levels of complexity. For example, at the organ level, petal-like structures, regardless of location in the flower (outer tepals, inner tepals, petals, staminodes), require B-function. At the level of the entire flower, restricted patterns of gene expression, particularly of B- and C-function homologs, results in whorled arrangements of distinct organs, regardless of the phylogenetic homology of these structures. Flowers of *Asimina* (Annonaceae, Magnoliales) and *Arabidopsis* (Brassicaceae, Malvales) exhibit similar floral whorls and patterns of gene expression despite obviously independent origins of their developmental genetic programs. Finally, comparison of gene expression patterns in close relatives may provide clues to the origin of novel structures, such as the calyptra of *Eupomatia*.

Despite exciting progress in understanding the origin and diversification of the floral genetic program, fundamental problems remain. Without adequate genetic models outside of core eudicots and grasses, we are left with inferences about the possible functions of floral regulators in basal angiosperms. But does expression necessarily imply function? Possible tests of this assumption include studies of protein production using antibodies—identifying protein localization is one step closer to function than expression patterns. Furthermore, the development of new model plants from among the basal lineages of angiosperms will permit true functional studies. Successful transformation of *Persea americana* (the avocado), in conjunction with special grafting techniques, permits functional studies of key genes, including floral regulators, in a tree species in a reasonable timeframe (Litz *et al.*, 2005). Likewise, *Aristolochia* is an experimentally tractable basal angiosperm, but with a relatively short generation time. These new models, coupled with increasing knowledge of the floral genome from comparative studies, will pave the way for new and comprehensive studies of the evolution of the flower and its genetic control.

REFERENCES

Albert, V. A., Gustafsson, M. H. G. and Di Laurenzio, L. (1998). Ontogenetic systematics, molecular developmental genetics, and the angiosperm petal. *In* "Molecular Systematics of Plants II" (P. S. Soltis, D. E. Soltis and J. J. Doyle, eds.), pp. 349–374. Kluwer Academic Publishers, Boston.

Albert, V. A., Soltis, D. E., Carlson, J. E., Farmerie, W. G., Wall, P. K., Ilut, D. C., Solow, T. M., Mueller, L. A., Landherr, L. L., Hu, Y., Buzgo, M., Kim, S., *et al.* (2005). Floral gene resources from basal angiosperms for comparative genomics research. *BMC Plant Biology* **5**, 1–15.

Ambrose, B. A., Lerner, D. R., Ciceri, P., Padilla, C. M., Yanofsky, M. F. and Schmidt, R. J. (2000). Molecular and genetic analyses of the *Silky1* gene reveal conservation in floral organ specification between eudicots and monocots. *Molecular Cell* **5**, 569–579.

Barkman, T., Chenery, J. G., Mcneal, J. R., Lyons-Weiler, J. and De Pamphilis, C. W. (2000). Independent and combined analyses of sequences from all three genomic compartments converge on the root of flowering plant phylogeny. *Proceedings of the National Academy of Sciences of the United States of America* **97**, 13166–13171.

Becker, A. and Theissen, G. (2003). The major clades of MADS-box genes and their role in the development and evolution of flowering plants. *Molecular Phylogenetics and Evolution* **29**, 464–489.

Bowers, J. E., Chapman, B. A., Rong, J. and Paterson, A. H. (2003). Unraveling angiosperm genome evolution by phylogenetic analysis of chromosomal duplication events. *Nature* **422**, 433–438.

Bowman, J. L. (1997). Evolutionary conservation of angiosperm flower development at the molecular and genetic levels. *Journal of Biosciences* **22**, 515–527.

Bowman, J. L., Alvarez, J., Weigel, D., Meyerowitz, E. M. and Smyth, D. R. (1993). Control of flower development in *Arabidopsis thaliana* by *APETALA1* and interacting genes. *Development* **119**, 721–743.

Buzgo, M., Soltis, D. E. and Soltis, P. S. (2004). Floral developmental morphology of *Amborella trichopoda* (Amborellaceae). *International Journal of Plant Sciences* **165**, 925–947.

Buzgo, M., Soltis, P. S., Kim, S. and Soltis, D. E. (2005). The making of the flower. *Biologist* **52**, 149–154.

Coen, E. S. and Meyerowitz, E. M. (1991). The war of the whorls: Genetic interactions controlling flower development. *Nature* **353**, 31–37.

Colombo, L., Franken, J., Koetje, E., Vanwent, J., Dons, H. J. M., Angenent, G. C. and Vantunen, A. J. (1995). The petunia MADS box gene *FBP11* determines ovule identity. *Plant Cell* **7**, 1859–1868.

Cronquist, A. (1988). "The Evolution and Classification of Flowering Plants," 2nd Edn. New York Botanical Garden, Bronx.

DeBodt, S., Maere, S. and Van de Peer, Y. (2005). Genome duplication and the origin of angiosperms. *Trends in Ecology and Evolution* **20**, 591–597.

Ditta, G., Pinyopich, A., Robles, P., Pelaz, S. and Yanofsky, M. F. (2004). The *SEP4* gene of *Arabidopsis thaliana* functions in floral organ and meristem identity. *Current Biology* **14**, 1935–1940.

Drinnan, A. N., Crane, P. R. and Hoot, S. B. (1994). Patterns of floral evolution in the early diversification of non-magnoliid dicotyledons (eudicots). *Plant Systematics and Evolution* **8** (Suppl.), 93–122.

Egea-Cortines, M., Saedler, H. and Sommer, H. (1999). Ternary complex formation between the MADS-box proteins SQUAMOSA, DEFICIENS GLOBOSA is involved in the control of floral architecture in *Antirrhinum majus*. *EMBO Journal* **18**, 5370–5379.

Endress, P. K. (2001). The flowers in extant basal angiosperms and inferences on ancestral flowers. *International Journal of Plant Sciences* **162**, 1111–1140.

Endress, P. K. (2003). Early floral development and nature of the calyptra in Eupomatiaceae (Magnoliales). *International Journal of Plant Sciences* **164**, 489–503.

Endress, P. K. and Igersheim, A. (2000). Reproductive structures of the basal angiosperm *Amborella trichopoda* (Amborellaceae). *International Journal of Plant Sciences* **161**, S237–S248.

Ferrandiz, C., Gu, Q., Martienssen, R. and Yanofsky, M. F. (2000). Redundant regulation of meristem identity and plant architecture by *FRUITFULL, APETALA1*, and *CAULIFLOWER*. *Development* **127**, 725–734.

Friis, E. M., Pedersen, K. R. and Crane, P. R. (2000). Reproductive structure and organization of basal angiosperms from the Early Cretaceous (Barremian or Aptian) of Western Portugal. *International Journal of Plant Sciences* **161**, S169–S182.

Gu, Q., Ferrandiz, C., Yanofsky, M. F. and Martienssen, R. (1998). The *FRUITFULL* MADS-box gene mediates cell differentiation during *Arabidopsis* fruit development. *Development* **125**, 1509–1517.

Honma, T. and Goto, K. (2001). Complexes of MADS-box proteins are sufficient to convert leaves into floral organs. *Nature* **409**, 525–529.

Huijser, P., Klein, J., Lonnig, W. E., Meijer, H., Saedler, H. and Sommer, H. (1992). Bracteomania, an inflorescence anomaly, is caused by the loss of function of the MADS-box gene *SQUAMOSA* in *Antirrhinum majus*. *EMBO Journal* **11**, 1239–1249.

Irish, V. F. (2003). The evolution of floral homeotic gene function. *BioEssays* **25**, 637–646.

Irish, V. F. and Sussex, I. M. (1990). Function of the *apetala-1* gene during *Arabidopsis* floral development. *Plant Cell* **2**, 741–753.

Jaramillo, M. A. and Kramer, E. M. (2004). *APETALA3* and *PISTILLATA* homologs exhibit novel expression patterns in the unique perianth in Aristolochia (Aristolochiaceae). *Evolution & Development* **6**, 449–458.

Kanno, A., Saeki, H., Kameya, T., Saedler, H. and Theissen, G. (2003). Heterotopic expression of class B floral homeotic genes supports a modified ABC model for tulip (*Tulipa gesneriana*). *Plant Molecular Biology* **52**, 831–841.

Keck, E., McSteen, P., Carpenter, R. and Coen, E. (2003). Separation of genetic functions controlling organ identity in flowers. *EMBO Journal* **22**, 1058–1066.

Kim, S., Yoo, M.-J., Albert, V. A., Farris, J. S., Soltis, P. S. and Soltis, D. E. (2004). Phylogeny and diversification of B-function MADS-box genes in angiosperms: Evolutionary and functional implications of a 260-million-year-old duplication. *American Journal of Botany* **91**, 2102–2118.

Kim, S., Koh, J., Yoo, M-J., Kong, H., Hu, Y., Ma, H., Soltis, P. S. and Soltis, D. E. (2005a). Expression of floral MADS-box genes in basal angiosperms: Implications for the evolution of floral regulators. *Plant Journal* **43**, 724–744.

Kim, S., Koh, J., Ma, H., Hu, Y., Endress, P. K., Buzgo, M., Hauser, B. A., Soltis, P. S. and Soltis, D. E. (2005b). Sequence and expression studies of A-, B-, and E-class MADS-box genes in *Eupomatia* (Eupomatiaceae): Support for the bractate origin of the calyptra. *International Journal of Plant Sciences* **166**, 185–198.

Kim, S., Soltis, P. S., Wall, K. and Soltis, D. E. (2006). Phylogeny and domain evolution in the *APETALA2*-like gene family. *Molecular Biology and Evolution* **23**, 107–120.

Kramer, E. M. and Hall, J. C. (2005). Evolutionary dynamics of genes controlling floral development. *Current Opinion in Plant Biology* **8**, 13–18.

Kramer, E. M., Dorit, R. L. and Irish, V. F. (1998). Molecular evolution of genes controlling petal and stamen development: Duplication and divergence within the *APETALA3* and *PISTILLATA* MADS-box gene lineages. *Genetics* **149**, 765–783.

Kramer, E. M., Di Stilio, V. S. and Schluter, P. M. (2003). Complex patterns of gene duplication in the *APETALA3* and *PISTILLATA* lineages of the Ranunculaceae. *International Journal of Plant Sciences* **164**, 1–11.

Kramer, E. M., Jaramillo, M. A. and Di Stilio, V. S. (2004). Patterns of gene duplication and functional evolution during the diversification of the *AGAMOUS* subfamily of MADS box genes in angiosperms. *Genetics* **166**, 1011–1023.

Li, G. S., Meng, Z., Kong, H., Chen, Z, Theissen, G. and Lu, A. M. (2005). Characterization of candidate class A, B and E floral homeotic genes from the perianthless basal angiosperm Chloranthus spicatus (Chloranthaceae). *Development Genes and Evolution* **215**, 437–449.

Litt, A. and Irish, V. F. (2003). Duplication and diversification in the *APETALA1/FRUITFULL* floral homeotic gene lineage: Implication for the evolution of floral development. *Genetics* **165**, 821–833.

Litz, R. E., Witjaksono, Raharjo, S., Efendi, D., Pliego Alfaro, F. and Barcelo-Munoz, A. (2005). *Persea americana* Avocado. *In* "Biotechnology of Fruit and Nut Crops" (R. E. Litz, ed.), pp. 326–349. CABI Publishing, Wallingford, UK.

Ma, H. and dePamphilis, C. (2000). The ABCs of floral evolution. *Cell* **101**, 5–8.

Park, J. H., Ishikawa, Y., Ochiai, T., Kanno, A. and Kameya, T. (2004). Two *GLOBOSA*-like genes are expressed in second and third whorls of homochlamydeous flowers in *Asparagus officinalis* L. *Plant Cell Physiology* **45**, 325–332.

Pelaz, S., Ditta, G. S., Baumann, E., Wisman, E. and Yanofsky, M. F. (2000). B and C floral organ identity functions require *SEPALLATA* MADS-box genes. *Nature* **405**, 200–203.

Pinyopich, A., Ditta, G. S., Savidge, B., Liljegren, S. J., Baumann, E., Wisman, E. and Yanofsky, M. F. (2003). Assessing the redundancy of MADS-box genes during carpel and ovlule development. *Nature* **424**, 85–88.

Posluszny, U. and Tomlinson, P. B. (2003). Aspects of inflorescence and floral development in the putative basal angiosperm *Amborella trichopoda* (Amborellaceae). *Canadian Journal of Botany* **81**, 28–39.

Qiu, Y.-L., Lee, J., Bernasconi-Quadroni, F., Soltis, D. E., Soltis, P. S., Zanis, M., Zimmer, E. A., Chen, Z., Savolainen, V. and Chase, M. W. (1999). The earliest angiosperms: Evidence from mitochondrial, plastid and nuclear genomes. *Nature* **402**, 404–407.

Riechmann, J. L., Krizek, B. A. and Meyerowitz, E. M. (1996a). Dimerization specificity of *Arabidopsis* MADS domain homeotic proteins APETALA1, APETALA3, PISTILLATA, and AGAMOUS. *Proceedings of the National Academy of Sciences of the United States of America* **93**, 4793–4798.

Riechmann, J. L., Wang, M. and Meyerowitz, E. M. (1996b). DNA-binding properties of Arabidopsis MADS domain homeotic proteins APETALA1, APETALA3, PISTILLATE, and AGAMOUS. *Nucleic Acids Research* **24**, 3134–3141.

Ronse De Craene, L. P. R., Soltis, P. S. and Soltis, D. E. (2003). Evolution of floral structures in basal angiosperms. *International Journal of Plant Sciences* **164**, S329–S363.

Rounsley, S. D., Ditta, G. S. and Yanofsky, M. F. (1995). Diverse roles for MADS box genes in *Arabidopsis* development. *Plant Cell* **7**, 1259–1269.

Soltis, D. E., Soltis, P. S., Chase, M. W., Mort, M. E., Albach, D. C., Zanis, M., Savolainen, V., Hahn, W. H., Hoot, S. B., Fay, M. F., Axtell, M. Swensen, S. M., *et al.* (2000). Angiosperm phylogeny inferred from a combined data set of 18S rDNA, *rbcL* and *atpB* sequences. *Botanical Journal of the Linnean Society* **133**, 381–461.

Soltis, D. E., Soltis, P. S., Endress, P. K. and Chase, M. W. (2005a). "Phylogeny and Evolution of Angiosperms." Sinauer, Sunderland, MA.

Soltis, D. E., Albert, V. A., Kim, S., Yoo, M-J., Soltis, P. S., Frohlich, M. W., Leebens-Mack, J., Kong, H., Wall, K., dePamphilis, C. and Ma, H. (2005b). Evolution of the flower. *In* "Plant Diversity and Evolution" (R. J. Henry, ed.), pp. 165–200. CABI Publishing, Wallingford, UK.

Soltis, P. S., Soltis, D. E. and Chase, M. W. (1999). Angiosperm phylogeny inferred from multiple genes as a tool for comparative biology. *Nature* **402**, 402–404.

Sundström, J., Carlsecker, A., Svensson, M. E., Svenson, M., Johanson, U., Theissen, G. and Engström, P. (1999). MADS-box genes active in developing pollen cones of Norway spruce (*Picea abies*) are homologous to the B-class floral homeotic genes in angiosperms. *Developmental Genetics* **25**, 253–266.

Taylor, S. A., Hofer, J. M. I., Murfet, I. C., Sollinger, J. D., Singer, S. R., Knox, M. R. and Noel Ellis, T. H. (2002). *PROLIFERATING INFLORESCENCE MERISTEM*, a MADS-box gene that regulates floral meristem identity in Pea. *Plant Physiology* **27**, 69–78.

Theissen, G. (2001). Development of floral organ identity: Stories from the MADS house. *Current Opinion in Plant Biology* **4**, 75–85.

Tzeng, T.-U., Liu, H.-D. and Yang, C.-H. (2004). The C-terminal sequences of LMADS1 is essential for the formation of homodimers for B function proteins. *Journal of Biological Chemistry* **279**, 10747–10755.

van Tunen, A. J., Eikelboom, W. and Angenent, G. C. (1993). Floral organogenesis in *Tulipa*. *Flowering Newsletter* **16**, 33–38.

Winter, K-U., Weiser, C., Kaufmann, K., Bohne, A., Kirchner, C., Kanno, A., Saedler, H. and Theissen, G. (2002). Evolution of class B floral homeotic proteins: Obligate heterodimerization originated from homodimerization. *Molecular Biology and Evolution* **19**, 587–596.

Yang, Y. and Jack, T. (2004). Defining subdomains of the K domain important for protein–protein interactions of plant MADS proteins. *Plant Molecular Biology* **55**, 45–59.

Yang, Y., Fanning, L. and Jack, T. (2003). The K domain mediates heterodimerization of the *Arabidopsis* floral organ identity proteins, APETALA3 AND PISTILLATA. *Plant Journal* **33**, 47–59.

Yu, H., Ito, T., Wellmer, F. and Meyerowitz, E. M. (2004). Repression of AGAMOUS-LIKE 24 is a crucial step in promoting flower development. *Nature Genetics* **36**, 157–161.

Zahn, L. M., Kong, H., Leebens-Mack, J. H., Kim, S., Soltis, P. S., Landherr, L. L., Soltis, D. E., dePamphilis, C. W. and Ma, H. (2005). The evolution of the *SEPALLATA* subfamily of MADS-box genes: A pre-angiosperm origin with multiple duplications throughout angiosperm history. *Genetics* **169**, 2209–2223.

Zanis, M. J., Soltis, D. E., Soltis, P. E., Mathews, S. and Donoghue, M. J. (2002). The root of the angiosperms revisited. *Proceedings of the National Academy of Sciences of the United States of America* **99**, 6848–6853.

The Molecular Evolutionary Ecology of Plant Development: Flowering Time in *Arabidopsis thaliana*

KATHLEEN ENGELMANN AND MICHAEL PURUGGANAN

Department of Biology and Center for Comparative and Functional Genomics, New York University, NewYork, New York 10003

ABSTRACT

Flowering time is a major fitness determinant of plants in seasonal habitats. In *Arabidopsis thaliana* flowering time is largely determined by photoperiod, vernalization, and ambient temperature, although foliar shade, water availability, and herbivory can also have an effect. There is selection on flowering time via both mortality and fruit production, and typically selection favors flowering time plasticity. Much of the variation in flowering time can be attributed to molecular variation in the genes that are responsible for sensing light and temperature, and many of these genes owe their discovery to these effects as determined by quantitative trait locus (QTL) mapping. Not surprisingly, many flowering time QTLs are environment-dependent. Molecular analyses of the genes underlying the response of flowering time to the environment provide further evidence that these genes have been repeated targets of natural selection.

Advances in Botanical Research, Vol. 44
Incorporating Advances in Plant Pathology

0065-2296/06 $35.00
DOI: 10.1016/S0065-2296(06)44013-1

I. INTRODUCTION

The onset of flowering, that is, the change from vegetative to reproductive development, is a major life history transition in flowering plants and is sensitive to various seasonal climatic signals (Koornneef *et al.*, 2004). Flowering phenology is critically tied to the reproductive ecology of flowering plants, and is a central feature in the evolutionary trajectory of many angiosperm species (Murfet, 1977). Moreover, the shift to flowering represents a major developmental transition that can reshape the architecture of the plant, and change its interaction with the biotic and abiotic environment. The study of flowering time provides an opportunity to investigate the diversification of a key developmental process in both molecular genetic and ecological contexts, and results in a synthetic view that encompasses the molecular evolutionary ecology of development.

II. EVOLUTIONARY ECOLOGY OF FLOWERING TIME

Seasonal habitats exhibit regular annual fluctuations in precipitation, temperature, day length, length of growing season, and potential disturbance due to storm activity or flooding. Some habitats may experience variation in more than one of these environmental parameters (e.g., a cool, dry season and a warm, wet season), and in some cases fluctuations can be extreme. Dry conditions can lead to desiccation whereas flooding can lead to anoxia, fungal growth, or can disrupt seed dispersal and seedling recruitment. Changes in daylength can reduce the amount and quality of photosynthetically active radiation available, therefore affecting growth rate. Plants in seasonal environments may also experience season-dependent fluctuations in predators and competitors.

Most plants living in seasonal environments have adapted to these changes in one of two ways: they have constitutive mechanisms that increase their range of environmental tolerances or they have plastic responses that are timed with seasonal changes (Alpert and Simms, 2002; Murfet, 1977). Plastic responses may be induced by direct exposure to the environmental condition (e.g., shade, predators) or may be timed to other cues that reliably predict seasonal shifts. In temperate environments, there are four major cues for anticipating seasonality: temperature, day length, change in temperature, and change in daylength. In other environments, water availability may also be an important seasonal cue.

Flowering time is an important determinant of fitness in a variable environment, and represents a discrete developmental transition in response to ecological cues. In outcrossing species, it is critical that flowering be timed such that both pollinators and other flowering individuals are present (Rathcke and Lacey, 1985). Even in selfing species, if the timing of flowering is correlated with the timing of seed set, timing of flowering can determine the conditions under which seeds will be dispersed and germinate. Not surprisingly, flowering time is a complex trait controlled by both internal states (developmental and physiologic status of the plant) as well as external conditions.

Arabidopsis thaliana (L.) Heynh. (family Brassicaceae) is a weedy annual plant, most often found in disturbed habitats such as the margins of agricultural fields. It has become a model system for the study of the molecular evolutionary ecology and genetics of plant adaptation (Mitchell-Olds, 2001; Pigliucci, 1998; Shimizu and Purugganan, 2005). *A. thaliana* is characterized by small size and rapid growth, able to complete its life cycle in less than six weeks depending on strain and conditions, and a low outcrossing rate, ~1% (compiled by Hoffmann *et al.*, 2003). *A. thaliana* is estimated to have diverged from other *Arabidopsis* species 5–6 million years ago (Hoffmann, 2002). Its native range extends across Eurasia and Northern Africa, although its naturalized range is much more extensive, including North America and Japan (Hoffmann, 2002). Genetic analysis performed by Sharbel *et al.* (2000) indicates that two major post-Pleistocene expansions in the species range occurred from glacial refugia in the Iberian Peninsula and Asia ~17,000 years ago. However, studies (Nordborg *et al.*, 2005; Schmid *et al.*, 2005) suggest that the post-Pleistocene expansion of *A. thaliana* may be more complex.

In *A. thaliana*, the early stage of the life cycle is a vegetative phase in which the shoot apical meristem produces rosette leaves. As the life cycle proceeds, external ecological cues or internal signals trigger the reproductive developmental transition, and the shoot apical meristem begins to produce the inflorescence, with the associated elongation or bolting of the main shoot. Life histories of *A. thaliana* ecotypes can be classified into three main flowering strategies: winter annual, summer annual, and rapid cycling. The winter annual strategy is prevalent, particularly at southern latitudes (Donohue, 2002; Weinig and Schmitt, 2004). Winter annuals germinate in the fall and overwinter as rosettes, where they experience both cold temperatures and short day lengths. Flowering is delayed until early spring. Fruits ripen and the plants die prior to the onset of the summer heat. The summer annual plants germinate in spring and grow quickly to maturity, dropping seeds in the fall of that same season. Rapid cycling plants germinate in fall, quickly mature, and drop their seeds prior to the onset of winter. Which life history a given ecotype will adopt, the timing of flowering within that life history, and therefore the reproductive success of that ecotype, depend on the interactions of photoperiod and temperature and, to a lesser extent, water availability with the genetic background of the population under study.

Both summer annual and rapid-cycling life histories have been observed in northern latitudes or at high altitudes (Donohue, 2002; Griffith *et al.*, 2004; Nordborg and Bergelson, 1999). Donohue (2002) has shown that when plants adopt an inappropriate strategy, mortality prior to initiation of flowering was 100% under natural conditions. In a similar study, Griffith *et al.* (2004) showed very poor survivorship and fruit production in spring germinating plants at a Kentucky field site. In fall-germinated plants, they showed selection both for earlier bolting and for bolting at a larger size. Winter annual plants often flower at a later developmental stage (Diggle, 1999; Weinig and Schmitt, 2004), that is, greater leaf number and typically larger size, leading to increased fruit production (Engelmann, K., and Purugganan, M., unpublished data; Ungerer *et al.*, 2003).

Flowering time variation is a classic quantitative trait, with phenotypes among individuals continuously distributed rather than qualitatively differentiated. The range of variation in flowering time can be large. Nordborg and Bergelson (1999) reported means ranging from 35 to 251 days, with some individuals not flowering at all under their experimental conditions. The broad sense heritability for bolting time among a set of recombinant inbred lines of *A. thaliana* ecotypes has been estimated to range from 0.07 to 0.744 (Ungerer *et al.*, 2002; Van Berloo and Stam, 1999). Rosette leaf number at bolting, widely regarded as a developmental surrogate for bolting time, is also highly variable, with means ranging from 4.2 to 70.9 leaves depending on conditions and strain (Karlsson *et al.*, 1993). Ungerer *et al.* (2002) also estimated the heritability for this trait in two sets of recombinant inbred lines and results ranged from 0.396 to 0.534.

Photoperiod or daylength is a major seasonal cue for flowering in *A. thaliana* and many studies have compared the fitnesses of plants grown under short vs long days. Studies in controlled environment growth chambers often show an increase in size and fruit production under short day lengths (Engelmann, K., and Purugganan, M., unpublished data; Ungerer *et al.*, 2003). Although it is important to bear in mind that in most field studies the effects of photoperiod are confounded with those of temperature, many studies have also showed increased fitness in winter annual plants. Donohue (2002) found that locally adapted strains of *A. thaliana* have increased survivorship and fruit production when grown in winter field conditions vs spring conditions. Weinig *et al.* (2003b) planted 98 *Ler* × Col recombinant inbred lines at field sites in Rhode Island and North Carolina and found spring germinants attained higher fitness at the northern Rhode

Island site, while winter annuals had higher fitness at the southern North Carolina site, perhaps due to the hotter, drier spring conditions at the southern field site. Studies done under 24 hr light are somewhat more difficult to interpret, as "long day" conditions can occur in both spring and fall. Nevertheless, Ungerer and Rieseberg (2003) showed there is selection for early flowering under long (24 hr) days. Collectively, these findings are consistent with the notion that a winter annual strategy is adaptive for *Arabidopsis* at a range of latitudes.

While selection for earlier bolting is common under most of the environments that have been studied, both photoperiod and light quality can affect optimal flowering time in a given environment. Callahan and Pigliucci (2002) found selection for earlier flowering in greenhouse studies done on local strains in Tennessee; however these seeds were planted in late January, making it difficult to compare with winter or summer annual life histories. They also found selection for earlier bolting, based on mortality, in two field plots, both planted in November. Selection for early bolting was more consistent in the shaded field site. Dorn *et al.* (2000) found that earlier reproduction at a smaller size under foliar shade (and long days) was consistent with selection under these conditions whereas later reproduction in full sun, where there are presumably fewer competitors, was associated with increased fitness.

Many plants in temperate habitats require a period of cold exposure in order to flower, a mechanism known as vernalization, to ensure that flowering takes place after winter has passed. In *A. thaliana*, vernalization is not required for flowering, but in many ecotypes vernalization substantially shortens the time to initiation of flowering. Stratification, that is, the exposure of seeds to cold period, has a similar effect on flowering time in many but not all ecotypes. Nordborg and Bergelson (1999) found that both treatments induced earlier flowering, both required about 30 days for a significant effect and they noted that combining seed and rosette cold treatment was not additive. Furthermore, all plants that responded to cold treatment as rosettes also responded to cold treatment as seeds, although the reverse was not always true.

Pigliucci and Marlow (2001) and Callahan *et al.* (2004) showed that under long days vernalization confers earlier flowering and a correlated increase in fitness, measured as fruit production. Callahan *et al.* (2004) also showed that nonplastic genotypes had a similar time to flowering and fitness as the plastic genotypes when fully vernalized.

Flowering time also responds plastically to ambient temperature, although the fitness consequences of this flexibility are not always clear. Westerman and Lawrence (1970) tested 33 inbred lines under long days and found that days to flowering and rosette leaf number at flowering decrease as ambient temperature increases. Stinchcombe *et al.* (2004a) reanalyzed the data of Westerman and Lawrence (1970) and showed that there was selection against flowering time plasticity, but only from 20 °C to 25 °C, which may only be representative of the extreme southern range of the distribution of *A. thaliana*. There was no such evidence of selection against plasticity from 15 °C to 20 °C.

In *Arabidopsis*, plants receiving limited water flower earlier than plants receiving liberal amounts of water (Engelmann and Schlichting, 2005), however, this response in water-limited plants did depend on the consistency of the water regime. Plants that received limited overall water in infrequent, large pulses bolted at the same time as plants receiving generous amounts of water, but these plants also suffered very high mortality and therefore a large decrement in fitness. The earlier flowering plants, those that received limited but consistent amounts of water, showed survivorship comparable to the generously watered plants.

It has also been noted that across genotypes later flowering strains tend to have higher water use efficiencies (Juenger *et al.*, 2005a; McKay *et al.*, 2003). These authors conducted a suite of studies to explore this correlation and found several lines of evidence suggesting likely pleiotropic effects of flowering time genes on water use efficiency, that is, amount of carbon fixed relative to amount of water transpired, as measured by carbon isotope ratios. They found that several known strains of flowering time mutants also showed changes in water use efficiencies, and several water use mutants, though not all, showed changes in flowering time. In a QTL study (Juenger *et al.*, 2005a), they found several loci that independently determined both flowering time and water use efficiency, but they also found a single locus on chromosome III that clearly regulated both, although the effects of this locus on each trait were negatively correlated.

The effect of nutrients on flowering time has not been extensively examined in *A. thaliana*. Pigliucci and Schlichting (1998) have addressed this question and found that the effect of nutrients on flowering varies greatly among genotypes and families. More generally, however, they have shown that differences in nutrient availability lead to differences in correlations between traits.

III. THE GENETIC BASIS OF ENVIRONMENTAL PERCEPTION IN FLOWERING TIME SIGNALING

Plants sense light via three known classes of photoreceptors: cryptochromes, phytochromes, and phototropins (Casal, 2002). Of these photoreceptors, both cryptochromes and phytochromes play important roles in plant development, including timing of flowering. Cryptochromes detect blue light and ultraviolet A radiation (UV-A), while the phytochromes detect red and far-red light. Numerous authors (Borevitz *et al.*, 2002; Casal and Smith, 1989a,b; Chory and Li, 1997; Johnson *et al.*, 1994; Pigliucci and Schmitt, 2004; Reed *et al.*, 1994; Schmitt *et al.*, 1999) have shown that these receptors mediate plant growth via gibberellin and brassinosteroid hormones, resulting in adaptively appropriate phenotypes for a given light quality. The ratio of red to far-red light, for example, is known to indicate the presence of both shade and twilight, and prolonged exposure to light with low red: far red ratios induces hypocotyl elongation, a mechanism that allows plants to overgrow the neighbors before initiating leaf expansion. It has been shown (Borevitz *et al.*, 2002; Maloof *et al.*, 2000, 2001; Stenoien *et al.*, 2002) that there is extensive variation to light sensitivity that is mediated by the cryptochrome and phytochrome pathways in natural accessions of *A. thaliana*.

Phytochromes are also thought to regulate the onset of flowering through a complex pathway that also involves cryptochrome photoreception, which defines the first steps in a photoperiod or daylength-dependent flowering time pathway (Schultz and Kay, 2003; Searle and Coupland, 2004; Valverde *et al.*, 2004). The mRNA of the gene *CONSTANS* (*CO*), a flowering time gene downstream of the phytochrome and cryptochrome loci, is expressed in a circadian pattern. Peak expression corresponds to what is early evening under short days, but what is late afternoon under long days. During dark periods, the CO protein cannot accumulate because it is degraded by the proteasome. However in the presence of both red and blue light, proteasome degradation of CO is inhibited by two of the phytochrome and cryptochrome proteins, PHYA and CRY2, respectively. Therefore under long days in full spectrum light, CO protein, a potent activator of the gene *FLOWERING TIME* (*FT*), can accumulate and flowering is induced. Under conditions of foliar shade, that is low red: far-red light ratios, PHYB activates expression of another gene, *PHYTOCHROME AND FLOWERING TIME 1* (*PFT1*), which upregulates *FT* expression (Cerdan and Chory, 2003).

The timing of flowering has also been shown to be sensitive to ambient growth temperature. The autonomous pathway, a series of photoperiod-independent genes and gene products necessary for flowering, may define a temperature-regulated flowering pathway. In wild-type plants, flowering occurs earlier at 23 °C than at 16 °C. This effect is mediated by two autonomous pathway genes, *FVE* and *FCA* (Blazquez *et al.*, 2003; Kim *et al.*, 2004). Furthermore this difference in flowering time is also enhanced in *CRY1* and *CRY2* mutants, suggesting there may be some interaction between light exposure and ambient temperature. Thingnase *et al.* (2003) did show that while increased ambient night temperatures and increased mean daily temperature both decrease the number leaves at bolting, their effects are statistically independent. Furthermore, they showed that increased ambient night temperature decreased the days to bolting whereas mean daily temperature did not.

One of the more dramatic temperature-regulated pathways in flowering is the vernalization pathway, which promotes flowering in response to a prolonged cold treatment such as that observed in winter conditions. Many plants have a vernalization requirement and in the Brassicaceae, there are two unique vernalization genes, *FRIGIDA* (*FRI*) and *FLOWERING LOCUS C* (*FLC*) (Koornneef *et al.*, 1994), and these are quite well characterized in *Arabidopsis*. The *FRI* gene upregulates *FLC* which expresses a MADS-box transcription factor that inhibits flowering in unvernalized plants. Vernalization, via *VERNALIZATION 1* (*VRN1*), *VRN2*, and *VERNALIZATION INSENSITIVE 3* (*VIN3*), irreversibly alters histone methylation at the *FLC* locus, permanently repressing *FLC* expression enabling photoperiod to induce flowering substantially earlier than in unvernalized plants (Sung and Amasino, 2005). *VIN3* is also required for *FLC*-independent vernalization via its action as a promoter of *LUMINIDEPENDENS* (*LD*) in the autonomous pathway (Sung and Amasino, 2004).

IV. QUANTITATIVE TRAIT LOCUS MAPPING OF FLOWERING TIME VARIATION

Molecular developmental genetic studies have elucidated many of the key pathways that plants utilize to sense seasonal cues, and allowed investigators to examine the molecular genetic basis of flowering time in an ecological context. Understanding the evolution and ecology of flowering time, however, requires us to understand not only what genes regulate this trait but also which specific genes are responsible for natural variation in flowering time and the role selection plays in defining this variation. The quantitative nature of flowering time variation in *A. thaliana* allows us to employ modern quantitative trait locus (QTL) mapping approaches in dissecting the genetic architecture of this trait. Flowering time has been the subject of the most intensive effort in QTL mapping in this species, with at least 18 QTL mapping studies published in the last 10 years (Bandaranayake *et al.*, 2004; Clarke *et al.*, 1995; El-Assal *et al.*, 2001; El-Lithy *et al.*, 2004; Jansen *et al.*, 1995; Juenger *et al.*, 2005a,b; Kuittinen *et al.*, 1997; Maloof, 2003; Mitchell-Olds, 1996; Remington and Purugganan, 2003; Stratton, 1998; Ungerer *et al.*, 2002, 2003; Weinig and Schmitt, 2004; Weinig *et al.*, 2002, 2003a; Werner *et al.*, 2005b). QTL mapping studies of flowering time have defined at least 28 loci that affect natural variation in flowering time among individual accessions of this species under different conditions (Weinig *et al.*, 2002). The effects of these individual QTL follow a long-tailed distribution, with one to three loci of moderate to large effect (>10% of variation explained) and a larger number of loci with smaller effects (Juenger *et al.*, 2005a,b; Kuittinen *et al.*, 1997; Ungerer *et al.*, 2002; Weinig *et al.*, 2003a). Epistatic effects among QTLs have also been observed, indicating that phenotypic variation could be explained in part by nonadditive multilocus interactions within genomes.

As discussed above, two ecological cues, daylength and exposure to a cold period associated with winter conditions (vernalization), have been extensively studied as environmental cues to flowering in *A. thaliana*. In two sets of recombinant inbred line mapping populations of *A. thaliana* (the Col × *Ler* and Cvi × *Ler* mapping populations), QTL × environment interactions, that is, the environment-dependent detection of loci, have been documented for three of the five identified loci between long and short day conditions (Ungerer *et al.*, 2003). Although still at the level of QTL identification, these studies have begun to address the genetic basis for differential response of *A. thaliana* accessions to ecological cues.

Studies on the genetic architecture of flowering time variation have largely been undertaken in controlled environmental conditions, and few studies have explored the extent to which the genetic basis for phenotypic variation in this trait differs under field settings. Studies by Weinig *et al.* (2002), however, have explored the genetic architecture of flowering time in the field under ecologically relevant conditions. A study of the Col × *Ler* recombinant inbred line mapping population at two field locations (Rhode Island and North Carolina) over winter and spring seasons revealed, not surprisingly, field- and season-dependent QTLs, suggesting strong genotype-by-environment interactions for this trait in natural conditions. The number and identity of QTLs differed in significant ways between controlled growth chamber and field conditions, indicating that our view of relevant genes underlying flowering time behavior may be skewed by reliance on controlled conditions in studying flowering time loci (Weinig *et al.*, 2002).

Intriguingly, and perhaps not surprisingly, variation in flowering time is genetically correlated with other developmental and physiological phenotypes in *A. thaliana*. One clear example is between bolting time, a life history trait, and rosette leaf number, a morphological trait. The genetic correlation, r_G, between these two traits is 0.94 in a collection of 21 accessions (Ungerer *et al.*, 2002). Other genetic correlations have been observed between bolting time and various features of shoot architecture, including lateral branch number and fruit production. Joint QTL analyses of flowering time and shoot architectural traits suggests that they share common QTLs, indicating the presence of trait suites underpinned by common loci (Engelmann, K., and Purugganan, M., unpublished data; Ungerer *et al.*, 2002, 2003).

One interesting correlation observed in an ecological context is that observed between flowering time and herbivory. Weinig *et al.* (2003a,c) examined resistance to herbivory and found that this depends significantly on flowering time. Early bolting plants are more susceptible to herbivores, in this study rabbits, but this susceptibility does not necessarily result in a fitness decrement. In fact, on average *Arabidopsis* tends to overcompensate for apical meristem damage such that herbivory can lead to an increase in fruit production via proliferation of basal branches. They also showed that three QTLs associated with flowering time are also associated with herbivore resistance, although not all flowering time QTLs showed this association.

Whether these correlations arise from pleiotropy or close linkage among specific genes must await fine mapping analyses and possibly isolation of relevant QTL genes. One approach has been to examine the mutational covariance associated with mutant alleles of known flowering time genes. A study of genetic correlations suggests a relationship between flowering time as a drought-escape mechanism and dehydration-avoidance mechanisms as measured by δ^{13} C isotope ratio. Near-isogenic lines of *FRI* and *FLC* flowering time alleles also affect δ^{13} C ratios, providing strong evidence for possible pleiotropic effects of these flowering time genes or very tight linkage to dehydration avoidance loci (Juenger *et al.*, 2005a; McKay *et al.*, 2003).

V. ISOLATION OF GENES UNDERLYING FLOWERING TIME VARIATION

Quantitative trait locus mapping studies have defined the genomic regions that harbor genetic polymorphisms associated with flowering time variation and have elucidated the genetic architecture of this trait. Further progress in examining the evolutionary ecology of flowering time, however, requires us to identify and isolate the specific genes that underlie natural variation in flowering time in this species. Genes underlying quantitative variation in flowering time have been isolated in recent years. Two approaches to isolating QTL genes have been pursued: (1) fine mapping and positional cloning of QTL genes and (2) candidate gene association studies.

One of the first genes demonstrated to underlie a flowering time QTL is *EARLY DAYLENGTH INSENSITIVE* (*EDI*), which was first identified in a QTL mapping study using the recombinant inbred Cvi × *Ler* mapping population (El-Assal *et al.*, 2001). Positional cloning of this QTL demonstrated that *EDI* is equivalent to the *CRY2* gene. The Cvi accession of *A. thaliana* has a *CRY2* allele with two amino acid changes that result in altered *CRY2* protein levels during the circadian cycle, and results in early flowering of plants under short day conditions. Although the *EDI* QTL is a large-effect allele, its cloning remains a landmark feat in the identification of the genetic basis for quantitative variation in *A. thaliana* (El-Assal *et al.*, 2001).

New genomics technologies have also advanced the ability to fine-map and isolate genes underlying QTL, including flowering time loci. A study identified flowering time QTLs in a recombinant inbred mapping population between the Nd and Col accessions (Werner *et al.*, 2005b). Microarray hybridization with genomic DNA identified an Nd-specific deletion of the *FLOWERING LOCUS M* (*FLM*) gene, which encodes a MADS-box transcription factor and is a duplicate of *FLC*. Like *FLC*, this gene appears to repress flowering and the effect of this gene deletion on flowering time was confirmed by transgenic complementation.

Candidate gene studies have also been useful in associating flowering time variation with particular *A. thaliana* haplotypes. Positional QTL cloning identified the *CRY2* Cvi allele as a flowering time QTL, but this allele has been observed only in the Cape Verde Islands (El-Assal *et al.*, 2001). It is thus unclear whether it represents an adaptation to local conditions or a rare, possibly slightly deleterious mutation maintained by relaxed selection in this inbreeding species. A study of nucleotide variation at this gene, undertaken as a candidate gene approach to identifying flowering time QTLs, revealed other major, moderate-frequency haplotypes that exist within the species range of *A. thaliana* (Olsen *et al.*, 2004). Two major *CRY2* haplotype groups exist within this species, one of which includes several alleles that feature a nonsynonymous glutamine (Q) substitution in an otherwise conserved serine (S). Candidate gene association studies reveal that these different haplotypes/haplotype groups are significantly associated with flowering time differences in the species (Olsen *et al.*, 2004).

An important determinant of standing genetic variation in flowering time in *A. thaliana* is the *FRI* gene (Johanson *et al.*, 2000). As noted above, the *FRI* gene appears to act by upregulating expression of another flowering time gene, *FLC*, which encodes a MADS-box transcriptional activator. Molecular analysis reveals that *FRI* harbors several large deletions that lead to loss-of-function alleles, at least two of which are found at moderate frequency (Hagenblad and Nordborg, 2002; Hagenblad *et al.*, 2004; Johanson *et al.*, 2000; Le Corre *et al.*, 2002; Stinchcombe *et al.*, 2004b).

A latitudinal cline in flowering time, measured in days to flowering, has been shown to be dependent on *FRI* genotypes in *A. thaliana* accessions.

This cline is observed only in accessions that do not carry any of the *FRI* deletion alleles, and when flowering time is assayed under field conditions where plants are vernalized by exposure to winter conditions or cold treatment (Lempe *et al.*, 2005; Stinchcombe *et al.*, 2004b). Shindo *et al.* (2005), however, did not find a statistically significant cline when several North American accessions were included.

The latitudinal cline observed in *A. thaliana* may be driven in part by an epistatic effect of *FRI* with *FLC* (Caicedo *et al.*, 2004; Lempe *et al.*, 2005; Michaels and Amasino, 2001). In the latter gene, two major haplotype groups have been detected in *A. thaliana* accessions, and there is significant flowering time variation associated with *FRI FLC* two-locus genotypes. *FLC* haplotypes also show a significant latitudinal distribution, but only in putatively functional *FRI* genotypic backgrounds. Finally, *FRI* and *FLC* show significant intergenic linkage disequilibrium, even though the two genes are found in two different *A. thaliana* chromosomes. Together, these results suggest that epistatic selection may underlie flowering time variation in winter annuals of this species (Caicedo *et al.*, 2004; Stinchcombe *et al.*, 2004b). Other *FLC* and *FRI* alleles associated with low levels of *FLC* expression have been found (Lempe *et al.*, 2005; Shindo *et al.*, 2005; Werner *et al.*, 2005a).

VI. MICROEVOLUTION OF FLOWERING TIME LOCI

QTL mapping studies and the subsequent isolation of genes provide crucial insights into the molecular genetic basis of natural flowering time variation. This, in turn, is complemented by molecular evolutionary studies, which determine the levels and patterns of nucleotide variation to infer evolutionary forces that have acted at specific loci. Molecular population genetic studies of six flowering time genes have been reported, including the *FRI* (Hagenblad and Nordborg, 2002; Le Corre *et al.*, 2002) and *FLC* (Caicedo *et al.*, 2004) genes associated with the vernalization response, the photoreceptor gene *CRY2* (Olsen *et al.*, 2004) in the photoperiod pathway, and the *LFY, AP1*, and *TFL1* (Olsen *et al.*, 2002) loci, which are either flowering time integrators and/or floral meristem identity genes. These genes show different patterns of nucleotide variation consistent with differing evolutionary forces than those acting on known flowering time genes.

From the viewpoint of molecular population genetics, the *FRI* gene is the most intensively studied flowering time gene in *A. thaliana*. Studies of this gene indicate the presence of several large independent deletions that remove portions of the coding region and presumably result in nonfunctional *FRI* alleles. At least three major deletions have been identified, each of which has

TABLE I
Levels and Patterns of Mean Nucleotide Variation for A. thaliana *Flowering Time Genes*

Gene	n[a]	π[b]	Tajima's *D* value
FRI	26	0.003	−1.225
CRY2	31	0.013	+0.348
FLC	14	0.004	+0.858
LFY	15	0.002	−1.572
AP1	15	0.005	−2.102[c]
TFL1 (coding region)	14	0.001	−2.032[c]
TFL1 (promoter/5′ UTR)	14	0.019	+1.532[c]

[a]Sample size.
[b]Nucleotide diversity per silent site.
[c]Significant at the $p < 0.05$ in a coalescent simulation of the neutral-equilibrium model under no recombination.
UTR, untranslated region.

arisen independently and two of which are present in moderate frequencies in *A. thaliana*. The level of silent site polymorphism in this gene, π, is 0.003 and is lower than the genomic average of $\pi = 0.009$ (Schmid *et al.*, 2005) (Table I), and this low value is due in part to the very low variation of the two major deletion alleles at this locus. Studies of the genomic region around *FRI* also appear to be consistent with positive selection, with linkage disequilibrium extending to ~250 kb and with the *FRI* deletion haplotypes extending unbroken across large genomic distances (Hagenblad *et al.*, 2004). Other nonsynonymous and premature stop codon mutations have been reported at this gene that may also have functional consequences, although these mutations are present at low frequency (Le Corre *et al.*, 2002; Shindo *et al.*, 2005). The results on *FRI* suggest recent and strong directional selection for the evolution of nonfunctional *FRI* alleles in this species.

The evolution of *FRI* appears closely linked to that of its downstream target gene *FLC* (see earlier discussion). A study indicates that *FLC* alleles are found in two major haplotype clades, and that epistatic interaction between *FRI* and these two *FLC* haplotypes may be associated with a latitudinal cline in flowering time (Caicedo *et al.*, 2004). Despite the presence of two differentiated allele groups at *FLC*, levels of variation are lower than the genomic average (Table I). The presence of these two haplotype groups, however, results in a positive Tajima's *D* value, which is associated with the presence of these two differentiated allele classes at moderate frequency.

Like *FLC, CRY2* also appears to have two major haplotype groups, and variation between and within these groups appears to be associated with flowering time variation (Olsen *et al.*, 2004). The level of variation at this gene is higher than the genomic mean (Table I) and Tajima's *D* value is also positive as a result of these differentiated haplotype groups, but these values are not significantly different from neutral-equilibrium expectations.

Among the three downstream genes of the flowering time pathway that have been explored, only the inflorescence developmental gene *TFL1* has a striking pattern of molecular variation. The silent site nucleotide diversity of the *TFL1*-coding region is the lowest of the six genes that have been studied at that time, and this variation is significantly reduced when compared to other inflorescence developmental genes (Olsen *et al.*, 2004). The value of Tajima's *D* value is also significantly negative for this gene. Both these results suggest that the coding region has been subjected to a recent bout of directional selection, and the reduced variation in the coding region is the result of a selective sweep. The promoter and 5′ untranslated region of *TFL1*, however, shows a contrasting pattern from the coding region. The level of silent site nucleotide diversity for the *TFL1* promoter is relatively high, and the Tajima's *D* value is significantly positive with respect to neutral-equilibrium expectations (Table I). Both the high level of nucleotide diversity and significantly positive Tajima's *D* value is associated with the presence of two moderate-frequency haplotype groups at the *TFL1* promoter, in contrast to the near absence of variation at the *TFL1* coding region. The variation at the *TFL1* promoter, with the observed differentiation into two distinct haplotype groups, is reminiscent to that observed for *FLC* and *CRY2*.

The two other floral integrator/meristem identity genes, *LFY* and *AP1*, do not appear exceptional with regard to their evolutionary dynamics (Olsen *et al.*, 2004). The levels of variation at these loci are both lower than the genomic mean for this species, and *AP1* has a significantly negative Tajima's *D* value with respect to the neutral-equilibrium model. However, the levels and patterns of nucleotide variation at these two loci do not show strong evidence of any nonneutral evolution in their recent history.

These studies provide important insights into the divergent types of evolutionary forces that act at specific flowering time loci. A comprehensive assessment of the microevolution of these genes in the context of the flowering time genetic network may permit us to draw conclusions on the diversification of regulatory gene networks in an explicitly evolutionary ecological context. By assessing the impact of evolutionary forces that have shaped variation at these loci, we may be able to determine to what extent network structures channel and constrain evolutionary trajectories of phenotypes.

VII. SUMMARY

The studies on the molecular, ecological, quantitative, and population genetics of flowering time in *A. thaliana* serve as a model for studying the evolution of development, particularly at the stage of microevolutionary changes that can ultimately lead to changes in fitness. By examining flowering time variation at various hierarchical levels, from the molecular to the organismal to the ecological, we can take a broad view of the evolution of a key developmental transition. Flowering time is a particularly appropriate trait in this regard, as it is a developmental transition that is sensitive to ecological cues to allow adaptive response to seasonal and other environmental variations. One of the features that sets plant development apart from animal development is the greater sensitivity of plant developmental processes to environmental cues. This is rooted in the sessile nature of plants, which relies in part on changing developmental patterns to react to environmental changes to optimize survival and reproductive success.

The study of the evolution of development in recent years has focused almost exclusively on understanding the molecular genetic mechanisms underlying developmental diversification at a macroevolutionary level. There have been few attempts to examine developmental diversification at microevolutionary levels, an approach that also allows us to examine the ecological context of the evolutionary process. A microevolutionary perspective on developmental diversification allows us to catch the origins of the evolutionary process which we can then strive to integrate with its observed endpoint, providing a fuller understanding of the nature of evolutionary change.

ACKNOWLEDGMENTS

This work was funded in part from grants from the Integrated Research Challenges in Environmental Biology, Frontiers in Biological Research, and Plant Genome Research Programs of the US National Science Foundation.

REFERENCES

Alpert, P. and Simms, E. L. (2002). The relative advantages of plasticity and fixity in different environments: When is it good for a plant to adjust? *Evolutionary Ecology* **16**, 285–297.

Bandaranayake, C. K., Koumproglou, R., Wang, X. Y., Wilkes, T. and Kearsey, M. J. (2004). QTL analysis of morphological and developmental traits in the Ler x Cvi population of *Arabidopsis thaliana*—QTL analysis in *Arabidopsis*. *Euphytica* **137**, 361–371.

Blazquez, M. A., Ahn, J. H. and Weigel, D. (2003). A thermosensory pathway controlling flowering time in *Arabidopsis thaliana*. *Nature Genetics* **33**, 168–171.
Borevitz, J. O., Maloof, J. N., Lutes, J., Dabi, T., Redfern, J. L., Trainer, G. T., Werner, J. D., Asami, T., Berry, C. C., Weigel, D. and Chory, J. (2002). Quantitative trait loci controlling light and hormone response in two accessions of *Arabidopsis thaliana*. *Genetics* **160**, 683–696.
Caicedo, A. L., Stinchcombe, J. R., Olsen, K. M., Schmitt, J. and Purugganan, M. D. (2004). Epistatic interaction between *Arabidopsis FRI* and *FLC* flowering time genes generates a latitudinal cline in a life history trait. *Proceedings of the National Academy of Sciences of the United States of America* **101**, 15670–15675.
Callahan, H. S. and Pigliucci, M. (2002). Shade-induced plasticity and its ecological significance in wild populations of *Arabidopsis thaliana*. *Ecology* **83**, 1965–1980.
Callahan, H. S., Dhanoolal, N. and Ungerer, M. C. (2004). Plasticity genes and plasticity costs: A new approach using an *Arabidopsis* recombinant inbred population. *New Phytologist* **166**, 129–139.
Casal, J. J. (2002). Environmental cues affecting development. *Current Opinion in Plant Biology* **5**, 37–42.
Casal, J. J. and Smith, H. (1989a). The function, action and adaptive significance of phytochrome in light-grown plants. *Plant Cell and Environment* **12**, 855–862.
Casal, J. J. and Smith, H. (1989b). The end-of-day phytochrome control of internode elongation in mustard—kinetics, interaction with the previous fluence rate, and ecological implications. *Plant Cell and Environment* **12**, 511–520.
Cerdan, P. D. and Chory, J. (2003). Regulation of flowering time by light quality. *Nature* **423**, 881–885.
Chory, J. and Li, J. (1997). Gibberellins, brassinosteroids and light-regulated development. *Plant Cell and Environment* **20**, 801–806.
Clarke, J. H., Mithen, R., Brown, J. K. M. and Dean, C. (1995). QTL analysis of flowering time in *Arabidopsis thaliana*. *Molecular & General Genetics* **248**, 278–286.
Diggle, P. K. (1999). Heteroblasty and the evolution of flowering phenologies. *International Journal of Plant Sciences* **160**, S123–S134.
Donohue, K. (2002). Germination timing influences natural selection on life-history characters in *Arabidopsis thaliana*. *Ecology* **83**, 1006–1016.
Dorn, L. A., Pyle, E. H. and Schmitt, J. (2000). Plasticity to light cues and resources in *Arabidopsis thaliana*: Testing for adaptive value and costs. *Evolution* **54**, 1982–1994.
El-Assal, S. E. D., Alonso-Blanco, C., Peeters, A. J. M., Raz, V. and Koornneef, M. (2001). A QTL for flowering time in *Arabidopsis* reveals a novel allele of *CRY2*. *Nature Genetics* **29**, 435–440.
El-Lithy, M. E., Clerkx, E. J. M., Ruys, G. J., Koornneef, M. and Vreugdenhil, D. (2004). Quantitative trait locus analysis of growth-related traits in a new *Arabidopsis* recombinant. *Plant Physiology* **135**, 444–458.
Engelmann, K. E. and Schlichting, C. D. (2005). Coarse- versus fine-grained water stress in *Arabidopsis thaliana* (Brassicaceae). *American Journal of Botany* **92**, 101–106.
Griffith, C., Kim, E. and Donohue, K. (2004). Life-history variation and adaptation in the historically mobile plant *Arabidopsis thaliana* (Brassicaceae) in North America. *American Journal of Botany* **91**, 837–849.
Hagenblad, J. and Nordborg, M. (2002). Sequence variation and haplotype structure surrounding the flowering time locus *FRI* in *Arabidopsis thaliana*. *Genetics* **161**, 289–298.

Hagenblad, J., Tang, C. L., Molitor, J., Werner, J., Zhao, K., Zheng, H. G., Marjoram, P., Weigel, D. and Nordborg, M. (2004). Haplotype structure and phenotypic associations in the chromosomal regions surrounding two *Arabidopsis thaliana* flowering time loci. *Genetics* **168**, 1627–1638.

Hoffmann, M. H. (2002). Biogeography of *Arabidopsis thaliana* (L.) Heynh. (Brassicaceae). *Journal of Biogeography* **29**, 125–134.

Hoffmann, M. H., Bremer, M., Schneider, K., Burger, F., Stolle, E. and Moritz, G. (2003). Flower visitors in a natural population of *Arabidopsis thaliana*. *Plant Biology* **5**, 491–494.

Jansen, R. C., Vanooijen, J. W., Stam, P., Lister, C. and Dean, C. (1995). Genotype-by-environment interaction in genetic-mapping of multiple quantitative trait loci. *Theoretical and Applied Genetics* **91**, 33–37.

Johnson, E., Bradley, M., Harberd, N. P. and Whitelam, G. C. (1994). Photoresponses of light-grown *phy*A mutants of *Arabidopsis*: Phytochrome-A is required for the perception of daylength extensions. *Plant Physiology* **105**, 141–149.

Johanson, U., West, J., Lister, C., Michaels, S., Amasino, R. and Dean, C. (2000). Molecular analysis of *FRIGIDA*, a major determinant of natural variation in *Arabidopsis* flowering time. *Science* **290**, 344–347.

Juenger, T. E., McKay, J. K., Hausmann, N. J., Joost, J. B., Saunak, S., Stowe, K. A., Dawson, T. E., Simms, E. L. and Richards, J. H. (2005a). Identification and characterization of QTL underlying whole plant physiology in *Arabidopsis thaliana*: δ13C, stomatal conductance and transpiration efficiency. *Plant Cell and Environment* **2005**, 1–12.

Juenger, T. E., Sen, S., Stowe, K. A. and Simms, E. L. (2005b). Epistasis and genotype-environment interaction for quantitative trait loci affecting flowering time in *Arabidopsis thaliana*. *Genetica* **123**, 87–105.

Karlsson, B. H., Sills, G. R. and Nienhuis, J. (1993). Effects of photoperiod and vernalization on the number of leaves at flowering in 32 *Arabidopsis thaliana* (Brassicaceae) ecotypes. *American Journal of Botany* **80**, 646–648.

Kim, H. J., Hyun, Y., Park, J. Y., Park, M. J., Park, M. K., Kim, M. D., Lee, M. H., Moon, J., Lee, I. and Kim, J. (2004). A genetic link between cold responses and flowering time through FVE in *Arabidopsis thaliana*. *Nature Genetics* **36**, 167–171.

Koornneef, M., Blankestijn-de Vries, H., Hanhart, C., Soppe, W. and Peeters, T. (1994). The phenotype of some late-flowering mutants is enhanced by a locus on chromosome 5 that is not effective in the Landsberg *erecta* wild-type. *The Plant Journal* **6**, 911–919.

Koornneef, M., Alonso-Blanco, C. and Vreugdenhil, D. (2004). Naturally occurring genetic variation in *Arabidopsis thaliana*. *Annual Review of Plant Biology* **55**, 141–172.

Kuittinen, H., Sillanpaa, M. J. and Savolainen, O. (1997). Genetic basis of adaptation: Flowering time in *Arabidopsis thaliana*. *Theoretical and Applied Genetics* **95**, 573–583.

Le Corre, V., Roux, F. and Reboud, X (2002). DNA polymorphism at the *FRIGIDA* gene in *Arabidopsis thaliana*: Extensive nonsynonymous variation is consistent with local selection for flowering time. *Molecular Biology and Evolution* **19**, 1261–1271.

Lempe, J., Balasubramanian, S., Sureshkumar, S., Singh, A., Schmid, M. and Weigel, D. (2005). Diversity of flowering responses in wild *Arabidopsis thaliana* strains. *PLoS Genetics* **1**, 109–118.

Maloof, J. N. (2003). QTL for plant growth and morphology. *Current Opinion in Plant Biology* **6**, 85–90.

Maloof, J. N., Borevitz, J. O., Weigel, D. and Chory, J. (2000). Natural variation in phytochrome signaling. *Seminars in Cell & Developmental Biology* **11**, 523–530.

Maloof, J. N., Borevitz, J. O., Dabi, T., Lutes, J., Nehring, R. B., Redfern, J. L., Trainer, G. T., Wilson, J. M., Asami, T., Berry, C. C., Weigel, D. and Chory, J. (2001). Natural variation in light sensitivity of *Arabidopsis*. *Nature Genetics* **29**, 441–446.

McKay, J. K., Richards, J. H. and Mitchell-Olds, T. (2003). Genetics of drought adaptation in *Arabidopsis thaliana*: I. Pleiotropy contributes to genetic correlations among ecological traits. *Molecular Ecology* **12**, 1137–1151.

Michaels, S. D. and Amasino, R. M. (2001). Loss of *FLOWERING LOCUS C* activity eliminates the late-flowering phenotype of *FRIGIDA* and autonomous pathway mutations but not responsiveness to vernalization. *The Plant Cell* **13**, 935–941.

Mitchell-Olds, T. (1996). Genetic constraints on life-history evolution: Quantitative-trait loci influencing growth and flowering in *Arabidopsis thaliana*. *Evolution* **50**, 140–145.

Mitchell-Olds, T. (2001). *Arabidopsis thaliana* and its wild relatives: A model system for ecology and evolution. *Trends in Ecology & Evolution* **16**, 693–700.

Murfet, I. C. (1977). Environmental interaction and genetics of flowering. *Annual Review of Plant Physiology and Plant Molecular Biology* **28**, 253–278.

Nordborg, M. and Bergelson, J. (1999). The effect of seed and rosette cold treatment on germination and flowering time in some *Arabidopsis thaliana* (Brassicaceae) ecotypes. *American Journal of Botany* **86**, 470–475.

Nordborg, M., Hu, T., Ishino, Y., Jhavei, J., Toomajian, C., Zheng, H., Bakker, E., Calabrese, P., Gladstone, J., Goyal, R., Jakobsson, S. Kim, S., *et al.* (2005). The pattern of polymorphism in *Arabidopsis thaliana*. *PLoS Biology* **3**, 1289–1299.

Olsen, K. M., Womack, A., Garrett, A. R., Suddith, J. I. and Purugganan, M. D. (2002). Contrasting evolutionary forces in the *Arabidopsis thaliana* floral developmental pathway. *Genetics* **160**, 1641–1650.

Olsen, K. M., Halldorsdottir, S. S., Stinchcombe, J. R., Weinig, C., Schmitt, J. and Purugganan, M. D. (2004). Linkage disequilibrium mapping of *Arabidopsis CRY2* flowering time alleles. *Genetics* **167**, 1361–1369.

Pigliucci, M. (1998). Ecological and evolutionary genetics of *Arabidopsis*. *Trends in Plant Sciences* **3**, 485–489.

Pigliucci, M. and Marlow, E. T. (2001). Differentiation for flowering time and phenotypic integration in *Arabidopsis thaliana* in response to season length and vernalization. *Oecologia* **127**, 501–508.

Pigliucci, M. and Schlichting, C. D. (1998). Reaction norms of *Arabidopsis*. V. Flowering time controls phenotypic architecture in response to nutrient stress. *Journal of Evolutionary Biology* **11**, 285–301.

Pigliucci, M. and Schmitt, J. (2004). Phenotypic plasticity in response to foliar and neutral shade in gibberellin mutants of *Arabidopsis thaliana*. *Evolutionary Ecology Research* **6**, 243–259.

Rathcke, B. and Lacey, E. P. (1985). Phenological patterns of terrestrial plants. *Annual Review of Ecology and Systematics* **16**, 179–214.

Reed, J. W., Nagatani, A., Elich, T. D., Fagan, M. and Chory, J. (1994). Phytochrome-A and phytochrome-B have overlapping but distinct functions in *Arabidopsis* development. *Plant Physiology* **104**, 1139–1149.

Remington, D. L. and Purugganan, M. D. (2003). Candidate genes, quantitative trait loci and functional trait evolution in plants. *International Journal of Plant Sciences* **164**, S7–S20.

Schmid, K. J., Ramos-Onsins, S., Ringys-Beckstein, H., Weisshaar, B. and Mitchell-Olds, T. (2005). A multilocus sequence survey in *Arabidopsis thaliana* reveals a genome-wide departure from a neutral model of DNA sequence polymorphism. *Genetics* **169**, 1601–1615.

Schmitt, J., Dudley, S. A. and Pigliucci, M. (1999). Manipulative approaches to testing adaptive plasticity: Phytochrome-mediated shade-avoidance responses in plants. *American Naturalist* **154**, S43–S54.

Schultz, T. F. and Kay, S. A. (2003). Circadian clocks in daily and seasonal control of development. *Science* **301**, 326–328.

Searle, I. and Coupland, G. (2004). Induction of flowering by seasonal changes in photoperiod. *EMBO Journal* **23**, 1217–1222.

Sharbel, T. F., Haubold, B. and Mitchell-Olds, T. (2000). Genetic isolation by distance in *Arabidopsis thaliana*: Biogeography and postglacial colonization of Europe. *Molecular Ecology* **9**, 2109–2118.

Shimizu, K. K. and Purugganan, M. D. (2005). Evolutionary and ecological genomics of *Arabidopsis*. *Plant Physiology* **138**, 578–584.

Shindo, C., Aranzana, M. J., Lister, C., Baxter, C., Nicholls, C., Nordborg, M. and Dean, C. (2005). Role of *FRIGIDA* and *FLOWERING LOCUS C* in determining variation in flowering time of *Arabidopsis*. *Plant Physiology* **138**, 1163–1173.

Stenoien, H. K., Fenster, C. B., Kuittinen, H. and Savolainen, O. (2002). Quantifying latitudinal clines to light responses in natural populations of *Arabidopsis thaliana* (Brassicaceae). *American Journal of Botany* **89**, 1604–1608.

Stinchcombe, J. R., Dorn, L. A. and Schmitt, J. (2004a). Flowering time plasticity in *Arabidopsis thaliana*: A reanalysis of Westerman & Lawrence (1970). *Journal of Evolutionary Biology* **17**, 197–207.

Stinchcombe, J. R., Weinig, C., Ungerer, M., Olsen, K. M., Mays, C., Halldorsdottir, S. S., Purugganan, M. D. and Schmitt, J. (2004b). A latitudinal cline in flowering time in *Arabidopsis thaliana* modulated by the flowering time gene *FRIGIDA*. *Proceedings of the National Academy of Sciences of the United States of America* **101**, 4712–4717.

Stratton, D. A. (1998). Reaction norm functions and QTL-environment interactions for flowering time in *Arabidopsis thaliana*. *Heredity* **81**, 144–155.

Sung, S. B. and Amasino, R. M. (2004). Vernalization in *Arabidopsis thaliana* is mediated by the PHD finger protein VIN3. *Nature* **427**, 159–164.

Sung, S. B. and Amasino, R. M. (2005). Remembering Winter: Towards a molecular understanding of vernalization. *Annual Review of Plant Biology* **56**, 491–508.

Thingnase, E., Torre, S., Ernstsen, A. and Moe, R. (2003). Day and night temperature responses in *Arabidopsis*: Effects on gibberellin and auxin content, cell size, morphology and flowering time. *Annals of Botany* **92**, 601–612.

Ungerer, M. C. and Rieseberg, L. H. (2003). Genetic architecture of a selection response in *Arabidopsis thaliana*. *Evolution* **57**, 2531–2539.

Ungerer, M. C., Halldorsdottir, S. S., Modliszewski, J. L., Mackay, T. F. C. and Purugganan, M. D. (2002). Quantitative trait loci for inflorescence development in *Arabidopsis thaliana*. *Genetics* **160**, 1133–1151.

Ungerer, M. C., Halldorsdottir, S. S., Purugganan, M. D. and Mackay, T. F. (2003). Genotype-environment interactions at quantitative trait loci affecting inflorescence development in *Arabidopsis thaliana*. *Genetics* **165**, 353–365.

Valverde, F., Mouradov, A., Soppe, W., Ravenscroft, D., Samach, A. and Coupland, G. (2004). Photoreceptor regulation of CONSTANS protein in photoperiodic flowering. *Science* **303**, 1003–1006.

Van Berloo, R. and Stam, P. (1999). Comparison between marker-assisted selection and phenotypical selection in a set of *Arabidopsis thaliana* recombinant inbred lines. *Theoretical and Applied Genetics* **98**, 113–118.

Weinig, C. and Schmitt, J. (2004). Environmental effects on the expression of quantitative trait loci and implications for phenotypic evolution. *Bioscience* **54**, 627–635.

Weinig, C., Ungerer, M. C., Dorn, L. A., Kane, N. C., Toyonaga, Y., Halldorsdottir, S. S., Mackay, T. F. C., Purugganan, M. D. and Schmitt, J. (2002). Novel loci control variation in reproductive timing in *Arabidopsis thaliana* in natural environments. *Genetics* **162**, 1875–1884.

Weinig, C., Stinchcombe, J. R. and Schmitt, J. (2003a). QTL architecture of resistance and tolerance traits in *Arabidopsis thaliana* in natural environments. *Molecular Ecology* **12**, 1153–1163.

Weinig, C., Dorn, L. A., Kane, N. C., German, Z. M., Hahdorsdottir, S. S., Ungerer, M. C., Toyonaga, Y., Mackay, T. F. C., Purugganan, M. D. and Schmitt, J. (2003b). Heterogeneous selection at specific loci in natural environments in *Arabidopsis thaliana*. *Genetics* **165**, 321–329.

Weinig, C., Stinchcombe, J. R. and Schmitt, J. (2003c). Evolutionary genetics of resistance and tolerance to natural herbivory in *Arabidopsis thaliana*. *Evolution* **57**, 1270–1280.

Werner, J. D., Borevitz, J. O., Uhlenhaut, N. H., Ecker, J. R., Chory, J. and Weigel, D. (2005a). *FRIGIDA*-independent variation in flowering time of natural *Arabidopsis thaliana* accessions. *Genetics* **170**, 1197–1207.

Werner, J. D., Borevitz, J. O., Warthmann, N., Trainer, G. T., Ecker, J. R., Chory, J. and Weigel, D. (2005b). Quantitative trait locus mapping and DNA array hybridization identify an *FLM* deletion as a cause for natural flowering-time variation. *Proceedings of the National Academy of Sciences of the United States of America* **102**, 2460–2465.

Westerman, J. M. and Lawrence, M. J. (1970). Genotype-environment interaction and developmental regulation in *Arabidopsis thaliana*.1. Inbred lines—description. *Heredity* **25**, 609–627.

A Genomics Approach to the Study of Ancient Polyploidy and Floral Developmental Genetics

JAMES H. LEEBENS-MACK,*,‡ KERR WALL,* JILL DUARTE,* ZHENGUI ZHENG,† DAVID OPPENHEIMER† AND CLAUDE DEPAMPHILIS*

*Department of Biology, Institute of Molecular Evolutionary Genetics, and Huck Institutes of Life Sciences, The Pennsylvania State University, University Park, Pennsylvania 16802
†Department of Botany and the Genetics Institute, University of Florida, Gainesville, Florida 32611

‡Present Address: Department of Plant Biology, University of Georgia, Athens, Georgia 30602.

Advances in Botanical Research, Vol. 44
Incorporating Advances in Plant Pathology

0065-2296/06 $35.00
DOI: 10.1016/S0065-2296(06)44014-3

ABSTRACT

Comparative genomics approaches are proving to be extremely valuable for the study of gene function, gene duplications, and genome evolution. In this chapter we discuss how cross-species comparisons of gene sequences and gene-expression patterns are elucidating the evolution of many plant processes including the regulation of reproduction. Emphasis is placed on the implications of gene and genome duplications for the evolution of genome structure and plant reproduction. In addition, we show that comparative analyses can both promote transfer of knowledge from model to non-model systems and inform our understanding of conserved processes in model species.

I. INTRODUCTION

As has been discussed in each chapter in this volume, much of our current understanding of flower development has been informed by cross-species comparative investigations (Albert *et al.*, 1998; Becker *et al.*, 2000; Coen and Meyerowitz, 1991; Ma and dePamphilis, 2000). This work is built on a strong foundation of forward genetics (see Davies *et al.*, Chapter 7; Irish, Chapter 3; Kramer and Zimmer, Chapter 9; Zahn *et al.*, Chapter 4,) and a growing understanding of the phylogenetic relationship among plant lineages with contrasting floral morphologies (Endress, Chapter 1; Soltis *et al.*, 2005; Zanis *et al.*, 2003). In recent years, genome and transcriptome analyses have also added to our understanding of genes involved in the regulation of flowering time (Schmid *et al.*, 2003) and floral development (Albert *et al.*, 2005; Laitinen *et al.*, 2005; Wellmer *et al.*, 2004; Zik and Irish, 2003). In this chapter, we discuss the comparative genomics approach as a useful way of identifying genes and noncoding sequences that may be involved in floral development. We also discuss the utility of comparative genomics for testing properly framed hypotheses. Finally, we consider new high-throughput technologies that promise to expand the scope and impact of comparative genomics.

A. PHYLOGENETIC CONTEXT

The improving resolution of phylogenetic relationships among plant lineages (Fig. 1) is providing the historical context necessary to understand events associated with the origin and diversification of seed plants (Burleigh and Mathews 2004), angiosperms (Davies *et al.*, 2004; Leebens-Mack *et al.*, 2005; Qiu *et al.*, 2005; Zanis *et al.*, 2002) and specific flowering plant lineages (Beilstein *et al.*, 2006 for the Brassicaceae; Malcomber *et al.*, Chapter 11 for the Poaceae). These advances have paved the way for comparative genomic studies aimed at understanding associations between organismal diversification and genome evolution (Bowers *et al.*, 2003; Buzgo *et al.*, 2005;

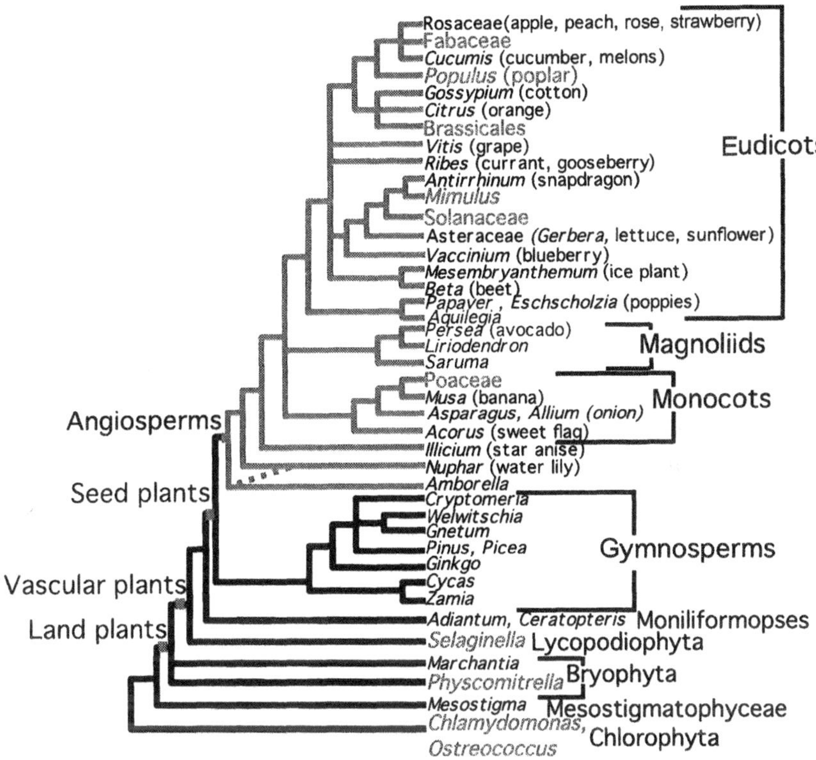

Fig. 1. Phylogenetic distribution of green plant taxa with large cDNA sequence sets (ESTs and unigenes) and genome sequencing projects (highlighted) complete or in progress. Topology and taxonomy are taken from Peter Stevens' angiosperm phylogeny website (http://www.mobot.org/MOBOT/research/APweb/), Pryer *et al.*, 2001 and Marin and Melkonian (1999).

Kellogg and Bennetzen, 2004; Paterson *et al.*, 2004; Soltis *et al.*, 2002; Vandepoele and Van de Peer, 2005; Zahn *et al.*, 2005a).

The analytical and conceptual tools of comparative genomics can be applied to questions pertaining to the entire continuum of evolutionary time scales. The precise question that can be addressed most effectively through comparative genomics varies depending on the degree of divergence among the taxa being compared (Fig. 2). Whereas comparisons of closely related species and intraspecific polymorphism can help identify genes or quantitative trait loci (QTL) associated with specific phenotypic differences (Aranzana *et al.*, 2005; Lexer *et al.*, 2005) including patterns of gene expression (eQTL; Doerge, 2002; Schadt *et al.*, 2003), facilitate positional cloning (Bortiri *et al.* 2006), and aid investigation of mechanisms responsible for speciation (Hey *et al.*, 2005 and other papers in this issue of PNAS devoted

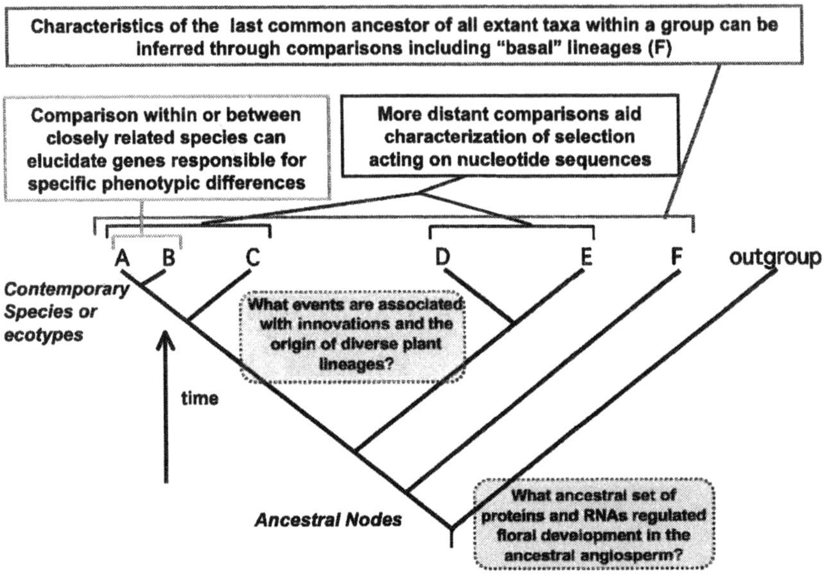

Fig. 2. Taxon sampling for comparative genomic analyses depends on the questions to be addressed. Sampling within species or among closely related species can elucidate the genetic basis of phenotypic differences, while sampling among more divergent species is necessary to investigate events associated with the origin and diversification of ancient groups such as the eudicots, monocots, angiosperms, or seed plants (after Hardison, 2003; dePamphilis, 1995).

to Ernst Mayr; Sweigart *et al.*, 2006) and domestication (Burke *et al.*, 2002; Clark *et al.*, 2004; Nesbitt and Tanksley, 2002; Yamasaki *et al.*, 2005), comparisons of more divergent genomes are useful for identifying conserved noncoding sequences that may have regulatory functions (Eddy, 2005; Hardison, 2003; Odenwald *et al.*, 2005; Siepel *et al.*, 2005). Understanding genetic events associated with the origin of ancient groups ranging from the grass family, or core eudicots, to all flowering plants, seed plants, or land plants also requires comparisons of increasingly divergent genomes (Soltis *et al.*, 2002). In this chapter, we describe a few examples of how comparative genomics research at each of the levels depicted in Fig. 2 has added greatly to our understanding of plant reproductive biology.

B. GENOMIC APPROACHES

"Genomic approaches" are typically high-throughput methods that provide a view of genetic variation across gene families, transcriptomes, genomic regions, or whole genomes. One may apply genomic approaches to test hypotheses and elucidate biological processes all along this continuum.

Transcriptome sequencing [e.g., expressed sequence tag (EST) sequencing], massively parallel signature sequencing (MPSS; Brenner *et al.*, 2000), microarray analyses, use of functional tools including targeting local lesions in genomes (TILLING; reviewed by Comai and Henikoff, 2006) or virus-induced gene silencing (VIGS; reviewed by Burch-Smith *et al.*, 2004), and genomic sequencing are just some of the high-throughput techniques that can yield data for comparative genomic analyses. In this chapter, we focus primarily on comparative genomic analyses of sequence and gene expression data aimed at understanding various aspects of plant reproduction.

Appropriate analyses of genomic data are developed in order to address questions of interest, and the field of bioinformatics has emerged and is growing in response to the demands that come with staggering increases in the amount of genomic data. Of particular importance for comparative genomics is the development of searchable databases with cDNA sequences (Albert *et al.*, 2005; Dong *et al.*, 2005; Lee *et al.*, 2005; Rudd, 2005) and repeat element sequences (Ouyang and Buell, 2004) for multiple species. Powerful phylogenomic analysis pipelines have been used to efficiently mine sequence databases such as these, construct alignments, and build gene family phylogenies (Chapman *et al.*, 2004; Hartmann *et al.*, 2006; Sjölander, 2004). Comparative analyses of data extracted from sequence and gene family databases are contributing to advances in plant reproductive biology, and this trend will continue as the volume of data rapidly increases and more investigators are trained how to build analysis pipelines and tailor them to specific research questions.

II. WIDESPREAD POLYPLOIDY IN ANGIOSPERM HISTORY

Botanists have long understood that polyploidy has been an important force in angiosperm history (Grant, 1981; Soltis, 2005; Soltis and Soltis, 1999; Stebbins, 1950). Analyses of chromosome numbers have suggested that many extant angiosperms are ancient polyploids (Grant, 1963; Otto and Whitton, 2000). Despite the small size of the *Arabidopsis thaliana* genome (157 Mb/C; Bennett *et al.*, 2003), a striking observation from early analyses of these data was that much of the genome consisted of large duplicated segments, suggesting a history of repeated rounds of ancient polyploidy (Blanc *et al.*, 2000; Bowers *et al.*, 2003; Simillion *et al.*, 2002; Vision *et al.*, 2000). The number and timing of genome duplication events came into better focus when the duplicated blocks were analyzed in the context of sequence data from pine species, monocots, asterids, and other rosids (Blanc *et al.*, 2003; Bowers *et al.*, 2003). Bowers *et al.* (2003) performed high-throughput phylogenetic analyses on genes found in duplicated blocks and inferred an ancient genome duplication event in the common ancestor of *Brassica* and *Arabidopsis*; a second event in the common ancestor of asterids and rosids, and possibly a third event predating the divergence of monocots and eudicots (Fig. 1). Ancient polyploidy is also evident as large duplicated blocks in the genome sequences of rice (*Oryza sativum*; Paterson *et al.*, 2004; Yu *et al.*, 2005) and *Populus trichocarpa* (Tuskan *et al.*, submitted for publication).

Analyses of EST data have implicated additional ancient polyploidization events throughout the angiosperms (Blanc and Wolfe, 2004; Cui *et al.*, 2006; Schlueter *et al.*, 2004). Following the earlier work of Lynch and Conery (2000), all-against-all BLAST searches (Altschul *et al.*, 1997) of large sets of coding sequences sampled from a species (e.g., EST or unigene sequences) can be used to identify putative paralog pairs, which can be aligned in coding frame. The number of synonymous changes per synonymous sites (K_s) is then estimated for each paralog pair alignment using yn00 or codeml in PAML (Yang 1997) or similar routines in HyPhy (Pond *et al.*, 2005), and the frequency distribution of K_s values can then be plotted. The underlying distribution of K_s plots is expected to reflect a background rate of gene duplication and extinction, with a peak near $K_s = 0$ and an exponentially decreasing frequency of duplicate gene pairs with increasing values of K_s (Fig. 3; Blanc and Wolfe, 2004). A secondary spike in the K_s distribution (Fig. 3) would only be expected if a large number of gene duplications occurred at the same time in the past. Therefore, a spike in the K_s distribution can be interpreted as indicating an ancient polyploidy event (including partial genome duplications; but see Hughes *et al.*, 2003) or a concerted increase in transposon activity (Hughes *et al.*, 2003). However, not all polyploidy events can be observed in K_s plots. Paralog pairs from polyploidy events may be indistinguishable from background single gene duplications or allelic variants ($K_s < 0.05$). Further, sampling error in the substitution process leads to increased variance in K_s with time (Fig. 3; Cui *et al.*, 2006). Finally, gene loss and incomplete sampling of a proteome may reduce the signal of ancient polyploidy in K_s plots. Therefore, whereas ancient polyploidy can be inferred from K_s plots, the absence of a spike in K_s, should not be interpreted as an absence of polyploidy in a species' ancestry.

Despite the limitations of K_s plots for inferring polyploidy, analyses of EST sequences from representatives of most major flowering plant lineages do provide evidence of frequent genome duplications throughout angiosperm history (Blanc and Wolfe, 2004; Cui *et al.*, 2006; Schlueter *et al.*, 2004). An understanding of ancient genome duplication in the basal-most angiosperm lineages is especially important for elucidating the role polyploidy may have played in the origin and early diversification of flowering plants (de Bodt *et al.*, 2005; Zahn *et al.*, 2005a,b). While K_s analyses of basal

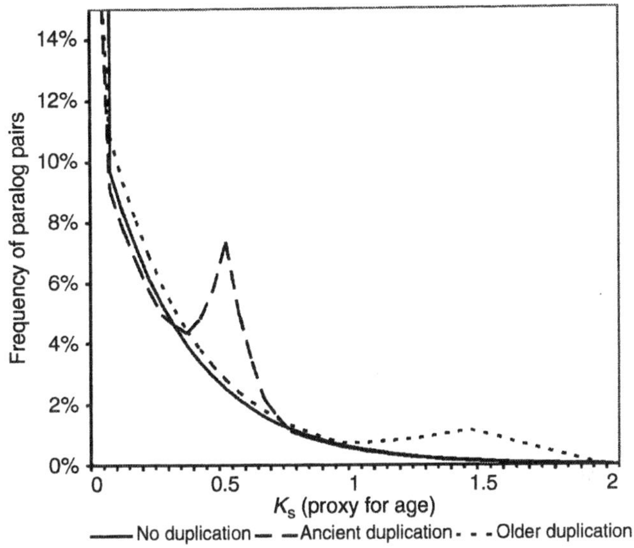

Fig. 3. The frequency distribution of K_s for paralog pairs can indicate ancient polyploidy events (e.g., secondary peaks shown for dashed lines), but not all genome duplications will be detected in K_s plots. Gene loss and increasing variance in K_s with time would both erode secondary peaks in these plots.

angiosperm lineages (Cui *et al.*, 2006) and phylogenomic analyses of duplicated blocks in the *Arabidopsis* genome (Bowers *et al.*, 2003; Sampedro *et al.*, 2005) provide evidence for paleopolyploidy in early angiosperm history, the precise timing of genome duplications relative to the origin of extant flowering plant lineages will require further investigation. At the same time, much research is now focused on the fate of duplicated genes (Adams *et al.*, 2003; Casneuf *et al.*, 2006; Chapman *et al.*, 2006; Maere *et al.*, 2005; Moore and Purugganan, 2005; others reviewed by Adams and Wendel 2005) and shifting function in retained duplicates (Duarte *et al.*, 2006; Force *et al.*, 1999; Lynch and Conery, 2000). Floral evolution has been linked to gene duplications (Irish, 2003, Chapter 3), and future research will investigate the apparent association between ancient polyploidy and innovations in plant reproduction.

III. IMPLICATIONS OF ANCIENT POLYPLOIDY FOR COMPARATIVE GENOMICS

A. ORTHOLOGS, HOMEOLOGS, AND PARALOGS

The discovery of widespread ancient polyploidy throughout the history of angiosperms complicates our understanding of orthology and paralogy in flowering plants. Orthologs and paralogs are defined as genes that originated from speciation or duplication events, respectively (Sonnhammer and Koonin, 2002; Theissen, 2002) and homeologs (or paleologs) are paralogs that originate from genome duplication. However, if genome duplication is a recurrent phenomenon in angiosperms, then any two distantly related angiosperms will be separated by one or more genome duplications. For example, according to our understanding of the polyploid histories of lineages leading to *Arabidopsis* and *Oryza*, at least three genome duplications have occurred since these species shared a common ancestor: one in the early history of the Brassicaceae, one before the diversification of the major core eudicot lineages, and one in the early history of the Poaceae. Therefore, even genes with what appear to be simple orthologous relationships in *Oryza* and *Arabidopsis*—for example *LEAFY* and its rice "ortholog" *RFL*—are in fact the survivors of a complex history of duplication and loss of duplicate copies. Given this dynamic nature of plant genome histories, phylogenetic analyses of gene families must be performed on a genomic scale in order to address issues ranging from the prediction of gene and protein function (Eisen, 1998; Engelhardt *et al.*, 2005; Sjölander, 2004) to the influences of polyploidy on genome content and structure (Bowers *et al.*, 2003; Rong *et al.*, 2005), as well as the evolution of regulatory networks influencing floral development (see later section).

B. CHARACTERIZING THE FATE OF DUPLICATED GENES

The polyploid histories of flowering plant genomes provide global opportunities for selective expansion of specific kinds of genes. For example, regulatory genes (Blanc and Wolfe, 2004; Maere *et al.*, 2005), and genes that encode long complex proteins (Chapman *et al.*, 2006), may be more likely to survive genome duplication. Analyses performed by Maere *et al.* (2005) suggest that duplicate regulatory genes are more likely to be retained following polyploidy events relative to single gene duplications. In contrast, many other genes may be particularly resistant to retention of duplicates. Such genes have existed over long periods of time as singletons or low-copy genes in the face of whole-genome duplications, implying that selection against duplicate copies is more intense for these genes.

Chapman *et al.* (2006) presented a slightly different interpretation of single-copy genes (singletons). Focusing on adaptive retention of functionally redundant duplicate genes that may buffer critical functions in developmentally and genetically unstable polyploids, the authors present evidence that single-copy genes may simply be genes for which duplicate copies offer no selective advantage. Analysis of intraspecific single nucleotide substitution polymorphisms (SNPs) revealed that genes retained in duplicate following the most recent polyploidizations in lineages leading to rice and *Arabidopsis* tended to have a lower ratio of amino acid replacement substitutions to nonreplacement substitutions (dN/ds) relative to singleton genes (Chapman *et al.*, 2006). This pattern implies that singletons evolve under *less* severe purifying selection than genes that have been retained as duplicates.

The intensity of purifying selection may not have anything to do with selection for retention or extinction of duplicate gene copies. Based on gene-clustering analyses (Enright *et al.*, 2002, 2003), we estimate that 727 strict ortholog sets exist as single-copy genes in the *Arabidopsis*, rice, and *Populus* genomes (Wall *et al.*, in preparation). This is a much larger number than would be expected if gene deaths were random following gene and genome duplications. The estimated frequencies of singletons in the *Arabidopsis*, rice, and *Populus* genomes are at most 15% (singletons/total gene number = 3862/26,207), 21% (11,954/57,915), and 12% (5396/45,555), respectively. The lineages leading to *Arabidopsis*, rice, and *Populus* have each experienced at least one genome duplication event that is independent of polyploidy events in the other lineages. Therefore, if gene extinctions were independent in these three lineages, we would expect the frequency of shared singletons to be the product of singleton frequencies in each genome multiplied by the number of genes in the smallest proteome (15% × 21% × 12% × 26,207 = 99). The expected value is less then one-seventh of the observed number of shared singletons, so gene deaths must not be random. This result could be explained, at least in part, by maintenance of duplicate genes (Chapman *et al.*, 2006), since the percentage of singletons among genes that were susceptible to extinction within each species would be higher if a fraction of duplicates were selectively maintained. However, even if one assumes that half of the genes are selectively maintained in duplicate, there are still many more shared singletons than expected. We surmise that a large fraction of shared singletons in the three sequenced angiosperm genomes are selectively maintained as such, and are not simply the survivors of random gene loss in the absence of selection for retention of duplicate gene copies. Selection for preservation of dosage balance could be an important force maintaining some genes in single-copy following gene and genome-duplication events.

C. A GENE FAMILY PERSPECTIVE ON GENOME DUPLICATIONS

Duplication patterns that are observed in phylogenetic studies of gene families provide a view of family history that can be interpreted in the context of known genome duplication events. For example, the *HUA* enhancer 2 family (*HEN2*) is part of a small family of putative DExH box RNA helicase enzymes (Western *et al.*, 2002). *HEN2* mutants display defects in petal number and position as well as phyllotaxy and floral number (Western *et al.*, 2002). Parsimony analysis of members of this family (Fig. 4) identified in *Arabidopsis*, rice, and basal angiosperm sequences mined from the Floral Genome Project PlantTribes database (http://fgp.huck.psu.edu/tribe.php; Albert *et al.*, 2005), suggest that three or four clades were established before the early diversification of angiosperm lineages. Surprisingly, there is a single *Arabidopsis* gene and between zero and two rice homologs in each of these clades. Thus, while none of the *HEN2* family genes were counted among the singletons described in an earlier section, most of the duplicate genes that have been generated through multiple rounds of polyploidy in angiosperm history have not survived.

The utility of gene family analyses placed in the context of genome duplication events is also illustrated in an analysis of the expansin gene family (Sampedro *et al.*, 2005). Expansins are cell wall loosening proteins that exist as a multigene family in all plants (Cosgrove, 2005; Sampedro and Cosgrove, 2005).

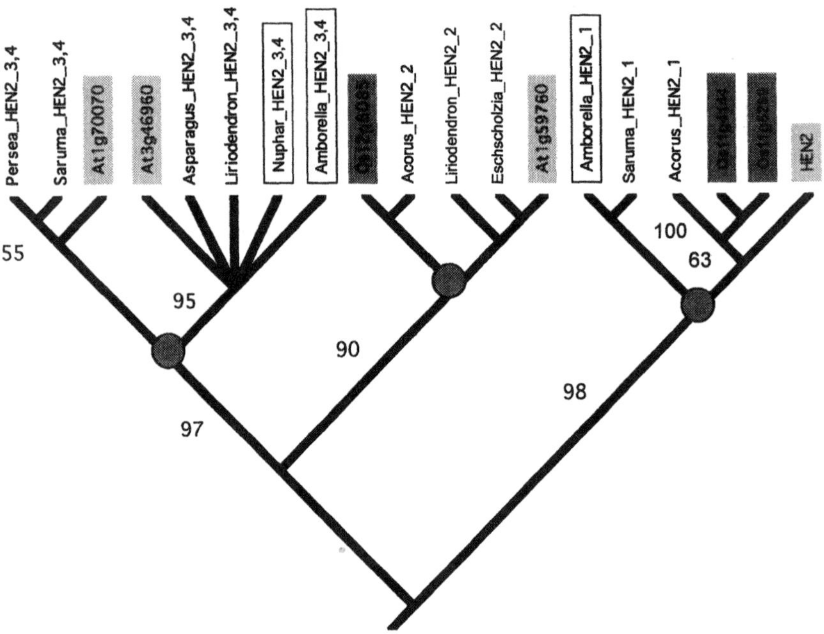

Fig. 4. A gene phylogeny for the *HEN2* RNA helicase gene family with single *Arabidopsis* genes in each of four clades. Each of the four clades include monocot, magnoliid, or basal-most angiosperm (outlined) genes, indicating at least two gene deaths in each *Arabidopsis* (light shading) gene lineage following the genome duplications in polyploid Brassicaceae and core eudicot ancestors. Gene deaths are also evident in the lineage leading to rice (dark shading) following a genome-wide duplication in the early history of the Poaceae. Bootstrap values are shown to the left or right of each branch when greater than 50%.

Although phylogenetic analysis was only partly able to resolve the history of the gene family, careful examination of duplicated blocks of genes in rice and *Arabidopsis* aided the resolution of the phylogeny and showed that nearly every member of the gene family in *Arabidopsis* could be accounted for as the product of genome duplications (Sampedro *et al.*, 2005). Thus, Sampedro *et al.* demonstrated that consideration of genomic context improves phylogenetic resolution of complex gene family histories.

As has been described in many other chapters in this volume, the MADS-box gene family has been intensively studied in terms of both gene duplication and functional diversification (Becker and Theissen, 2003; Irish, 2003, Chapter 3; Kim *et al.*, 2005; Kramer and Hall, 2005; Soltis *et al.*, Chapter 12; Zahn *et al.*, 2005a,b, 2006). Given the role of many MADS-box genes in the regulation of flowering time and floral organ specification, it has been hypothesized that gene duplications and subsequent functional shifts have been a driving force behind reproductive innovations (e.g., preceding references). Duplications in multiple MADS-box gene subfamilies coincide with major events in angiosperm history, most notably the earliest diversification of extant flowering plants and the diversification in the major core eudicot lineages (Fig. 1). Whole genome duplications have also been hypothesized for these nodes of the angiosperm phylogeny (Bowers *et al.*, 2003; Buzgo *et al.*, 2005; Cui *et al.*, 2006; Zahn *et al.*, 2005a), but it has not been shown conclusively whether or not ancient polyploidy events spawned the diversification of MADS-box genes in the common ancestors of all core eudicots or all extant flowering plant lineages.

D. SHIFTS IN SELECTIVE CONSTRAINT

Mechanistic hypotheses are required for genome-wide investigations of the relationship between gene and genome duplications and the evolution of plant reproduction. Most studies of functional evolution following gene duplication have built on a model of evolution wherein selection may be relaxed as the result of functional redundancy immediately following duplication events. However, evolutionary constraint is eventually restored after one duplicate becomes a pseudogene, ancestral gene function is split between the two duplicates (subfunctionalization), or one of the duplicates takes on new function (neofunctionalization) (Lynch and Conery, 2000; Force *et al.*, 1999; Ohno, 1970). Whereas these models predict that functional redundancy between duplicate genes is a temporary (nonequilibrium) condition, the notion of an adaptive "error-buffering" role for functional redundancy mentioned in an earlier section, provides an alternative explanation for the maintenance of duplicate genes (Chapman *et al.*, 2006; Hileman and Baum, 2003; Nowak *et al.*, 1997; Moore *et al.*, 2005). Under these adaptive redundancy models, developmental instability in gene expression would have to be so deleterious that natural selection would favor individuals with functionally redundant gene copies over those with single copies of some genes. This hypothesis could be tested in populations of synthetic hybrids (Wang *et al.*, 2006), although subtle differences in fitness may be difficult to detect. Alternatively, a phylogenetically based retrospective approach may (or may not) detect even subtle changes in selective constraint following duplication events (see later section).

Hileman and Baum (2003) also proposed that duplicate genes may be retained if expression levels for both gene copies were reduced such that both genes would be required to maintain ancestral gene product dosage. This additive dosage model, described by Force *et al.* (1999) as "quantative subfunctionalization," is distinct from the more commonly hypothesized form of subfunctionalization in that there is no differential tissue or stage-specific compartmentalization of gene expression (Hileman and Baum, 2003). Duarte *et al.* (2006) identified instances of "hypofunctionalization" in their analysis of microarray expression profiles for duplicate genes where one duplicate was expressed at much lower levels than the other, but the degree of expression level divergence between duplicates was rarely constant across all organs. In addition, Duarte *et al.* (2006) identified significant gene by organ interactions (divergence of gene-expression patterns) in the majority of their ANOVA-based comparisons of expression levels for duplicate gene pairs. This result is consistent with models of regulatory sub- and neofunctionalization following gene duplication.

Changes in the mode of selection on protein-coding regions of gene sequences are often diagnosable through analyses of substitution rates (see reviews in Nielsen, 2005; Yang, 2002). Maximum likelihood estimates of per site synonymous (dS [$=K_s$]) and nonsynonymous (dN) nucleotide substitution frequencies, and the ratio of these substitution types (dN/dS $= \omega$) can be estimated from nucleotide alignments in a pair-wise fashion as described in Section II, or within the context of a gene phylogeny using codon-based models of sequence evolution (Goldman and Yang, 1994; Muse and Gaut, 1994; Nielsen and Yang, 1998; Yang *et al.*, 2005) as implemented in the codeml program of PAML (Yang, 1997) or HyPhy (Pond *et al.*, 2005). Bayesian estimates of these parameters can be obtained using MrBayes as described by Huelsenbeck and Dyer (2004), and Bayesian estimates of site-specific rate ratios (ω) are provided in the codeml output (Yang *et al.*, 2005).

There has been an explosion of interest in analyses of dN/dS ratios aimed at detecting adaptive amino acid changes (positive selection) associated with changing gene function within gene families (Barkman, 2003; Yang, 1998) or across the whole genome (Bustamante *et al.*, 2005; Clark *et al.*, 2003; Nielsen *et al.*, 2005). Shifts in gene expression and function may be driven by changes in noncoding regulatory elements rather than protein-coding sequences (Doebley and Lukens, 1998), and adaptive divergence in protein-coding sequences is not necessary under the *functional divergence model* for retention of duplicated genes (Lynch and Conery, 2000; Force *et al.*, 1999; Ohno, 1970). The model does, however, predict that selection would be relaxed immediately following gene duplication. In contrast, the *developmental instability-buffering model* (Chapman *et al.*, 2006; Hileman and Baum, 2003; Moore *et al.*, 2005; Nowak *et al.*, 1997) predicts that purifying selection would not be relaxed following duplication events. Tree-based analysis of dN/dS ratios could test the null hypothesis that selective constraint averaged across the coding sequence does not change following duplication events (see the "branch" model of Yang, 1998; Barkman, 2003). While rejection of this null hypothesis might favor the functional divergence model, failure to reject would not necessarily favor the developmental instability-buffering model. A short period of relaxed selection following duplication may be difficult to detect, and a power analysis (Leebens-Mack and dePamphilis, 2002) would be required in order to interpret failure to reject the null hypothesis of equal selective constraint (dN/dS) before and after duplication. Further, if dN/dS is quite variable across a coding sequence, a "branch X sites" model (Yang and Nielsen, 2002) would provide more power for detecting relaxed selection in the portions of a gene that were more highly conserved before duplication. Nam *et al.* (2005) found significant variation in amino acid substitution rates across regions of duplicate MIKC-type MADS genes, suggesting that dN/dS does vary across coding sequences. Moreover, analyses that characterize this variation can identify specific domains that contribute to functional divergence (Nam *et al.*, 2005).

IV. COMPARATIVE ANALYSES OF DISTANTLY RELATED TAXA ELUCIDATE GENE FUNCTION IN *ARABIDOPSIS*

Comparative analyses are also providing insights into gene function in *Arabidopsis* and other model systems. Investigations of single-copy gene families are especially interesting and straightforward because duplicates arising from repeated polyploidy events (see Section II) may have been selectively culled due to dosage constraints, and functional studies employing reverse genetics are less likely to be confounded by redundancy. At the same time, a high proportion of these genes have been annotated as "hypothetical" or "expressed" proteins because unlike members of larger gene families, the annotation process is not aided by similarity to functionally characterized genes.

We have been combining comparative and functional approaches to investigate an interesting class of single-copy genes that are found in a wide array of plant species, but seem to have been lost in the grasses (Poaceae). Using the search tools in PlantTribes (http://fgp.huck.psu.edu/tribe.php) we identified approximately 1500 single-copy genes in *Arabidopsis* that have no orthologs in rice. Of these, a subset of about 500 genes had "hypothetical" or "expressed" protein annotations. The predicted protein sequences of these genes were used to search the FGP Unigene database, the Plant Genome Database (PlantGDB http://www.plantgdb.org), The Institute for Genomic Research maize genome database (TIGR, http://tigrblast.tigr.org/tgi_maize/index.cgi), the moss database (COSMOS, http://www.cosmoss.org) and the *Chlamydomonas* genome database (http://www.chlamy.org) using TBLASTN (Altschul *et al.*, 1997). Those sequences that showed hits in other plants, but no hits in grasses, were chosen for further study. One of these genes, Di05 (At5g48480), has orthologs in a moss, a fern (*Ceratopteris*), gymnosperms, basal angiosperms, eudicots, and the nongrass monocots *Asparagus* and *Yucca*, but no orthologs were found in any members of the Poaceae. This is remarkable given the large number of expressed gene sequences available for multiple species in the grass family.

An alignment of Di05 orthologs was constructed using CLUSTALW (Thompson *et al.*, 1994) and refined using Se-AL (http://evolve.zoo.ox.ac.uk/software.html?id = seal). A phylogenetic tree was derived from parsimony analysis in PAUP* (Swofford, 2003). The resulting tree is consistent with known organismal relationships (Fig. 1), and supports the hypothesis that Di05 homologs existed in the common ancestor of mosses, ferns, gymnosperms, and angiosperms, but the gene was lost on the monocot branch leading to the Poaceae.

In *Arabidopsis* seedlings, Di05 is strongly expressed in the shoot apical meristem and leaf primordia. In reproductive structures, transcripts could be detected in the floral apical meristems, and floral primordia, but only in developing pollen and ovules after stage 9 (Fig. 5). During seed development, Di05 is strongly expressed in the developing embryo. After seed germination, it is only expressed in the shoot and root tips. These results suggest that Di05 may be involved in cell or tissue differentiation. In *Eschscholzia* and *Persea* flowers, the Di05 expression pattern is very similar to that seen in *Arabidopsis* (Fig. 5). Interestingly, in the *Ceratopteris* sporophyte, Di05 is only expressed in the shoot and root tips, but not in the developing or

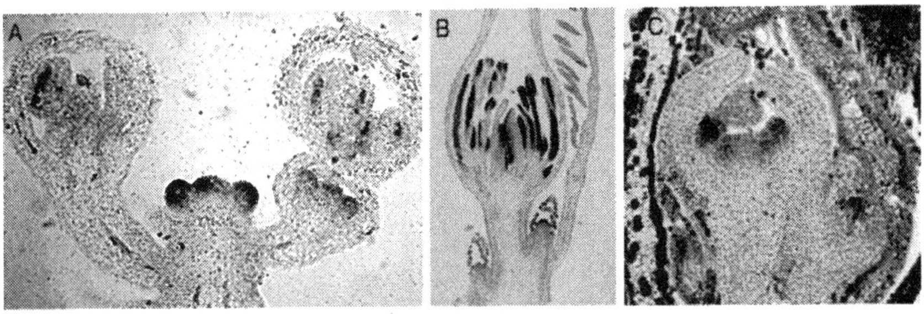

Fig. 5. *In situ* hybridizations show similar expression of an uncharacterized single-copy gene in reproductive meristems, pollen and ovules in *Arabidopsis* (A), *Eschscholzia* (B), and *Persea* (C).

mature spores. These results suggest that Di05 expression (and perhaps function) is conserved in eudicots, basal angiosperms, and perhaps ferns. Further investigation is required to understand the function of Di05 and the consequence of its loss some time after the divergence of the Asparagales and commelinid lineages within the monocots.

To further define the function of single-copy genes, we are examining available *Arabidopsis* T-DNA insertion lines. By focusing on those organs and tissues in which expression was detected by *in situ* hybridization, we can quickly identify phenotypes associated with the T-DNA insertion. Using this strategy, we are successfully identifying phenotypes for several genes, and thus elucidating gene function.

V. FUTURE PROSPECTS: DEVELOPING A GENE FAMILY FRAMEWORK TO CHARACTERIZE PLANT GENE AND GENOME EVOLUTION

This is an exciting time in plant genomics. The number of plant genome sequencing projects is expanding (Fig. 1), and this trend will continue with technological advances (Margulies *et al.*, 2005). Increasingly powerful analytical tools are being developed to allow more questions to be addressed through comparative analyses. The high frequency of genome duplications and complicated gene birth-and-death process in plants relative to animals pose challenges to phylogenomic analyses aimed at transferring understanding of gene function from model to nonmodel systems (Eisen, 1998; Engelhardt *et al.*, 2005; Sjölander, 2004), but much can be gained through analyses of sequence evolution and variation in gene expression within gene families. What is more, inferences concerning the evolution of genome structure are being drawn from analyses of gene family phylogenies placed in the context of the chromosomal positions of duplicated genes (or gene blocks) (Bowers *et al.*, 2003; Mudge *et al.*, 2005; Paterson *et al.*, 2004; Sampedro *et al.*, 2005). A number of research groups are now independently developing databases for plant gene families (Albert *et al.*, 2005; Cannon *et al.*, 2004; Hartmann *et al.*, 2006), and these efforts are laying the foundation for evolutionary analyses of changing gene function and genome structure that may be associated with innovations in plant reproduction.

ACKNOWLEDGMENTS

We would like to thank all participants in the Plant Reproductive Genomics Workshop held at the Plant and Animal Genome Conference in 2005. We also thank all those involved in the Floral Genome Project for many stimulating discussions on the points raised in this chapter, especially André Chanderbali, Sangtae Kim, Doug Soltis, and Pam Soltis for their thoughtful comments on an early version of our manuscript. Finally, we thank the National Science Foundation Plant Genome Research Program for support of the Floral Genome Project (DBI-0115684).

REFERENCES

Adams, K. L. and Wendel, J. F. (2005). Polyploidy and genome evolution in plants. *Current Opinion in Plant Biology* **8**, 135–141.

Adams, K. L., Cronn, R., Percifield, R. and Wendel, J. F. (2003). Genes duplicated by polyploidy show unequal contributions to the transcriptome and organ-specific reciprocal silencing. *Proceedings of the National Academy of Sciences of the United States of America* **100**, 4649–4654.

Albert, V. A., Soltis, D. E., Carlson, J. E., Farmerie, W. G., Wall, P. K., Ilut, D. C., Solow, T. M., Mueller, L. A., Landherr, L. L., Hu, Y., Buzgo, M., Kim, S., *et al.* (2005). Floral gene resources from basal angiosperms for comparative genomics research. *BMC Plant Biology* **5**, 5.

Albert, V. A., Gustafsson, M. H. G. and Di Laurenzio, L. (1998). Ontogenetic systematics, molecular developmental genetics, and the angiosperm petal. *In* "Molecular Systematics of Plants II DNA Sequencing" (D. E. Soltis, P. S. Soltis and J. J. Doyle, eds.), pp. 349–374. Kluwer Academic Publishers, Boston.

Altschul, S. F., Madden, T. L., Schaffer, A. A., Zhang, J., Zhang, Z., Miller, W. and Lipman, D. J. (1997). Gapped BLAST and PSI-BLAST: A new generation of protein database search programs. *Nucleic Acids Research.* **25**, 3389–3402.

Aranzana, M. J., Kim, S., Zhao, K., Bakker, E., Horton, M., Jakob, K., Lister, C., Molitor, J., Shindo, C., Tang, C., Toomajian, C., Traw, B., *et al.* (2005). Genome-wide association mapping in *Arabidopsis* identifies previously known flowering time and pathogen resistance Genes. *PLoS Genetics* **1**, e60.

Barkman, T. J. (2003). Evidence for positive selection on the floral scent gene isoeugenol-O-methyltransferase. *Molecular Biology and Evolution* **20**, 168–172.

Becker, A. and Theissen, G. (2003). The major clades of MADS-box genes and their role in the development and evolution of flowering plants. *Molecular Biology and Evolution* **29**, 464–489.

Becker, A., Winter, K. U., Meyer, B., Saedler, H. and Theissen, G. (2000). MADS-Box gene diversity in seed plants 300 million years ago. *Molecular Biology and Evolution* **17**, 1425–1434.

Beilstein, M. A., Al-Shehbaz, I. A. and Kellogg, E. A. (2006). Brassicaceae phylogeny and trichome evolution. *American Journal of Botany* **93**, 607–619.

Bennett, M. D., Leitch, I. J., Price, H. J. and Johnston, J. S. (2003). Comparisons with *Caenorhabditis* (approximately 100 Mb) and *Drosophila* (approximately 175 Mb) using flow cytometry show genome size in *Arabidopsis* to be approximately 157 Mb and thus approximately 25% larger than the *Arabidopsis* genome initiative estimate of approximately 125 Mb. *Annals of Botany (London)* **91**, 547–557.

Blanc, G. and Wolfe, K. H. (2004). Widespread paleopolyploidy in model plant species inferred from age distributions of duplicate genes. *Plant Cell* **16**, 1667–1678.

Blanc, G., Barakat, A., Guyot, R., Cooke, R. and Delseny, M. (2000). Extensive duplication and reshuffling in the *Arabidopsis* genome. *Plant Cell* **12**, 1093–1101.

Blanc, G., Hokamp, K. and Wolfe, K. H. (2003). A recent polyploidy superimposed on older large-scale duplications in the *Arabidopsis* genome. *Genome Research* **13**, 137–144.

Bortiri, E., Jackson, D. and Hake, S. (2006). Advances in maize genomics: The emergence of positional cloning. *Current Opinion in Plant Biology* **9**, 164–171.

Bowers, J. E., Chapman, B. A., Rong, J. and Paterson, A. H. (2003). Unravelling angiosperm genome evolution by phylogenetic analysis of chromosomal duplication events. *Nature* **422**, 433–438.

Brenner, S., Johnson, M., Bridgham, J., Golda, G., Lloyd, D. H., Johnson, D., Luo, S., McCurdy, S., Foy, M., Ewan, M., Roth, R., George, D., *et al.* (2000). Gene expression analysis by massively parallel signature sequencing (MPSS) on microbead arrays. *Nature of Biotechnology* **18**, 630–634.

Burch-Smith, T. M., Anderson, J. C., Martin, G. B. and Dinesh-Kumar, S. P. (2004). Applications and advantages of virus-induced gene silencing for gene function studies in plants. *Plant Journal* **39**, 734–746.

Burke, J. M., Tang, S., Knapp, S. J. and Rieseberg, L. H. (2002). Genetic analysis of sunflower domestication. *Genetics* **161**, 1257–1267.

Burleigh, J. G. and Mathews, S. (2004). Phylogenetic signal in nucleotide data from seed plants: Implications for resolving the seed plant tree of life. *American Journal of Botany* **91**, 1599–1613.

Bustamante, C. D., Fledel-Alon, A., Williamson, S., Nielsen, R., Hubisz, M. T., Glanowski, S., Tanenbaum, D. M., White, T. J., Sninsky, J. J., Hernandez, R. D., Civello, D., Adams, M. D., *et al.* (2005). Natural selection on protein-coding genes in the human genome. *Nature* **437**, 1153–1157.

Buzgo, M., Soltis, P. S., Kim, S. and Soltis, D. E. (2005). The making of a flower. *The Biologist* **52**, 149–154.

Cannon, S. B., Mitra, A., Baumgarten, A., Young, N. D. and May, G. (2004). The roles of segmental and tandem gene duplication in the evolution of large gene families in *Arabidopsis thaliana*. *BMC Plant Biology* **4**, 10.

Casneuf, T., De Bodt, S., Raes, J., Maere, S. and Van de Peer, Y. (2006). Nonrandom divergence of gene expression following gene and genome duplications in the flowering plant *Arabidopsis thaliana*. *Genome Biology* **7**, R13.

Chapman, B. A., Bowers, J. E., Schulze, S. R. and Paterson, A. H. (2004). A comparative phylogenetic approach for dating whole genome duplication events. *Bioinformatics* **20**, 180–185.

Chapman, B. A., Bowers, J. E., Feltus, F. A. and Paterson, A. H. (2006). Buffering of crucial functions by paleologous duplicated genes may contribute cyclicality to angiosperm genome duplication. *Proceedings of the National Academy of Sciences of the United States of America*.

Clark, A. G., Glanowski, S., Nielsen, R., Thomas, P. D., Kejariwal, A., Todd, M. A., Tanenbaum, D. M., Civello, D., Lu, F., Murphy, B., Ferriera, S., Wang, G., *et al.* (2003). Inferring nonneutral evolution from human-chimp-mouse orthologous gene trios. *Science* **302**, 1960–1963.

Clark, R. M., Linton, E., Messing, J. and Doebley, J. F. (2004). Pattern of diversity in the genomic region near the maize domestication gene tb1. *Proceedings of the National Academy of Sciences of the United States of America* **101**, 700–707.

Coen, E. S. and Meyerowitz, E. M. (1991). The war of the whorls: Genetic interactions controlling flower development. *Nature* **353**, 31–37.

Comai, L. and Henikoff, S. (2006). TILLING: Practical single-nucleotide mutation discovery. *Plant Journal* **45**, 684–694.

Cosgrove, D. J. (2005). Growth of the plant cell wall. *Nature Reviews Molecular Cell Biology* **6**, 850–861.

Cui, L., Wall, P. K., Leebens-Mack, J. H., Lindsay, B. G., Soltis, D. E., Doyle, J. J., Soltis, P. S., Carlson, J. E., Arumuganathan, K., Barakat, A., Albert, V. A. Ma, H., *et al.* (2006). Widespread genome duplications throughout the history of flowering plants. *Genome Research* in press.

Davies, T. J., Barraclough, T. G., Chase, M. W., Soltis, P. S., Soltis, D. E. and Savolainen, V. (2004). Darwin's abominable mystery: Insights from a supertree of the angiosperms. *Proceedings of the National Academy of Sciences of the United States of America* **101**, 1904–1909.

De Bodt, S., Maere, S. and van de Peer, Y. (2005). Gene duplication and the evolution of angiosperms. *Trends in Ecology and Evolution* **20**, 591–597.

dePamphilis, C. W. (1995). Genes and genomes. *In* "Parasitic Plants" (M. C. Press and J. D. Graves, eds.), pp. 177–205. Chapman and Hall, London.

Doebley, J. and Lukens, L. (1998). Transcriptional regulators and the evolution of plant form. *Plant Cell* **10**, 1075–1082.

Doerge, R. W. (2002). Mapping and analysis of quantitative trait loci in experimental populations. *Nature Reviews Genetics* **3**, 43–52.

Dong, Q., Lawrence, C. J., Schlueter, S. D., Wilkerson, M. D., Kurtz, S., Lushbough, C. and Brendel, V. (2005). Comparative plant genomics resources at PlantGDB. *Plant Physiology* **139**, 610–618.

Duarte, J. M., Cui, L., Wall, P. K., Zhang, Q., Zhang, X., Leebens-Mack, J., Ma, H., Altman, N. and dePamphilis, C. W. (2006). Expression pattern shifts following duplication indicative of subfunctionalization and neofunctionalization in regulatory genes of *Arabidopsis*. *Molecular Biology and Evolution* **23**, 469–478.

Eddy, S. R. (2005). A model of the statistical power of comparative genome sequence analysis. *PLoS Biology* **3**, e10.

Eisen, J. A. (1998). Phylogenomics: Improving functional predictions for uncharacterized genes by evolutionary analysis. *Genome Research* **8**, 163–167.

Engelhardt, B. E., Jordan, M. I., Muratore, K. E. and Brenner, S. E. (2005). Protein Molecular Function Prediction by Bayesian Phylogenomics. *PLoS Computational Biology* **1**, e45.
Enright, A. J., Van Dongen, S. and Ouzounis, C. A. (2002). An efficient algorithm for large-scale detection of protein families. *Nucleic Acids Research* **30**, 1575–1584.
Enright, A. J., Kunin, V. and Ouzounis, C. A. (2003). Protein families and TRIBES in genome sequence space. *Nucleic Acids Research* **31**, 4632–4638.
Force, A., Lynch, M., Pickett, F. B., Amores, A., Yan, Y. L. and Postlethwait, J. (1999). Preservation of duplicate genes by complementary, degenerative mutations. *Genetics* **151**, 1531–1545.
Goldman, N. and Yang, Z. (1994). A codon-based model of nucleotide substitution for protein-coding DNA sequences. *Molecular Biology and Evolution* **11**, 725–736.
Grant, V. (1963). "The Origin of Adaptations." Columbia University Press, New York.
Grant, V. (1981). "Plant Speciation." Columbia University Press, New York.
Hardison, R. C. (2003). Primer on Comparative Genomics. *PLoS Biology* **3**, e58.
Hartmann, S., Lu, D., Phillips, J. and Vision, T. J. (2006). Phytome: A platform for plant comparative genomics. *Nucleic Acids Res.* **34**, D724–D730.
Hey, J., Fitch, W. M. and Ayala, F. J. (2005). Systematics and the origin of species: An introduction. *Proceedings of the National Academy of Sciences of the United States of America* **102** (Suppl. 1), 6515–6519.
Hileman, L. C. and Baum, D. A. (2003). Why do paralogs persist? Molecular evolution of CYCLOIDEA and related floral symmetry genes in Antirrhineae (Veronicaceae). *Molecular Biology and Evolution* **20**, 591–600.
Huelsenbeck, J. P. and Dyer, K. A. (2004). Bayesian estimation of positively selected sites. *Journal of Molecular Evolution* **58**, 661–672.
Hughes, A. L., Friedman, R., Ekollu, V. and Rose, J. R. (2003). Non-random association of transposable elements with duplicated genomic blocks in *Arabidopsis thaliana*. *Molecular Phylogenetics and Evolution* **29**, 410–416.
Irish, V. F. (2003). The evolution of floral homeotic gene function. *Bioessays* **25**, 637–646.
Kellogg, E. A. and Bennetzen, J. L. (2004). The evolution of nuclear genome structure in seed plants. *American Journal of Botany* **91**, 1709–1725.
Kim, S., Koh, J., Yoo, M. J., Kong, H., Hu, Y., Ma, H., Soltis, P. S. and Soltis, D. E. (2005). Expression of floral MADS-box genes in basal angiosperms: Implications for the evolution of floral regulators. *Plant Journal* **43**, 724–744.
Kramer, E. M. and Hall, J. C. (2005). Evolutionary dynamics of genes controlling floral development. *Current Opinion in Plant Biology* **8**, 13–18.
Laitinen, R. A., Immanen, J., Auvinen, P., Rudd, S., Alatalo, E., Paulin, L., Ainasoja, M., Kotilainen, M., Koskela, S., Teeri, T. H. and Elomaa, P. (2005). Analysis of the floral transcriptome uncovers new regulators of organ determination and gene families related to flower organ differentiation in *Gerbera hybrida* (Asteraceae). *Genome Research* **15**, 475–486.
Lee, Y., Tsai, J., Sunkara, S., Karamycheva, S., Pertea, G., Sultana, R., Antonescu, V., Chan, A., Cheung, F. and Quackenbush, J. (2005). The TIGR Gene Indices: Clustering and assembling EST and known genes and integration with eukaryotic genomes. *Nucleic Acids Research* **33**, D71–D74.
Leebens-Mack, J. and dePamphilis, C. (2002). Power analysis of tests for loss of selective constraint in cave crayfish and nonphotosynthetic plant lineages. *Molecular Biology and Evolution* **19**, 1292–1302.

Leebens-Mack, J., Raubeson, L. A., Cui, L., Kuehl, J. V., Fourcade, M. H., Chumley, T. W., Boore, J. L., Jansen, R. K. and Depamphilis, C. W. (2005). Identifying the basal angiosperm node in chloroplast genome phylogenies: Sampling one's way out of the Felsenstein zone. *Molecular Biology and Evolution* **22**, 1948–1963.

Lexer, C., Rosenthal, D. M., Raymond, O., Donovan, L. A. and Rieseberg, L. H. (2005). Genetics of species differences in the wild annual sunflowers, *Helianthus annuus* and *H. petiolaris. Genetics* **169**, 2225–2239.

Lynch, M. and Conery, J. S. (2000). The evolutionary fate and consequences of duplicate genes. *Science* **290**, 1151–1155.

Ma, H. and dePamphilis, C. (2000). The ABCs of floral evolution. *Cell* **101**, 5–8.

Maere, S., De Bodt, S., Raes, J., Casneuf, T., Van Montagu, M., Kuiper, M. and Van de Peer, Y. (2005). Modeling gene and genome duplications in eukaryotes. *Proceedings of the National Academy of Sciences of the United States of America* **102**, 5454–5459.

Marin, B. and Melkonian, M. (1999). Mesostigmatophyceae, a new class of streptophyte green algae revealed by SSU rRNA sequence comparisons. *Protist* **150**, 399–417.

Margulies, M., Egholm, M., Altman, W. E., Attiya, S., Bader, J. S., Bemben, L. A., Berka, J., Braverman, M. S., Chen, Y. J., Chen, Z., Dewell, S. B., Du, L., *et al.* (2005). Genome sequencing in microfabricated high-density picolitre reactors. *Nature* **437**, 376–380.

Moore, R. C. and Purugganan, M. D. (2005). The evolutionary dynamics of plant duplicate genes. *Current Opinion in Plant Biology* **8**, 122–128.

Moore, R. C., Grant, S. R. and Purugganan, M. D. (2005). Molecular population genetics of redundant floral-regulatory genes in *Arabidopsis thaliana. Molecular Biology and Evolution* **22**, 91–103.

Mudge, J., Cannon, S. B., Kalo, P., Oldroyd, G. E., Roe, B. A., Town, C. D. and Young, N. D. (2005). Highly syntenic regions in the genomes of soybean, Medicago truncatula, and *Arabidopsis thaliana*. BMC *Plant Biology* **5**, 15.

Muse, S. V. and Gaut, B. S. (1994). A likelihood approach for comparing synonymous and nonsynonymous nucleotide substitution rates, with application to the chloroplast genome. *Molecular Biology and Evolution* **11**, 715–774.

Nam, J., Kaufmann, K., Theissen, G. and Nei, M. (2005). A simple method for predicting the functional differentiation of duplicate genes and its application to MIKC-type MADS-box genes. *Nucleic Acids Research* **34**, e12.

Nesbitt, T. C. and Tanksley, S. D. (2002). Comparative sequencing in the genus Lycopersicon. Implications for the evolution of fruit size in the domestication of cultivated tomatoes. *Genetics* **162**, 365–379.

Nielsen, R. (ed.) (2005). "Statistical Methods In Molecular Evolution" (Statistics for Biology and Health Series). Springer-Verlag, New York.

Nielsen, R. and Yang, Z. (1998). Likelihood models for detecting positively selected amino acid sites and applications to the HIV-1 envelope gene. *Genetics* **148**, 929–936.

Nielsen, R., Bustamante, C., Clark, A. G., Glanowski, S., Sackton, T. B., Hubisz, M. J., Fledel-Alon, A., Tanenbaum, D. M., Civello, D., White, T. J., J., J. S., Adams, M. D. and Cargill, M. (2005). A scan for positively selected genes in the genomes of humans and chimpanzees. *PLoS Biology* **3**, e170.

Nowak, M. A., Boerlijst, M. C., Cooke, J. and Smith, J. M. (1997). Evolution of genetic redundancy. *Nature* **388**, 167–171.

Odenwald, W. F., Rasband, W., Kuzin, A. and Brody, T. (2005). EVOPRINTER, a multigenomic comparative tool for rapid identification of functionally important DNA. *Proceedings of the National Academy of Sciences of the United States of America* **102**, 14700–14705.

Ohno, S. (1970). "Evolution by Gene Duplication." Springer-Verlag, New York.

Otto, S. P. and Whitton, J. (2000). Polyploid incidence and evolution. *Annual Review of Genetics* **34**, 401–437.

Ouyang, S. and Buell, C. R. (2004). The TIGR Plant Repeat Databases: A collective resource for the identification of repetitive sequences in plants. *Nucleic Acids Research* **32**, D360–D363.

Paterson, A. H., Bowers, J. E. and Chapman, B. A. (2004). Ancient polyploidization predating divergence of the cereals, and its consequences for comparative genomics. *Proceedings of the National Academy of Sciences of the United States of America* **101**, 9903–9908.

Pond, S. L., Frost, S. D. and Muse, S. V. (2005). HyPhy: Hypothesis testing using phylogenies. *Bioinformatics* **21**, 676–679.

Pryer, K. M., Schneider, H., Smith, A. R., Cranfill, R., Wolf, P. G., Hunt, J. S. and Sipes, S. D. (2001). Horsetails and ferns are a monophyletic group and the closest living relatives to seed plants. *Nature* **409**, 618–622.

Qiu, Y.-L., Dombrovska, O., Lee, J., Li, L., Whitlock, B., Bernasconi-Quadroni, F., Rest, J., Borsch, T., Hilu, K. W., Renner, S. S., Soltis, D. E., Soltis, P. S., *et al.* (2005). Phylogenetic analysis of basal angiosperms based on nine plastid, mitochondrial, and nuclear genes. *International Journal of Plant Science* **166**, 815–842.

Rong, J., Bowers, J. E., Schulze, S. R., Waghmare, V. N., Rogers, C. J., Pierce, G. J., Zhang, H., Estill, J. C. and Paterson, A. H. (2005). Comparative genomics of *Gossypium* and *Arabidopsis*: Unraveling the consequences of both ancient and recent polyploidy. *Genome Research* **15**, 1198–1210.

Rudd, S. (2005). openSputnik—a database to ESTablish comparative plant genomics using unsaturated sequence collections. *Nucleic Acids Research* **33**, D622–D627.

Sampedro, J. and Cosgrove, D. J. (2005). The expansin superfamily. *Genome Biology* **6**, 242.

Sampedro, J., Lee, Y., Carey, R. E., dePamphilis, C. and Cosgrove, D. J. (2005). Use of genomic history to improve phylogeny and understanding of births and deaths in a gene family. *Plant Journal* **44**, 409–419.

Schadt, E. E., Monks, S. A., Drake, T. A., Lusis, A. J., Che, N., Colinayo, V., Ruff, T. G., Milligan, S. B., Lamb, J. R., Cavet, G., Linsley, P. S. Mao, M., *et al.* (2003). Genetics of gene expression surveyed in maize, mouse and man. *Nature* **422**, 297–302.

Schlueter, J. A., Dixon, P., Granger, C., Grant, D., Clark, L., Doyle, J. J. and Shoemaker, R. C. (2004). Mining EST databases to resolve evolutionary events in major crop species. *Genome* **47**, 868–876.

Schmid, M., Uhlenhaut, N. H., Godard, F., Demar, M., Bressan, R., Weigel, D. and Lohmann, J. U. (2003). Dissection of floral induction pathways using global expression analysis. *Development* **130**, 6001–6012.

Siepel, A., Bejerano, G., Pedersen, J. S., Hinrichs, A. S., Hou, M., Rosenbloom, K., Clawson, H., Spieth, J., Hillier, L. W., Richards, S., Weinstock, G. M., Wilson, R. K., *et al.* (2005). Evolutionarily conserved elements in vertebrate, insect, worm, and yeast genomes. *Genome Research* **15**, 1034–1050.

Simillion, C., Vandepoele, K., Van Montagu, M. C., Zabeau, M. and Van de Peer, Y. (2002). The hidden duplication past of *Arabidopsis thaliana*. *Proceedings of the National Academy of Sciences of the United States of America* **99**, 13627–13632.

Sjölander, K. (2004). Phylogenomic inference of protein molecular function: Advances and challenges. *Bioinformatics* **20**, 170–179.

Soltis, D. E. and Soltis, P. S. (1999). Polyploidy: Recurrent formation and genome evolution. *Trends in Ecology and Evolution* **14**, 348–352.

Soltis, D. E., Soltis, P. S., Albert, V. A., Oppenheimer, D. G., dePamphilis, C. W., Ma, H., Frohlich, M. W. and Theissen, G. (2002). Missing links: The genetic architecture of flowers floral diversification. *Trends Plant Science* **7**, 22–31; discussion 31–34.

Soltis, D. E., Soltis, P. S., Chase, M. W. and Endress, P. (2005). "Phylogeny, Evolution, and Classification of Flowering Plants." Sinauer Associates, Sunderland, MA.

Soltis, P. S. (2005). Ancient and recent polyploidy in angiosperms. *New Phytologist* **166**, 5–8.

Sonnhammer, E. L. and Koonin, E. V. (2002). Orthology, paralogy and proposed classification for paralog subtypes. *Trends in Genetics* **18**, 619–620.

Stebbins, G. L. (1950). "Variation and Evolution in Plants." Columbia University Press, New York.

Swofford, D. L. (2003). PAUP*. Phylogenetic Analysis Using Parsimony (* and other methods). Version 4, Sinauer Associates, Sunderland, Massachusetts.

Sweigart, A., Fishman, L. and Willis, J. (2006). A simple genetic incompatibility causes hybrid male sterility in Mimulus. *Genetics* **172** (4), 2465–2479.

Theissen, G. (2002). Secret life of genes. *Nature* **415**, 741.

Thompson, J. D., Higgins, D. G. and Gibson, T. J. (1994). CLUSTAL W: Improving the sensitivity of progressive multiple sequence alignment through sequence weighting, position-specific gap penalties and weight matrix choice. *Nucleic Acids Research* **22**, 4673–4680.

Vandepoele, K. and Van de Peer, Y. (2005). Exploring the plant transcriptome through phylogenetic profiling. *Plant Physiology* **137**, 31–42.

Vision, T. J., Brown, D. G. and Tanksley, S. D. (2000). The origins of genomic duplications in *Arabidopsis*. *Science* **290**, 2114–2117.

Wang, J., Tian, L., Lee, H. S., Wei, N. E., Jiang, H., Watson, B., Madlung, A., Osborn, T. C., Doerge, R. W., Comai, L. and Chen, Z. J. (2006). Genome-wide nonadditive gene regulation in *Arabidopsis allotetraploids*. *Genetics* **172**, 507–517.

Wellmer, F., Riechmann, J. L., Alves-Ferreira, M. and Meyerowitz, E. M. (2004). Genome-wide analysis of spatial gene expression in *Arabidopsis* flowers. *Plant Cell* **16**, 1314–1326.

Western, T. L., Cheng, Y., Liu, J. and Chen, X. (2002). HUA ENHANCER2, a putative DExH-box RNA helicase, maintains homeotic B and C gene expression in *Arabidopsis*. *Development* **129**, 1569–1581.

Yamasaki, M., Tenaillon, M. I., Bi, I. V., Schroeder, S. G., Sanchez-Villeda, H., Doebley, J. F., Gaut, B. S. and McMullen, M. D. (2005). A large-scale screen for artificial selection in maize identifies candidate agronomic loci for domestication and crop improvement. *Plant Cell* **17**, 2859–2872.

Yang, Z. (1997). PAML: A program package for phylogenetic analysis by maximum likelihood. *Computer Applications in the Biosciences* **13**, 555–556.

Yang, Z. (1998). Likelihood ratio tests for detecting positive selection and application to primate lysozyme evolution. *Molecular Biology and Evolution* **15**, 568–573.

Yang, Z. (2002). Inference of selection from multiple species alignments. *Current Opinion in Genetics and Development* **12**, 688–694.

Yang, Z. and Nielsen, R. (2002). Codon-substitution models for detecting molecular adaptation at individual sites along specific lineages. *Molecular Biology and Evolution* **19**, 908–917.

Yang, Z., Wong, W. S. and Nielsen, R. (2005). Bayes empirical bayes inference of amino acid sites under positive selection. *Molecular Biology and Evolution* **22**, 1107–1118.

Yu, J., Wang, J., Lin, W., Li, S., Li, H., Zhou, J., Ni, P., Dong, W., Hu, S., Zeng, C., Zhang, J., Zhang, Y., *et al.* (2005). The Genomes of *Oryza sativa*: A history of duplications. *PLoS Biology* **3**, e38.

Zahn, L. M., Kong, H., Leebens-Mack, J. H., Kim, S., Soltis, P. S., Landherr, L. L., Soltis, D. E., Depamphilis, C. W. and Ma, H. (2005a). The evolution of the *SEPALLATA* subfamily of MADS-box genes: A preangiosperm origin with multiple duplications throughout angiosperm history. *Genetics* **169**, 2209–2223.

Zahn, L. M., Leebens-Mack, J., DePamphilis, C. W., Ma, H. and Theissen, G. (2005b). To B or Not to B a flower: The role of *DEFICIENS* and *GLOBOSA* orthologs in the evolution of the angiosperms. *Journal of Heredity* **96**, 225–240.

Zahn, L. M., Leebens-Mack, J. H., Arrington, J. M., Hu, Y., Landherr, L. L., dePamphilis, C. W., Becker, A., Theissen, G. and Ma, H. (2006). Conservation and divergence in the AGAMOUS subfamily of MADS-box genes: Evidence of independent sub- and neofunctionalization events. *Evolution & Development* **8**, 30–45.

Zanis, M. J., Soltis, D. E., Soltis, P. S., Mathews, S. and Donoghue, M. J. (2002). The root of the angiosperms revisited. *Proceedings of the National Academy of Sciences of the United States of America* **99**, 6848–6853.

Zanis, M. J., Soltis, P. S., Qiu, Y.-L., Zimmer, E. and Soltis, D. E. (2003). Phylogenetic analyses and perianth evolution in basal angiosperms. *Annals of the Missouri Botanical Garden* **90**, 129–150.

Zik, M. and Irish, V. F. (2003). Global identification of target genes regulated by APETALA3 and PISTILLATA floral homeotic gene action. *Plant Cell* **15**, 207–222.

AUTHOR INDEX

Numbers in **bold** refer to page on which full references are listed.

C

D

G

H

K

M

O

P

U

V

X

Y

Z

SUBJECT INDEX

D

H

FSC
www.fsc.org
MIX
Papier | Fördert
gute Waldnutzung
FSC® C083411

Kolibri 360 GmbH
Ferdinand-Jühlke-Straße 7
99095 Erfurt, Deutschland
produktsicherheit@kolibri360.de